YELLOW SEA
HONSHU
SHIKOKU
KYUSHU
C H I N A
CHUNGKING
SHANGHAI
EAST CHINA SEA
RYUKYU RETTO
BONIN IS
FORMOSA
HONGKONG
CANTON
PHILIPPINE SEA
HAINAN I
LUZON
MANILA
PHILIPPINE IS
MARIANAS IS
THAILAND
BANGKOK
INDO CHINA
SAIGON
GULF OF SIAM
MINDORO
SAMAR
PANAY
CEBU
LEYTE
PALAWAN
NEGROS
BOHOL
SULU SEA
MINDANAO
M I C R
CAROLINE IS
NORTH BORNEO
MALAYA
CELEBES SEA
HALMAHERA
SINGAPORE
M E L A N E
S O U T H C H I N A S E A
BORNEO
MOLUCCA SEA
SUMATRA
CELEBES
NETHERLANDS INDIES
NEW GUINEA
BISMARCK ARCHIPELAGO
PAPUA
BATAVIA
JAVA SEA
BANDA SEA
SOERABAJA
JAVA
FLORES SEA
FLORES I
TIMOR
ARAFURA SEA
PORT MORESBY
TIMOR SEA
INDIAN OCEAN
DARWIN
GULF OF CARPENTARIA
GREAT BARRIER REEF
A U S T R A L I A

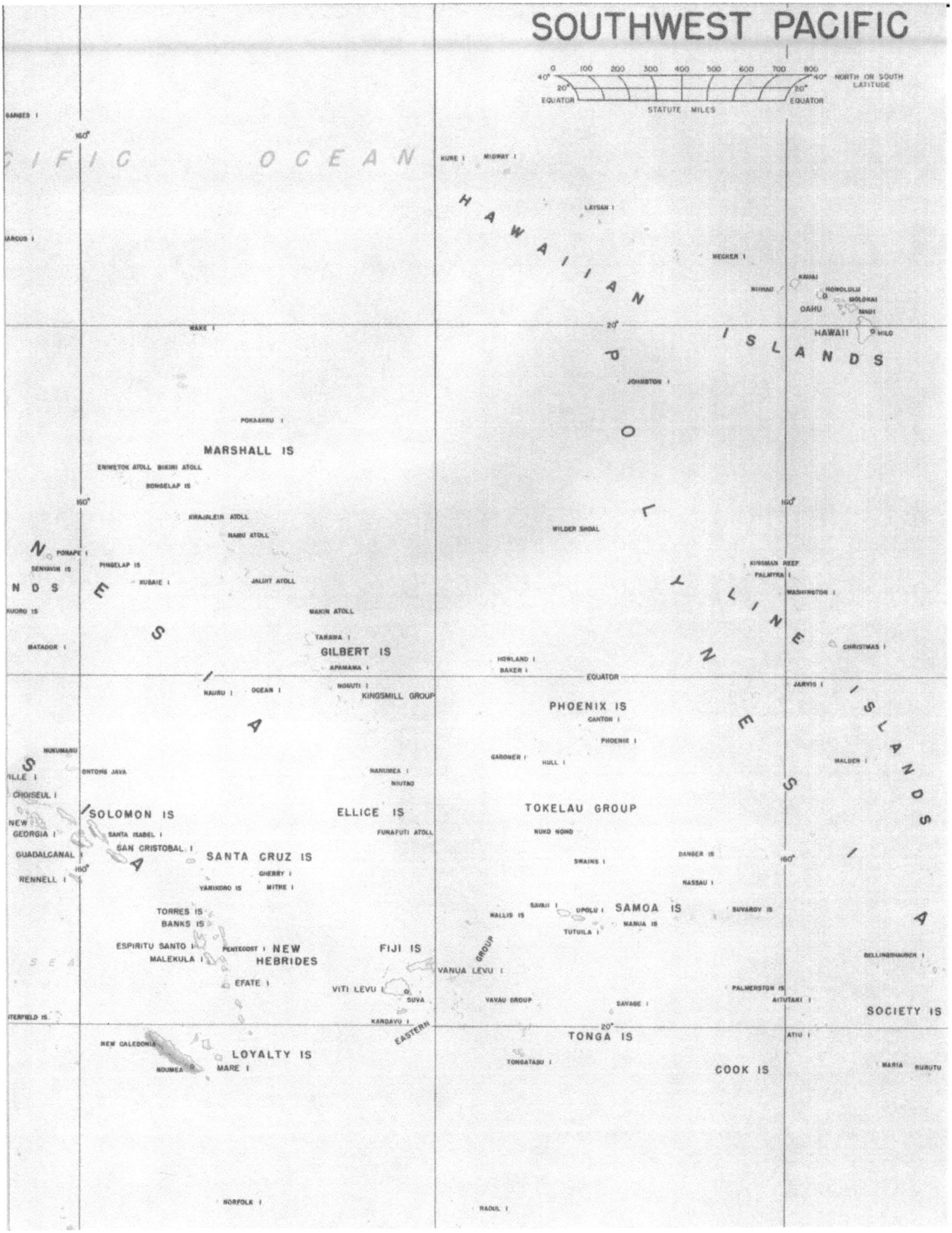

SOUTHWEST PACIFIC
0 100 200 300 400 500 600 700 800
NORTH OR SOUTH LATITUDE
EQUATOR
STATUTE MILES
OCEAN
KURE I
MIDWAY I
HAWAIIAN ISLANDS
LAYSAN I
NECKER I
NIIHAU
KAUAI
HONOLULU
MOLOKAI
OAHU
MAUI
HAWAII
HILO
WAKE I
JOHNSTON I
POLYNESIA
MARSHALL IS
ENIWETOK ATOLL
BIKINI ATOLL
RONGELAP IS
KWAJALEIN ATOLL
WILDER SHOAL
PONAPE I
PINGELAP IS
KUSAIE I
JALUIT ATOLL
KINGMAN REEF
PALMYRA I
WASHINGTON I
LINE ISLANDS
MAKIN ATOLL
TARAWA I
GILBERT IS
APAMAMA I
CHRISTMAS I
MATADOR I
HOWLAND I
BAKER I
EQUATOR
JARVIS I
NAURU I
OCEAN I
KINGSMILL GROUP
PHOENIX IS
CANTON I
PHOENIX I
GARDNER I
HULL I
MALDEN I
ONTONG JAVA
NANUMEA I
NIUTAO
CHOISEUL I
SOLOMON IS
ELLICE IS
TOKELAU GROUP
NEW GEORGIA I
SANTA ISABEL I
FUNAFUTI ATOLL
NUKO NONO
GUADALCANAL I
SAN CRISTOBAL I
SANTA CRUZ IS
SWAINS I
DANGER IS
RENNELL I
CHERRY I
NASSAU I
VANIKORO IS
MITRE I
TORRES IS
BANKS IS
SAVAII
UPOLU I
SAMOA IS
SUVAROV IS
WALLIS IS
MANUA IS
TUTUILA I
ESPIRITU SANTO I
PENTECOST I
NEW HEBRIDES
MALEKULA I
FIJI IS
BELLINGSHAUSEN I
EFATE I
VANUA LEVU I
PALMERSTON IS
VITI LEVU I
SUVA
VAVAU GROUP
SAVAGE I
AITUTAKI I
SOCIETY IS
KANDAVU I
EASTERN GROUP
TONGA IS
ATIU I
NEW CALEDONIA
LOYALTY IS
NOUMEA
MARE I
TONGATABU I
COOK IS
MARIA
RURUTU
NORFOLK I
RAOUL I

Reports of General MacArthur

THE CAMPAIGNS OF MACARTHUR IN THE PACIFIC

VOLUME I

PREPARED BY HIS GENERAL STAFF

Library of Congress Catalog Card Number: 66–60005

Facsimile Reprint, 1994

CMH Pub 13–3

FOREWORD

The *Reports of General MacArthur* include two volumes being published by the Department of the Army in four books reproduced exactly as they were printed by General MacArthur's Tokyo headquarters in 1950, except for the addition of this foreword and indexes. Since they were Government property, the general turned over to the Department in 1953 these volumes and related source materials. In Army and National Archives custody these materials have been available for research although they have not been easily accessible. While he lived, General MacArthur was unwilling to approve the reproduction and dissemination of the *Reports,* because he believed they needed further editing and correction of some inaccuracies. His passing permits publication but not the correction he deemed desirable. In publishing them, the Department of the Army must therefore disclaim any responsibility for their accuracy. But the Army also recognizes that these volumes have substantial and enduring value, and it believes the American people are entitled to have them made widely available through government publication.

The preliminary work for compiling the MacArthur volumes began in 1943 within the G–3 Section of his General Staff, and was carried forward after the war by members of the G–2 Section, headed by Maj. Gen. Charles A. Willoughby with Professor Gordon W. Prange, on leave from the University of Maryland, as his principal professional assistant. Volume II of the *Reports* represents the contributions of Japanese officers employed to tell their story of operations against MacArthur's forces. The very large number of individuals, American and Japanese, who participated in the compilation and editing of the *Reports* would make a complete listing of contributors relatively meaningless.

Volume I narrates the operations of forces under General MacArthur's command from the Japanese attack on Luzon in 1941 through the surrender in 1945. While service histories have covered much of the same ground in separate volumes, no single detailed narrative of General MacArthur's leadership as commander of the Southwest Pacific Area has yet appeared. Chapters dealing with the reconquest of Borneo, plans for the invasion of Japan, and the Japanese surrender make a distinctly new contribution. Volume I Supplement describes the military phase of the occupation through December 1948, reporting events not treated elsewhere in American publications. Volume II on Japanese operations brings together a mass of information on the enemy now only partially available in many separate works. Collectively, the *Reports* should be of wide interest and value to the American people generally, as well as to students of military affairs. They are an illuminating record of momentous events influenced in large measure by a distinguished American soldier.

Washington, D.C.
January 1966

HAROLD K. JOHNSON
General, United States Army
Chief of Staff

FOREWORD TO THE 1994 EDITION

I determined for several reasons to republish General MacArthur's reports to commemorate the fiftieth anniversary of World War II. First, the *Reports of General MacArthur* still stand as a detailed account from MacArthur's perspective of his operations against the Japanese in the Southwest Pacific Area. Second, the *Reports* offer a unique Japanese version of their operations in the Southwest Pacific that remains one of the few English-language descriptions of Imperial Army campaigns during World War II. Third, excellent illustrations, many of them original artwork commissioned for the *Reports*, plus superb maps give these volumes an enduring value for military historians and the American public. Finally, while General MacArthur remains a towering figure in American historiography, the passage of fifty years has dimmed the contributions of the U.S. Army units that first checked the Japanese southward advance in Papua New Guinea, then spearheaded the counteroffensive along the north New Guinea coastline that enabled MacArthur to make good his promise to return to the Philippines. The veterans of these campaigns, both men and women, deserve to be remembered for their contributions to the Nation in its time of greatest peril. These are General MacArthur's *Reports*, but they are also his testament to the American soldiers who served under his command.

Washington, D.C.
31 January 1994

HAROLD W. NELSON
Brigadier General, USA
Chief of Military History

PREFACE

This report has been prepared by the General Staff of GHQ to serve as a background for, and introduction to the detailed operational histories of the various tactical commands involved.

The pressure of other duties having prevented my personal participation in its preparation, it has been entrusted by me to that magnificent staff group which actually conducted the staff work during the progress of the campaigns. They speak with that sincere and accurate knowledge which is possessed only by those who have personally participated in the operations which they record.

DOUGLAS MACARTHUR

TABLE OF CONTENTS

ILLUSTRATIONS

Page

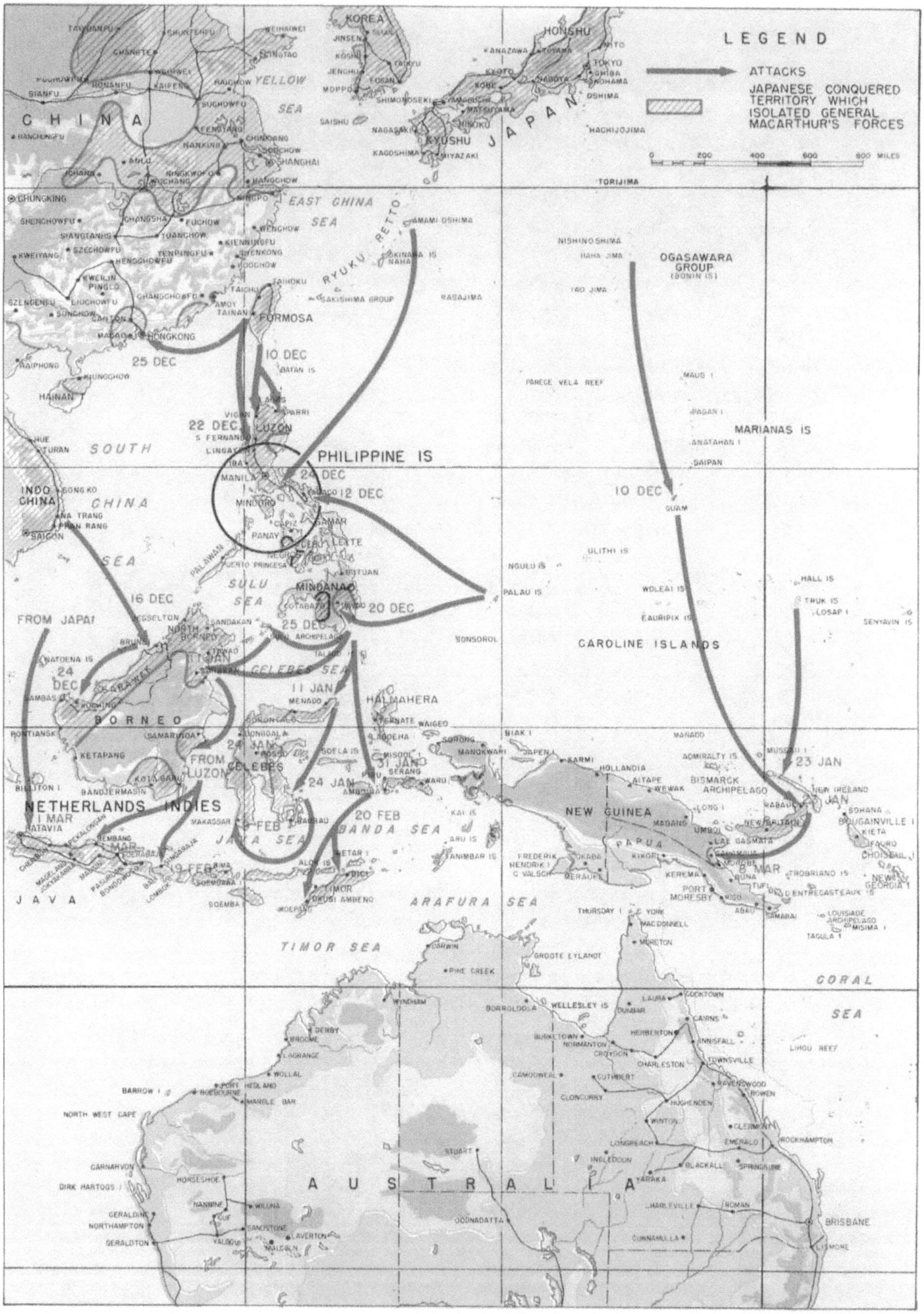

PLATE NO. 1
The Japanese Conquests which Isolated General MacArthur's Forces in the Philippines

CHAPTER I

THE JAPANESE OFFENSIVE IN THE PACIFIC

Pearl Harbor

The devastating attack against Pearl Harbor on 7 December 1941[1] and the subsequent Japanese thrusts in Asia left only one important obstacle in the path of the Japanese onslaught in the Southwest Pacific—General MacArthur and his small forces, isolated in the Philippines. (Plate No. 1) Their tenacious defense against tremendous odds completely upset the Japanese military timetable and enabled the Allies to gain precious months for the organization of the defense of Australia and the vital eastern areas of the Southwest Pacific. Their desperate stand on Bataan and Corregidor became the universal symbol of resistance against the Japanese and an inspiration for the Filipinos to carry on the struggle until the Allied forces should fight their way back from New Guinea and Australia, liberate the Philippines, and then press on to the Japanese homeland itself.[2]

By the end of November 1941 the Japanese had completed their over-all preparations for war against the United States, Great Britain, and the Netherlands. The Japanese Combined Fleet was directed to attack the naval forces of those nations and to combine with the Japanese Army and Air Force in a many-pronged offensive against Allied territories. The target date for the commencement of operations was 8 December 1941, East Longitude Time.[3]

Japanese Strategic Objectives

Japan's ultimate aim was complete hegemony in Asia and unchallenged supremacy in the western Pacific. Her strategic objectives were the subjugation of the Philippines and the capture of the immense natural resources of the Netherlands East Indies and Malaya. The conquest of the Philippines became an immediate military necessity. The Islands represented America's single hope of effective resistance in Southeast Asia, and, given the time and resources, General MacArthur would

1 Unless otherwise indicated, hours and dates used throughout this volume refer to the local time of the specific area under discussion.

2 The Japanese themselves realized the important effect of the protracted resistance in the Philippines. "Politically it stood as a symbol to the Filipinos and encouraged them to continue their resistance even after the fall of Corregidor," said Maj. Moriya Wada of the Fourteenth Army Staff. Lt. Col. Yoshio Nakajima, Intelligence Staff, Fourteenth Army, expressed the same opinion: "There was an influence, a spiritual influence, exerted by the American resistance on Bataan. Not only did the Japanese at home worry about the length of the period of American resistance on Bataan, but it served as a symbol to the Filipinos that the Americans had not deserted them and would continue to try to assist them." Interrogation Files, G–2 Historical Section, GHQ, FEC.

3 *Navy Operation Plans and Orders, 1941–1944, Recovered from CA NACHI*, ATIS, G–2, GHQ, SWPA, Limited Distribution Translation No. 39, Part VIII.

accomplish his long-range plan of making the islands impregnable.[4] He once called the Philippines "the key that unlocks the door to the Pacific." The Japanese understood this completely, for the islands lay directly athwart their path of future aggression. Close to South China and the island stronghold of Formosa, they were not only an obstacle to Japan's international ambitions, but they could be made into a powerful strategic springboard for their drive south and eastward. Flanking the vital sea routes to the south, they were the hub of the transportation system to Southeast Asia and the Southwest Pacific; from the Philippines, lines of communication radiated to Java, Malaya, Borneo, and New Guinea. Economically too, they were necessary to Japan's grandiose scheme of the Greater East Asia Co-Prosperity Sphere. As a thriving democracy, the Philippines were a living symbol of American political success in Asia and a direct negation of the national and moral principles represented by Japan. The Japanese were convinced that the Philippines must be conquered.

The abundant supply of oil, rubber, and other essential products in the Netherlands East Indies and Malaya which Japan needed for her vast war machine was another lucrative prize. The Japanese planned to isolate this region by destroying Allied naval power in the Pacific and Far Eastern waters, thus severing British and American lines of communication with the Orient. The unsupported garrisons of the Far East would then be overwhelmed and the areas marked for conquest quickly seized. Air attacks launched from progressively advanced airfields would prepare the way for amphibious assaults.

The first major operations would be directed against the Philippines and Malaya, with the invasion of British Borneo following as soon as possible. In the early stages of these campaigns, other striking forces were to seize objectives in Celebes, Dutch Borneo, and southern Sumatra, enabling the forward concentration of aircraft to support the invasion of Java. After the fall of Singapore, northern Sumatra would be occupied; operations would also be carried out against Burma at an appropriate time to cut the Allied supply routes to China. Singapore, Soerabaja, and Manila were expected to become major bases.[5]

The Japanese also planned to capture other strategic areas where they could establish advance posts and raise an outer barrier against an Allied counteroffensive. Their scheme of conquest envisaged control of the Aleutians, Midway, Fiji and Samoa, New Britain, eastern New Guinea, points in the

---

4 Lt. Gen. Torashiro Kawabe, Deputy Chief of the Army General Staff, stated that an important factor in Japan's decision to go to war and to invade the Philippines was the fear on the part of the Japanese General Staff of General MacArthur's ten-year plan for the defense of the Philippines. The plan was in its sixth year and a potential menace to Japan's ambitions. The Japanese had to intervene before it was too late. Lt. Gen. Akira Muto, Director of the Military Affairs Bureau of the War Ministry, voiced virtually the same opinion: "General MacArthur's program among the Filipinos was a potential obstacle to the Japanese plan of expansion in Asia. . . . If the Philippines were fortified and the defense strengthened by additional troops, Japan could not have undertaken war with the United States." Lt. Col. Hikaru Haba, Intelligence Staff, Fourteenth Army, said: "If there had been 50,000 additional men in the Philippines, and had the defenses been completed, we would have had to reconsider carefully the consequences of going to war." Interrogation Files, G–2 Historical Section, GHQ, FEC.

5 *Navy Operation Plans and Orders, 1941–1944, Recovered from CA NACHI,* ATIS, G–2, GHQ, SWPA, Limited Distribution Translation No. 39, Part VIII.

Australian area, and the Andaman Islands in the Bay of Bengal. All these would be seized or neutralized when operational conditions permitted.[6]

If the offensive succeeded, the United States would be forced back to Pearl Harbor, the British to India, and China's life line would be cut. With this eminently favorable strategic situation and control of the raw materials which they required, the Japanese felt they would be in a position to prosecute the war to a successful conclusion and to realize their ambition to dominate the Far East.

Initial American Dispositions

Japan's war potential, her probable action and plans of invasion had been brilliantly anticipated many years before by Homer Lea in his amazing book, *The Valor of Ignorance*, published in 1911.[7] Although this extraordinary publication had created a sensation among the general staffs of the world, the growing Japanese menace was not fully appreciated. Among the few, General MacArthur had clearly recognized the danger signals. His long and close association with the Philippines and the Orient had given him a rich background of knowledge and experience with which to judge the situation in the Far East. His grasp of the Japanese character and psychology and his understanding of Japanese military policy and aggressive intentions had induced him to voice repeated warnings of the shape of things to come. In a desperate race against time, he had attempted to stem the tide by initiating preparations for the defense of the Philippines. Working against almost insuperable political and administrative obstacles, he had commenced in 1935 to create a modern Philippine Army of ten divisions to counter the Japanese attack that he knew would soon come from the north, swiftly, fiercely, and without warning.

As the signs of impending conflict became unmistakably clear, General MacArthur prepared his meager forces in the Philippines for the inevitable storm. He grouped them into three principal commands, the Northern Luzon Force, the Southern Luzon Force, and the Visayan-Mindanao Force. The major unit was assigned to northern Luzon, for General MacArthur expected the principal enemy attack to be launched at the entrance to the great Central Plain, the natural corridor from Lingayen Gulf to Manila.[8]

The forces at General MacArthur's disposal included regular United States Army

6 Ibid.

7 Homer Lea's accurate forecast of the basic strategy which the Japanese would use in their invasion of the Philippines reads: "As the conquest of Cuba was accomplished by landing forces distant from any fortified port, so will the Philippines fall. Lingayen Gulf on the north coast of Luzon, or Polillo Bight on the east coast, will form the Guantanamo Bays of the Japanese.... Japan, by landing simultaneously one column of twenty thousand men at Dagupan and another column of the same size at Polillo Bight, would, strategically, render the American position untenable. These points of debarkation are almost equidistant from Manila, and are connected with it by military roads, while a railroad also connects Dagupan with the capital. The impossibility of defending Manila with the force now stationed on the islands is seen in the strategic advantages inherent in Japan's convergent attack. These two columns, more than double the strength of the American force, converge on Manila at right angles....If the American forces, on the other hand, should remain behind their lines at Manila, they would, in two weeks after the declaration of war, be surrounded by overwhelming numbers. The lines about Manila, as was demonstrated during the Spanish-American War, are incapable of prolonged defense. An aggressive enemy in control of the surrounding country can render them untenable in a short period of time." (Harper's 1942 edition) pp. 174–176.

8 This strategy followed in general War Plan Orange III which envisaged holding the entrance to Manila Bay.

units, regiments of the Philippine Scouts which were a part of the Army of the United States, the Philippine Army, and the Philippine Constabulary units. From 1 September 1941, orders were progressively issued to various units of the Philippine Army calling them to active duty. The over-all command was designated the United States Army Forces in the Far East (USAFFE) which had been created by the War Department on 26 July 1941. The air arm of USAFFE, designated the Far East Air Force (FEAF), was headed by Maj. Gen. Lewis R. Brereton.

The Japanese operations which plunged the United States into total war occurred in rapid sequence and were well timed. The surprise attacks against Pearl Harbor and the Philippines were followed at once by the invasion of Malaya, the seizure of Guam, and the capture of other American and British areas in the Pacific and the Far East. When the first blows fell, Nazi Germany, Japan's ally, had already conquered most of Europe. The German armies were deep in Russia on a broad front, and in the Middle East Rommel's armored divisions were attacking the British troops defending Egypt.

General MacArthur recommended that the Soviet Union strike Japan from the north. Such pressure, he felt, supplemented by United States air concentrations in Siberia, would limit the range of Japan's striking power, counter her initial successes, and gain time to strengthen the Philippines and the Netherlands East Indies. He believed it would throw Japan from the offensive to the defensive and that it would save the enormous outlay in blood, money, and effort necessary to regain lost ground. The Soviet Union, however, did not elect to engage in hostilities with Japan, and the British were unable to supply the reinforcements needed to protect their outposts in the Far East. The burden of stopping the Japanese rested largely upon the Allied forces already in Southeast Asia and the Pacific region and such additional strength as the United States could provide.[9]

Allied Strategy after Pearl Harbor

The United States was not prepared for war, and the effort which could be exerted against the Japanese was immediately and sharply limited by the global strategy adopted. President Roosevelt and Prime Minister Churchill, in a Washington conference after the Japanese attack on Pearl Harbor, confirmed a previous decision to concentrate first on the defeat of Germany.[10] Until victory was won in Europe, operations in the Pacific would be directed toward containing the Japanese and maintaining pressure upon them by conducting such offensive action as was possible with the limited resources available.[11]

The prospects for the Allies in the Pacific were highly discouraging. Eight battleships of the United States Pacific Fleet had been sunk or damaged at Pearl Harbor. Two days later the British battleship, *Prince of Wales,* and the battle cruiser, *Repulse,* were sent to the bottom off the east coast of Malaya. There remained in Netherlands East Indies waters some United States, Dutch and British cruisers, destroyers and submarines, but Japa-

---

9 Military Intelligence Division (MID), War Department (WD), *The World at War, 1939–1944* (Washington, 1945), pp. 109–110.

10 The general plan to support the Atlantic Theater was previously crystallized in the Anglo-American staff conferences that had begun in Washington in the spring of 1941.

11 Ibid pp. 105–107. Prime Minister Curtin, Ltr to CINCSWPA, 30 May 42, AG GHQ 381, Aust Req B SWPA, (MS).

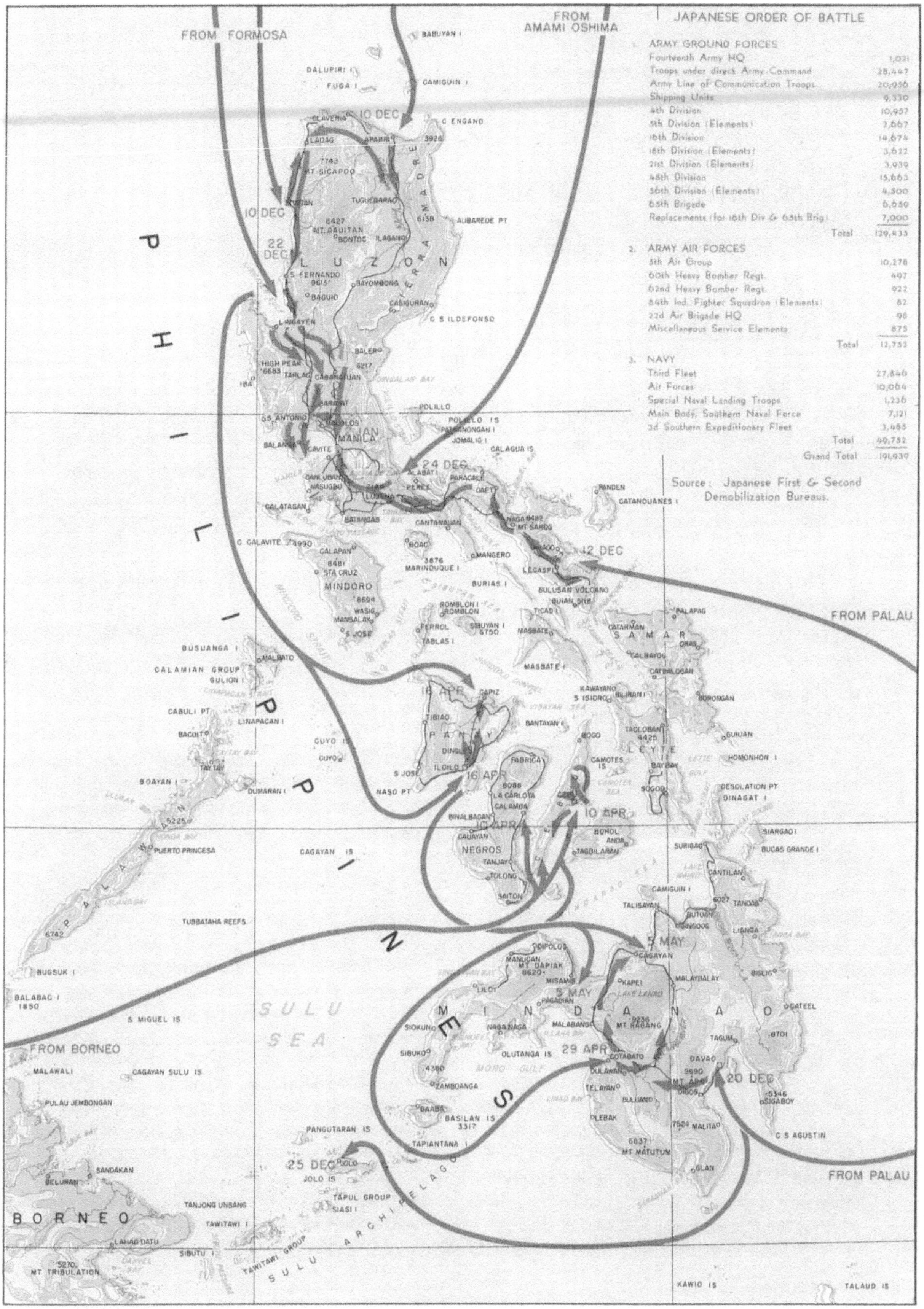

1. ARMY GROUND FORCES	
Fourteenth Army HQ	1,021
Troops under direct Army Command	28,447
Army Line of Communication Troops	20,956
Shipping Units	9,330
4th Division	10,957
5th Division (Elements)	2,667
16th Division	14,674
18th Division (Elements)	3,622
21st Division (Elements)	3,939
48th Division	15,663
56th Division (Elements)	4,300
65th Brigade	6,659
Replacements (for 16th Div & 65th Brig)	7,000
Total	129,433
2. ARMY AIR FORCES	
5th Air Group	10,278
60th Heavy Bomber Regt.	497
62nd Heavy Bomber Regt.	922
84th Ind. Fighter Squadron (Elements)	82
22d Air Brigade HQ	96
Miscellaneous Service Elements	875
Total	12,752
3. NAVY	
Third Fleet	27,846
Air Forces	10,064
Special Naval Landing Troops	1,236
Main Body, Southern Naval Force	7,121
3d Southern Expeditionary Fleet	3,485
Total	49,752
Grand Total	191,939

PLATE NO. 2

The Japanese Invasion of the Philippines and the Forces Employed nese naval supremacy was temporarily assured. The enemy already possessed overwhelming superiority in air and ground forces available at well-developed bases extending from French Indo-China to Formosa and the Mandated Islands.

The principal obstacles to Japanese expansion in Southeast Asia were the United States and Filipino forces under the command of General MacArthur in the Philippines, the British Imperial III Corps defending Singapore, and the concentration of Dutch forces in Java. Without substantial reinforcement these forces were not capable of more than delaying action.[12]

The Attack against the Philippines

The Japanese strike against the Philippines on 8 December was anticipated by General MacArthur several days in advance. On 4 December his pursuit interceptor planes began night patrols. Each night, they located hostile Japanese bombers from twenty to fifty miles out at sea, but the enemy planes turned back before actual contact was made. The last of these night flights was intercepted and turned back at the exact time of the attack on Pearl Harbor.[13]

When the enemy attacked the Philippines in overwhelming strength, the relatively weak American air force of fewer than 150 planes suited to combat service was totally inadequate to turn back the attackers.[14] Enemy bombers were guided in by sympathizers or espionage agents located near military objectives. Complete reports on American airfields and troop dispositions, procured by an extensive espionage net just prior to hostilities, enabled the Japanese to concentrate their attacks accurately on the most important objectives. Serious damage was inflicted on American planes and airdrome installations in the central Luzon area. On 10 December the naval base at Cavite was heavily bombed, and simultaneously the Japanese began their ground

12 MID, WD, op. cit., pp. 109–110.

13 GHQ, USAFFE, Press Release, 23 Jan 42. There has been some speculation as to what might have happened if our bombers had duplicated these pre-hostilities flights and attacked the Formosan airfields. The answer is that they would in all probability have run into a hornet's nest of 300-700 planes. See footnote 14.

14 In the scattered notes which Lt. Gen. Masaharu Homma wrote during the subsequent fighting on Bataan, the following statement is pertinent: "One of our greatest advantages is that we have complete control of the air." ("General Homma's Notes During the Battle of Bataan,") Interrogation Files, G–2 Historical Section, GHQ, FEC.) Against the weak Far East Air Force the Japanese used 307 first-line army planes in the Philippine operation and 444 navy planes (land-based and carrier)—a total of 751 aircraft. (Japanese First (Army) and Second (Navy) Demobilization Bureau Reports, G–2 Historical Section, GHQ, FEC.) The American air strength in the Philippines had been closely watched by the Japanese, and it figured heavily in their military calculations before Pearl Harbor. Rear Adm. Sadatoshi Tomioka, formerly of the Navy General Staff, offered a significant opinion on this subject: "The Japanese through long experience learned that they must have a 3–1 ratio in the air to attain supremacy; if General MacArthur had had an air force which exceeded 500 planes, Japan would never have been able to strike the Philippines." The statement of Col. Monjiro Akiyama, Organization and Order of Battle Department, Imperial General Headquarters, is also of interest: "In my opinion, the presence of a well-equipped air force in the Philippines would have had a great effect on the decision to attack Pearl Harbor and to begin a war with the United States." Important too, is the statement of Lt. Col. Tokutaro Sato, Fourteenth Army Staff Operations: "We had estimated that there were 200 planes available to General MacArthur in the Philippines before the opening of hostilities. Had there been twice this amount, I doubt that it would have been possible to attack the Philippines successfully. Had General MacArthur had this additional air power, the Japanese might not have been able to attack the Philippines at all, and possibly would have been unable to open hostilities elsewhere." Interrogation Files, G–2 Historical Section, GHQ, FEC.

offensive against the island of Luzon; twelve transports escorted by two light cruisers, thirteen destroyers, and numerous auxiliary craft landed troops at Aparri in the north and Vigan on the west coast. (Plate No. 2)

Reports issued that day covered these landing actions briefly:

> *7:30 A.M All indications point to a heavy enemy attack with land troops supported by naval elements and air forces on North Luzon.*
>
> *10:00 A.M. The enemy is in heavy force off the north coast of Luzon, extending from Vigan to Aparri.*
>
> *Large Japanese navy elements are escorting transports with Japanese air support. At Vigan at about 7:30 six transports were engaged in landing operations. At that time our bombardment attack on these ships created grave damage. Three were directly hit, one immediately capsizing and bombs were observed hitting close to the other three.*
>
> *At Aparri and perhaps at other contiguous points, landings were effected but exact strengths are unknown.*
>
> *4:00 P.M. Situation in North Luzon remains unchanged.*
>
> *Military objectives in the vicinity of Manila were bombed early this afternoon. While not yet verified, it is reported that the Japanese did not escape unscathed. Several planes have been reported shot down. There is no evidence of any other landings than those reported this morning at Vigan and Aparri.*[15]

General MacArthur realized that these landings were only preliminary. The enemy had not yet shown himself in sufficient force for these two operations to be anything more than diversionary attacks or security actions to protect the flanks of the main effort that would soon take place. When the Japanese landed in the south at Legaspi on 12 December under strong naval escort,[16] General MacArthur again held back his main forces, explaining his strategy as follows:

> *With the scant forces...at my disposal, they could not be employed to defend all the beaches. The basic principle of handling my troops is to hold them intact until the enemy has committed himself in force....*[17]

Future operations throughout this widespread theater were characterized by his tenacious adherence to this single strategic concept and the tactical measures to implement it.

Although small units were sent out to contain these relatively weak Japanese elements, General MacArthur held back his main force, waiting for the principal Japanese thrust.

At the end of the first week of war, the General reported many widely scattered actions but revealed that the all-out attack had not yet come:

> *The situation, both on the ground and in the air, was well in hand as the first week of military operations came to a close. A resume of the operations during the week follows: The enemy*

---

15 GHQ, USAFFE, Press Release, 10 Dec 41.

16 The troop transports for the Legaspi invasion left Palau on the morning of 8 December. The attack force which accompanied the transports consisted of one cruiser, six destroyers, and numerous auxiliary craft. Indirect support was provided by the seaplane carriers (twenty-plane capacity) *Chitose* and *Mizuho*. Additional units of the Japanese fleet cruised in the waters east of Legaspi. They consisted of the three heavy cruisers *Myoku, Haguro,* and *Nachi* (each with three catapult planes), the aircraft carrier *Ryujo* (thirty-six planes), and a number of destroyers. (Combined Fleet Headquarters Report, *Operational Study of the Philippines No. 1, Philippine Invasion Operation, 8 December—29 December,* G–2 Historical Section, GHQ, FEC.) The plane capacity of the Japanese ships in the Legaspi landing and the adjacent area alone was over half the number of the entire operational air force which General MacArthur had had at his disposal when the Japanese first struck the Philippines.

17 GHQ, USAFFE, Press Release, 12 Dec 41.

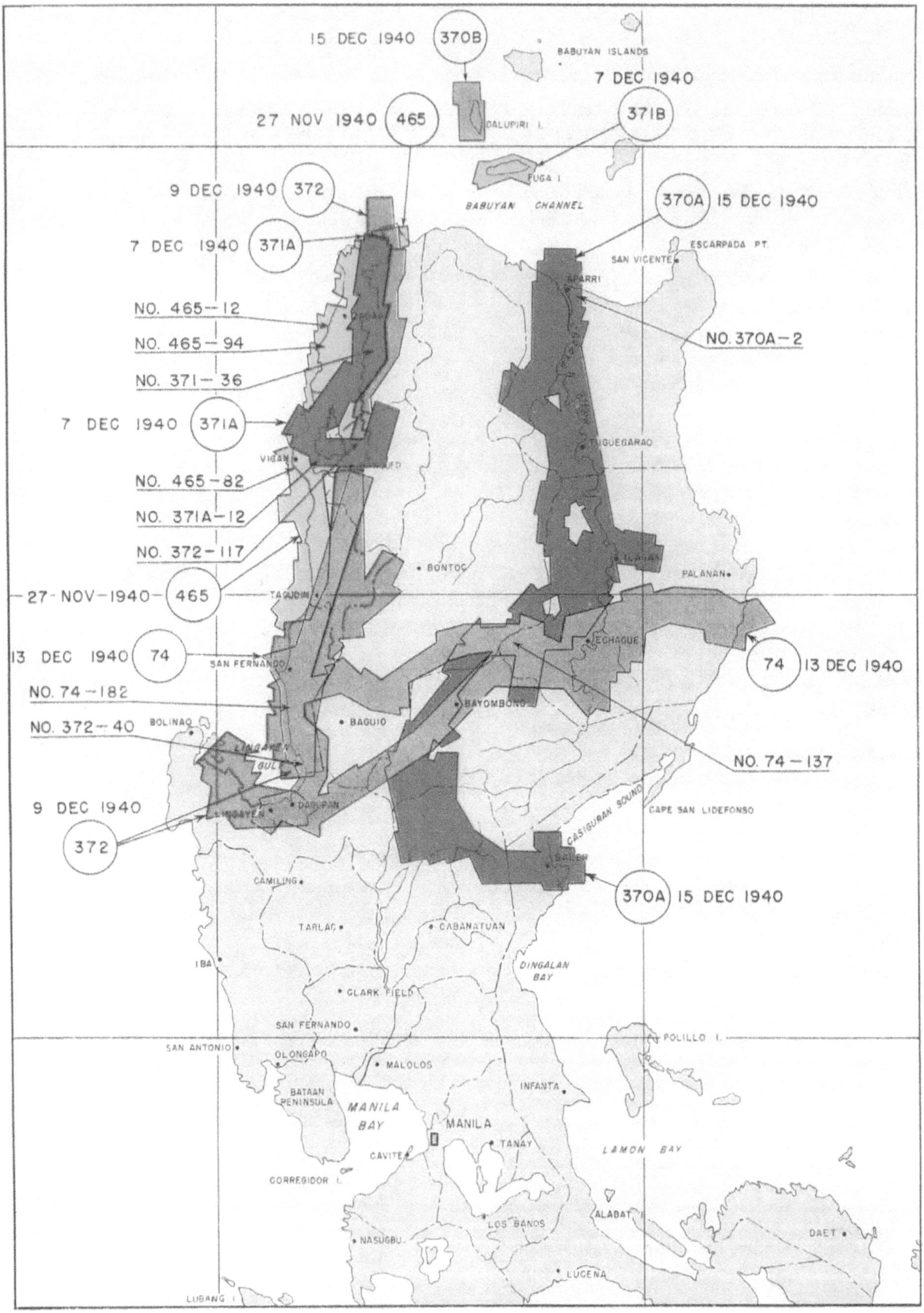

PLATE NO. 3

Aerial Reconnaissance of Luzon

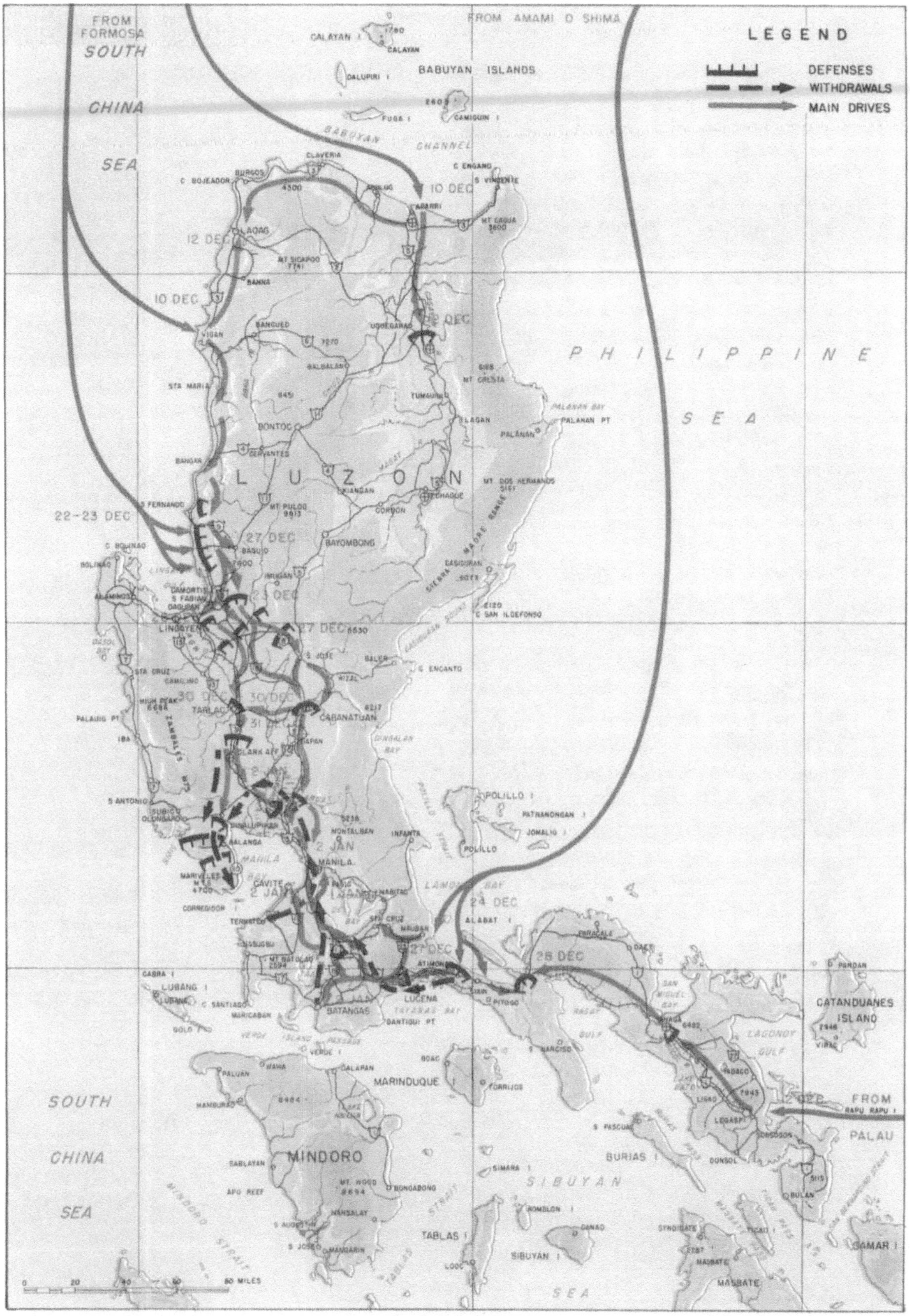

PLATE NO. 4
Operations on Luzon
December 1941

carried out 14 major air raids on military objectives in the Philippines, but paid dearly in the loss of transports, planes and troops and at least two battleships badly damaged as a result of action by our air and ground forces. An enemy landing was attempted in the Lingayen area, but was repulsed by a Philippine Army division. The enemy effected unopposed landings in limited numbers at Vigan, Legaspi and Aparri, but there is only local activity in those areas. Enemy naval units, troops and materiel on the ground were bombed effectively in the Vigan and Aparri areas, hampering landing operations. Four enemy transports are known to have been sunk and three others seriously damaged by our air force off northern Luzon. Individual deeds of heroism and bravery on the part of American and Filipino ground troops and air units marked the week's operations....The total of enemy air losses from all causes during the week is not less than 40 actually accounted for and probably many more which could not be verified....[18]

In the midst of his last-minute preparations for the big blow, General MacArthur did not forget the needs of the stricken civilian population. He had strongly endorsed a request of President Manuel Quezon to the United States for funds for public relief and the protection of the civilian population. On 14 December General MacArthur announced that, pending legislative action, he had been authorized by President Roosevelt to make available the sum of 20,000,000 pesos for these purposes. He took immediate action to turn this amount over to President Quezon.[19]

The second week of war came to a close without the major attack which General MacArthur was anticipating and for which he was holding his reserves in readiness. He did not commit his forces because the whole pattern of the enemy's plans and the direction of his main thrust had not been revealed. Except for a Japanese landing in force at Davao,[20] no large-scale operations had taken place, and aerial activity still predominated. General MacArthur's report of 21 December indicated, however, that the fighting was growing in intensity:

Increasing activity on the northern and southern fronts marked the second week of the war in which Japan is seeking to gain a foothold in the Philippine Archipelago.

Ground and air commands of the United States Army Forces in the Far East struck back with increasing fury at the invaders, routing enemy patrols in the Vigan and Legaspi sectors.

The enemy launched 12 major air raids during the week, but for the most part damage was not serious and casualties were light.

Our air force landed telling blows on the invaders at Legaspi, where it bombed and seriously damaged two transports and shot down a number of enemy planes. At Vigan, on the northern front, our airmen destroyed at least 25 planes on the ground and set fuel supplies afire.

In Davao on the extreme southern front the enemy landed in force and as the week came to a close, heavy fighting was in progress.[21]

General MacArthur in the meantime kept

18 GHQ, USAFFE, Press Release, 14 Dec 41.

19 GHQ, USAFFE, Press Release, 18 Dec 41.

20 The Davao invasion force consisted of fourteen transports which sailed from Palau on 17 December. The transports were escorted by a cruiser, six destroyers, and numerous auxiliary craft. The actual invasion was supported by the seaplane carriers *Chitose* and *Mizuho*. Additional units of the Japanese fleet stood by in the waters east of Davao. These consisted of the three heavy cruisers, *Myoko, Haguro*, and *Nachi*, the aircraft carrier *Ryujo*, the destroyer *Shiokaze*, and the patrol boat *Shirataka*. These units had also supported the Legaspi landings on 9 December. Combined Fleet Headquarters Report, *Operational Study of the Philippines No. 1, Philippine Invasion Operation, 8 December–29 December*, G–2 Historical Section, GHQ, FEC.

21 GHQ, USAFFE, Press Release, 21 Dec 41.

a close watch on the low, sloping beaches of Lingayen Gulf where, weeks before the initial blows, he had stationed the 11th and 21st Divisions of the Philippine Army at beach defenses. In the deceptive quiet of the gray dawn that spread over Lingayen Gulf on 22 December 1941, the great enemy blow fell at last. The communique that morning made no attempt to minimize the ominous situation. General MacArthur reported tersely:

> *Nichols Field was bombed at about 8:00 A.M. today.*
>
> *There was sighted this morning off Lingayen Gulf a huge enemy fleet estimated at 80 transports. Undoubtedly this is a major expeditionary drive being aimed at the Philippines.*[22]

The Japanese knew where to strike. Strength, location, and routes of the American main groupings were accurately plotted on the intelligence maps of the Japanese Headquarters. More than a year before their invasion of the Philippines—November and December 1940—the Japanese had made extensive aerial surveys of northern Luzon, selecting the areas of strategic importance and the points they could best attack. (Plate No. 3) Their objective was the prompt and complete annihilation of the defending forces so as to clear without delay the way to the rich areas that lay to the south.

Only two days after the mass landings at Lingayen Gulf, where the greatly outnumbered defenders on the beaches were ultimately overwhelmed by the invaders, another large Japanese force landed at Atimonan on Lamon Bay, on the east coast of southern Luzon.[23] This was much closer to Manila and central Luzon than the original Japanese elements that had landed far to the south at Legaspi.

With these landings the strategy of the Japanese commander, Lt. Gen. Masaharu Homma, became immediately apparent. It was obvious that he sought to swing shut the jaws of a great military pincers, one prong being the main force that had landed at Lingayen, and the other the smaller but nonetheless powerful units that had landed at Atimonan. (Plate No. 4) If these two forces could effect a speedy juncture, General MacArthur's main body of troops would have to fight in the comparatively open terrain of central Luzon, with the enemy to the front and to the rear. The Japanese strategy in arrogant optimism envisaged complete annihila-

22 GHQ, USAFFE, Press Release, 22 Dec 41. The huge invasion force which entered Lingayen Gulf on 22 December consisted of three transport echelons. The first echelon was composed of twenty-seven transports from Takao under the command of Rear Adm. Kensaburo Hara; the second echelon of twenty-eight transports from Mako was under the command of Rear Adm. Yoji Nishimura; the third echelon of twenty-one transports from Keelung was under the command of Rear Adm. Sueto Hirose. This force of seventy-six transports was supported by three escort units of cruisers, destroyers, and auxiliary craft which had been previously used in the landings at Vigan, Aparri, and Batan Island. The American estimate of eighty transports was within four of the actual number used by the Japanese in Lingayen Gulf. War Department Intelligence (WDI), G–2, GHQ, AFPAC, Document 19692C (WDI–113), *Operations of the Japanese Navy in the Invasion of the Philippines.* Combined Fleet Headquarters Report, *Operational Study of the Philippines No. 1, Philippine Invasion Operation, 8 December–29 December*, G–2 Historical Section, GHQ, FEC.

23 The force of twenty-four transports which made the landing at Atimonan left Amami Oshima on 17 December. It was escorted by the light cruiser *Nagara*, the 24th Destroyer Division, the 1st Section of the 16th Destroyer Squadron, subchasers, gunboats, minesweepers, and other minor vessels. Combined Fleet Headquarters Report, *Operational Study of the Philippines No. 1, Philippine Invasion Operation, 8 December–29 December*, G–2 Historical Section, GHQ, FEC.

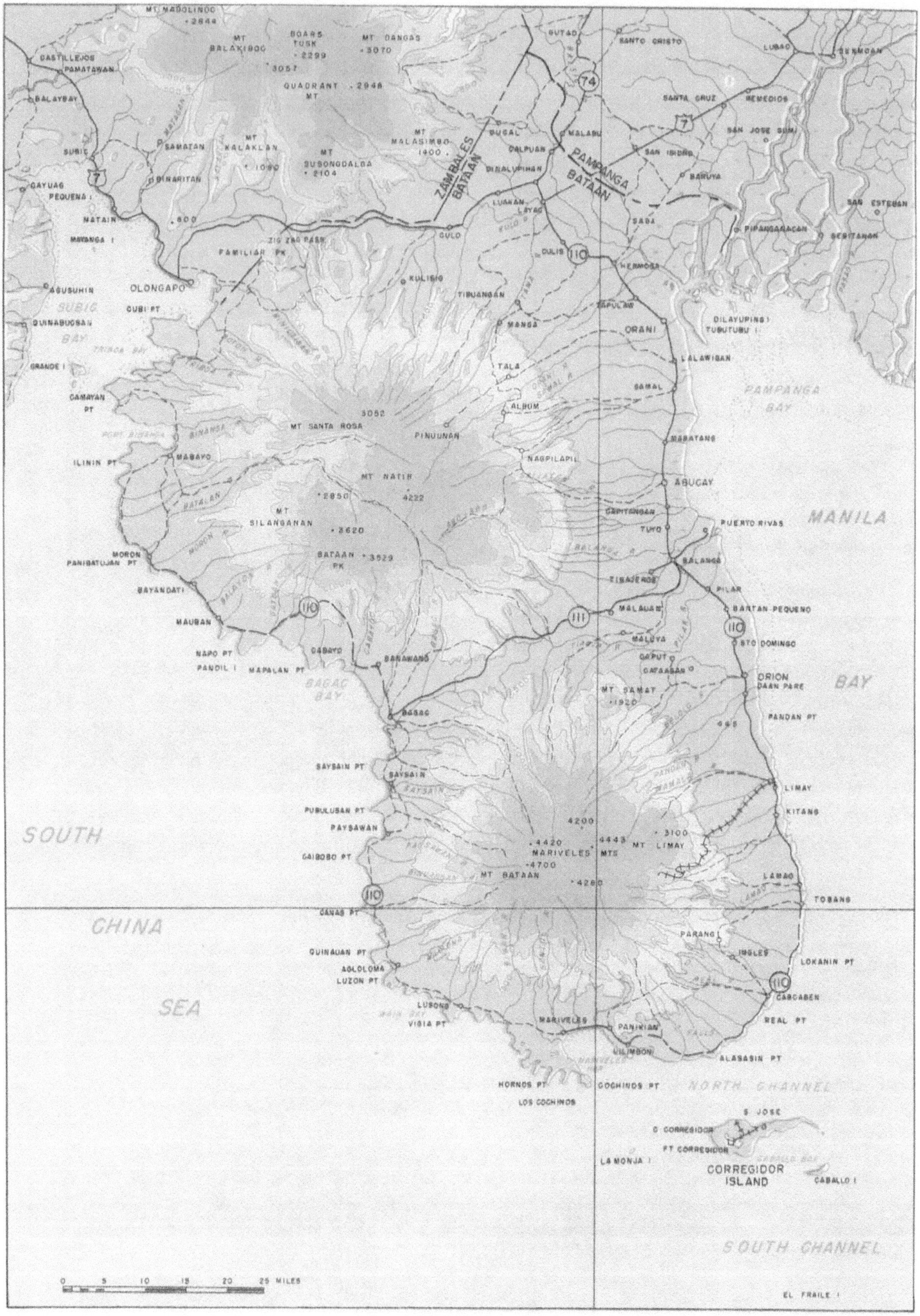

PLATE NO. 5

Terrain Features of Bataan

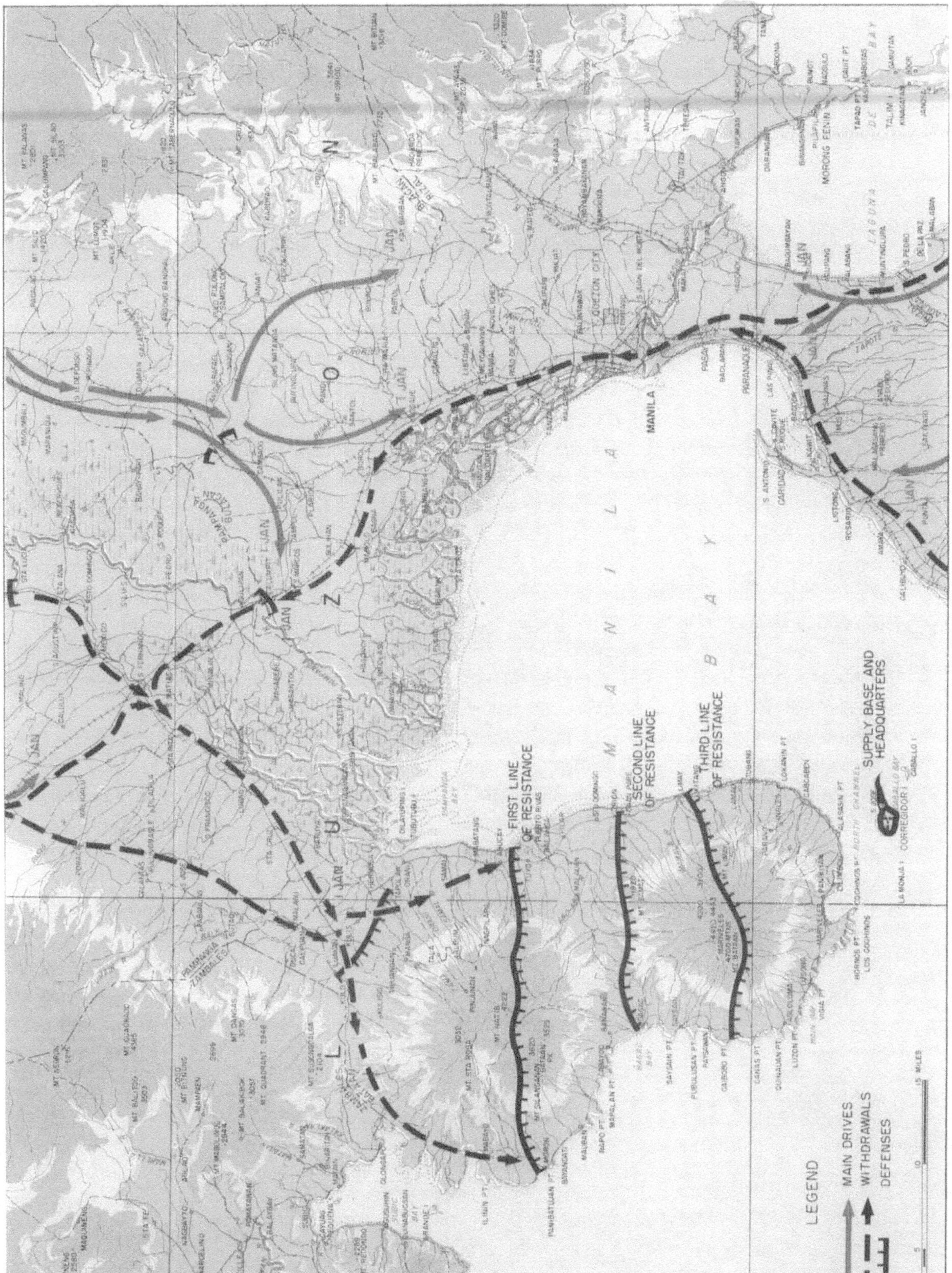

PLATE NO. 6

Route of Strategic Withdrawal to Bataan and Concept of Defense tion of the Luzon defense forces within a very short period.[24] With the principal island of the archipelago under their control, the Japanese could look forward to an easy conquest of the remainder of the islands.

Strategic Withdrawal to Bataan

General MacArthur's concept of defense against a superior force landing in northern Luzon was one of delaying actions on successive lines drawn across the great Central Plain from Lingayen Gulf in the north to the neck of the Bataan Peninsula in the south. While these delaying actions were retarding the enemy advance, General MacArthur would withdraw his troops from Manila, from the south, and from the Central Plain into Bataan where he could pit his intimate knowledge of the terrain against the Japanese superiority in air power, tanks, artillery, and men. To allow his main bodies to be compressed into the Central Plain by the Japanese advancing from both directions could only mean early and total destruction. By retiring into the peninsula, he could exploit the maneuverability of his forces to the limit, for Bataan's forests and ravines gave him his only chance of survival. (Plate No. 5)

On Bataan, the main line of resistance would run from Moron, on the coast of the China Sea, to Abucay, on the shore of Manila Bay. Should it become necessary, the defenders could drop back to a reserve line some six or eight miles to the rear. A third projected line crossed the Mariveles Mountains, the highest part of the peninsula. Still farther to the rear, Corregidor, separated from Bataan by two miles of water, would serve as the supply base for the Bataan defense and deny the Japanese the use of Manila Harbor, even though they should take the city of Manila itself.

The growing menace of encirclement by superior numbers induced General MacArthur to act quickly. The difficult problem he faced was to sideslip his troops westward in a series of rapid maneuvers and holding actions to the rocky peninsula and to the island forts in Manila Bay before superior forces of the enemy could cut off their path from the north. The crux of the problem was the successful passage of a famous tactical defile—the bridge at Calumpit, just south of San Fernando in Pampanga, where Highway No. 3 from northern Luzon to Manila joined with Highway No. 7 leading into Bataan. (Plate No. 6) The movement of the Southern Luzon Force, already complicated through the passage of Manila, was inevitably canalized at Calumpit, and once across the bridge, this force would also have to pass through San Fernando before it was safely on the road to Bataan.

As a corollary, the hard-pressed Northern Luzon Force would have to hold the enemy back from San Fernando and the Calumpit bridge no matter what the price until the Southern Luzon Force had cleared the critical

---

24 Japanese estimates on the length of time it would require to destroy or capture General MacArthur's forces varied from one to two months. Lt. Gen. Moriji Kawagoe, former Chief of Staff, 48th Division, Fourteenth Army, stated: "Imperial General Headquarters, Fourteenth Army, and I estimated that Luzon could be taken in one and one-half months after landing." Col. Kotoshi Nakayama, Senior Operations Officer, Fourteenth Army Staff, said: "Prior to the campaign, we estimated that we could annihilate the greater part of the American forces in one or two months." Lower estimates came from officers of the Fifth Army Air Force. "It was estimated by the Fourteenth Army," said Maj. Tsutomu Mizutani, "that the Philippines could be taken in one month." Maj. Koroshi Doba reported that the Japanese believed that General MacArthur's forces "could be annihilated within twenty days after the landings at Lingayen." Interrogation Files, G-2 Historical Section, GHQ, FEC.

defile, or General MacArthur stood to lose nearly half the forces with which he expected to defend Bataan and Corregidor.

To effectuate his strategy for protecting San Fernando until the movement of the Southern Luzon Force was completed, General MacArthur designated several lines for delaying action across the great Central Plain. The line directly north of San Fernando was to be held at all costs until the beginning of 1942, by which time, according to the schedule for the withdrawal of the Southern Luzon Force, the latter would be safe.

The lack of military transportation was solved by using scores of commercial buses which were commandeered to speed up the movement of the Southern Luzon Force.[25] At the same time both commercial and military trucks were employed day and night in endless columns to move goods, ammunition, equipment, and medical supplies from Manila to Bataan. In anticipation of the needs of war, General MacArthur had initiated the construction of a general hospital at Limay on the east coast of Bataan, and now, with the outbreak of the conflict, additional supplies were taken to the peninsula for the establishment of a second hospital.[26]

As the complicated but precisely timed withdrawal operation was being carried out, General MacArthur on 27 December declared Manila an open city to save the population from unnecessary suffering. His proclamation read:

> *In order to spare the metropolitan area from possible ravages of attack either by air or ground, Manila is hereby declared an open city without the characteristics of a military objective. In order that no excuse may be given for a possible mistake, the American High Commissioner, the Commonwealth government, and all combatant military installations will be withdrawn from its environs as rapidly as possible.*
>
> *The Municipal government will continue to function with its police force, reinforced by constabulary troops, so that the normal protection of life and property may be preserved. Citizens are requested to maintain obedience to constituted authorities and continue the normal processes of business.*[27]

The Japanese discovered too late the strategy underlying the movement of General MacArthur's forces behind the curtain of the rear guard actions in Pampanga and Batangas. Driving with reckless fury, they strove desperately to cut the vulnerable road junction at San Fernando and the bridge defile at Calumpit. The year ended with the Japanese applying increasing pressure to the beleaguered American and Filipino forces.

---

25 In peace-time maneuvers, the great commercial bus transportation companies operating in the major provinces of Luzon had already been organized as provisional motor transport battalions, utilizing the company personnel as officers and men. It is interesting to note that General MacArthur's G-2 in this period had previously served as G-4 of the Philippine Department, 1939-1940, and developed the motorization program.

26 Bataan had been quickly organized for a protracted defense through the hasty construction of depot areas in the primeval forests west of Lamao, the development of docks at Cabcaben, Limay, and Lamao, and the improvement of the road net, especially along the west coast; this organization of Bataan also came under G-4 Philippine Department.

27 GHQ, USAFFE, Press Release, 27 Dec 41. The Japanese nevertheless continued to bomb the city and on the 28th a press release reported: "Nothing new since this morning from the fronts. Until Manila was declared an open city it was noticeable that the Japanese did not attempt to attack civil installations from the air, but as soon as the army, including antiaircraft protection, withdrew, they immediately raided hitting all types of civilian premises including churches, convents, the cathedral, business houses and residences. Manila will no longer be blacked out. Tonight and all nights in the future Manila will be lighted."

The enemy is driving in great force from both the north and the south. His dive bombers practically control the roads from the air. The Japanese are using great quantities of tanks and armored units. Our lines are being pushed back.[28]

At this critical point General MacArthur threw in his small light tank force. In a series of costly counterattacks this force held the enemy long enough for the withdrawal into Bataan to be successfully completed. The hazardous timing of this movement was its most notable feature; one slip in the coordinated maneuver and the motor columns from southern Luzon would have been cut off and cornered in Manila. The success of this split-second operation enabled the assembly and reorganization on Bataan of the bulk of General MacArthur's forces and contributed to the subsequent brilliant defense of the peninsula.[29] No trained veteran divisions could have executed the withdrawal movement more admirably than did the heterogeneous force of Filipinos and Americans.

On 1 January General MacArthur announced the successful completion of the historic withdrawal:

In order to prevent the enemy's infiltration from the east from separating the forces in southern Luzon from those in northern Luzon, the Southern Luzon Force for several days has been moving north and has now successfully completed junction with the Northern Luzon Force. This movement will uncover the free city of Manila which, because of complete evacuation by our forces previously, has no practical military value. The entrance to Manila Bay is completely covered by our forces and its use thereby denied the enemy.[30]

The Battle for Bataan

General MacArthur regrouped his forces on Bataan into two corps, with I Corps under General Wainwright on the left and II Corps under Brig. Gen George Parker on the right.[31] As reorganization took place, the thousands of troops on the Bataan Peninsula began to dig in. New field artillery regiments were improvised in the field from miscellaneous elements to offset the enemy's great artillery superiority. While final delaying actions took place in the vicinity of Guagua, work went ahead on two runways near the southern tip of the peninsula to serve as a base for the

---

28 GHQ, USAFFE, Press Release, 31 Dec 41.

29 Most of the Japanese commanders who opposed General MacArthur in the Philippines generally agreed that the withdrawal into Bataan was an excellent strategical maneuver. "The fact that the Americans entered Bataan where there were well-prepared positions was a brilliant move strategically. American resistance was very fierce," said Lt. Gen. Susumu Morioka, Commanding General, 16th Division. While some of the Japanese commanders thought that General MacArthur would move into Bataan, the operation caught the bulk of them unawares, for they did not anticipate that it would be done so soon or so efficiently. The following statements are typical of their reactions: "We were completely surprised by General MacArthur's withdrawal to Bataan. We thought the Americans were cowards at the time. However, later studying the move objectively, I have come to believe that it was a great strategic move," said Colonel Akiyama, Organization and Order of Battle Department, Imperial General Headquarters. "The Japanese had never planned for or expected a withdrawal to Bataan," said Colonel Sato, Fourteenth Army Staff Operations. "It had been anticipated that the decisive battle would be fought in Manila. The Japanese commanders could not adjust to the new situation caused by the withdrawal of General MacArthur's forces into Bataan which they learned about from wireless, intelligence, and aerial reconnaissance around 28 December." Interrogation Files, G–2 Historical Section, GHQ, FEC.

30 GHQ, USAFFE, Press Release, 1 Jan 42.

31 Jonathan M. Wright, *General Wainwright's Story* (Garden City, N. Y., 1946), p. 48.

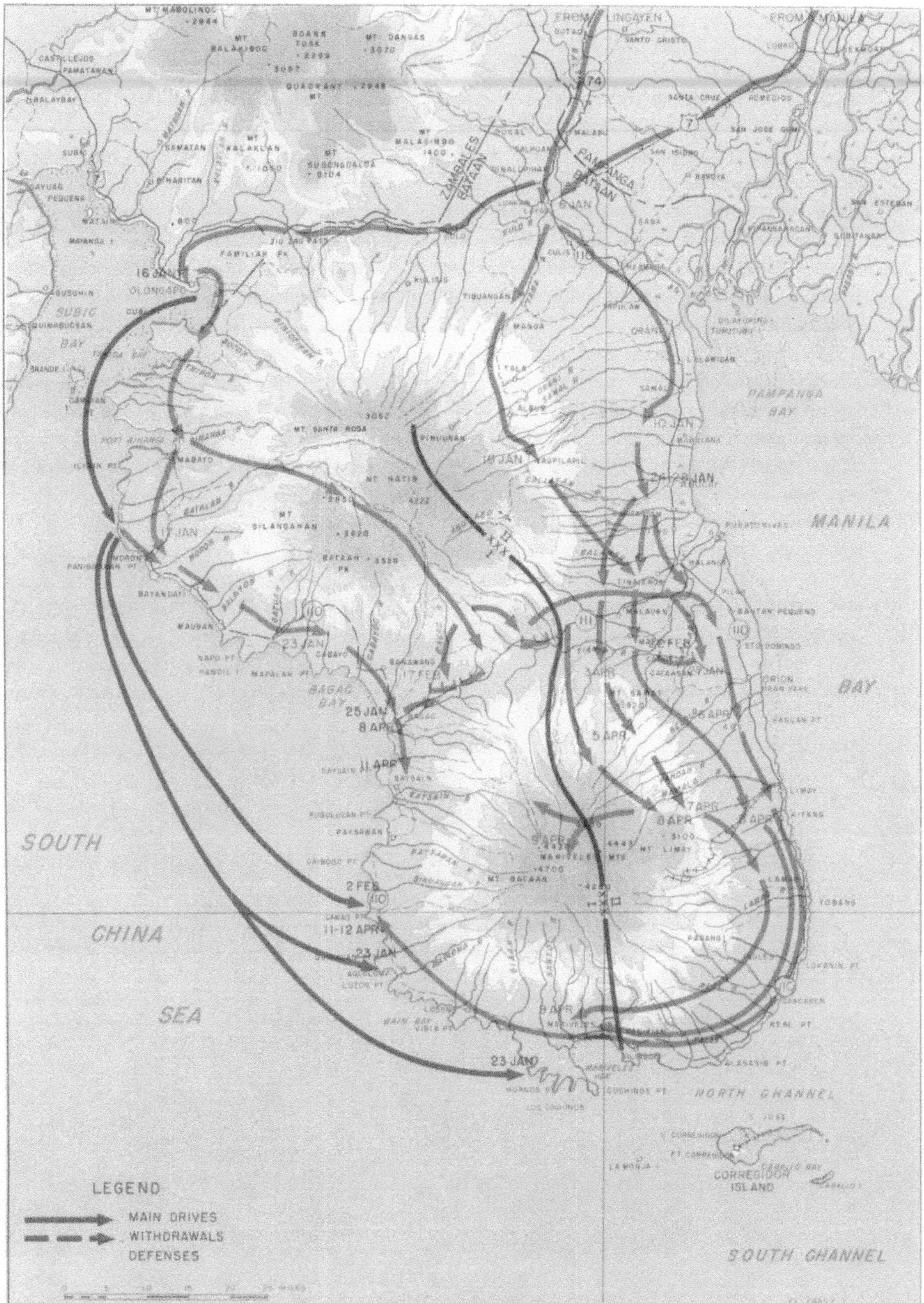

PLATE NO. 7
Action on Bataan
January-April 1942

remaining P-40's of the Far East Air Force. A hospital was set up near Cabcaben, and, because of the great number of civilians that had fled into Bataan with the army forces, refugee camps had to be established to keep them out of the way of military activities.

General Homma entered the city of Manila on 2 January 1942, but he had failed in his primary purpose, the annihilation of General MacArthur's field forces.

The baffled Japanese immediately launched headlong attacks against the first organized line of resistance, which ran from Moron to Abucay on the twenty-mile wide peninsula. They struck first on 12 January in the vicinity of Abucay, the eastern anchor. Initially successful, they were thrown back a few days later, only to strike again on 17 January in the Abucay Hacienda area, a little farther to the west. When unsuccessful there, they struck a third time on 20 January, still farther west in the II Corps sector. At the same time another Japanese force attacked Moron, the left anchor of the whole line, only to be repulsed. (Plate No. 7)

The hard-pressed defenders of this initial defense line on Bataan were finally forced to give ground. They were already beginning to suffer from malaria, and their daily rations, which had been sharply reduced, were to become even more meager as time went by.[32] They were unable to prevent the Japanese from infiltrating across the steep jungle-covered slopes of Mt. Natib in the center of the general defense line. At the same time a fresh, coordinated Japanese attack on the left flank had succeeded in taking Moron. The constant pressure against I Corps was so heavy that the only possibility of preserving alignment was to effect a partial withdrawal and reorganization. II Corps also withdrew in order to maintain a continuity of front on the line from Bagac to Orion.

Although successful in forcing this first line on Bataan, the Japanese had suffered so many casualties that their initial efforts against the new Bagac-Orion line were ineffective. By the end of January General Homma in turn practically had to cease operations while awaiting substantial reinforcements.[33] During this waiting period minor penetrations were made into the I Corps sector, but the raiding forces were pinched off and annihilated. A serious Japanese attempt to land from the

---

32 The shrewd enemy applied pressure and accelerated the influx of these refugees. This influx became an important factor in the logistics of defense. Under existing war plans, General MacArthur undertook to hold the entrance of Manila Bay—Bataan—for six months, presumably sufficient time to organize a relief expedition from the United States. Consequently, supply stockage, especially food, was calculated for 180 days only. Obviously, the defending forces could not stand heavy inroads into their limited supplies by thousands of homeless refugees.

33 "At first, General Homma thought that the 65th Brigade could occupy Bataan by continuing the pursuit of the American forces which had retreated to Bataan," said General Morioka, Commanding General, 16th Division. "However, they met fierce resistance at Hermosa and were stalled. General Homma studied documents captured in Manila and realized that Bataan was very strongly defended and so changed his plans. General Sugiyama, Chief of Staff, Imperial General Headquarters, personally came to Bataan in the middle of February to investigate. It was realized that Bataan could not be taken with what troops were present. As a result of General Sugiyama's visit, Imperial General Headquarters transferred the 4th Division and one brigade of another division from China plus the Kitajima Artillery Group (about 150 guns) from Hong Kong to Bataan. Several aerial bombardment groups from Burma and Malaya were also recalled to Bataan for the subsequent operations. These units arrived in the first part of March and underwent about three weeks of jungle training. On 3 April, the attack was resumed." Interrogation Files, G–2 Historical Section, GHQ, FEC.

China Sea behind the American lines on the west coast of Bataan was likewise frustrated. These attacking forces were completely destroyed.[34]

The brilliant use of artillery, superbly placed for interdiction fire, and carefully timed counter-attacks had held the baffled and infuriated enemy at bay in a series of increasingly costly attacks. Attack and counter-attack had serrated the opposing lines; Abucay had been lost and retaken; Moron had been lost and retaken, and lost again. For three bloody weeks, the line from Moron to Abucay had held, only to be swept back at last by the overwhelming Japanese tide. For many additional weeks, the line from Bagac to Orion had held steady.[35] The enemy had begun to avoid bloody frontal attacks and to resort to landings on the west coast behind the American-Filipino lines, or to attempted penetration raids, to stab deep into the battle line, only to be sealed off and liquidated. If only help could have reached the Philippines, even in small force, if only limited reinforcement could have been supplied, the end could not have failed to be victory. It was Japan's ability to bring in fresh forces continually and America's inability to do so that finally settled the issue.

During this period of fighting President Quezon continued to carry on the government of the Philippines from General MacArthur's Headquarters on Corregidor. On 28 February he addressed a message to his people, praising their resistance and urging continued support of their Allies:

Almost three months have passed since the enemy first ravaged our sacred soil. At the cost of many lives and immeasurable human suffering we have been resisting his advance with all our might. He has taken our capital and occupied several of our provinces, but we are neither beaten nor subdued....

Our soldiers in the field and the civilians behind the lines are animated by one determination and one aim—to fight the invader until death and to expel him from our land....

For the last month, the enemy has failed to make any advance. Every attack he has launched against us has been repulsed and his losses have been mounting every day. Under the leadership of General MacArthur our men are valiantly doing their duty. We are holding our ground against overwhelming odds....

Already the gallantry of our soldiers has aroused the admiration of the whole world....The most glorious chapter in the history of our country is being written in Bataan and Corregidor on the epic stand of our armies....

I urge every Filipino to be of good cheer, to have faith in the patriotism and valor of our soldiers in the field, but above all, to trust America....

The United Nations will win this war. America is too great and too powerful to be vanquished in this conflict. I know she will not fail us.[36]

The resistance afforded by the American and Filipino forces against successive concentrations of superior strength in all components and branches had developed into an epic of courage and tenacity. The Philippines were

34 General MacArthur's G-2, on a staff visit to I Corps, became involved in this landing attack in the sector of the 71st Division. When all United States officers had become casualties, he took command of the 1st Philippine Constabulary, defending Agloloma Pt. and reestablished the position by a sharp counterattack.

35 GHQ, USAFFE, Press Release, 28 Feb 42. The situation was summarized as follows: "The enemy has adopted a defensive attitude. This is definite indication that his heavy losses have shaken him. He begins to show signs of exhaustion. We are probably entering upon a phase of positional warfare of indecisive character. In North Luzon our mountain troops have forced him to evacuate the Abra Valley from Cervantes to Bangued . . . ."

36 GHQ, USAFFE, Press Release, 28 Feb 42.

now entirely surrounded by the Japanese, who were north in Formosa and on the China coast, west in Indo-China, south in Borneo and the Netherlands East Indies, and east in the Marianas, Carolines, and Marshall Islands. The United States Navy, charged with the vital mission of maintaining and keeping open sea-borne supply lines to the Philippines, was unable to do so. This spelled the doom of the forces fighting there. The fall of Singapore on 15 February enabled the enemy to reinforce his units on Luzon, but the continuing resistance on Bataan denied the Japanese an essential base for the projected thrust to the South Pacific. The enemy was forced to retain large army and navy forces in the Philippines,[37] which otherwise could have been employed against Allied shipping of men and materials to Australia and New Caledonia from the United States and the Middle East. As it finally developed, this enabled the barely sufficient preparations to be accomplished which saved those areas from what appeared inevitable invasion.

General MacArthur was not permitted to remain on Bataan or Corregidor to direct the last defense. Instead, he was ordered by President Roosevelt on 22 February to proceed to Australia to organize a new headquarters. On 11 March, in compliance with repeated orders, the General and selected members of his staff began the hazardous six-day journey by PT boat and plane through enemy-controlled areas to Australia from whence he was to launch the campaigns that would carry him not only back to Manila but on to the heart of the enemy's homeland—Tokyo.

General Wainwright assumed command of some 60,000 men remaining on Bataan and 10,000 more on Corregidor. Detachments cut off in northern Luzon, numbering about 2,000, continued to fight the enemy in guerrilla fashion.

General MacArthur's order of battle when he left for Australia was as follows:[38]

Bataan

1. *From Bagac to Mt. Samat: Philippine I Corps (including the 11th Division of the Philippine Army; the 1st Regular Division and the 71st and 91st Divisions of the Philippine Army, each less one regiment; and the 45th Infantry and 26th Cavalry Regiments of the Philippine Scouts.)*
2. *From Mt. Samat to Orion: Philippine II Corps (including the 21st, 31st, and 41st Divisions and the 51st Combat Team of the Philippine Army; the Provisional Air Corps Regiment of the United States Army; and the Philippine Army Provisional Air Corps Battalion and Training Cadre Battalion.)*

Corregidor and the forts at the entrance to Manila Bay

1. *The Harbor Defense Command under Maj. Gen. George F. Moore.*

---

37 The combined American and Filipino troops, partially trained, inexperienced, and ill-equipped as they were, numbered less than half the Japanese forces, many of whom were battle-tried veterans from China. These handicaps, together with the complete control of the air enjoyed by the Japanese and the supremacy of their Navy which kept their all-important supply lines open and America's closed, spelled the final word of doom for General MacArthur's forces in the Philippines.

38 CINCSWPA Radio No. 3, 21 Mar 42, to WARCOS, 323. 36 AG GHQ (S). General MacArthur's order of battle was purposely designed to meet the varied and special problems involved in the Philippine situation. In clear recognition of the geographical division of the Philippines and the tactical isolation of Luzon, General MacArthur carefully designated three separate command entities, with a view toward protracted warfare in each, if one should be lost. By this arrangement he hoped to prevent the Japanese from forcing a surrender of all the islands by the capture of any single headquarters.

Visayas

1. The Visayan Forces under Brig. Gen. Bradford G. Chynoweth.

Mindanao

1. The Mindanao Force under General Sharp.

His plans already formulated for the future conduct of operations on Bataan, were aimed at a fight to the end; they included attempts at opening supply lines from the southern portion of the Philippines and also a counter-attack to capture the enemy base at Subic Bay.[39] There was still the possibility that the defense could be protracted until the United States had time to recover from the first blows of the war and begin to send in supplies and reinforcements.

United States Army Forces in Australia

While the fighting in the Philippines was retarding the Japanese general advance, Australia was being prepared as a major Allied and United States base. A convoy en route to the Philippines from the United States at the outbreak of hostilities had been diverted to Australia. While at sea on 21 December, the troops were formed into Task Force South Pacific, and after debarking at Brisbane they were designated, on 5 January 1942, United States Army Forces in Australia (USAFIA). This expeditionary force was placed under the command of Maj. Gen. George H. Brett who was directed to establish a service of supply in support of the Philippines.[39]

Plans had been formulated by the War Department for building an air force in Australia for the defense of critical points in the Southwest Pacific Area. The remaining heavy bombers—fewer than twenty—from General MacArthur's Far East Air Force had been moved south to Darwin and were already operating from Australia. General Brereton, in command of this force, was under instructions to protect lines of communication to the Philippines and furnish support for that operation, as well as to cooperate with the Navy and Allied forces in Australia and the Netherlands East Indies.

---

39 General MacArthur's estimates and plans were outlined to General Marshall in a radio sent from Australia on 4 April 1942. "In an endeavor to permit passage of supplies from Cebu to Corregidor, I prepared prior to my departure detailed plans for an air attack by B-17 from here to Mindanao.... I believe there is a chance for blockade runners from the United States to reach their destination if they approach by the route around the north of Luzon.... When I left on March 11, I estimated that serious shortage would not develop before May 1 at the earliest, allowing sufficient time for the arrival of blockade runners from the United States....

I am utterly opposed, under any circumstances or conditions, to the ultimate capitulation of this command. If it is to be destroyed, it should be on the field of battle in order to exact full toll from the enemy. To this end I had long ago prepared a comprehensive plan for cutting a way out if food or ammunition failed. This plan contemplated an ostentatious artillery preparation on the left by the I Corps as a feint and a sudden surprise attack on the right by the II Corps...taking the enemy's Subic Bay positions in reverse simultaneously with a frontal attack by the I Corps. If successful, the supplies seized at this base might well rectify the situation. This would permit them to operate in Central Luzon where food supplies could be obtained and where they could still protect Bataan and the northern approach to Corregidor. If the movement is not successful and our forces defeated, many increments thereof, after inflicting important losses upon the enemy, could escape through the Zambales Mountains and continue guerrilla warfare in conjunction with forces now operating in the north....

I would be very glad if you believe it advisable for me to attempt to rejoin this command temporaily and take charge of this movement. The pressure on this situation could be immeasurably relieved if a naval task force with its own air protection could make some kind of threat in that general direction...." CINCSWPA Radio, 4 Apr 42, to WARCOS, Rec Sec, GHQ, 384.3 420404-C (S).

The ABDA Command

Allied strategy envisaged holding the Malay barrier (Malaya, Sumatra, Java, and northern Australia) while re-establishing communications with Luzon through the Netherlands East Indies. Burma and Australia were essential supporting positions. The former constituted the only remaining route for supplying China. The latter was to be the staging area for the reserves of manpower and materiel which would have to be shipped from the United States if the Japanese push were to be stopped. To carry out this strategic policy, the American, British, Dutch, and Australian Governments established a unified control over the area. The ABDA Command, as it was called, included Burma, Malaya, the Netherlands East Indies, and the Philippines within its jurisdiction. General Sir Archibald P. Wavell of the British Army was appointed Supreme Commander with headquarters in Java. All land, sea, and air forces of the participating governments in the area were directly under him as were also the supporting forces in Australia.

The mission of the United States forces in Australia became twofold: to get supplies through the Japanese blockade into the Philippines and to continue logistic support of the United States Army units in the Nertherlands East Indies. These were largely Air Corps units, including the heavy bombers of the Far East Air Force which had been shifted from Darwin to Java for their base of operations.[40] General Brett left Australia to become Deputy Commander under General Wavell, and General Brereton was appointed Commander of the United States Operating Air Forces in the ABDA area. Maj. Gen. Julian F. Barnes assumed command of United States troops and facilities in Australia.

It was evident that Japanese seizure of the entire region from Burma to Australia could not be long delayed;[41] at every point the Allied forces were inferior in strength. General Wavell's Headquarters were dissolved on 28 February, and he departed to organize the defenses of India. General Brett returned to direct the United States Army Forces in Australia, and the command in Java passed to the Dutch who prepared for a final stand.[42]

A desperate but futile effort to halt the Japanese was made in the Java Sea on 27 February. An Allied force of five cruisers and nine destroyers under the command of Rear Adm. Karel W. F. Doorman of the Royal Netherlands Navy moved out to intercept the approaching invasion forces. In the resulting action, the Allied fleet suffered a crushing defeat. The enemy immediately proceeded with landings on Java and overcame major resistance by 9 March. Tokyo thereupon announced the completion of the conquest of the Netherlands East Indies. The announcement followed by one day the fall of Rangoon and the cutting of the Burma supply line to China.[43]

The Threat to Australia

To the east the Japanese were soon established in the southern Solomons, at Rabaul and Gasmata in New Britain, and at Kavieng

40 *Report of Organization and Activities, United States Army Forces in Australia,* AG GHQ 314.7, USAFIA.

41 Ibid. See also *Operation Journal of 19th Bomb Grp, 8 Dec 41–2 Feb 42, G–2 Historical Section, GHQ, FEC.*

42 WD, *The World at War, 1939–1944* (Washington 1945), pp. 109–123. WARCOS Radio, 23 Feb 42, to CG USAFIA, AG USAFFE.

43 WD, *The World at War, 1939–1944* (Washington 1945), pp. 122–123. COMINCH, *Our Navy at War* (Washington 1944), pp. 28–29.

in New Ireland. They had extended their control of the Central Pacific by the occupation of the Gilbert Islands. On 8 March enemy forces seized Lae and Salamaua on the northeast coast of New Guinea. From these positions and from the Netherlands East Indies, the Japanese directly menaced the security of Australia and the sea lanes through the South Pacific.[44] (Plate No. 8)

The Allied commanders in Australia considered that the next enemy moves would be directed against Darwin and Port Moresby, both already under air attack. The occupation of Darwin might be designed merely to deny it to the Allied forces or as one stage in a general offensive against northwest Australia and the Gulf of Carpentaria. An attack on Port Moresby would eliminate it as a threat to the flank of an advance from Rabaul southward against the east coast of Australia or New Caledonia and the sea and air routes from the United States.[45]

Intelligence reports indicated that there were at least nine Japanese divisions in Dutch New Guinea, Ambon, Makassar, Timor, Celebes, and Java. Two or more were available to seize Darwin. The attack could be supported by about four aircraft carriers as well as shore-based planes from Koepang, Ambon, and Namlea. A force estimated at one division, with two or three aircraft carriers and land-based planes from New Britain, was ready for operations against Port Moresby.[46]

The forces available in Australia were inadequate to meet the Japanese threat. Air and naval strength were particularly weak, and these deficiencies had proved disastrous to the Allied cause elsewhere in the Pacific. The Commonwealth's first line troops consisted of three divisions of the Australian Imperial Forces which were in the Middle East when the Pacific War began. This fact and the rapidity of the Japanese offensive dictated the return of the 6th and 7th Australian Divisions. They were scheduled to arrive in late March, while the 9th Division remained in Egypt. In the Southwest Pacific there were the 1st Armored Division of the Australian Imperial Forces and several divisions of militia troops conscripted for home defense. These units were only partially trained and inadequately equipped. The Royal Australian Air Force had approximately sixteen squadrons classed as first-line, but there was a serious shortage of planes and trained crews in some of these squadrons. They were equipped, in general, with inferior types of aircraft which limited their capabilities to small-scale bombing and reconnaissance.[47] The principal units of the Royal Australian Navy were two heavy and

---

44 WD, *The World at War, 1939–1944* (Washington 1945), pp. 123–125.

45 Australian Chiefs of Staff, "Probable Immediate Japanese Moves in Proposed New Anzac Area," 5 Mar 42, G–3, GHQ, SWPA *Journal* File prior to 5 Apr 42 (MS).

46 Allied estimates of enemy intentions were accurate. As early as 2 February 1942, Imperial General Headquarters had ordered the South Seas Detachment to prepare for the invasion of Port Moresby in coordination with the Fourth Fleet. Japanese First Demobilization Bureau Report, *Southeastern Area Operations Record,* Part III, "Operations of the Eighteenth Army," Vol I, pp. 4–6, G–2 Historical Section, GHQ, FEC. The reports of the Japanese First and Second Demobilization Bureaus are operational histories prepared by former Japanese Army and Navy officers, all of whom participated in some phase of the war. The purpose of these reports is to document fully the history of the Japanese Armed Forces in World War II.

47 Australian Chiefs of Staff, "Defense of Australia," 27 Feb 42, G–3, GHQ, SWPA *Journal* File prior to 5 Apr 42 (MS). USASOS, *Report of Organization and Activities, United States Army Forces in Australia,* 9 Jul 43, AG GHQ 314. 7.

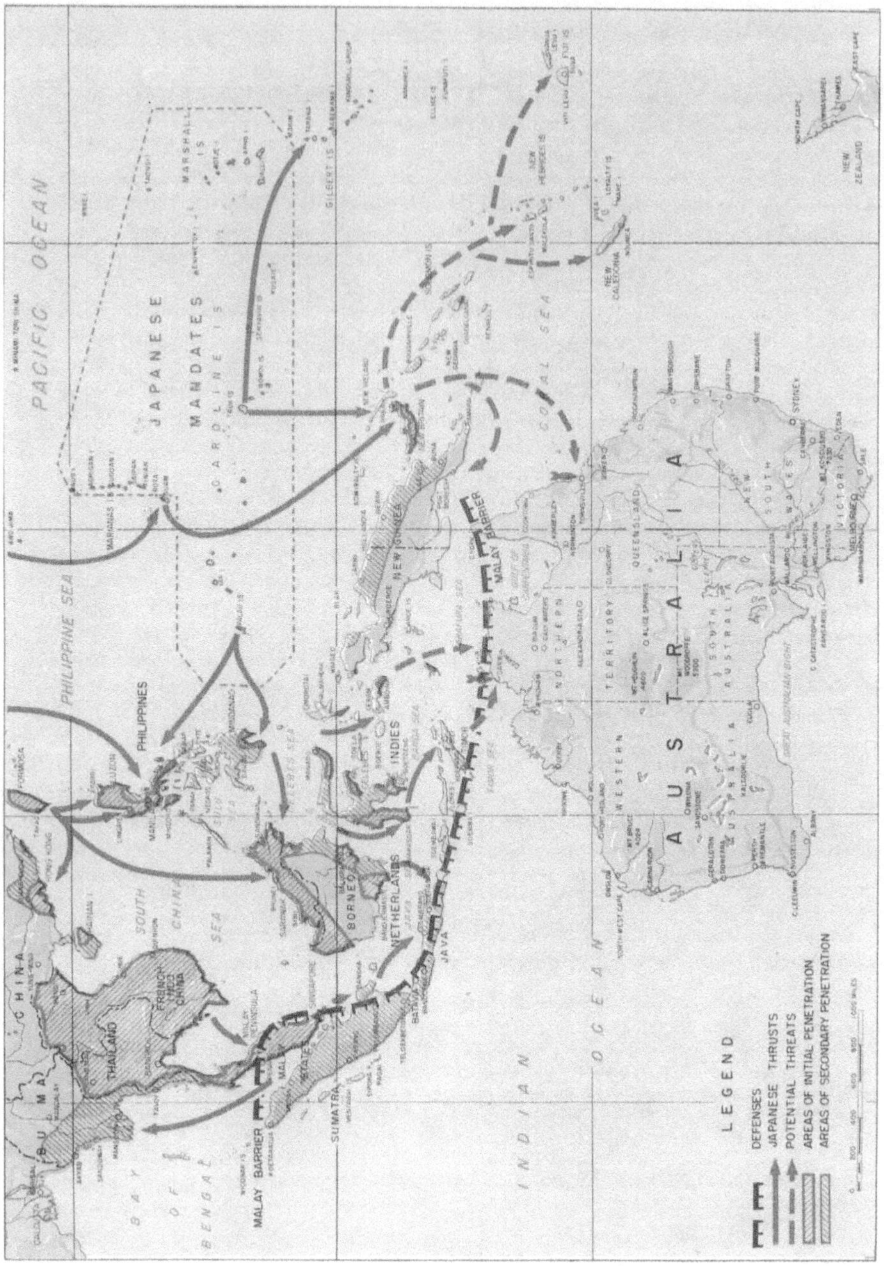

PLATE NO. 8

Disintegration of the Malay Barrier and the Threat to Australia

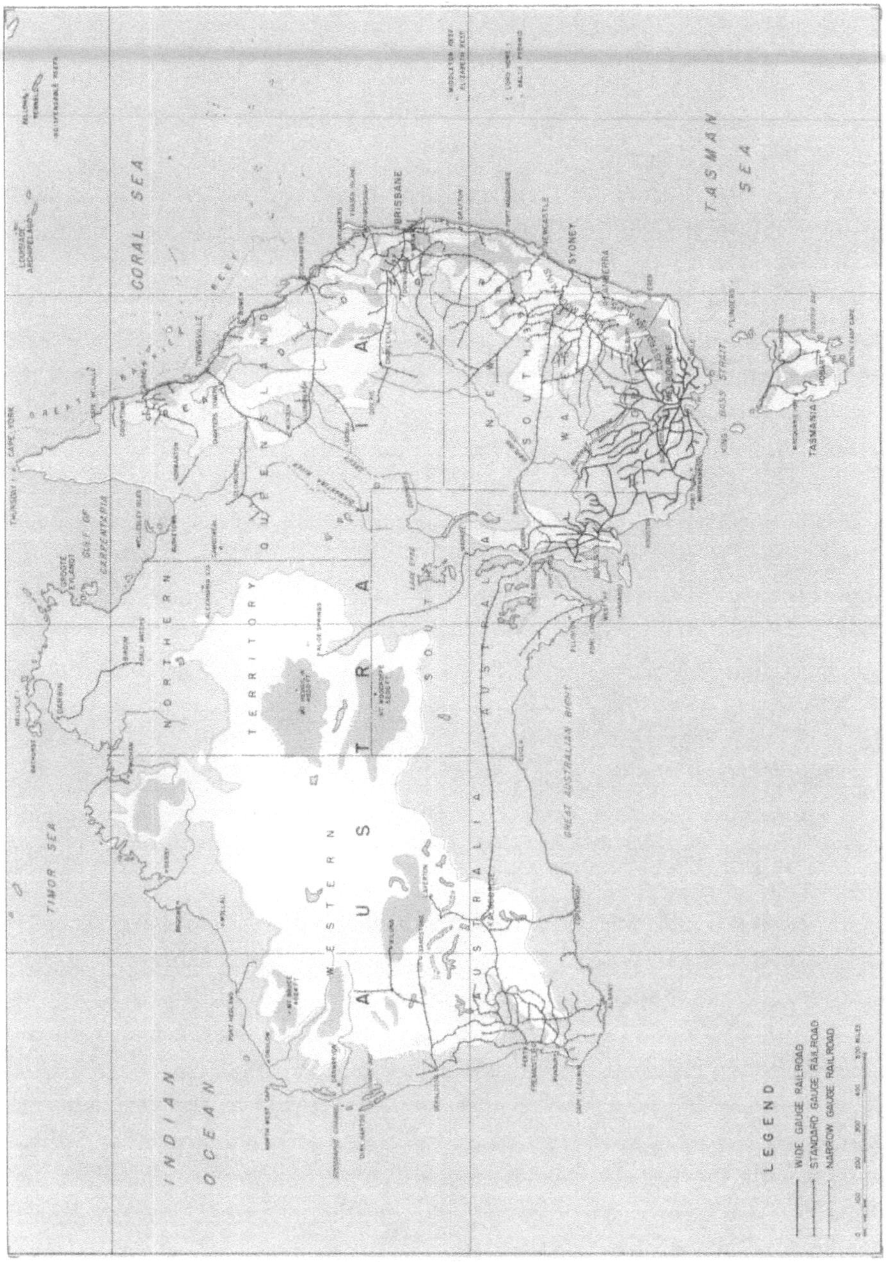

PLATE NO. 9

Orientation Map Showing the Lack of Rail Transportation along the Vulnerable Northern Coast two light cruisers.

Limitations of manpower and productive capacity severely restricted expansion of the Australian armed forces and made support from overseas essential.[48] The War Department in Washington planned to bring to full strength the two heavy bomber groups, two medium bomber groups, one light bomber group, and three pursuit squadrons already in Australia and to provide additional fighter units. With the exception of field artillery and antiaircraft elements, there were no United States ground combat forces, but the 41st Division was to arrive in April, and another division had been promised if the Australians permitted one of their divisions to remain in the Middle East.[49]

The combined United States and Australian forces were obviously insufficient for the protection of the whole of a vast continent, almost equal in size to the United States and with a coast line 12,000 miles in length. Concentration in any area to oppose Japanese landings was exceedingly difficult because of the great distances to be covered and the inadequate transportation facilities. The road and rail network of Australia was of limited capacity and of varying gauges and did not permit the rapid movement of troops. Principal routes frequently paralleled the coast and were vulnerable to amphibious attack. Darwin possessed no rail connection with the rest of the country. Port Moresby and Tasmania depended solely upon sea lines of communication and their control by friendly air and naval forces.[50] (Plate No. 9)

Disposition of Forces

The Australians disposed the bulk of their forces in the general region around Brisbane and Melbourne where most of the industries, the principal food producing centers, and the best ports, were located. The area was all-important to the country's war effort, and its defense was the prime consideration of its military authorities.[51]

Small forces were stationed in Tasmania and Western Australia and at Darwin, Port Moresby, Thursday Island, and Townsville. Because of their relative isolation, the retention of these points on the outer perimeter depended largely upon their garrisons, none of which was strong enough to oppose a major assault successfully. Reinforcing them was impossible, as additional troops could be drawn only from the vital southeastern region which, in the opinion of the Australian Chiefs of Staff, was itself inadequately held. Withdrawal from the outlying areas was equally out of the question. Their importance warranted an effort to hold them as long as possible, and the destructive effect of such an evacuation on public morale also had to be considered. The return of the 6th and 7th Divisions and the expected arrival of a division from the United States would allow the strengthening of the forces at Darwin, Western Australia, and Tasmania. The Australians, however, planned to retain two of these units

---

48 "Appreciation by Australian Chiefs of Staff," 27 Feb 42, G–3, GHQ, SWPA *Journal* File prior to 5 Apr 42 (MS).

49 WARCOS Radio No. 739, 18 Mar 42, to CG USAFFE, AG GHQ 323, 26 Supr Com (S).

50 "Appreciation by Australian Chiefs of Staff," 27 Feb 42; GOC Home Forces Memo, for Minister for the Army, 4 Feb 42, G–3, GHQ, SWPA *Journal* File prior to 5 Apr 42. *Report of Organization and Activities, United States Army Forces in Australia,* AG GHQ 314.7, USAFIA.

51 "Appreciation by Australian Chiefs of Staff," 27 Feb 42 and 5 Mar 42, G–3, GHQ, SWPA *Journal* File prior to 5 Apr 42 (MS).

to increase the reserve in the Brisbane-Melbourne zone.[52]

Coordination of the United States and Australian forces which manned the defense was achieved through cooperative effort and a system of joint committees established early in January.[53] So long as the continent had not been directly threatened this was adequate, but by the middle of March the Japanese were approaching the doorstep of Australia. Darwin, the only port on the northwest coast, had already been severely bombed. Townsville had also been subjected to air raids. With each passing day the Japanese were forging new links in their chain of encirclement and preparing new strikes against Australia and her life line to the outside world. The organization of a combined Allied defense under a single directing authority had become a matter of urgent necessity. Proposals concerning the organization of a new area of command had been made early in March by the Australian and New Zealand Governments. A decision concerning these proposals was the next important problem facing the governments engaged in this theater of the war. Final action awaited the safe arrival of General MacArthur from the Philippines.

52 Ibid.

53 *Report of Organization and Activities, United States Army Forces in Australia,* AG GHQ 314.7, USAFIA.

CHAPTER II

ESTABLISHMENT OF THE SOUTHWEST PACIFIC AREA

General MacArthur Arrives in Australia

General MacArthur arrived in Australia on 17 March, after a hazardous trip by motor torpedo boat from Corregidor, and by B-17 flight from Mindanao. His report contains a graphic account of the event:

Departed from Corregidor at dark on the 12th making the trip with four U. S. Navy motor torpedo boats. Air reconnaissance revealed one hostile cruiser and one destroyer off the west coast of Mindoro but we slipped by them in the darkness. The following day was passed in the shelter of an uninhabited island but we risked discovery by air and started several hours before dark in order to approach Mindanao at dawn. Sighted an enemy destroyer at 15,000 yards but escaped unseen, making the scheduled run despite heavy seas and severe buffeting. Upon arrival at Mindanao we learned that of four planes dispatched only one had arrived and that without brakes or superchargers and being unfit for its mission it had already departed. Brett selected three more planes for the trip of which one developed mechanical trouble and two arrived safely taking the entire party out. Safe arrival and departure forced us to pass the latitude of Ambon at dawn but our course was set somewhat to the eastward enabling the party to escape interception. We landed at Batchellor Field while Darwin was under air raid....[1]

---

1 CINCSWPA Radio, 17 Mar 42, to WARCOS, AG, GHQ, 370.05 No. 1 (S). Note comments by Lt. Gen. George H. Brett, A. C., then Commander of United States Army Forces in Australia:

"I received a radiogram from Gen. Marshall informing me that Gen. MacArthur would call on me to send a flight of long-range bombers to Mindanao before March 15th.... MacArthur had been ordered out.

From the wreckage of Java I had brought a dozen B-17s of the 19th Group. They were in pretty bad shape. In a fully equipped air force they all would have been scrapped.... But they were all we had, and we had to keep them flying, if we tied them together with chewing gum and baling wire.... I looked over my B–17s.... There wasn't a bomber in the lot fit for the Philippines trip.... There was only one way out. Twelve new Flying Fortresses had just arrived in Australia, but they were assigned to the Navy.... I went to Admiral Herbert Fairfax Leary, and told him I had to get MacArthur out of the Philippines.... Could I borrow three of his planes to bring him to Australia? Leary had the reputation of saying no to all requests, unless he could see that the Navy would benefit by his acquiescence. 'I'd like to help you, Brett,' he said, 'but it is quite impossible. We need those planes here, and can't spare them for a ferry job, no matter how important it is....'

I had no jurisdiction over the Navy, and could not commandeer those planes.... Leary was...determined to hang on to his B–17s come hell or high water. There was nothing else to do but send the best of our combat-shattered group. We did all we could with them mechanically, which, I'll be the first to admit, wasn't too much.... Only one of the planes...was able to get through.... Of the other three I had dispatched, two experienced engine trouble, and turned back. The third fell into the sea, but the crew was picked up.... The problem remained exactly the same as it had been in the first place.... Back I went to Adm. Leary. I expected the same answer I'd had before, but was prepared to get tougher. But Leary didn't give me a single 'no'. Perhaps he had heard directly from Washington...the Admiral loaned me four beautiful new bombers...." George H. Brett, "The MacArthur I Knew," *True* (October 1947), pp. 139–140.

General MacArthur's successful completion of this perilous journey caused enormous popular enthusiasm in the United States and Australia. Expressing his deep appreciation for the cordiality extended to him by the Australians upon his arrival, the General pledged the full co-operation of the United States in the struggle against Japan in an address at Canberra on 26 March:

I am deeply moved by the warmth of the greeting extended to me by all of Australia.... There is a link that binds our countries together which does not depend upon written protocol, upon treaties of alliance or upon diplomatic doctrine.... It is that indescribable consanguinity of race which causes us to have the same aspirations, the same hopes and desires, the same ideals and the same dreams of future destiny.

My presence here is tangible evidence of our unity. I have come as a soldier in a great crusade of personal liberty as opposed to perpetual slavery. My faith in our ultimate victory is invincible and I bring to you tonight the unbreakable spirit of the free man's military code in support of our just cause.... Under its banner the free men of the world are united today. There can be no compromise. We shall win or we shall die, and to this end I pledge you the full resources of all the mighty power of my country and all the blood of my countrymen.

Hailed for his achievements in the Philippines, his safe arrival in Australia was a good omen for the future. The Allied nations acclaimed this new development and looked forward with renewed hope to stemming the Japanese tide. The controlled Japanese press, however, tried to exploit for propagandist ends General MacArthur's transfer to Australia. Although it attempted to hide the true meaning of this important event, it unwittingly expressed what in reality proved to be true —that one name meant so much to the Allied cause in the Southwest Pacific.[2]

It had taken repeated orders to transfer General MacArthur from the Philippines to Australia. He had been loath to leave because he knew that the Japanese had lost heavily on Bataan and he had planned to organize an early counterattack against them. He was without reinforcements, however, and realized that it would be impossible to get enough supplies through the Japanese blockade to drive them out of Bataan. He was finally convinced that the counteroffensive could not be organized in the Philippines and that he would have to use Australia as his new base of operations. The conviction that he and his staff would find the means for a general counteroffensive in Australia constituted the real reason for his acquiescence in his transfer. With this in mind, he voiced his famous words when he left the Philippines: "I shall return."[3]

Acting upon instructions from President Roosevelt, General Brett informed Prime Minister John Curtin that General MacArthur had

---

2 The following editorial is typical of the reactions of the Japanese press: "The fact that the general public in the Allied countries is hailing MacArthur as the only man now capable of leading the Allied forces is a clear evidence of the pitiful dearth of Allied leadership. That one general can mean so much to the Allied cause is a damaging confession that the rest of their leaders are unworthy of confidence. Hysterical emphasis laid on the supposedly miraculous qualities of this one man reveals how worried the British and Americans are over their continued reverses and how frantically they are clutching at this one straw represented by MacArthur." *Japan Times and Advertiser*, Tokyo, March 20, 1942.

3 Radio from MacArthur to Beebe, 21 Mar 42, 000.75, AG, GHQ (S): "The President of the United States ordered me to break through the Japanese lines and proceed from Corregidor to Australia for the purpose, as I understand it, of organizing the American offensive against Japan, a primary object of which is the relief of the Philippines. I came through and I shall return."

arrived in Australia to set up his new headquarters. The Australian Government immediately nominated General MacArthur for the post of Supreme Commander in a proposed new Allied command, the Southwest Pacific Area (SWPA). In this capacity he would have command of all Allied forces in the general region of Australia and New Zealand as far north as the equator and would be responsible to the British and American Combined Chiefs of Staff. The governments concerned were in agreement on the main points of the Australian-New Zealand proposals for a unified command in the Southwest Pacific and concurred in the selection of the Commander-in-Chief. Within one month the Southwest Pacific Area was established under General MacArthur's command.

Pending completion of the formal arrangements, steps were taken to integrate the United States and Australian forces and to develop the defensive strength of the continent on a purely co-operative basis.[4] The Australian Army was reorganized and regrouped and more intensive efforts were made to train the new recruits. The combined Australian and United States air forces were placed under the command of General Brett and Vice Adm. Herbert F. Leary, USN, assumed command of the naval forces.

The Australian authorities readily adopted suggestions made by General MacArthur. The most complete co-operation existed throughout the war between him and the other nationalities within his command—Australians, Filipinos, Dutch, British, and New Zealanders. Not only was there an almost complete lack of friction and misunderstanding, but the ties of mutual respect, good will, and admiration among the commanders, staffs, and troops might well serve as a model for a mixed international force. General MacArthur's ability to gain and maintain the full confidence of these nations and their forces, of such marked national variance, was an important factor in the success of the Pacific War.

Organization of the Pacific Theater

Revision of the machinery required for the over-all direction of the war in the Pacific progressed rapidly in Washington. By agreement among the Governments of Australia, New Zealand, the United Kingdom, the Netherlands, and the United States, the Pacific Theater was designated an area of primary United States strategic responsibility. While the British and American Combined Chiefs of Staff retained jurisdiction over the allocation of forces and materials, the Pacific commands were put directly under the United States Joint Chiefs of Staff.[5]

The Pacific Theater was reorganized and divided into three separate areas of responsibility, only two of which, the Southwest Pacific Area and the Pacific Ocean Areas, were directly concerned with operations against the Japanese. The Southwest Pacific Area, under

---

4 In a letter to the Secretary of the Defense Committee, Australian Department of Defense Coordination, on 1 April 1942 (323.36, AG GHQ), General MacArthur explained his position in the following words: "At the present time I am operating not as Commander of SWPA which is to be established, but as the Commanding General of all United States troops in the Far East to which have been attached Admiral Leary's naval forces. With the Australian forces, I am functioning by coordination and cooperation."

The tactful assistance and discreet intervention of the Australian Defense Secretary, Sir Frederick G. Shedden, one of the ablest of Australian civil servants, should be noted in this connection. Sir Frederick maintained a liaison office directly with GHQ which aided greatly in the co-operative effort.

5 JCS Directive to CINCSWPA, 30 Mar 42, G-3, GHQ JCS/CCS 42-43 (S).

General MacArthur, included the Philippines, New Guinea, the Bismarck Archipelago, the Solomon Islands to 159° East Longitude, Australia, and the Netherlands East Indies less Sumatra. (Plate No. 10) The Pacific Ocean Areas under Adm. Chester W. Nimitz, were subdivided into the North, Central, and South Pacific Areas. Admiral Nimitz was to retain direct command in the first two of these, but was to appoint a Commander, South Pacific Area, who, acting under his authority, would have control of its combined forces.[6]

The strategic policy approved for the Pacific Theater was initially defensive. The Joint Chiefs of Staff directed General MacArthur to hold the key military regions of Australia as bases for a future offensive and to check the Japanese southward advance by destroying enemy shipping, aircraft, and bases in the Netherlands East Indies, New Guinea, and the Solomon Islands. General MacArthur was also to attempt to maintain the position in the Philippines, protect communications, route shipping, and support the operations of Allied forces in the South Pacific and Indian Theaters. Economic pressure was to be exerted against the Japanese through attacks on vessels transporting raw materials from the conquered territories back to the Homeland.[7]

Admiral Nimitz was directed to hold the islands between the United States and the Southwest Pacific which would be necessary to provide security for lines of communication and support for operations against the Japanese. Other tasks assigned his command were to contain the enemy, aid in the defense of North America, protect essential sea and air routes, and "support the operations of SWPA forces." Both General MacArthur and Admiral Nimitz were to "prepare for the execution of major amphibious offensives," the initial stages of which would be launched from the South Pacific and Southwest Pacific Areas.[8]

Establishment of General Headquarters, Southwest Pacific Area

The Southwest Pacific Area was constituted on 18 April 1942 by agreement among the Governments of Australia, the United Kingdom, the Netherlands, and the United States. On that date General MacArthur assumed command and proceeded to establish his General Headquarters at Melbourne. The forces assigned to him were organized into five subordinate commands designated as Allied Land Forces (ALF), Allied Air Forces (AAF), Allied Naval Forces (ANF), United States Army Forces in Australia (USAFIA), and United States Forces in the Philippines (USFIP).[9]

The principal Staff Officers of General Headquarters included the following:[10]

Chief of Staff:
Maj. Gen. Richard K. Sutherland
Deputy Chief of Staff:
Brig. Gen. Richard J. Marshall
Asst. Chief of Staff, G–1:
Col. Charles P. Stivers, GSC
Asst. Chief of Staff, G–2:
Col. Charles A. Willoughby, GSC

6 Ibid. Also JCS Directive to CINCPOA, 30 Mar 42, G–3, GHQ JCS/CCS 42–43 (S).

7 JCS Directive to CINCSWPA, 30 Mar 42, G–3, GHQ JCS/CCS 42–43 (S).

8 Ibid. Also JCS Directive to CINCPOA, 30 Mar 42, G–3, GHQ JCS/CCS 42–43 (S).

9 GHQ General Order No. 1, 18 Apr 42, AG, GHQ 300.4 GO (S); GHQ General Order No. 2, 19 Apr 42, AG, GHQ 300.4 GO (S).

10 Except for General Chamberlin and Cols. Whitlock and Fitch who were members of the Staff of USAFIA, these officers were with General MacArthur in the Philippines before the outbreak of the war.

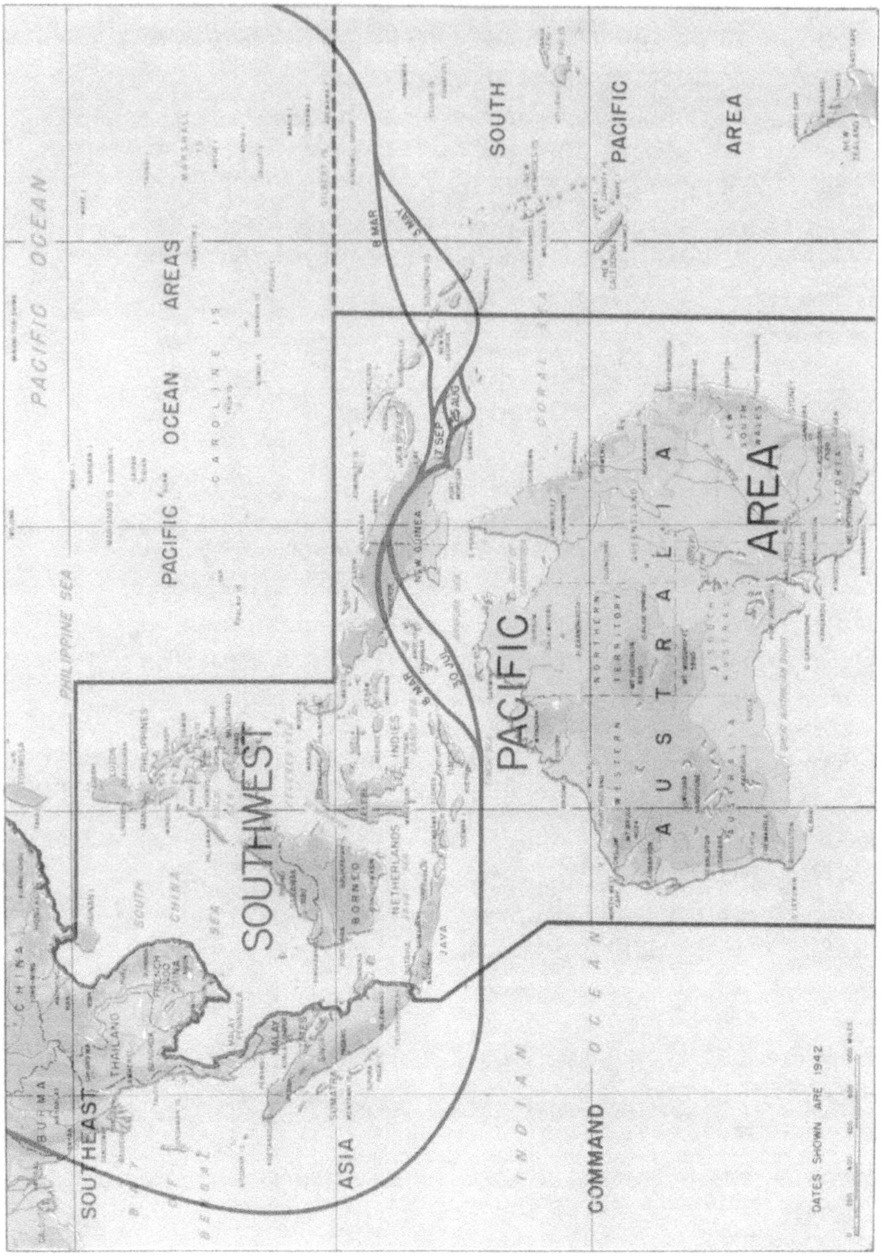

PLATE NO 10

The Boundaries of the Southwest Pacific Area and the Extent of the Japanese Advance

Asst. Chief of Staff, G–3:
Brig. Gen. Stephen J. Chamberlin
Asst. Chief of Staff, G–4:
Col. Lester J. Whitlock, GSC
Chief Signal Officer:
Brig. Gen. Spencer B. Akin
Antiaircraft Officer:
Brig. Gen. William F. Marquat
Chief Engineer:
Brig. Gen. Hugh J. Casey
Adjutant General:
Col. Burdette M. Fitch, AGD
Public Relations Officer:
Col. LeGrande A. Diller, Inf.

The Organization of Forces

The Allied forces in the Southwest Pacific Area were formed into three components for the purposes of operational control. The ground troops, designated Allied Land Forces, were placed under the command of Gen. Sir Thomas A. Blamey, Commander-in-Chief of the Australian Army. They consisted of several territorial divisions in various stages of development and training; the command echelons were represented by the Australian First and Second Armies, and III Corps; the combat elements consisted of the 6th Division (less two brigades diverted to Ceylon), the 7th Division (both divisions having been returned from the Middle East), and the U. S. 41st Division.[11] In addition, the 32nd Division was due to reach Australia from the United States in May.[12]

Command of United States Army Air Forces tactical and service units, together with operational control of the combat elements of the Royal Australian and Netherlands East Indies Army Air Forces, was assigned to Allied Air Forces under General Brett. American air units ultimately to be allotted to the Southwest Pacific Area included 2 heavy bomber groups, 2 medium bomber groups (less 2 squadrons), and 3 fighter groups. The Australian elements totaled 17 squadrons. One partially equipped Dutch bomber squadron was also attached to Allied Air Forces.[13]

The naval elements of the United States in the Southwest Pacific Area were assigned to Allied Naval Forces under the command of Admiral Leary. Operational control of the combat sections of the Royal Australian and Royal Netherlands Navies was also included within Admiral Leary's command. The principal units of Allied Naval Forces consisted of 2 Australian heavy cruisers, 1 Australian light cruiser and 1 United States heavy cruiser, supplemented by a number of destroy-

---

11 Radio No. AG 381 from MacArthur to AGWAR (for Marshall), 20 Apr 42, WD (28) C/S. Of the 11 divisions of Australian forces, only about 2 divisions and 1 brigade could be considered available for operations. General MacArthur reported: "The Australian Army has 1 division and 1 division less 2 brigades which have returned from the Middle East which are...effective troops. One additional division in the home force is approaching combat condition. The remainder of the Army is composed of a militia in a very indifferent state of training and equipment which can be prepared for combat only by prolonged and intensive efforts...." CINCSWPA Radio No. 558 to WARCOS, 1 May 42, AG GHQ 384 No. 1 (S).

12 GHQ General Order No. 1, 18 Apr 42, AG GHQ 300.4 GO (S); WARCOS Radio No. 1167 to CG USAFIA, 8 Apr 42, C/S GHQ WD 10 (S). Forces of the Royal Netherlands East Indies Army in Australia then consisted chiefly of one infantry company which was assigned to Allied Land Forces in May.

13 GHQ General Order No. 1, 18 Apr 42, AG GHQ 300.4 GO (S). WARCOS Radio No. 1188 to CG USAFFE, 9 Apr 42, C/S GHQ WD 10A (S). CINCSWPA Radio to WARCOS, 24 Apr 42, AG 5AF 322.99 Dutch Units (S). CINCSWPA Radio No. 558 to WARCOS, 1 May 42, AG GHQ 384 No. 1 (S). It would take many months for this allocation to become effective. See n. 18.

ers, submarines, and escort and auxiliary craft.[14]

Responsibility for the administration and supply of their own units remained with the Australian branches of the services and with the units of the U.S. Navy and the Royal Netherlands Navy assigned to General MacArthur's control. The administration of Netherlands Army and Air Force elements was handled through national channels, but logistical support was provided by American agencies. Administration and supply of the U. S. Ground and Air Forces, except for certain activities charged to the Air Force, were the responsibility of United States Army Forces in Australia, operating in accordance with policies set forth by General Headquarters. General Barnes, in command of United States Army Forces in Australia, was responsible for all U. S. Army Forces (other than air corps elements) except for operational control of the units assigned to Allied Land Forces.[15]

Decision to Take the Offensive

The immediate and imperative problem which confronted General MacArthur was the defense of Australia. The forces available, however, were completely inadequate to cover such an extensive coastline. The Australian Chiefs of Staff would be virtually compelled to yield the northern part of the continent to the Japanese should they attempt an invasion; they recommended that the defense be made in the Brisbane area and had disposed their available forces accordingly. (Plate No. 11) The Second Army, which had approximately two divisions under its control, although primarily a training command, was assigned the tactical mission of providing local protection for Melbourne. The First Army, with an allocation plan of seven training divisions, was directed to defend the coast from Brisbane to the Victoria border, inclusive, and Townsville with a garrison to be built up to one division. Defense by major units north of Townsville was not contemplated because the lack of communication facilities would not permit effective maneuver. Sufficient forces to secure Fremantle and Darwin against determined enemy assault were not available. The principal concentration areas of the air force were Melbourne-Sydney, Brisbane, Townsville-Cloncurry, and the main naval bases at Sydney and Brisbane.[16]

---

14 GHQ General Order No. 1, 18 Apr 42, AG GHQ 300.4 GO (S). Samuel Eliot Morison, *History of United States Naval Operations in World War II*, Vol. III, "The Rising Sun in the Pacific, 1931—April 1942" (Boston, 1948), p. 261. and Vol. IV, "*Coral Sea, Midway and Submarine Actions, May 1942—August 1942*" (Boston, 1950), p. 15. General MacArthur repeatedly urged that an aircraft carrier be provided, "even of the smallest type", to give these naval units offensive power. He stated that "the surface element of our naval force is unbalanced because of lack of integral air units, in consequence of which its value as a striking force is nullified, reducing it to the execution of minor missions." If the necessary carrier protection were provided, it "could transform the force from an ineffective element into a powerful offensive weapon.... The enemy's lines are partially uncovered and carelessly defended due to the fact that he has been practically unchallenged. This opportunity will not continue indefinitely; when he consolidates his position, the situation will be much more difficult. I consider it a waste of our potentialities to operate such an unbalanced force at the point of immediate contact." CINCSWPA Radio No. 453 to WARCOS, 24 Apr 42, WD 35, C/S GHQ (S).

15 JCS Directive to CINCSWPA, 30 Mar 42, G-3, GHQ JCS/CCS 42-43 (S). *Report of Organization and Activities, United States Army Forces in Australia*, AG GHQ 314.7, USAFIA. CINCSWPA Radio to WARCOS, 24 Apr 42, AG 5AF 322.99 Dutch Units (S).

16 GHQ (Aust) Opn Instr No. 1, 10 Apr 42, and No. 3, 12 Apr 42, G-3, GHQ, SWPA *Journal* and G-3, GHQ, SWPA Admin 370 (S). AHQ Opn Instr No. 50, 9 Apr 42, G-3, GHQ, SWPA Admin 370 (S).

General MacArthur felt strongly that passive defense was strategically unsound. He decided to move forward more then a thousand miles into eastern Papua and beat the Japanese to the punch. By making the first move, he could force them to fight on his terms—across the barrier of the Owen Stanley Range.

This range of mountains, almost 14,000 feet high, covered with dense jungle, formed a natural barrier which ran the entire length of eastern New Guinea. Winding foot trails were the only means of crossing the range. Port Moresby, to its rear, was well protected and could be used as a supply and air base. The only other feasible approach by which the enemy could attack was the sea route through Milne Bay at the eastern tip of New Guinea. Such an attack might be blocked if adequate air bases could be constructed in time. General MacArthur's problem was to move his center of gravity forward fifteen hundred miles and secure this line before the enemy could seize it. The operation involved almost insuperable obstacles of time and space. It necessitated the construction of airfields for fighter planes and bases for heavy bombers, first within range of Papua and then on Papua itself. It involved provision for supply and reinforcement of advanced areas from rear bases in Australia which were in large part merely ports for the reshipping of materiel from the distant west coast of the United States. It necessitated the exploration and mapping of unfamiliar areas, the charting of unknown waters, and demanded friendly co-operation from the New Guinea natives. It required the training of our forces for combat against an experienced and numerically superior enemy. And it risked carrying out these moves by sea into areas dominated by the unchallenged naval power of the Japanese.

General MacArthur's confidence in the ultimate success of his plan was not shared in all quarters; many believed that the difficulties made it impracticable. It is a credit to the officers and men involved, however, that they justified the confidence placed in them by their Commander and that they successfully overcame these difficulties. The results achieved completely vindicated his judgment. This bold and imaginative decision was one of the most crucial and decisive of the war and the final successful culmination of the Papuan operations undoubtedly saved Australia.[17]

None of the three elements of his command —naval, air, or ground—was considered by General MacArthur to be adequate to carry out the missions assigned. The naval force

17 "In the spring of 1942 the Japanese did not think that General MacArthur would establish himself in New Guinea and defend Australia from that position," said Captain Toshikazu Ohmae, IJN, Senior Staff Officer, Southeast Area Fleet at Rabaul, June 1942—December 1943. "They also did not believe that he would be able to use New Guinea as a base for offensive operations against them. The Japanese felt that General MacArthur could not establish himself in Port Moresby because he did not have sufficient forces to maintain himself there and because the Japanese Navy was confident that it could control the Coral Sea and keep him out of New Guinea. In view of the successful air attacks against Darwin and Townsville, the Japanese reasoned that General MacArthur's forces were weak or they would have staved off the attacks. Such was the attitude of the Japanese Army and Navy High Command before the battle of Midway." This opinion was corroborated by other Japanese commanders who fought in the New Guinea area. For example, Colonel Jinmatsu Morifuji, Eighth Area Army Staff Officer, stated: "The Japanese did not anticipate that General MacArthur would attempt to defend Australia by establishing his forces in New Guinea. For this reason the Japanese did not stress the importance of New Guinea at the beginning of the campaign." Interrogation Files, G–2 Historical Section, GHQ, FEC.

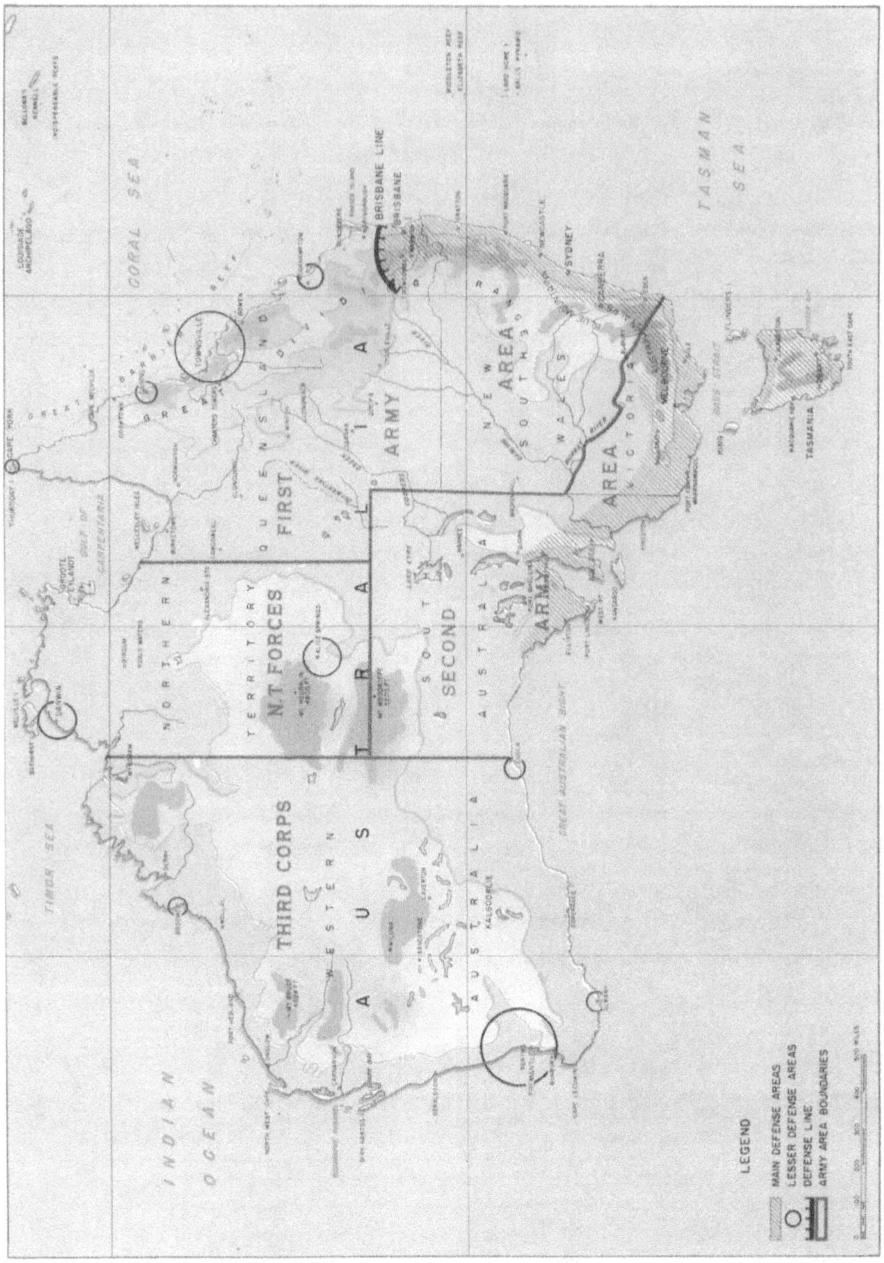

PLATE NO. 11

The Main Australian Defense Areas

was small and unbalanced. It lacked direct air support because of the absence of carriers and was therefore suitable " only for operations of a minor and subsidiary nature." The Royal Australian Air Force would require many months for its development. The organization and training of the American air component was below the required standard; months of intensive effort would be needed for it to reach a satisfactory condition.[18] The ground troops, too, were considered inadequate for the tremendous task ahead; not only were they too few in number, but they lacked the equipment and the strenuous training necessary for combat.[19]

The Japanese, on the other hand, had many veteran troops, well supplied and equipped, with which to continue their offensive. They pressed on into the South Pacific in an attempt to disrupt the sea lanes between Australia and the United States. The offensive in the Solomons began on 30 March with landings on Buka Island and on the northern tip of Bougainville Island. A simultaneous landing was made in the Shortland Islands just south of Bougainville. Kieta was occupied the following day and Buin was taken on 7 April, completing the conquest of the island. New Georgia was next in the line of enemy advance and fell to the Japanese on 16 April. Tulagi, on Florida Island, was taken on 3 May.

The continuing struggle on Luzon and at other points in the Philippines, which had forced the Japanese to retain large numbers of troops in those areas, was slowly drawing

---

18 General Brett, Allied Air Force Commander in Australia, commented on these problems as follows: "We had flown what was left of our planes out of Java, just as Brereton had flown his out of the Philippines. I took over as commander of the American forces. The situation was, to put it mildly, muddled and unhappy. Australia's defenses were weak, and Australia expected an invasion. There are no better fighting men in the world than the hard-bitten soldiers of the island continent, but there were too few of them. The Royal Australian Air Force was equipped with almost obsolete planes and was lacking in engines and spare parts, as well as personnel. We had only one American infantry division, and that was incompletely trained. When Gen. MacArthur arrived, he was extremely disappointed in what he found. He had not wanted to leave the Philippines.... However, his better judgment prevailed, influenced, possibly, by reports that a great American army was being gathered in Australia for him to lead. It did not take long for him to find out how erroneous these reports had been. There was no great army, and the air force consisted of a few battered planes, and combat-weary men....

I knew MacArthur was thoroughly dissatisfied with what he was getting, and he could not be blamed for that.... We...were fighting in the air over New Guinea, on a starvation ration of planes and ammunition and men.... We were working sixteen and eighteen hours a day.... The airmen found it difficult to understand why their country, the greatest industrial nation on earth, could not give them the tools with which to fight. What seemed almost as bad was the fact that our country seemed more concerned with the German phase of the global conflict than with the Japanese. There was only a comparative handful of Americans in the Pacific theater. We didn't have much to fight with, but we were hopeful that plenty would come through sooner or later. But, nothing much came through in those dreary months, and this unescapable fact, as well as the danger and monotony and bad living conditions, had our boys very badly down.... On my way back Stateside, everywhere I went I saw bombers and fighters stacked up waiting to move to Australia. Many had been waiting for a long time.... Our effectiveness was curtailed, our losses higher than they should have been, because those men and planes were held back. I was compelled to send into combat fighter pilots with less than ten hours' experience on the type of plane assigned them, when there were available men with hundreds of hours of flying time in the same type of aircraft...." Brett, Op. Cit. pp. 26, 27, 139, 149.

19 CINCSWPA Radio No. 588 to WARCOS, 1 May 42, AG GHQ 384 No. 1 (S).

to a close. The inevitable fall of Bataan came on 9 April and the stubborn defense of Corregidor, which was maintained for almost another month, ended on 6 May when General Wainwright was forced to surrender his command. With the collapse of major resistance in the Philippines, many of the enemy forces in the islands were released for new drives in other directions.[20] General MacArthur analyzed the uses to which the available enemy forces might be put in his radio of 8 May 1942 in reply to a query from President Roosevelt:[21]

The fall of Corregidor and the collapse of resistance in the Philippines, with the defeat in Burma, brings about a new situation in this theater. At least two enemy divisions and all the air force in the Philippines will be released for other missions. Japanese troops in Malaya and the Netherlands East Indies are susceptible of being regrouped for an offensive effort elsewhere since large garrisons will not be required because of the complacency of the native population. The Japanese Navy is as yet unchallenged and is disposed for further offensive effort. A preliminary move is now under way probably initially against New Guinea and the line of communications between the United States and Australia.[22] *The series of events releases an enormously dangerous enemy potential in the Western Pacific. That the situation will remain static is most improbable. I am of the opinion that the Japanese will not undertake large operations against India at this time. That area is undoubtedly within the scope of their military ambitions but it would be strategically advisable for them to defer it until a later date. On the other hand, the enemy advance toward the south has been supported by the establishment of a series of bases while his left is covered from the Mandated Islands. He is thus*

---

20 In order to delay the Japanese even longer, General MacArthur had planned to continue resistance as long as possible in the southern islands of the Philippines after the fall of Bataan and Corregidor. His plan is illustrated by a radio sent to General Sharp, CG, Mindanao Force on 9 May: "Since his surrender, orders emanating from General Wainwright are no longer valid. Insofar as possible, separate your force into small elements and initiate guerrilla operations. You, of course, have full authority to make any decisions required by immediate emergency. Keep in communication with me as much as possible. You are a gallant and resourceful Commander and I am proud of what you have done."

21 CINCSWPA Radio No. 176 to WARCOS, 8 May 42, WD C/S (S).

22 General MacArthur's estimate of the strategic situation was correct; the Japanese were making plans to complete the isolation of Australia. "In the spring of 1942, the Japanese Navy, eager to capitalize on Japan's initial success in the war, hoped to invade Australia. It was thought that such key areas as Darwin in the north and Townsville, Brisbane, and Sydney on the east coast should be occupied," said Captain Ohmae. "The Navy was responsible for defending New Guinea, New Britain and the Solomons, so Australia figured heavily in its plans.... The Navy realized that Australia would become not only the base from which counterattacks would be launched against Japanese forces, but the steppingstone for an invasion of Japan itself.... By invading Australia the supply of war materials, particularly airplanes, gas and oil which had already begun to flow from the United States would be stopped. The Navy suggested the idea of invading Australia to the Army in March 1942. The Army estimated that it would require at least ten or twelve divisions to carry out such a large-scale operation. The Army, however, felt that it could not adopt the Navy's suggestion because it did not wish to move that many troops from Manchuria and other occupied areas at that time. Adequate transportation and supplies were also lacking. The Navy then proposed a countermeasure because it was determined to isolate Australia and prevent the shipment of American war materials. A master plan was accordingly prepared during April and May 1942 which provided for the occupation of Fiji, Samoa and New Caledonia. Submarine and air bases were to be set up to cut the supply lines from America, isolate Australia, and force her out of the war.... At the insistence of Adm. Isoroku Yamamoto, Commander-in-Chief of the Combined Fleet, however, the invasion of Midway received priority and as a result the plan to occupy New Caledonia, Fiji, and Samoa was postponed." Interrogation Files, G-2 Historical Section, GHQ, FEC.

prepared to continue in that direction. Moreover, operations in these waters will permit the regrouping of his naval and air forces to meet a threat from the East. Such is not the case in a movement towards India. He must thrust into the Indian Ocean without adequate supporting bases, relinquishing the possibility of concentrating his naval strength in either ocean. The military requirements for a decisive Indian campaign are so heavy that it cannot be undertaken under those conditions. On the other hand, a continuation of his southern movement at this time will give added safety for his eventual move to the west. In view of this situation I deem it of the utmost importance to provide adequate security for Australia and the Pacific Area, thus maintaining a constant frontal defense and a flank threat against further movement to the southward. This should be followed at the earliest possible moment by offensive action or at least by a sufficiently dangerous initial threat of offensive action to affect the enemy plans and dispositions....[23]

The first step in the execution of this conception is the strengthening of the position in this area. At this time there are present all the elements to produce another disaster. If serious enemy pressure were applied against Australia prior to the development of adequate and balanced land, sea, and air forces, the situation would be extremely precarious. The extent of territory to be defended is so vast and the communication facilities are so poor that the enemy, moving freely by water, has a preponderant advantage.... In view of the enemy potentialities I consider it essential for the security of this country that it be reinforced as follows...two aircraft carriers in order to provide a balanced sea force and a reasonable coverage of the adjacent sea areas; an increase from 500 to 1000 front line planes in U. S. Air Forces with an adequate flow of replacement personnel and material to maintain Table of Organization strength; one U. S. Army Corps of three first-class divisions capable of executing a tactical offensive movement. Such a force will give reasonable assurance of a successful defense of Australia and will provide an adequate base for counter offensive action. I cannot too strongly represent that the defensive force here must be built up before hostile direct pressure is applied for it would then be too late. We must anticipate the future or we will find ourselves once more completely outnumbered....[24]

---

23 General MacArthur's insistence on offensive action was contrary to Japanese hopes and expectations. They planned to secure their positions, occupy other key areas, and force the Allies into the hopeless Maginot psychology of the defensive. In March 1942, Imperial General Headquarters published the following principles for the conduct of future operations: "Mopping up in the areas already occupied will be completed as soon as possible, and our combat troops in those areas will become garrison troops.... The strategic initiative obtained from the operations at the very beginning of the war will be maintained. Positions will be established which will withstand an extended period of enemy attacks and which will force the United States and British forces into a negative defensive position. In order then to hasten the termination of the war, the necessary operations will be prepared and carried out at key points along the outer perimeter of the occupied areas." Japanese First Demobilization Bureau Report, *Southeast Area Operations Record,* Part II, "Seventeenth Army Operations," Vol I, p. 2, G–2 Historical Section, GHQ, FEC.

24 The strategic importance of Australia and the Southwest Pacific Area was fully realized by Japanese Imperial General Headquarters as revealed in their estimate of the situation in October 1942: "The South Pacific seems to be the most likely position from which the enemy may carry out an offensive counterattack. The reason for this is that Australia and its surrounding islands are connected to the United States by a chain of islands, thus making the South Pacific a very potential position. The enemy will be able to threaten our command of the sea in the West Pacific from this area, recapture our southern occupational area, and occupy our South Sea Islands. Air raids can be carried out easily against our areas of important resources. After a thorough study of the situation we have decided that the enemy will probably attack this South Pacific area and will enforce regular counterattack measures after deploying his strength. The decisive battle between Japan and the United States will be the occupation of this strategic region." Japanese First Demobilization Bureau Report, *Imperial General Headquarters Army High Command Record,* p. 64, G–2 Historical Section, GHQ, FEC.

Prime Minister Curtin, through the Australian representatives in Washington and London, also pressed for increased forces to safeguard his country. The Commonwealth Government had approved a program for expanding the Royal Australian Air Force to seventy-three squadrons, the aircraft for which would have to be provided mainly from American and British resources. The Government agreed that the 9th Division remain temporarily in the Middle East, but urged that two British divisions be sent to Australia until such time as the 9th Division and the two brigades at Ceylon could be returned. The Prime Minister also sought to obtain a British aircraft carrier and other naval support.[25]

The additional forces requested were not forthcoming. All United States and British carriers were employed elsewhere and for the present no substantial increase in the air forces was assured. No additional United States divisions were allocated, and the British planned to send ground forces to Australia only in case of dire necessity. The execution of the tasks assigned to the Southwest Pacific Area rested with the limited forces at hand.[26]

The Magnitude of the Task

The magnitude of the Southwest Pacific Theater can best be appreciated against a background of comparative geographical distances. If a map of the United States is superimposed on one of the Southwest Pacific, the continental area of the United States will fit roughly between Australia and the Philippines. (Plate No. 12) The distance from New York to San Francisco is approximately the same as from Rabaul to Java. With Miami located at the coastal city of Townsville in Australia, Hollandia will fall on Milwaukee. Biak will be south of Minneapolis and Morotai will be near Bismarck, North Dakota. Leyte will be across the Canadian border approximately 300 miles northwest of Regina and Manila will fall some 700 miles north of the United States—Canadian border in Alberta Province.

The inherent tactical and strategical problems of a land and sea advance from Australia and Papua, directed from Headquarters at Port Moresby and Brisbane, through New Guinea and the Moluccas to Manila, are comparable to those of a hypothetical American Headquarters stationed in the Carribean Sea 750 miles southeast of Miami, with its advance echelon in the vicinity of Roanoke, Virginia, charged with air strikes on Boston, New York, and Washington to points west of Denver, and an advance by land and sea from Norfolk (Milne Bay) via Milwaukee (Hollandia), Minneapolis (Biak), Bismarck (Morotai), and northwest across the middle of Saskatchewan, Canada (Leyte) to a point northwest of Lake Claire in northern Alberta (Manila).

Against this geographical background, it is evident that the logistical difficulties of the Southwest Pacific Theater in the conduct of the war were tremendous. Not only was the line of communications from the United States to the scene of operations one of the longest the world has ever seen, but the entire route was by water at a time when the Japanese Navy was undefeated and roaming the Pacific almost at will. The time factor

25 "Appreciation by Australian Chiefs of Staff," 27 Feb 42, G-3, GHQ, SWPA *Journal* prior to 5 Apr 42, (MS). PM Curtin Ltr to CINCSWPA, 28 Apr 42, AG GHQ 381 Aust Req B (MS). CINCSWPA Ltr to PM Curtin, 10 May 42, AG GHQ 381 Aust Req B (S).

26 WARCOS Radios Nos. 1188 and 1499 to CINCSWPA, 9 and 24 Apr 42, C/S GHQ WD 10A, 35 et seq (S). PM Curtin Ltr to CINCSWPA, 4 May 42, C/S GHQ Aust 14. PM Curtin Ltr to CINCSWPA, 30 May 42, AG GHQ 381 Aust Req B (MS).

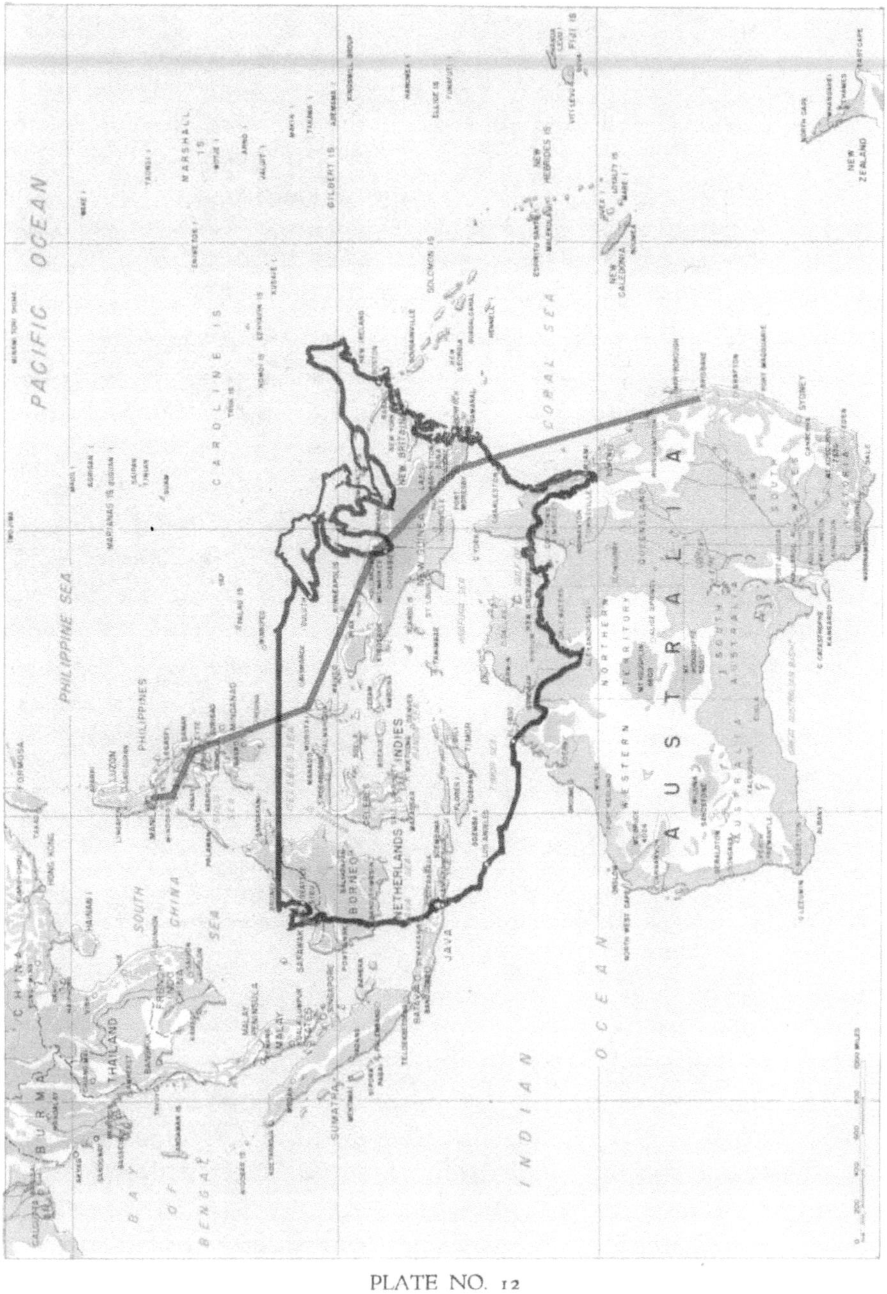

PLATE NO. 12

The United States Superimposed on the Southwest Pacific Area alone demanded more than usual foresight in logistic planning and required the application of methods not taught in any of the American Service Schools.

The shortage of water transportation for supplies and equipment, as well as for troops, was probably the most difficult problem of all. There was not sufficient shipping available to mount operations adequately and then to support the troops in widely dispersed locations. Allowances of supplies and equipment accompanying troops into battle and also the amounts to be sent later for their maintenance had to be curtailed—at times to the danger point. Factors used for the computation of these requirements in the Southwest Pacific were less than those used in any other theater or area in the world; at times they were less than half the usual figure.

Tremendous logistic risks were taken in making decisions for planning and launching operations. These were "forced risks" which had to be taken if the war was to be pursued aggressively and if the Allies were to seize the initiative and hold it. Shortages of every description necessitated "crisis management" to meet the needs of each individual operation. Other important logistic features which adversely affected the campaigns and operations in the Southwest Pacific Area included shortages of combat equipment and insufficient service troops. There was also a scarcity of equipment and troops for construction of ports, bases, and air facilities which often had to be created out of virgin jungle before successive operations could be launched. In certain decisive operations, there was even a lack of critical calibers of ammunition.

In addition to all these difficulties there was New Guinea itself—as tough and tenacious an enemy as the Japanese. Few areas in the world presented such a formidable variety of terrain obstacles to military operations. (Plate No. 13) The great mountain ranges with their high peaks and deep gorges, the dense jungles which cover almost all of the huge island, the reeking nipa and mangrove swamps—"a stinking jumble of twisted, slime-covered roots and muddy 'soup'"—the hazardous jungle trails, the vast patches of kunai grass, with its sharp-edged blades growing to a height of six or seven feet, the swollen streams, the ever-present mud, the dangerous off-shore reefs, most of them uncharted, the poor harbors—these terrain characteristics exerted a constant and adverse influence on troops and military tactics.[27]

The problems of climate and health were no less severe. The penetrating, energy-sapping heat was accompanied by intense humidity and frequent torrential rains that defy description. Health conditions were among the worst in the world. The incidence of malaria could only be reduced by the most rigid and irksome discipline and even then the dreaded disease took a heavy toll. Dengue fever was common while the deadly blackwater fever, though not so prevalent, was no less an adversary. Bacillary and amoebic dysentery were both forbidding possibilities, and tropical ulcers, easily formed from the slightest scratch, were

27 The psychological effect of the New Guinea terrain on combat troops was well described by Lt. Gen. Robert L. Eichelberger, Commanding General of the Buna Forces: "The psychological factors resulting from the terrain were...tremendously important. After a man had lain for days in a wet slit trench, or in the swamp, his physical stamina was reduced materially. This reduction served to make him extremely nervous and to attribute to the unfamiliar noises of the jungle, spectres of Japanese activities. These reactions preyed on his mind until he was reduced often to a pitifully abject state, incapable of aggressive action." *Report of the Commanding General, Buna Forces on the Buna Campaign, Dec 1, 1942—Jan 25, 1943,* p. 64.

difficult to cure. Scrub typhus, ringworm, hookworm, and yaws all awaited the careless soldier. Millions of insects abounded everywhere. Clouds of mosquitoes, flies, leeches, chiggers, ants, fleas, and other parasites pestered man night and day. Disease was an unrelenting foe. New Guinea provided a background in which almost every threat of nature combined with the sudden and unforeseeable dangers of modern war to provide a microcosm of the vast struggle in the Southwest Pacific.

New Guinea was the last obstacle between the Japanese and Australia, and lodgements had already been made by enemy forces at points along the northern New Guinea coast. The Japanese held the initiative and their command of the sea enabled them to concentrate on a particular objective and overwhelm the defenders with superior forces. Countermeasures had to be devised before the Japanese should choose their next point of attack and strike at the vulnerable Allied defenses.

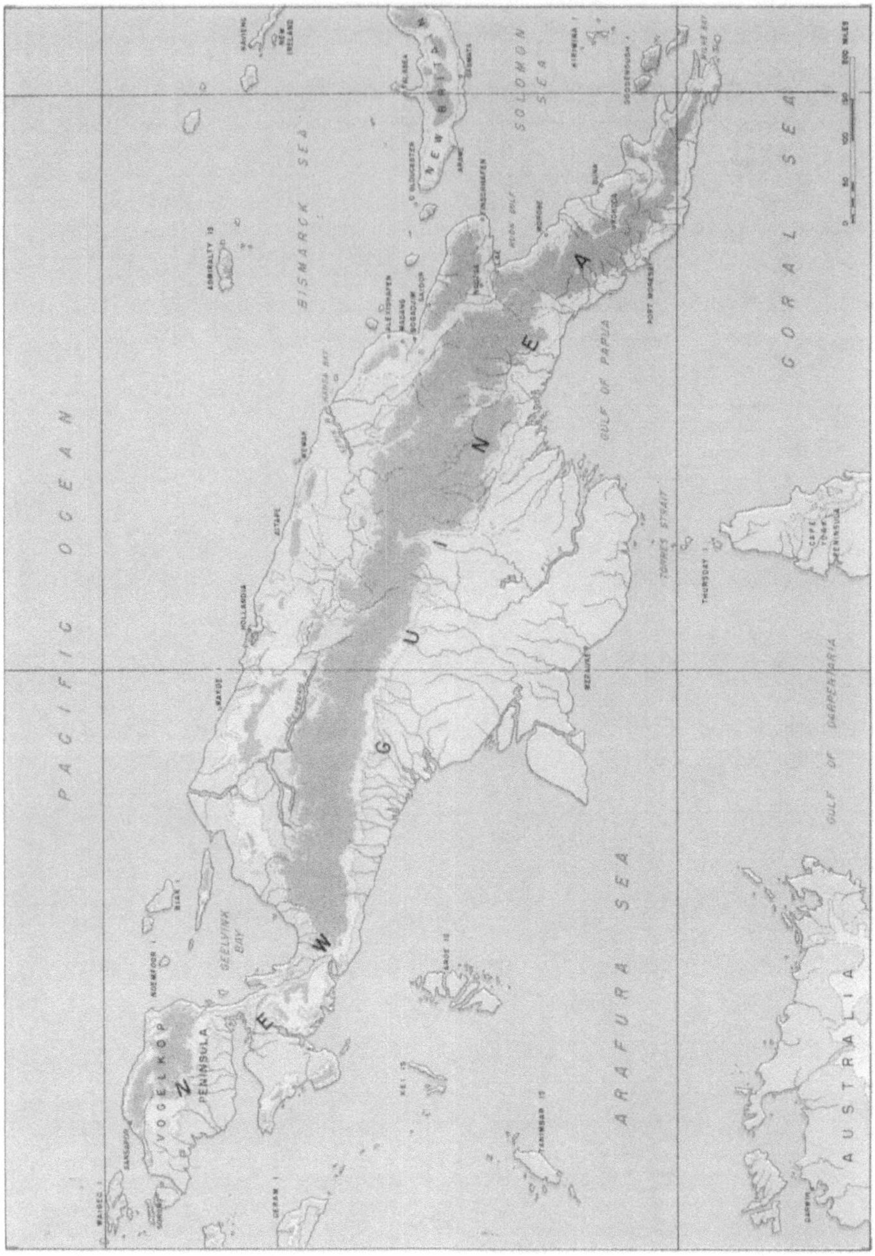

PLATE NO. 13
Relief Map of New Guinea

CHAPTER III
HALTING THE JAPANESE

Concept of the Offensive

Following the capture of Lae and Salamaua by the Japanese, General MacArthur unhesitatingly discarded the prevailing concept of a purely passive defense of Australia and decided to achieve his mission by offensive operations in New Guinea.

The meager resources at his disposal necessarily limited the scope of such actions, but by striking boldly at the enemy's weak points and avoiding frontal engagements in force, he expected to upset the enemy's timetable and frustrate his plan to occupy southeastern New Guinea. Relying on tactical maneuver, the exploitation of terrain factors, and an increasing flow of intelligence, the General planned to operate with maximum striking power in order to reduce the loss of life and materiel to a minimum.

When the Japanese temporarily halted their advance toward Australia, they controlled the entire chain of islands which lie across its northern sea approaches, with the single exception of southeastern New Guinea. These forward positions formed a front three thousand miles long, extending from Java to the Solomons. Interior sea and air lines of communication, radiating southward from Japan through intermediate supporting bases from Singapore to Formosa and to Truk, permitted easy concentrations for further offensive moves. The western and northern coasts of Australia to the Torres Straits were directly exposed to Japanese operations from the Netherlands East Indies. In addition to these geographic advantages, the Japanese held the initiative as to the point of attack.

In April reinforcements dispatched to III Corps in Western Australia and to Northern Territory Force at Darwin raised each to division strength, plus certain auxiliaries. Small air force elements were maintained in each locality and airdrome facilities were expanded in order to accommodate major striking forces for projected operations as the center of gravity of the Allied air forces moved forward. Naval protection at this time was limited to two light cruisers and a number of destroyers and submarines based at Fremantle.

Strengthening Port Moresby

After the Japanese occupation of Rabaul and the New Guinea coast, the most pressing danger facing General MacArthur was a potential enemy thrust to Port Moresby in preparation for a drive against Australia. Consistent with his strategy of defending Australia in New Guinea, General MacArthur decided to strengthen immediately the Allied position in Papua and develop Port Moresby as a major air and land base.

Port Moresby, flanked on the east by Milne Bay and protected on the north by the formidable 14,000-foot Owen Stanley Range, was well placed strategically. Except for its vulnerability to amphibious assault, it was ideally situated for future air operations against enemy positions to the north and the northeast. A strong base in this region would not only serve to protect Australia from hostile raids but also provide the starting point on the road back to the Philippines—the heart of the enemy's supply and communications network in his newly conquered empire.

Constant enemy air raids and the lack of adequate airdrome facilities made Port Moresby virtually untenable for big bombers except as a forward fueling point. The nearest supporting airfields were at Townsville, 700 miles to the south. To strengthen his Port Moresby base, General MacArthur planned to move his bombers into more forward positions. A program of airdrome construction was undertaken in Australia, first expanding installations in the Townsville—Cloncurry area, then building north along the York Peninsula. This development was in its early stages when, at the end of April, the presence of a large Japanese naval force in the Mandated Islands and concentrations of aircraft in the New Britain—Solomons region indicated that the enemy was about to renew his offensive. To meet this threat, General MacArthur ordered the Allied Air Forces to intensify their reconnaissance, to assemble a maximum striking force on the Townsville—Cloncurry airfields, and to bomb enemy positions. During the first week in May the important stronghold of Rabaul was attacked repeatedly. Missions were also sent against Lae, Buka and Deboyne Islands, and enemy convoys in adjacent waters. Garrison commanders in northeastern Australia and at Port Moresby were alerted to the possibility of landings in their respective areas.[1] Allied Naval Forces (SWPA) dispatched three cruisers to augment the Pacific fleet units which already included the aircraft carriers *Lexington* and *Yorktown*. Every measure was taken to check the threatened enemy advance.

Battle of the Coral Sea

The Allied countermeasures were soon justified. Eight days after General MacArthur had issued orders for the all-out alert, the Japanese made their first move. On 3 May an enemy naval force made an unopposed landing at Tulagi Harbor in the southern Solomons.[2] The next morning, however, the *Yorktown*, which had been refueling at Espiritu Santo, made a high speed run north and launched its planes in three strikes against harbor shipping, crippling a destroyer and three or four small boats. It then returned southward to rejoin the *Lexington*.

While the occupation of Tulagi was being carried out, the main force of the Japanese Fourth Fleet was completing its final preparations for a full scale amphibious invasion of Port Moresby. The decision to take Port Moresby had been made on 29 January and specific orders to the 5th Carrier Division were issued in the middle of April. The landings were to take place on 10 May.[3]

On 1 May, a strong fleet of Japanese warships, including the carriers *Zuikaku* and *Sho-*

1 GHQ, SWPA, Opn Instr No. 2, 25 April 42, G–3 Admin, 370 (S) and CINCSWPA Radio No. 719 to WARCOS, 13 May 42, C/S GHQ, WD No. 68 (S).

2 The Japanese landings were carried out under orders of the Commander in Chief of the Fourth Fleet, Vice Adm. Shigeyoshi Inouye, who wanted an advanced seaplane base established at Tulagi to cover the forthcoming operations against Port Moresby. Japanese Second Demobilization Bureau, *Southeast Area Naval Operations*, Part I, p. 1, G–2 Historical Section, GHQ, FEC.

3 The exact composition of the Japanese Fleet Units participating in this operation may be found in Document No. 18865 F (WDI 56), ATIS, GHQ, SCAP, *Full Translation of the Port Moresby Operation*, Vol 5, 22 May 46.

kaku of the 5th Carrier Division, had left Truk to follow a route through the Mandated Islands around San Cristobal and into the Coral Sea. (Plate No. 14) The main missions of this task force were to protect the Port Moresby landing group and to engage the Allied Fleet which was expected to appear on the scene.

Meanwhile, the Japanese South Seas Detachment, a veteran unit which had already taken Guam, Rabaul, and Salamaua, had been designated as the invasion landing force under Maj. Gen. Tomitaro Horii. This group, loaded on approximately 12 transports, departed from Rabaul on 4 May escorted by 1 light cruiser and 6 destroyers and also supported by the carrier *Shoho* and four heavy cruisers.

Long-range, land-based bombers from General MacArthur's Command were combing the area for the Japanese convoy but failed to locate it either that day or the next.[4] On the 5th, however, Allied intelligence reported that Port Moresby was the main enemy objective and that landings could be expected any time between 5 May and 10 May.[5] B–17's and B–26's of the Southwest Pacific Area stood by on the alert to attack while other planes carried out neutralizing raids to keep enemy land-based air power from participating in the coming battle. It was not until late on the 6th, however, that three B–17's on reconnaissance duty finally located the Japanese invasion force headed for the Jomard Passage and the Louisiade Islands. Rear Adm. Frank J. Fletcher, Commander of the Allied Fleet, dispatching a group of cruisers and destroyers to cover the Jomard Passage, moved north with his carrier force to contact and close with the main enemy fleet.

On the morning of 7 May, army scout planes reported sighting an enemy carrier, which proved to be the *Shoho*, off Misima Island. Ten B–17's immediately were sent to the attack. They were generally unsuccessful in their high level bombing, but were able to start fires on one cruiser and, by throwing the Japanese formation into complete disorder, they caused the carrier to reverse its course. Flyers from the *Lexington* and the *Yorktown* arrived shortly thereafter catching the Japanese in a state of unpreparedness, with but few planes in the air and with their carrier headed away from the wind.[6] Nine bomb hits and four torpedoes sank the *Shoho* within five minutes after the first blow was scored. A second strike at the retiring enemy force was readied but not ordered aloft because the other Japanese carriers had not yet been located.

The undiscovered *Shokaku* and *Zuikaku* were at this time to the northeast busily searching for the United States carriers. Reconnaissance planes from these two enemy ships misdirected a heavy attack against a group of oil tankers, mistaking them for a carrier task force. Admiral Fletcher did not learn of this attack until dusk, too late to take any effective counteraction.

The Japanese learned of the sinking of the *Shoho* as their planes were returning from the attack on the tanker group, and at dusk about twenty-seven bombers and torpedo planes again left the *Shokaku* and *Zuikaku* in another effort to locate and sink the *Lexington* and the *Yorktown.* After a fruitless search of almost 300 miles, the planes were forced to jettison all bombs and torpedoes and swing back to their ships. During the return flight, these planes passed over the United States carriers at night and some landings were actually attempted before the Japanese pilots realized their mistake.

Both sides initiated all-out attacks on 8 May.

---

4 United States Strategic Bombing Survey (USSBS), Naval Analysis Division, *The Campaigns of the Pacific War*, p. 52.

5 GHQ, SWPA, *G–2 Information Bulletin*, Special Bulletin, 5 May 42.

6 Gilbert Cant, *America's Navy in World War II* (Rev. ed., New York, 1944), p. 200.

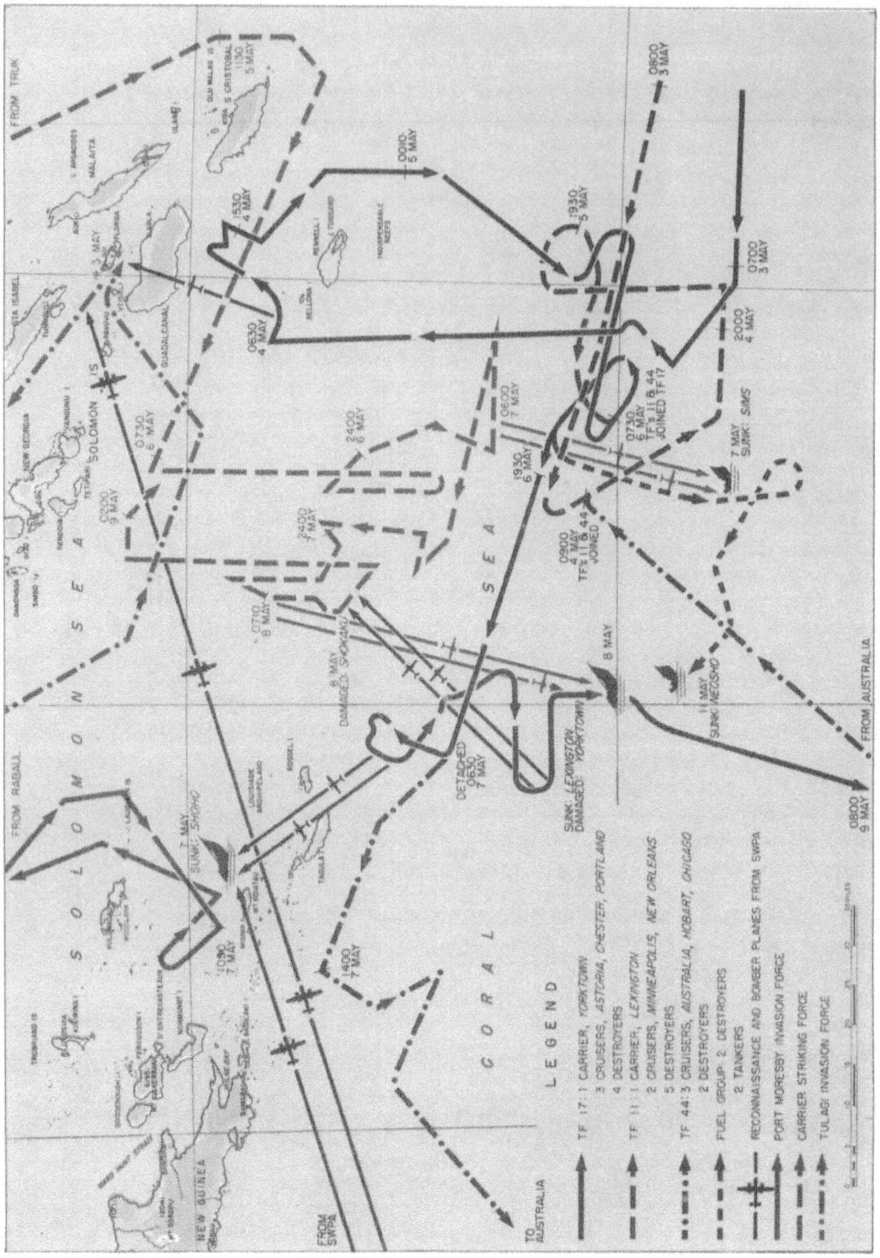

PLATE NO. 14

Battle of the Coral Sea

About mid-morning, United States carrier planes scored three hits on the *Shokaku* which was forced to retire. At the same time that the *Shokaku* was undergoing attack, planes from the *Shokaku* and *Zuikaku* were attacking the *Lexington* and *Yorktown*. Early in the afternoon the *Lexington*, put out of control by enemy attacks, was abandoned and sunk by its own destroyer escorts. The *Yorktown* was damaged but remained operational. This exchange terminated one of the most unusual battles in the annals of naval warfare. Not only was it the first engagement between carrier forces in history but surface ships did not exchange a single shot throughout the entire battle.[7]

In addition to reconnaissance and preparatory raids against enemy air installations, land-based aircraft from the SWPA supported the action of the naval forces by flying some forty-five sorties against the enemy fleet. Bad weather intervened, however, and frustrated all attempts to bomb the crippled *Shokaku*, which succeeded in escaping to the sanctuary of Rabaul.[8]

The Battle of the Coral Sea prevented the Japanese from occupying Port Moresby by sea and temporarily delayed their plans to capture Guadalcanal and occupy the Solomons.[9] The race against time by the Allies had now become a split-second effort to develop the northeastern Australia—New Guinea area. The construction of airdromes at Cairns, Cooktown, Coen, Horn Island, and Port Moresby was pushed rapidly, and small garrisons were provided for the airdromes on the York Peninsula and at Port Moresby. On the other hand, the Allied victory was a purely defensive one. Allied forces in the Southwest Pacific were still unable to launch a major offensive. The Japanese had lost an important battle, but the strategic initiative still remained in their hands.[10]

Kanga Force

Kanga Force was formed in April 1942 to harass and interrupt the development of the Japanese bases at Lae and Salamaua. This force consisted of a small formation of local Europeans known as New Guinea Volunteer Rifles, and a reinforced Australian Independent Company.[11] A part of the force was already at Wau, a prewar gold mining town in the Owen Stanleys, from which trails led to Lae and Salamaua. Additional troops were flown in from Port Moresby during the latter part of May.

At the beginning of May, General MacArthur suggested to General Blamey that ground raids be initiated against Lae and Salamaua to destroy enemy installations and, if possible, to occupy the airfields.[12] If Lae

---

7 Adm. Ernest J. King, USN, *Our Navy at War, A Report to the Secretary of the Navy*, 27 Mar 44, p. 30.

8 CINCSWPA Radio No. 719 to WARCOS, 13 May 42, C/S, GHQ, WD No. 68 (S). AAF DOI, *Brief Appreciation of Coral Sea Battle, 5 to 8 May 42*, G–2 Historical Section, GHQ, FEC.

9 "The loss of the Battle of the Coral Sea affected our strategy in New Guinea to a great extent," said Lt. Col. Shiro Hara, Eighth Area Army Staff Officer, Operations Section. "After the loss, we realized that our offensive assault against Port Moresby by sea was blocked. This made it obvious that if we were to attack our objective it would have to be made overland via the Owen Stanleys." Interrogation Files, G–2 Historical Section, GHQ, FEC.

10 The Japanese assault troops en route to Port Moresby were not affected by the action in the Coral Sea but returned to Rabaul. Japanese strategists by no means abandoned the plan to seize Port Moresby but only decided to postpone the next attempt until additional preparations could be completed.

11 New Guinea Force, Opn Instr No. 7, 23 Apr 42, G–3, GHQ Admin NGF (S).

12 CINCSWPA Ltr to General Blamey, 1 May 42, 385 (Plan 3) G–3, GHQ Admin (S).

and Salamaua could be denied to the enemy for a limited period and the airfields used by Allied pursuit planes, valuable time would be gained for strengthening the defenses at Port Moresby.

Toward the end of May, as Kanga Force was making preparations to launch raids on Lae and Salamaua, scout reports indicated that the Japanese themselves were about to move into the Bulolo Valley against Wau. The Kanga Force Commander, Maj. N. L. Fleay, was committed to the defense of the important air installations at Wau, but with the small force available he was unable to protect the airfields and at the same time carry out offensive operations against Lae and Salamaua. The raids, therefore, were postponed for a later date.

The planned attacks were finally made at the end of June and although some damage and casualties were inflicted on the enemy, no decisive results were obtained. Kanga Force was maintained at Wau after these raids under almost prohibitive conditions to continue the protection of air facilities and to secure the crest of the Owen Stanleys in that area.

Battle of Midway

While the Southwest Pacific Area was thus engaged, another naval battle was being fought in the Central Pacific that considerably altered the general strategic situation. Soon after the retirement of the Japanese from the Coral Sea, a powerful concentration of enemy naval forces in home waters was reported. The United States Pacific Fleet prepared to meet an attack eastward. On 3 June the enemy was sighted southwest of Midway Island. Beginning that afternoon and continuing for the next three days, land-based planes from Midway and carrier aircraft from the *Yorktown, Enterprise,* and *Hornet* repeatedly battered the Japanese ships, sinking four aircraft carriers and one heavy cruiser. One heavy cruiser was severely damaged in addition to two destroyers. American losses were light, consisting of the *Yorktown* and one destroyer. Again, as in the Battle of the Coral Sea, the opposing warships never once sighted each other and not a single direct shot was exchanged. The dominant role of air power in naval warfare was now clearly demonstrated.

This decisive victory restored the balance of naval power in the Pacific and automatically removed the threat to Hawaii and the west coast of the United States. Thereafter, except for the Aleutians where the Japanese had landed on Attu and Kiska Islands, enemy operations were confined to the South and Southwest Pacific Areas.

Milne Bay and Merauke

Whereas the Battle of Midway compelled the Japanese to concentrate their attention on the New Guinea area, the Battle of the Coral Sea had pointed to the strategic importance of Milne Bay. It convinced the Japanese that a base at that location would be invaluable for the air support of any future convoy rounding the southeastern tip of Papua. It also disclosed another line of approach to Port Moresby—the coastal trails along which ground troops could infiltrate to divert Allied strength from its defenses and to reinforce Japanese troops attacking over the Owen Stanley mountains.

General MacArthur was equally aware of the strategic potentialities of Milne Bay. He fully realized not only its importance to the enemy but also its value to his own scheme of offense. While affording a powerful barrier against any Japanese attack, the Owen Stanley mountains also presented an obstacle to Allied advance. An Allied base at Milne Bay would pave the way for a move around the end of New Guinea and along the Papuan coast to Buna and thence to the main objectives of Lae and Salamaua, avoiding the mountains entirely. With Milne Bay in Allied hands, the east flank of Port Moresby would be secured.

Accordingly, in June General MacArthur ordered the secret construction of an air, land, and naval base at Gili Gili located at the head of Milne Bay.[13] The move along the coast to the site of the projected base, known as Operation *Fall River*, was made secretly and skillfully so that the initial execution of the plan was unhampered by enemy opposition. Had the Japanese been aware of General MacArthur's intention to fortify Milne Bay, they would have made every effort to prevent its fulfillment. This was clearly demonstrated by the fury and determination of their counter-attacks as soon as they learned that he had beaten them to that all-important position.

In addition to the cover afforded by the air forces at Darwin and on the York Peninsula, General MacArthur directed the construction of an airdrome at Merauke on the southern coast of Dutch New Guinea as a further protection to his west flank.[14] Although the project had to be temporarily suspended in August because the engineer effort was required elsewhere, a garrison of one infantry company and auxiliaries was kept at Merauke to deny its use to the Japanese.

Kokoda Trail

General MacArthur also anticipated enemy efforts along the Kokoda Trail, the only passable route across the mountains from Buna on the east Papuan coast to Port Moresby. (Plate No. 15) He was convinced that the Japanese intended to employ this route either to assault Port Moresby directly or as a supply base for a possible amphibious force attacking by sea through the Louisiade Archipelago. To forestall such a move, he asked General Blamey to advise him of the plan of New Guinea Force for the protection of Kokoda and the points along the trail.[15] General MacArthur's estimate of the enemy's intention was strikingly correct. Three days after this query to General Blamey, Imperial Japanese Headquarters ordered the commander of the Japanese Seventeenth Army to co-operate with the Navy and " immediately make land attack plans against Port Moresby."[16]

On 21 June, the Australian Maroubra Force,

---

13 GHQ, SWPA, Memo to Comdrs. ALF, AAF, ANF and CG USAFIA, 12 Jun 42, 385 (4) (Fall River), G-3, GHQ Admin (S). Also New Guinea Force, Opn Instr No. 17, 15 Jun 42, G-3, GHQ, SWPA *Journal* (S).

14 GHQ, SWPA, Memo to Commanders AAF, ALF, ANF and CG USAFIA, 22 Jun 42, G-3, GHQ, SWPA *Journal* (S).

15 CINCSWPA Ltr to Comdr. ALF, 9 Jun 42, AG, GHQ, 384, No. 1 (S).

16 To carry on their offensive in the Southwest Pacific after the Battle of the Coral Sea, Imperial General Headquarters rushed plans for the organization of the Seventeenth Army under the command of Lt. Gen. Harukichi Hyakutake. In spite of the Japanese losses at Midway, which postponed other operations in the Southwest Pacific, the Seventeenth Army renewed preparations to seize Port Moresby. On 1 July the Commander of the Seventeenth Army ordered the 15th Independent Engineer Regiment and an infantry battalion of the South Seas Detachment to land at Buna and conduct reconnaissance to determine the feasibility of an overland assault on Port Moresby. On 11 July the Seventeenth Army was given final orders to attack and occupy Port Moresby. General Hyakutake intended to carry out this order by having the South Seas Detachment infiltrate over the Owen Stanleys to Port Moresby while the Aoba Detachment carried out simultaneous amphibious operations. Japanese First Demobilization Bureau, *Southeast Area Operations Record*, Part II, " Seventeenth Army Operations," Vol. I pp. 1-20, G-2 Historical Section, GHQ, FEC. This detailed study of Japanese strategy after the Coral Sea and Midway battles reveals the emphasis placed by the Japanese on isolating and defeating the Allies in the Southwest Pacific.

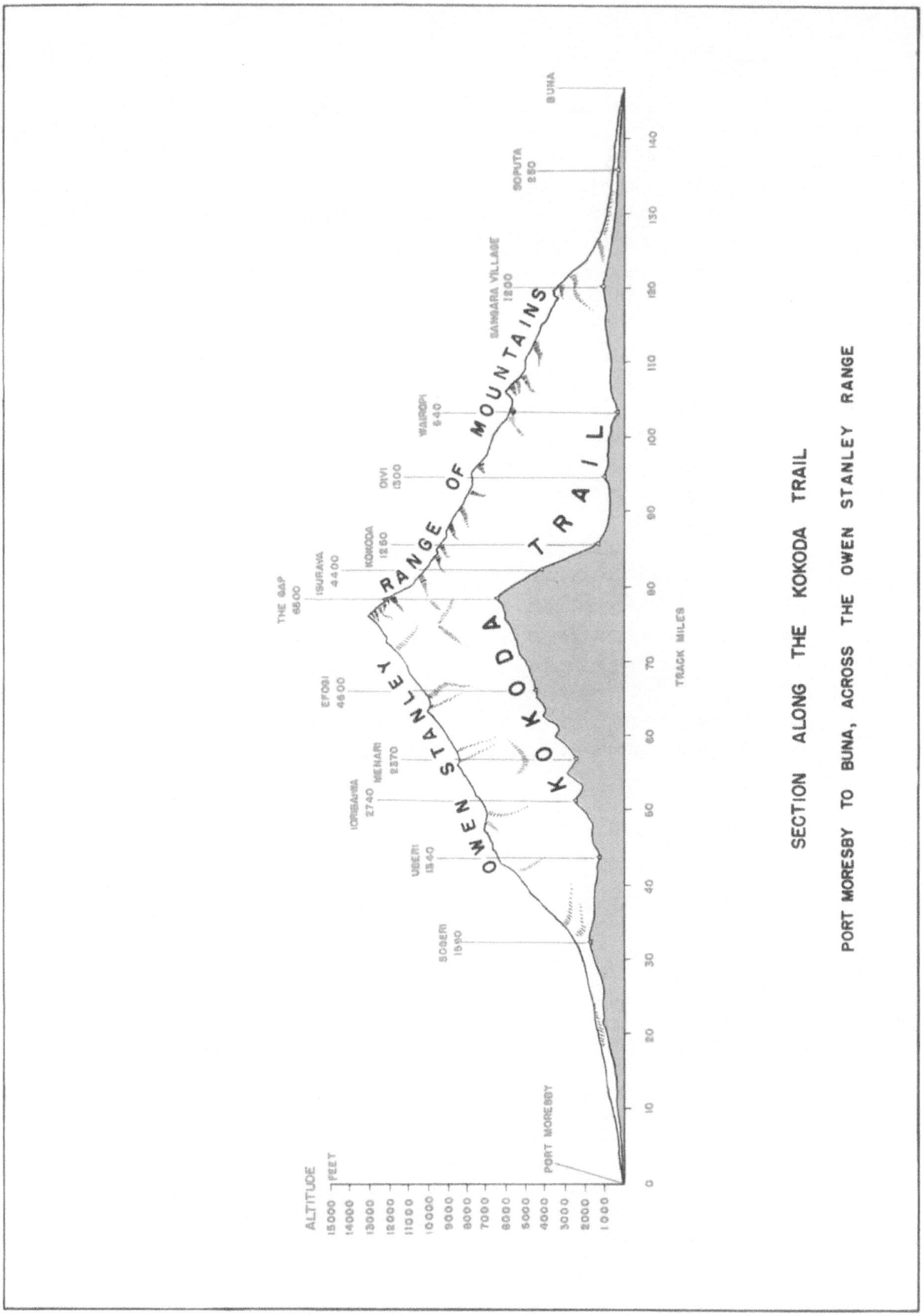

PLATE NO. 15

Section along the Kokoda Trail

consisting of the Papuan Infantry Battalion and a battalion of the 30th Brigade, was organized for concentration at Kokoda on the northern slopes of the Owen Stanleys northeast of Port Moresby. It was directed to prevent an enemy penetration of the mountains along the route of the Kokoda Trail.[17] The Papuan Infantry Battalion, about 280 native troops with white officers, was already north of the range and a company of the brigade started forward overland in early July. The remainder stayed at Port Moresby to assist in the strengthening of the southern terminus of the Kokoda Trail, and to reinforce the advance detachment if necessary.

In addition to defending the passes along the Owen Stanleys, General MacArthur worked out a plan for the occupation of key points along the Kokoda Trail from Port Moresby to Buna. This plan, designated as Operation *Providence*, called for an overland force to proceed from Port Moresby to Buna. Upon arrival at Buna this force would meet a light convoy proceeding up the coast from Milne Bay. The commander of the overland expedition, utilizing the supplies and equipment of this light convoy, would then form a beachhead to cover the landing of a second and main convoy proceeding from Milne Bay. According to this plan, the main convoy would contain the personnel for airdrome construction and garrison units and also necessary supplies.[18]

Intelligence in Operations

General MacArthur's plan to penetrate deeply into the difficult and largely unknown terrain held by the Japanese required special preparation and the collection of new information. New Guinea was a wilderness compared with Western Europe where professional armies in being for over a century had left a rich heritage of military and topographical information. The amount of material on the geography and topography of New Guinea and adjacent areas, however, was inadequate; available hydrographic charts were old and faulty, and data on health and climatic conditions peculiar to these regions were meager. Information concerning the activity of the enemy, his strength and dispositions, his combat methods and current equipment, and his actual relationships with the natives was imperative. Such intelligence would have to be obtained by secret operations behind the enemy lines and from other sources.

General MacArthur directed that special agencies be developed to provide him with this information.[19] On 19 July 1942 the Allied Geographical Section (AGS) was formed to assemble and evaluate all geographic information regarding the Southwest Pacific Area. The primary task of the new section was to prepare and publish reports and locality studies on the areas of immediate tactical interest in

17 New Guinea Force, Opn Instr No. 18, 21 Jun 42, G–3, GHQ, SWPA *Journal* (S).

18 CINCSWPA Ltr to Comdrs. AAF, ALF, ANF and CG USAFIA, 15 Jul 42, G–4, GHQ, Admin (S).

19 The formation and administration of these agencies were the responsibility of the G–2 Section which had begun their organization, in Melbourne, in May. The principal agencies and their fields of coverage were: (1) Allied Translator and Interpreter Section (ATIS) which trained, organized, and sent into the field linguist detachments to interrogate prisoners of war and translate captured documents; (2) Allied Intelligence Bureau (AIB) which conducted clandestine operations, sabotage, and espionage behind the enemy lines and in enemy-held territories; its European counterpart was the Office of Strategic Services (OSS) which was never employed in the SWPA; (3) Allied Geographical Section (AGS) which gathered and published geographical information on areas within the SWPA; its Washington counterpart was Joint Army Navy Intelligence Service (JANIS); (4) Central Bureau (CB) which was an inter-Allied cryptanalytical service, co-ordinated with British and United States establishments.

New Guinea, New Britain, New Ireland, and the Solomons. Subsequently, more extensive "Terrain Studies" were issued to all planning staffs, supplemented by pocket-size "Handbooks" supplied to troops in the field, down to platoon commanders, as a sort of individual "Baedeker" for the assault echelons when they hit the landing beaches.[20]

The Allied Intelligence Bureau (AIB) was established in July 1942 to collect intelligence through clandestine operations behind enemy lines. A secondary function was to conduct sabotage operations and to secure co-operation and aid from the natives in fighting the Japanese. The Bureau incorporated several previously existing agencies and absorbed the Royal Australian Navy's "coast watching" system. It dispatched agents and radio transmitters by submarine and established stations ranging from Bougainville to Borneo, from Papua to Palawan. This organization performed invaluable service in giving immediate reports by radio on Japanese plane and ship movements in the area from the Solomons to New Ireland, New Britain, and the mainland of New Guinea. (Plate No. 16) This information enabled Allied planes to be ready and in the air at a time when the Allies were relatively weak in air power. Operational areas of New Guinea were explored by parties sent in by AIB to ascertain Japanese strength, activities, and movements. Intelligence was radioed for strategical use directly to local force commanders, Air Force units, and the Navy, as well as to AIB Headquarters in Brisbane and Port Moresby. The coastal waters between Milne Bay and Buna were sounded, charted, and marked by secret buoys to guide the transports of proposed landing operations.[21]

Experience on Bataan with a handful of Nisei interpreters had clearly shown the potentialities of a competent interrogation and

---

20 By the end of the war, AGS had published 110 Terrain Studies, 62 Terrain Handbooks and 101 Special Reports on virtually every phase of the geographic features of New Guinea, the Philippines, and adjacent areas. Indicative of the wide variety of topics covered by AGS are such typical publications as "The Native Carrier," "Getting About in New Guinea," "Vegetation Study, Eastern New Guinea," and "Sailing Directions, New Guinea Areas." These studies were designed especially for use in Allied operations and every effort was made to insure widest dissemination. The demand was so great that by 15 September 1945 almost 200,000 copies had been distributed to the different services of the Allied Command as shown in the following table:

	Ground Forces	Naval Forces	Air Forces	GHQ etc.	Totals
Terrain Studies	21,349	16,674	14,817	12,988	65,828
Terrain Handbooks	50,345	21,487	25,432	17,092	114,356
Special Reports	5,416	1,737	918	5,300	13,371
Totals	77,110	39,898	41,167	35,380	193,555

AGS, GHQ, SWPA, Final Progress Report, 15 Sep 45.

21 Part of the success of Allied operations was directly attributable to the fact that the Australian and American soldiers were well informed while their Japanese adversary was not. AIB and the other intelligence agencies provided information which the Japanese commanders chronically complained was never furnished to them in important combat areas. Maj. Gen. Kazuo Tanikawa, Staff Officer of the Japanese Eighth Area Army, stated, "We did not have information about the terrain in New Guinea.... Allied intelligence was far above what we could attain." Colonel Hara, Eighth Army Staff Officer, Operations Section said, "Allied intelligence activities were responsible in great part for our losses in New Guinea. They were very effective and the Allies seemed to know our strength and dispositions beforehand for it always seemed to me that they continually attacked our weak points and by avoiding our concentrated strong points managed to obtain their objectives with minimum losses. I always felt that Allied intelligence gained its great effectiveness through the failure of our own intelligence to combat it." Interrogation Files, G-2 Historical Section, GHQ, FEC.

translation service. Out of this experience grew one of General MacArthur's most important single intelligence agencies—the Allied Translator and Interpreter Section (ATIS) which was organized on 19 September 1942. Invaluable results were achieved by ATIS personnel in neutralizing one of the greatest advantages possessed by the Japanese—a language which was almost as effective as a secret code. The Japanese had found early in the war that they could label their mine fields, carry personal diaries, use their spoken language freely, and even handle military documents with little regard for security.

With complete confidence in the Nisei, G-2 employed hundreds of second-generation Japanese from Hawaii and California in linguist detachments, to be sent into the field with the combat forces. ATIS intelligence teams accompanied the troops in all initial landing operations. Captured maps and orders processed by ATIS revealed enemy strength and dispositions and plans of attack. Diaries contained excellent clues to the psychology and the state of morale of the Japanese troops. Other documents indicated the enemy's problems of food and supply, his order of battle, the effect of our air attacks, his relations with the natives, the relative effectiveness of Allied and Japanese weapons, and other equally important data. Spot interrogations of prisoners taken in battle were at times of such importance that they caused a shift in Allied plans of attack. ATIS provided information of immediate operational as well as over-all strategic value.[22]

These intelligence agencies combined to furnish the SWPA with its own Office of Strategic Services. As the war progressed they were enlarged and co-ordinated to meet the immediate needs of each new tactical and strategical situation. By using this weapon of superior intelligence, General MacArthur was in a position to gain the greatest strategic advantage in the shortest time with minimum losses.

General Headquarters Transferred to Brisbane

On 20 July, General MacArthur transferred his entire headquarters from Melbourne to Brisbane, and drew with it forward echelons of all component forces—Ground, Air and Navy. Brig. Gen. Richard J. Marshal replaced General Barnes in command of United States Army Forces in Australia, which was redesignated United States Army Services of Supply, and Maj. Gen. George C. Kenney relieved General Brett as Commander, Allied Air Forces.

Pursuing the Counteroffensive

The Allied successes in the Coral Sea and at Midway created a new situation which, General MacArthur urged, should be exploited by immediate offensive action. In a radio to the War Department on 8 June, he outlined his plans and recommendations and emphasized

---

22 During the war ATIS issued thousands of printed documents, falling into 13 different main classes or categories. They included ATIS "Publications," "Bulletins," "Inventories," "Interrogation Spot Reports," "Current Translations," "Enemy Publications," "Interrogation Reports," "Research Reports," "Philippine Series Bulletins," and "Philippine Series Translations."

ATIS average monthly production was as follows:	Nov 1942	Sept. 1945
Documents received, examined, and listed	1,000	33,500
Documents translated	750	19,000
Documents printed and published	0	17,000

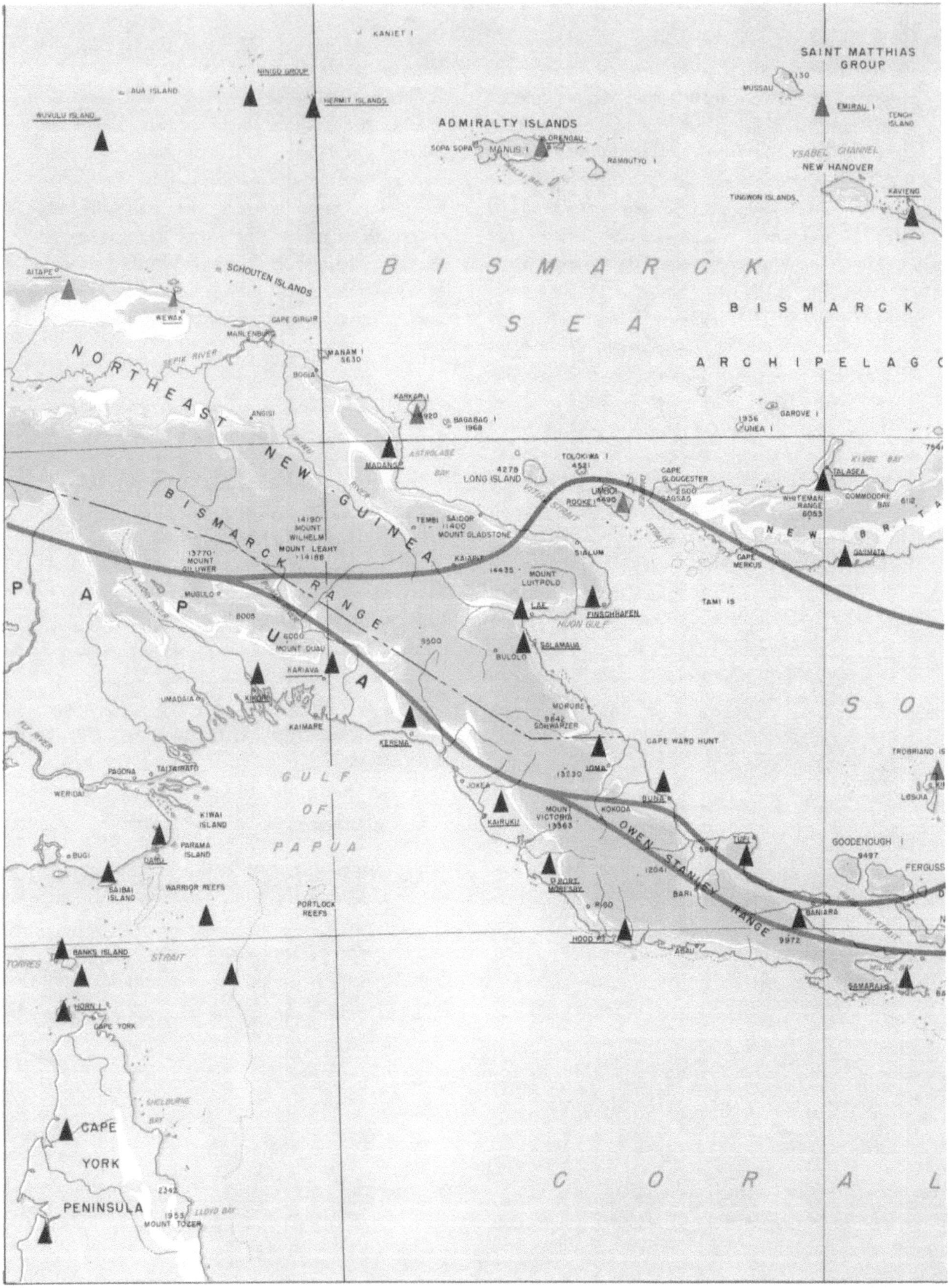
SAINT MATTHIAS GROUP
ADMIRALTY ISLANDS
NEW HANOVER
BISMARCK SEA
BISMARCK ARCHIPELAGO
NORTHEAST NEW GUINEA
BISMARCK RANGE
PAPUA
OWEN STANLEY RANGE
GULF OF PAPUA
CAPE YORK PENINSULA
CORAL
SCHOUTEN ISLANDS
LONG ISLAND
HUON GULF
TORRES STRAIT
GOODENOUGH I
NEW BRITAIN

PLATE NO. 16

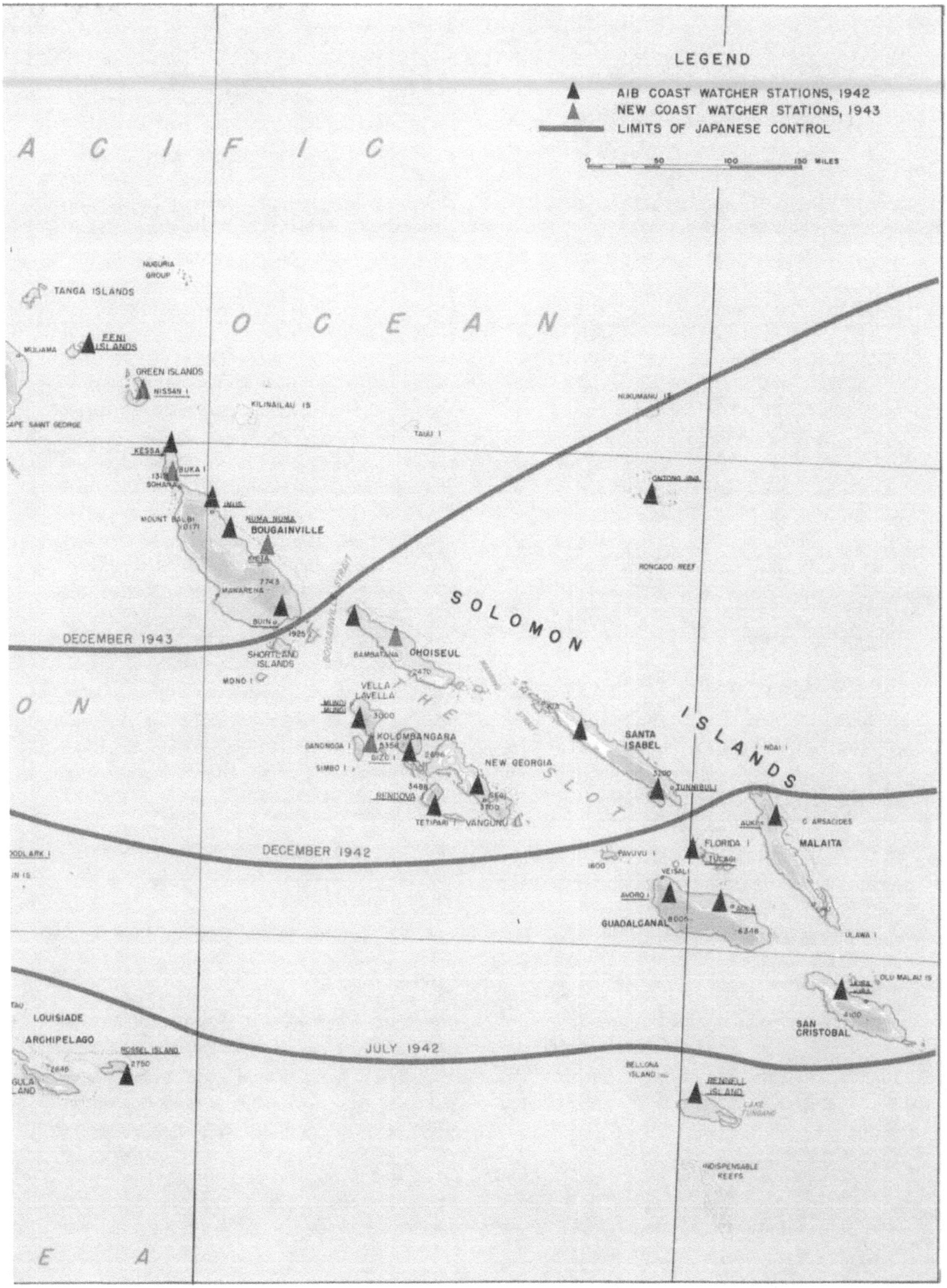

Coast Watching Teleradio Stations

the necessity of speed in their execution:

The first objective should be the New Britain—New Ireland area, against which I would move immediately if the means were available; have the 32nd and 41st Infantry Divisions and the 7th Australian Division which can be used in support of a landing force, but which cannot be employed in initial attack due to lack of specialized equipment and training; naval component excellent, but lack integral air elements; recommend 1 division trained and equipped for amphibious operations and a task force, including 2 carriers, be made available to me at earliest practicable date; with such a force I could retake that important area. . . . speed is vital; not possible for me to act quickly if I must build equipment and train my divisions; cannot urge too strongly that the time has arrived to use the force, or a portion of it, which you have informed me to be 40,000 men on the west coast (USA) trained for amphibious operations, and could be employed in conjunction with the forces available to me in the initiation of offensive operations in the SWPA.[23]

On 24 June General MacArthur elaborated on his plan for an offensive along the northeast coast of New Guinea and in the New Britain-New Ireland area. He stressed the important strategic principles of concentration of forces and the selection of objectives which could be held once they were won. An operation against Timor had been proposed, but General MacArthur doubted whether Timor could be held in proximity to strong Japanese bases in the Netherlands East Indies and with the Japanese in complete control of the adjacent sea areas. He was opposed to any division of the Allied effort and, rather than dissipate limited strength in various areas, he proposed a single unified drive under united command:

Impracticable to attempt the capture of Rabaul by direct assault supported by the amount of land-based aviation now available; my plan contemplates a progressive movement, involving primary action against the Solomons and the north coast of New Guinea, in order to protect the naval surface forces and to secure airfields from which essential support can be given to the forces participating in the final phase of the operation; necessary that all ground operations in this area be under my direction; to bring in land forces from other areas with a view to their operation under naval direction exercised from distant points can result in nothing but complete confusion; the very purpose of the establishment of the Southwest Pacific Area was to obtain unity of command. . . . the action projected northwest of Australia, [attack on Timor] even if successful, cannot be supported as is the case in the New Britain area; doubtful whether Timor could be held under present circumstances in view of nearby Japanese bases and their control of the sea; its capture should not be undertaken unless we are prepared to support it fully which includes the continuous control of the sea lane between there and Australia; the situation is entirely different in the New Britain—New Ireland area, which if captured can unquestionably be held; no effort from the northwest coast of Australia should be made until naval and air forces are built up which will insure retention of objectives. . . .

If correct [command] procedure is not followed it makes a mockery of the unity of command theory, which was the basis of the organizational plan applied to the Pacific Theater.[24]

On 2 July, the Joint Chiefs of Staff issued a directive under which Admiral Ghormley was assigned the conquest and garrisoning of the Santa Cruz Islands, Tulagi, and adjacent positions.[25] General MacArthur's task was to capture the remaining Solomon Islands, and also Lae, Salamaua, and the northeast coast of New Guinea. The target date for

23 CINCSWPA Radio No. 913 to WARCOS, 8 Jun 42, C/S GHQ, WD No. 101 (S).

24 CINCSWPA Radio No. 248 to WARCOS, 24 Jun 42, C/S GHQ, WD No. 118 (S).

25 JCS Radio to CINCPAC, CINCSWPA and COMSOPAC, 2 Jul 42, C/S SOPAC No. 33 (S).

the first operation against the Solomon Islands was set as 7 August.

The task would not be easy. The enemy was bringing large reinforcements into the region of the projected offensive and was intensifying his efforts to complete airdromes at the key points of Kavieng, Rabaul, Lae, Salamaua, Buka, and Guadalcanal.[26] The Allies, on the other hand, lacked trained amphibious forces; they were short on shipping facilities and airfields; and they needed experienced pilots and more planes to counter Japanese power.[27]

Plans for the offensive, however, were rushed. Operation *Tulsa Two B* envisioned two drives. (Plate No. 17) One force would proceed generally along the north coast of New Guinea securing in succession Lae—Salamaua—Gasmata, Cape Gloucester—Talasea—Madang, and Wewak—Lorengau. The other force would follow along the Solomons to New Ireland seizing Faisi—Kieta—Buka and Kavieng. In addition to his operations in New Guinea, General MacArthur prepared to support the attack on Guadalcanal by interdicting hostile air and naval activities in the area west of the Solomons.[28]

Air force units were meanwhile being moved into Cape York Peninsula and New Guinea as rapidly as the completion of airdromes and limited shipping allowed. By 19 August four airfields had been completed at Port Moresby, one for heavy bombers, one for medium bombers, and two for fighters. Two more heavy bomber fields and one for medium bombers were expected to be ready by 5 September. At Milne Bay, one field, already in operation for pursuit planes, was to be ready for heavy bombers by 25 August, and two other heavy bomber strips were under construction. On Cape York Peninsula two fields for heavy bombers and one for fighters were completed, with three additional heavy bomber fields planned for September. Adequate service facilities had been completed in the Townsville—Cloncurry area.

The acceleration of air movements and the construction of air establishments is an early characteristic of General MacArthur's strategy in three-dimensional warfare. His complete air mindedness is demonstrated by the principle of moving his bomber line forward: (1) advancing in successive bounds, (2) establishing forward airfields, and (3) making the next move under cover of his own air umbrella.[29]

The United States 32nd and 41st Divisions

---

26 CINCSWPA Radio to WARCOS, 27 Jun 42, C/S GHQ, WD No. 121 (S). Morison, *Coral Sea, Midway and Submarine Actions May 1942—August 1942*, p. 256.

27 General MacArthur pointed out that Japanese policy was to maintain front-line air elements at peak efficiency by constantly bringing in their ablest pilots developed in 2nd and 3rd line echelons. This was one of the reasons why the Japanese air force had consistently made such a good showing since the beginning of the war. General MacArthur, therefore, firmly opposed any exchange plan which would replace his experienced pilots with partially trained personnel, thus using the Southwest Pacific Area as an indoctrination school and training ground. Such a plan, he insisted, would only result in pitting inexperienced personnel against the best that Japan had to offer. The only outcome would be heavy losses in personnel and equipment by the American air forces. CINCSWPA Radio to WARCOS, 27 Jun 42, 321 AC Personnel, AG, GHQ, (S).

28 GHQ, SWPA, Opn Instr No. 14, 26 Jul 42, G–3 Admin 370 (S).

29 "This was the type of strategy we hated most," said Lt. Col. Matsuichi Iino, Senior Intelligence Officer, Eighth Area Army Staff. "The Americans, with minimum losses, attacked and seized a relatively weak area, constructed airfields and then proceeded to cut the supply lines to troops in that area. Without engaging in a large scale operation, our strong points were gradually starved out. The Japanese Army preferred direct assault, after the German fashion, but the Americans flowed into our weaker points and submerged us, just as water seeks the weakest entry to sink a ship.... We respected this type of strategy for its brilliance because it gained the most while losing the least." Interrogation Files, G–2 Historical Section, GHQ, FEC.

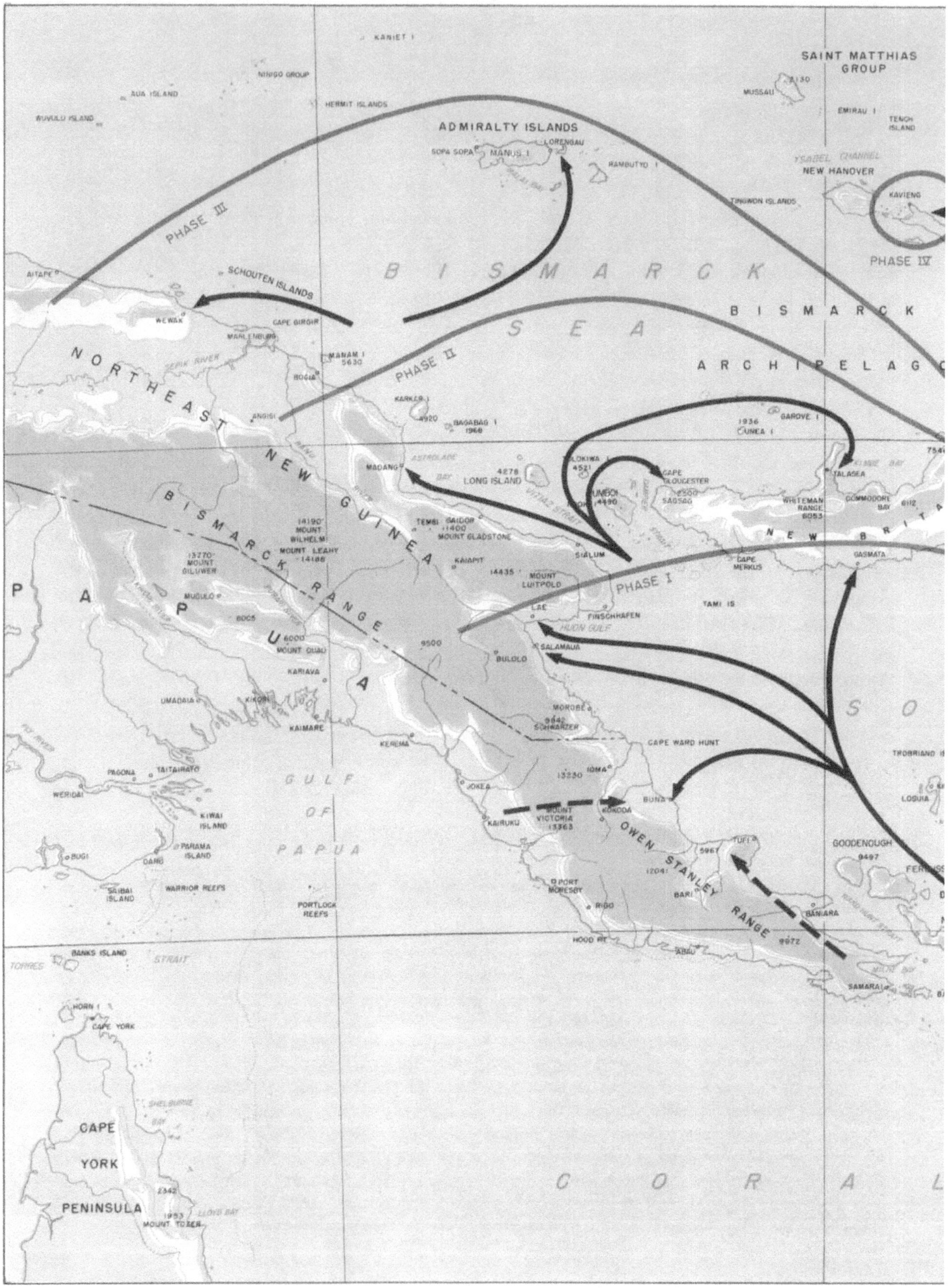

PLATE NO. 17

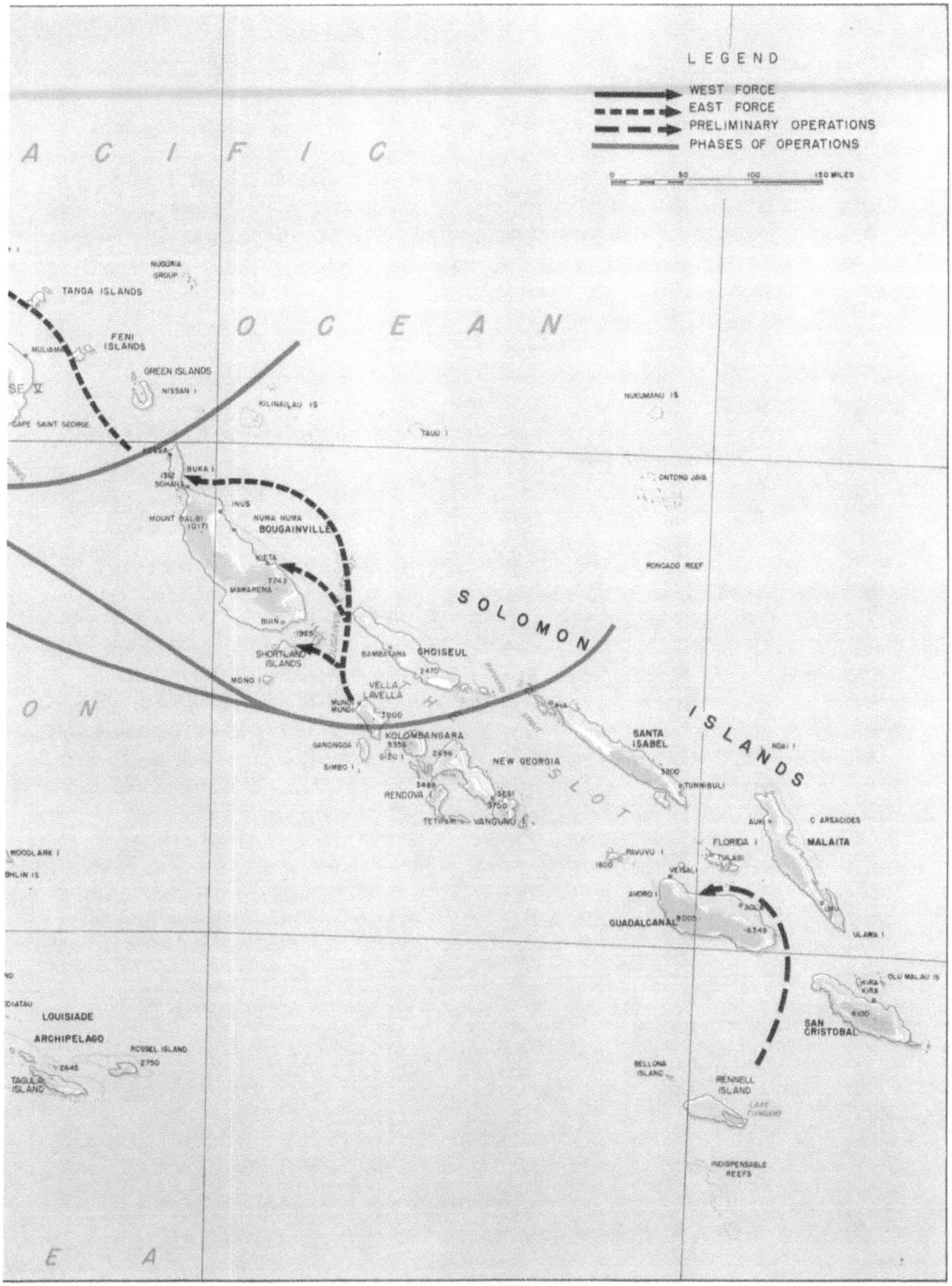
LEGEND
WEST FORCE
EAST FORCE
PRELIMINARY OPERATIONS
PHASES OF OPERATIONS
0 50 100 150 MILES
P A C I F I C
O C E A N
SOLOMON
ISLANDS
TANGA ISLANDS
FENI ISLANDS
GREEN ISLANDS
NISSAN I
BUKA I
BOUGAINVILLE
KIETA
SHORTLAND ISLANDS
CHOISEUL
VELLA LAVELLA
KOLOMBANGARA
NEW GEORGIA
RENDOVA I
SANTA ISABEL
FLORIDA I
TULAGI
GUADALCANAL
MALAITA
SAN CRISTOBAL
RENNELL ISLAND
BELLONA ISLAND
LOUISIADE ARCHIPELAGO
ROSSEL ISLAND
TAGULA ISLAND
ONTONG JAVA
RONCADO REEF
NUKUMANU IS
INDISPENSABLE REEFS
E A

Tulsa Two B

were shifted to Brisbane and Rockhampton respectively, to train for offensive operations. These divisions, formed into a corps, were to be placed under the command of Lt. Gen. Robert L. Eichelberger, Commanding General, I Corps, en route from the United States.[30]

By early August, plans and preparations for the offensive had proceeded to such a point that General MacArthur was ready to execute the first phase—the capture of Buna, Lae, Salamaua, and Gasmata—upon the successful conclusion of Admiral Ghormley's operation at Guadalcanal.[31]

Japanese Landing near Buna

While these preparations were under way, the Japanese were pursuing their own plans for taking Port Moresby. After the Midway debacle, Imperial General Headquarters cancelled the plans for operations against Samoa, New Caledonia, and the Fiji Islands and ordered the commander of the Japanese Seventeenth Army to concentrate his efforts on securing eastern New Guinea.[32]

As a result of these orders, the Allied move to execute Operation *Providence* and seize Buna was checked. On 21 July, a strong convoy of Japanese transports and warships landed near Basabua northwest of Buna. Although the convoy was sighted and attacked by Allied planes, General MacArthur's air force was as yet too small to effect any serious damage. Japanese troops occupied the Buna area immediately and dispatched a spearhead inland towards Kokoda—the first elements in the overland drive to Port Moresby. Their only opposition was a small advance detachment of Australians, carrying out Operation *Providence*, which had just completed an exhausting march from Port Moresby to the outskirts of Buna. The Australians, heavily outnumbered, and operating against adverse terrain factors, were forced to begin a slow retreat which did not stop until the enemy had reached a point within twenty airline miles of Port Moresby. (Plate No. 18)

Enemy Advance along Kokoda Trail

On 23 July, the first ground action between the opposing forces in New Guinea took place at Awala. The Australians were forced to give way after it became obvious that they could not hope to withstand increasing enemy pressure from the beachhead area. On 25 July, after breaking out of an enemy encirclement near Oivi, the Australians fell back to Deniki, a few miles south of Kokoda airdrome. When the Japanese at first failed to occupy Kokoda itself, the Australians on 28 July moved back to the airfield. That night, however, the Japanese, also recognizing the importance of the airstrip for supply, launched a heavy attack which drove the Australians once more to Deniki. A counterattack on 10–11 August proved futile, and the Australians were forced to evacuate Deniki and retreat to Isurava, 3 miles south of Kokoda.

Both sides were making every effort to in-

30 CINCSWPA Radio No. C–7 to WARCOS, 2 Jul 42, AG, GHQ 332 No. 2.

31 CINCSWPA Radio No. Q–147 to WARCOS, 2 Aug 42, C/S GHQ, WD No. 173 (S).

32 General Hyakutake, Commander of the Seventeenth Army, was notified of the cancellation of this offensive on 14 June while in Manila. At the same time, he was informed that Imperial General Headquarters desired a land offensive against Port Moresby. Accordingly, he ordered Lt. Gen. Tomitaro Horii, Commander of the South Seas (Nankai) Detachment, at Rabaul to make the required preparations for such an assault. Initial units left Rabaul on 20 July bound for Basabua, 6 miles northwest of Buna, to make a reconnaissance in force. The detailed composition of the Japanese task force may be found in ATIS, GHQ, SWPA, Current Translations No. 5, 3 Dec 42, pp. 58–59.

crease their strength at this stage. The Australians planned to build Isurava into a base from which they could attack and reoccupy Kokoda. Once in possession of the airstrip on the Kokoda plateau, they could receive air reinforcements from Port Moresby and drive the Japanese back to the coast. The Japanese likewise planned to build Kokoda into a large forward supply base. As companies of Australian infantry were trickling through the Owen Stanleys to bolster the battle-weary forces at Isurava, the Japanese were dispatching troops from Rabaul for the final all-out drive on Port Moresby. The main strength of the South Seas Detachment had arrived at Basabua, near Buna, during 13–21 August under orders to proceed to Port Moresby. On 18 August, Lt. Gen. Sidney F. Rowell replaced Maj. Gen. Basil M. Morris as Commander of New Guinea Force. Maj. Gen. Arthur S. Allen was put in charge of the Australian 7th Division and made responsible for operations on the Kokoda Trail. The troops at Isurava had been reinforced by the 2nd Battalion of the 30th Brigade and two battalions of the 21st Brigade.

By 26 August, the date on which the Australians were to have launched their counterattack, the enemy himself was applying heavy pressure on Isurava. Outflanked and outnumbered, the Australians were pushed back along the trail. Attempts to supply them by air were abortive as supplies were either smashed or lost in the air drops. The reinforcements which were to have ousted the Japanese from Kokoda were now forced to assume the defensive and to help extricate the battered remnants of the forces retreating from Isurava. General MacArthur was faced with the grim task of halting a Japanese drive which, if successful, would not only disrupt the Allied timetable in the southwest Pacific but could mean an eventual struggle for Australia itself.

Action in the Solomons

Admiral Ghormley launched his scheduled attack against the Solomons on 7 August when the 1st Marine Division, reinforced, went ashore at Tulagi and on Guadalcanal. Three aircraft carriers were among the participating naval units, which also included several cruisers and destroyers temporarily assigned to Admiral Ghormley's control from General MacArthur's naval forces. General MacArthur did everything in his power to assist the landing operation and to support the Solomons offensive. Submarines were dispatched to operate in the vicinity of Rabaul, reconnaissance was intensified, and air attacks against enemy airfields and shipping were pressed vigorously. All planes capable of the range were directed against Rabaul, Buka, and Kieta while others struck at Lae and Salamaua. Land operations in New Guinea were also intensified to relieve the pressure on the Solomons.[33]

The Marine landings at Tulagi and Guadalcanal took the enemy garrisons by surprise. At the end of the second day, the Marines were in complete control of Tulagi and held the airfield on Guadalcanal, the immediate objectives of the operation. The Japanese, however, struck back viciously by sea and air from their bases in the northern Solomons and the Bismarck Archipelago. On the night of 8 August, hostile surface units entered the area between Guadalcanal and Florida Island undetected and sank the cruisers *Canberra*, *Quincy*, *Vincennes*, and *Astoria*, and badly damaged the *Chicago* and two destroyers. Enemy land-based planes began bombing Marine positions during daylight hours and submarines were active against the American lines

33 GHQ, SWPA, Opn Instr No. 14, 26 Jul 42, G–3 Admin 370 (S).

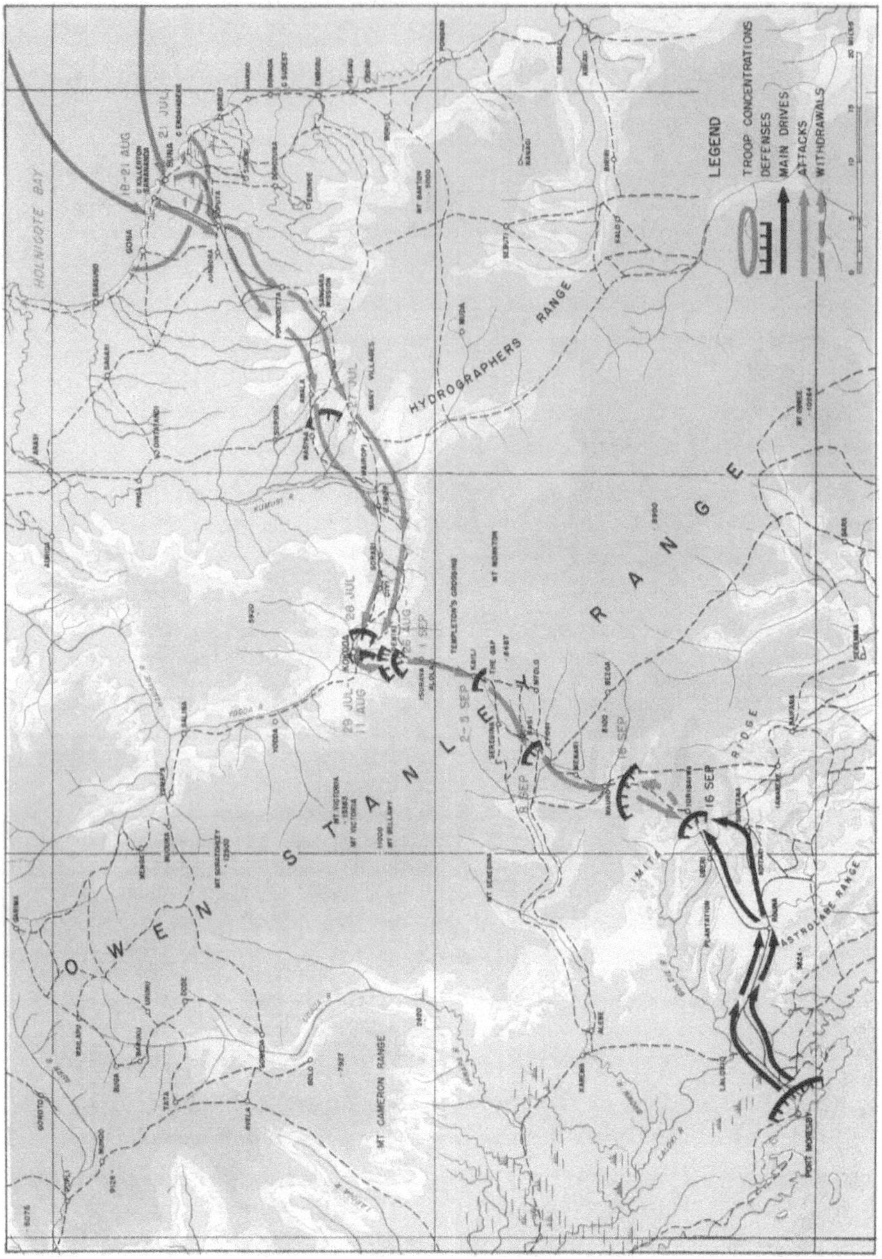

PLATE NO. 18
Japanese Thrust toward Port Moresby

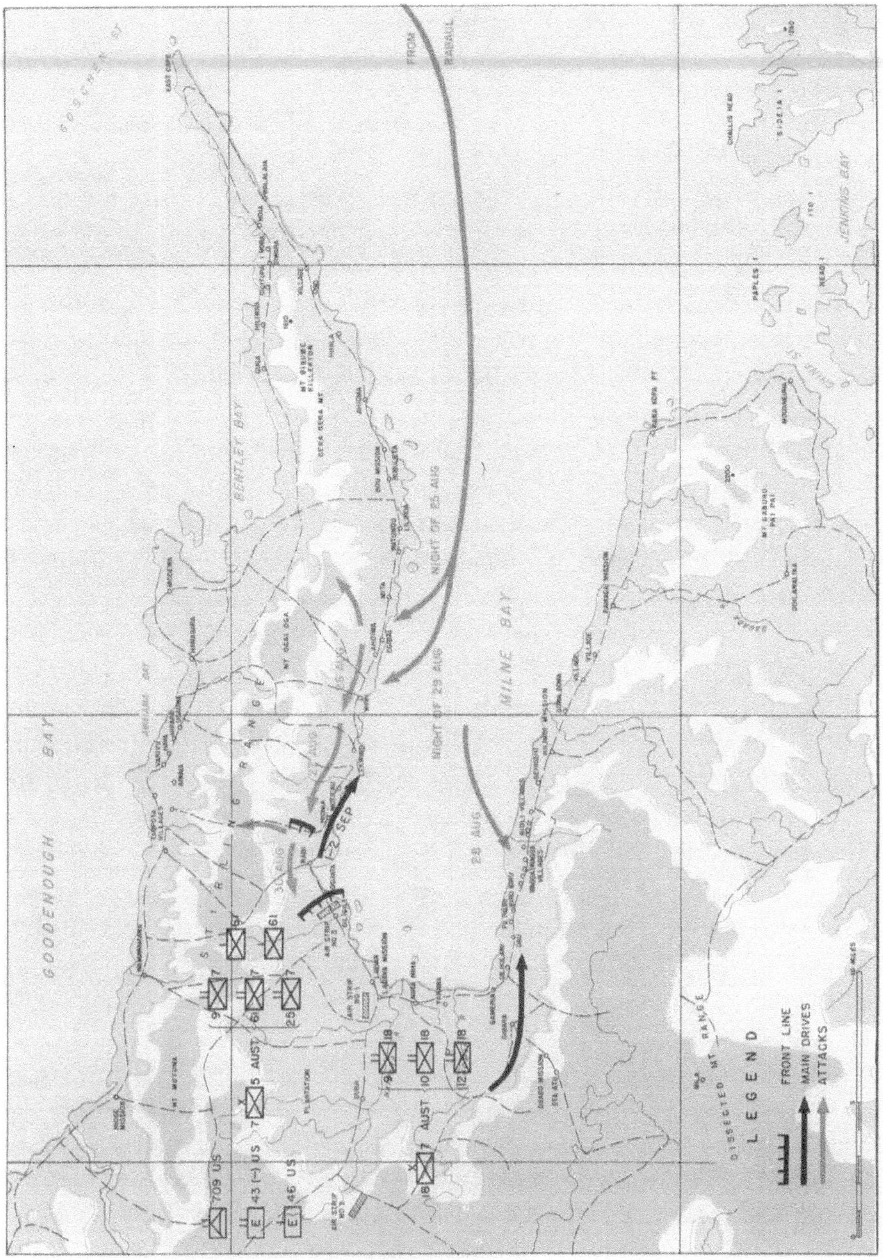

PLATE NO. 19

Enemy Landings at Milne Bay, August 1942.

of communication. At night enemy warships bombarded shore installations virtually at will.

On 23 August, a heavily guarded convoy of transports was sighted moving south of Rabaul. The Allied fleet took partial revenge for its earlier losses as carrier aircraft combined with land-based planes from Guadalcanal and the Southwest Pacific Area succeeded in inflicting damage that turned the enemy back. The Japanese, however, were able to land additional troops on Guadalcanal by means of the "Tokyo Express"—fast night runs by destroyers from the northern Solomons—and the Marines defending the airfield faced increasing pressure. The campaign became a race to see which side could build up the greater number of reinforcements and supplies, the outcome depending upon control of the sea and air.

Battle of Milne Bay

Intelligence meanwhile indicated that the Japanese had discovered the Allied occupation of Milne Bay.[34] General MacArthur anticipated that the enemy, incensed at having been forestalled, would soon attempt retaliatory measures. Preparing for such a move, he secretly dispatched the Australian 18th Brigade to reinforce the 7th Brigade which had been sent in July to garrison the airfield. In addition to these two brigades, the forces at Milne Bay included some 1300 United States combat and service troops.[35] They were under the command of Maj. Gen. Cyril A. Clowes, who arrived on 13 August to take charge of the forces in that area. Air power consisted of two fighter squadrons and part of a Hudson bomber squadron based on the Milne airdrome.

The first indications of a Japanese attack became evident on 24 August when seven enemy barges were sighted moving down the Papuan coast midway between Buna and Milne Bay. An air raid on Milne Bay by about twelve Japanese fighters was carried out on the same day. On 25 August, the enemy barges were discovered beached off Cape Varietta on Goodenough Island. They were immediately strafed and destroyed by Allied planes.[36]

A more dangerous threat, however, appeared on the 25th in the form of an enemy convoy of nine vessels moving southward from the direction of Rabaul past the Trobriands and Normanby Island. Bad weather hampered efforts to bomb the convoy which subsequently succeeded in landing over 1,100 troops by power barge near Waga Waga and Wanadala. (Plate No. 19) Japanese warships moved into the bay under cover of darkness to shell Allied positions and then, covered by heavy mists, retreated to open waters by dawn. Enemy troops put ashore initially and later as reinforcements numbered slightly over 1,900 men.

The Japanese proceeded with the operations, unaware of the secret Allied reinforcement of their Milne Bay garrison. The first important clash of the opposing forces took place near KB Mission, a few miles west of Waga Waga. The enemy, bringing ashore light tanks in addition to mortars and 25-pounders pushed the

---

34 G–2, GHQ, SWPA, *Daily Summary of Enemy Intelligence* No. 112, 13 Jul 42.

35 CINCSWPA Radio No. C–382 to WARCOS, 30 Aug 42, C/S GHQ, WD No. 215 (S).

36 Captured Japanese documents reveal that the Japanese on these barges had intended to land at Taupota, cross the Stirling Range, and attack Milne Bay from the rear in conjunction with the amphibious assault. The barges landed at Goodenough Island at 1130 on 25 August, intending to set sail for Taupota. At about 1230, ten fighters (RAAF) made a slashing attack which sunk all the barges and left over fifty casualties among the helpless Japanese. Their mission had resulted in complete failure. ATIS, GHQ, SWPA, Current Translations No. 14, 18 Jan 43, pp. 3–4.

defenders back beyond the Mission to the west bank of the Gama River.

By dawn of the 27th, the fighting had moved still further back toward No. 3 Airstrip. Here the Allies had drawn up a strong defense line hidden in the wooded area on the edge of the strip. For three days, the enemy did his utmost to advance but was unable to pass the strip defenses. Finally, on 30 August, having been hastily reinforced on the previous night by the 3rd Kure Special Landing Party, the Japanese launched a determined attack in force.[37] They came forward in repeated and furious charges backed by field pieces and heavy machine guns but a solid wall of Allied fire prevented a single Japanese from crossing the strip. The attack was repulsed and the enemy suffered heavily. That night marked the turning point of the battle.

With their tanks bogged in mud, their task force commander and most of their officers killed, their rations almost exhausted, and their equipment unsuited for the terrain and climate around Milne Bay, the Japanese began to crumble rapidly. The Allies, seizing the advantage, started their counterattack. Moving rapidly across No. 3 Airstrip, past the Gama River, they reoccupied KB Mission. In the early morning of 2 September the advance was continued. On 5 September the enemy was driven from his position at Waga Waga, thus putting an end to organized resistance. Several pockets farther east were cleaned out by air strafing and artillery.

The Japanese managed to evacuate about 1300 of their troops on the evening of 5 September, abandoning the remainder to the jungle. Some of these hapless troops tried to make their way over the Stirling Range and along the coast to rejoin their forces at Buna. Some were caught in the lower foothills and in the native villages near Ahioma, at the eastern end of the Bay. Others who had succeeded in crossing the range were found and killed near Taupota on the north shore. The rest died of starvation and disease along the coast. Despite their incompetent leadership, the Japanese troops at Milne Bay fought bravely and tenaciously to the bitter end. A grim foretaste of the future was to be found in the fact that at one phase of the operations, nearly 700 Japanese were killed and only three, (two of whom later died) were taken prisoner. Wounded Japanese who could not be evacuated were shot by their own men.

General MacArthur analyzed the signifi-

37 The difference between Allied and Japanese staff work was clearly demonstrated by the fighting at Milne Bay, as related by Comdr. Minoru Yano, IJN, Commanding Officer of the Kure 3rd Special Landing Party (SLP) and the senior Japanese officer at Milne Bay when the original task force commander, Comdr. Masajiro Hayashi, was killed. A substantial part of the Japanese troops used at Milne Bay was at Kavieng when the first Japanese landing took place. These troops were the 3rd Kure SLP, which had been formed at Kure in January 1942 for the purpose of invading Port Moresby. After the Battle of the Coral Sea these troops were sent back to Kavieng, where they stayed until sent to Milne Bay. When the Japanese troops who landed at Milne Bay on 25 August ran into totally unexpected opposition, the 3rd Kure SLP was sent to Milne Bay without the slightest intelligence about either the opposition, terrain, or climate. The Commanding Officer of the 3rd Kure SLP declared that when his unit was sent, it was supposed to get information which should have been known beforehand for a successful invasion. When the 3rd Kure SLP landed, the men were to attack immediately but they were too tired so the attack was postponed until 30 August. There was need for reconnaissance but no one was able to carry it out. The Japanese nevertheless proceeded with their operation, completely unaware of Allied ground strength at Milne Bay. Interrogation Files, G-2 Historical Section, GHQ, FEC. This account is in keeping with General MacArthur's early analysis of the victory at Milne Bay in which he pointed out that "the decisive factor was the complete surprise obtained over him [the enemy] by our preliminary concentration of superior forces." CINCSWPA Radio to C/S WD, 6 Sep 42, G-3, GHQ, SWPA *Journal* (S).

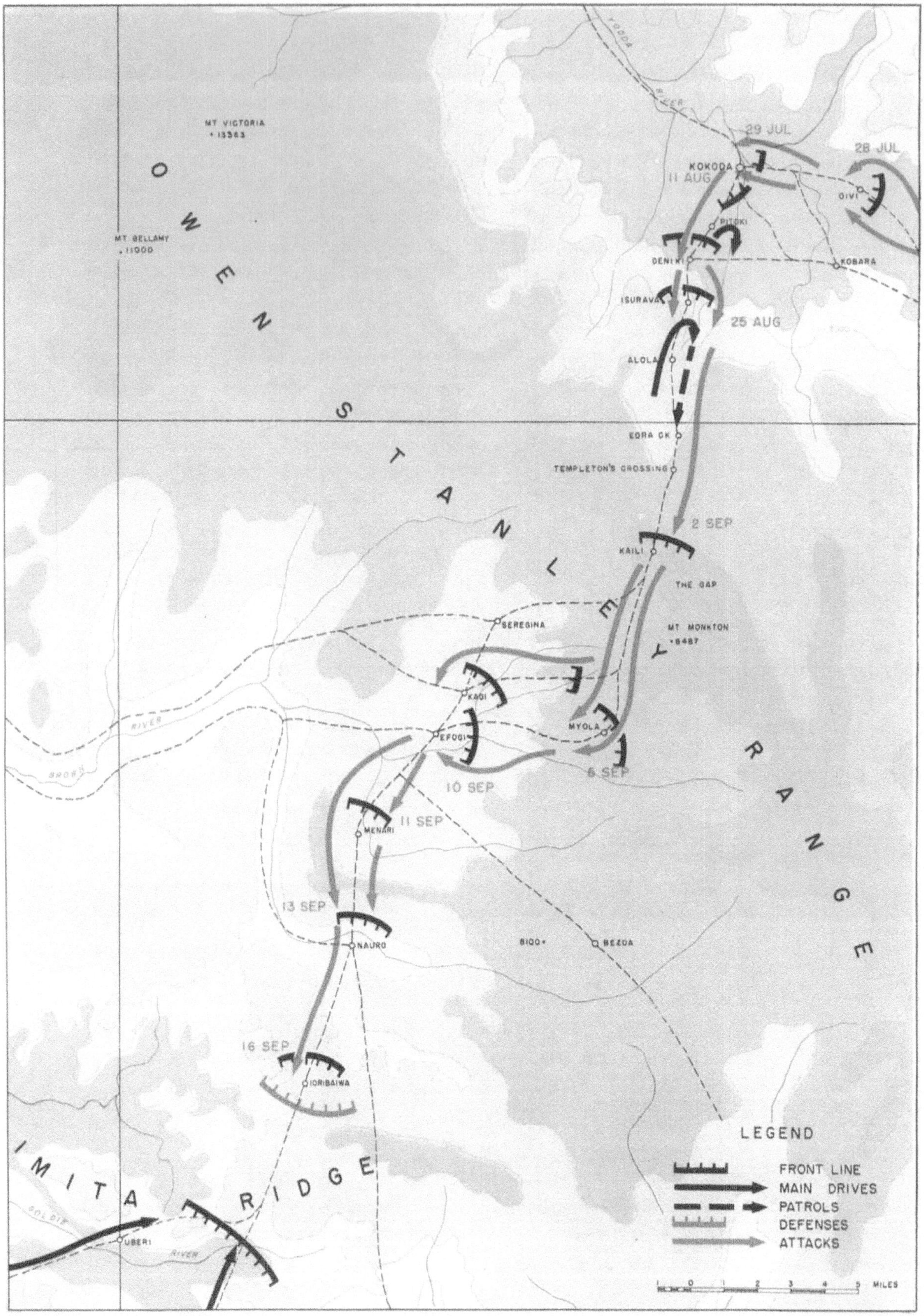

PLATE NO. 20

Action from Oivi to Imita Ridge, July—September 1942.

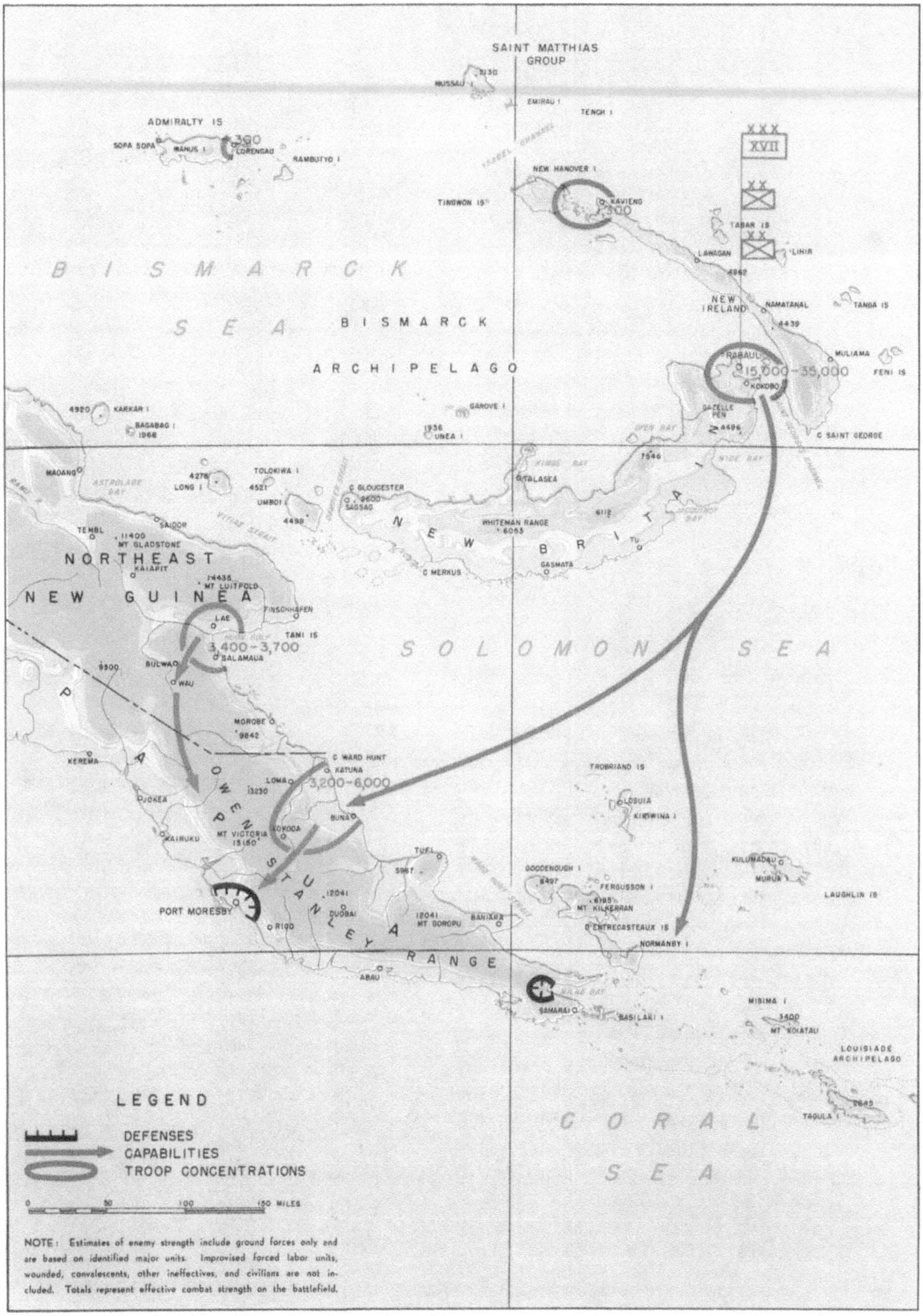

PLATE NO. 21

Japanese Dispositions and Capabilities, September 1942.

cance of the early operations in New Guinea as follows:

This operation represents another phase in the pattern of the enemy's plans to capture Port Moresby. This citadel is guarded by the natural defense line of the Owen Stanley Range. The first effort was to turn its left flank from Lae and Salamaua which proved impracticable. He then launched an attack in large convoy force against its rear. This was repulsed and dissipated by air and sea action in the Coral Sea. He then tried to pierce the center in a weak attempt by way of Buna—Gona—Kokoda subjecting himself to extraordinary air losses because of the extreme vulnerability of his exposed position. His latest effort was to turn the right flank by a surprise attack at Milne Bay. The move was anticipated, however, and prepared for with great care. With complete secrecy the position was occupied by our forces and converted into a strong point. The enemy fell into the trap with disastrous results to him.[38]

Enemy Checked near Port Moresby

The enemy in the meantime had not diminished his pressure along the Kokoda Trail. Australian units arriving from Port Moresby were immediately committed to stemming the steady advance of determined Japanese troops. There was no opportunity to choose advantageous positions or to build elaborate defenses; fighting took place wherever the enemy was encountered.

During September the Australians continued to withdraw, still "fighting tenaciously and gallantly under conditions of extraordinary hardship and difficulty," as General MacArthur phrased it.[39] From Isurava they fell back across The Gap but were again outflanked by Japanese infiltration tactics, despite concentrated Allied bombing and strafing attacks from the air. (Plate No. 20) On 14 September the Australians took up a last desperate stand at Imita Ridge. This ridge, within view of the sea on the Australian side, was the final remaining hurdle before Port Moresby.

Here the line held. The Japanese, having left the comparatively smooth slopes of their initial advance from Buna, had begun to experience increasing difficulties with the terrain and their extended, tenuous lines of supply. Casualties from bombing, sickness, and shortages of food had exacted a heavy toll. A mere twenty airline miles from his long-sought goal, the difficulties of the situation forced the Japanese commander to halt and consolidate his worn forces and await reinforcements of men and supplies. His grim story of the battle for the Kokoda Trail, told a short time before his death, is revealed by General Horii in a proclamation to his troops:

1. *Repeatedly we were in hot pursuit of the enemy. We smashed his final resistance in the fierce fighting at Ioribaiwa, and today we firmly hold the heights of that area, the most important point for the advance on Port Moresby.*

2. *For more than three weeks during that period, every unit forced its way through deep forests and ravines, and climbed scores of peaks in pursuit of the enemy. Traversing knee deep mud, clambering up steep precipices, bearing uncomplainingly the heavy weight of artillery ammunition, our men overcame the shortage of our supplies, and we succeeded in surmounting the Stanley Range. No pen or word can depict adequately the magnitude of the hardships suffered. From the bottom of our hearts we appreciate these many hardships and deeply sympathize with the great numbers killed and wounded. . . .*

3. *We will strike a hammer blow at the stronghold of Moresby. However, ahead of us the enemy still crawls about. It is difficult to judge the direction of his movement, and many of you*

38 GHQ, SWPA, Communique No. 140, 31 Aug 42.

39 GHQ, SWPA, Communique No. 151, 10 Sep 42.

have not yet fully recovered your strength. I feel keenly that it is increasingly important during the present period while we are waiting for an opportunity to strike, to strengthen our positions, reorganize our forces, replenish our stores, and recover our physical fitness. . . .

4. *When next we go into action, this unit will throw in its fighting power unreservedly.*[40]

In spite of the victory at Milne Bay and the checking of the enemy on the Kokoda Trail, General MacArthur felt that the overall situation was becoming increasingly precarious. During July and August the Japanese had been steadily pouring planes and ships into the Truk—Rabaul region and after their defeat at Midway the main battleground was now definitely in the South and Southwest Pacific Areas. Estimates on Japanese dispositions and capabilities indicated that they would soon be in a position to renew their drive for control of New Guinea. (Plate No. 21) General MacArthur reported to Washington that, unless steps were taken to match the heavy ground, air, and naval forces which the Japanese were assembling, a situation might develop similar to that which had successively overwhelmed Malaya, the Netherlands East Indies, and the Philippines early in the war.[41]

40 ATIS, GHQ, SWPA, Current Translations No. 2, "Message of Instruction to South Seas Detachment," 20 Sep 42.

41 CINCSWPA Radio No. C-381 to WARCOS, 30 Aug 42, C/S WD No. 216 (S).

CHAPTER IV

CLEARING OF PAPUA

Plans and Preparations

General MacArthur's immediate problems were to secure and strengthen his forward bases in New Guinea and to begin a counteroffensive that would drive the Japanese from Papua. Operations would have to be carried out over some of the most difficult terrain of any area in the world. (Plate No. 22) Except for Port Moresby and Milne Bay, the Japanese possessed control of the mountain ranges and occupied the narrow coastline of all northeastern Papua. Immediate action was imperative before they could fully exploit the geographic features of New Guinea and transform their holdings into impregnable positions supported by their stronghold at Rabaul.

The presence of General Horii's troops in command of the Kokoda Trail was a threat to Port Moresby that had to be removed. The security of this important strategic base demanded a counterdrive to regain the crest of the Owen Stanleys. General Horii's units could be reinforced from such key bases as Buna, Sananda, and Gona which the Japanese were constantly fortifying by shipments from Rabaul and Kavieng. The Japanese could also threaten Milne Bay by advancing down the northeastern coast of New Guinea. Port Moresby and Milne Bay were absolutely essential to support any Allied counteroffensive in New Guinea; their loss would constitute an irreparable blow to the Allied cause in the Southwest Pacific. Without these two forward bases, all New Guinea would soon fall to the enemy and Australia would be open to air attack from Japanese bases less than 350 miles from Cape York Peninsula. It was a sober realization of this fact that caused General MacArthur to warn at the time: "If New Guinea goes, the results will be disastrous."[1]

General MacArthur had studied carefully the whole picture of the enemy's operations in New Guinea and decided that General Horii had overreached himself along the Kokoda Trail. He felt that the moment was opportune for an immediate counterdrive that would catch the Japanese before they could find their second wind. "In view of the changed situation," he stated, "an attack to clear the north coast of New Guinea should be undertaken as soon as possible."[2] Anticipating that the Japanese drive would abate and spend itself as it ran ahead of its supply line, General MacArthur had already started the execution of his plans for a counteroffensive before General Horii halted at Imita Ridge on 14 September. I Corps, under General Eichelberger, was undergoing intensive training and reequipment for seaborne operations and jungle warfare. On 4 September, General MacArthur had instructed General Eichelberger to be prepared to carry out amphibious operations along the northeastern coast of Papua.[3] He strengthened the

1 CINCSWPA Radio to C/S WD, 6 Sep 42, G-3, GHQ, SWPA *Journal* (S).

2 Ibid.

3 CINCSWPA Ltr to CG, I Army Corps, 4 Sep 42, G-3, GHQ, SWPA *Journal* (S).

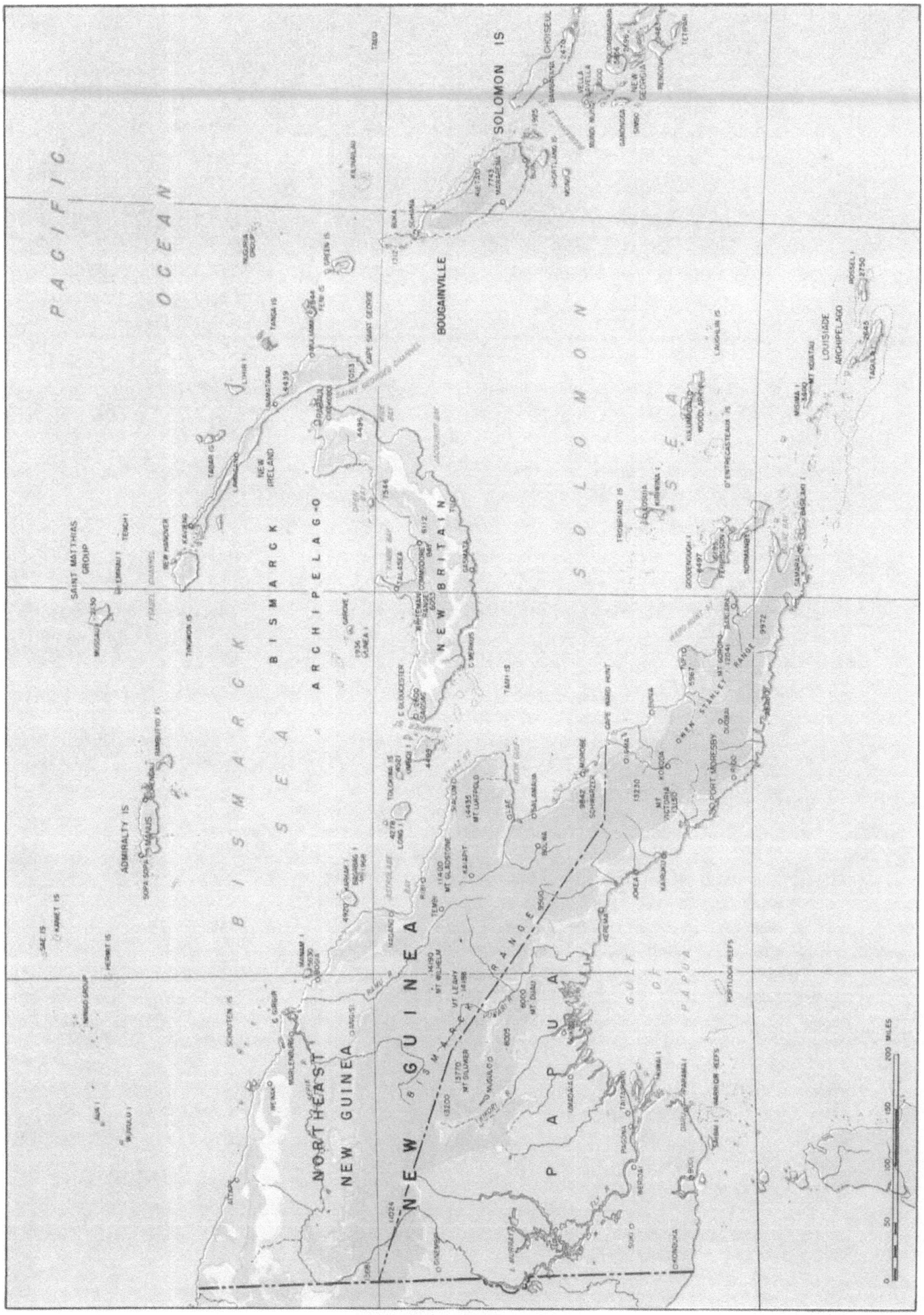

PLATE NO. 22

Relief Map of Eastern Papua

Corps in order to provide a strong spearhead for a forward thrust and placed it under the operational control of Allied Land Forces.[4]

In order to seize the initiative while the Japanese were attempting to regroup their forces, General MacArthur rushed all available troops into New Guinea. General Blamey moved Allied Land Forces Headquarters forward to Port Moresby and Lt. Gen. Sir Edmund F. Herring replaced General Rowell as Commander, New Guinea Force.[5] The Australian 16th Brigade and auxiliary troops of the Australian 6th Division were sent from Brisbane on 10 September. To insure that no time was lost, General MacArthur directed that the 126th and 128th Regimental Combat Teams of the United States 32nd Division be transported to Port Moresby by air. The movement of these two regiments, ordered on 11 September, was completed in record time by 28 September.[6] This air transfer from Australia to New Guinea was "the first large-scale airborne troop movement by United States forces in a theater of operations."[7] By the end of October, General MacArthur had augmented his ground forces in Papua to comprise a total of ten brigades and regiments together with supporting and attached troops.

Shipping resources and supply facilities were taxed to the limit to transport and maintain a force of this size. Docks for handling ships, roads for the distribution of supplies to depots, and buildings for hospitals and the protection of materiel against the heat and torrential rains were inadequate at Port Moresby, almost nonexistent at Milne Bay, and entirely lacking in the forward areas. These facilities had to be developed at the same time that the troops were moved into position.

It was in this phase of operations that General MacArthur felt most keenly the lack of service troops.[8] Under terrain conditions which demanded an abnormally large proportion of service elements,[9] his barest requirements were not being filled. Construction engineers and other supporting personnel were forced to work

4 The U.S. 32nd and 41st Divisions, the 147th Field Artillery, and the 1st Battalion of the 148th were assigned to I Corps. General Order No. 30, 5 Sep 42, GHQ, SWPA Special File No. 3, Folder 4 (S).

5 CINCSWPA Ltr to PM Curtin, 28 Sep 42, C/S Aust 161.

6 MID, WD, *Papuan Campaign, The Buna-Sanananda Operation*, p. 4.

7 Ibid., p. 20.

8 During 1942, only fifteen percent of all Allied military resources were channelled into the war against Japan. This pitifully small allotment had to be parcelled out to the Southeast Asia Command, China, and to the North, Central, South, and Southwest Pacific areas.

In September 1942, General MacArthur requested enlisted filler and loss replacements at a rate of 3,500 a month for the succeeding six months. This amount was necessary in order to bring his units to table of organization strength, provide for estimated losses other than battle casualties, and establish a replacement pool equal to ten percent of unit strength. The African invasion preparations, however, hungry for manpower, swallowed the entire output of the replacement training centers in the United States during the months of September, October, and November. It was not until the end of January 1943 that troops could be spared for shipment to Australia.

9 The extreme difficulties encountered in supplying Allied forces in New Guinea are well illustrated by the ingenious means employed by the Australian 7th Division in supporting its units just prior to their drive across the Owen Stanleys. Shipments were loaded on six 3-ton trucks, and transported from Port Moresby to the roadhead at Koitaki. A transfer was then made to ten 1/4-ton jeeps and carried to Owers Corner, a distance of only three miles. Here another transfer was made and the supplies, distributed on 60 pack animals and 1,000 native carriers, were moved over steep, treacherous trails to Uberi, three miles on, and thence by porters to the front line units. These 16 motor vehicles, 60 animals, and 1,000 porters, delivering at maximum capacity over a comparatively short line of supply, were still insufficient to move enough materiel to support a force of 3,500 troops. G-2, GHQ, Gen. Willoughby Pers, Circ 22 Sept 42 (S).

to the point of exhaustion. The shortage of these troops was so critical that combat troops were continually called upon to unload ships, transport supplies, and perform other service duties which diverted their energies from the fighting fronts. Instead of being accorded a period for recuperation and an opportunity for reequipment after an area was captured, battle-weary troops had to develop their own facilities for the support of future operations.

In an effort to unravel the knotty logistic problem confronting him, General MacArthur directed the formation of the Combined Operational Service Command under Brig. Gen. Dwight F. Johns.[10] This organization was established in Port Moresby and functioned directly under New Guinea Force as a co-ordinating agency between the American and Australian supply services. The consolidation of activities resulted in the increase of the cargo handling capacities of Port Moresby from under 2,000 tons to an average of over 6,000 tons daily.[11] It also unified the utilization of shipping under the United States Army Transportation Corps and the Australian transportation units so that their combined facilities could be employed with maximum effect.

Clearing the Owen Stanleys

General MacArthur's preparations were designed to clear the Japanese from Papua by carrying out a plan of attack along three axes of advance. (Plate No. 23) One axis would engage the enemy in a frontal action along the Kokoda Trail; the second would involve a wide flanking movement over the Owen Stanleys east of Port Moresby against the enemy lines of communication and supply; and the third axis of advance would consist of large-scale infiltration from Milne Bay along the northeastern coast of Papua. The immediate objectives of the first two drives were to regain the Nauro and Kokoda airfields and to secure the line of the Kumusi River from Owalama Divide to the crossing of the Kokoda—Buna trail at Wairopi; the third drive aimed at seizing selected points south of Cape Nelson and occupying Goodenough Island to deny the enemy these areas. All three axes of advance were to converge upon the Buna—Sananada—Gona area for a final simultaneous attack against this enemy coastal stronghold.[12]

The first move in the planned counteroffensive was made by the 7th Division of the Australian I Corps under Maj. Gen. A. S. Allen. With the mission of regaining control of the Isurava—Deniki—Kokoda route, it launched its drive on 26 September. The 25th Brigade, reinforced, spearheaded the advance while the 16th Brigade moved up in support. The 21st and 30th Brigades, which had borne the brunt of earlier fighting along the Kokoda Trail, and the 14th Brigade were placed under command of the Australian 6th Division Headquarters and retained in Port Moresby to defend the area against a possible seaborne attack. Ioribaiwa fell to the Australians on 28 September after the main enemy body evacuated its defenses and withdrew rapidly northward.

General Allen continued to press his advance, seizing Nauro on 30 September. Thus General MacArthur could report:

> *Ioribaiwa Ridge has been captured by our troops. The enemy is retiring in the direction of Nauro, ten miles to the north. Quantities of supplies and equip-*

---

10 Establishment of Combined Operational Service Command in New Guinea, 5 Oct 42, GHQ, SWPA, AGO 323. 36.

11 OCE, GHQ, AFPAC, *Engineers of the Southwest Pacific, 1941–1945*, Vol VII, "Engineer Supply," p. 55.

12 GHQ, SWPA Opn Instr No. 19, 1 Oct 42, G-3 Admin 370 (S).

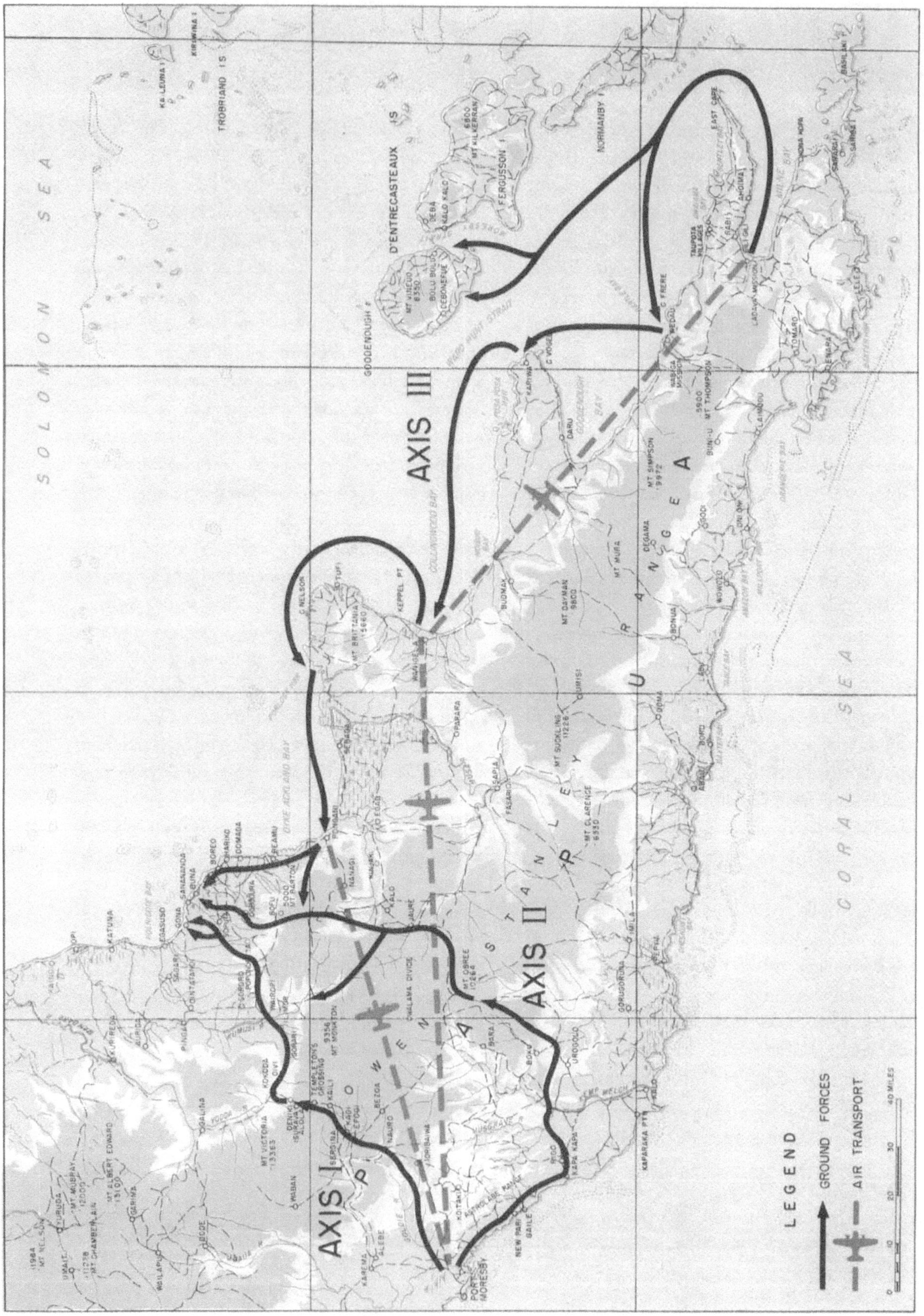

PLATE NO. 23

Axes of Advance, Papuan Campaign

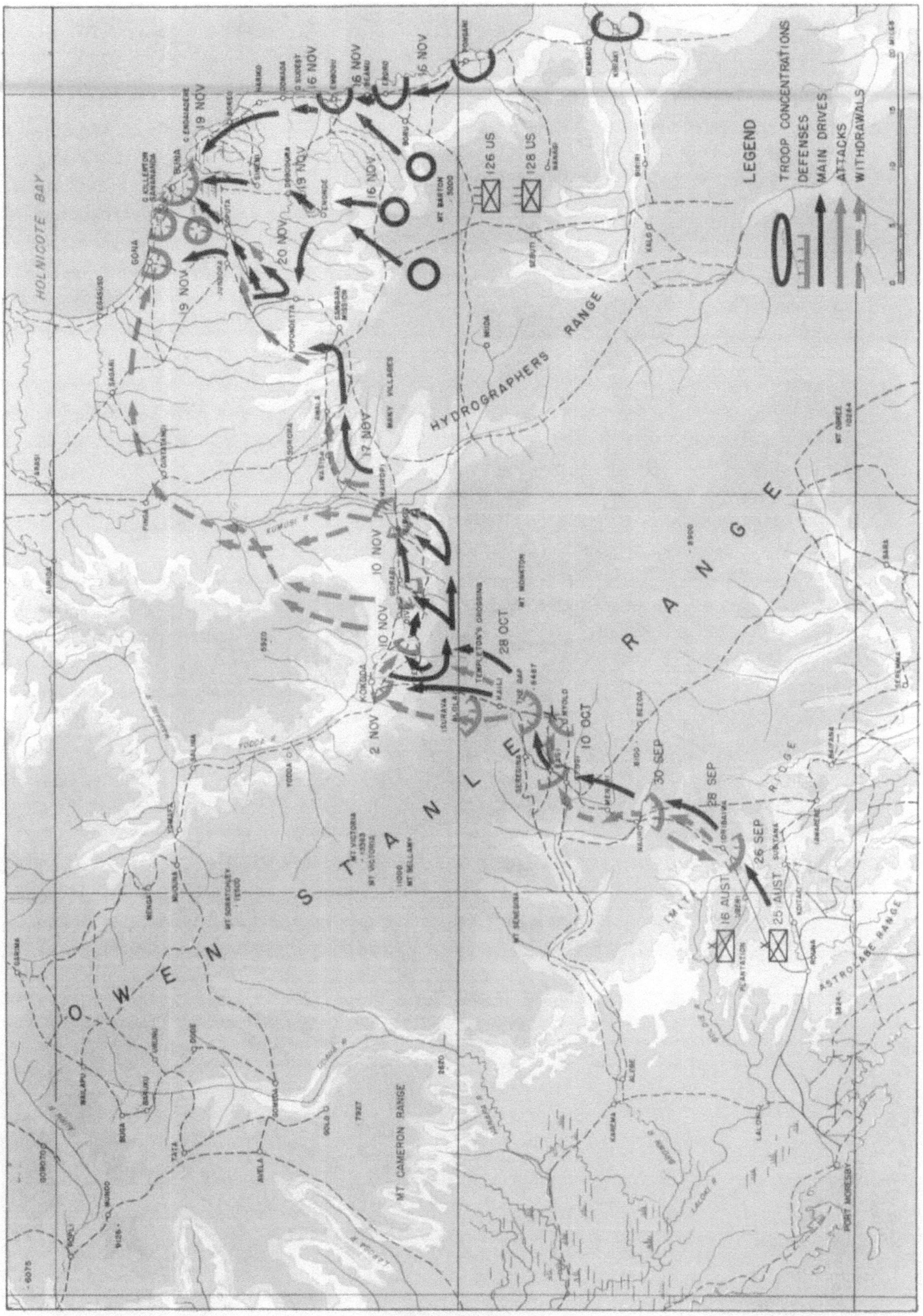

PLATE NO. 24

Owen Stanley Offensive

ment were abandoned by him and are in our hands. The position was one of natural strength and had been prepared for defense with barricades, obstructions and field trenches. . . .

Our forward elements have captured Nauro and are continuing the advance. There was no contact with the main enemy body yesterday, which is still retiring. Additional supplies and equipment abandoned by his retreating troops have fallen into our hands. Strong formations of our attack bombers, with fighter escort, harried enemy lines of supply in the Menari area.[13]

Pursuit of the enemy was delayed temporarily by the bad state of the trail and adverse weather conditions which hampered the air dropping of supplies. By 10 October, however, forward elements again met the enemy on top of the range in the area of The Gap. There the Japanese fought tenaciously from a series of well-prepared positions on the high ground dominating the trail.[14] The last of these was not cleared until the end of the month when the enemy retreated toward Oivi. (Plate No. 24)

Meanwhile, troops of the United States 32nd Division under General Harding had been assigned the task of operating over the mountains east of Port Moresby along the second axis of advance. Two routes were considered : one from Rigo and the other from Abau, both of which passed through Jaure to Wairopi. While the United States 128th Infantry upon its arrival in New Guinea was temporarily allocated to the Australian 6th Division for the defense of Port Moresby, the 126th Infantry at Kapa Kapa dispatched its 2nd Battalion over the rugged trail rising more than 8,000 feet above Rigo. The difficulties of supply prevented the movement of a larger force until the route could be further developed and supply dumps established. The United States 127th Infantry was to be shipped from Australia and, as suggested by General MacArthur, utilized along the second trail. Later, conditions at Abau made rapid port development impractical, and the plan to establish it as a port of embarkation was cancelled on 14 October.[15]

Airlift to Wanigela

General MacArthur realized the importance of starting the third phase of his Papuan plan of attack at the earliest moment. He was handicapped, however, by a lack of trained amphibious forces and an acute shortage of landing craft with which to carry out his projected advance along the eastern Papuan coast. He accordingly asked the War Department for additional naval facilities and Marine troops and stressed the fact that the time was opportune for such an advance :

There are growing indications that the enemy intends to throw large ground reinforcements into New Guinea from the north by means of numerous small increments

13 GHQ, SWPA, Communiques Nos. 171 and 172, 30 Sep 42 and 1 Oct 42.

14 In September General Horii, Commander South Seas Detachment, after receiving orders from the Army Commander to assemble his main force near Isurava and Kokoda, redistributed his troops. On 25 September the detachment withdrew from Ioribaiwa and an element of the unit, the Stanley Detachment consisting of the 3rd Battalion of 144th Infantry as its main force, was posted in commanding positions in the area of The Gap. As the Stanley Detachment had been under attack since the early part of October, the main force of the South Seas Detachment had been sent forward from Kokoda to give it assistance. The enemy continued to send more units to the scene of battle. Allied air attacks on his supply routes, however, in addition to heavy rains which made overland transportation impossible for the Japanese, left the Stanley Detachment without the means to carry on. On 28 October it retreated to Kokoda. Japanese First Demobilization Bureau Report, *Southeast Area Operations Record, Part III, Eighteenth Army Operations,* Vol I, p. 14, G-2 Historical Section, GHQ, FEC.

15 CINCSWPA Radio No. XC853 to General Blamey, 14 Oct 42, G-3, GHQ, SWPA *Journal* (S).

using usual Marine infiltration technique. A golden opportunity is being lost in not anticipating the enemy through clearing the coast as far as Buna by trained amphibious combat teams operating in conjunction with projected enveloping movement by infantry over the trails of the Owen Stanleys. Earnestly request reconsideration of previous decision not to make available such forces until later developments. Small craft for landing and maintenance of supply are a necessary adjunct of the combat teams.[16]

He was informed, however, that the type and number of landing craft required for immediate operations in New Guinea along with trained amphibious combat teams were not available at the time. The first shipment of landing craft would not be sent from the United States until sometime in October and other shipments would not be ready until November and December, too late to take part in the campaign to capture Buna.[17]

Determined nevertheless to seize the initiative, General MacArthur sought to establish his forces on the northeastern coast of Papua by overland march and by air transport. At Wanigela, south of Cape Nelson, an airstrip was located which proved to be in sufficiently good order for the landing of transport aircraft. Instructions for its occupation were issued immediately and landings were successfully carried out on 5 October without the enemy's knowledge or opposition.

During the middle of October, plans to move the remainder of the Australian 18th Brigade to Wanigela were changed to speed the operation; instead, Regimental Headquarters and two battalions of the United States 128th Infantry, the Australian 2/6th Independent Company, and miscellaneous artillery units were substituted and flown from Port Moresby. The United States troops, together with the Australian Independent Company, started to move overland from Wanigela to Pongani, 23 miles from Buna, but found the Musa River impassable. The overland plan was consequently abandoned in favor of a coastwise ferry service. The first elements arrived at Pongani on 20 October.[18]

By 2 November the 128th Infantry, except for half a battalion at Port Moresby, had completed its move to the Pongani area. The 2/6th Independent Company covering the advance, reconnoitered the forward routes through Hydrographers Range, and an airstrip was cleared at Pongani for transport planes. The 2nd Battalion of the 126th Infantry which had marched overland from Rigo was deployed in the Jaure area. The remainder of the two United States regiments was concentrated north of the Owen Stanleys ready to attack the Japanese in conjunction with the Australian 7th Division.

Landings on Goodenough Island

Concurrently with the advance from Wanigela, a reconnaissance in force was made on Goodenough Island. This force was to locate suitable approaches for supply purposes, to examine the terrain for possible airfield sites, and to occupy the island thus denying it to the enemy. Drake Force, composed of a battalion of the Australian 18th Brigade and auxiliary troops, accordingly boarded two destroyers on the night of 22 October. The battalion, less one company, landed at Mud

16 CINCSWPA Radio to C/S WD, 27 Sep 42, G-3, GHQ, SWPA *Journal* (S).

17 WARCOS Radio No. 2283 to CINCSWPA, 30 Sep 42, WD No. 245 C/S GHQ (S).

18 The movement would have been long delayed except for General MacArthur's foresight and advance planning. Early in October he had directed an aerial engineer reconnaissance of the Pongani area to select small boat landing points for a move along the northeast coast of Papua by sea and to determine possible beachhead lines and inshore routes north to Buna. CINCSWPA Radio No. XC731 to Matthews, 3 Oct 42, G-3, GHQ, SWPA *Journal* (S).

Bay while the detached company landed at Taleba Bay. The Australians discovered a small force of the enemy on Goodenough Island near Kilia Mission which had been stranded there since August when Allied aircraft had destroyed the barges in which they were moving from Buna in an attempt to join the assault on Milne Bay. The Australians attacked the Japanese at Kilia Mission from the two landing points and forced them to withdraw eastward to Fergusson Island. Drake Force reassembled at Mud Bay on 26 October and remained in occupation of Goodenough Island.[19] By 20 November, it had constructed an emergency airfield and had begun the preparation of an advance air base with radar stations. The lack of garrison troops, however, coupled with other developments in the New Guinea theater caused a temporary postponement of the project.[20] Meanwhile, the troops stayed on as a "ghost force" and by elaborate camouflage and deceptive measures made a show of strength which deterred the enemy from any attempt to retake the island.

Although enemy reaction to the operations along the north Papuan coast was limited to minor resistance at Goodenough Island, grave risks were involved. The Japanese Navy, controlling the sea north of New Guinea, could cut the Allied line of communications almost at will and bring overwhelming force to bear against the newly won bases. The advance had to proceed with caution, since withdrawal would be necessary should the supply lines fail. Airstrips were cleared inland at Sapia and Kinjaki primarily for this purpose and supply dumps were established to permit the retirement overland of troops from the coast should the enemy attack in force.

Guadalcanal

Following the naval engagement in the eastern Solomons on 24 August, no decisive action took place in the South Pacific Area for a period of about six weeks. During this time, however, the supply lines to Guadalcanal had to be kept open. Japanese submarines and aircraft were active in the vicinity, and there were numerous scattered actions which sunk the carrier *Wasp* and several destroyers and damaged a number of other ships.

Despite offensive operations directed against his ground troops and naval forces by the Allies, the enemy had succeeded in putting almost the equivalent of a division on Guadalcanal by early September. Additional Japanese fleet units had been assembled to the north and the situation again became serious. Reinforcements for the Marines had now become imperative even though they had to be made in the face of enemy land and naval superiority.

On the night of 11 October, a United States cruiser and destroyer group surprised an enemy force off Cape Esperance, sinking and damaging a number of vessels and causing the remainder to withdraw. Although the engagement was a victory for the Allied forces, attributable both to the enemy's surprise and confusion and to the accuracy of Allied gunfire, the situation remained critical.

General MacArthur continued to do everything possible to support the operation in the Solomons and relieve the pressure on Guadalcanal. His drive in New Guinea was in part a diversion for that purpose, involving great risks which would not normally be justified.[21]

19 Adv LHQ Report, "Operations, Goodenough Island, 22-26 October 1942," 23 Nov 42, AG, GHQ 370. 2. (S).

20 CINCSWPA Ltr to General Blamey, 10 Dec 42, ALF No. 43, C/S GHQ (S).

21 CINCSWPA Radio to SECWAR, 18 Oct 42, C/S GHQ WD 257 (S).

It served to divide the enemy effort, however, siphoning strength which would otherwise have been concentrated in the Solomons.[22] The Japanese, confused in their estimate of the relative importance of the two theaters of operations and considering each a separate problem, frittered away their strength in unco-ordinated and spasmodic reinforcement efforts. General MacArthur, on the other hand, realized that these operations were interdependent and that the utmost mutual co-operation was necessary to insure their success. He saw the picture in clear perspective and was prepared to do everything in his power to aid another operation even though he scarcely had enough manpower and equipment to pursue his own operational plans. In a radio to the War Department on 17 October, he stressed the extreme gravity of the struggle in the Solomons, reported on the contributions which his forces were making to the battle on Guadalcanal, and emphasized its importance to the over-all operational strategy against the Japanese:

From its inception I have been acutely aware of the critical situation in the Solomous and in fact anticipated it and reported it with Ghormley in July and in August. . . . I invited attention to the acute danger developing and begged review of the question by the President and the Chiefs of Staff lest it become too late.

Action in SOPAC has been supported to maximum capacity of my Air Force, using all planes that can reach targets; Ghormley three times has radioed his appreciation; Turner has also communicated to that effect; while it is impossible to assess all damage it is believed that our bombardment missions have had a vital effect upon the success of operations to date by pounding hostile air and supply installations; the major effort has been expended in Rabaul area, because it can be reached with full bomb loads, while planes going to Solomons area must carry bomb bay tanks, reducing bomb load by 50 %; am in constant communication with Ghormley coordinating my reconnaissance with his and have given immediate priority to his requests for reconnaissance and attack; planes capable of affecting his situation have been used exclusively to that end; three times within the last week I have ordered missions on Ghormley's request, using every bomber available during the period; my own operations in New Guinea have been supported only by short-range aircraft. . . .

My operation to capture the north coast of New Guinea in full swing; am greatly hampered by the total lack of light shipping, landing boats, and barges, which I have previously requested; in their absence am moving overland and by air; supply is the controlling factor and must be accomplished by native carrier and by air; improvised landing fields have been and are being prepared; . . . supply difficulties incredible and limit speed of movement and size of forces and are of course multiplied by lack of shipping and shortage of transport planes; the possibility of success of the movement and the time factor are still unpredictable; under severe pressure from this Headquarters, much progress has been made beginning with the development of Port Moresby, which had initially a garrison of a regiment of militia, one landing strip without even parking space for planes and a port capacity of only 500 tons a day. . . .

If we are defeated in the Solomons, as we must be unless the Navy accepts successfully the challenge of the enemy surface fleet, the entire Southwest Pacific Area will be in the gravest danger; information derived from enemy sources indicates that an attack on Milne Bay and possibly elsewhere in New Guinea is contemplated for mid-November; urge that the entire resources of the United States be used temporarily to meet the critical situation; that shipping be made available from

---

22 An example of the effect of General MacArthur's campaign in New Guinea on Japanese operations in Guadalcanal is shown by a statement of Colonel Kazuyoshi Obata, Supply Staff Officer of the Japanese Eighteenth Army: "Due to the Allied advance in New Guinea, engineer, air, and anti-aircraft units meant for Guadalcanal were deployed to New Guinea. At this time also submarines and sea trucks were bringing supplies to New Guinea instead of Guadalcanal as it was impossible to use them there. One example of a unit deployed to New Guinea that was meant for Guadalcanal was the 51st Division." Interrogation Files, G-2 Historical Section, GHQ, FEC.

any source, that one corps be dispatched immediately; that all available heavy bombers be ferried here at once and that urgent action be taken to increase the air strength at least to the full complement allotted for this area; that immediate action be taken to prepare bases for naval operations on the east coast of Australia; that the British Eastern Fleet be moved to the west coast of Australia.[23]

An important factor in the Guadalcanal Campaign made itself felt at this time—AIB "coast watchers" sent out by General MacArthur's G-2 to operate secret radio stations behind enemy lines and report on Japanese troop, plane and ship movements. Carefully placed at strategic locations in the Solomon Islands, these agents were particularly effective in sending radio spot reports on imminent Japanese aerial attacks.

The main Japanese air bases for operations against Guadalcanal were at Rabaul on New Britain, Buin on Bougainville, and Buka Island, with Kavieng on New Ireland as a supporting base. AIB agents located at the key points of Buin and Buka Passage were ideally situated for observation purposes. (Plate No. 25) They had perfected a network by which they were able to give three successive warning signals of Japanese bombers en route to Tulagi and Guadalcanal. United States forces at Tulagi and at Henderson Field had ample notice of impending air attacks and were able to gain a decided advantage by having their planes aloft and ready to strike at the most opportune time.[24]

AIB "coast watchers" also reported on Japanese harbor activity in the waters adjacent to the Solomon Islands. One party in the hills overlooking Bougainville Island sent daily reports on enemy harbor activity to the Allied Fleet off Guadalcanal's shore. Another party gave details of sea and air arrivals and departures at Buka Passage, an important anchorage for ships operating against Guadalcanal. Other agents at Gold Ridge near Lunga and in northwest Guadalcanal formed an interlocking and efficient intelligence and radio communication net.[25]

Despite the reverses suffered in the Battle of Cape Esperance, the Japanese during the succeeding days continued to press their attacks on Guadalcanal. Although sustaining heavy losses, the enemy succeeded in getting a number of transports through and landed the major portion of another division. Meanwhile, enemy submarines and aircraft renewed their efforts to interrupt Allied lines of communications, and it became increasingly clear that the

---

23 CINCSWPA Radio No. C731 to WARCOS, 17 Oct 42, WD No. 255 C/S GHQ (S).

24 The effectiveness of these "coast watcher" parties is well illustrated by the results during the first days of the Allied landings. Four hours after United States forces launched their attack on Guadalcanal, AIB agents warned that twenty-four enemy torpedo bombers were speeding to counterattack. The bombers arrived and were met by Allied fighter planes in position. Only one enemy plane escaped. Early the next morning, the AIB agent at Buka Passage reported forty-five Japanese bombers and fighters heading southeast. Within thirty minutes preparations were underway at Tulagi to intercept the expected attack. The same afternoon the AIB agent at Buin reported more aircraft. Again the American fighters met them and again the intended blow was smashed. An enemy attack the next day met disaster in like manner. Crippled by losses in the air, the enemy was unable to mount another strike for several days, giving the Allied forces precious time in which to consolidate their beachhead gains. In the afternoon of 20 August, the first Grumman "Wildcats" flew into Henderson Field. The following day, warned by "coast watchers" that Japanese planes were on the way, they soared aloft, intercepted at the most favorable altitude, and shot down a number of enemy craft. The same routine was carried out the following day with equally satisfactory results. Thereafter, almost daily for a month, forewarned "Wildcats" intercepted the oncoming enemy, shooting them out of the sky in large numbers. Eric Feldt, *The Coast Watchers* (New York, 1946), pp. 58–103.

25 Ibid. See also: G–2, GHQ, FEC, *Intelligence Series*, Vol. IV, Operations of the Allied Intelligence Bureau, GHQ, SWPA.

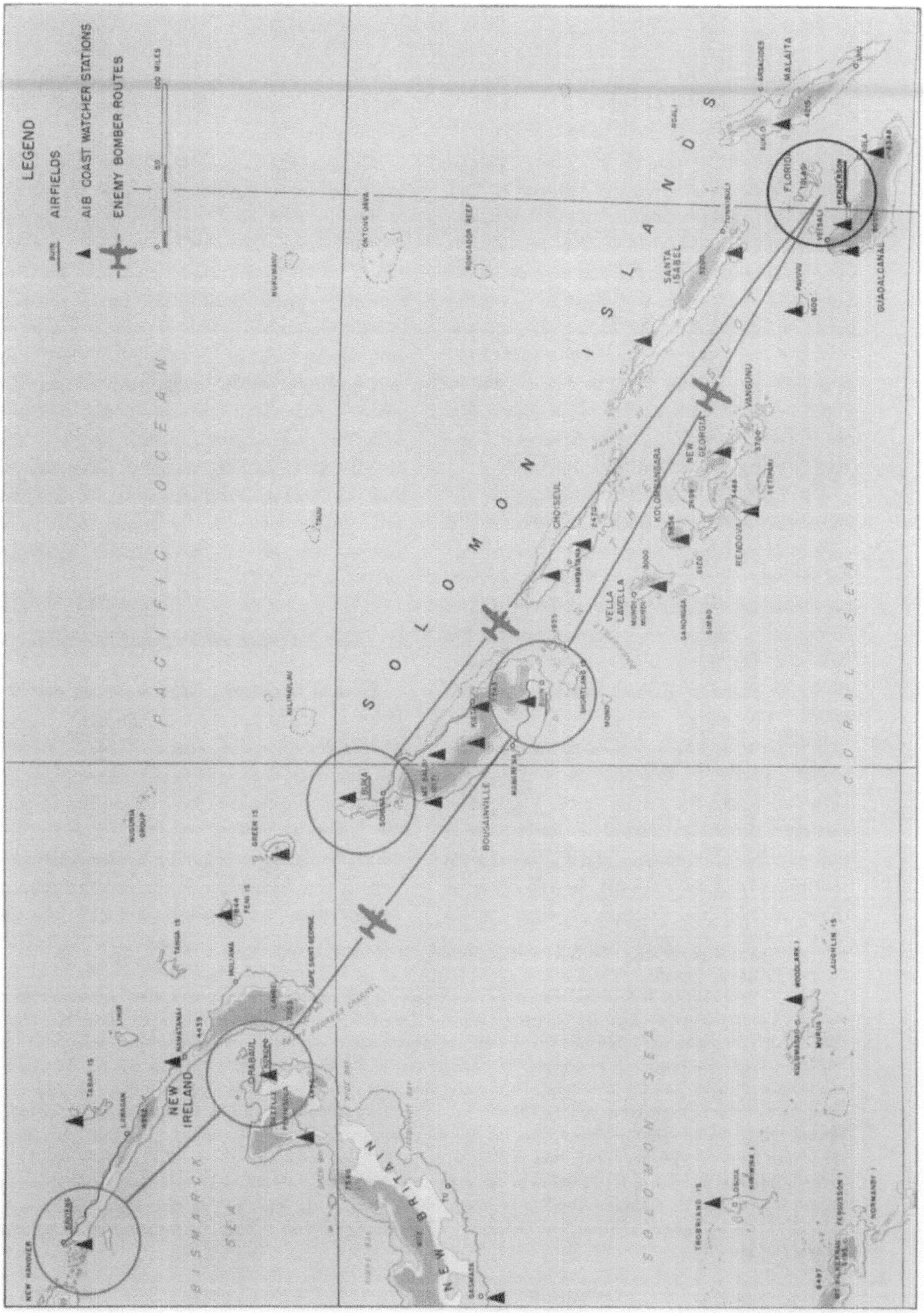

PLATE NO. 25
Coast Watcher Stations (AIB) along the Japanese Bomber Routes

next Japanese move would be supported by powerful surface and air units.

On 18 October, Vice Admiral William F. Halsey relieved Admiral Ghormley in the South Pacific Area. On the night of 23 October, the Japanese began a ground offensive against the Guadalcanal airfield and at the same time moved their naval units east of the Solomons to support the offensive. In the battle which followed on 26 October, aircraft from the American carriers *Hornet* and *Enterprise* put two enemy carriers out of action. Four enemy air groups were cut to pieces, but the *Hornet* was sunk and the *Enterprise* damaged, forcing the Allied fleet to retire.[26]

On shore that same day, following an all-out enemy attack, the question of whether the Allied forces could retain the airfield hung in the balance. A counterattack, however, restored the situation, and initiated a series of Allied successes. Except for a minor assault the following day, this constituted the last serious threat by Japanese ground forces on Guadalcanal. The critical period was safely past.

The Japanese still exercised control over the waters adjacent to Guadalcanal, and for the next two weeks the Allied forces engaged in scattered actions calculated to interfere with that control. Submarines and PT boats were continuously active against hostile lines of communication and launched repeated attacks against enemy attempts to land during the night from surface craft dispatched from neighboring islands.

Allied troops on Guadalcanal had been reinforced on 6 November and a second group of ships arrived on 11–12 November. Simultaneously, the Japanese, in another invasion effort, concentrated their naval forces and transports in the Rabaul-Buin area and again moved southward. In a three-day battle beginning the night of 12 November, the Allied forces, despite heavy losses, gained a decisive victory, and their position in the southern Solomons was secured. Except for the "Tokyo Express," which succeeded from time to time in landing small quantities of supplies and reinforcements to the enemy, control of the sea and air in the southern Solomons passed to the Allies.[27]

The Advance in Papua Continues

General MacArthur was convinced that the Japanese would not deviate from their original plan to complete the seizure of New Guinea. He felt that their operational strategy would call for a firm hold in both New Guinea and the Solomons if they were to protect the key points necessary to maintain a dominant position in the Southwest Pacific and thereby in eastern Asia.[28] In order to forestall Japanese

26 Adm. Ernest J. King, *Our Navy at War, A Report to the Secretary of the Navy,* 27 Mar 44, pp. 35-36.

27 Ibid., pp. 37-38.

28 The correctness of General MacArthur's assumption is demonstrated by a message which Lt. Gen. Hatazo Adachi, Commander of the Japanese Eighteenth Army in New Guinea, gave to his troops on 26 November 1942: "The East New Guinea and Solomon Island areas are very important not only for the immediate protection of the important southern areas, which we occupied at the beginning of the Great East Asia War, but also for the security and defense of Japan Proper. Therefore, it is necessary for us to occupy these areas as the first line. Furthermore, these are the most strategic areas, and the absolute control of them as offensive bases is necessary in order to cut the communication line between the United States and Australia and thus disrupt the enemy's plans. For this very reason, the United States and Britain . . . have been making a serious full-scale counteroffensive for the past four months in order to recapture these bases. . . . The area of operations in which we are participating [New Guinea] is as extremely important as that of the Solomons Islands. In view of the situation of the whole war, the first objective of our army is to secure the strongholds in East New Guinea." ATIS, GHQ, SWPA, Current Translations No. 13, Spot Report No. 45.

efforts to strengthen their positions and renew their assault on the Allies in Papua,[29] General MacArthur decided to execute the final phase of his offensive—the simultaneous advance on Buna—Sanananda. On 6 November he moved forward into New Guinea and established an advance echelon of General Headquarters at Government House, Port Moresby. Losing no time, he set 15 November as a tentative date on which his three-pronged drive would stab at Buna—Sanananda.[30]

On 2 November the Australian 7th Division, now commanded by Gen. George A. Vasey who had relieved General Allen on 29 October, captured the airfield at Kokoda. The Australians quickly readied the landing strip to ease the supply problem, and then continued to push forward. The Japanese made a resolute stand on the high ground at Oivi and Gorari. General Vasey attacked with the 16th Brigade assaulting frontally, while the 25th Brigade moved against the enemy flank and rear. (Plate No. 26) On 10 November the enemy was driven from his positions in full retreat after suffering heavy casualties.[31]

General MacArthur described this envelopment and rout of the Japanese forces:

> *The enemy was forced from his main positions near Oivi with heavy loss. His retreat has been blocked by our enveloping troops astride the main track and he is endeavoring to cut his way through the rear. Simultaneously our forces enveloped and destroyed enemy forces trapped south of Gorari.*[32]
>
> *Our ground troops have reached the vicinity of Ilimor. Here the enemy defending force has been surrounded. Of the enemy detachment which was encircled and destroyed at Gorari, bodies of five officers and more than five hundred men have already been counted in the jungle. In the Oivi pocket several hundred additional dead have been found. Allied fighters cooperating with ground units strafed and silenced enemy positions in the rear areas.*[33]

The surviving Japanese set out for Buna. The main body moved toward the coast by plodding laboriously along the flooded west bank of the Kumusi River. A few attempted the hazardous trip down the river itself on hastily improvised rafts. Hearing the heavy mortar fire in the direction of Buna and feeling that his immediate presence on the coast was

29 To cope with the imminent threat to their holdings, Imperial General Headquarters on 16 November 1942 had established the Eighth Area Army under which defense of each geographical region was entrusted to a separate army. The new Eighteenth Army was to take over all operations in New Guinea while the Seventeenth Army was directed to concentrate its resources entirely on combating the Allied attack in the Solomons. At the same time plans were made to increase their air power and augment the shipment of troops and supplies to the Southwest Pacific Area. Japanese First Demobilization Bureau Report, *Southeast Area Operations Record, Part III, Eighteenth Army Operations,* Vol I, pp. 16–18, G–2 Historical Section, GHQ, FEC.

30 The Australian 7th Division was to strike from Kokoda, one part of the United States 32nd Division from Kinjaki, and the other part from Pongani. The date was dependent upon developments in the Solomons struggle and upon the speed with which supplies could be placed behind each column of advance. Memo AC/S G-3, 3 Nov 42, G-3, GHQ, SWPA *Journal.* (S).

31 This frontal attack and double envelopment by the Australian 16th and 25th Brigades is described in the official Japanese war records as follows: "Since the latter part of October, the main force of the South Seas Detachment occupied positions in the Oivi Sector (southeast of Kokoda) and resisted enemy attacks. The enemy not only came in contact with our defensive lines but also cut off our retreating route north of Ilimor and occupied Ilimor Hamlet. This disrupted the line of communications between the frontal and reserve units of the South Seas Detachment. The commander of the Detachment decided to retreat to the right of the Kumusi River, and on 10 November the Detachment left the battlefield for Papaki." Japanese First Demobilization Bureau Report, *Southeast Area Operations Record, Part III, Eighteenth Army Operations,* Vol I, pp. 14–15, G-2 Historical Section, GHQ, FEC.

32 GHQ, SWPA, Communique No. 214, 12 Nov 42.

33 GHQ, SWPA, Communique No. 216, 14 Nov 42.

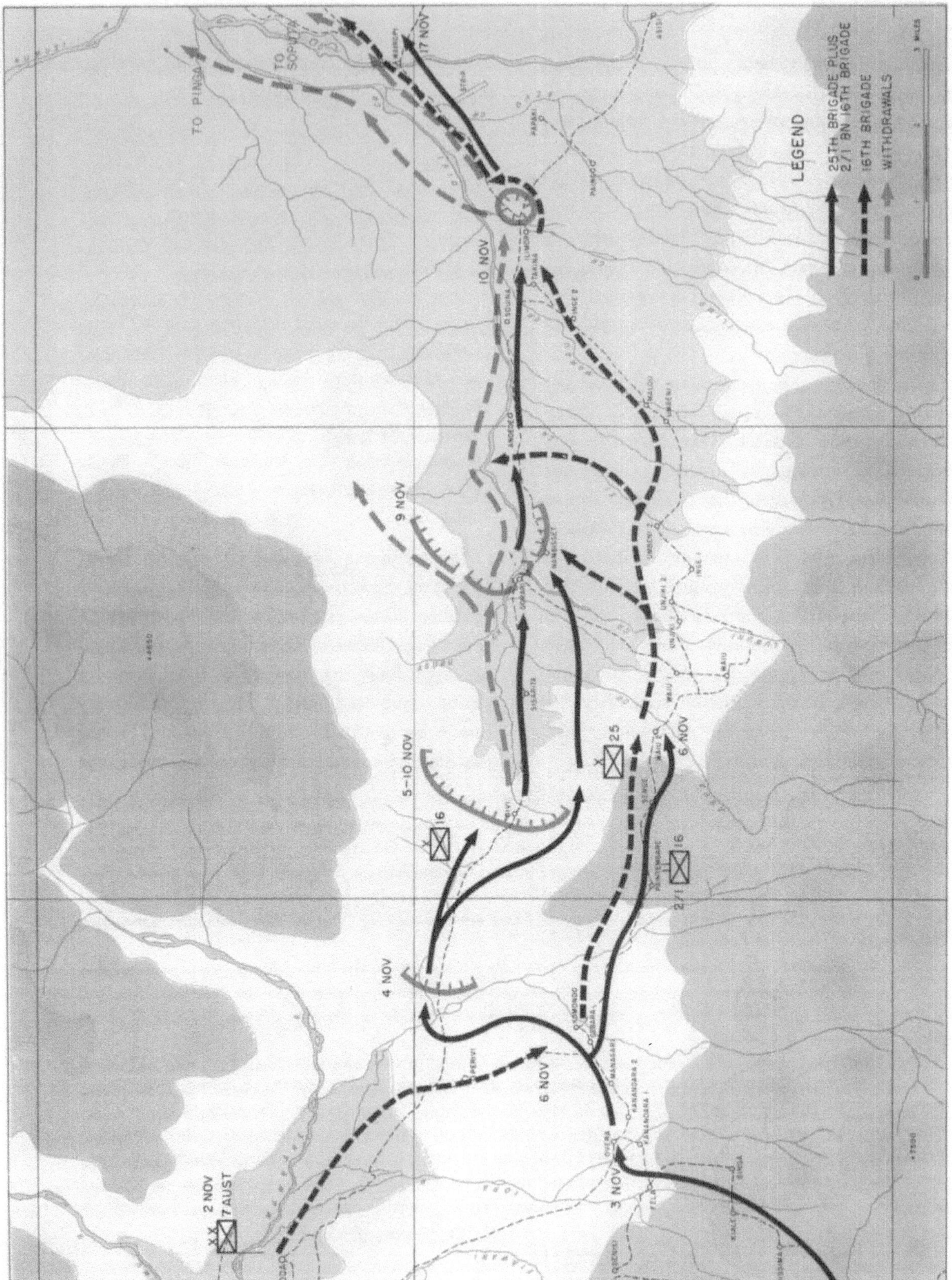

PLATE NO. 26

Action in Kokoda—Oivi—Wairopi

November 1942

imperative, General Horii decided to speed his progress by riding one of these rafts to the mouth of the Kumusi River and then paddling along the shore to the Buna area. He succeeded in reaching the sea with only one mishap, but during the short trip from the river mouth to the Japanese headquarters by native canoe the General and his staff officer were drowned during a squall. Many of his troops, however, succeeded in escaping down the Kumusi to the coast and eventually reached Gona to join in its defense.

On 14 November, New Guinea Force issued instructions for the Australian 7th and the United States 32nd Divisions to advance toward the sea from their positions on Hydrographers Range. The 7th Division was ordered to move from the vicinity of Hagenahambo toward Sanananda—Gona, and the 32nd Division, up the coast from the vicinity of Oro Bay and Bofu to Buna.[34] The boundary between the two forces was formed for a distance by the Girua River and then followed a stream running southwest between Inonde and Popondetta.

The combat zone in which the final Papuan operations took place is a low, flat coastal plain stretching inland from Gona and Buna to the foothills of the Owen Stanley Range. It is covered with dense jungle, strips of forest, and large kunai grass patches with extensive sage and mangrove swamps bordering the coast. The area is divided by the Girua River which empties into a broad delta stretching between Sananda Point and Buna Village. The overflow from the river runs into the low-lying ground south of the Buna Government Station and the Buna airstrips, forming with the delta a thick, impassable swamp. Movement was generally limited to the few native tracks which, although following the higher ground, were muddy at all times and hopelessly boggy in wet weather. In the Buna—Gona area there were five native trails: (1) Soputa to Sanananda, (2) Junbora to Gona, (3) Soputa to Buna Village, (4) Dobodura to the Buna airstrips, and (5) the coastal track north through Cape Endaiadere.[35]

The Australians crossed the Kumusi River on 14 November and quickly constructed a bridge at Wairopi; by 17 November the two Australian brigades were across the bridge in pursuit of the Japanese. The 25th Brigade, moving along the Junbora—Gona track, was held up about a mile south of Gona by strongly resisting enemy forces. On the following day the 16th Brigade, on the Soputa—Sanananda track, encountered well-prepared positions a mile and a half north of Soputa.

East of the Girua River the movement of the United States 126th and 128th Infantry Regiments from Port Moresby was completed by air transport to the Wanigela, Pongani, and Sapia landing strips. The use of air transport not only speeded the placement of combat troops on the fighting lines but also served to avoid the extreme physical exhaustion of a march through the New Guinea jungle, a factor which had proved so costly to the Japanese in their initial advance.

By the time the Australians had defeated the Japanese at Oivi and Gorari and pushed across the Kumusi River, the American forces under General Harding had started forward.[36] The rapidity of their advance and the devious paths of the difficult terrain divided the units of the 32nd Division into two main groups.

34 New Guinea Force, Opn Instr No. 42, 14 Nov 42, G-3, GHQ, SWPA *Journal* (S).

35 *Report of the Commanding General, Buna Forces on the Buna Campaign, Dec 1, 1942-Jan 25, 1943,* pp. 2–9.

36 The 126th Infantry at Pongani, Bofu, and Natunga moved forward on Embogu and Inonde. The 128th Infantry at Embogu, Eroro, and Embi advanced on Cape Sudest and Dobodura. New Guinea Force, Situation Report 280–281, 17 Nov 42, G-3, GHQ, SWPA *Journal.*

Urbana Force was deployed along the Ango—Buna track and contacted the enemy about a mile south of Buna. Warren Force was situated in the coastal area south of Cape Endaiadere, on a line running inland to Sinemi Creek. One battalion of the 126th Infantry had crossed the Girua River at Inonde and, joining forces with the Australian 7th Division, had remained to participate in the defense of Soputa and the subsequent operation on Sananda.

Thus, at the end of November the Japanese were pinned down in the narrow coastal strip from Gona to Buna and, with the sea at their backs, faced the Allied forces to the west and south.

Attack Stalemated

It was hoped that a co-ordinated drive of the three Allied forces, poised before the last remaining enemy strongholds in Papua, would be able to gain an early victory. The Japanese, however, realizing that they could no longer retreat without forfeiting the vital air bases along the northern coast, were determined to hold their positions no matter what the cost.[37] They accordingly took every advantage of the terrain, prepared strong defenses in depth, and fought with great courage and tenacity. The final phase of the struggle for Papua and the reduction of the enemy pockets on the northern coast witnessed some of the most gruelling and savage fighting of the entire campaign.

During the operations along the Kokoda Trail, enemy troops remaining in the Buna—Gona area had methodically prepared a series of well-designed and extremely strong defenses. Every contour of the terrain was exploited and the driest stretches of land were carefully chosen to be occupied and fortified, making it impossible for the Allies to execute any lateral movement without becoming mired in swamp.[38] All potential roads of approach were cleverly channelled into narrow corridors along tracks which led into a murderous crossfire of well-hidden machine guns. The western flank of the Buna defenses was protected by the sea and the unnavigable bog of the Girua River. (Plate No. 27) The worst swampland in the Buna area, between Entrance and Sinemi Creeks, formed a wide, impassable stretch joining with the thick jungle to protect the middle of the line. The eastern flank was anchored on the seacoast south of Cape Endaiadere. While any movement of the Allies from one flank to the other would have

37 The "Old Strip" at Buna, for example, which was enlarged by the Japanese to 1300 x 90 yards and protected by blast bays, was a valuable military asset for either side. If the Japanese could defend and secure Buna, the "Old Strip" would provide a base from which they could attack nearby Allied installations with fighter cover. Conversely, if it were in the hands of the Allies, it would give them an air base unhampered by the limitations of the Port Moresby bases, which were subject to the hazards of the Owen Stanleys and their unpredictable weather. *Report of the Commanding General, Buna Forces on the Buna Campaign, Dec 1, 1942-Jan 25, 1943*, p. 3.

38 The occupation of the limited dry areas was a great advantage to the enemy when the weather in the Buna area is considered. During the closing stages of the Papuan Campaign, eight inches of rain were recorded on one day (11 January 1943) in the Buna area. *Report of the Commanding General, Buna Forces on the Buna Campaign, Dec 1, 1942-Jan 25, 1943*, p. 5. The heavy rains may be cited as an example of climatic conditions which constantly hampered combat operations and which were utterly foreign to the xperience of all troops entering the New Guinea combat areas. The mean annual rainfall in the Buna area, for example, is about 120 inches with an average of 163 rain days a year; even these figures are exceeded by those of several other key points in New Guinea. AGS, GHQ, SWPA, Terrain Study No. 27, *Buna and Plains of Northern Division*, pp. 23-25. The heavy rainfall of New Guinea can best be appreciated by a comparison with the United States where about 40 percent of the land area receives less than 20 inches of rain annually, about 35 percent receives from 20 to 40 inches a year, and 25 percent receives from 40 to 60 inches a year.

entailed an arduous two-day trek via Ango and Sinemi, the Japanese could move by fast motor transport inside the perimeter of their defenses to bolster any weakening point almost immediately.

West of the Girua River, the enemy position protecting Sananada constituted a deep beachhead roughly triangular in shape, the apex three and one-half miles inland on the Soputa—Sananada track and the base resting on strongpoints along the coast between Cape Killerton and Tarakena. Gona, strongly fortified, was a flank position to the northwest.

To complicate the problem of piercing the enemy defenses, the already difficult logistical situation had become even more critical. Allied lines of communication were stretched 1,700 miles from Australia to the landing strips and supply dumps along the coast of New Guinea. Rations, ammunition, and equipment for Urbana Force and the Australian 7th Division were generally transported by the Fifth Air Force from Port Moresby to Dobodura and Popondetta.[39] Flying weather was usually bad and for days at a time supply planes were grounded, unable to penetrate the thick, low-hanging clouds which veiled The Gap. This break in the supply line at so crucial a period caused the failure of the first Allied thrust at the enemy's coastal defenses. When elements of the 25th Brigade attacked Japanese positions at Gona on 19 November, they were forced to withdraw because of a shortage of ammunition.[40]

Warren Force was supported by light vessels creeping along the coast from Port Moresby to Porlock Harbor and thence to Oro Bay where the Combined Operational Service Command had established advance bases. At Oro Bay the supplies were transferred to small craft for movement to Hariko. Jeeps, trailers, and native porters then took over the task of transportation to the forward areas. This limited means of conveyance under constant threat of enemy air and sea attacks restricted the flow of supplies and equipment to the barest essentials.[41]

As an additional aggravation, the New Guinea jungle and swamp were unrelenting in their toll of disease and sickness.[42] The troops in the Buna area were beginning to suffer from an increasing malaise and lassitude engendered by the stifling climate and an insufficient food ration. The incidence of illness climbed rapidly as man after man was stricken and hospitalized with at least one of the multitude of ever-present

---

39 General MacArthur's reliance upon air transport for the conduct of operations is shown in his radio of 24 November: "My campaign in northern Papua is being supported and supplied entirely by air; in view of enemy reinforcement and increased resistance the situation of my air transport is serious. Do not feel I should call for assistance upon SOPAC, which is also hard pressed; believe assistance should be supplied from theaters not engaged in active operations; request two groups each of 3 troop carrier squadrons be dispatched here as soon as possible; advise if this can be done." CINCSWPA Radio No. P-289 to WARCOS, 24 Nov 42, WD No. 296, C/S GHQ (S).

40 ALF Operations Report No. 222, 19 Nov 42, G-3, GHQ, SWPA *Journal*.

41 Attacks by enemy planes on 16–17 November put this seaborne supply line out of operation for about three weeks. MID, WD, *Papuan Campaign, The Buna-Sanananda Operation*, p. 22.

42 Disease riddled the Allied ranks throughout the entire New Guinea campaign. In the Buna operation, for example, the total number of casualties of all kinds in the U.S. 32nd Division was 10,960. Of this total, 8,286 were disease casualties, 5,358 of them caused by fever, principally malaria. *Report of the Commanding General, Buna Forces on the Buna Campaign, Dec 1, 1942-Jan 25, 1943*, p. 105. The same story is told by an Australian report: "The sickness rate at least doubled the battle casualties, proving that, in some respects, the country was an enemy as formidable as...the Japanese. In this phase of the first New Guinea campaign, of every nine men put out of action, three were battle casualties—one killed, two wounded—and six were sickness casualties. Any reserve force was, therefore, suffering two-thirds the wastage of manpower of those engaged in actual battle." *The Jap was Thrashed, An Official Story of the Australian Soldier*, (Director General of Public Relations, Melbourne), p. 85.

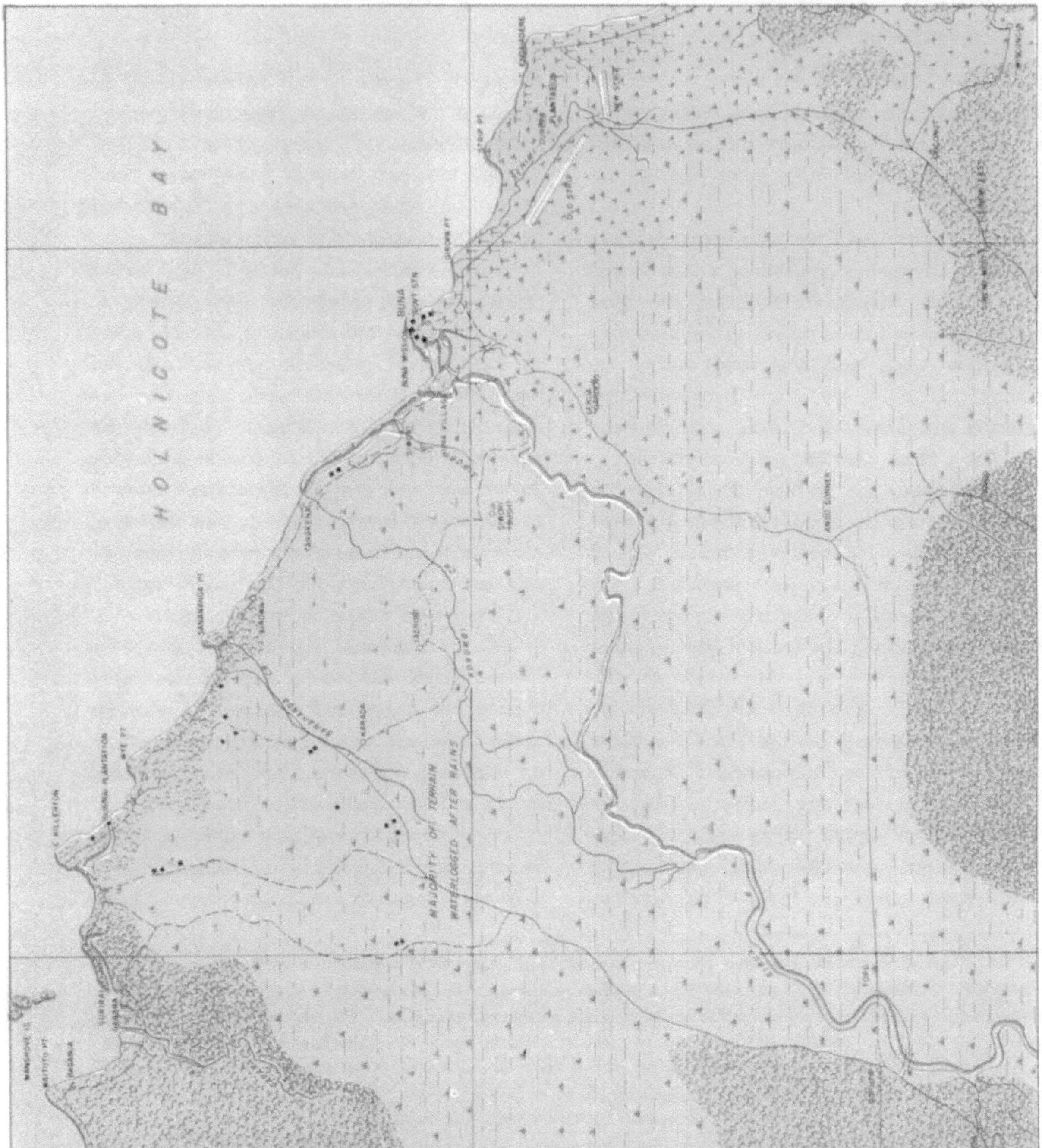

PLATE NO. 27

Terrain Features, Buna—Gona Area

fevers which infested the entire region.[43]

Initial assaults against the enemy entrenchments failed to achieve any appreciable gains. General Blamey, feeling that his covering artillery fire was inadequate to support a frontal attack, proposed that an amphibious landing be made at the rear of the enemy's positions at Buna.[44] The lack of small landing craft, however, and the fact that the Navy was unwilling to risk destroyers and corvettes in the shallow, uncharted, and dangerous waters off Collingswood Bay prevented his plan from being carried out.[45]

The Allied offensive, therefore, unable to pierce the enemy's powerful positions by frontal assault and incapable of mounting an amphibious flanking maneuver, had to give way to painstaking infiltration tactics. The situation developed into a virtual stalemate with activity on both sides restricted to sporadic thrusts by small groups. In fact, the entire Buna campaign was characterized by feats of the individual soldier, constantly isolated from his main unit and left to his own resources in the fight against the enemy.

The Fall of Buna

In order to break the existing deadlock, General MacArthur sent a new commander, General Eichelberger, to direct the Allied forces attacking Buna. He arrived with his staff at Dobodura on 1 December and took charge of the operations east of the Girua River. General Eichelberger found that the units of the 32nd division had become badly disorganized during the fighting, seriously weakening the chain of command. He proceeded immediately to regroup the scattered units and at the same time reorganize the supply system which had become disrupted and unreliable. Corps and division staffs were merged to form Headquarters Buna Force. The combined United States and Australian troops north of the Owen Stanley Range were placed under General Herring, Commander of Advanced New Guinea Force, whose headquarters were established at Popondetta on 28 November. General Blamey acted in the dual capacity of Commander, New Guinea Force, and Commander, Allied Land

---

43 The casualty rates for the Japanese forces in New Guinea also present a grim picture. The Papuan Campaign destroyed the myth, born during the Malayan operation, that the Japanese were unbeatable jungle fighters. Referring to New Guinea, a Japanese Army report stated: "Epidemics are numerous and the climate is bad.... The large number of patients is caused by malaria, diarrhea, skin disease, beri-beri, and malnutrition.... It is preferable to replace personnel at least every five or six months.... In the first month and a half, malaria manifested itself and by the end of the second month and a half the strength of the active force engaging in the operations decreased to half After three and a half months the strength had decreased to a third, and by the fifth and sixth months, the fraction of men physically fit became less than a seventh...." Imperial Japanese Headquarters, Army Section, *Lessons from Actual Experiences in Eastern New Guinea Operations, July 1942-April 1943,* published 18 Jan 42 as *Enemy Publication No. 285,* ATIS, GHQ, SWPA.

44 General Blamey Ltr to CINCSWPA, 26 Nov 42, ALF No. 32, C/S GHQ, SWPA (S).

45 Vice Adm. Arthur S. Carpender, COMSOWESPACFOR reported: "I do not favor sending destroyers north of Milne Bay for protection of small ships proceeding to Buna via Ward Hunt Strait and Cape Nelson, and strongly recommend against their being so employed. The area involved is poorly surveyed and charts unreliable. Shallow draft vessels [only] can proceed via coastal routes from Cape Nelson—this entire area is filled with reefs to such an extent that there is little or no sea room available for destroyers to maneuver. On the other hand, via the northern approach route, from Gasmata, the enemy can readily move destroyers and cruisers via deep water areas against the Buna area. To put a minor surface force in the Buna area would be of no value against a heavy force which the enemy might easily send in." COMSOWESPACFOR Indorsement, 10 Nov 42, to Comdr ALF Ltr to CINCSWPA, 7 Nov 42, ANF No. 5, C/S GHQ, (S).

Forces until after the fall of Buna when he returned to Australia. Brig. Gen. Albert W. Waldron replaced General Harding as commander of the United States 32nd Division[46]

On 5 December General Eichelberger's reorganized forces made a co-ordinated attack on both fronts in another effort to dislodge the Japanese. (Plate No. 28) They resisted stubbornly, however, determined not to yield a foot of ground. Warren Force made repeated assaults, only to be thrown back exhausted and disheartened. The efforts of Urbana Force proved more successful. By resourceful leadership and daring exploits, a platoon of the 126th Infantry managed to pierce the enemy network of interlacing bunkers and break through to the sea. The enemy counterattacked immediately but the units on the Urbana front held doggedly to their corridor which wedged into the enemy defenses, isolating Buna Village from Buna Mission. For the first time since the beginning of the campaign against Buna, the Japanese line had been breached.[47]

General MacArthur insisted that immediate advantage should be taken of the gains on the Buna front and directed that maximum strength be concentrated in that sector, while the Japanese were contained west of the Girua River with a minumum force. The United States 127th Infantry was rushed by air to bolster Urbana Force. The Australian 18th Brigade and a number of tanks were moved by water to the Warren front below Cape Endaiadere near Cape Sudest.[48] To avoid the necessity of diverting the artillery elements of the 32nd Division for reinforcement of its depleted infantry strength, General MacArthur ordered the 163rd Infantry of the 41st Division to be sent forward from Australia.

The presence of fresh troops in the Buna area made itself felt immediately. On 14 December, Urbana Force launched a heavy attack against Buna Village. Despite fierce resistance and repeated attempts by the enemy to bring in water-borne reinforcements, the village was at last wrested from the Japanese. General MacArthur, describing the first major victory in his coastal offensive, reported as follows :[49]

---

46 Col. Clarence A. Martin became the commander of Warren Force and Col. John E. Grose, of Urbana Force. The 1st Battalion, 126th Infantry, took over the lines on the Sinemi-Buna trail south of the bridge ; the 1st Battalion, 128th Infantry occupied the Plantation position ; the 3rd Battalion 128th Infantry went into reserve on the coast behind the right flank. MID, WD, *Papuan Campaign, The Buna-Sananda Operation*, p. 37.

47 The Division Commander, General Waldron, was wounded during the attack of 5 December and was replaced by Brig. Gen. Clovis E. Byers.

48 Brig. Wootten of the Australian 18th Brigade, took over command of Warren Force upon his arrival on 14 December.

49 GHQ, SWPA, Communique No. 247, 15 Dec 42. The conclusion in this report was later confirmed from Japanese sources. Two Japanese landing forces had each made two attempts to bring aid to their besieged comrades in Papua. This second landing force was composed of the 25th Field Machine Gun Company and the 1st Battalion (less two companies), the 9th Company, and the Regimental Gun Company of the 170th Infantry. It was also accompanied by personnel from all sections of Eighteenth Army Headquarters including Staff Officer Col. Aotsu. On its first attempt the landing party was scheduled to land at Gona on 8 December but was so pounded by Allied planes that it was forced to return to Rabaul. It left again on 12 December in another largely unsuccessful effort to bring in reinforcements. The elements that succeeded in escaping Allied bombs landed about forty miles to the north of Gona, too distant and too late to render effective assistance. Gona had fallen to the Allies on 9 December, one day after the original landing force of enemy reinforcements was scheduled to arrive. Japanese First Demobilization Bureau Report, *Southeast Area Operations Record, Part III, Eighteenth Army Operations*, Vol I, pp. 25-28. G-2 Historical Section, GHQ, FEC.

Buna Village has been taken. It was occupied by our troops at ten o'clock this morning, December 14. . . . In another attempt by the enemy's naval forces to reinforce their ground troops in the Buna area, the enemy launched a convoy of two cruisers and three destroyers for a landing at the Mambare and Kumusi estuaries. Our air force intercepted this convoy. In heavy bombing and strafing attacks the enemy's landing barges were sunk or disabled. Survivors attempted to reach land by swimming, suffering heavy casualties. Supplies were set afire. Several hits and many near misses were made on the war vessels. The enemy's air force intervened unsuccessfully. . . . It is believed that a major enemy effort was largely parried.

Heartened by its success, Urbana Force pushed ahead. After the capture of Buna Village, it attacked on its right flank in an effort to eliminate the strong enemy positions at Coconut Grove and The Triangle. By 17 December, the first of these positions had been reduced and cleared. During this attack on the Grove, General Byers was wounded and General Eichelberger assumed direct control of the 32nd Division. When The Triangle resisted all attacks, it was contained and Urbana Force prepared to cross Entrance Creek and seize Buna Mission.[50]

On the Warren front, the stalemate was finally broken. On 18 December, Brigadier Wootten's 18th Brigade moved forward with tank support and secured the Cape Endaiadere area. United States units joined the advance and together the Allies pushed slowly westward along the coast and along the two airstrips toward Giropa Point. Sinemi Creek was crossed and the entire Warren Force moved on against weakening enemy resistance.

The final phase of the struggle for Buna was at hand. Urbana Force crossed Entrance Creek and, duplicating its feat at Buna Village, drove a wedge to the sea isolating Buna Mission from Giropa Point. The enemy, thus cut off in the Mission, was hammered mercilessly by a series of heavy Allied assaults. The Japanese troops, severed from all reinforcement, their leaders lost in action or by suicide, their bodies worn with starvation and sickness, and their morale shattered by the unrelenting blows of their attackers, were finally forced to yield their positions.[51] On 2 January, Buna Mission fell. Watching this successful conclusion to forty-five days of bitter and continuous fighting, General Eichelberger wrote:

At 4:30 PM, I crossed the bridge (from The Island) after "C" Company had passed and I saw American troops with their bellies out of the mud and their eyes in the sun circling unafraid around the bunkers. It was one of the grandest sights I have ever seen. . . . the 127th Infantry found its soul. . . .[52]

Warren Force, meanwhile, had cleared and occupied Giropa Point. The first week of January saw the entire coast of Papua east of the Girua River in Allied hands and the enemy force in the Buna area completely destroyed. It was a fitting culmination to a year of hardship and a propitious beginning for a New Year

50 When the Triangle was finally evacuated by the enemy and occupied by units of the 32nd Division on 28 December, it was found to be honeycombed with no less than 18 bunkers. These defensive strongholds were powerfully fortified, mutually supported by connecting trenches and almost impervious to artillery bombardment. MID, WD, *Papuan Campaign, The Buna-Sananda Operation*, p. 57.

51 As the Papuan campaign came to a close, the Japanese made several desperate attempts to aid their sorely pressed force at Buna, but their efforts to land reinforcements at Lae for transfer south were repeatedly thwarted by the alert Allied Air Forces. The forced march of a hastily assembled rescue force from Gona and Giruwa failed to arrive in time to be of any effect. Japanese First Demobilization Bureau Report, *Southeast Area Operations Record, Part III, Eighteenth Army Operations*, Vol I, pp. 27–29. G-2 Historical Section, GHQ, FEC.

52 *Report of the Commanding General, Buna Forces on the Buna Campaign, Dec 1, 1942-Jan 25, 1943*, p. 36.

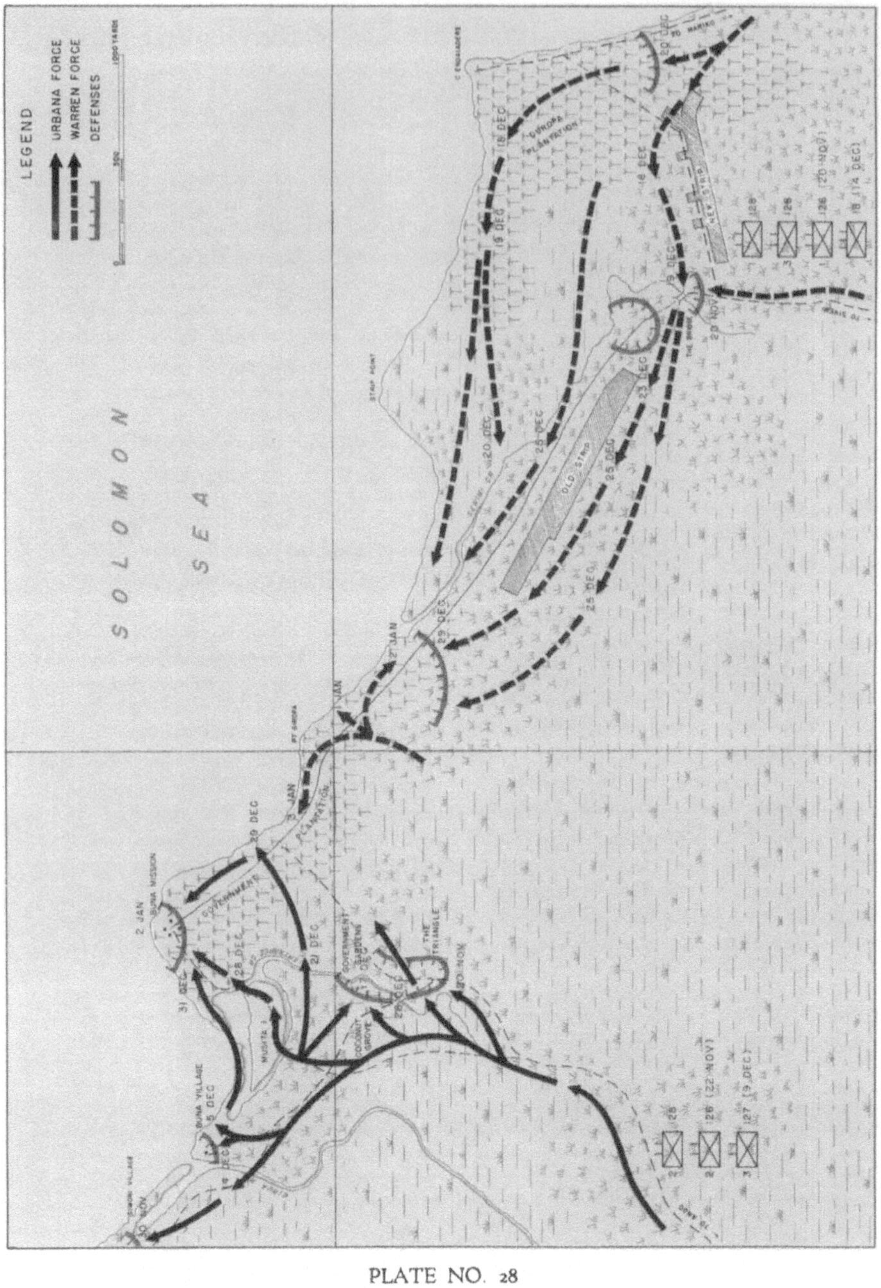

PLATE NO. 28
The Capture of Buna

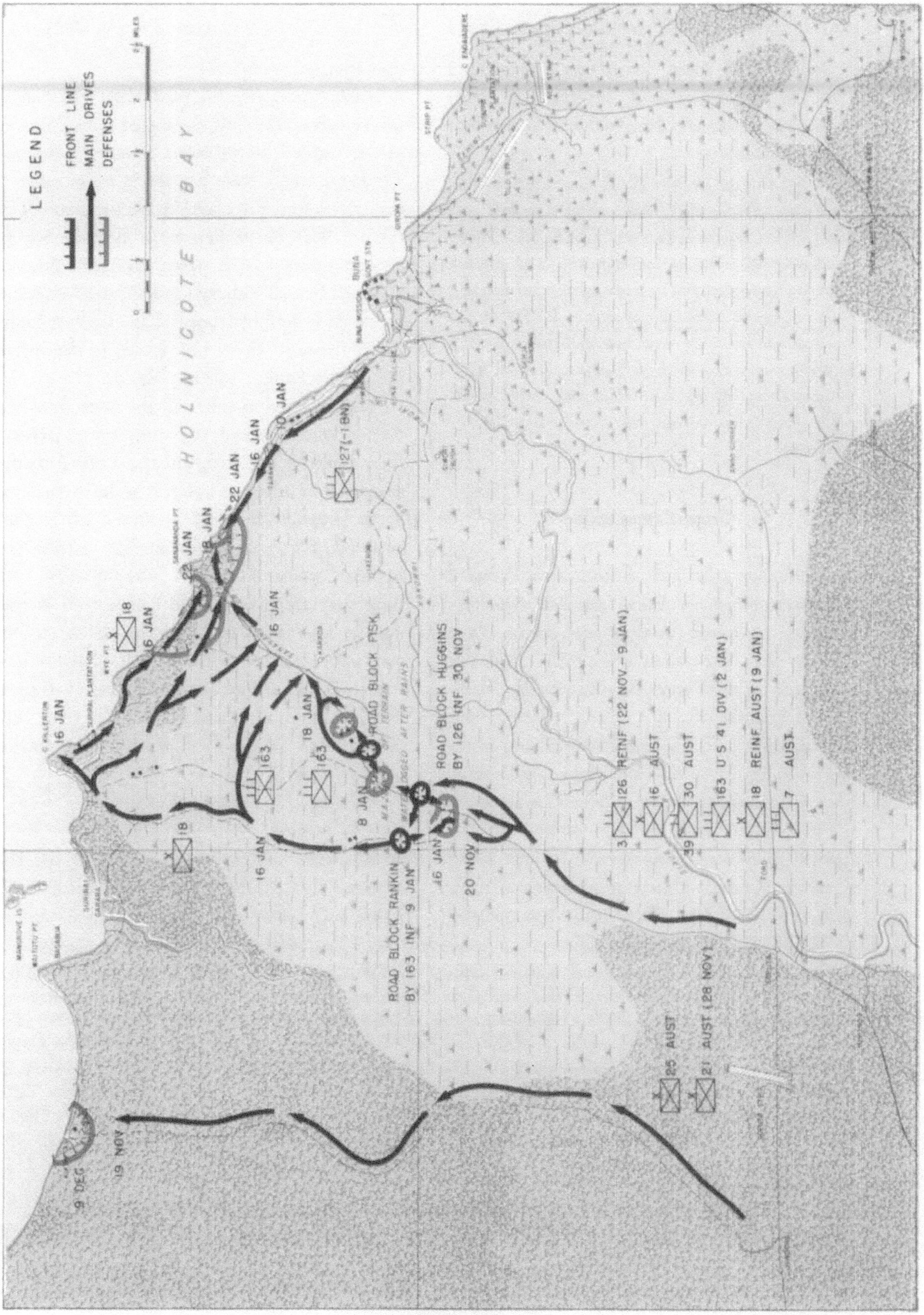

PLATE NO. 29

Sananande Front, 22 November 1942–22 January 1943

of hope. The Buna victory was a heartening tonic to the Allies and Secretary of War Stimson sent General MacArthur a warm letter of congratulations:

The coming of the New Year, coinciding as it does with your success at Buna impels me again to send you my warmest congratulations and good wishes. Have followed your masterly campaign with close interest and much gratification. . . . it is a tremendous satisfaction to feel that American fortunes in SWPA are in such skillful hands; am in constant touch with President Quezon here and we are both beginning to think with encouragement of the time, which now really seems approaching, when we shall redeem our promise to the Filipinos. . . .[53]

Gona-Sanananda

While the right arm of the Allied offensive was hammering at Buna, the left arm was pressing its attack west of the Girua River against the Sanananda triangular defensive position. It was from these two points, Buna and Sanananda, that the enemy had originally launched his land drive on Port Moresby along the winding twin trails which joined at Soputa. The Allied counterdrive, prevented from lateral movement by the deep swamps of the coastal plain, was compelled to follow along these very same trails and was consequently split by the Girua River into two parallel operations.

On 22 November, the Australian 7th Division under General Vasey, combined with elements of the United States 126th Infantry, launched a general attack from Soputa in an attempt to encircle the enemy positions on the Sanananda track. The Australian 25th Brigade moved against Gona to the west; the 16th Brigade attacked frontally along the road itself supported on each side by troops of the United States 126th Infantry whose task it was to drive into the enemy flanks. (Plate No. 29)

The frontal assaults of the 25th and the 16th Brigades lacked the numbers to achieve their objective but troops of the 126th Infantry managed to advance about a mile to the rear of the enemy's forward positions where they established a road block squarely astride the Soputa-Sanananda track.[54] The Japanese, realizing that this block would choke off a vital supply line to their forward defenses, immediately counterattacked with characteristic ferocity. The Americans however, would not be dislodged but held grimly to their entrenchments in the face of incessant assault from all sides.

Although the Allied forces struggled desperately to improve their positions around Sanananda, they were too exhausted and too few in number to do more than hold their ground. Their major effort was directed at maintaining

53 SECWAR Ltr to CINCSWPA, 6 Jan 43, WD No. 328, C/S GHQ (S). Prime Minister Curtin also expressed his gratitude in a letter of commendation dated 11 January 1943 which General MacArthur published to his troops: "Dear General MacArthur: The Order of the Day issued by you at the concluding stage of operations in South-Eastern New Guinea marks a notable historic stage of the war in the Pacific.... I would express to you, your commanders and all ranks of Australian and American Forces, the thanks and admiration of the Australian people and the Government for their magnificent services. The campaign has been fought under most trying conditions in one of the most difficult regions in the world. The forces under your command have not only overcome these immense natural difficulties but have decisively defeated a tenacious and stubborn foe. The campaign has been a demonstration of comradeship in arms and cooperation between the forces of the United States and Australia which I am sure will continue until the common foe is totally defeated." GHQ General Order No. 6, 13 Jan 42, AG, GHQ, SWPA No. 430113.

54 This block, established on 30 November, became known as "Huggins Block" after Captain Huggins who commanded the United States group which held it under ceaseless enemy attack until relieved by the Australian 39th Infantry Battalion on 22 December.

supplies to the all-important road block. In this they were greatly assisted by Papuan natives who risked their lives in daily infiltrations through watchful enemy lines, helping to carry in needed food and ammunition and to evacuate the wounded.

While the 126th Infantry was thus engaged, the Australian 21st Brigade was flown from Popondetta to join the exhausted 25th Brigade for another attempt to seize Gona. This time their drive was successful. Despite a fanatical, last-ditch defense by the Japanese who fought until they dropped amid the putrefying bodies of their dead comrades, Gona fell to the Allies on 9 December.[55]

It was not until after the fall of Buna, however, that the Allies were able to make any real progress toward Sanananda. Until the end of December it was a bitter see-saw struggle measured in yards, rather than in miles, of steaming, swampy road. Only after the close of the operations against Buna were the Allies permitted to shift their strength to the campaign west of the Girua River. The 163rd Infantry of the 41st Division, flown from Port Moresby to Popondetta and Dobodura, and Brigadier Wootten's 18th Brigade, transferred from Buna, were brought in to reinforce the hard-pressed forces on the Sanananda front.

General MacArthur advised General Blamey that, for strategic and logistic reasons, it was essential that an all-out attack on Sanananda be initiated immediately to close the campaign.[56] On 13 January, General Eichelberger was placed in command of Advance New Guinea Force which included all Allied troops north of the Owen Stanleys.

The Australian 18th Brigade pushed vigorously along a branch trail to Cape Killerton, reaching the coast on 16 January. It then turned southeast to approach Sanananda and Giruwa on the flank. The United States 163rd Infantry advanced frontally along the Sanananda track to assault the main enemy defenses before Sanananda.[57] The enemy's left flank was being threatened by units of the United States 127th Infantry which had crossed the Girua River after the capture of Buna and advanced northwest to take Tarakena. In the face of these three converging columns the battered enemy defenses began to crumble rapidly[58] and General MacArthur was able to

55 The Japanese defending Gona were under specific orders to fight to the bitter end. A bulletin of 19 November 1942 from the Commander of the Yazawa Unit at Gona Village read: "It is not permissable to retreat even a step from each unit's original defensive position. I demand that each man fight until the last. As previously instructed, those without firearms or sabers must be prepared to fight with sharp weapons such as knives or bayonets tied to sticks, or with clubs." ATIS, GHQ, SWPA, Current Translations, No. 15, p. 40.

56 CINCSWPA Radio to Gen. Blamey, 4 Jan 43, ALF No. 61, C/S GHQ, SWPA (S).

57 Alert handling and rapid interrogation of a captured Japanese soldier disclosed that a withdrawal had been ordered from the enemy's forward positions along the Soputa-Sanananda track. This valuable information enabled the Allies to catch the Japanese at a decided disadvantage as they were about to begin their evacuation. MID, WD, *Papuan Campaign, The Buna-Sanananda Operation*, pp. 73–74. G-2, GHQ, SWPA, *Daily Summary* No. 298, 14/15 Jan 43.

58 The extreme plight of the enemy at this point is vividly described by Maj. Gen. Kensaku Oda, who had replaced General Horii as commander of the South Seas Detachment, in his report to the Chief of Staff, Eighteenth Army; "Enemy aircraft flying above all day and bombing and strafing everything in sight; entire enemy artillery in action and fire being concentrated on us.... Communications disrupted between our first line and central positions since yesterday. Strong enemy force has penetrated into our central positions. Coast also being battered by enemy gunfire. Fresh enemy force from Buna area very active. Our short supply line leading from the coast already on verge of collapse. Most of the men stricken with dysentery. Those not kept in bed with illness are without food and too weak for hand-to-hand fighting. As the days go by, starvation is taking many lives, and it is weakening the already extended lines. We are doomed...." Japanese First Demobilization Bureau Report, *Southeast Area Operations Record, Part III, Eighteenth Army Operations*, Vol I, pp. 29–30. G-2 Historical Section, GHQ, FEC.

report the situation as follows:

The Papuan campaign is in its final closing phase. The Sananada position has now been completely enveloped. A remnant of the enemy's forces is entrenched there and faces certain destruction. With its elimination, Papua will be entirely cleared of the enemy. One of the primary objects of the campaign was the annihilation of the Japanese Papuan army under Lieutenant General Horii. This can now be regarded as accomplished. This army was approximately 15,000 strong.[59]

The End of the Papuan Campaign

Sanananda Point and Sanananda Village fell on 22 January and Giruwa was taken the same day. The only Japanese left in Papua were isolated groups fleeing or swimming along the coast to Salamaua. The Papuan campaign was finished.

During the entire struggle, the enormous flexibility of modern air power was constantly exploited. The calculated advance of bomber lines through seizure of forward bases meant that a relatively small force of bombers could attack under cover of an equally limited fighter force, operating at short and medium ranges. Each phase of advance had as its objective an airfield which could serve as a steppingstone to the next advance. In addition, as this airline moved forward, naval forces under newly established air coverage began to regain the sea-lanes which hitherto had been the undisputed arteries of the enemy's far-flung positions. Ground, air and sea operations were thoroughly co-ordinated.

In announcing the conclusion to six months of bitter, ceaseless struggle, General MacArthur said:

The destruction of the remnants of the enemy forces in the Sananada area concludes the Papuan Campaign. The Horii Army has been annihilated. The outstanding military lesson of this campaign was the continuous calculated application of air power, inherent in the potentialities of every component of the Air Forces, employed in the most intimate tactical and logistical union with ground troops. The effect of this modern instrumentality was sharply accentuated by the geographical limitations of this theater. For months on end, air transport with constant fighter coverage moved complete infantry regiments and artillery battalions across the almost impenetrable mountains and jungles of Papua, and the reaches of the sea; transported field hospitals and other base installations to the front; supplied the troops and evacuated casualties. For hundreds of miles bombers provided all-around reconnaissance, protected the coast from hostile naval intervention, and blasted the way for the infantry as it drove forward.[60] *A new form of campaign was tested which points the way to the ultimate defeat of the enemy in the Pacific. The offensive and defensive power of the air and the adaptability, range and capacity of its transport in an effective combination with ground forces, represent tactical and strategical elements of a broadened conception of warfare that will permit the application of offensive power in swift, massive strokes, rather than the dilatory and costly island-to-island advance that some have assumed to be necessary in a theater where the enemy's far-flung strongholds are dispersed throughout a vast expanse of archipelagos. Air forces and ground forces were welded together in Papua and when in sufficient strength with proper naval support, their indissoluble union points the way to victory through new and broadened strategic and tactical conceptions.*[61]

59 GHQ, SWPA, Communique No. 271, 8 Jan 43.

60 Enemy aircraft identified as lost or destroyed since 23 July, the opening date of the Papuan Campaign, were as follows: 333 destroyed, 89 probably destroyed, 117 damaged; total 539. Naval losses inflicted by Allied aircraft were as follows: sunk, destroyed, or seriously damaged: 6 cruisers, 13 destroyers, 1 destroyer tender, 2 seaplane tenders, 2 gunboats, 44 large to medium merchant ships, 39 small to medium merchant ships, 150-200 landing barges. GHQ, SWPA, Communique No. 271, 8 Jan 43.

61 GHQ, SWPA, Press Release, 24 Jan 43.

The close of the Papuan Campaign marked the first victorious operation of the Allied ground forces against the Japanese and the end of the first phase of operations in the Southwest Pacific Area.[62] In the northwestern sector, with the exception of the occupation of the Kai, Aroe, and Tanimbar Islands on 30 July and intermittent air raids on Darwin, the Japanese had been engaged in consolidation. In the critical New Guinea—Bismarcks—Solomons sector, their deepest penetrations had been halted and thrown back. Australia and the supply lines from the United States were secure, and the firm establishment of a base in Papua formed an initial and vital step in the Allied counteroffensive which would eventually reduce Rabaul to impotence and drive the Japanese from their forward bases in the Southwest Pacific Area.

62 General MacArthur cited a number of officers for their conduct of the Papuan Campaign in his Order of the Day on 9 January 1943: "It is my high honor to cite to the Order of the Day, for extraordinary courage, marked efficiency and precise execution of operation during the Papuan Campaign, the following officers: General Sir Thomas Blamey; Lt. General Kenney; Lt. General Herring; Lt. General Eichelberger; Major General Sutherland; Major General Vasey; Brigadier General Willoughby; Brigadier General Whitehead; Brigadier General Walker; Brigadier Wootten; Brigadier Eather; Group Captain Garing. The victory which has been achieved would have been impossible of accomplishment without the invincible leadership which they have provided. I have directed that each be awarded the Distinguished Service Cross of the United States, the highest decoration at my disposal, with appropriate individual citation. This award will serve for all time and for all eyes as the outward symbol of the devotion and gallantry with which they have performed their dangerous and difficult duty. The magnificent conduct of the troops and elements of this command, operating under difficulties rarely, if ever, surpassed in campaign, has earned my highest praise and commendation. In spite of inadequate means in many categories, their resourcefulness, their ingenuity, their adaptability, have produced a self-reliance that has overcome all handicaps and deficiencies. Through skill and courage and an indomitable will for victory they have defeated a bold and aggressive enemy possessing a marked superiority of resources and potentialities in the areas of campaign and combat. . . . To Almighty God I give thanks for that guidance which has brought us to this success in our great Crusade. His is the honor, the power and the glory forever, Amen." GHQ, SWPA, Press Release, 9 Jan 43.

CHAPTER V

UP FROM PAPUA

Concept of Strategy

The successful culmination of the Papuan Campaign opened the way for a drive up the New Guinea coast and laid the groundwork for long-range offensive planning which would disrupt Japanese strategy and destroy their war machine in the Southwest Pacific. The bitterness of the struggle in Papua, however, indicated the formidable task that lay ahead. The Japanese still occupied most of New Guinea, maintaining strong bases at Salamaua and Lae. On New Britain, Rabaul remained the focal point for the protection and reinforcement of their holdings in New Ireland, the northern Solomons, the Bismarck Archipelago, and the whole northeast area. Allied victories in Papua and Guadalcanal had temporarily contained the Japanese but did not threaten their main centers of power. These victories, however, provided invaluable bases for further assault and made possible a broader concept of strategy and offensive operations. (Plate No. 30)

To push back the Japanese perimeter of conquest by direct pressure against the mass of enemy occupied islands would be a long and costly effort. General MacArthur lacked the forces necessary to carry out such a scheme of frontal attack even if he was so minded for the Papuan Campaign had exhausted many of his troops and much equipment. Replacements trickled in slowly, providing only the minimum essentials with which to conduct immediate operations.

General MacArthur, however, envisioned a strategy of an entirely different nature:

> *My strategic conception for the Pacific Theater, which I outlined after the Papuan Campaign and have since consistently advocated, contemplates massive strokes against only main strategic objectives, utilizing surprise and air-ground striking power supported and assisted by the fleet. This is the very opposite of what is termed 'island hopping' which is the gradual pushing back of the enemy by direct frontal pressure with the consequent heavy casualties which will certainly be involved. Key points must of course be taken but a wise choice of such will obviate the need for storming the mass of islands now in enemy possession. 'Island hopping' with extravagant losses and slow progress ... is not my idea of how to end the war as soon and as cheaply as possible. New conditions require for solution and new weapons require for maximum application new and imaginative methods. Wars are never won in the past.*[1]

The successful employment of this type of strategy called for the wise selection of key points as objectives and the careful choosing of the most opportune moment to strike. General MacArthur accordingly applied his major efforts to the seizure of areas which were suitable for airfield and base development but which were only lightly defended by the enemy. In this way he could move his bomber line forward and yet avoid the bloody losses and drain on his resources which would result from frontal assaults on positions where the Japanese were

1 GHQ, SWPA, Press Release, 21 Sep 43.

concentrated in force. Thus, by daring forward strikes, by neutralizing and by-passing enemy centers of strength, and by judicious use of his air forces to cover each movement, General MacArthur intended to destroy Japanese power in New Guinea and adjacent islands and clear the way for a drive to the Philippines.[2]

The Struggle for Wau

When the ultimate fall of Guadalcanal and Papua became a certainty, the Japanese decided to consolidate their positions in the Southwest Pacific Area and retract their first line of defense. The key points along this new defensive perimeter included northern New Guinea, New Britain, and the northern Solomons. New Guinea in particular assumed special significance. Not only was it a strategic point on the right flank of the new defense line but its loss would provide the Allies with an ideal springboard for a thrust into the heart of the Japanese inner zones of operations.[3]

Accordingly, at the end of 1942, the Commander of the Japanese Eighteenth Army began to occupy and fortify points along the northern coast of New Guinea, landing fresh forces at Wewak and Madang. Airfields on Lae and Salamaua on the east coast were further developed and put to use. Plans were also made to send the entire Japanese 51st Division to secure and hold established positions in New Guinea and at the same time to exploit all possibilities for offensive action.[4]

The town of Wau, which had been occupied by a small contingent of Australians since March 1942, was particularly desirable for the fulfillment of Japanese plans. (Plate No. 31) Strategically situated in the Bulolo Valley, it had access to the key inland trails leading northward to Lae and Salamaua, and southward to Mambare and Kokoda. In addition it had a small airfield already constructed and sites suitable for new strips. In Allied hands, it was a valuable outpost for the defense of Port Moresby, and a constant threat to the security of the Japanese positions at Lae and Salamaua. Conversely, if it were controlled by the Japanese, it would provide additional protection for their New Guinea positions and at the same time serve as a strategic intermediate base for a new drive to the south toward Port Moresby should the opportunity arise.

Once the Japanese decided to take Wau, they moved swiftly. While their troops were still fighting in Papua, other forces moved southward from Salamaua along the trails leading into Wau. On 27 January, just five days after the fall of Giruwa and Sananda, they launched an attack against the

2 Lt. Col. Masaru Shinohara, Senior Intelligence Officer of the Eighth Area Army, commenting on General MacArthur's tactics, said: "I believe that, more or less, all of the Allied operations depended on deception by landing in places where we thought a landing and the building of airfields impossible." Interrogation Files, G-2 Historical Section, GHQ, FEC.

3 In emphasizing the particular importance of New Guinea to their new defensive positions, the Japanese said: "New Guinea especially, was the strategic point on the right flank of the defensive line, and if it should fall into the hands of the enemy, who had already secured firm operational bases in Australia and in one corner of the Solomon Islands, it would be a case of giving the enemy the best possible route to penetrate into the Philippines and any part of the South Co-Prosperity Sphere. This would be a great menace to the foundation of our general defense system. It was clear that the northern coastal area of New Guinea Island was of great importance to the enemy for his offensive bases and for conducting his operations, and also because of the difference in the terrain of the northern and southern areas of New Guinea. The strategic value of the Lae and Salamaua areas in the present stage of the operation was of immense importance." Japanese First Demobilization Bureau Report, *Southeast Area Operations Record, Part III: Eighteenth Army Operations*, Vol. I, p. 54, G-2 Historical Section, GHQ, FEC.

4 Ibid., pp. 54–56.

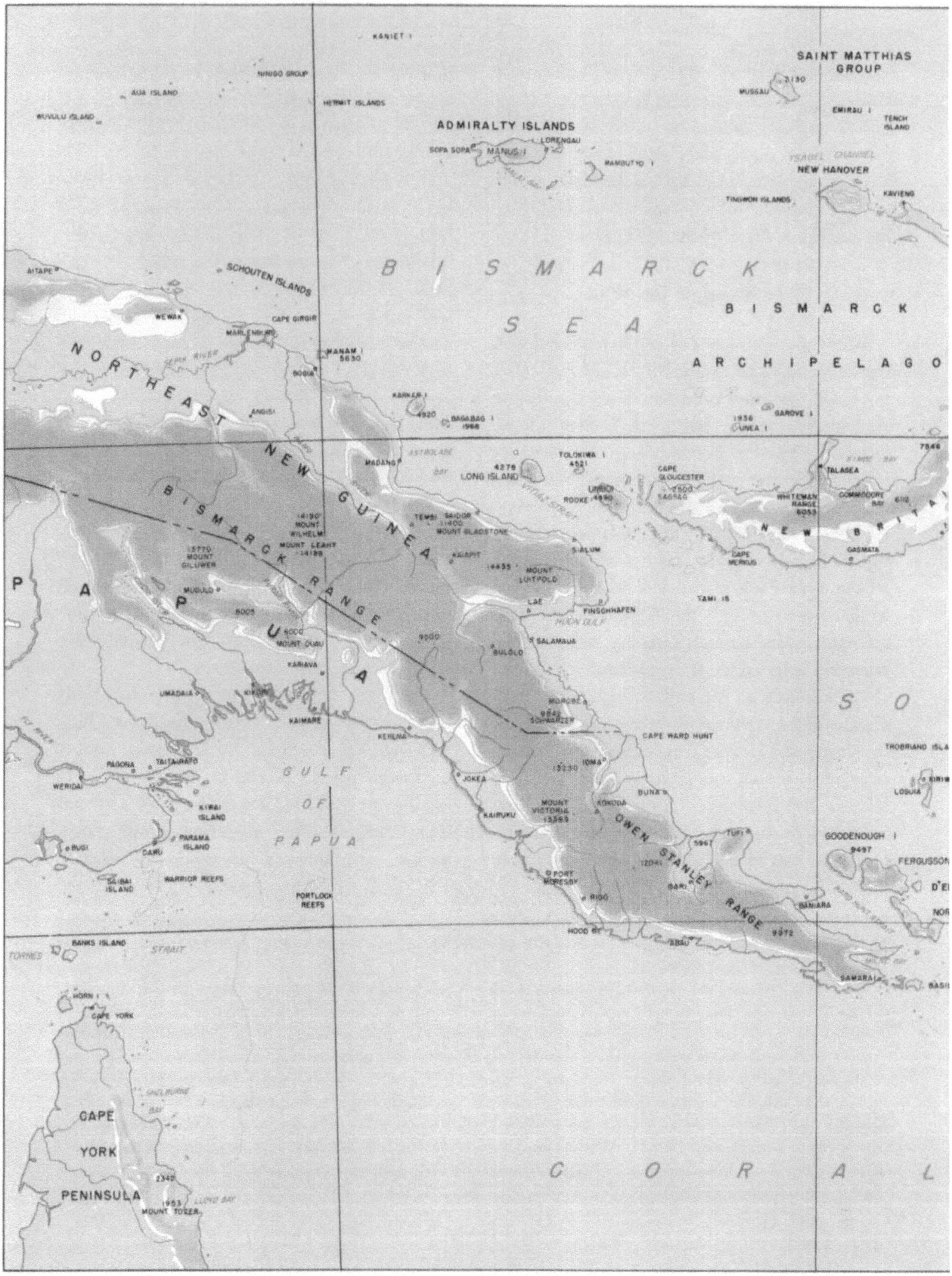
SAINT MATTHIAS GROUP
ADMIRALTY ISLANDS
BISMARCK SEA
BISMARCK ARCHIPELAGO
NORTHEAST NEW GUINEA
BISMARCK RANGE
PAPUA
OWEN STANLEY RANGE
GULF OF PAPUA
CAPE YORK PENINSULA
CORAL

PLATE NO. 30

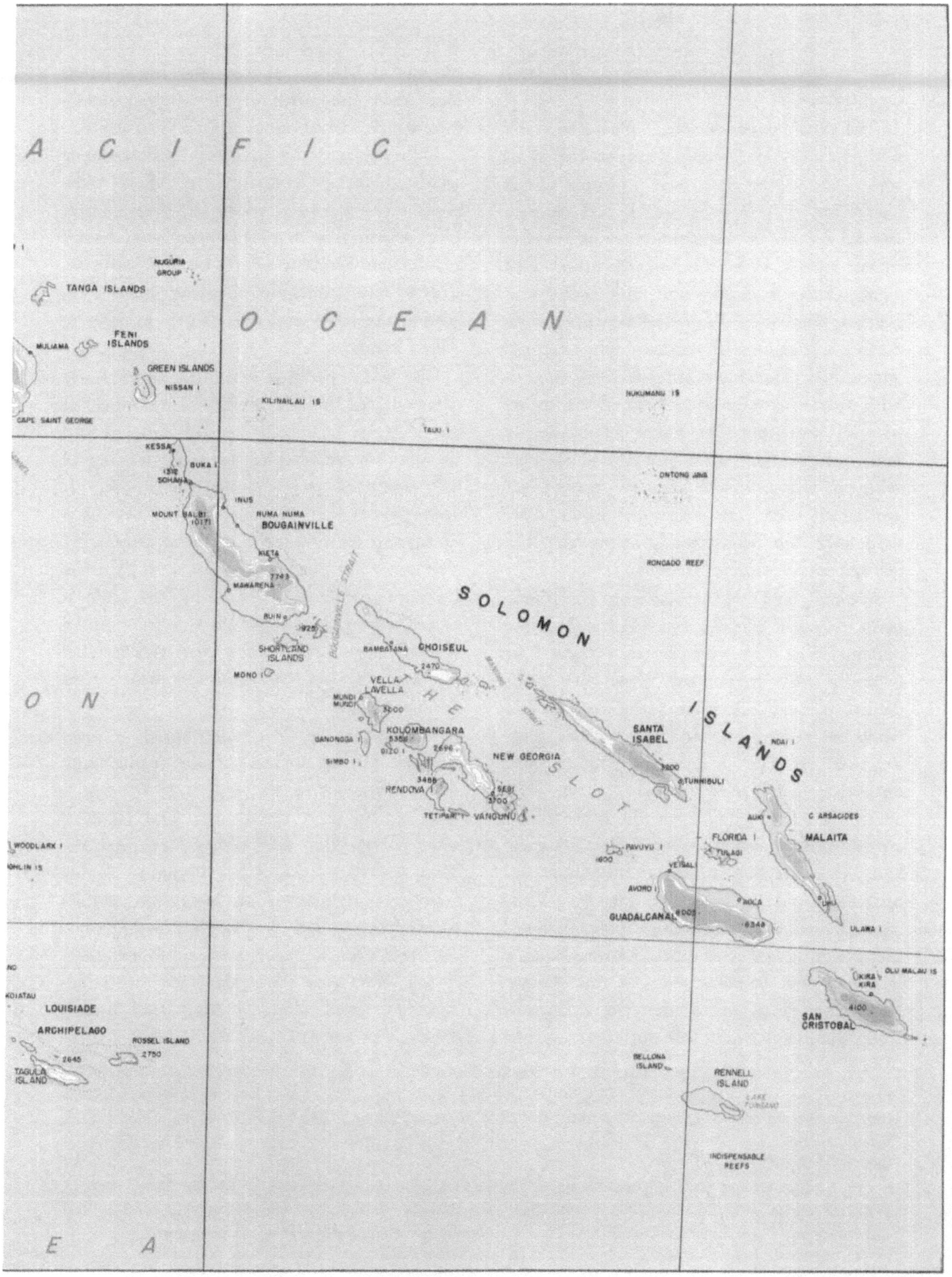

New Guinea—Solomons Area

weak Australian garrison occupying this strategic outpost.[5]

The situation was serious. Only the small Kanga Force, which had occupied the Wau area since March 1942, and a portion of the Australian 17th Brigade, which had recently arrived by air in anticipation of a possible enemy assault, were protecting the town. Bad flying weather had prevented the further dispatching of airborne troops and the thick jungle trails made any rapid overland reinforcement impossible. The Australian defenders, numerically inferior, attempted to hold off the enemy advance until more reinforcements could be sent, but they were gradually forced back toward the airstrip. The Japanese pushed forward until they were within 400 yards of the strip itself and had virtually surrounded the beleaguered Australian garrison.

A timely break in the weather at this critical point, however, enabled the Allies to fly in the remainder of the 17th Brigade from Port Moresby. On 29 January, plane after plane landed in rapid succession on the small airfield while the fighting was still in progress. During the first day of airborne reinforcement, fifty-seven landings were made on the Wau airstrip and for several days thereafter Allied air transports shuttled between Port Moresby and Wau carrying additional personnel and materiel to the sorely pressed garrison. Thus strengthened, the Australians soon shattered the enemy attack and moved forward. By 4 February the Japanese were in full retreat. Beaten and disorganized, they fled in disorder along the jungle trails, seeking refuge in their outpost at Mubo. It was later learned that more than one-fourth of the original enemy force which had set out to capture Wau was lost.

On 6 February the Japanese Air Force made a belated attempt to disrupt the Allied transportation system by attacking the Wau airstrip. Disproportionate losses, however, soon halted any further bombing efforts by the Japanese. This air attack concluded the struggle for Wau, leaving the town and the airstrip securely in Allied hands.

The Wau operation had once more clearly demonstrated the adaptability of air transport not only for conveying needed supplies and ammunition but also for the rapid landing of fully equipped troops under enemy fire. It again proved that air transport had become a strong and trusty weapon of the armed forces.[6] The battle also marked the last attempt by the Japanese to seek new territory in New Guinea. From then on, in anticipation of a major Allied advance, their entire efforts were concentrated on strengthening the positions which they already occupied.

The action at Wau contributed in great measure to the success of the forthcoming Allied offensive in the region of the Huon Gulf and the drive against the vital Japanese bases at Salamaua and Lae. Soon after the Wau operation, the Allies began to deploy their forces in preparation for future assaults against the enemy. In March, Headquarters, 3rd Australian Division took over operations in the Wau area while the 15th Australian Brigade moved along another trail toward Bobdubi on the Francisco River above Mubo.

5 The main enemy force sent against Wau was the Okabe Detachment of the Japanese 51st Division. This unit, consisting of 7 infantry companies, 1 artillery battalion, and 1 engineer company, had assembled at Salamaua from Lae, and was ordered to begin its advance against Wau on 14 January. Japanese First Demobilization Bureau Report, *Southeast Area Operations Record, Part III: Eighteenth Army Operations*, Vol I, pp. 59, 65–67, G-2 Historical Section, GHQ, FEC.

6 "Although there had been some experience in reinforcement of ground troops by air previously, the risk calculated and assumed in this operation demonstrated that such employment of air transport provided a mature and potent weapon." Air Evaluation Board, SWPA, *Air Transport Operations, Battle of Wau*, 10 Jun 45, p. 2.

In April the Australians held positions in front of both Bobdubi and Mubo where they waited until other troops could be readied for a full-scale attack on Salamaua and Lae.

General Situation in Early 1943

General MacArthur, by skillful maneuver of his meager forces, had resisted and thrown back each Japanese attempt to press their initial advantages of surprise and power. Not satisfied with merely parrying the enemy's thrusts, he had, in addition, seized upon every opening to strike forward with adroit ripostes of his own. The Allies had accomplished much; in one year their strategic position had vastly improved and at the end of February 1943, General MacArthur was able to report:

> *Allied forces control Australia—Papua—Southeast Solomons—Fijis and the land and sea areas south of the line Buna—Goodenough—Fijis. Advanced airdromes are along the line Darwin—Moresby—Buna—Guadalcanal and these, with further development, will support initial operations toward the Bismarcks.*[7]

Although the Allied successes in Papua and the Solomons were encouraging in themselves, they were only preliminary steps in the major effort necessary to destroy the enemy's military power.[8] These successes did show, however, that the Japanese were being forced to relinquish the tactical initiative. In spite of the fact that General MacArthur still lacked the resources required for a full-scale offensive, he was now able to plan along lines which would bring his forces to decisive grips with the full strength of the Japanese war machine.

The problem presented was unique. Throughout the entire Southwest Pacific there was not a single concentration of Japanese strength in any particular area which could be brought to battle as in ordinary warfare. Japanese forces were placed to protect strategically vital ports and airdromes throughout an archipelago which extended over 900 miles from Manus in the Admiralties to New Georgia in the Solomons, and 400 miles from Kavieng on New Ireland to Lae on New Guinea. In this way the Japanese could command large expanses of territory by merely controlling certain key approaches. If these specific points, however, could be deprived of naval and air protection, they would become extremely vulnerable to amphibious assault and the Allies could, in turn, dominate and exploit these areas to their own advantage.

General MacArthur had long recognized the necessity of amphibious training for his troops. He knew that the Japanese were already trained and equipped for seaborne operations. Their campaigns into the Southwest Pacific had been planned far in advance, and their veteran troops had been given previous experience in amphibious maneuvers. By comparison, his own ground troops, including those employed in the Papuan Campaign, needed additional amphibious training before they could be effectively used in an over-water operation. His training program, however, had been held up by a chronic shortage of landing craft and other special equipment. In early 1943 this shortage was partially relieved by the assignment of the 7th Amphibious Force of the U.S. Navy[9] and the 2nd Engineer Special Brigade of the U.S. Army to the Southwest Pacific Area. Each of these units was designated for a specific task. The 7th Amphibious Force,

---

7 G-3, GHQ, SWPA, *Elkton II*, 28 Feb 43, Obsolete Plans, G-3, GHQ Planning (S).

8 The Japanese Eighth Area Army, with headquarters at Rabaul, was commanded by Lt. Gen. Hitoshi Imamura and consisted of General Hyakutake's Seventeenth Army in the Solomons and Lt. Gen. Hatazo Adachi's Eighteenth Army in New Guinea and New Britain.

9 The 7th Amphibious Force was a component of the Seventh Fleet, which prior to 15 March 1943 had been designated Southwest Pacific Force.

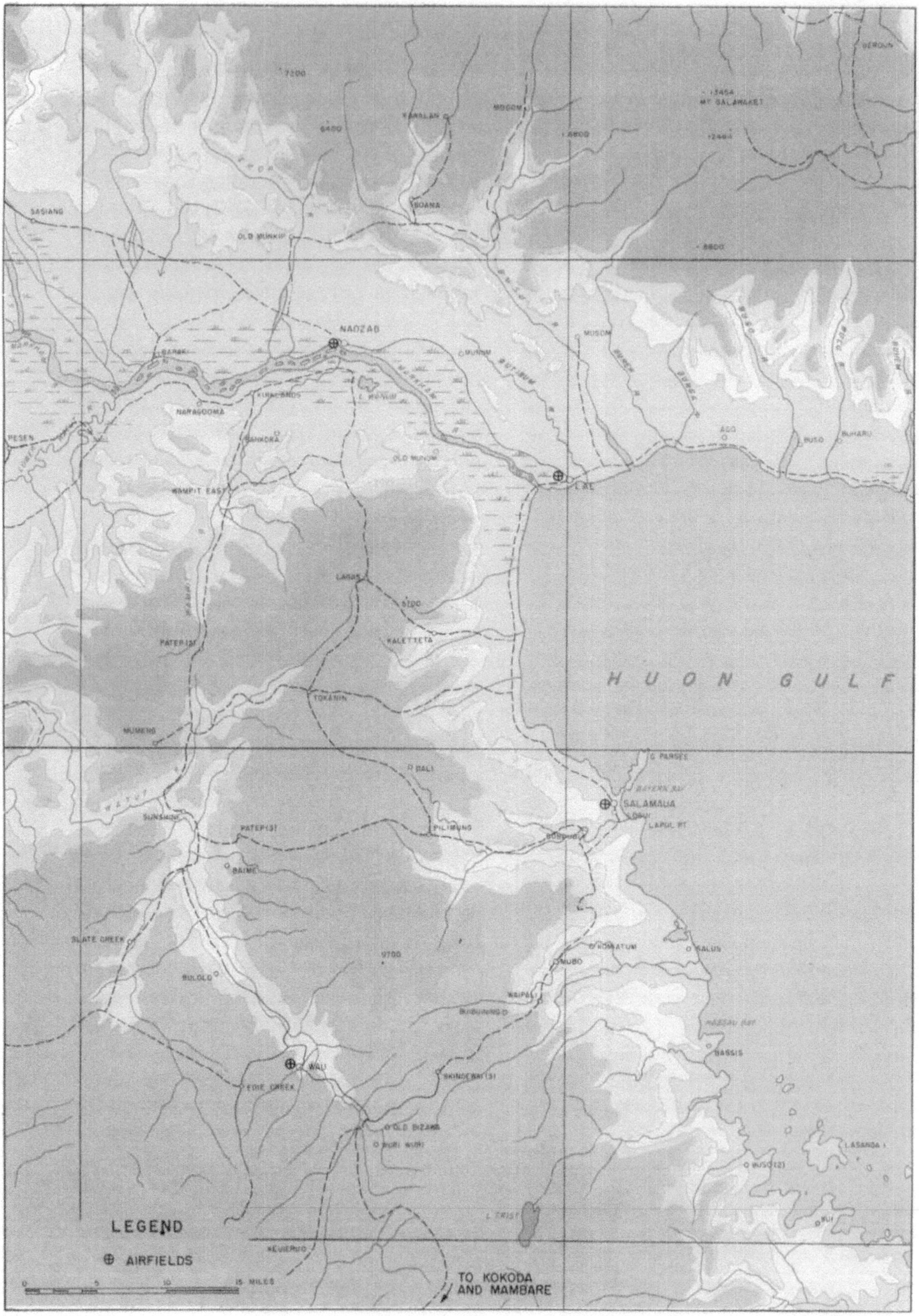

PLATE NO. 31.

Strategic Location of Wau

equipped with transports, cargo vessels, and landing craft of all types, was to be employed for major amphibious movements, while the engineer brigade with small boats limited to a range of sixty miles was to be used for shore-to-shore operations.

Responsibility for amphibious training was given to Rear Adm. Daniel E. Barbey, Commander of the 7th Amphibious Force, in January 1943. The program provided four weeks of intensive training for the battalion, regimental, brigade, and divisional combat teams, during which time they were taught to embark, proceed overseas, and land on hostile shores. A complete rehearsal was given each task force assigned to a particular assault mission immediately before the operation was launched.[10] Ground, air, and naval units utilized the interval between campaigns to complete their standard training. Weaknesses brought out in combat were corrected, and newly arrived replacements were taught the techniques peculiar to tropical and jungle warfare. With a smoothly running amphibious training program, General MacArthur felt that his troops would soon be ready for a broadening of the offensive.

Southwest Pacific Area Command

In order to derive the maximum striking power from the different nations united in the fight against Japan, a rather complex command organization was worked out for the Southwest Pacific Area. (Plate No. 32) Originally the various national components were united for training and tactical employment but remained separate and distinct for all other purposes. The Australian First and Second Armies and the United States Sixth Army[11] were under the operational control of Allied Land Forces which established separate task forces for each operation. New Guinea Force, which conducted the Papuan Campaign, was such a command and directed both Australian and American troops. In the spring of 1943, however, Alamo Force, composed entirely of United States units, was set up as an independent organization directly under General Headquarters and without an intermediate echelon of command. It actually comprised the Sixth Army. New Guinea Force remained under the Allied Land Forces, which by this action was relieved of the control of United States troops except as they were assigned for specific operations.

The United States Fifth Air Force and the Royal Australian Air Force Command were assigned to Allied Air Forces, while the United States Seventh Fleet and units of the Royal Australian and Royal Netherlands Navies were grouped under Allied Naval Forces. These two commands remained united throughout all operations of the Southwest Pacific Area, but the naval units were divided into separate task

---

10 CINCSWPA Ltr to ALF, AAF, ANF, 1st Mar Div, USASOS, SWP Amph F, 8 Feb 43 AG, GHQ, SWPA (S). An integrated, collateral feature of this amphibious training was the increasing development, for issue to staffs and troops of "Terrain Studies" and pocket size "Handbooks" of the landing areas and beaches of contemplated operations.

11 General MacArthur, having felt early the need for a United States Army in the Southwest Pacific Area, sent the following message to General Marshall: "Experience indicates the necessity for a tactical organization of an American Army. In the absence of such an echelon the burden has been carried by GHQ. I recommend the U.S. Third Army under General Krueger, which would provide an able commander and an efficient operating organization. I am especially anxious to have Krueger because of my long and intimate association with him." CINCSWPA Radio to C/S, WD, 11 Jan 43, WD No. 321, C/S GHQ (S); CINCSWPA Radio No. C–149 to C/S, WD, 16 Jan 43, WD No. 324, C/S GHQ (S). In response to General MacArthur's request, Headquarters Sixth Army, commanded by Lt. Gen. Walter Krueger, arrived in the Southwest Pacific Area and all American combat units were assigned to it on 16 February 1943. GHQ General Order No. 17, 16 Feb 43, AG, GHQ No. 322 (S).

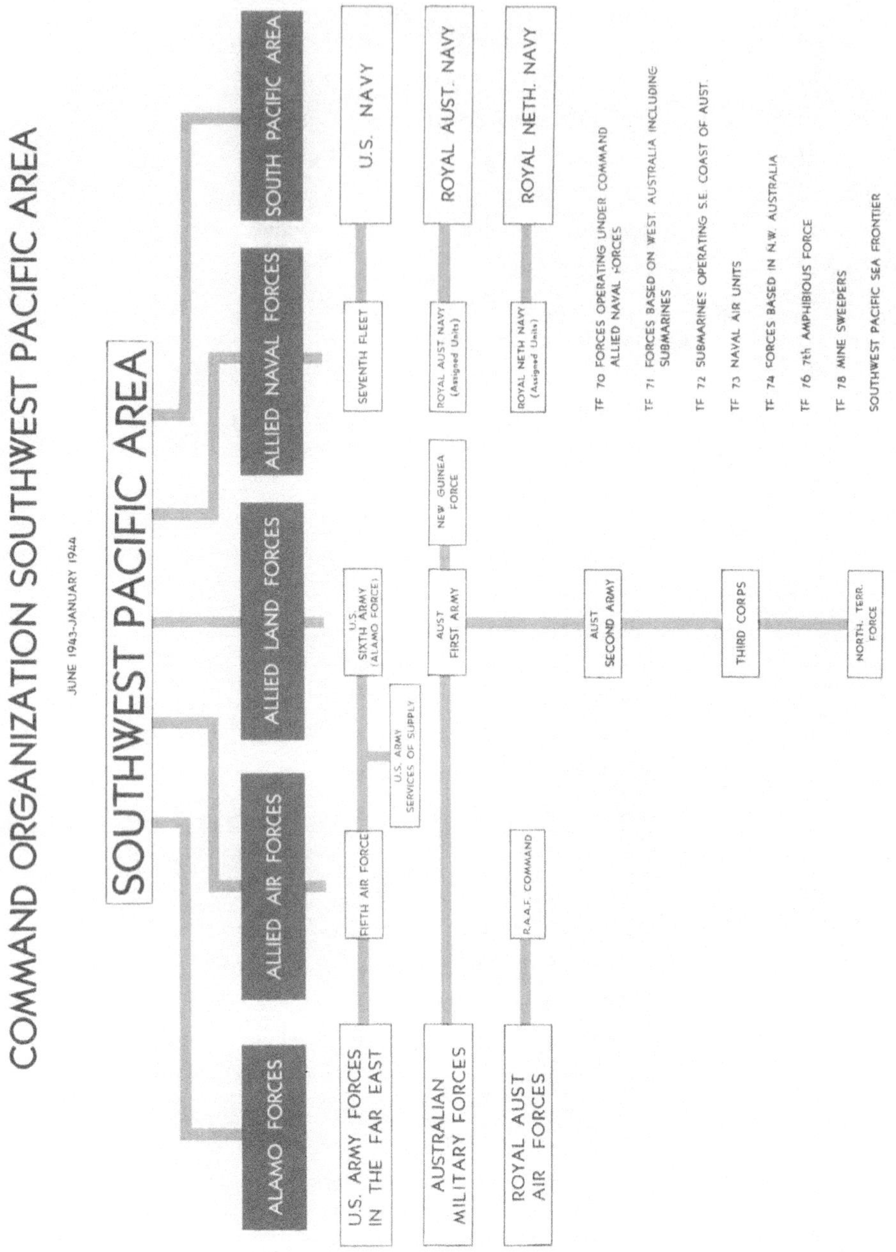

PLATE NO. 32

Command Organization, Southwest Pacific Area

forces according to the specific mission assigned. The Australian Military Forces, the Royal Australian Air Forces, and the United States, the Royal Australian, and the Royal Netherlands Navies were each responsible for the administration and supply of its own nationals. The United States Army Services of Supply, placed under United States Army Forces in the Far East, provided logistical support for the United States ground and air forces. The Sixth Army and the Fifth Air Force were also placed under the administrative control of USAFFE Headquarters which performed all staff duties incident to command, except those relating to strategic or tactical operations. General MacArthur, as Commander in Chief, Southwest Pacific Area, directed the combat employment and training of the combined armies, navies, and air forces, but at the same time, as Commanding General, United States Army Forces in the Far East, he was the direct commander of United States ground and air units.[12]

General Blamey was commander of the Australian Military Forces and Allied Land Forces; Admiral Carpender, of Allied Naval Forces and the Seventh Fleet; General Kenney, of the Allied Air Force and the Fifth Air Force; and General Krueger, of the Sixth Army and Alamo Force. In spite of an apparent complexity, the command channels were clear and distinct. Each subordinate echelon had definite and specific duties and responsibilities, and was co-ordinated and controlled by General MacArthur and his staff. In addition, the forces of Admiral Halsey, Commander South Pacific Area, were to operate under general directives of the Commander in Chief, SWPA.[13]

General MacArthur described his headquarters in July 1943 as follows:

Complete and thorough integration of ground, air, and naval Headquarters within General Headquarters is the method followed with marked success in the Southwest Pacific Area rather than the assembly of an equal number of officers from those components into a General Headquarters staff. Land, air and naval forces each operate under a commander with a completely organized staff. Naval and air commanders and their staffs are in the same building with General Headquarters. The land commander and his staff are nearby.

These commanders confer frequently with the Commander-in-Chief and principal members of General Headquarters. In addition to their complete functions as commanders, they operate, in effect, as a planning staff to the Commander-in-Chief.

When operating in forward areas the same conditions exist.

The personal relationships established and the physical location of subordinate Headquarters make possible a constant daily participation of the staffs in all details of planning and operations. Appropriate members of General Headquarters are in intimate daily contact with members of the three lower Headquarters.

Air officers and naval officers are detailed as members of General Headquarters staff and function both in planning and operations on exactly the same basis as army officers similarly detailed. The problem in this area is complicated by the fact that it is an Allied effort. Australian and Dutch naval, air, and army officers have been assigned to General Headquarters.

General Headquarters is, in spirit, a Headquarters for planning and executing operations, each of which demands effective combinations of land, sea, and air power. General Headquarters has successfully developed an attitude that is without service bias. Although the physical location and staff procedures of all four Headquarters are of the utmost importance, it is only the determination that General Headquarters shall act as a General Headquarters rather than as the Headquarters of a single service that will produce the unanimity of action and singleness of purpose that is essential for the successful conduct of combined

12 HQ USAFFE, General Order No. 1, 26 Feb 43, AG AAF 323. 3 (S).

13 JCS Radio No. 2407 to GHQ, SWPA, 29 May 43, G-3, GHQ, SWPA *Journal*, 30 May 43 (S).

operations.[14]

Battle of the Bismarck Sea

The defeat of the Japanese at Wau increased the danger to their general position in eastern New Guinea. Realizing that the enemy would become apprehensive about the security of his defenses below the Huon Peninsula, General MacArthur and his planning staff anticipated a major effort by the Japanese to reinforce their garrisons at Salamaua and Lae. Their assumption was further substantiated by numerous intelligence reports of a growing concentration of shipping in Rabaul harbor and of a noticeable increase in activity on enemy airfields along the probable convoy route.[15]

General MacArthur, accordingly, alerted his forces to be ready for a large-scale effort by the enemy to transport troops from Rabaul to eastern New Guinea. So certain were the Allies of an imminent convoy movement that General Kenney's air forces carried out actual practice maneuvers under conditions similar to those expected, reconnoitering the most advantageous routes of attack so that superiority could be attained at the point of combat. In addition, special preparations were made to carry out a new technique of skip bombing in the event of unfavorable weather and low cloud formations. Extensive rehearsals were carried out in February 1943 until the new method was perfected for the anticipated task.

Allied precautions were well taken. As expected, on 28 February a strong enemy convoy of 8 transports and 8 destroyers carrying the remainder of the 51st Division, certain key personnel, and various vital supplies for the New Guinea front, left Rabaul harbor. The convoy was spotted off Cape Gloucester on 1 March but unusually bad weather prevented an effective Allied strike. The next day, in spite of haze, rain, and thick clouds, the attack was launched according to rehearsed plans. Skip bombing practice had not been wasted. Driving in at low altitudes through heavy flak, General Kenney's planes skimmed over the water to drop their bombs as close to the target as possible. That morning the bombers left one transport sinking and scored several other hits.[16] Adverse weather during the afternoon limited the effect of further strikes to minor damage. The battle was resumed on 3 March, about 30 miles southeast of Finschhafen, with an all-out attack by the Fifth Air Force bombers.[17] In less than an hour after the main assault all the remaining transports were in a sinking condition and several

14 CINCSWPA Radio No. C-4369 to WARCOS, 31 Jul 43, AG, GHQ 323. 36 (S).

15 G-2, GHQ, SWPA, *Daily Summary* No. 337, 22/23 Feb 43 stated the following : "Merchant shipping at Rabaul has reached a new high. Fifty-nine vessels of this type were in the harbor 22 February, totalling 299,000 tons. About 200,000 tons of merchant shipping appears normal for Rabaul. In view of the comparative inactivity in the Solomons, this increase over the normal assumes serious aspects, in relation to possible employment against New Guinea." The following was also reported : "Enemy interest in the Gasmata and Cape Gloucester bases at this time strengthens other indications of impending landing operations, possibly simultaneously, at Wewak, Madang and Lae. The airdromes at Cape Gloucester and Gasmata provide staging and refueling points for air cover over the Vitiaz Straits." G-2, GHQ, SWPA, *Daily Summary* No. 335, 20/21 Feb 43.

16 On 2 March, two destroyers of the convoy picked up the survivors of the sinking ship and carried them at forced draft to Lae. After disembarking the rescued troops, these two destroyers then returned to the scene of action on the morning of 3 March.

17 Referring to the Battle of the Bismarck Sea, Vice Adm. Gunichi Mikawa, Commander of the Japanese Eighth Fleet, said: "The percentage of hits of the American Air Force's low-level bombing was excellent. Our fleet mistook it for a torpedo attack and suitable evasive action was not taken." ATIS, GHQ, SWPA, Document No. 16269B, "Report on the Naval Battle of the Bismarck Sea," 10 Apr 46.

of the escorting destroyers were heavily damaged. The battle continued throughout the day and dawn of 4 March revealed a lone destroyer in the battle area which was soon sent to the bottom. It was a unique sea battle; not a single Allied vessel was involved.

The rain of bombs from the skies had been most destructive. Of the original convoy of 16 ships, all transports were sunk and only four destroyers survived.[18] Captured documents later disclosed that more than half of the approximately 7,000 troops loaded on the convoy were lost. Of the survivors, only about 800 had managed to reach Lae. The entire load of provisions and materiel, including a large amount of airplane fuel and a four months' supply of food for 20,000 men, was totally destroyed.[19] The forces in eastern New Guinea were thus deprived of supplies and reinforcements necessary to withstand the forthcoming Allied blows at Salamaua and Lae.

The destruction of the Bismarck convoy was a devastating blow to the Japanese[20] which General MacArthur analyzed as follows:

> *We have achieved a victory of such completeness as to assume the proportions of a major disaster to the enemy. Our decisive success cannot fail to have most important results on the enemy's strategic and tactical plans. His campaign, for the time being at least is completely dislocated.*[21]

Along with the loss of critical troops, supplies, and ships, the battle of the Bismarck Sea conclusively demonstrated that the Japanese could no longer reinforce the Salamaua—Lae area by cargo vessels or fast destroyer convoys. After this disaster, all attempts at running large transports into Lae were abandoned and the hungry enemy garrisons in eastern New Guinea had to be satisfied with the thin trickle of supplies and replacements carried in by destroyers, barges, or submarines. It was not until the battle for Leyte that the Japanese again attempted to bring in large reinforcements and supplies to beleaguered units within range of aerial bombardment. Control of the air over the sea had been definitely lost by the Japanese air forces.[22]

---

18 The ships sunk were the following: the transports *Aiyo Maru, Kembu Maru, Kyokusei Maru, Oigawa Maru, Shinai Maru, Taimei Maru, Teiyo Maru, Nojima* and the destroyers *Shirayuki, Arashio, Asashio,* and *Tokitsukaze.* The surviving destroyers *Shikinami, Uranami, Asagumo* and *Yukikaze,* carried the bulk of the rescued personnel to Rabaul. The details of the Battle of the Bismarck Sea are based on Japanese sources and all available documents and interrogations of the Fifth Air Force and ATIS, GHQ, FEC.

19 Japanese First Demobilization Bureau Report, *Southeast Area Operations Record, Part III: Eighteenth Army Operations,* Vol. I, p. 161, G-2 Historical Section, GHQ, FEC.

20 A fortuitous incident of major importance in the wake of the Battle of the Bismarck Sea brilliantly illustrated the value and efficiency of ATIS. A ship's captain, seeking refuge on Goodenough Island carried with him navigational charts, ship's records, and the Japanese Army List and Directory for 1942/43, a document of some 2700 pages containing the name and unit of every Japanese officer then in the service. Up to this time, information on Japanese Order of Battle and identification of tactical commands was based on outdated Chinese reports. It naturally was inconclusive for other Asiatic areas. The patrol which captured the ship's captain sent back the precious intelligence material to higher headquarters. The significance of the documents was immediately realized and all work in ATIS from the forward field units to the base detachment in Brisbane was completely devoted to the translation of this great find. With the newly discovered information, every Japanese unit in the field could be reconstructed, from company through division, corps, and army. The work was finished in a few weeks. The translated book (See Plate: *Alphabetical List of Japanese Army Officers,* ATIS Publication No. 2, May 43.) was printed by the Australian Government Printer on emergency order and distributed quickly to intelligence units from Alaska to India.

21 GHQ, SWPA, Press Release, 4 Mar 43.

22 Comdr. Yasumi Doi, member of the staff of the Southeast Area Fleet at Rabaul from 1943 to 1945, stated that after the Battle of the Bismarck Sea it was realized that control of the air was lost and, consequently, supplies to New Guinea were shipped by destroyer and submarine only. Finschhafen was thus supplied until captured by the Allies, but supplies in the New Guinea area were totally inadequate. United States Strategic Bombing Survey *Interrogations of Japanese Officials,* Vol. II, p. 397.

ALLIED TRANSLATOR AND INTERPRETER SECTION

SOUTHWEST PACIFIC AREA

A. T. I. S. PUBLICATION

No. 2 MAY, 1943

ALPHABETICAL LIST
OF
JAPANESE ARMY OFFICERS

INTRODUCTION.

This list is compiled from the "Register of Officers", and the "Register of Reserve Officers on the Active List" issued by the Japanese War Department on 15 October, 1942.

Both Registers are compiled as at 1 September, 1942.

An Asterisk immediately preceding the name of an officer indicates that he is a Reserve (Special Volunteer) Officer.

The following particulars are given in this list against the name of each officer: Rank and date of appointment thereto; arm of service; appointment and date thereof; the number of the page in the original Register of Officers or Register of Reserve Officers on which appear all entries relating to the officer concerned.

Subjoined are a glossary of words, phrases and place names the translation of which is considered to require some explanation, and a list of the abbreviations used throughout the work. The glossary shows Romaji, Kanji and the English translation adopted.

It should be remembered that, in Japanese, the letters "G" and "K" are almost interchangeable, as are the syllables "SHI" and "JI"; also that names commencing with the letters "Kami" will often be found written commencing with the letters "Ue." For example, KOSHI will often be found to be written GOSHI, and KAMIMURA to be written UEMURA. These points should be borne in mind when using this list.

A PW has stated that the names of certain Reserve Officers do not appear in the original register.

SIDNEY F. MASHBIR,
Colonel, S. C.,
Co-ordinator.

WJM/BM

ABAIE, Shotaro : Capt (8/39) : Inf : 1 Anchorage Inspectorate (9/41) : 802.
ABE, Akio : lst Lt (8/41) : Inf : Coy Comdr 226 Inf Regt (8/41) : 1312.
ABE, Akira : lst Lt (10/41) : Tpt : Adjt 5 Tpt Regt (4/41) : 1440.
ABE, Akira : lst Lt (3/42) : Medical : Attd Reinf TAI 79 Inf Regt (3/42) : 2772.
ABE, Bumpachi : 2nd Lt (10/41) : Air : Attd 30 Airfd Bn (10/41) : 1839.
ABE, Chiharu : lst Lt (10/41) : Air : Attd 55 Ind Air Sqn (4/42) : 1435.
***ABE, Chikao** : lst Lt (8/40) : Rly (Engr) : Attd 2 Rly Regt (2/40) : 485.
ABE, Chi,o'i : Maj (12/41) : Air Tech : Army Air Tech Research Inst (12/41) : 2155.
ABE, Chuhachi : lst Lt (10/41) : Intend : Attd 59 Inf Regt (8/40) : 2405.
ABE, Chu'i : lst Lt (3/42) : Tech : Attd Army Fuel Depot (3/42) : 2093.
ABE, Chutako : Maj (11/37) : Vet : Chief, Vet Dept 32 Div (8/42) : 2999.
***ABE, Doichiro** : Col (8/36) : Inf : Attd KWANTUNG Army HQ; Sch Instr (8/40) : 3.
ABE, Eiichiro : lst Lt (8/40) : Vet : Vet Dept 18 Div (9/41) : 3028.
ABE, Eji : Lt (10/41) : Intend : Intend, 1 Air Army (6/42) : 2395.
ABE, Eisaku : lst Lt (12/41) : Air : Attd Tng TAI KUMAGAYA Army Flying Sch (11/40) : 1622.
***ABE, Eisaku** : lst Lt (8/40) : Inf : Adjt Reinf TAI 16 Inf Regt (10/41) : 453.
ABE, Fukusaburo : Col (10/41) : Intend : Attd Army Clothing Factory (10/41) : 2241.
***ABE, Gisaburo** : lst Lt (8/41) : Fd Mtn Arty : Comdr Reinf TAI, 25 Mtn Arty Regt (8/41) : 579.
ABE, Hachi'iro : Capt (8/36) : Mtn Arty : 732.
ABE, Hakuhiko : Col (3/39) Medical : Comdr Medical Dept 1 Air Army (6/42) : 2463.
ABE, Haruo : 2nd Lt (6/41) : Medical : Attd 48 Airfield Bn (12/41) : 2848.
ABE, Haruo : 2nd Lt (4/42) : Medical : Attd 92 Inf Bn (4/42) : 2895.
ABE, Hayami : Maj (10/41) : Fd Mtn Arty : Bn Comdr 33 Mtn Arty Regt (10/41) : 691.
ABE, Heisuke : Lt-Gen (8/41) : Inf : Comdr 41 Div (7/42) : 19.
ABE, Hideaki : Col (8/41) : Intend : Chief Int Dept 9 Div (4/42) : 2240.
ABE, Hideharu : Maj (10/41) : Medical : Attd DAIREN Army Hosp (3/42) : 2547.
***ABE, Hidemi** : lst Lt (8/41) : Air : Attd 41 Airfield Bn (3/41) : 599.
***ABE, Hidenosuke** : lst Lt (9/38) : Fd Mtn Arty : Attd 9 M n Arty Regt (9/39) : 246.
ABE, Hideo : 2nd Lt (4/42) : Medical : Attd 239 Inf Regt (4/42) : 2898.
ABE, Hikaru : lst Lt (4/42) : Medical : Attd 19 Air Group HQ (8/42) : 2783.
ABE, Hirokichi : lst Lt (8/39) : Inf : Attd Reinf TAI 4 Inf Regt Imp Gds (8/42) : 1104.
ABE, Hiroshi : lst Lt (12/40) : Intend : Intend Dept 23 Army (10/41) : 2370.
ABE, Hiroshi : lst Lt (8/41) : Inf : Coy Comdr 27 Inf Regt (8/41) : 1302.
ABE, Hiroshi : 2nd Lt (3/42) : Air Tech : Attd Army Fuel Depot (3/42) : 2220.
ABE, Hisakichi : Capt (11/37) : Medical : Attd 3 Southern Area Army Hosp (8/42) : 2976.
ABE, Hisao : Capt (10/41) : Air Tech : Superintendent, Army Air HQ (10/41) : 2165.
ABE, Hisashi : lst Lt (8/41) : Inf : Coy Comdr 5 Inf Regt (11/41) : 1327.
ABE, Ichiji : lst Lt (7/41) : Inf : Attd GS HQ; Attd Imp GHQ Army Dept (10/40) : 1253.
***ABE, Ichiro** : Capt (12/41) : Air Attd 109 Air Tng Regt (4/42) : 157.
***ABE, Isao** : Capt (8/42) : Field Mtn Arty : Attd to N CHINA Area Army HQ (8/39) : 196.
ABE, Junichi : Capt (3/41) : Medical : Chief MO, Medical TAI, 33 Div (3/41) : 2602.
ABE, Jun i : 2nd Lt (3/42) : Air : Attd 3 Air Regt (3/42) : 1884.
***ABE, Kaname** : lst Lt (3/40) : Mech : Attd 26 Cav Regt (2/42) : 438.
ABE Kane hige : Capt (3/40) : Engrs : Comdr 1 Fd Well Construction Coy (9/41) : 821.
***ABE, Kaoru** : Capt (3/42) : Inf : 26 Ind Garrison Inf Bn (4/42) : 177.
ABE, Katsuya : lst Lt (12/39) : Engr : Comdr 5 Fd Well Const Coy (9/40) : 1114.
ABE, Ka umi : Capt (8/39) : Vet : Vet Sec 2 Army (7/42) : 3014.
ABE, Keiichi : Lt-Col (10/41) : Inf : Sec Comdr KUSHIRO Regt Dist HQ (8/42) : 401.
ABE, Keigo : 2nd Lt (10/41) : Inf : Attd Reinf TAI 34 Inf Regt (10/41) : 1799.
ABE, Ken o : Maj (3/39) : Fd Mtn Arty : Instr Mil Acad (8/42) : 486.
ABE Kikuo : lst Lt (8/41) : Fd Mtn Arty : Attd 10 Ind Air TAI (9/41) : 1297.
***ABE, Kikuo** : lst Lt (12/39) : Inf : Coy Comdr 1 Raid Tng Regt (3/42) : 321.
ABE, Kiyomine : Capt (4/42) : Judicial : Judicial Dept, 52 Div; Judge Advocate, 52 Div Court Martial (4/42) : 3014.
ABE, Kiyoshi : lst Lt (12/41) : Inf : Attd 75 Inf Regt (10/40) : 1622.
ABE, Kiyoshi : lst Lt (12/40) : Mech (Cav) : Attd Maint TAI 1 Tk Gp (8/41) : 1243.
ABE, Kiyoshi : Lt-Col (8/41) : Gend : Comdr KWANTUNG Gend TAI (7/41) : 1950.
ABE, Kiyosuke : Capt (8/42) : Medical : Attd HSINKING Army Hosp (10/38) : 2980.

Henceforth, except for desperate counterattacks by isolated units, the Japanese were compelled to abandon all plans for the offensive in the Southwest Pacific.

Final Plans

The final plan for the 1943 offensive was a carefully evolved synthesis of previously drawn plans, necessarily modified and adjusted to fit the resources available to the theater at the time.[23] Briefly, the plan envisioned simultaneous operations along two lines of advance from Guadalcanal and from Papua, securing northeast New Guinea and the Solomons group, and converging to pinch off the Japanese strongholds at Rabaul and Kavieng. The immediate objective was the seizure of airfields along these routes from which to whittle down the enemy's strength and at the same time provide cover for Allied assaults.

To furnish air support for operations in the two sectors of intended advance, it was planned to occupy Woodlark and Kiriwina Islands and begin immediate development of airfields there. In New Guinea, meanwhile, an initial feint would be made at Salamaua to divert enemy forces to its protection. The main drive, however, would be made against Lae, seizing its valuable airstrips by a combined assault of an airborne force operating through the Markham Valley and an amphibious force moving along the coast from Milne Bay and Buna. Then Finschhafen and other bases in the Huon Gulf—Vitiaz Strait area would be taken by shore-to-shore movements.

In the Solomons, the forces of the South Pacific Area under the immediate command of Admiral Halsey were assigned the capture of the New Georgia island group. After attaining these objectives, both parts of General MacArthur's command, the Southwest Pacific and the South Pacific, covered and supported by the newly won bases, would push on to strike simultaneous blows against New Britain to the west and Bougainville to the east. Operations could then be undertaken to deprive Rabaul of naval support and airborne supply and to eliminate it as a threat to the Allied flank.[24] (Plate No. 33)

Each subordinate command was delegated the responsibility of developing its own part of the designated operations. Liaison officers kept the various headquarters informed of what the others were doing and gave technical aid and advice. In addition, conferences between the planning staffs were held frequently. The rare problem that could not be solved by these methods was taken up by the commanders involved, or, if necessary, referred to General MacArthur for final decision.

These plans required extensive and careful preparation. Existing airdromes at Port Moresby and Buna were incapable of furnishing adequate fighter protection for an airborne assault in the Markham Valley, so additional airfields in the interior were developed at Bena Bena, Tsili Tsili, and in the Bulolo Valley. Bases on the coast through such points as the mouth of the Mambare River and Morobe Bay were constructed, and work was rushed on the road from the Lakekamu River to Wau. Further study of the facilities of the U. S. 2nd Engineer Special Brigade indicated that it could transport only one stripped brigade, approximately 3,000 men, which was insufficient for the planned assault upon Lae. It was decided, therefore, to move the entire Australian 9th Division from its staging area at Milne Bay in the larger craft of the U. S. 7th Amphibious

23 This final plan, Elkton III (G-3, GHQ, SWPA, Planning File), superceding previous plans and directives, became the basic plan for 1943 operations on 26 April 43.

24 Ibid.

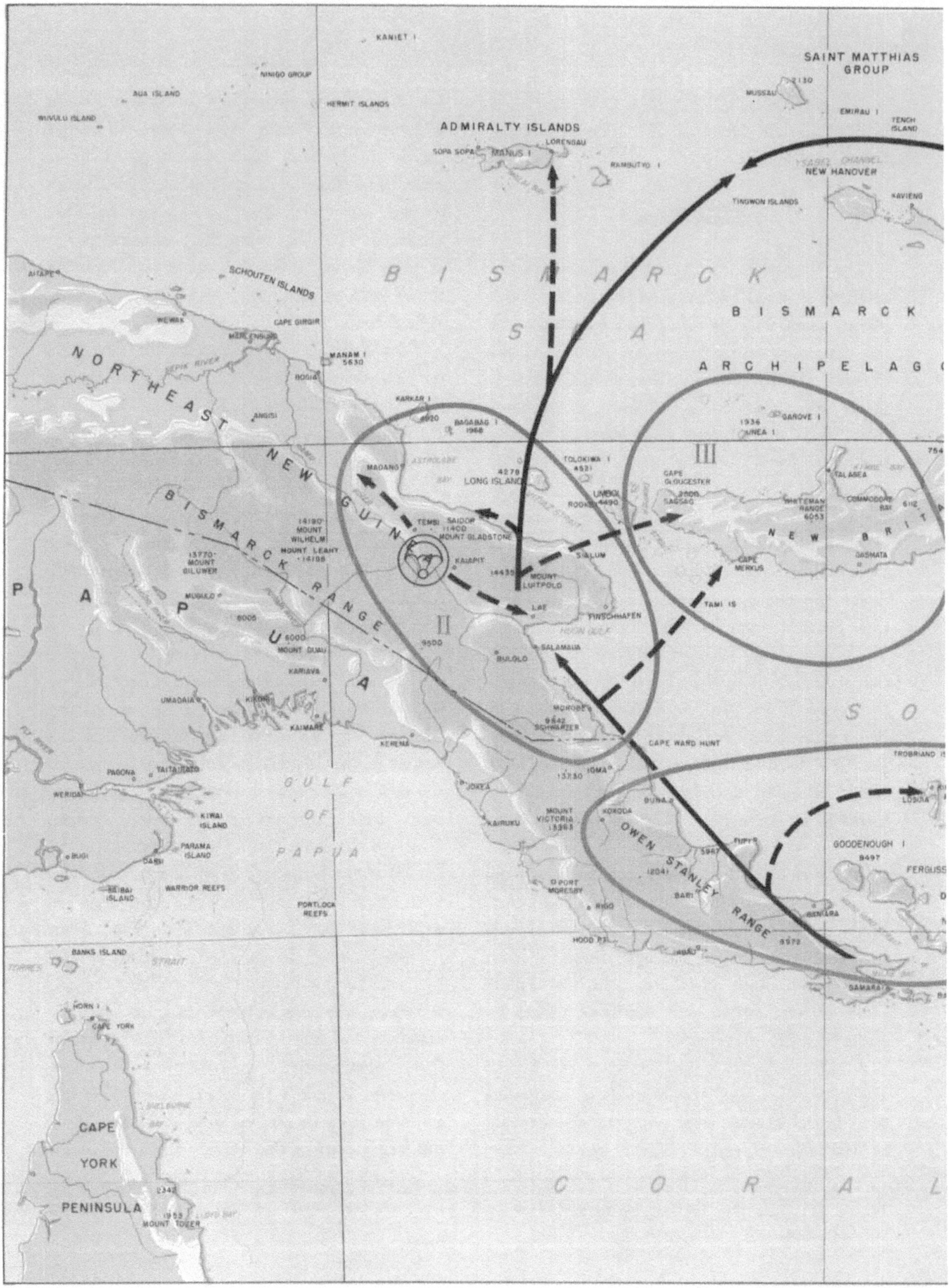

SAINT MATTHIAS GROUP
ADMIRALTY ISLANDS
NEW HANOVER
BISMARCK SEA
BISMARCK ARCHIPELAGO
NORTHEAST NEW GUINEA
BISMARCK RANGE
PAPUA
LONG ISLAND
NEW BRITAIN
GULF OF PAPUA
OWEN STANLEY RANGE
CAPE YORK PENINSULA
CORAL
II
III

PLATE NO. 33

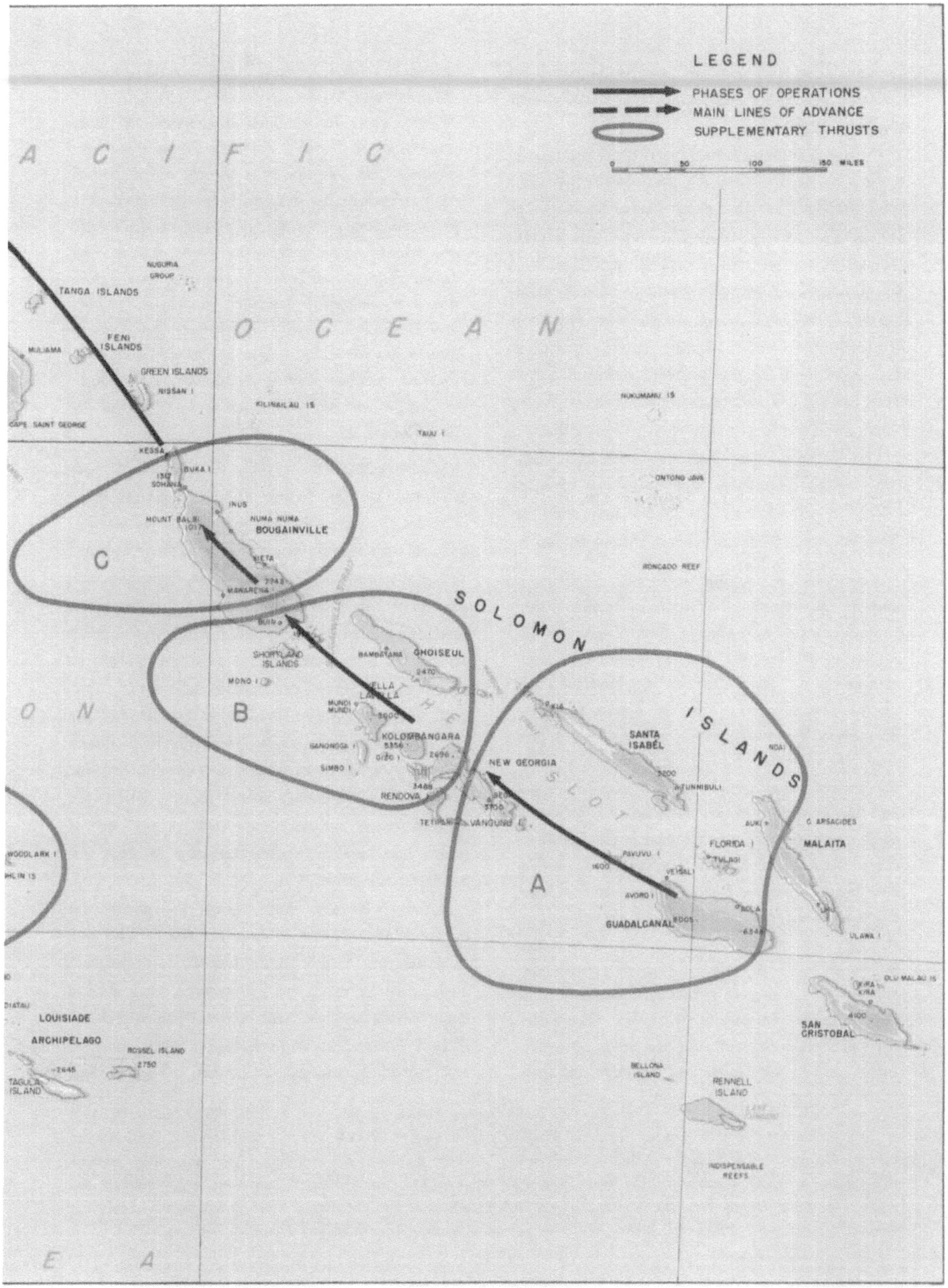

Operations Chart "Elkton Plan," New Britain—New Ireland—New Guinea Area

Force. The 2nd Brigade was to be used by the boat and shore parties and for local shore-to-shore supply.

As the task forces completed their plans and assembled their troops, the engineers and other units of the United States Army Services of Supply and of the Australian Military Forces developed and improved airfields at Milne Bay, Buna, and Goodenough Island to provide close fighter cover and support for the projected amphibious movements and assaults. Supplies and equipment for the separate national forces were brought forward to the advance bases from the Zone of the Interior in Australia.

The logistical problem was extremely difficult, primarily because of the shortage of shipping. Some idea of its magnitude may be obtained from General MacArthur's report:

Operations are getting under way, involving a monthly dispatch from Australia to islands where combat is expected, including New Guinea, of an average of 600,000 ship tons for troops and cargo movements. . . . Much of this tonnage handled from the mainland of Australia to New Guinea and other islands must be transshipped on vessels of from 100–300 feet long, of shallow draught, and sometimes again to trawlers and similar small boats for island to island and coastal creeping supply operations. . . . Water is the principal means of transportation in the Southwest Pacific Area. Roads between bases are non-existent in New Guinea; road and railroad capacity between bases on the mainland is negligible. As a result 95 per cent of supply and troop movements must be made by water. In considering the Southwest Pacific Area transportation problem, it must be remembered that it is 1500 miles from Port Moresby to Milne Bay. Our operations are already scattered along 300 miles of coastline on the north of New Guinea and will more and more extend to numbers of small islands from the advanced bases in New Guinea, frequently involving runs of from 100, 200 and 300 miles across water. All floating equipment found in Australia has been sent into this area, but Australia is a sparsely settled, under-equipped country that has already exceeded expectations in contributing as she has 40 coastal vessels (cargo) and several scores of trawlers, fishing smacks, and similar craft. . . . Attention is invited to the fact that the enemy frequently assembles 250,000 or 300,000 tons of merchant shipping, exclusive of small boats in Rabaul Harbor alone. If we are to amass sufficient force to overcome a force maintained by this much tonnage, it can well be expected that our tonnage be far greater. . . .[25]

Reconnaissance parties, aircraft, and submarines intensified their efforts to gather information concerning landing beaches, terrain, ground and aerial concentrations, shipping and defensive installations. As at Guadalcanal, patrols of the Allied Intelligence Bureau operated deep behind the enemy lines, and "coast watchers" were stationed at strategic points to observe and report on hostile activities. Aircraft and submarines conducted regular searches of the Bismarck and Solomon Seas to check shipping, airdromes, and troop concentrations. Each task force had trained scouts, who were landed from submarines and motor patrol boats for tactical reconnaissance of the immediate objectives.[26]

Every measure was taken to insure the completeness of final preparations. The 7th Amphibious Force concentrated its transports and landing craft at Townsville and Milne Bay; other surface and sub-surface vessels of Allied Naval Forces protected supply convoys and attacked enemy shipping. The Allied

25 CINCSWPA Radio No. C-3107 to CG Army Service Forces, 12 Jun 43, C/S GHQ, WD No. 421 (S).

26 "We knew that the enemy was using scouting patrols sent in by submarines," said General Tanikawa, at that time Colonel on the Staff of the Eighth Area Army. "These patrols did a wonderful job in getting detailed information on supply, convoys and the like. Both the Americans and Australians sent in patrols. . . . Whenever our supply ships came into port, no matter where it was and regardless of foul weather, the enemy raiders would come to bombard these ships. . . . The only answer was these patrols." Interrogation Files, G-2 Historical Section, GHQ, FEC.

Air Force stepped up its attacks on airdromes from the northern Solomons through New Ireland and New Britain to Wewak on New Guinea, destroying or damaging hundreds of Japanese planes in combat and on the ground. Heavy air raids on Koepang, Ambon, Timor, Tanimbar, and other bases in the Netherlands East Indies were carried out to deceive the enemy as to the main direction of the offensive. Increased dummy signal traffic in code through radios at Darwin, Perth, and Merauke helped to convince the Japanese that at least a diversionary attack was to be mounted from northwestern Australia. No step was overlooked which might aid the forthcoming operations.

Woodlark and Kiriwina

The 1943 offensive was initiated by General Krueger's Alamo Force which was charged with the task of occupying Woodlark and Kiriwina Islands. According to the over-all plan, these two islands were to be developed as air bases to cover the advance of both the Southwest Pacific and South Pacific Forces.[27]

It was the first amphibious landing movement assigned to the Southwest Pacific Area and, although neither of the islands was held by the enemy, much advance preparation was necessary. The garrison and construction troops for Woodlark were furnished by the South Pacific Command and arrived at Townsville in northeastern Australia between 21 May and 4 June. The units for Kiriwina were scattered along the Australian mainland and at Port Moresby. Both forces were transported to Milne Bay for staging purposes and movements were carefully co-ordinated to avoid interference with essential supply activities. The difficulties presented by a poor road net and the inadequate loading facilities at Milne Bay further complicated the problem. The concentration of troops for the amphibious operation was accomplished in good time, however, and the target date for the main landing was set for 30 June.

Operations were carried out as scheduled. On 23 June, advance engineer and survey construction parties were landed on both islands to prepare for the arrival of the main force. On 30 June, while New Guinea Force was holding the enemy's attention by landings along the New Guinea coast near Nassau Bay and Admiral Halsey's forces were moving toward Rendova Island on New Georgia, Alamo Force carried out the complete occupation of Woodlark and Kiriwina Islands.

Enemy reaction was limited to aerial reconnaissance and one or two feeble bombing raids. Intense attacks by the Allied Air Force on hostile bases during previous months had weakened the enemy's power to organize effectual opposition to the troops of Alamo Force. Besides, the Japanese could not divert their attention from areas of the concurrent Allied operations in New Guinea and New Georgia.

Ground and antiaircraft defenses on Woodlark and Kiriwina were quickly installed. Work on the airstrips proceeded rapidly, and by 24 July the 67th Fighter Squadron of the South Pacific Command was ready for missions based at Woodlark. On 18 August, the 79th RAAF Fighter Squadron was ready for action at Kiriwina.

New Georgia

On the eastern crescent of the Allied offensive, the South Pacific Force was carrying out its attack on New Georgia Island in the northern Solomons. Areas which were only lightly

27 HQ Alamo Force, "History of Chronicle Operation," 23 Aug 43, AG, GHQ, 370.22 (S).

defended by the enemy had been selected for the initial landings. Enemy ground troops were strongly entrenched around Munda on New Georgia and at Vila on Kolombangara; other enemy outposts were stretched from Vanganu Island to Vella Lavella. The minor concentrations of Japanese forces, except for those at Vella Lavella, were to be by-passed.

On 21 June, a small group of Marines made a preliminary landing at Segi Point on the southeast tip of New Georgia. On 30 June, forces of the 43rd Division went ashore at Wickham Anchorage on southeast Vanganu Island and at Rendova Harbor. The Japanese, apparently, were not expecting Allied landings at these points since the first incursions were virtually unopposed on land.[28]

Enemy reaction in the air and on the sea, however, was strong and rapid. During the first five days of the Allied operations, the Japanese attacked shipping and shore positions with over 315 aircraft of which approximately 155 were destroyed. Hostile naval forces were sent in with supplies and reinforcements. In two naval engagements in the Kula Gulf on 6 July and 12 July, several Japanese cruisers and destroyers were put out of action.[29]

Despite these attempts to check Allied progress, the 43rd Division on Rendova drove ahead to eliminate all pockets of enemy resistance. Artillery was set up on the Rendova coast in preparation for the assault on the main Japanese stronghold at Munda across the Blanche Channel. On 2 July, units of the Division crossed over from Rendova to secure beaches east of Munda airfield and six days later a coordinated attack was made on Munda itself.

The enemy defended his positions stubbornly, however, and Allied reinforcements from the 25th and 37th Divisions were brought into the struggle. After severe fighting and heavy casualties on both sides, Munda fell to the Allies on 5 August. The northern portion of New Georgia was cleared by Marine units which had landed at Rice Anchorage on 5 July.

The Japanese evacuated their forces on Kolombangara by barge and destroyer between 28 September and 4 October. By 10 October, New Georgia and its adjacent islands, including the airdrome at Vila, were in Allied hands. Offensive operations against the enemy were progressing smoothly and after the capture of Munda General MacArthur stated:

We are doing what we can with what we have. Our resources are still very limited, but the results of our modest but continuous successes in campaign have been cumulative to the point of being vital. A measure of their potentiality can be obtained by imagining the picture to have been reversed, with the enemy capturing Guadalcanal and besieging Port Moresby rather than we in possession of Munda and at the gates of Salamaua. Such a contrast might well have meant defeat for us in the war for the Pacific.

The margin was close but it was conclusive. Although for many reasons our victories may have lacked in glamorous focus, they have been decisive of the final result in the Pacific. I make no predictions as to time or detail, but Japan on the Pacific fronts has exhausted the fullest resources of concentrated attack of which she was capable, has failed, and is now on a defensive which will yield just in proportion as we gather force and definition. When that will be I

28 Referring to the New Georgia operations, Colonel Tanikawa, Staff Officer of the Eighth Area Army said: "We had estimated an Allied landing somewhere in the New Georgia group but did not anticipate a landing on Rendova on 30 June 1943. We rather expected them to land on Munda where we were ready for such an attack. Also, the time of the actual landing was about a month earlier than we had anticipated." Interrogation Files, G-2, Historical Section, GHQ, FEC.

29 G-2, GHQ, SWPA, *Daily Summary* 4/5 Jul No. 469, 11/12 Jul No. 476 and 18/19 Jul No. 483, Appendix "A," Weekly Review of Enemy Activity.

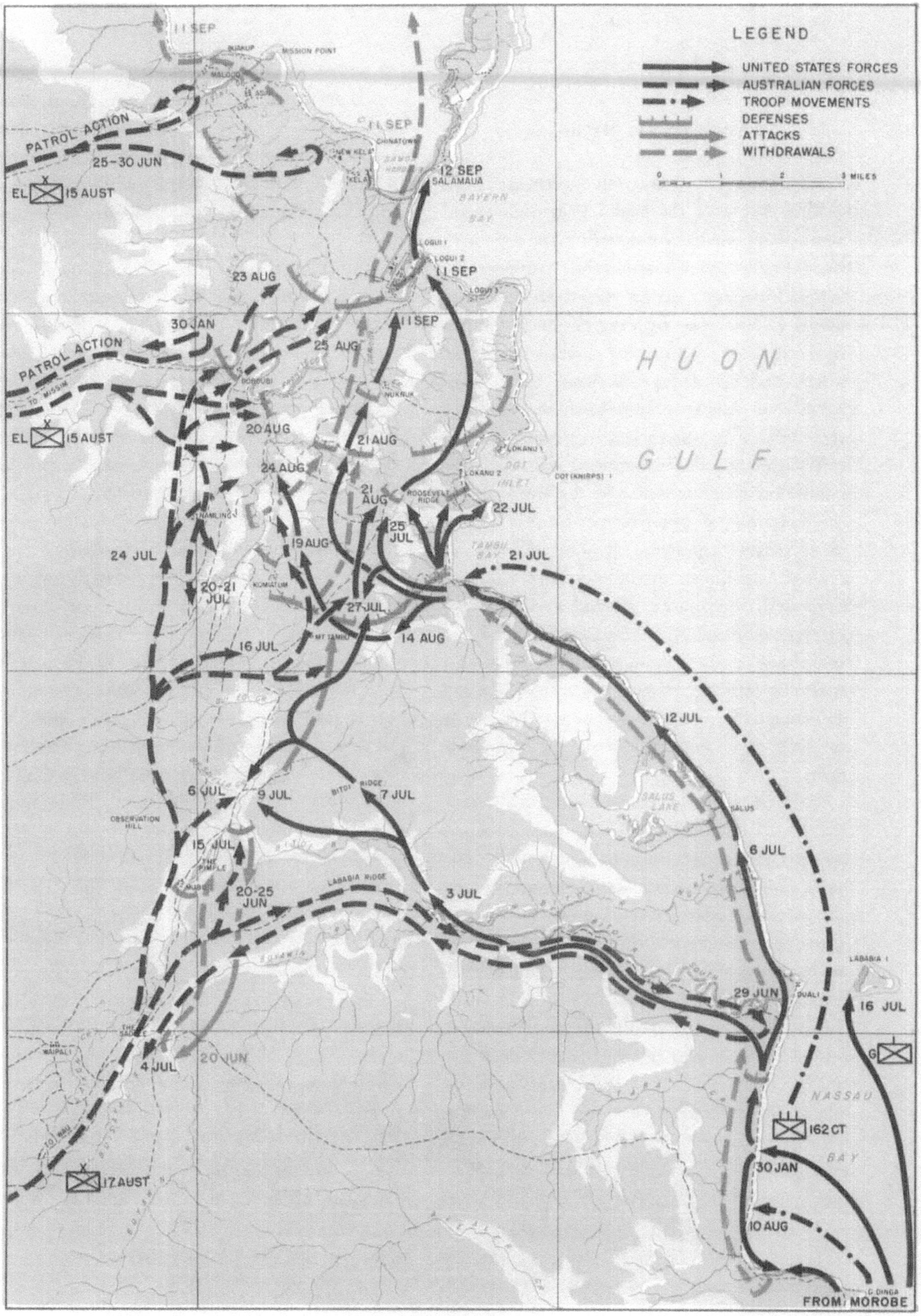

PLATE NO. 34

Operations, Nassau Bay to Salamaua

do not know, but it is certain."[30]

Nassau Bay to Salamaua

The plan to advance in northeast New Guinea and seize the Huon Peninsula necessitated comprehensive preparations to cope with the many complex terrain and hydrographic factors involved. Large ship-to-shore movements in the seas adjacent to New Guinea were impractical because of shallow, restricted waters and the danger of major losses from enemy air attack. Shore-to-shore advances were limited by the shortage of small landing craft and overland operations through the mountains were hampered by the impossibility of maintaining a strong supply line. To surmount these problems, it was planned to conduct a combined amphibious, airborne, and overland drive from the east and west, utilizing each type of maneuver where most practical and co-ordinating the over-all operation to obtain maximum striking power. Again the keynote of General MacArthur's policy was the use of his land, naval and air forces as a composite team.

Early in May, he issued Warning Instructions No. 2 which directed New Guinea Force to seize and occupy the area containing Salamaua, Lae, Finschhafen, and Madang. Under the warning order, New Guinea Force was given the code name " Phosphorous " and put under the command of General Blamey with General Herring named as his deputy.[31]

Lae, the gateway to the Huon Peninsula, was the first main objective for this new advance in New Guinea. This enemy stronghold was protected not only by well-prepared defenses of its own but also by a cordon of fortified positions at Mubo, Bobdubi, Komiatum, and Salamaua which stood guard over its approaches. (Plate No. 34) It was decided that a United States amphibious force would be sent along the eastern coast of New Guinea to effect a landing which would permit a junction with the Australians already operating along the outskirts of Mubo. The boats of the 2nd Engineer Special Brigade, selected to transport the troops and equipment, were limited in range, making it necessary to debark within sixty miles of Morobe, a port of departure secured by the Allies early in April. Nassau Bay was chosen as the point of the proposed amphibious landing. The United States troops were to join the Australian 17th Brigade in a combined attack on Mubo while the 15th Brigade moved forward from the Missim area to take Bobdubi Ridge to the north between Mubo and Salamaua.

On 30 June, while other Allied forces were striking in the Trobriands and at New Georgia, a reinforced battalion of the United States 162nd Infantry landed at Nassau Bay. The area was unoccupied by the enemy and subsequent resistance was light. While the 15th Brigade drew enemy attention to its attack on Bobdubi, United States troops, according to plan, drove west from the coast to link forces with the Australians at Mubo on 14 July. With the aid of heavy artillery bombardment and effective air support from the 5th Air Force, the Allies succeeded in occupying Mubo and pushing the surviving Japanese back toward Salamaua.

The Japanese took up new positions along the line Bobdubi—Mt. Tambu—Komiatum, where the Allied advance temporarily halted. Meanwhile, two more battalions of the 162nd Infantry landed at Nassau Bay and moved up to Tambu Bay where they encountered difficult terrain and powerful enemy defenses around Roosevelt Ridge. The Allies soon maneuvered

30 GHQ, SWPA, Press Release, 9 Aug 43.

31 G-3, GHQ, SWPA Warning Instructions No. 2, 6 May 43, G-3, GHQ, Opns 370.5 (S).

to cut the Japanese line of communications, however, and began to encircle their positions from the flanks and to the rear.

The Japanese tried vainly to prevent the steady closing of the Allied trap by desperate counterattacks from Mt. Tambu and Roosevelt Ridge. They knew that if any of their outer defenses were pierced the Allies would command dominating positions from which to bombard Salamaua itself with direct fire. The Allied forces continued to close in, nevertheless, and on 19 August the Japanese abandoned Mt. Tambu and Komiatum Ridge and retreated to a line along the Francisco River. Salamaua now lay open to assault.

General MacArthur, however, had not planned to take Salamaua immediately. Its airfield had been rendered useless and the town proper was of little importance to operations. The main purpose of the Allied attack on this small isthmus on the east coast of New Guinea was to siphon off enemy strength from his Lae defenses and lure his troops and supplies southward to be cut to pieces on the Salamaua front. It was intended to deceive the Japanese into believing that Salamaua was the prime objective of the Allied advance and this strategy was later to prove most successful.[32] Meanwhile, under cover of the Salamaua operations, the Allies were preparing for the principal drive to capture the strategically important town of Lae. Accordingly, Salamaua was not to be taken until the assault on Lae was actually underway.

Air Attack on Wewak

When the spotlight of the Pacific Theater focused on operations in the New Guinea area, Japanese Imperial General Headquarters felt that air power in that region had to be considerably strengthend. Consequently, on 28 July, orders were issued for the formation of the Japanese Fourth Air Force.[33]

The Allied Fifth Air Force, however, suspecting some such move by the enemy, intensified its assaults on key Japanese air bases. On 17 and 18 August, while the Japanese were consolidating their forces on the northeast New Guinea coast to carry out their assigned mission, a strong formation of Allied planes struck suddenly at Wewak. Heavy attacks on its major airdromes destroyed large numbers of planes caught on the ground.

As a result of this successful surprise attack, the sailing of enemy convoys intended to reinforce the Wewak and Hansa Bay garrisons, was rendered next to impossible owing to lack of air cover.[34] In commenting on the Wewak raid, General MacArthur said :

32 In the final drive on Salamaua, 2,000 Japanese troops were killed and large quantities of material, provisions, and barges were lost in futile efforts to hold a comparatively useless position. Japanese First Demobilization Bureau Report, *Southeast Area Operations Record, Part III : Eighteenth Army Operations*, Vol. I, pp. 104–108, G-2 Historical Section, GHQ, FEC. See also : *The History of the Lae-Salamaua Garrison*, GS (Int) Adv LHQ, SWPA, p. 7.

33 Japanese First Demobilization Bureau Report, *Southeast Area Operations Record, Part III : Eighteenth Army Operations*, Vol II, p. 27, G-2 Historical Section, GHQ, FEC.

34 The air attack on Wewak is described in the report of the Japanese First Demobilization Bureau as follows : "On 17 August 1943, Wewak airfield was attacked by a large formation of enemy bombers and fighters and the damage was great. In one attack, we lost more than 100 planes. Therefore, the seventh transport movement which was expected to arrive at Wewak on 25 August and its following transportation plans were postponed to the next month because of insufficient air escort which was caused by deterioration and frequent and continuous attacks by the enemy planes. Ever since 17 August a large convoy movement to New Guinea to increase shipping strength was expected, but the above described events affected this military operation to a large extent." Japanese First Demobilization Bureau Report, *Southeast Area Operations Record, Part III : Eighteenth Army Operations*, Vol I, p. 156, G-2 Historical Section, GHQ, FEC.

It was a crippling blow at an opportune moment. Numerically the opposing forces were about equal in strength, but one was in the air and the other was not. Nothing is so helpless as an airplane on the ground. In war, surprise is decisive.[35]

Smaller raids upon this base, Hansa Bay, and other positions in New Guinea were continued daily. Enemy interception grew lighter and the quality of his pilots deteriorated steadily after the heavy losses suffered in this sector and in the Solomons.

Nadzab and Lae

Preparations for the Lae operation were intensified during this period. On 20 August General Blamey arrived from Australia to take personal command of New Guinea Force. The amphibious and airborne training of the troops was completed, full-dress rehearsals were held, and the necessary equipment brought forward for loading. A strong United States carrier task force in the South Pacific sortied from Espiritu Santo and Efate in the New Hebrides and began round-the-clock raids on southern Bougainville to divert enemy attention to possible Allied assaults in that area. At the same time, PT boats and submarines struck at enemy barges along the coast and protected Allied waterborne movements against surface and subsurface attack.

The development of the hard-won Allied port at Buna had proceeded rapidly. It was to be the base of operations for the Australian 9th Division in its attack on the enemy's left flank. Actually, the timing of the Lae operation depended to a great extent upon the speed with which Buna could be built up as an intermediate supply base and staging area on the road from Milne Bay to the Lae beaches.[36]

On 1 September, the Australian 9th Division embarked in the transports and assault craft of the 7th Amphibious Force at Milne Bay. After a stop at Buna for reinforcements, fueling, and a final check, the landing force moved forward. On 4 September, under cover of continuous Allied air strikes at Wewak, Hansa Bay, Alexishafen, Madang, and New Britain, the main attack on Lae was launched. The assault troops hit the beaches at Bulu Plantation and at the mouth of the Busu River, less than twenty miles from Lae. (Plate No. 35) Minor opposition from enemy snipers was quickly eliminated and the troops drove on up the coast toward Lae itself.

At the same time that the Allied amphibious assault was being carried out east of Lae, a bold scheme was in progress to strike the enemy simultaneously on his right flank. It was planned to fly the Australian 7th Division directly from Port Moresby and land it in the Markham Valley to attack Lae from the west. The success of this maneuver depended upon the seizure of a pre-war emergency landing field located at Nadzab across the Markham River. The previously developed airstrips at Tsili Tsili and Wau were to provide the necessary fighter cover for such an airborne maneuver.

On 5 September, the United States 503rd Parachute Regiment, accompanied by General MacArthur himself, took off from Port Moresby on the first major jump of United States paratroopers in the Pacific War. "I did not want our paratroops to enter their first combat fraught with such hazard," said General MacArthur, "without such comfort as my presence might bring to them."[37] Trans-

35 GHQ, SWPA, Press Release, 18 Aug 43.

36 HQ 1st Aust Corps, "Report on Operations of New Guinea Force and 1st Australian Corps in New Guinea from 22 Jan 43 to 8 Oct 43," 17 Jan 44.

37 GHQ, SWPA, Press Release, 5 Sep 43.

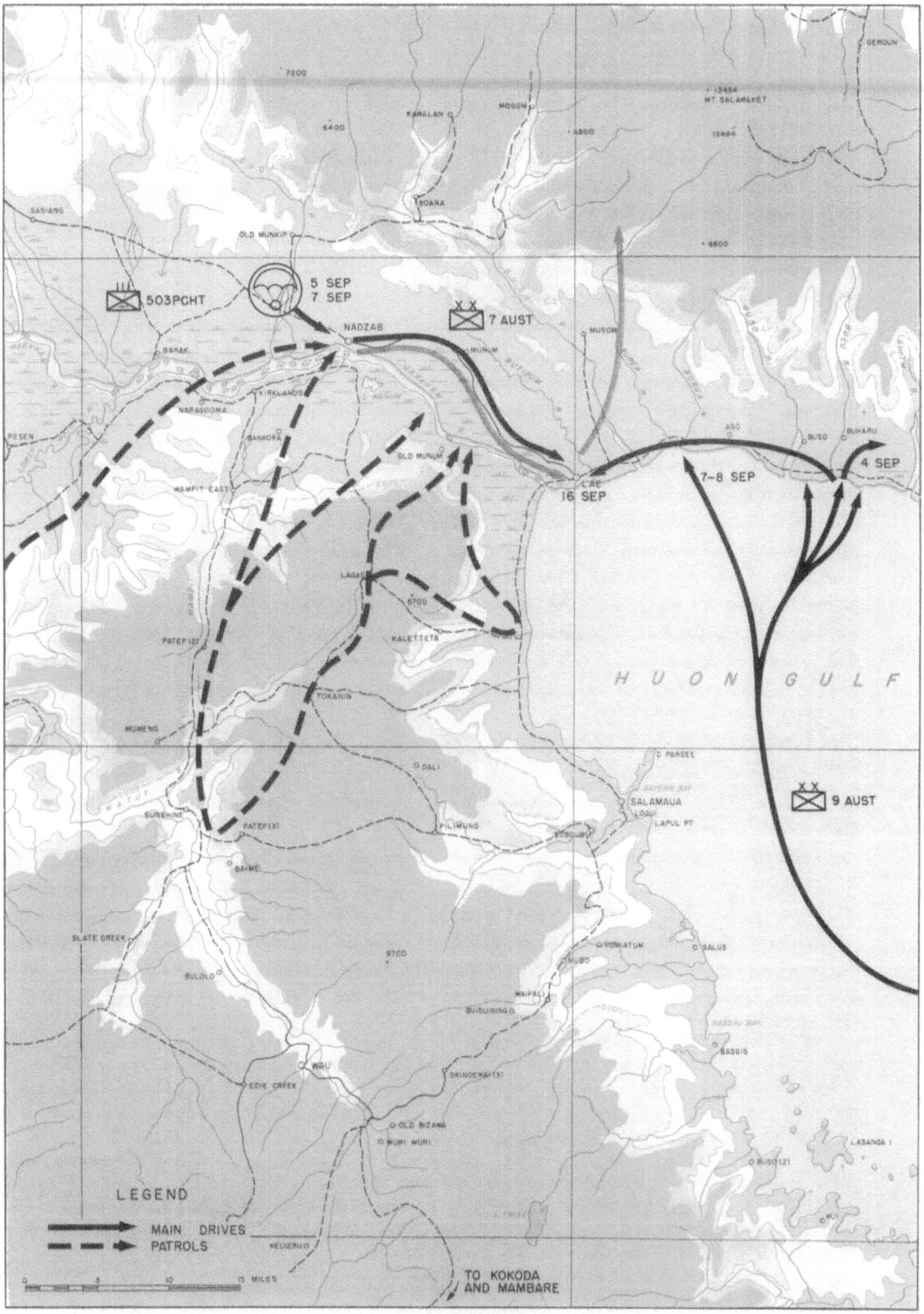

PLATE NO. 35

Nadzab and Lae

port after transport poured out its cargo of fully equipped paratroopers upon the vital airstrip. Watching the multi-colored parachutes spread themselves over the valley, General MacArthur felt the satisfaction of seeing this daring operation carried out with smooth precision. Even before he left the scene, the ground was being prepared for the big transports as flame throwers began to eat away the large patches of tall kunai grass. On the next day, the first of the planes bearing elements of the 7th Division landed on the runway to discharge its precious load of troops and equipment.[38]

The Allied circle around the enemy at Lae began to tighten rapidly. Advance troops of the 7th Division joined the paratroopers and drove along the Markham Road to approach Lae from the west. The 9th Division hammered in from the coast while the Allies increased the pressure from the air and on the Salamaua front to the south. On 11 September the Francisco River line was breached and the grim defense of Salamaua was smashed. The Allies closed in on Lae from all sides.

The Japanese were surrounded except for one narrow route of escape northward through the dense jungles and almost impassable mountain trails of the Huon Peninsula. As the Allied noose gradually choked them off from all hope of aid, the Japanese yielded their positions and, discarding almost all equipment, began a precipitous flight through jungle and mountains towards Kiari in a desperate effort to escape complete annihilation. On 16 September, Lae was occupied by the Allies. Another valuable link was forged in the chain of airbases that would eventually encircle and render helpless the powerful Japanese war machine.

Securing the Huon Peninsula

While the airstrip at Lae was being developed as a new forward base for Allied transport planes, New Guinea Force moved onward. Its next objective was Finschhafen—a busy enemy port on the tip of the Huon Peninsula. A keystone in the arch of the Japanese defenses guarding the western side of the strategic Vitiaz Straits, it was a valuable prize to be plucked from enemy hands.

Plans for the assault of Finschhafen had been mapped well in advance. Immediately after the fall of Lae, the 20th Brigade of the Australian 9th Division which had made the initial Lae landing, started on its next mission. On 22 September it rounded the jutting Huon peninsula and landed just north of Finschhafen. (Plate No. 36) The beachhead was secured against stiff opposition and the Allied troops, reinforced by another battalion, advanced toward their objective. Moving steadily on against stubborn but hastily prepared enemy counterattacks, Allied forces occupied Finschhafen on 2 October.[39] The Japanese retreated northwest to Satelberg where, taking up strong defensive positions, they contested the Allied

38 Colonel Shinohara, Intelligence Officer of the Eighth Area Army, commenting on the Nadzab operations, said: "We were retreating from the Salamaua area over the Finisterre Mountains toward Reiss Point when the Allied paratroopers landed at Nadzab which was one place where we thought the enemy would never attack. The remaining elements of the retreating 51st Division were virtually cut in half by this surprise pincer movement." Interrogation Files, G-2 Historical Section, GHQ, FEC.

39 The Allied landing at Finschhafen came at a timely moment. The Japanese, confused by the simultaneous Allied actions in New Guinea, had started to transfer their 20th Division from Madang in order to protect Finschhafen. The Allies caught the unit in the middle of its coastwise movement and the Japanese were forced into an insufficiently prepared defensive struggle. Japanese First Demobilization Bureau Report, *Southeast Area Operations Record, Part III: Eighteenth Army Operations*, Vol. I, p. 115, G-2 Historical Section, GHQ, FEC.

advance. Enemy lines of supply and communication were maintained from bases at Sio and Gusika and reinforcements were brought in during the month of October to check the Allied drive.

As one part of New Guinea Force carried out amphibious operations along the coast of the Huon Peninsula, another part was cutting inland below the enemy along the Markham Valley. Elements of the Australian 7th Division moving up from Nadzab seized Kaiapit, west of the Leron River, on 18–19 September. From there they went on to Gusap and thence to Dumpu which was taken on 4 October. These latest conquests provided new sites for the development of advanced airfields which would give the Allies an additional string of important operational airbases stretching from Tsili Tsili through Nadzab, Kaiapit, and Gusap to Dumpu.

On 25 November, New Guinea Force resumed its drive on the Huon Peninsula and pressed forward against continuing bitter resistance to occupy Satelberg. Moving north from Satelberg, it pushed on to take Wareo on 8 December.

The capture of Finschhafen and the subsequent drive up the New Guinea coast together with the simultaneous air-ground movement of 200 miles up the Markham Valley through the center of New Guinea gave the Allies control of the entire Huon Peninsula. The Allied maneuver outflanked and contained all important enemy centers on the Peninsula and rendered impotent his numerous positions and installations along the northeast coast of New Guinea. The speed of this double envelopment apparently caught the enemy unprepared and resulted not only in the serious dislocation of his grip on New Guinea but caused him enormous losses of men and material which he found impossible to replace.

Bougainville

Admiral Halsey's South Pacific Force, meanwhile, was engaged along the other crescent of advance in the northern Solomons. Originally, it had been planned to capture and consolidate the airfields near Faisi in the Shortland Islands and near Buin on southern Bougainville to cover a further advance to Kieta on the east coast. (Plate No. 37) Heavy concentrations of enemy strength at these two places, however, later made it advisable to amend the plan in favor of by-passing these enemy strongholds and making an incursion farther north at Torokina along Empress Augusta Bay. Aircraft could then neutralize Buka to the north while naval forces could close the sea route through St. George's channel to Rabaul and at the same time cover the projected assault on western New Britain.

To assist the operations of the South Pacific Force, bombers from the 5th and 13th Air Forces, escorted by fighters from Kiriwina and Woodlark, were to attack airdromes and shipping at Rabaul and Buka from the middle of October to 6 November. The South Pacific Force, shielded by these raids, would then occupy Treasury Island and northern Choiseul about 20 October to establish radar and motor torpedo boat bases for the subsequent attack against Empress Augusta Bay.[40]

The daylight air raids on Rabaul by General MacArthur's air forces commenced with the great strike on 12 October. To inflict maximum damage upon the enemy, the attack was timed to take place when observation photographs disclosed the largest number of enemy planes on the ground. Just as the right wing of the Japanese air force had been smashed at Wewak,

40 G-3 Memo of Record, 10 Sep 43, G-3, GHQ, SWPA *Journal* (S).

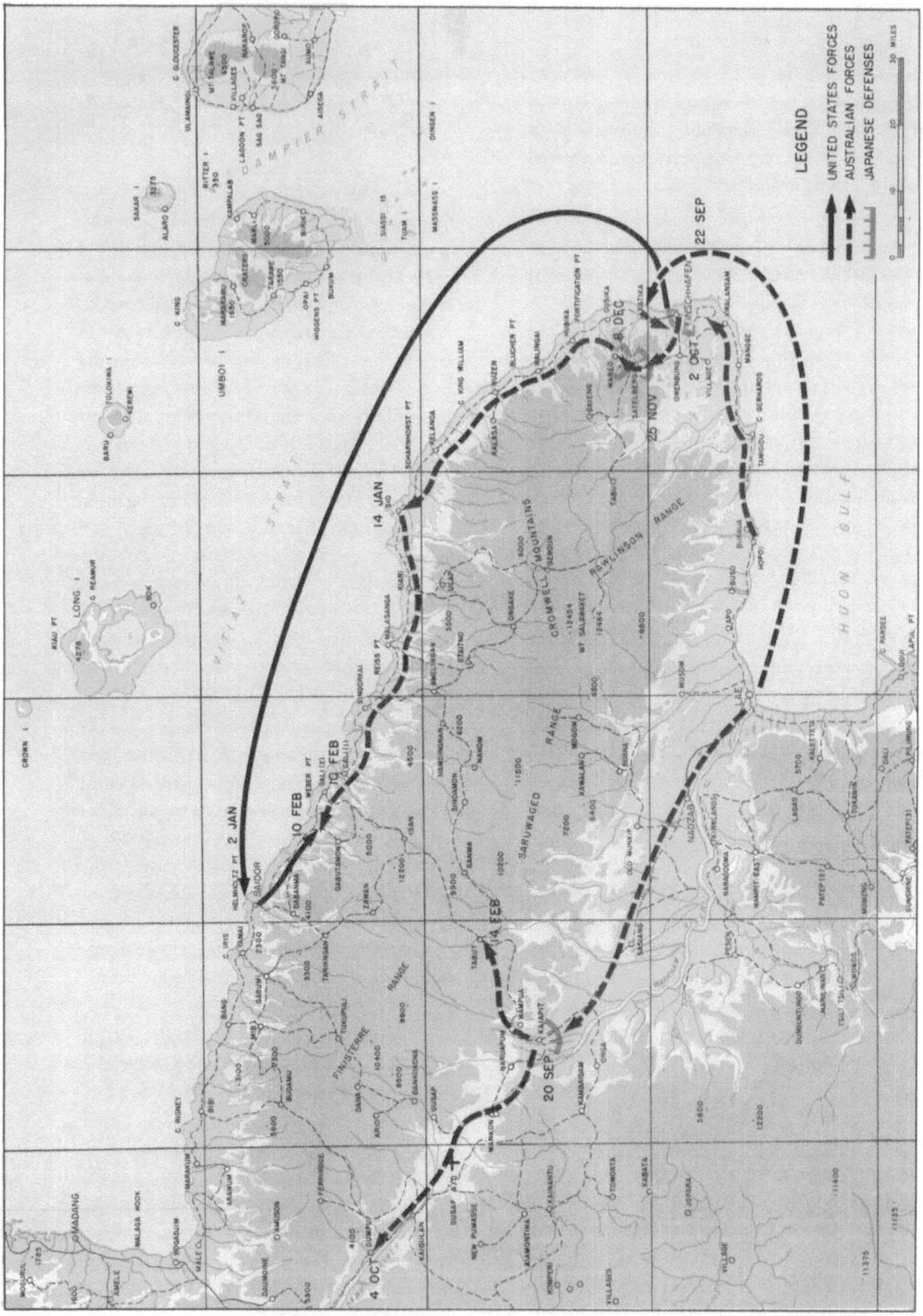

PLATE NO. 36

The Envelopment of the Huon Peninsula

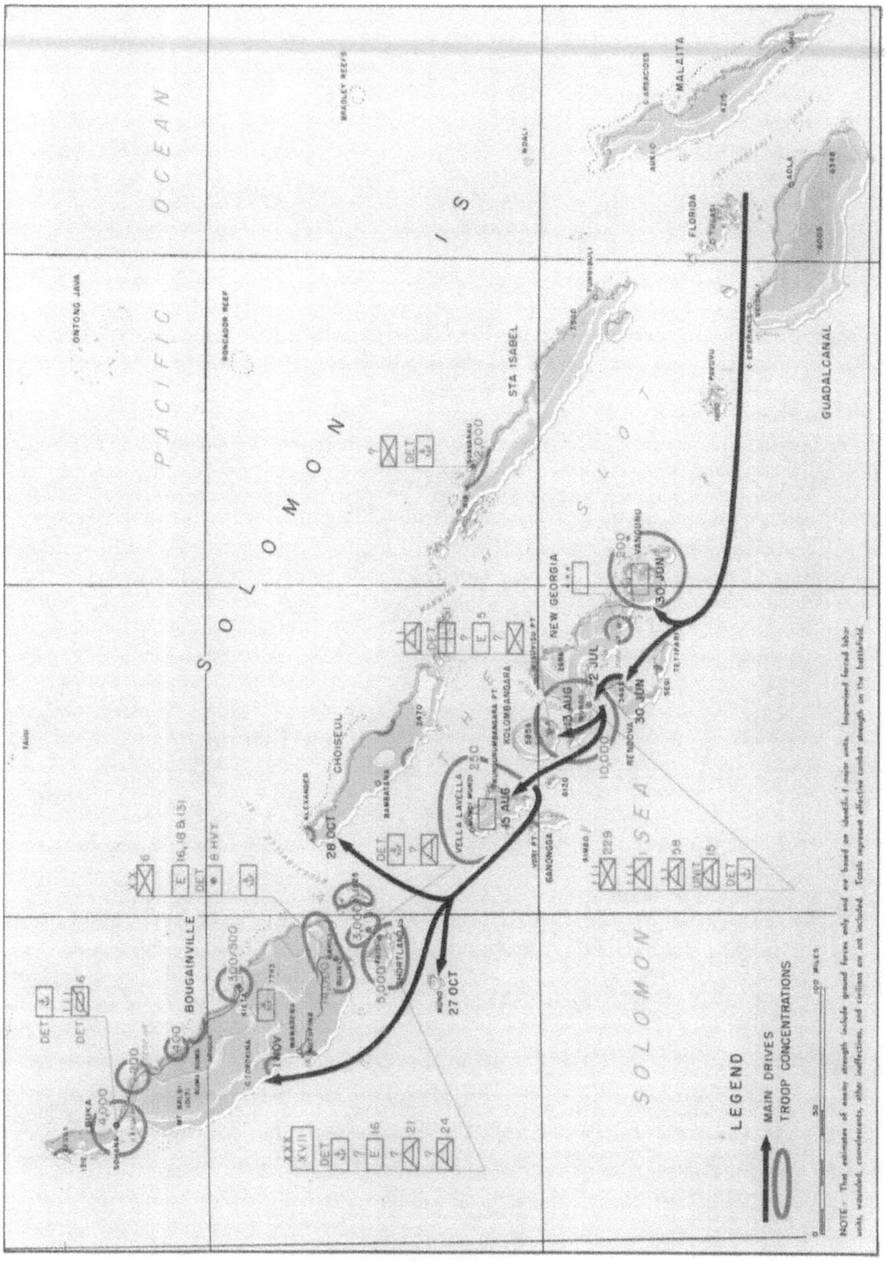

PLATE NO. 37

Allied Operations and Estimated Enemy Dispositions, Solomon Islands, 30 June 1943

General MacArthur hoped this time to destroy their left wing at Rabaul. The division of enemy air power into groups operating from these two widely separated bases made it possible for him to throw his entire available air strength alternately against one flank and then the other, thereby acquiring in each case superiority at the point of attack.

The Allied strike at Rabaul like the one at Wewak again caught Japanese planes on the ground, severely crippling the enemy's air strength and greatly damaging his merchant and naval units in the harbor. Continuous subsequent raids upon Lakunai, Vunakunau, Tobera, and Rapopo, the four major fields at Rabaul, greatly reduced Japanese air strength and contributed to the success of Admiral Halsey's attack in the northern Solomons.

According to the revised plan, the Allies moved forward 200 miles from New Georgia to carry out the projected landings in central Bougainville. On 1 November the United States 3rd Marine Division hit the beaches at Torokina on the northern shore of Empress Augusta Bay and pushed inland against relatively light opposition. The Japanese apparently did not expect an attack at this point. Their main forces concentrated at Buin and Buka were cut off by mountains, swamp, and jungle from the Allies at Torokina, and were consequently unable to counterattack effectively on land.[41] The Allies by their by-passing maneuver thus outflanked the Japanese on southern Bougainville and stood squarely athwart their supply line to that area.

Scattered hostile naval and air attacks failed to retard the rapid development of the Torokina beachhead. During the month of November, three enemy naval task forces were intercepted and driven back with substantial losses. The Marines, reinforced by the 37th Division, formed a defensive cordon around the airfield sites as a safeguard against probable counterattack. Enemy counterefforts, however, hampered by the difficulties of the intervening terrain, could not be mounted until March. When his anticipated attacks did materialize they were completely crushed. After that, the Japanese contented themselves with small, ineffective raids which continued sporadically until the end of the war.

The strategic aim of the Bougainville operation had been accomplished with the investment of Empress Augusta Bay. The Allies now possessed new airfields within 250 miles of Rabaul to help complete its isolation and neutralization.

New Britain

With the ousting of the Japanese from the Huon Peninsula, General MacArthur had secured domination over the western approaches to the Vitiaz Straits. To remove completely all threats of enemy interference with his operations in those strategic waters, he decided to lose no time in completing his conquest of the Straits and seizing control of its eastern shores. He accordingly directed Alamo Force to occupy Cape Gloucester on the southwestern tip of New Britain, setting the target date initially as 20 November. This date was based upon the assumption that the airdromes in the Lae—Finschhafen—Markham Valley area would be fully developed and capable of providing close support and cover. On 14 November, one regimental combat team was to seize an airfield

41 "The first real surprise maneuver, after I had arrived in Rabaul, occurred when the enemy landed on Cape Torokina on western Bougainville, during the latter part of 1943," said Lt. Col. Matsuichi Iino. "Because we thought the poor topographical features of this area would hamper enemy landing operations, we did not anticipate a direct landing here and consequently were not adequately prepared. This Allied operation proved very annoying in that we could not launch an immediate counterthrust in this area because of the poor network of roads. We knew an airstrip was being constructed but we were helpless." Interrogation Files, G-2 Historical Section, GHQ, FEC.

site at Lindenhafen Plantation near Gasmata on the south coast and then prepare to neutralize the surrounding region and prevent its use by the enemy. This operation would protect the southeast flank of the forces involved in the main effort against Cape Gloucester which was to be launched six days later by a combined amphibious and airborne attack on the other side of New Britain.[42]

Unforeseen developments, however, necessitated some changes in the original plans. The engineers reported in late October that the supporting airdromes could not be completed on schedule. The attack upon Lindenhafen, therefore, was postponed until 2 December and that upon Cape Gloucester until 26 December.[43] Intelligence reports received in November indicated that the Japanese garrison in the Gasmata area had been reinforced and was expecting an attack but that other areas, particulary Arawe, were weakly held. (Plate No. 38) In addition, it was decided that air facilities at Gasmata were not essential to the success of the operation. Under these circumstances, the attack was cancelled and instead it was planned to capture Arawe on Cape Merkus to the west in order to provide the necessary motor torpedo boat base for the protection of the main convoy movement to Cape Gloucester. The seizure of this port was therefore substituted for Lindenhafen, and a target date of 15 December was assigned to allow Alamo Force adequate time to provide for the change in plan.[44]

In preparing for the New Britain landings, the heavy concentration of enemy air strength on the Rabaul airfields constituted a serious menace to Allied operations. In spite of repeated raids on their airdromes, the Japanese continued to send planes aloft in considerable numbers. During the month of November, particularly, their efforts in the air reached a new peak of activity, indicating that plane reinforcements in substantial strength were arriving at Rabaul, Kavieng, Wewak, Hollandia, and Madang.[45]

This resurgence of enemy air power presented an added problem in the guarantee of adequate air coverage for the New Britain operation. There was no illusion that the naval convoy would escape heavy Japanese bombing attacks. To enable the fighter planes of the Allied Air Force, which were based on widely scattered fields, to rendezvous over Vitiaz Straits and maintain a continuous air umbrella over the convoy, General Kenney needed advance reports of Japanese air sorties against Allied landing forces at Arawe and Gloucester. General MacArthur's G-2 had anticipated this need and had taken steps to develop a comprehensive radio warning net of intelligence agents behind the Japanese lines. Three months prior to the Allied invasion of New Britain the U. S. Submarine *Grouper* had landed 16 Allied Intelligence Bureau operatives and 27 specially trained natives on the island.[46] These operatives had slowly infiltrated inland, and on the day of the

42 CG Alamo Force Ltr to CINCSWPA, 28 Sep 43, G-3, GHQ, SWPA *Journal.*

43 G-3 Memo to C/S, 26 Oct 43, G-3, GHQ, SWPA *Journal.*

44 G-3 Memo to CINCSWPA, 21 Nov 43, G-3, GHQ, SWPA *Journal.* GHQ, SWPA, Opn Instr No. 38/15, 22 Nov 43, G-3, GHQ Admin 370 (S).

45 G-2, GHQ, SWPA, *Daily Summary* Nos. 590, 591, 593, 597, 599, 605 and 607, Nov 43.

46 Air opposition to the New Britain operations was anticipated primarily from Rabaul. It was almost certain that the Japanese fighters and bombers would fly a direct route from Rabaul to the invasion points in order to attack Allied forces in the act of landing. Therefore, the Allied Intelligence Bureau established a chain of air watcher radio stations across the neck of the Gazelle Peninsula while other AIB agents took up assigned positions at Wide Bay, Open Bay, Gasmata, and Cape Orford. Pre-operational reconnaissance parties were also inserted on Rooke Island in the Vitiaz Straits to cover enemy flights originating from New Guinea. Eric Feldt, *The Coast Watchers* (New York, 1946), p. 218.

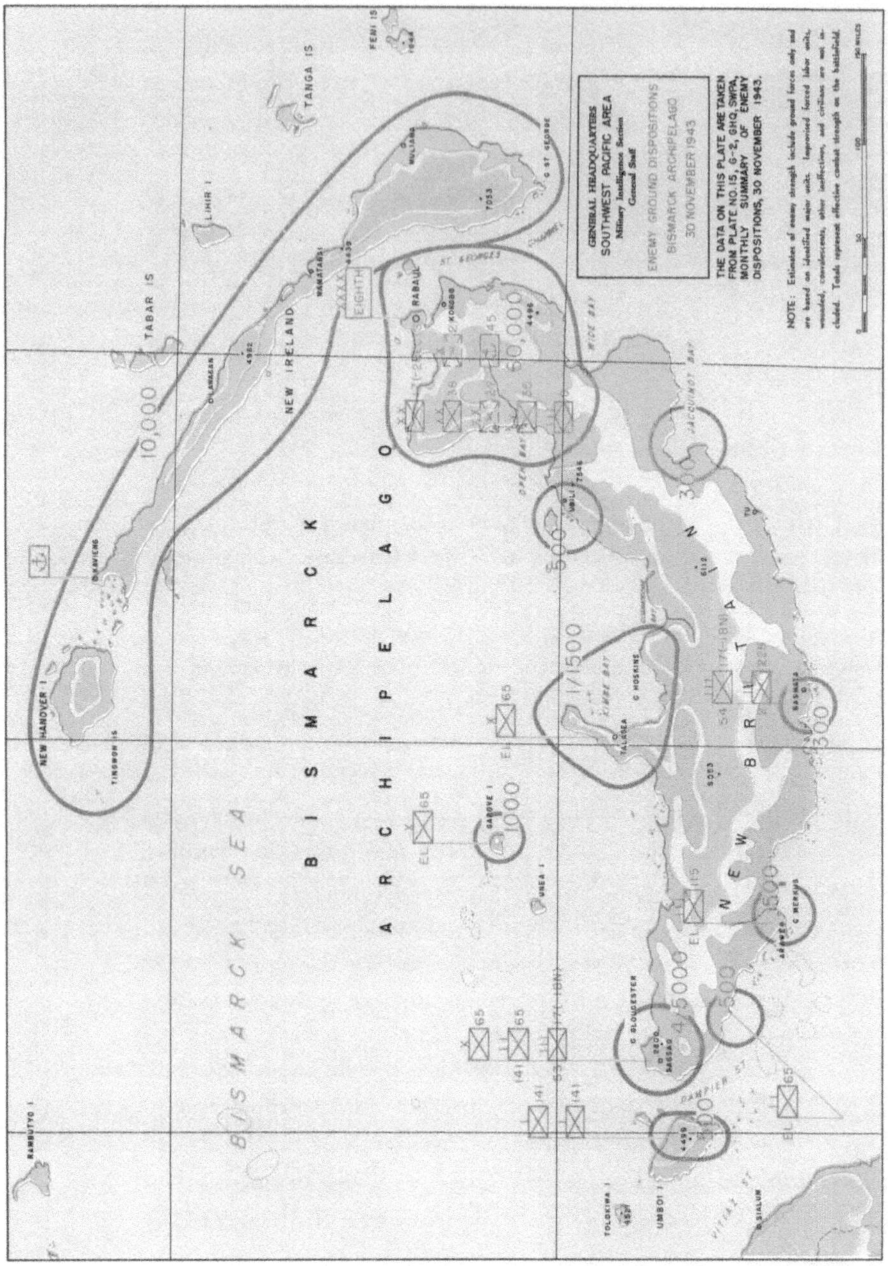

PLATE NO. 38

Enemy Ground Dispositions, Bismarck Archipelago, 30 November 1943

Allied landings they were in strategic positions to report on enemy activity and to give ample notice of oncoming enemy air formations.

The requisite troops and amphibious craft of Alamo Force were concentrated at Milne Bay, Buna, and Goodenough Island, while Allied planes began preparatory bombardment. Over 1,500 sorties were flown against Cape Gloucester alone during December and 3,700 tons of bombs were dropped. Gasmata and other positions on the south coast of New Britain were raided frequently, but to avoid alerting the enemy Arawe was not attacked until the day preceding the landing.

On 15 December, the U. S. 112th Cavalry Regiment, with strong air and naval support, went ashore at Arawe. Ground resistance was ineffectual and beachhead positions were rapidly consolidated. As predicted by initial intelligence forecasts, however, the Japanese attempted to strike back swiftly and forcibly in the air. An estimated 100 planes appeared over Arawe within three hours after the landing and from the 15th through the 17th approximately 350 enemy sorties were made in one of the largest defensive air operations which the Japanese had thus far engaged upon in the Southwest Pacific.[47] Allied fighters, however, forewarned of the impending strikes, scored heavily against each raid.[48] The Japanese nevertheless continued to expend their plane strength against Arawe until it became apparent that the prime interest of the Allies lay elsewhere.

Persistent Allied raids upon Cape Gloucester marked it clearly as General MacArthur's next objective. Even after the Japanese became aware of the Allies' true intention they could do little to protect this vital position which guarded the Vitiaz Strait and the lines of communication between Rabaul and New Guinea. The Allied aerial and submarine blockade restricted troop movements to small craft operating under cover of darkness and prevented the concentration of air and naval forces in sufficient strength to resist an attack. The Japanese attempted to reinforce their garrison from the ground troops at Rabaul, but this reinforcement was inadequate to counter the force pitted against it.

The mounting evidence of enemy weakness caused Alamo Force to cancel the airborne phase of the Gloucester operation and to rely entirely upon amphibious assault. On 26 December, two combat teams of the U. S. 1st Marine Division, commanded by Maj. Gen. William H. Rupertus, covered and supported by a heavy naval and air bombardment, landed on the beaches southeast and southwest of Cape Gloucester. The bombardment had proved very effective and the beachheads were secured against minor initial opposition. The main difficulties were presented by the narrow beach, which heavy rains had converted into a sea of mud, and the thick swamp and jungle beyond. At the end of the first day the Marines had established a perimeter approximately 900 yards in from the shore against stiffening resistance. As at Arawe, air reaction was initially strong and determined but after continued successful interception by Allied fighters enemy air activity virtually ceased by the end of December.[49] Both

47 G-2, GHQ, SWPA, *Daily Summary* No. 634, 16/17 Dec 43.

48 According to plan, Fighter Command Headquarters (Nadzab) received warnings of all approaching enemy formations directly from the coast watchers previously located on New Britain. On each message from the AIB operatives, Allied fighters took to the air and intercepted the raiders.

49 Again AIB agents gave warning of oncoming formations thirty to sixty minutes in advance, enabling Allied fighters to meet the enemy at the most advantageous altitude. Four raids were intercepted over the beaches the first day with disproportionate losses to the enemy. On 26-27 December in two missions of 70 to 90 fighters and bombers, the enemy lost over 75 planes as compared with 5 of the Allies. It was estimated that over 180 enemy planes were destroyed from 23–27 December. G-2, GHQ, SWPA *Daily Summary* No. 644, 26/27 Dec 43.

flanks of the vital Vitiaz Straits had been secured, opening the waters of this strategic highway between the Solomon and Bismarck Seas for unrestricted use in the furtherance of Allied plans.

Saidor

After the recapture of Satelberg and Wareo, New Guinea Force continued to advance along the Huon Peninsula against a steadily retreating enemy. The Japanese, with their supply lines rapidly disintegrating under sweeping Allied assaults, picked their way westward toward Sio and Madang. They were given no opportunity for respite or consolidation; their weary troops were compelled to resume their retreat once more as the Australians moved into Sio on 14 January 1944.

General MacArthur intended that this withdrawal should cost the enemy as high a price as possible. To cut off the main route of retreat and at the same time speed the Allied advance, he directed Alamo Force to make an amphibious landing at Saidor, midway between Sio and Madang.[50]

The landing was made on 2 January and the United States troops turned eastward to meet the fleeing enemy. General MacArther described the predicament of the Japanese when he reported:

We have seized Saidor on the north coast of New Guinea. In a combined operation of ground, sea and air forces, elements of the Sixth Army landed at three beaches under cover of heavy air and naval bombardment. The enemy was surprised both strategically and tactically and the landings were accomplished without loss. The harbor and airfields are in our firm grasp. Enemy forces on the north coast between the Sixth Army and the advancing Australians are trapped with no source of supply and face disintegration and destruction.[51]

The Japanese, caught between the closing pincers of the two advancing forces, abandoned the main route of retreat and, trying desperately to reach Madang, scattered in chaotic flight into the jungle[52] The hundreds of exhausted and starved bodies of Japanese discovered later along the ridges and mountain trails attested to the fact that the jungle had done its part in completing the work of the Allies. The last enemy soldier was squeezed off the Sio-Saidor road when elements of the Alamo and New Guinea Forces made contact on 10 February.[53]

Saidor was developed as an advance air and naval base to assist in the further conquest of New Guinea and for penetration into the Bismarck Sea, soon to be the next zone of Allied operations.

Objectives of 1943 Achieved

The year 1943 had fulfilled its early promise. The propitious victory at Buna had presaged the Allied conquests to come. In marked contrast to the slow, uphill struggle of the preceding year, 1943 witnessed long strides by the Allies deep into enemy territory. The goals set at the beginning of the year had been successfully achieved. On his left flank, General MacArthur had moved on from Buna according to plan, seizing in succession Salamaua,

50 GHQ, SWPA, Opn Instr No. 38/19, 31 Dec 43, G-3, GHQ, SWPA Admin 370 (S).

51 GHQ, SWPA, Communique No. 633, 3 Jan 44.

52 The Allied amphibious operation at Saidor split the Japanese Eighteenth Army. The landing constituted a wedge which isolated the main force of the Japanese 20th and 51st Divisions, approximately 15,000–20,000 troops, from the operating base of the Army at Madang. Japanese First Demobilization Bureau Report, *Southeast Area Operations Record, Part III: Eighteenth Army Operations*, Vol. II, pp. 167–8.

53 HQ Michaelmas Task Force, "Report of Michaelmas Operation," 8 Mar 44, G-3, GHQ, SWPA *Journal*, 10 Feb 44 (S).

Lae, Finschhafen, and the entire Huon Peninsula. His forces were rolling back General Adachi's Eighteenth Army at an accelerating pace all along the coast of New Guinea. On his right flank, the capture of Arawe and Gloucester Bay had placed the southern end of New Britain in his hands. Farther to the east, the Allies controlled the entire stretch of the Solomon Islands and the waters of the Solomon Sea.

Rabaul was being steadily emasculated by a growing Allied air arm which slashed constantly at its vital airfields and harbor installations. Enemy thrusts from that once powerful stronghold were becoming weak and ineffectual and by the end of February 1944, Rabaul had " no air support whatsoever." In spite of replacements and reinforcements the once-powerful " Japanese air force in this area had been driven to the point of extinction."[54] The gradual decimation of the enemy's land-based air power by the Fifth and Thirteenth Air Forces sharply decreased his ability to defend his vital sea lanes and opened the way for Allied naval craft to advance in increasing strength. It was these unrelenting and punishing attacks against major Japanese airfields and ground installations that won the battle in the skies and gradually destroyed the enemy air force in the New Guinea—Solomons area.

The Japanese at the opening of the year had held an advanced line running from Salamaua and Lae across southern New Britain and the Solomon Sea to New Georgia. Now their ground forces were pushed back to Madang, Rabaul, and Kavieng. In the air, the power of the Allied Air Force was increasing rapidly and could drive the enemy from the sky at almost any chosen point. On the sea, Allied Naval units had demonstrated that they could cope with any threat to amphibious operations and defeat the enemy wherever he chose to give battle.

The Japanese had suffered heavily in men and materiel. Huge amounts of discarded and abandoned supplies, guns, and ammunition marked the veering paths of their confused retreat as they were forced from one position after another. Thousands of their dead, killed in battle or stricken by starvation and disease, were left strewn on the battlefields and along the jungle trails. Additional thousands of troops were pocketed on Bougainville, Buka, Choiseul, and the Shortlands, their sting removed by isolation and their whole attention occupied with maintaining their own sustenance. Their transport and supply ships were sunk in increasing numbers as Allied planes and submarines exacted an ever-rising toll for their passage. Clearly, the Japanese had lost the power to conduct any futher large-scale offensives in New Guinea.

As supplies kept coming into the Southwest Pacific Area and the shortages which shackled his every plan were gradually being removed, General MacArthur prepared to launch heavier blows against the Japanese.

54 United States Strategic Bombing Survey, *The Allied Campaign Against Rabaul*. p. 24.

CHAPTER VI

THE WESTWARD DRIVE ALONG NEW GUINEA

Preliminary Plans

The route of General MacArthur's return to the Philippines lay straight before him—westward along the coast of New Guinea to the Vogelkop Peninsula and the Moluccas. (Plate No. 39) The successive Allied blows in New Guinea, New Britain, and the Solomons had seriously breached the Japanese outer wall of island defenses. General MacArthur was still about 1,600 miles from the Philippines and 2,100 miles from Manila, but he was now in a position to carry out with increasing speed the massive strokes against the enemy which he had envisioned since the beginning of his campaigns in the Southwest Pacific Area.

With the Japanese confused and thrown off balance by the recent series of Allied successes, General MacArthur urged that the situation be exploited immediately:

> *There are now large forces available in the Pacific which, with accretions scheduled for the current year, would permit the execution of an offensive which would place us in the Philippines in December if the forces were employed in effective combination. . . . All available ground, air, and assault forces in the Pacific should be combined in a drive along the New Guinea—Mindanao axis supported by the main fleet based at Manus Island [planned for a later operation] and other facilities readily available in these waters. I propose that on completion of operations in the Marshalls, the maximum force from all sources in the Pacific be concentrated in my drive up the New Guinea coast, to be coordinated with a Central Pacific operation against the Palaus and the support by combatant elements of the Pacific Fleet with orders to contain or destroy the Japanese Fleet. Time presses. . . .*[1]

General MacArthur planned to advance through western New Guinea because that route would provide the best opportunity for the complete utilization of the Allied ground-air-navy team. Such a drive, penetrating Japan's defense perimeter along the New Guinea line, would permit the by-passing of heavily defended areas and, at the same time, follow his tried and proved pattern of operations.[2] The land-based bomber line would again be moved westward by the successive occupation of new air sites; ground forces would be rapidly deployed forward by air transport and amphibious movements; additional plane and ship bases would be established as each objective was taken. Enemy naval

1 CINCSWPA Radio No. C-1217 to WARCOS, 2 Feb 44, WD C/S, JCS and CCS Papers No. 2, G-3, GHQ Exec Files (S).

2 Lt. Gen. Ryozo Sakuma, who became Chief of Staff of the Second Area Army in December 1944, expressed his opinion concerning General MacArthur's tactics in the following words: "I think that they were excellent tactics. I say this without prejudice. If any other plans had been used, the Americans would have had a very difficult time. . . . What the Americans did, as a whole, in the entire operation was good. . . . When General MacArthur retreated from our advance in the Philippines, he was not relieved as Commander-in-Chief of the area. . . . The fact that General MacArthur was kept at his post made it possible for him to conduct this campaign of retaking the Philippines as he saw best." Interrogations Files, G-2 Historical Section, GHQ, FEC.

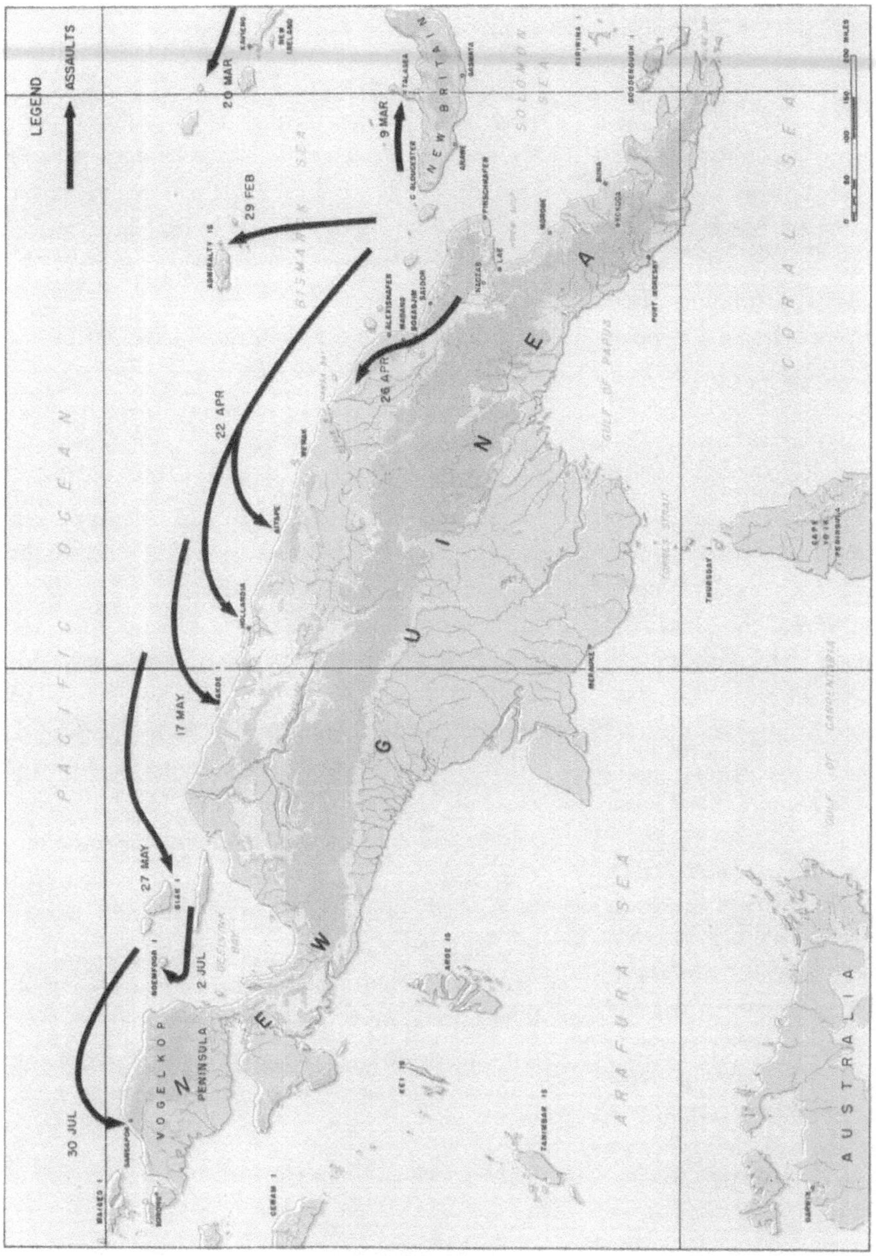

PLATE NO. 39

Westward Drive along New Guinea

forces and shipping would be eliminated along the line of advance to prevent reinforcement; then the same pattern would be repeated, neutralizing and pocketing hostile concentrations until the Allied forces were in position to make a direct attack against the Japanese in the Philippines.

As a preliminary to the coming drive, General MacArthur directed the seizure of the Green Islands, north of Bougainville. On 15 February, Allied troops carried out a weakly opposed amphibious landing and captured the islands to cut off an estimated 22,000 enemy troops to the south. Although the occupation of the Green Islands was a relatively minor operation, it completed the envelopment of the Solomons and rung down the curtain on that campaign. General MacArthur summarized the operation as follows:

We have seized the northern end of the Solomons Archipelago. New Zealand and American ground troops, covered by naval and air forces, landed and occupied the Green Islands. Enemy ground resistance was negligible and his air reaction weak. This culminates the successful series of flank movements commencing in the New Georgia group which has gradually enveloped all enemy forces in the Solomons. These forces, estimated at 22,000 strong, dispersed through Choiseul, Shortland, Bougainville and Buka Islands, are now isolated from their source of supply at Rabaul. Starvation and disease, certain to ensue from military blockade, render their position hopeless. With their airfields destroyed and their barge traffic paralyzed, the relief of these scattered garrisons is no longer practicable and their ultimate fate is sealed. For all strategic military purposes this completes the campaign for the Solomon Islands.[3]

During this same period, Admiral Nimitz's forces struck powerful blows in the Central Pacific. After successful attacks against Japan's strategic island bases in the Marshalls, which included the capture of Kwajalein in early February, they carried out a series of heavy raids westward on Truk in the Carolines. On 17 and 18 February two powerful carrier air strikes were directed against this Japanese naval stronghold where an estimated 325 enemy planes were destroyed and over 30 ships sunk. These damaging attacks against Japan's two key defenses in the Central Pacific opened a new horizon of activity for the U. S. fleet, and forced the Japanese to withdraw their navy to more secure bases.

Another indication of Japan's declining naval and air strength in the Central and Southwest Pacific at this time was reflected in the moves of other U. S. fleet units during 22 and 23 February. A South Pacific destroyer team, on a raiding and reconnaissance mission, made a two-day sweep of the waters in and around the Bismarck Sea without opposition. Although the destroyers' course led them through regions hitherto patrolled by enemy planes based in New Ireland and the Admiralties, and although they made several successful strikes on enemy merchant vessels and barges encountered along the way, their progress was never contested by Japanese air or naval units. The waters, once under complete enemy domination, could, it appeared, be traversed by Allied craft with comparative security.[4]

The Admiralties

Before launching a full-scale attack to the west, General MacArthur needed an additional base near enough for staging purposes and with a harbor of sufficient size to accommodate a large amphibious striking force. At the same time, he wished to insure the protection of his right flank, and to prevent reinforcements

3 GHQ, SWPA, Communique No. 677, 16 Feb 44.

4 G-2, GHQ, SWPA, *Daily Summary* No. 702, 23 Feb 44.

from reaching enemy troops bottled up in the Bismarcks—Solomons area.

The Admiralty Islands in the Bismarck Archipelago filled these requirements. They possessed ideal natural harbors and airfield sites which could support subsequent operations along the New Guinea coast and against the Carolines and Marianas to the north. In addition, their strategic location would provide another base from which to complete the encirclement of the Japanese cut off on New Britain and in the Solomons. Occupation of the Admiralties by the Allies would put the final seal on the isolation of Rabaul.

The original plan called for an amphibious operation on 1 April against the Admiralty Islands by Southwest Pacific Forces in conjunction with an attack on Kavieng by Admiral Nimitz's fleet units. In February, however, General MacArthur saw reflected in the lack of opposition to Allied air and naval craft in the Bismarck area, a temporary confusion and weakness of the enemy which he decided to capitalize upon without delay. He felt that the situation presented an ideal opportunity for a *coup de main* which, if successful, could advance the Allied timetable in the Pacific by several months and deal another powerful blow to Japan's crumbling defense structure in the New Guinea area.

The principal enemy positions in the Admiralties were on Los Negros and Manus Islands, both of which were in a precarious position at this time. Allied air power blocked the shipment of sufficient supplies to maintain a strong defense and seriously reduced the possibility of bringing in reinforcements in time to withstand an Allied attack. The feeble reaction to Allied raids offered by Japanese naval and air units demonstrated that their resistance would probably be minor and easily overcome.

The question of enemy ground dispositions in the Admiralties, however, posed a different problem. Opinions differed sharply on the number of Japanese troops in the islands. Repeated air observation flights over Manus and Los Negros at the end of February reported a total lack of enemy activity. Momote airfield was observed to be entirely unused; even the bomb-craters in the runway had been left unfilled and grass was growing along the air strip. Surrounding buildings and installations were seemingly unattended and in a bad state of disrepair. The continued lack of antiaircraft fire and apparent desertion of the air strips reported by air reconnaissance supported the inference that the Japanese had left the islands altogether.[5]

General MacArthur's G-2, however, held firmly to a different opinion. On the basis of intercepted information and ground reports over a long period of time, the conclusion reached was that "cumulative intelligence does not support air observer reports that the islands have been evacuated."[6] On 15 February,

5 Typical of air reconnaissance reports are the following: "Momote and Lorengau strips appeared unserviceable. Nil activity. Nil new aircraft. Nil unusual signs of activity in entire Admiralty Islands." GHQ, SWPA, *Situation Report* No. 54/44, 23 Feb 44. "Aircraft flew low but nil A/A fire encountered. Nil signs of enemy activity. The island [Lorengau] appears deserted." Ibid., No. 55/44 24 Feb 44. "No signs of enemy activity on Manus and Los Negros Islands. All crews claim these islands have been evacuated. Grass growing thickly on Momote and Lorengau strips. Runways unserviceable, and badly pitted. No A/A fire, even at low altitude. (The B-25's flew over Momote strip at 20 feet)." Ibid., No. 56/44, 25 Feb 44. "Both Lorengau and Momote strips are unserviceable. The wrecked aircraft and trucks on Momote are untouched and bomb craters still unfilled. Villages on Los Negros Island appeared deserted and roads have not been used lately. Damage in Lorengau town has not been repaired. No activity of any kind observed." Ibid., No. 57/44, 26 Feb 44.

6 G-2, GHQ, SWPA, *Daily Summary* No. 706, 26/27 Feb 44. AIB operatives had picked up natives from the Admiralties and had pieced together a fairly accurate picture.

the probable enemy units still on the islands were listed and their total strength was estimated to be approximately 3,250 men.[7] On 24 February, these figures were revised and the estimate of Japanese troops in the Admiralties sector was raised to 4,000.[8] G-2 analysis discounted the significance of the lack of antiaircraft oppositions pointing out that "prior to the Allied landing at Cape Gloucester a similar situation was reported but resistance was encountered to the landing and to the subsequent offensive moves." It attributed the lack of antiaircraft fire to the deteriorated state of the enemy's logistical situation and predicted that "the enemy will hold his fire until the final defense of the Admiralties is imminent."[9]

On 24 February General MacArthur directed that an immediate reconnaissance in force be made to probe the island defenses by landing at Hyane Harbor on eastern Los Negros, where enemy positions apparently were weakly held. His aim essentially was to strike swiftly, achieve surprise, and thus avoid bitter fighting and heavy casualties at the beachhead. If an initial foothold could be established without undue losses, the reconnaissance force would then advance and seize Momote airstrip near the harbor entrance; if unforeseen enemy strength should be encountered at the beaches, however, and an unfavorable situation should develop, a speedy withdrawal would be carried out. In cognizance of his G-2 estimates of enemy strength in the Admiralties, General MacArthur also took precautions against the eventuality of a delayed counterattack in strength. A strong auxiliary force was readied at Finschhafen which would be sent two days later as reinforcements to enlarge the perimeter, if established, and to guard against any powerful Japanese reaction that might develop after the Allied landing was secured.

On the morning of 29 February, the daring strike was launched. The reconnaissance force was light, consisting of approximately 1,000 troops of the 1st Cavalry Division supported by two cruisers, a dozen destroyers, and the necessary air cover. The troop units were made up mainly of a reinforced squadron of the 5th Cavalry under Brig. Gen. William C. Chase; the naval forces were under the command of Admiral Kinkaid. General MacArthur was relying almost entirely upon surprise for success and, because of the delicate nature of the operation and the immediate overall decision required, he accompanied the force in person aboard Admiral Kinkaid's flagship,

7 G-2, GHQ, SWPA, *Daily Summary* No. 695, 15/16 Feb 44 contained the following precise listing of enemy units, almost two weeks in advance of the actual landing; compare these data with the negative reports of flight reconnaissance on 23–25 February. (See footnote 5 supra).

Admiralty Islands	*I.*	*Mobile Combat*	*Base Defense*	*Service*	*Total*
El 17th Div	*1/44*	*1000*			*1000*
1 Bn 1st Ind Mixed Regt	*1/44*	*600 ?*			*600*
El 38th Div	*1/44*	*300*			*300*
Oita S. Sea Det	*1/44*	*300 ?*			*300*
U/i AA	*?*		*200*		*200*
51st Tpt	*4/43*			*800*	*800*
El 51st Div Fd Hosp	*1/44*			*50*	*50*
		2200	*200*	*850*	*3250*

8 Ibid., No. 704, 24/25 Feb 44.

9 Ibid., No. 706, 26/27 Feb 44. Several days previously G-2 had also reported on the Admiralties: "A situation similar to Madang is encountered here in that no enemy activity is apparent. This is regarded as a case of passive antiaircraft defense necessitated by dwindling reserve ammunition. Other intelligence indicates that the enemy plans to defend the Admiralties with the forces at present located there." Ibid., No. 704, 24/25 Feb 44.

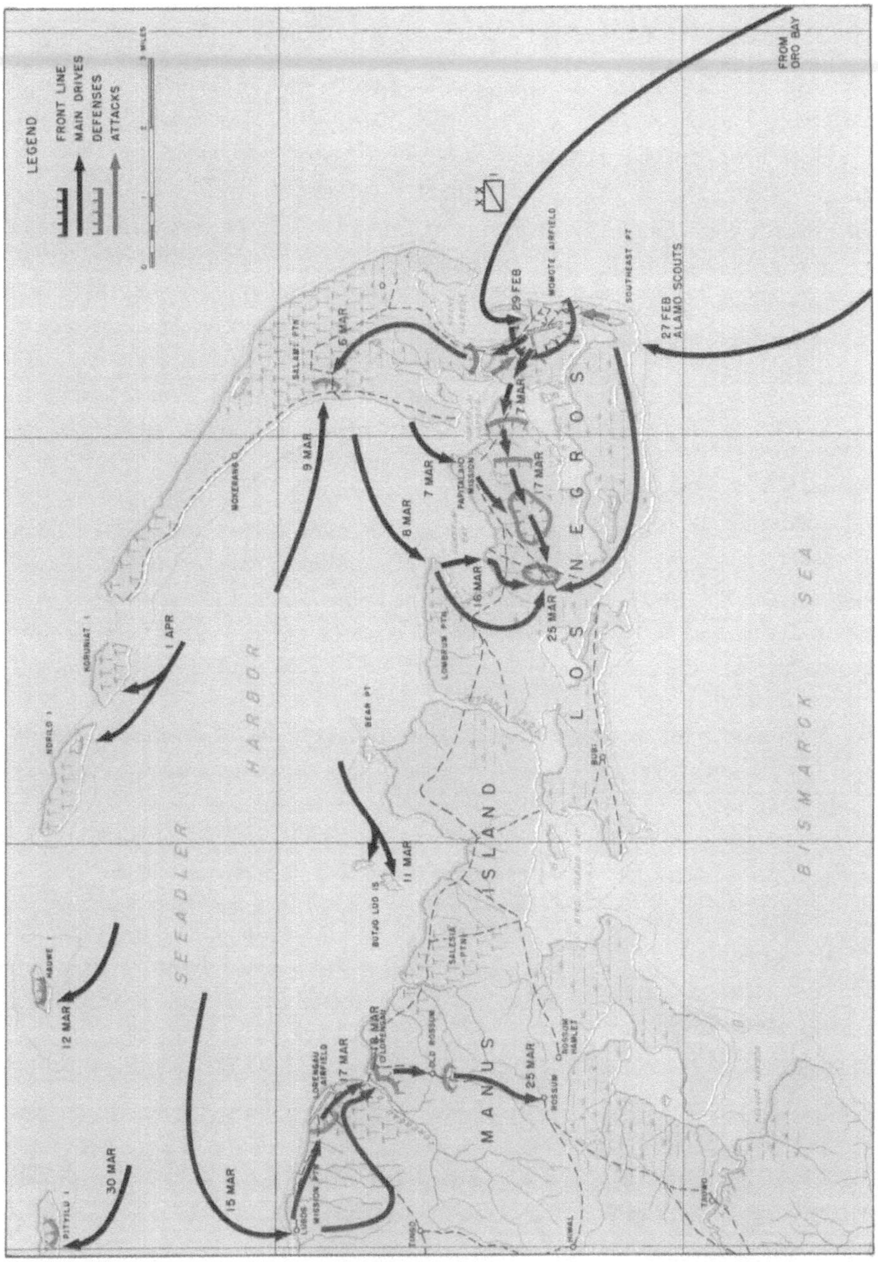

PLATE NO. 40

Admiralty Islands Campaign

the *Phoenix*.

After a preliminary air and naval bombardment, the initial landing was made on the southwest shore of Hyane Harbor under a completely overcast sky. (Plate No. 40) A reconnaissance group of Alamo Scouts, who had secretly explored the vicinity of the intended landing two days previously, provided valuable information on the dispositions of enemy concentrations near the harbor.[10] The first assault waves met little opposition. Before the Japanese could recover from their surprise and swing their heavy guns, which were facing Seeadler Harbor, into position, most of the troops were ashore. Working in a heavy rain against sporadic and weak enemy fire they established a wide perimeter along the airstrip and started the preparation of defensive positions.

The calculated risk had been justified. General MacArthur, coming ashore just six hours after the first assault wave to survey the situation, said to General Chase: "Hold what you have taken, no matter against what odds. You have your teeth in him now—don't let go."[11] With the scheduled arrival of the auxiliary force from Finschhafen on 2 March, the perimeter could be enlarged and the airstrip secured and prepared for incoming Allied planes.

Assured of the ultimate success of the operation, General MacArthur returned to New Guinea and his Headquarters issued the following report:

> *We have landed in the Admiralty Islands which stand at the northern entrance to the Bismarck Sea almost due south of Guam and 1,300 miles from the Philippines. . . . This marks a final stage in the great swinging move pivoting on New Guinea which has been the basic purpose of the operations initiated on June 29, 1943, when the Southwest Pacific Area and South Pacific Area were united under General MacArthur's command. The axis of advance has thereby been changed from the north to the west. This relieves our supply line of the constant threat of flank attack which has been present since the beginning of the Papuan Campaign. This line, previously so precariously exposed, is now firmly secured not only by air coverage, but by our own front to which it is perpendicular. The operation has been a delicate one and its final success lays a strategically firm foundation for the future. Tactically it tightens the blockade of the enemy's remaining bases. Their supply lines are definitely and conclusively severed and only a minimum of blockade running, by submarine or individual surface craft, is now possible. In addition to the troops trapped in the Solomons, some fifty thousand of the enemy, largely in New Britain and at Rabaul, are now inclosed. Their situation has become precarious and their ultimate fate is certain under blockade, bombardment and the increasing pressure of besieging ground forces. The end of the Bismarck Campaign is now clearly in sight with a minimum of loss to ourselves.*[12]

After two days of hard fighting in which determined Japanese infiltration attempts and counterattacks were beaten off, the original landing force was joined on 2 March by General MacArthur's auxiliary forces which had arrived from Finschhafen. These reinforcements came at a timely moment. An enemy document captured early the next morning disclosed that a Japanese effort to regain Momote airfield would be made on the night of 3 March. Hurried but thorough preparations were made for the forthcoming attack, which was expected to include most of the

---

10 This is an interesting example of clandestine operations, in conjunction with operatives of the AIB, a theater unit, under the operational control of G-2.

11 Historical Division, War Department, *The Admiralties, Operations of the 1st Cavalry Division (29 February—18 May 1944)* p. 31. Hereinafter cited as: *The Admiralties.*

12 GHQ, SWPA, Communique No. 691, 1 Mar 44.

Japanese remaining on western Los Negros and whatever troops could be brought from nearby Manus Island.

The scheduled attack came as promised. Wave after wave of tough Japanese infantry came pouring through the darkness against the Allied defenses.[13] In fierce but foolhardy charges, they sought to overwhelm each Allied position by sheer weight of numbers. When those in front were mowed down by mines and machine gun fire, the others rushed on undaunted over the bodies of their comrades only to be cut down in turn.

During the course of these suicidal "banzai" charges, the enemy resorted to wily and fanciful tactics. Japanese, who had somehow managed to learn the names of Allied platoon leaders, tried to trick them into misdirecting or ceasing their fire by interpolating false orders in perfect English into tapped telephone wires. One column of enemy troops came marching forward singing, for no apparent reason, "Deep in the Heart of Texas." Bayonets were affixed to five-foot poles and used as spears by the Japanese when their ammunition gave out. Bandages were discovered tied around their arms at pressure points, presumably to provide a ready tourniquet which would permit them to continue fighting should they lose the use of either hand.[14]

It was a bizarre and bloody night but the efforts of the Japanese, however brave, were erratic and unco-ordinated. Their attacks tapered off as their dead began to litter the field of battle and by morning the main enemy counterdrive had been decisively crushed. The cavalry forces awaited the arrival of more troops before clearing the rest of the islands. Further reinforcement echelons arrived on the 4th, 6th and 9th of March. The Allies struck out from the Momote perimeter westward to take Papitalai and northward to occupy Salami Plantation. From the latter point they moved on across Seeadler Harbor to seize Papitalai Mission and Lombrum Plantation. By 30 March all organized enemy resistance on Los Negros had been overcome.

Meanwhile operations against the larger island of Manus had begun. On 9 March, the 2nd Brigade Combat Team of the 1st Cavalry Division, under Brig. Gen. Verne D. Mudge, disembarked at Salami Plantation. On 15 March, after units of the 7th Cavalry had secured the two small islands of Hauwei and Butjo Luo off the north coast of Manus, assault troops of the 2nd Brigade landed at Lugos Mission, two miles west of Lorengau. The Japanese forces, depleted by the Los Negros counterattacks, put up a stubborn but useless struggle. Lorengau Airfield was captured on the 17th and the Allied units moved on to take Rossum on the 25th. This brought to an end the campaign for the Admiralties.

The success of the operation was complete. The Japanese had been powerless to present any naval opposition to the Allied landings and their efforts in the air were sporadic and easily overcome. On the ground, G-2 estimates had been remarkably accurate. It was calculated, after a count of enemy killed and captured, that the Japanese had used approximately 4,300 troops to defend the islands—substantially the force predicted in G-2 estimates of enemy strength.[15]

The occupation of the Admiralties had tied the last knot in the noose around Rabaul. More than 100,000 Japanese were now choked off in the Bismarcks—New Britain—Solomons

13 A desperate night counterattack of this kind had been forecast as one of the enemy's capabilities in the G-2 Daily Summary of 28/29 February 1944, viz: "Enemy capabilities are a) surreptitious withdrawal from the Admiralties via small craft, b) typical desperate counter-attack, probably at night."

14 *The Admiralties*, p. 47.

15 Ibid., p. 148.

area. Henceforth the enemy was compelled to rely almost exclusively on risky submarine and destroyer runs for carrying in bare essentials. In addition the supply route to his outer defenses was pushed farther west and the whole sea region between Truk and New Guinea was placed within Allied air range. With the threat to his right flank removed, General MacArthur could now concentrate his entire attention on the task of moving westward along the coast of New Guinea.

Supplementary Operations

Concurrently with the landings in the Admiralties, several lesser operations were undertaken to supplement the main Allied drives in the Bismarck and Solomon Seas areas. On 6 March, the 1st Marine Division on New Britain pushed on to take Talasea on the Willaumez Peninsula and then occupied the Cape Hoskins area. Most of the Japanese south of Rabaul had by this time retreated northeastward into the Gazelle Peninsula.

On 20 March, Admiral Halsey's forces captured Emirau Island in the St. Matthias group. The enemy troops had been withdrawn before the landing and the objective was gained without cost. This operation further strengthened the Allied grip on the Bismarck Archipelago and, in addition, provided an excellent air base within short range of Kavieng and Rabaul.

Meanwhile, the Australians, who had been fighting vigorously on the Huon Peninsula and in the Ramu Valley, pressed westward along the New Guinea coast. Progress had been held up considerably by the steep ranges of the Finisterre Mountains but, working steadily across this natural barrier, New Guinea Force moved into Bogadjim on Astrolabe Bay on 13 April. Madang was captured eleven days later, and on 26 April the Australians occupied Alexishafen as the Japanese carried out a mass retreat to the west toward Wewak and Aitape.

Plan to By-Pass Hansa Bay

The sudden seizure of the Admiralties put the Allied timetable well ahead of schedule. According to the original plan Hansa Bay was the next objective on the New Guinea coast but in view of the spectacular success of the Admiralties landing, General MacArthur was anxious to take a longer step forward. The proposed capture of Hansa Bay would mean an advance of only 120 miles from the Allied position at Saidor, and, even if the Allies occupied that base they would still have to face the large Japanese concentrations around Wewak and Aitape.

With the outcome of his operations into the Admiralties assured, General MacArthur planned a bold maneuver which, at one stroke, would move his forces almost 500 miles forward and at the same time squeeze some 40,000 Japanese troops between the jaws of a strong Allied vise. He would by-pass Hansa Bay, by-pass the enemy stronghold at Wewak, and strike well to the rear at Aitape and Hollandia.

On 5 March, just four days after his return from Los Negros, General MacArthur submitted his daring plan to the Joint Chiefs of Staff in Washington:

Recent seizure of a foothold in the Admiralties which will shortly be followed by complete occupation presents an immediate opportunity for rapid exploitation along the north coast of New Guinea. To this end I propose to make the Hollandia area instead of Hansa Bay my next objective, capturing the airfields and base in that vicinity for further support of operations toward the Philippines. The enemy has concentrated the mass of his ground forces forward in the Madang—Wewak area, leaving relatively weak forces in the Hollandia Bay area. He is attempting to concentrate land-based air forces in the area of western New Guinea and is developing additional fields to consolidate this area into a bulwark of air defense. SWPA land and air forces are disposed favorably to launch a two division reinforced assault against the objective. Preparations are being completed for the Kavieng operation and this stronghold should fall quickly into our hands. Naval forces of SOPAC and SWPA have established surface control of the Bismarck Sea and have been augmented for the Kavieng—Admiralties and Hansa Bay operations, the latter tentatively scheduled for late April. I propose that the Kavieng operation, proceeding in accordance with schedule, be followed between 15 and 24 April by an assault on Hollandia, using for the Kavieng—Hollandia operation the means of SOPAC and SWPA as augmented for the Kavieng—Admiralties and Hansa Bay operations including coverage. For this purpose it will be necessary to retain that part of the naval reinforcements allocated from CENPAC to SOPAC and SWPA for the Kavieng—Admiralties but which were not included in the Hansa Bay allocation for a short additional period. It will also be necessary to continue the operation of SOPAC Force in the theater of SWPA until Hollandia is secured. I request prompt approval. Admiral Halsey concurs. Establishment of our forces in this area will have wide implications. It will hopelessly isolate some 40,000 of the enemy ground forces along the New Guinea coast, will place in our hands airdromes from which our land-based air forces can dominate the Vogelkop, and hasten the advance westward by several months.[16]

It was later decided that the rapid development of the Admiralties' extensive harbor and airfield facilities, coupled with the occupation of Emirau Island, made the capture of Kavieng unnecessary. The plan to invade New Ireland was consequently cancelled.[17] The naval forces originally allocated for the Kavieng operation could now be shifted to the New Guinea theater to be employed against Hollandia.

Admiral Nimitz agreed to support fully the proposed assault on Hollandia. The Pacific Fleet would lend all possible assistance by furnishing the large naval forces required for such an ambitious undertaking. Carrier planes would provide close cover for the initial assaults of the landing parties since the projected point of attack was too distant from the forward bases held at the time for adequate protection by land-based fighters. The navy would also carry and protect the tremendous stores of supplies, munitions, and special equipment which were needed to maintain a major invasion operation. In addition, carrier task groups would begin the immediate neutralization of enemy positions in the western Carolines from which possible interference to the Allied operation might be forthcoming.

General MacArthur's plan called for landings on both sides of Hollandia, at Humboldt and Tanahmerah Bays, with a simultaneous landing by a smaller force at Aitape, midway between Hollandia and Wewak. The Hollandia forces would converge on the flanks of the three airstrips inland while the forces to the east would seize the fighter field at Aitape and prevent any junction of the enemy's Wewak and Hollandia troops. The target date was set for 22 April.[18]

During March and the first part of April, intensive air raids were launched by the Fifth Air Force against the intervening Japanese airfields on New Guinea. Strong formations of B-25's with fighter escort blasted continuously at Hansa Bay, Wewak, and Hollandia. General MacArthur described the punishing raid of 3 April against Hollandia as follows:

In the heaviest attack so far made on this enemy base, our escorted heavy units followed by medium and attack planes at minimum altitude, dropped 400

16 CINCSWPA Radio C-2473 to C/S WD, 5 Mar 44, G-3, GHQ, SWPA *Journal* (S).

17 JCS Radio to CINCPOA and CINCSWPA, 10 Mar 44, JCS and CCS Papers No. 2, G-3, GHQ, Exec Files.

18 HQ Sixth Army, Field Order No. 12, 23 Mar 44, G-3, GHQ, SWPA *Journal*, 8 Apr 44.

tons of bombs and expended 275,000 rounds of ammunition on the three airdromes. Our heavies started the attack by destroying or silencing the antiaircraft defenses; our medium and light bombers followed and in wave after wave swept the dispersal bays, revetments and landing strips. The runways were left unserviceable, and equipment and installations heavily damaged. Smoke from burning planes and fuel dumps totally obscured the target. In the air, we destroyed twenty-six intercepting enemy fighters for the loss of one plane. The enemy's strong air reserves built up in this area have been destroyed at negligible cost to us. Of the 288 planes present in the area on 30 March all have been demolished or irreparably damaged.[19]

During the same period navy carrier planes struck in force at the Palau Islands and at Yap, Ulithi, and Woleai to interdict any reinforcement efforts by the enemy from the western Carolines. Large numbers of Japanese planes were destroyed on the ground and in the air, and shipping concentrations were heavily hit. Any threat of opposition from the northwest was virtually removed. The Thirteenth Air Force also sortied from the Admiralties to continue pounding Rabaul and Kavieng, and the Royal Australian Air Force, operating from northwestern Australia, attacked enemy airdromes in the regions of Arafura Sea, Geelvink Bay, and the Netherlands East Indies.

Intelligence reports during March indicated that the Japanese were hurriedly strengthening their bases at Hansa Bay and Wewak.[20] Information on their defense dispositions indicated that an Allied attack was expected at either or both of these places. General MacArthur's G-2 thought that the opportunity to encourage this belief was too attractive to be overlooked, and on 7 March submitted a comprehensive deception plan to strengthen the enemy's conviction and at the same time lure him into diverting his forces from the Hollandia—Aitape area.

General MacArthur was in accord with this suggestion, and ordered his forces to carry out definite deceptive measures along the New Guinea coast.[21] The air force was directed to

---

19 GHQ, SWPA, Communique No. 726, 5 Apr 44. The Japanese described the effect of the Allied raids in their official report: "From the end of March, the enemy air attacks gradually intensified. Our air power was rapidly diminished as a result of the enemy attacks on 30–31 March and 3 April. The rise and fall of the fighting power of our fighter units exerted a decisive influence upon the future operations in the New Guinea Area. It also played a decisive role in the supply of the areas east of Hollandia and the operation of the bases in Western New Guinea." Japanese First Demobilization Bureau Report, *Southeast Area Operations Record, Part III: Eighteenth Army Operations*, Vol III, p. 36.

20 G-2, GHQ, SWPA, *Daily Summary* No. 735, 26 Mar 44 and No. 740, 31 Mar 44.

21 The effectiveness of these deceptive measures was clearly proved by later evidence. As late as 21 April, one day before the Allied landings at Hollandia and Aitape, the Japanese estimated enemy intentions as follows: "The signs of an enemy plan to make a new landing in the New Guinea area . . . are clear. The probability of a landing between Madang and Hansa or on the Karkar Islands is estimated to be greatest.

"According to the general situation a landing in the Wewak sector is next in probability. In the light of the recent bombings of Hansa, of reconnaissance and naval bombardment of Wewak, and the dropping of pamphlets by the enemy stating that he would land on Wewak on 24 April, precautions must be taken in the Wewak sector.

"It is also possible that the enemy will land in the Hollandia sector. . . . However, since there was no reconnaissance carried out in this region by submarines, destroyers or other means, and since air attacks were of a purely destructive nature, no signs of the usual pre-landing operations are discernible. Furthermore, the enemy has no air base at present from which to neutralize our airdromes west of Sarmi. Therefore, the probability of a landing in this sector is thought to be minor. . . ." Japanese First Demobilization Bureau Report, *Southeast Area Operations Record, Part III: Eighteenth Army Operations*, Vol III, p. 39.

continue the air attacks on Madang and Wewak and to drop dummy parachutes in the Hansa Bay area. Frequent reconnaissance flights were sent on ostensible mapping and photographic missions. The navy was directed to make suitable demonstrations along the same lines. PT boats were ordered to conduct operations against certain points along the coast in the Hansa Bay region, and empty rubber landing boats, indicative of disembarked scouting parties, were to be left at various spots along the shores east of Wewak.

Hollandia—Aitape Invasion

Hollandia was the principal Japanese rear supply base in New Guinea. A sheltered but undeveloped harbor located on Humboldt Bay, it provided the only protected anchorage of any size between Wewak and Geelvink Bay. The airfield area, shielded from the sea by the high Cyclops mountain range, was near Lake Sentani, about twelve miles from Humboldt Bay and midway to Tanahmerah Bay to the west. The bulk of the enemy's remaining New Guinea air strength was based on three large airdromes in this area. Hollandia also served the Japanese as an important transshipment point for the unloading and transfer of personnel and cargo from large transports to smaller coastal vessels. A considerable backlog of supplies was observed on the beaches and in the vicinity of the airfields. Intelligence estimated the number of Japanese troops in the Hollandia—Aitape area at the end of March to be 15,000; at Wewak—Hansa Bay, 30,000; at Wakde—Sarmi, 5,500; and in the Manokwari—Geelvink Bay area, 11,000.[22]

A large Allied task force was gathered in expectation of a difficult campaign. The closest teamwork of all participating components would be required to accomplish the largest operation and the longest amphibious move yet attempted in the Southwest Pacific Area. The projected operation involved a distance of 985 miles from Goodenough Island, the principal staging area, and over 480 miles from Cape Cretin, south of Finschhafen, the advance staging point.

General Krueger, Commanding General of Alamo Force, was made responsible for the co-ordination and planning of the ground, air, and naval forces. General Eichelberger, commanded the main landing group comprising the 24th and 41st Divisions, which was to seize the Hollandia airdromes by invasions at Humboldt Bay and Tanahmerah Bay. Brig. Gen. Jens Doe, commanded the 163rd Regimental Combat Team which was to land in the Aitape—Tadji area and capture the Tadji airfield. Another combat team would be kept in reserve. The naval forces were to provide air cover and escort protection. Admiral Barbey was in charge of the naval amphibious forces while the carrier forces of the Fifth Fleet were under Rear Adm. Marc A. Mitscher.

The entire task force for the invasion rendezvoused north of the Admiralties and then proceeded in a northwesterly direction toward Palau. Although this course was 200 miles longer than the direct route, it was intended to mislead the Japanese and prevent them from determining the exact objective in case of discovery by aerial reconnaissance. Swinging suddenly southward the huge convoy approached the New Guinea coast. On 21 April, fast carrier planes struck the Wakde—Sarmi, and Hollandia airfields while the Fifth and Thirteenth Air Forces concentrated on Wewak and Hansa Bay.

On 22 April, after heavy preliminary naval and air bombardment, the invasion troops went ashore according to plan. Complete tactical and strategic surprise was achieved. The con-

22 G-2, GHQ, SWPA, *Daily Summary* No. 730, 21/22 Mar 44, and No. 761, 21/22 Apr 44.

voys had sailed within striking distance of their objectives apparently without detection. The ease of the landings exceeded even the most sanguine hopes. No more than token resistance was met at any point and there was no interference from the enemy's air or naval forces. The painstaking deception measures had been remarkably effective.

In the Hollandia area, United States troops made a virtually unopposed advance. At both Humboldt Bay and Tanahmerah Bay opposition to the landings was ineffective and the two jaws of the giant invasion pincers clamped down rapidly on both sides of Mt. Cyclops. (Plate No. 41) So stunned was the enemy by the unexpected landings, that in the beach area complete radar sets and other valuable equipment, still uncrated, were left behind to be captured.[23] Progress was limited only by the problems of supply and the difficult terrain since the Japanese had failed to exploit the natural defensive positions offered by the narrow mountain defiles. On 26 April the Sentani, Cyclops and Hollandia airdromes were captured. In the meantime the Tami airstrip, lying on the coastal flat east of Humboldt Bay, was secured. With all objectives achieved, only mopping up operations and consolidation remained for the combat troops. On 6 June the Hollandia phase of the operation was officially closed.[24]

At Aitape the story was much the same. On D-Day, the Tadji airfield was seized and a perimeter defense was consolidated. Although there were many contacts with scattered Japanese groups, resistance on the whole was light.

General MacArthur emphasized the strategic importance of the Hollandia—Aitape operation in his official communique:

We have seized the Humboldt Bay area on the northern coast of Dutch New Guinea, approximately 500 miles west of Saidor. Our ground troops have landed at Aitape, Hollandia, and Tanahmerah Bay, covering a front of about 150 miles. The landings were made under cover of naval and air bombardment and followed neutralizing attacks by our air forces, and planes from carriers of the Pacific Fleet. Complete surprise and effective support, both surface and air, secured our initial landings with slight losses. We are pushing forward to secure the local airfields. Our feints over the past weeks towards Madang and Wewak apparently deceived the enemy into concentrating the mass of his forces forward into those areas thus leaving the vital sector of Hollandia vulnerable and making possible our surprise movement to his rear. The operation throws a loop of envelopment around the enemy's Eighteenth Army, dispersed along the coast of New Guinea in the Madang, Alexishafen,

23 The importance of the time element as a decisive factor in the Hollandia invasion is emphasized in the following statements by Japanese commanders who served in western New Guinea. "The Allied invasion of Hollandia and Aitape was a complete surprise to us," said Lt. Gen. Jo Iimura, Commander of the Second Area Army. "Although, after considering the past operational tactics of the enemy, we were confident that the Allies would eventually attack Hollandia, we rather believed they would attempt to acquire an important position somewhere east of Aitape, prior to an invasion of either Aitape or Hollandia. Because we misjudged the time of the Allied invasion on Hollandia and Aitape, we were neither able to reinforce nor send war supplies to their defending units." According to Lt. Col. Nobuo Kitamori, Staff Officer of the Second Area Army, the attack on Hollandia "was not a complete surprise in that we expected the enemy to come some time or other because it was such an important place. However, we did not think that the attack would come when it did. The morning that we found out that the Allies were going to come to Hollandia, they were already in the harbor with their transports and battleships. In that sense it certainly was a surprise." Colonel Kazuo Horiba, Staff Officer of the Southern Army, offered the following opinion: "It was a surprise attack as far as operations go, but not so strategically. We had planned on the fact the enemy was coming, but it was a surprise when the enemy came when he did, far before the time we expected and our defense preparations were not completed." Interrogation Files, G-2 Historical Section, GHQ, FEC.

24 HQ Sixth Army, *Report of the Hollandia—Aitape Operation*, G-3, GHQ, SWPA *Journal*, 25 Aug 44.

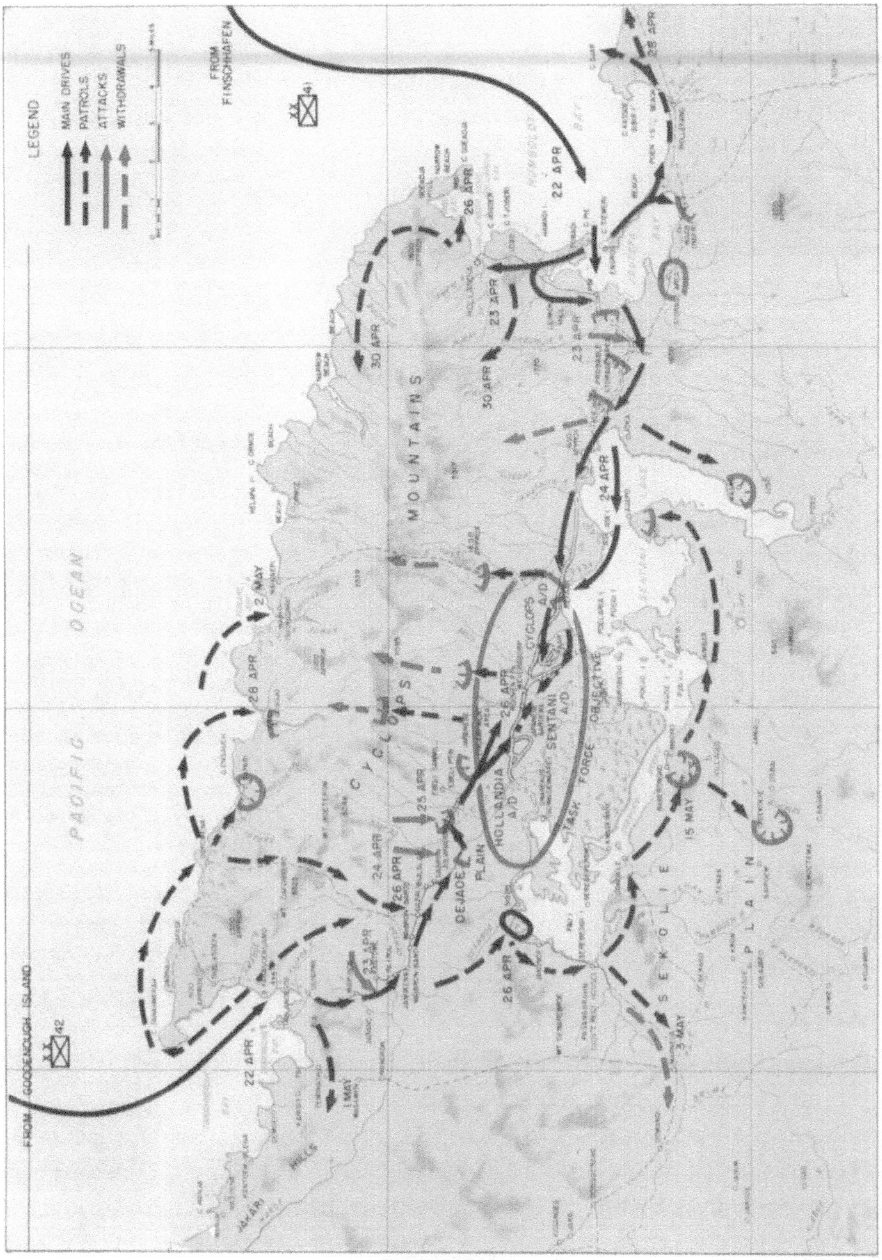

PLATE NO. 41

Hollandia Operation, 22 April—6 June 1944

Hansa Bay, Wewak sectors, similar to the Solomons and Bismarck loops of envelopment. To the east are the Australians and Americans; to the west the Americans; to the north the sea controlled by our Allied naval forces; to the south untraversed jungle mountain ranges; and over all our Allied air mastery. This enemy army is now completely isolated. With its communication and supply lines severed, its condition becomes similar to that of the beleaguered Seventeenth and Eighteenth Armies in the Bismarck and Solomon Archipelagos. Its present strength is estimated at 60,000. The total remaining forces of these two armies, which are now themselves surrounded are estimated at 140,000: 50,000 in New Britain, 10,000 in New Ireland, 20,000 in Bougainville, and 60,000 in New Guinea. Since the start of the campaign they have lost 110,000 men, 44 percent of their original strength of a quarter of a million and the remainder is now neutralized and strategically impotent. The enemy's maritime and air losses have been proportionately heavy. His invested garrisons can be expected to strike desperately to free themselves and time and combat will be required to accomplish the annihilation, but their ultimate fate is now certain. Their situation reverses Bataan. The present operation when completed frees British New Guinea from enemy control, and is the first recapture of Dutch territory in the war.[25]

Later evidence revealed that the garrison strength in the Hollandia area closely paralleled the G-2 estimate of 15,000 troops. This garrison, however, was comprised of a heterogeneous mixture of lightly armed and ill-trained service units and failed to respond with the last-ditch type of resistance heretofore so characteristic of the Japanese.[26] The enemy had intended to build Hollandia into a well defended strongpoint. The service units were forerunners of the heavy fortifications and combat troops that were slated to meet an anticipated Allied drive from Hansa and Wewak. If the Allied attack had not come as soon as it did the results might well have been different.

One of the most urgent tasks after the initial assault waves were ashore was the building of dispersal and storage areas and the forcing of exits from the congested beaches. Engineer units worked rapidly and efficiently to complete high priority construction projects. The supply track from Depapre on Tanahmerah Bay to the airdromes was improved to permit jeep traffic, and a road from Pim on Humboldt Bay was open for traffic as far as the Cyclops strip by the time it was captured. For major construction, the entire Hollandia area was divided into the Hollandia, Sentani, Tanahmerah, Tami, and Dock sectors. Permanent facilities included docks, a 900-foot channel through a coral reef at Depapre, the completion and improvement of the road net, the rapid repair and improvement of captured airdromes, water supply facilities, and fuel pipe lines. As roads improved and building materials became available, emphasis shifted to the construction of hospitals, camps, headquarters areas, and warehouses. All this was necessary to make Hollandia a supply base and staging area for future operations, and to establish it as the operating headquarters of the Southwest Pacific Area.

The Hollandia—Aitape operation had neutralized enemy forces to the east, increased the range and effectiveness of the Army Air Forces, and gained major harbor facilities. The initial success made possible the advancement of the schedule for subsequent operations to the west by releasing troops, ships, and planes for attacks from the newly developed air and naval facilities in the Hollandia—Aitape area. The

25 GHQ, SWPA, Communique No. 745, 24 Apr 44.

26 According to the Japanese official report on the war, there were approximately 14,600 personnel in the Hollandia Area: 6,600 Eighteenth Army; 7,000 Fourth Air Army; 1,000 Navy. About 90 percent of these troops were rear area service units including 1,000 hospital patients. Japanese First Demobilization Bureau Report, *Southeast Area Operations Record, Part III: Eighteenth Army Operations*, Vol III, pp. 47–49.

seizure of Hollandia forced the Japanese to withdraw their line of defense westward to Sarmi and Biak and reduced their communications in the New Guinea area practically to zero.[27]

Regrouping of Command Functions

The expanding nature of the tasks ahead, involving a substantial increase in the United States forces allocated to the Southwest Pacific Theater, caused General MacArthur to reorganize his command. He recommended that, excepting the units required for garrison operations, the forces of the South Pacific Command be regrouped and apportioned to the areas in forward contact with the enemy.

Accordingly, on 15 June, the Thirteenth Air Force, the XIV Corps, the 25th, 37th, 40th, 43rd, 93rd, and the Americal Divisions in addition to the 1st and 2nd Philippine Regiments, were assigned to the Southwest Pacific Area. On the same day, the Far Eastern Air Force, under General Kenney, was established to direct the operations of the Fifth and Thirteenth Air Forces. The XIV Corps, under Maj. Gen. Oscar W. Griswold, was given the responsibility of defending air and naval installations along the Solomons—New Britain—Emirau axis. New Guinea Force was given the same mission for all of New Guinea east of the Ramu River. The advance echelon of a new Eighth Army Headquarters was brought in to take over part of the administration and tactical duties of the Sixth Army; General Eichelberger assumed command on 7 September when the Eighth Army was formally established.[28]

The Seventh Fleet, after 5 May, comprised 3 light cruisers, 27 destroyers, 30 submarines, and supporting vessels. The deficiencies were filled by Admiral Nimitz from the Third Fleet which was transferred to the Central Pacific along with the 15th Marine Amphibious Corps and most of the naval air forces of the South Pacific Area. Since Admiral Halsey remained with the Third Fleet, he relinquished his command of the South Pacific Force on 14 June.[29]

As the Allied advance speared westward into Dutch territory, General MacArthur directed that civil affairs in that region be administered by Netherlands East Indies officials accompanying the troops. This policy of immediate transfer of civil control without a preliminary military government was instituted by General MacArthur at the start of the war. Rather

27 "The Allied seizure of Hollandia indicated the battle of New Guinea was rapidly drawing to a close," said Colonel Kitamori. "We knew that it would only be a matter of time before the Allies would control the air and the waters of New Guinea, thus paving the way for their expected counterlanding on the shores of the Philippines. Its seizure only indicated that defensive preparations in the Philippines would have to be accelerated, and to facilitate this our troops were to stall the enemy in northwestern New Guinea, as well as in the Halmahera group, as long as possible.

"From the relationship of supplies, the seizure of Hollandia meant that heavy sea transportation could not be effected in the surrounding waters of New Guinea, with the possible exception of Wasile, on the mouth of Kau Bay in Halmahera, and to Amboina, on Amboina Island. However, a little while later, the waters between the Philippines and the Halmaheras were infested with Allied aircraft and warships, which necessitated heavy sea transportation to be routed directly to Amboina by way of Java. From here, small craft carrying considerably less load than a 3,000 or 5,000-ton ship would be able to carry, made their way to the Halmaheras and into the McCluer Gulf, from where supplies were to be sent to Manokwari and the Sarmi areas, although most of these attempts ended in failure because of enemy PT boat activities, the rugged terrain of the countryside, and our own lack of shipping." Interrogation Files, G-2 Historical Section, GHQ, FEC.

28 Joint Staff Planners Report No. 713/5 to JCS, 17 Mar 44, G-3, GHQ, Exec JCS and CCS Papers (S).

29 Ibid.

than establish a military governor, he retained and supported the constitutional authority of the existing government. This policy was continued during the occupation of the islands north of Australia, when Australian civil officials assumed authority promptly in the wake of the operations. The success of this method was reflected in the complete lack of friction between the various governments concerned.

Capture of Wakde Islands

The Hollandia invasion initiated a marked change in the tempo of the Allied advance westward. Subsequent assaults against Wakde, Biak, Noemfoor, and Sansapor were mounted in quick succession and, in contrast to previous campaigns, no attempt was made to complete all phases of one operation before moving on to the next objective. General MacArthur was determined to reach the Philippines before December and consequently concentrated on the immediate utilization of each seized position to spark the succeeding advance.

Even while the first waves of troops were pouring on to the Hollandia beaches, he radioed his Chief of Staff to complete preparations for the Wakde Islands operation with the least possible delay.[30] On 29 April, he informed the War Department that he would attack positions in the Wakde region about 15 May with the primary objective of obtaining " more airdrome space for displacement of the air forces forward." He also wished to secure light naval bases to prevent hostile interference with the development of the Hollandia area and to support subsequent assaults upon the Vogelkop Peninsula.[31]

The Wakde—Sarmi region, beginning approximately 140 miles west of Hollandia, had been developed by the Japanese into a ground and air position of considerable strategic importance. There were good airfields at Maffin and Sawar on the mainland and on Wakde Island itself just off the New Guinea shore opposite Toem. The Japanese had established numerous bivouac and storage areas along the entire coastal road from Maffin Bay to Sarmi. Intelligence sources indicated that the enemy was concentrated in strength in the Wakde—Sarmi region.[32] It was therefore decided to employ a division less one regimental combat team at Sarmi and use the regimental team for the seizure of Wakde Island.[33]

Accordingly, the 163rd Regimental Combat Team of the 41st Division was directed to secure a beachhead in the Toem—Arara area, occupy Wakde Island, and protect the development of the required base construction. Wakde was too small an island to permit the direct landing of all required combat and service troops without serious congestion. Since Toem was in close range of any enemy coastal guns possibly on Wakde, it was planned to land first at Arara and from there maneuver into position for the Wakde invasion.

On 17 May, after a heavy air and naval bombardment, the 163rd Regimental Combat Team landed at Arara and established a beachhead flanked on the west by the Tor River and on the east by the Tementoe River. (Plate No. 42) At the same time a reinforced company landed without opposition on tiny Insoemanai Island, south of Wakde, to lend mortar and machine gun support. The following day, the task force went into Wakde in a shore-to-shore

---

30 CINCSWPA Radio to C/S GHQ, SWPA, 22 Apr 44, 385 Plan 10/11, G-3, GHQ, Admin (TS).

31 CINCSWPA Radio No. CA-11199 to C/S WD, 29 Apr 44, 385 Plan 10/13, G-3, GHQ, Admin (TS).

32 G-2, GHQ, SWPA, *Daily Summary* No. 786, 16/17 May 44.

33 HQ Sixth Army, *Report on the Wakde—Biak Operation, 17 May—2 Sep 44*, 25 Feb 45, G-3, GHQ, SWPA *Journal*.

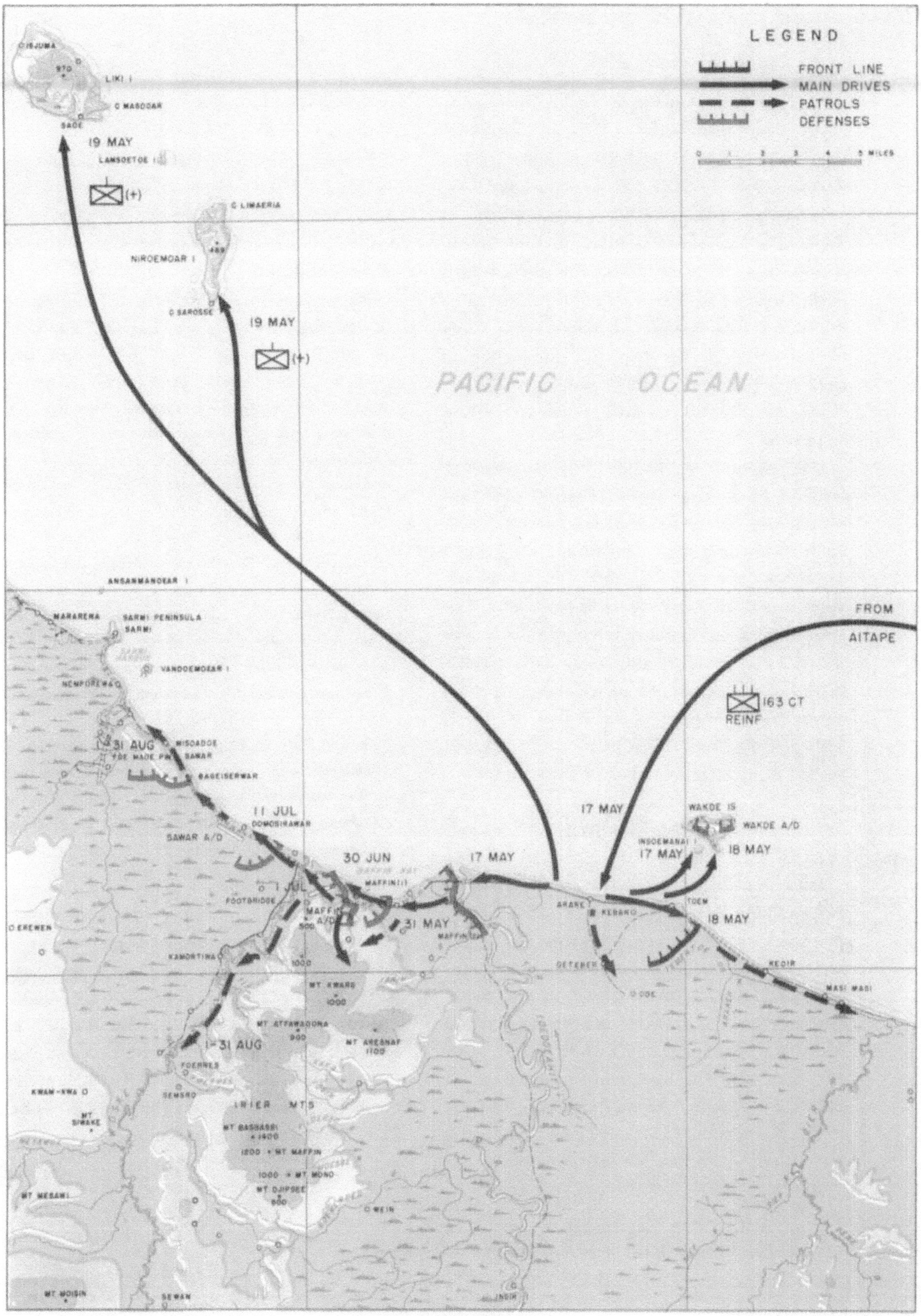

PLATE NO. 42

Wakde—Sarmi Operation, 17 May—2 September 1944

movement from the Toem area. Stubborn resistance from a strongly entrenched enemy was encountered but by 20 May all opposition had been overcome and the airdrome cleared for use. On 19 May, supplementary landings were made on Liki and Niroemoar Islands to the west where radar installations were rapidly established. Two days later, the drive toward Sarmi was initiated at Toem by a landing of the 158th Regimental Combat Team. This force crossed the Tor River and, augmented by later reinforcements during June, moved forward to Maffin airfield against strong opposition.[34]

The capture of Wakde Islands provided General MacArthur with a new advance base within easy fighter range of his forward objectives, deepening his penetration of Japan's perimeter defenses. All remaining enemy airdromes and harbors in New Guinea were now subject to Allied medium bomber raids. The enemy's rear areas, already dislocated and disrupted by the seizure of Hollandia, were thrown into further confusion as the Allied air, naval, and ground forces continued to pound the pocketed Japanese in northern New Guinea.

Struggle for Biak

Biak Island lay next in the line of General MacArthur's sweeping advance. This island, 200 miles west of Wakde, possessed one of the two remaining major groups of enemy airdromes between Hollandia and Halmahera and therefore constituted a valuable military asset to the Allied scheme of operations. On 27 May, just one week after the last enemy guns were silenced on Wakde, the 41st Division, less the 163rd Regimental Combat Team, made the first assault at Bosnek, on southern Biak. Its immediate objectives were the three airdromes at Mokmer, Borokoe, and Sorido near the southwest coast. (Plate No. 43)

Initial opposition was relatively light and the landing force began an advance inland toward the airfields. General MacArthur described the landing operation and its strategic value as follows:

> *We have landed on Biak Island. . . . The capture of this stronghold will give us command domination of Dutch New Guinea except for isolated enemy positions. For strategic purposes this marks the practical end of the New Guinea campaign. The final stage has also been reached in the offensive initiated in this theater on 29 June 1943, by the combined forces of the Southwest Pacific and South Pacific Areas. It has resulted in the reconquest or neutralization of the Solomons, the Bismarcks, the Admiralties, and New Guinea. From the forward point reached by the Japanese we have advanced our front approximately 1,800 statute miles westward and approximately 700 miles to the north. . . . Compared with the enemy our offensive employed only modest forces and through the maximum use of maneuver and surprise has incurred only light losses. The operations have effected a strategic penetration of the conquered empire Japan was attempting to consolidate in the Southwest Pacific and have secured bases of departure for the advance to its vital areas in the Philippines and the Netherlands East Indies.*[35]

The slight opposition to the beach landings, however, was little indication of the bitter struggle to follow, for "the Japanese defense of Biak," as General Eichelberger's report described it, "was based on brilliant appreciation and use of the terrain."[36] In this instance, the enemy had purposely withheld his main forces until the United States troops had advanced to the rugged terrain beyond the beaches. Then, from the dominating cliffs and caves overlooking the moving Allied columns, the Japanese

34 Ibid.

35 GHQ, SWPA, Communique No. 780, 28 May 44.

36 HQ I Corps, *History of the Biak Operation 15–27 June 1944*, p. 2.

launched a savage counterattack and, aided by 5-ton tanks, succeeded in driving a block between the beachhead and invading forces. Enemy frontal pressure increased considerably and a temporary retirement and regrouping was necessary.

The situation remained critical until the remainder of the 41st Division could be brought in from Wakde to bolster the United States positions.[37] Thus strengthened, the 41st Division renewed its drive toward Mokmer at dawn on 2 June and, after several days of severe and violent fighting, seized the important airdrome at Mokmer on 7 June. Even then, the struggle continued unabated as the Japanese poured a heavy fire into the newly established airfield positions. It was another week before the strip could be brought into use. Additional United States reinforcements were sent in and, despite the enemy's fierce and valiant attempts to hold the remaining airfields at Sorido and Borokoe, the two strips were finally wrested from the Japanese. Major fighting on Biak was over by 22 July, although the Japanese continued sporadic resistance throughout the rest of the year. Mopping-up operations continued with amphibious shore-to-shore landings in August at Korim Bay, Wardo, and Warsa Bay. Other landings on Soepiori Island in September brought the campaign to a close. The figure of over 7,200 enemy killed by 20 January 1945 indicates the tenacity with which the Japanese fought to keep Biak from falling to the Allies.[38]

Noemfoor Island

With hardly a pause, General MacArthur's forces swept forward to invade nearby Noemfoor Island, west of Biak. Since strategic considerations dictated a landing on the northwest end of the island where the enemy was judged to be strongest, the preceding air and naval bombardment was more intense than usual. Prior to and during the entire operation, naval forces maintained patrols, preventing any reinforcement of the Japanese garrisons. The Fifth Air Force, now merged with the Thirteenth Air Force under the Far Eastern Air Force, was strong enough to send 150 bombers on a single mission. For three weeks, ending with a record assault the day prior to the landings, the air bases at Noemfoor and other fields nearby were continuously and heavily pounded.

On 2 July, troops of the 158th Regimental Combat Team landed near Kamiri airdrome. (Plate No. 44) Wading ashore through the surf, the landing force moved swiftly to the east, northeast and southwest and formed a strong perimeter. By evening Kamiri airstrip was securely in Allied possession and on the next day paratroopers of the 503rd Parachute Infantry were flown in to help destroy enemy resistance and to capture the other airfields. Thus reinforced, the United States troops attacked to the northeast and seized Kornasoren airfield on 4 July. Two days later a shore-to-shore landing was made at Roemboi Bay to capture the airdrome at Namber and

37 According to Colonel Kitamori, a plan had been formed immediately after the Allied landing on Biak to regain that important area, utilizing Army-Navy co-operation for the first time in the western New Guinea area. The plan called for a strong naval task force, led by the battleship *Musashi* and composed of six cruisers and destroyers loaded to the brim with reinforcements, to make a landing behind the Allied positions. "I personally ordered this unit to be sent to Biak," said Col. Kazuo Horiba, former Senior Operations Officer, Second Area Army. "It was the 2nd Amphibious Brigade fresh from Japan. The Fleet got to the Sorong area but could not reach Biak because of the presence of enemy air and naval power, and it had to land at Sorong." Interrogation Files, G-2 Historical Section, GHQ, FEC.

38 G-3, GHQ, SWPA, *Monthly Summary of Operations*, Jul 44; 41 Div, Addenda to "41st Infantry Division Report on the Biak Campaign, 21 Aug 44—20 Jan 45," G-3, GHQ, SWPA *Journal*.

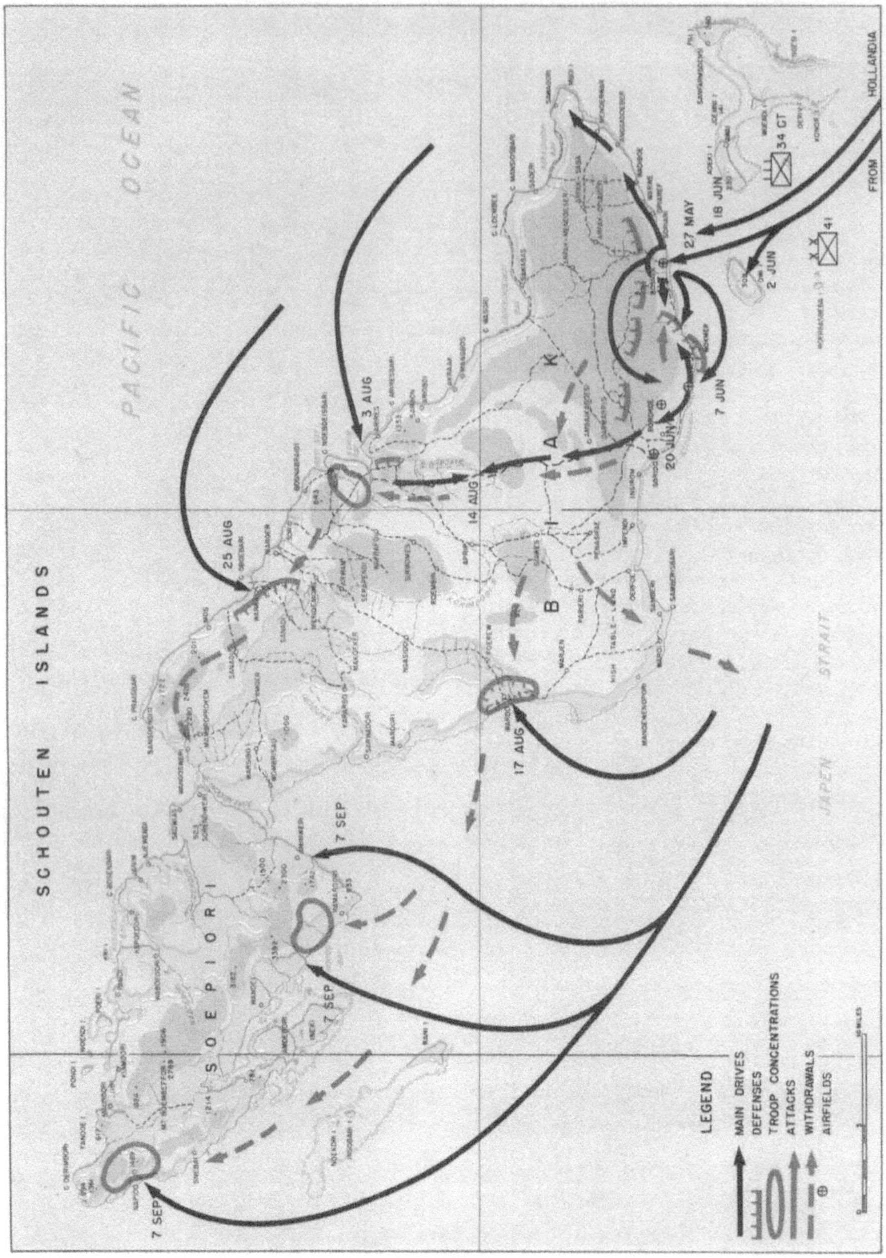

PLATE NO. 43

Biak Island Operation, 27 May—20 August 1944

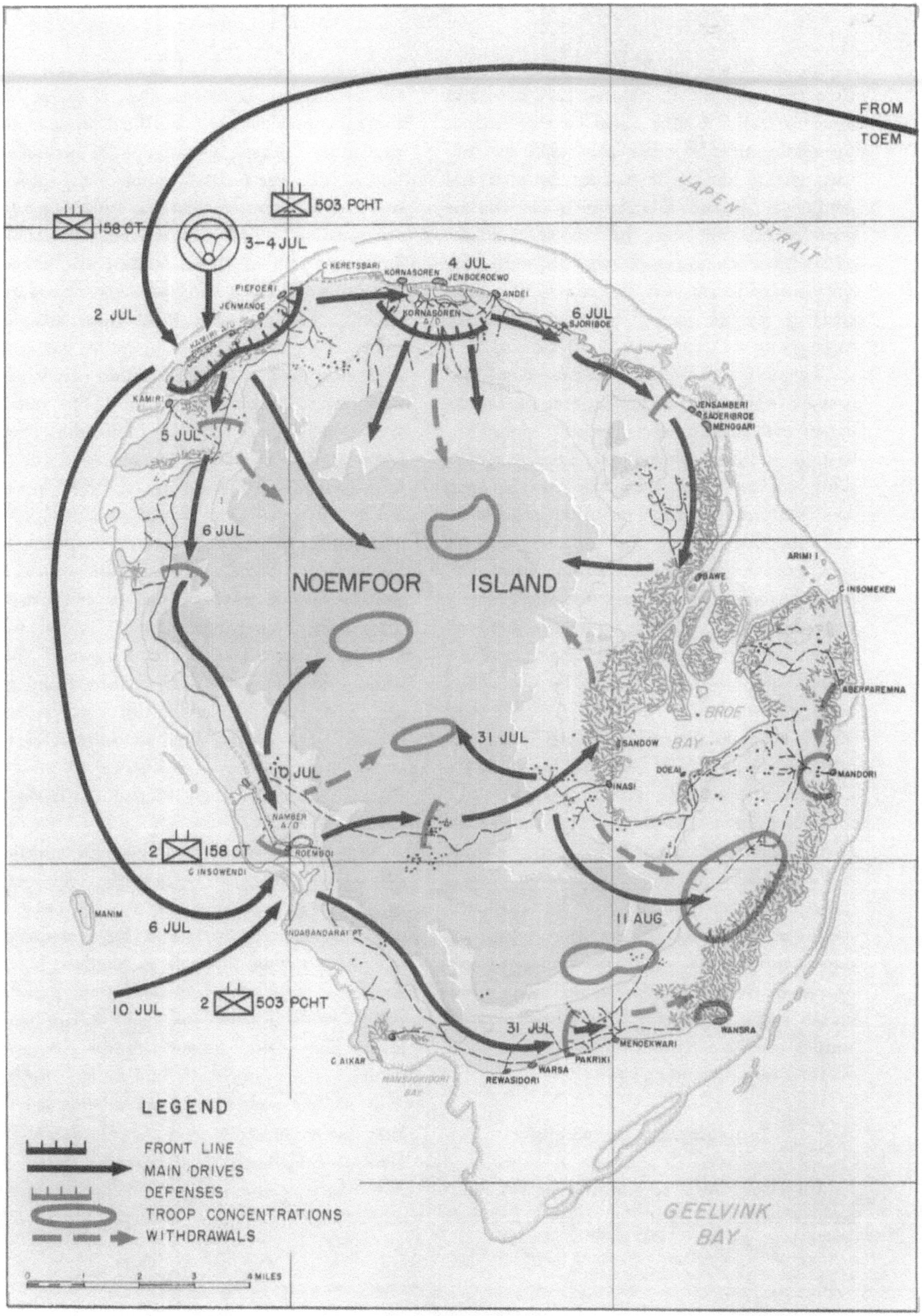

PLATE NO. 44

Noemfoor Island Operation, 2 July—31 August 1944

within one hour Allied planes were operating from the airstrip. It required only five days after the initial assault waves hit the beaches for all the airfields on the island to be put into use, giving almost immediate breadth and depth to General MacArthur's air deployment operations.[39] By 7 July the main phases of the Noemfoor operation were accomplished. Subsequent action was devoted to the final clearing up of enemy remnants scattered throughout the island and along the coast.

Beginning with the Hollandia invasion, Japanese air and naval resistance had been unco-ordinated and intermittent. Ground opposition, though stubborn, had been equally spotty. Only on Biak Island and in the Toem—Sarmi area on the mainland had major battles developed. At these points, the isolated Japanese garrisons put up a strong and resolute resistance, necessitating the employment of comparatively large numbers of Allied ground forces supported by planes, tanks, naval gunfire, and artillery. Sarmi itself was never occupied by the Allies although between 17 May and 6 October units of the 158th Regimental Combat Team, the 6th Division, the 31st Division, and the 33rd Division relieved each other in that area. As the Allied line moved forward after the seizure of the Maffin and Sawar airfields, the position at Sarmi lost its importance. Wakde and Toem, however, served admirably as an advanced staging area for the assaults on Noemfoor, Sansapor, and Morotai. To have mounted these operations from rear areas would have put an undue strain on General MacArthur's limited amphibious facilities and slowed down the pace of his advance considerably.

Counterattack at Aitape

While the Allied forces leaped forward from one success to another, the trapped Japanese Eighteenth Army below Hollandia was writing the last chapters in its epic story of defeat. As the situation developed, the final phase of the Hollandia—Aitape operation—the neutralization of the large isolated enemy units—proved to be more contested than the initial stages of the invasion. The Japanese near Hollandia and a portion of those sandwiched between Aitape and Wewak had fled westward towards Sarmi. The greater part of these escaping troops were destroyed en route by starvation and sickness; a small fraction eventually managed to trickle into Sarmi. The rest of the enemy troops east of Hollandia joined forces at Wewak where they remained cut off from all routes of escape to Dutch New Guinea. There, while the Allies poured in more supplies and continued to consolidate their positions, the pocketed Japanese waited helplessly and in vain for outside relief or some definite instructions from higher headquarters. After two months of isolated confinement, however, they could wait no longer. With their food running out, their nerves frazzled by frustrating inactivity, and with fresh Allied successes at Sarmi and Wakde making a withdrawal to western New Guinea increasingly difficult, the Japanese decided to act.

General Adachi, commanding the remnants of the once powerful Eighteenth Army, cast aside all logic and pretense at strategy and gave orders to make preparations for a desperate attempt to break through at Aitape. It was an almost hopeless enterprise, for any gains at Aitape would have left his troops still isolated, unless they could detour around the even stronger Allied force at Hollandia. Rather than submit passively to a situation which gave every promise of eventual starvation, General Adachi chose to chance a course of positive action which might somehow constitute an impediment to Allied strategy. In an im-

39 HQ Sixth Army, Report on Noemfoor Operation, G-3, GHQ, SWPA *Journal*, 31 Aug 44

passioned address, he exhorted his troops virtually to destroy themselves in a blind mass attack:

I cannot find any means nor method which will solve this situation strategically or technically. Therefore, I intend to overcome this by relying on our Japanese 'Bushido'. . . . I am determined to destroy the enemy in Aitape by attacking him ruthlessly with the concentration of our entire force in that area. . . . This will be our final opportunity to employ our entire strength to annihilate the enemy. . . . Make the supreme sacrifice, display the spirit of the Imperial Army.[40]

The preliminary movements leading to General Adachi's final attack against the Aitape positions were picked up by General MacArthur's G-2 Section. Cumulative intelligence from native sources, aerial reconnaissance, prisoner of war interrogations, intercepts, perusal of captured documents, and information from ground and PT boat patrols built up so unmistakable a picture of enemy intentions that Allied forces were able to take complete countermeasures well in advance of the actual attack.

During the period 15 June to 3 July, special intelligence studies were made available to the staffs and the troops concerned, predicting 5–10 July as the date of attack and describing enemy dispositions, including the strength and identity of the forces involved.[41] It was forecast that the Japanese contemplated employing the 20th and 41st Divisions assisted by garrison units. The probable plan of enemy formation would consist of two regiments abreast, a third regiment in the rear, and advance elements of the 41st Division available in the forward areas. The attack itself was to be divided into two main phases: Phase One, the seizure of the Driniumor River line; Phase Two, the general assault on Aitape.[42] In expectation of the coming attack, the Allied forces at Aitape were heavily reinforced.

The initial effort of the main enemy offensive took place soon after midnight on 11 July as the Japanese plunged forward in a mad suicidal rush against the Allied line south of the mouth of the Driniumor River. (Plate No. 45) Allied machine gunners mowed down wave after wave of oncoming Japanese troops, but by sheer weight of numbers, the enemy managed to effect a small breakthrough at the point of attack. Since the ensuing assaults followed substantially the course outlined in the intelligence reports, however, the Allies were fully prepared. Artillery, already placed in position and ranged beforehand, wiped out enemy assembly points while Allied planes kept up a continuous and accurate bombardment of supply points and routes of attack.

When the Japanese were frustrated in their frontal attempt near the mouth of the Driniumor River, they tried to pierce the Allied right flank to the south through the foothills of the Torricelli Mountains. For over a month, they battered violently against the Allied lines in the Afua region in fruitless efforts to overrun the Aitape defenses but every drive was repulsed as the toll of enemy dead mounted sharply. Finally, on 31 July, the Allied forces launched a double enveloping counteroffensive, advancing eastward to Niumen Creek and then pivoting south to encircle the enemy's rear at Afua. This counterdrive cut the Japanese into three main segments, leaving scattered groups west of Afua, east of the Driniumor River, and in the Harech River area. The encircling Allied forces then hunted down and

40 The full text of General Adachi's speech may be found in the Japanese First Demobilization Bureau Report, *Southeast Area Operations Record, Part III: Eighteenth Army Operations*, Vol III, pp. 100–101.

41 G-2, GHQ, SWPA, *Daily Summary* No. 843, 12/13 Jul 44.

42 Ibid.

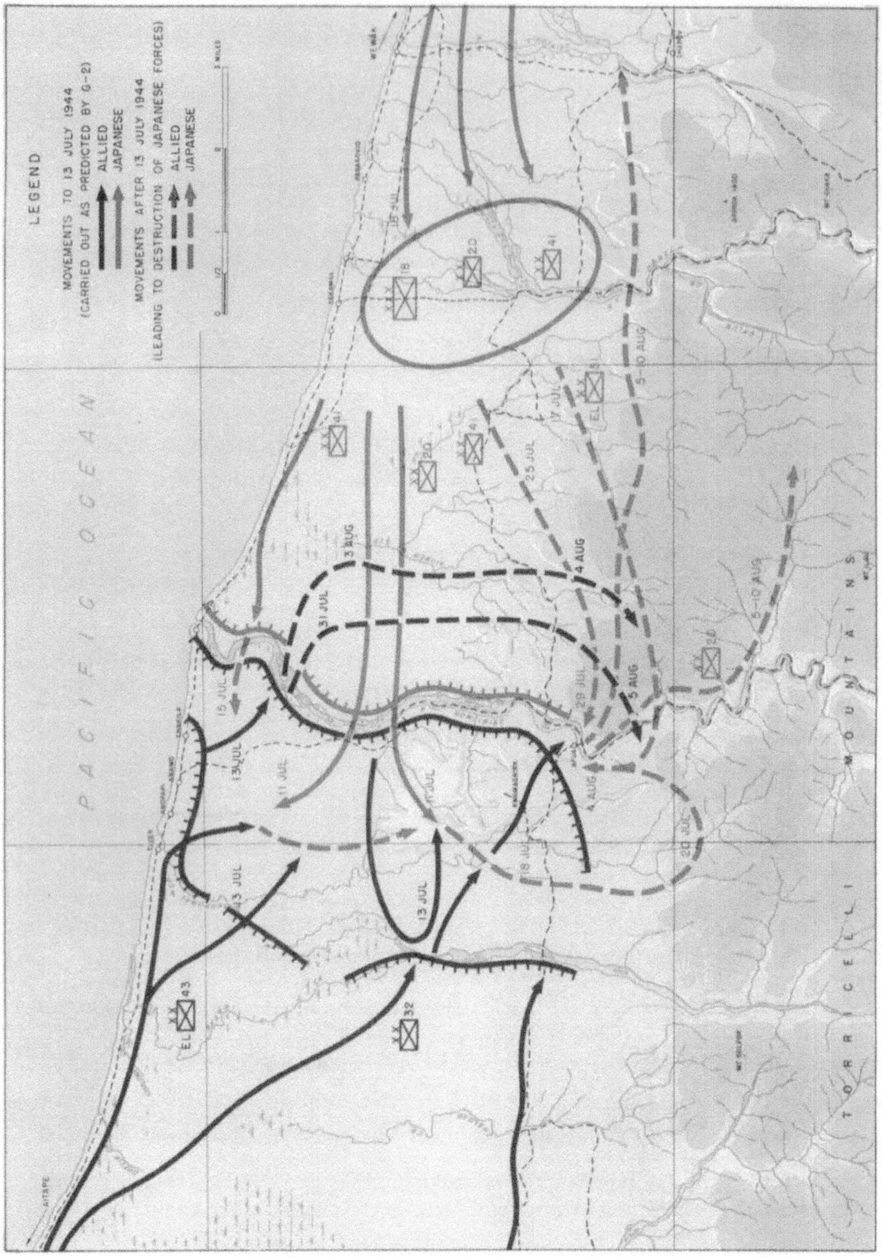

PLATE NO. 45

Japanese Counterattack East of Aitape

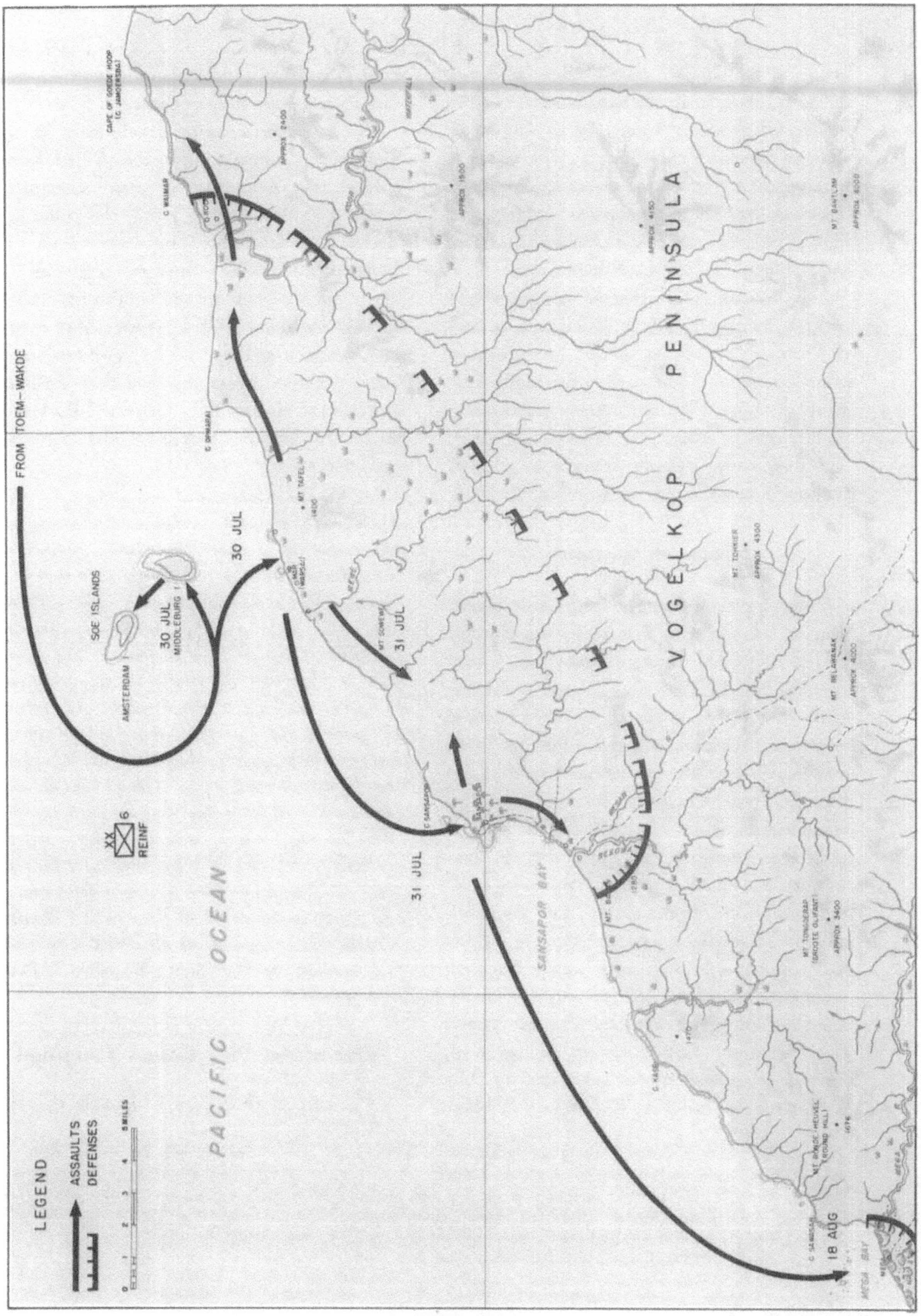

PLATE NO. 46

Sansapor—Cape Opmarai Operation, 30 July—31 August 1944

destroyed the divided enemy units in detail, virtually annihilating the remainder of General Adachi's Eighteenth Army. By 10 August all effective resistance had ceased. The scattered remnants of an utterly defeated enemy force dispersed into the mountains to congregate at Wewak and But where they remained isolated and helpless until the close of the war.

The reckless and abortive campaign had cost the Japanese heavily in killed and wounded. As General Adachi, himself, later gloomily acknowledged: "The story of the Eighteenth Army is tragic. We lost 10,000 killed when we decided to attack the Allies at Aitape."[43] This final costly episode brought to an end the last Japanese hopes in New Guinea.

Seizure of Sansapor

Following the capture of Noemfoor Island, General MacArthur's next strike was directed at the Vogelkop Peninsula, the last enemy stronghold in the New Guinea area. Adhering again to the principle of avoiding massed enemy concentrations where feasible, his forces attacked the western end of the Peninsula, by-passing several thousand Japanese defending Manokwari. (Plate No. 46) On 30 July, a task force comprising elements of the 6th Division made simultaneous and unopposed landings near Cape Opmarai on the mainland, and on Middelburg and Amsterdam Islands to the northwest. Only a few enemy stragglers were encountered. The next day a shore-to-shore landing from Cape Opmarai was carried out at Sansapor. As in other New Guinea areas, airdrome construction proceeded rapidly. In a short time airfields at Mar and on Middelburg Island and a float plane base at Amsterdam Island were fully operative.

The operation to seize airfield sites on the Vogelkop Peninsula advanced Allied forces another 200 miles to the west. Groups of Japanese troops moving along the coast and inland were intercepted by units of the 6th Division in the Mar—Sansapor region and at Mega, twenty miles to the southwest. Only Sorong remained as the westernmost enemy base in the Vogelkop and its significance as a threat to Allied operations had been nullified by constant air assault. General MacArthur issued an official account of the Sansapor operation stating:

> *We have seized the coastal area of Sansapor, and the nearby islands of Amsterdam and Middelburg, at the western tip of Dutch New Guinea, nearly 200 miles beyond Noemfoor and slightly more than 600 miles southeast of the Philippines. . . . The operation was an amphibious one, the ground troops acting in coordination with naval and air forces. The enemy base at Manokwari, the pivot of the defenses of the Vogelkop Peninsula is now by-passed and useless. Its garrison, 15,000 strong. . . is now isolated with its only possible escape route to the south over hazardous terrain of swamp and jungle. Our air bases are now established from Milne Bay along the entire coast of New Guinea. The enemy is no longer able to operate in this area, either by air or sea, beyond the Halmahera—Philippine line which is the main defense cover for his conquered empire in the Southwest Pacific. Should this line go, all of his conquests south of China will be imperiled and in grave danger of flank envelopment.*[44]

End of the New Guinea Campaign

The capture of Sansapor marked the suc-

43 Testimony of General Adachi during his trial in Rabaul, AP, World Service, Item 33, Sydney, April 14, 47. With reference to the casualties at Aitape, General Adachi's figures are in close agreement with G-2 reports. *Daily Summary* No. 887 for 26/27 August 1944 gives a total of 8,502 killed and taken prisoner to that date. This compares favorably with General Adachi's estimate of 10,000 killed, since the conditions of jungle fighting and malnutrition probably accounted for deaths not included in the G-2 report of identified killed.

44 GHQ, SWPA, Communique No. 845, 1 Aug 44.

cessful termination of the long and hard-fought New Guinea Campaign. Control of the entire stretch of coastline from Milne Bay to the Vogelkop Peninsula was now firmly in Allied hands. In less than thirteen months, General MacArthur's forces, boring through layer after layer of Japan's outer defense perimeter, had moved 1,300 miles closer to the core of her island empire. The thousands of Japanese troops they had pocketed and cut off from outside aid had lost all ability to interfere seriously with Allied operational plans. When queried as to the immediate disposition of these isolated enemy segments, General MacArthur recommended that they be ignored until the main task was accomplished:

> *The enemy garrisons which have been bypassed in the Solomons and New Guinea represent no menace to current or future operations. Their capacity for organized offensive effort has passed. The various processes of attrition will eventually account for their final disposition. The actual time of their destruction is of little or no importance and their influence as a contributing factor to the war is already negligible. The actual process of their immediate destruction by assault methods would unquestionably involve heavy loss of life without adequate compensating strategic advantages. The present allotment of shipping and assault craft would not permit such operations except at the expense of those which are now scheduled. . . .*"[45]

The Allies quickly developed their newly captured territory for immediate use. A major heavy bomber base was constructed on Biak, and fighters and medium bombers were stationed on Wakde, Noemfoor, and at Sansapor. From these bases the Far Eastern Air Force was to strike during the next few months at Japanese positions in Ceram, Celebes, Halmahera, Dutch Borneo, Java and the Palaus in preparation for the assault against the Philippines. In the meantime, the Royal Australian Air Force Command, was to interdict the Arafura, Banda and Ceram Seas and adjacent land areas thus protecting General MacArthur's southern flank.

The New Guinea Campaign, waged through broiling sun and drenching rain amid tangled jungle and impassable mountain trails, had been a difficult and gruelling struggle. In combating the extraordinary problems of climate and terrain, however, many valuable lessons were learned which were to prove of great benefit in pursuing the tasks ahead. This ability to cope with crises and profit from experience was a distinguishing characteristic of the Allied conduct of the war in the Southwest Pacific.

One of the important victories won by General MacArthur's forces was their triumph over the anopheles mosquito. It was a battle involving science and discipline, waged by the troops, both officers and men, under the guidance of the Medical Corps. During the first stages of the New Guinea Campaign, malaria had been as bitter and deadly a foe as the enemy. On the Papuan front, it had been responsible for more non-effectives than any other single factor. By the time General MacArthur was ready to go into the Philippines, however, it was reduced to secondary importance as a cause of disablement and no longer deserved serious consideration in planning tactical operations.

This remarkable achievement, comparable with Goethals' and Gorgas' triumph over yellow fever, was accomplished by the cooperation of everyone who served in New

---

45 CINCSWPA Radio No. C-15910, to C/S WD, 9 Aug 44, C/S, GHQ, WD 803 (S). This is characteristic of General MacArthur's unvaried concern with the health and safety of his troops. His abhorrence of bloody frontal assaults, of reckless plunging were predicated on his deep wish to reduce or avoid losses. The record of his campaigns is exceptional in this respect.

Guinea. The Medical Department surveyed, researched, lectured, demonstrated and recommended, and General Headquarters issued the necessary directives to insure the success of the struggle against malaria. General MacArthur appointed a special committee of representatives, authorities from both the Australian and American medical services, to formulate the general principles under which the campaign would be carried out. Medical officers and Malaria Control Units, specially trained to cope with the menace to the Allied forces in infested areas, waged a bitter war against the mosquitos, frequently within sound of the enemy's guns.[46] Malariologists provided expert advice to unit commanders. Troops were educated to the dangers and prevention of malaria with posters and pamphlets and every man was urged to wage his own personal war against the mosquitos. The result was complete success.

Japanese efforts along these same lines were not nearly so effective. Even as late as the Hollandia operation their malaria casualties continued to assume enormous proportions, despite the fact that they had captured huge quantities of quinine in the Netherlands East Indies.[47] As General MacArthur stated: "Nature is neutral in war but, if you beat it and the enemy does not, it becomes a powerful ally."

Shortages of equipment and the lack of adequate shipping presented another problem which was solved by the Southwest Pacific Command in its seizure of New Guinea from the Japanese. In addition to the fact that many important items of equipment for both combat and service troops were in short supply right up to the end of operations, the difficulty of handling the necessary tonnage of materiel was increased by the limited port facilities in New Guinea. Construction of bases had to be initiated almost from scratch in the native villages and missionary settlements that constituted the "cities" of New Guinea. Whatever facilities were needed had to be built or brought in as the forces advanced. The jungles and swamps afforded poor foundations upon which to erect the supply and operational bases necessary to carry on a war of rapid maneuver over long distances. Major construction projects were carried out in a primitive area devoid of railroads, highways, shipping facilities, and air transport.[48]

In many cases, shipment of unit equipment from the United States was lacking in organization and completeness. The initial equipment of a single organization was often transported on many different ships, destined for discharge at various ports. This situation caused much rehandling and reshipping, contributing to the congestion which invariably occurred in all new ports during their more critical periods of development. Equipment shortages of those units about to be employed in forthcoming operations were generally overcome by "crisis management." Equipment in the hands of units not scheduled for immediate combat was transferred to those troops marked for earlier action and loaded on

46 Col. Howard Smith, Public Health Service, on the Staff of USAFFE deserves great credit in this preventive campaign. An expert in tropical medicine and Chief of the Philippine Quarantine Service for many years prior to to the war, Colonel Smith was well known to General MacArthur. The latter's far-sighted planning brought Colonel Smith to his staff at an early date.

47 "The worst enemy of the Japanese Army in the withdrawal from Hollandia was malaria, and it was the major factor in causing much suffering and death during the march. Conditions were such that after a rain storm the men were forced to sleep in wet clothing without the benefit of mosquito netting. Because of the lack of medicine to counteract malaria, large numbers of stragglers fell by the wayside." Japanese First Demobilization Bureau Report, *Southeast Area Operations Record, Part III: Eighteenth Army Operations*, Vol III, p. 55.

48 Staff Report by Brig. Gen. H. E. Eastwood, AC of S. G-4.

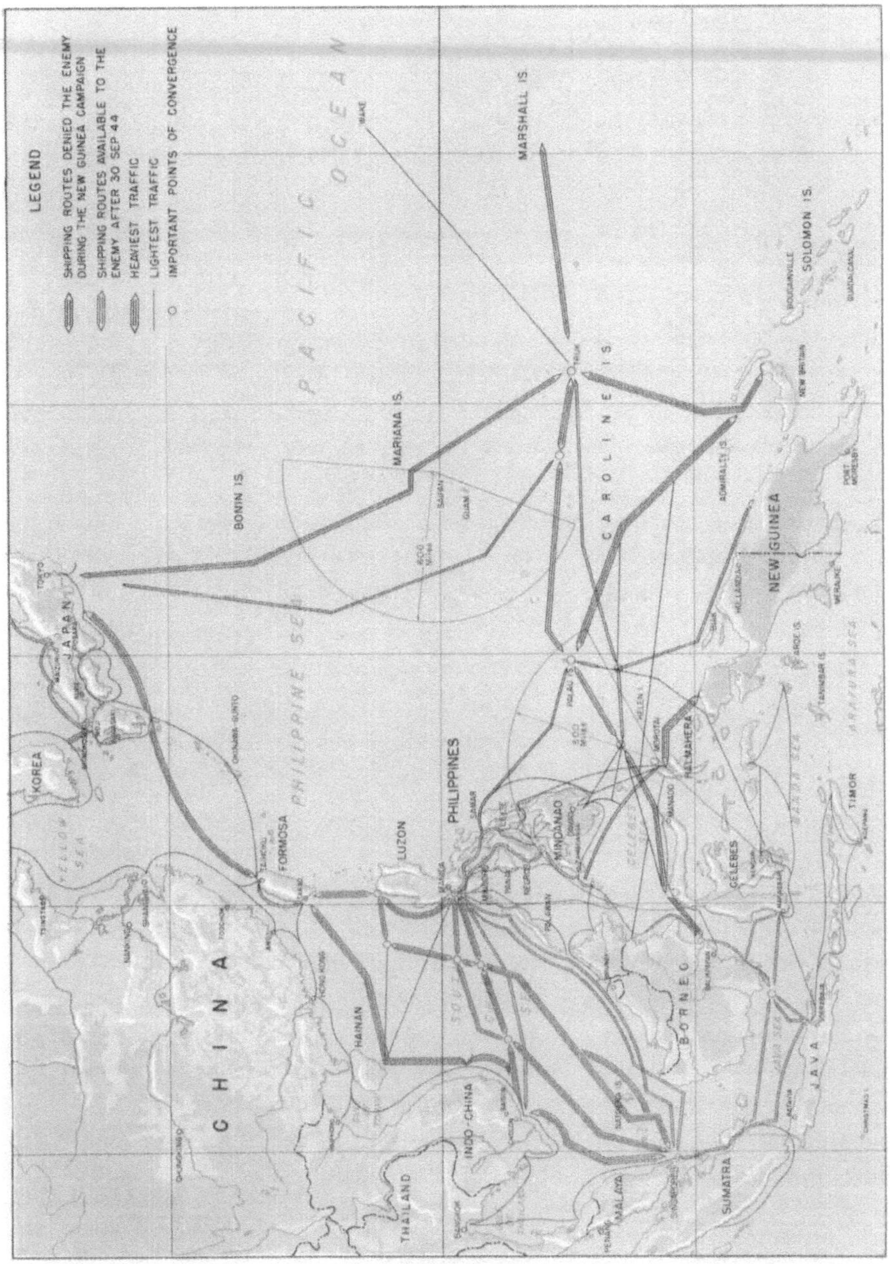

PLATE NO. 47

Enemy Shipping Routes Destroyed during the New Guinea Campaign special ships where water movement was necessary. In some instances, the troops had but little time to familiarize themselves with the weapons they were to take into battle.[49]

The lack of adequate and suitable water transportation with which to concentrate troops in staging areas and take them into combat at times considerably hampered the planning of operations. Amphibious landing craft were always critically short and liberty type ships had to be used as troop transports, both for assembling troops and for moving supporting elements into objective areas. Such cargo vessels were altered to add necessary facilities by the construction of crude accomodations on deck. Movements of troops in the tropical areas of the Southwest Pacific were accomplished in a manner which would have been utterly impossible in less temperate climates.[50]

The Battle against Enemy Coastal Shipping

The battle against enemy shipping was another significant feature of the New Guinea Campaign. The wholesale destruction by our planes, submarines, and PT boats of enemy coastal vessels, transports, barges, schooners, and sailing craft in the Southwest Pacific Area gradually paralyzed enemy efforts to supply, reinforce, or evacuate the remnants of his armies cut off in New Guinea, New Britain, New Ireland, and the Solomons. (Plate No. 47) More than 5,000 of these craft were destroyed. After the conclusion of the fighting in the Buna—Lae area and the Solomons, the Japanese were reluctant to risk major naval units, and as a result of their heavy losses in cargo ships and transports they were forced to devise a new supply technique. Without additional food, ammunition, and reinforcements, the Japanese forces would be unable to prevent the Allies from seizing all of New Guinea and the adjacent areas. Submarines were too small and unmaneuverable to be of more than minor aid and, moreover, were too few in number. Accordingly, the enemy's most ambitious efforts along these lines were directed to a greatly expanded use of barge traffic. These barges were manufactured and collected in Japan, China, the Philippines, and various island ports and then routed to the New Guinea area. They had a troop capacity of from 35 to 60 men and a cargo carrying capacity of up to 20 tons, and were generally of excellent workmanship.

When the enemy's use of small craft for transportation and supply purposes began to assume serious dimensions, the Allies found it imperative to develop effective counter tactics. The answer lay in a coordinated, intensive employment of PT boats, Catalinas, and low-flying planes. This combination destroyed the enemy's small craft much faster than they could be rebuilt. The Japanese reacted by emplacing scores of heavy caliber shore batteries to cover their lugger and barge movements. These measures proved ineffective against the fast striking PT boats and planes, however, and the Japanese were forced to abandon daylight traffic almost entirely. Attempts were thereafter confined to movement by night as their barges, restricted by the darkness, crept furtively from cove to cove under cover of elaborate camouflage and security measures. This almost complete interdiction of all water-borne reinforcement was a major factor in the enemy's defeat in New Guinea and demonstrated again the resourceful capabilities of the Allied forces in countering each new threat as it arose.

---

49 Ibid.

50 Ibid.

New Strategic Situation

With the end of operations in western New Guinea the Japanese were in an extremely vulnerable position throughout the Pacific. The powerful spearhead of General MacArthur's forces now pointed menacingly only 600 miles below Mindanao. Two whole armies had been destroyed or paralyzed—the Seventeenth Army in the Solomons and the Eighteenth Army in New Guinea—and the remainder of the once powerful Eighth Area Army was scattered and isolated in New Britain and New Ireland. Over a quarter of a million Japanese troops were withering on the vine, virtually erased as a military fighting force.

The unhappy strategic situation to which the Japanese had been reduced was further aggravated by Admiral Nimitz's forces which had by-passed Truk and invaded Tinian, Guam, and Saipan in the Marianas, thus tightening the Allied ring which encircled Japan's inner zone of defense. In southwest New Guinea and the Netherlands East Indies, the Allied Air Forces had so successfully neutralized the areas around the Arafura, Banda, and Ceram Seas that it became unnecessary to carry out ground actions in those regions.

Japan's air forces beyond the Philippines had been swept from the skies. The constant and implacable pressure exerted by the Allied Air Forces gradually eliminated the enemy's air resources in the Southwest Pacific. Thousands of planes had been destroyed along with experienced pilots and crewmen. Huge stocks of valuable equipment, supplies, and ammunition which the Japanese could not replace, had been completely lost.

General MacArthur's conduct of operations along the New Guinea axis had not only yielded large stretches of enemy holdings but had also kept Allied losses in the Southwest Pacific to an extraordinarily low level. By adhering closely throughout his campaign to the four basic principles of war—the principles of mass, surprise, economy, and objective (adherence to a master plan)—he had achieved maximum gains in territory with minimum losses in combat.

While Japan's whole defense structure was being rocked to its very foundations, General MacArthur was preparing even stronger blows. The vexing problems of logistics had been largely solved. His newly captured bases, with steadily increasing stock piles of materiel, now assured adequate supplies for his projected drive forward into the heart of the enemy's Pacific possessions—the Philippines.

CHAPTER VII

THE PHILIPPINES: STRATEGIC OBJECTIVE

Basic Strategy

The Philippine Islands constituted the main objective of General MacArthur's planning from the time of his departure from Corregidor in March 1942 until his dramatic return to Leyte two and one half years later.[1] From the very outset, this strategic archipelago formed the keystone of Japan's captured island empire and therefore became the ultimate goal of the plan of operations in the Southwest Pacific Area. (Plate No. 48) After the Philippines were liberated, they would form the main base for the final assault against the Japanese Homeland.

As the Allies advanced westward along New Guinea and across the Central Pacific, a wide divergence of opinion developed among international planners and military strategists on the methods of defeating Japan, but General MacArthur never changed his basic plan of a steady advance along the New Guinea—Philippine axis, from Port Moresby to Manila. This plan was conceived as a forward movement of ground, sea, and air forces, fully co-ordinated for mutual support, operating along a single axis with the aim of isolating large Japanese forces that could be attacked at leisure or slowly starved into surrender. By choosing

1 The Japanese also gave the Philippines high priority in their over-all strategic war plans. "The Philippines were the east wing of the so-called 'Southern Sphere' in the Japanese operations in the southern regions," said Lt. Gen. Shuichi Miyazaki, Chief of the Operations Section, Imperial General Headquarters. "They took the shape of the main line of defense against American counterattacks. The western wing was Burma and Malaya, and together the two wings protected Japanese access to the southern regions. Viewed from the standpoint of political and operational strategy, holding the Philippines was the one essential for the execution of the war against America and Britain. With the loss of these islands, not only would Japanese communications with the southern regions be severely threatened, but the prosecution of strategic policies within the southern regions as far as supply and reinforcements were concerned would be a paramount difficulty. The loss of the Philippines would greatly affect civilian morale in Japan. The islands were also essential and appropriate strategic bases for the enemy advance on Japan. After their capture the advantage would be two to one in favor of the enemy and the prosecution of the war would suddenly take a great leap forward." Maj. Gen. Naokata Utsunomiya, Assistant Chief of Staff, Fourteenth Area Army, indicated the important role which the Philippines played in Japanese strategic plans when he suggested that they were looked upon as the final stepping stones to Japan itself. "The main value of the Philippines lay in their role as a bulwark against the invasion of the Homeland. They were to absorb as much of the enemy attack as they could and to delay the advance as long as possible. Defense preparations in Japan were in the meantime being hastened." This opinion was corroborated by the testimony of Col. Shujiro Kobayashi, Tactical Staff Officer of the Fourteenth Area Army: "The Philippines formed the first line of defense of the Japanese Homeland. For this reason the Combined Fleet was committed to the battle off Leyte. It was considered necessary that the decisive battle be fought in the Philippines. This was the opinion of Imperial General Headquarters." Lt. Gen. Seizo Arisue, Chief of G-2, General Staff Headquarters added the following comment: "To shatter American war plans, the Japanese Army held it necessary to maintain the Philippines to the end and to fight a decisive battle with the Americans who planned to recapture the Philippines. Furthermore, the Philippines were of absolute necessity for the security of traffic between Japan Proper and the southern areas." Interrogation Files, G-2 Historical Section, GHQ, FEC.

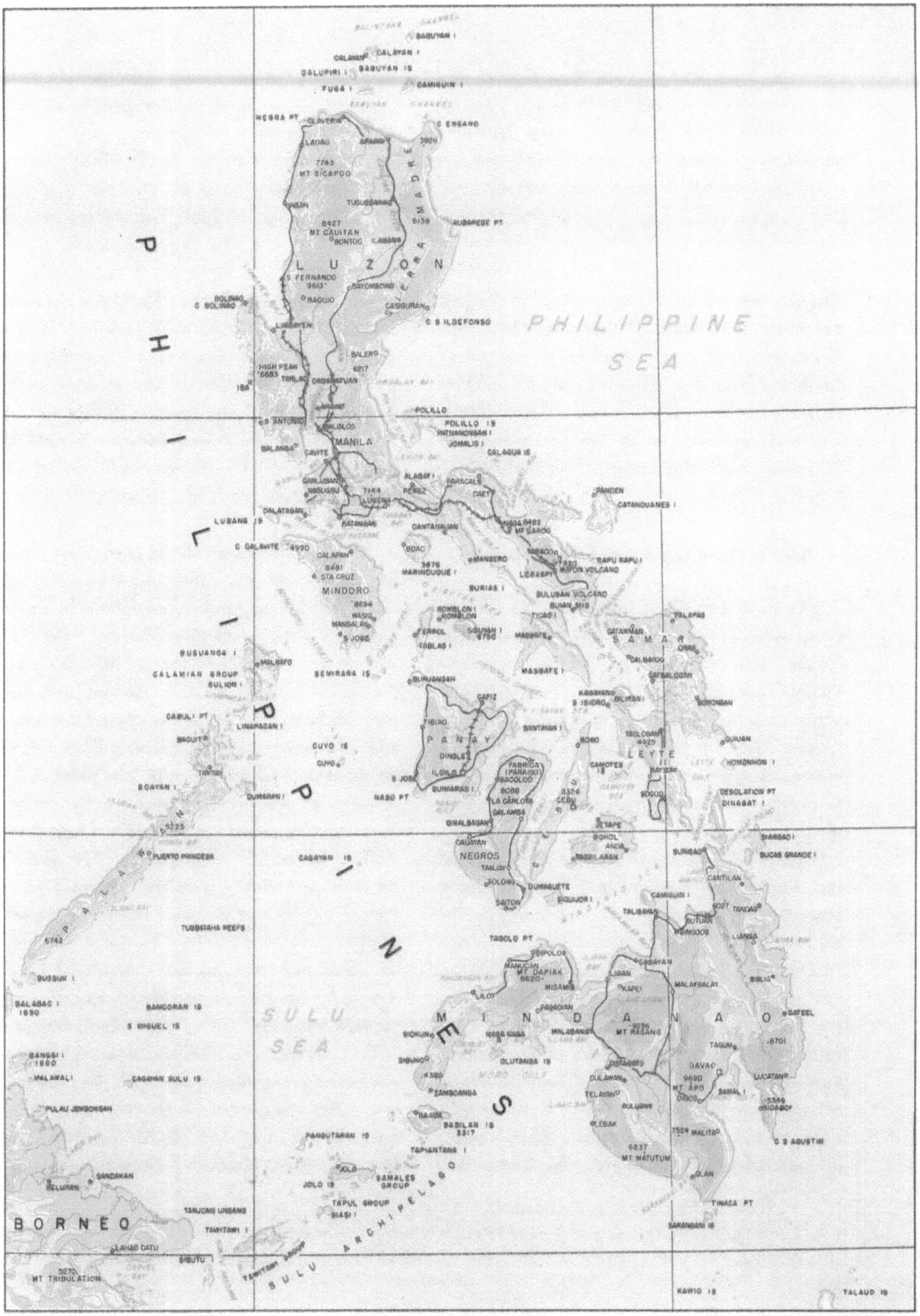

PLATE NO. 48

The Philippines

the route from Australia via New Guinea and the Halmaheras to the Philippines, General MacArthur could constantly keep his lines protected, pushing his own land-based air cover progressively forward with each advance. His plan insured control of the air and sea during major amphibious operations and was so designed that land-based air power with its inherent tactical advantages could be used to the maximum extent. This remained his fundamental concept of operations as he moved his forces from Port Moresby and Milne Bay to Buna and Lae, through the Vitiaz Straits to the Admiralties, on to Hollandia and the Vogelkop, until they reached their final springboard at Morotai.

Initial Planning for the Philippines

The first over-all plan naming the Philippines as a final objective for the Southwest Pacific Area was prepared by General MacArthur's G-3 planning staff at the conclusion of the Buna Campaign early in 1943. Called "Reno Plan," it formed the basis for ensuing operations against the Japanese although it was to undergo several changes during the course of the Pacific War.

"Reno I" was based on the premise that the Philippine Archipelago, lying directly athwart the main sea routes from Japan to the sources of her vital raw materials and oil in the Netherlands Indies, Malaya, and Indo-China, was the most important strategic objective in the Southwest Pacific Area.[2] Whoever controlled the air and naval bases in the Philippine Islands logically controlled the main artery of supply to Japan's factories. If this artery were severed, Japan's resources would soon dry up, and her ability to maintain her war potential against the advancing Allies would deteriorate to the point where her main bases would become vulnerable to capture.

The choice of routes to the Philippines was narrowed down until an advance along the northwest coast of New Guinea was selected as providing the best opportunities for the full coordination of General MacArthur's air, ground, and naval forces. Extensive use would be made of airborne and parachute troops, seizing certain key bases and "leap-frogging" past others. Mindanao was selected as the tactical objective area in the Philippines and the flanks of the Allied advance beyond the western tip of New Guinea were to be safeguarded by air and naval neutralization of Palau and Ambon.

During the course of 1943, the fast changing combat situation necessitated several alterations in the original "Reno Plan." In August, "Reno I" was succeeded by "Reno II" and in October still further modifications were published in "Reno III."[3] At that time, General MacArthur had driven past Finschhafen and had penetrated the enemy's New Guinea defenses to a depth of over 300 miles. The strategy of his advance was briefly outlined in a memorandum to the War Department. "This advance," he stated, "is along a decisive operational axis that drives a wedge into the [Japanese defense] perimeter toward a central core—the Philippines—that dominates all aerial and shipping lanes employed by the enemy for his current reinforcement and maintenance program. . . ."[4] The establishment of Allied forces in the Philippines would not only cut Japan's communications with the islands on which she was dependent for the means to run her war machine, but would also provide an ideal base from which to prepare the final blows

2 GHQ, SWPA, *Estimate of the Situation and Rough Draft, Reno Plan*, 25 Feb 43 (S).

3 GHQ, SWPA, *Reno II*, 3 Aug 43; GHQ, SWPA, *Reno III*, 20 Oct 43.

4 CINCSWPA memo to WARCOS, 31 Oct 43, JCS-CCS Papers No. 1, G-3, GHQ, Exec Files (S).

against Tokyo itself.

General MacArthur continued his analysis with an appraisal of the various approaches to the Philippines. He considered that an attack from the Southeast Asia Command would be frontal in nature, pushing the Japanese back upon successive lines of defense which they could readily keep supplied. "A major effort along this line of action," he concluded, "is undesirable both tactically and logistically." An attack across the Pacific would also have to be delivered against a position organized in great depth. It would have to be supported entirely by carrier-based aircraft as opposed to land-based air cover, and it would not sufficiently cut the enemy's lines of communication nor seriously curtail his war potential.[5]

A drive from the Southwest Pacific Area, on the other hand, possessed several advantages which General MacArthur explained as follows:

> *The attack from the Southwest Pacific Area departs from a base that is closest to the objective and advances against the most lightly organized portion of the enemy's defenses, effecting a decisive penetration. It is the only plan that permits an effective combination of land, sea and air power. The advance can be made by a combination of airborne and seaborne operations, always supported by the full power of land-based aviation and assisted by the fleet operating in the open reaches of the Pacific. A penetration of the defensive perimeter along this line results in by-passing heavily defended areas of great extent that will fall, practically of their own weight, to mopping-up operations with a minimum of loss.*[6]

He agreed that if the forces were available it would be advantageous to attack in force along each line of advance, but he maintained that the paucity of means in the initial stages, particularly in amphibious craft, should limit the Allies to a single drive in full strength. "To attempt a major effort along each axis," he stated, "would result in weakness everywhere in violation of cardinal principles of war, and would result in failure to reach the vital strategic objective at the earliest possible date, thus prolonging the war." Concentration on the most advantageous route of advance, with a powerful drive forward, was in General MacArthur's view the surest way to an early victory.[7]

By March of 1944, SWPA forces had landed in the Admiralties and Central Pacific forces had won positions in both the Gilberts and the Marshalls. Parallel drives converging upon the Philippines and Formosa from both the Southwest Pacific and Pacific Ocean Areas appeared to be possible. The strategic objectives, set forth under a new "Reno IV" plan, were expanded to include the securing of land, naval, and air bases in the southern Philippines from which to launch an attack upon Luzon in the north. Bases in Luzon would in turn be used for the support of subsequent Allied assaults against the Formosa—China coast area.[8]

The spring and summer of 1944 saw accelerated action in the war against Japan as the Allied advance gathered greater momentum. In the latter part of March, naval task force strikes were carried out against Palau, Yap, and Woleai, and at the end of April a heavy and damaging carrier raid was made on Truk. SWPA forces meanwhile invaded Aitape and Hollandia and during May assaulted Wakde and Biak. On 15 June, Central Pacific forces invaded Saipan in the Marianas to begin construction of a base for the mighty B-29 bombers. The Allied forces were fast reaching tactical positions from which to launch a powerful offensive into Japan's main lines of defense.

5 Ibid.

6 Ibid.

7 Ibid.

8 GHQ, SWPA, *Reno IV*, 6 Mar 44.

Again the question arose as to the best points of attack and the most advantageous routes of approach.

Various proposals were put forth for by-passing certain previously planned objectives or for by-passing all objectives short of the Japanese Home Islands themselves. Other plans advocated the use of long-range bombing from the Marianas to force the Japanese into surrender. Still others considered bases in China necessary to accomplish the task, with landings to be made along the China coast from positions in the Philippines. The suggestion was made that western New Guinea, Halmahera, and the Palaus be by-passed for an immediate blow at the Philippines; another plan advocated the capture of airfields in the southern Philippines and then striking directly at Formosa; and finally, a proposal was put forth to take the Philippines as a base for an assault straight into Kyushu, the southernmost of the four Japanese Home Islands.

Certain of these proposals were discarded upon further study as entailing too great a risk, with the possibility that failure would delay the conclusion of the war by many months. Other suggestions appeared to be more time-consuming and costly in men and materials than substitute objectives which might be found. The plan to force Japan to surrender by strategic bombing would undoubtedly require supplementing with actual territorial conquest of numerous vital areas.

Throughout the strategic planning on ways and means to defeat Japan, General MacArthur held firmly to his original concept. In view of the over-all situation in the summer of 1944, he considered that the route through Halmahera and the Palaus into the southern Philippines, and thence to Luzon was the most feasible approach to the Japanese Homeland. It permitted an advance that could move under the cover of land-based air power from start to finish, with a minimum of risk and with every assurance of success. This route would also take advantage of the large Philippines guerrilla forces which were already furnishing valuable intelligence and which were waiting impatiently to assist in ousting the enemy. Aside from military considerations, moreover, the liberation of the Philippines was a duty that General MacArthur felt had to be fulfilled as soon as possible. He outlined his views in a radio to the War Department a few days after completing the final draft of his "Reno V" plan:

> *In my opinion purely military considerations demand the reoccupation of the Philippines in order to cut the enemy's communications to the south and to secure a base for our further advance; even if this were not the case and unless military factors demanded another line of action, it would in my opinion be necessary to reoccupy the Philippines. It is American territory, where our unsupported forces were destroyed by the enemy; practically all of the seventeen million Filipinos remain loyal to the United States and are undergoing the greatest privation and suffering because we have not been able to support or succor them; we have a great national obligation to discharge. . . . I feel also that a decision to eliminate the campaign for the relief of the Philippines, even under appreciable military considerations, would cause extremely adverse reactions among the citizens of the United States; the American people, I am sure, would acknowledge the obligation.*[9]

The "Musketeer" Plans

General MacArthur's views regarding the approach to the Philippines met with general acceptance, and early in July 1944, joint staff conferences were held with representatives of the Central Pacific Area to prepare detailed

9 CINCSWPA Radio No. CX-13891 to WARCOS, 18 Jun 44, WD No. 761, C/S GHQ, SWPA (S).

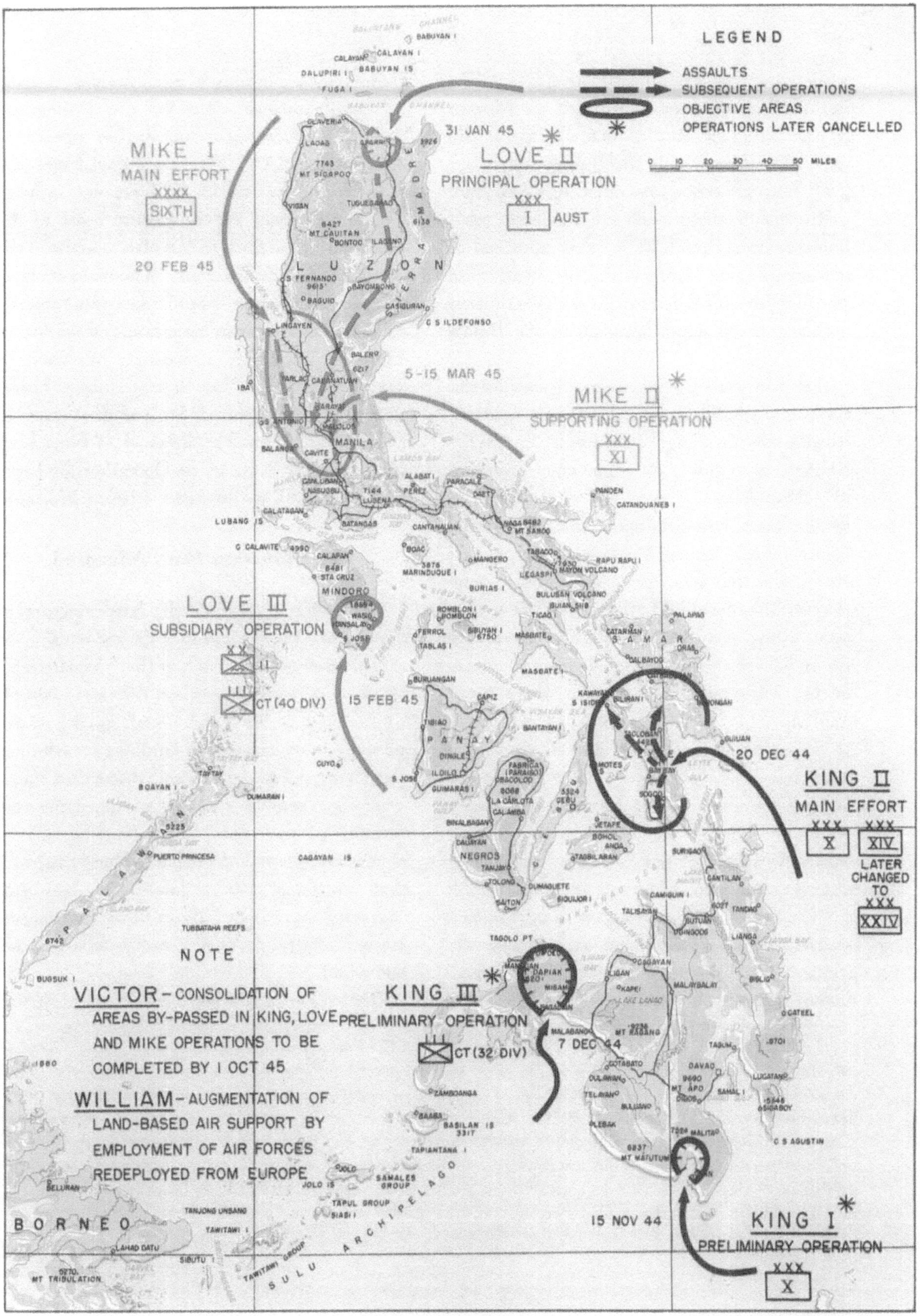

PLATE NO. 49

"Musketeer II" Plan

studies for the preliminary operations.[10] Coordinated assaults were scheduled for Morotai in the Moluccas and certain islands in the western Carolines. In the former operation, SWPA forces were to be aided by naval units of the Pacific Fleet while, in the latter task, bombers from the SWPA were to assist in attacks on Palau, Yap, and Ulithi. With these positions in Allied hands, the way would then be clear for the actual invasion of the Philippines.

The first version of the over-all plan for the conduct of the Philippines Campaign was published under the name " Musketeer " by GHQ, SWPA, on 10 July 1944. The chief objectives of " Musketeer I " were the destruction of hostile forces in the Philippines and the prompt seizure of the central Luzon area to provide air support and naval bases for possible operation of POA forces in the China coast–Formosa area. This plan, like " Reno IV, " provided for an Allied advance along the eastern shores of the Philippines to establish bases for a final attack on Luzon. Initial lodgements were to be made at Sarangani Bay in southern Mindanao on 15 November and at Leyte Gulf on 20 December. Except for these preliminary operations, however, Mindanao and the Visayas were to be by-passed and not consolidated until after the occupation of Luzon was completed.[11]

"Musketeer II" of 29 August 1944, enlarged on the original plan and had as its major objective "the prompt seizure of the Central Luzon area to destroy the principal garrison, command organization and logistic support of hostile defense forces in the Philippines and to provide bases for further operations against Japan."[12] The plan envisioned the full support of the U. S. Fleet not only to secure a foothold on the eastern coast of the Philippine Archipelago but also to assist in the invasion of central Luzon. The main effort in the Central Plains—Manila area was integrated into the Lingayen operation and set for 20 February 1945. A supporting operation to land at Dingalen Bay in the eastern Luzon area was contemplated for the first part of March. The invasion dates of 15 November for Sarangani Bay and 20 December for Leyte Gulf remained unchanged.[13] (Plate No. 49)

Leyte Invasion Date Advanced

A sudden change in the battle picture of the Pacific Theater led to a drastic revision of Allied strategy as set forth in the " Musketeer " plans. On 9–10 September, Admiral Halsey's Third Fleet carrier-borne aircraft, giving strategic support to impending landings on Morotai and Palau, hit Mindanao and discovered unexpected and serious weakness in the enemy's air defenses of that area. Few Japanese planes were encountered and further probing disclosed that Southwest Pacific land-based bombers, operating out of New Guinea fields, had caused severe damage to enemy air installations on the island.

Immediate orders were issued to capitalize

---

10 CINCPOA Radio to COMINCH, 4 Jul 44, JCS-CCS Papers No. 2, G-3, GHQ Exec Files. Admiral Nimitz in this radio expressed the opinion that short cuts and prompt exploitation of favorable situations were prime considerations, but that operations should be so conducted as to insure control of the sea and air during major amphibious assaults. He advocated the use of shore-based aircraft to the maximum extent possible and said that "tactical situations in which the fast carrier task forces are more or less immobilized in support of protracted fighting on shore should be avoided. In my opinion," he continued, "the basic concept of operations proposed by CINCSWPA in which shore-based air forces, ground troops, and naval forces are advanced together is sound." Ibid.

11 GHQ, SWPA, *Basic Outline Plan for Musketeer Operations*, 10 Jul 44, G-3, GHQ Plans File (TS).

12 GHQ, SWPA, *Basic Outline Plan for Musketeer II*, 29 Aug 44, G-3, GHQ Plans File (TS).

13 Ibid.

on the enemy's apparent aerial weakness. The Allies moved on beyond Mindanao and carrier task groups of the Third Fleet sortied northward to hit the Visayas on 12 and 13 September. Again enemy air reaction was surprisingly meager and heavy loss was inflicted upon Japanese planes and ground installations. It became more and more apparent that the bulk of the once mighty Japanese Air Force had been destroyed in the costly war of attrition incidental to the New Guinea operations.

The disclosure of such great vulnerability in the enemy's air shield over the Philippines caused an immediate reassessment of the situation to ascertain whether an acceleration of the existing schedule would be possible by omitting certain previously planned operations designed mainly for air support. Intelligence sources indicated that the Japanese had been increasing their ground forces in the Philippines.[14] With the enemy thus strengthening his positions to meet the anticipated Allied assaults, each month or week that could be cut from the Allied timetable for the Philippines would accordingly reduce the over-all cost of the campaign and help ensure its rapid accomplishment.

On 13 September Admiral Halsey advised Admiral Nimitz and General MacArthur that he believed the seizure of the western Carolines, including Palau, was no longer essential to the occupation of the Philippines.[15] He suggested that Leyte could be seized immediately if all projected operations in the Carolines, with the exception of Ulithi, were cancelled and the landings were covered by carrier aircraft. Admiral Nimitz concurred in the proposal to by-pass Yap, but directed that the Palau and Ulithi operations be carried out as scheduled, the former being needed as an air base and the latter as a fleet anchorage.[16] Admiral Halsey's recommendation was relayed to the Joint Chiefs of Staff who were then participating in the Quebec Conference between President Roosevelt and Prime Minister Churchill. General MacArthur's views were requested on the proposed change of the invasion date for Leyte and the reply came back as follows:

> *"In view of COM3rdFLT's latest report on carrier operations in the Philippines Islands area, I am prepared to move immediately to execution of King II [Leyte] with target date of 20 Oct 44."*[17]

The advance planning and the preparation of alternative solutions which were standard operational procedure for General MacArthur's staff, again, as in the case of Hollandia, permitted the necessary flexibility for rapid change. Accordingly, with the assurance that the SWPA forces could conform to the proposed change in schedule, the target date for the Leyte landing was advanced fully two months ahead of the original schedule.[18]

14 "An appraisal of information, over the period 15 May to 15 June, discloses a contemporary trend of events, indicating definitely that the Japanese are massing troops, bolstering dispositions, and shifting units in the Philippine area. This trend was not indicated prior to 1 April." G-2, GHQ, SWPA, *Philippine Monthly Combined Situation Report*, 15 Jun 44. Japanese reinforcement in the Philippines continued throughout the summer and autumn of 1944. G-2, GHQ, SWPA, *Philippine Monthly Combined Situation Report*, 15 Jul and 15 Aug, 44. See also G-2, GHQ, SWPA, *Monthly Summary of Enemy Dispositions*, 30 Sep 44.

15 COM3rdFLT Radio to CINCPOA, CINCSWPA, COMINCH, 13 Sep 44, G-3, GHQ, Admin 385 (TS).

16 CINCPOA Radio to COM3rdFLT, 13 Sep 44, C/S, GHQ, SOPAC No. 522 (S).

17 CINCSWPA Radio No. C-17744 to JCS, CINCPOA, 14 Sep 44, G-3, GHQ, Admin 385 (TS).

18 General Marshall described the dramatic sequence of events leading up to the orders for the Leyte invasion in his report to the Secretary of War: "...General MacArthur's views were requested and 2 days later he advised us that he was already prepared to shift his plans to land on Leyte 20 October, instead of 20 December, as previously intended. It was a remarkable administrative achievement.

"The message from MacArthur arrived at Quebec at night and Admiral Leahy (Chief of Staff to the President),

Other important decisions were made in rapid order. Preliminary operations, such as the seizure of Talaud, Yap, Misamis Occidental, and Sarangani, were cancelled. The Palau and Morotai operations, still considered necessary for air support and flank protection against thrusts from the Mandated Islands or from Halmahera and the Celebes, remained scheduled for 15 September. The landing on Ulithi, to secure a forward area logistic support base for the fleet, was planned for 23 September.

Morotai

The sustained air assaults of the Far Eastern Air Force against hostile bases in the Netherlands East Indies, together with Allied landings on Biak, Noemfoor, and Sansapor, had greatly weakened the enemy stronghold at Halmahera. His airfields there and at contiguous intermediate bases were almost wholly neutralized, his maritime forces largely interdicted, and his ground troops immobilized in almost static helplessness. The flexibility of the Halmahera base, from which he once funneled his forces to strategic outposts, was now completely destroyed. If he failed to retrieve his position, one of the main lines of defense protecting the Philippines and the Netherlands East Indies was gravely threatened.[19]

Intelligence sources revealed that, beginning in June 1944, the Japanese had displayed considerable interest in the defense of Morotai at the northern tip of Halmahera, but that little had been accomplished, presumably because other areas were placed higher on the list of probable Allied objectives. With its small garrison, Morotai offered an excellent opportunity for securing a base without prolonged fighting or heavy losses. It fitted well into General MacArthur's scheme of winning important positions by attacking weakly held but stratagically valuable points.

Additional information indicated that, while the Japanese strength on Morotai was relatively weak, the island could be substantially reinforced by barge and transport traffic with troops from Halmahera within forty-eight hours and from the Philippines within a week.[20]

The full weight of Allied air power was thrown against the Japanese to forestall any reinforcement efforts. Medium and heavy bombers of the Southwest Pacific Area struck airdromes on Mindanao, Halmahera, and the Netherlands East Indies, while fighters attacked and strafed defense installations, shipping, and troop concentrations. The fast carriers of the Third Fleet hit Chichi Jima and Iwo Jima in the Bonins, and Mindanao and the Visayas in the Philippines, with repeated raids between 31 August and 14 September. Hostile aircraft were confined to the northern Philippines and

Admiral King, General Arnold, and I were being entertained at a formal dinner by Canadian officers. It was read by the appropriate staff officers who suggested an immediate affirmative answer. The message, with their recommendations, was rushed to us and we left the table for a conference. Having the utmost confidence in General MacArthur, Admiral Nimitz, and Admiral Halsey, it was not a difficult decision to make. Within 90 minutes after the signal had been received in Quebec, General MacArthur and Admiral Nimitz had received their instructions to execute the Leyte operation on 20 October, abandoning the three previously approved intermediary landings." *Biennial Report of the Chief of Staff of the United States Army to the Secretary of War, July 1, 1943 to June 30, 1945.* In his "State of the Union Message" on 6 January 1945, President Roosevelt commented on the advanced date for the invasion of the Philippines as follows: "Within the space of twenty-four hours, a major change of plans was accomplished which involved Army and Navy forces from two different theaters of operations—a change which hastened the liberation of the Philippines and the final day of victory—a change which saved lives which would have been expended in the capture of islands which are now neutralized far behind our lines."

19 GHQ, SWPA, Press Release, 15 Aug 44.

20 HQ Alamo Force, *G-2 Estimate of Enemy Situation, Morotai Island,* 1 Aug 44. (R).

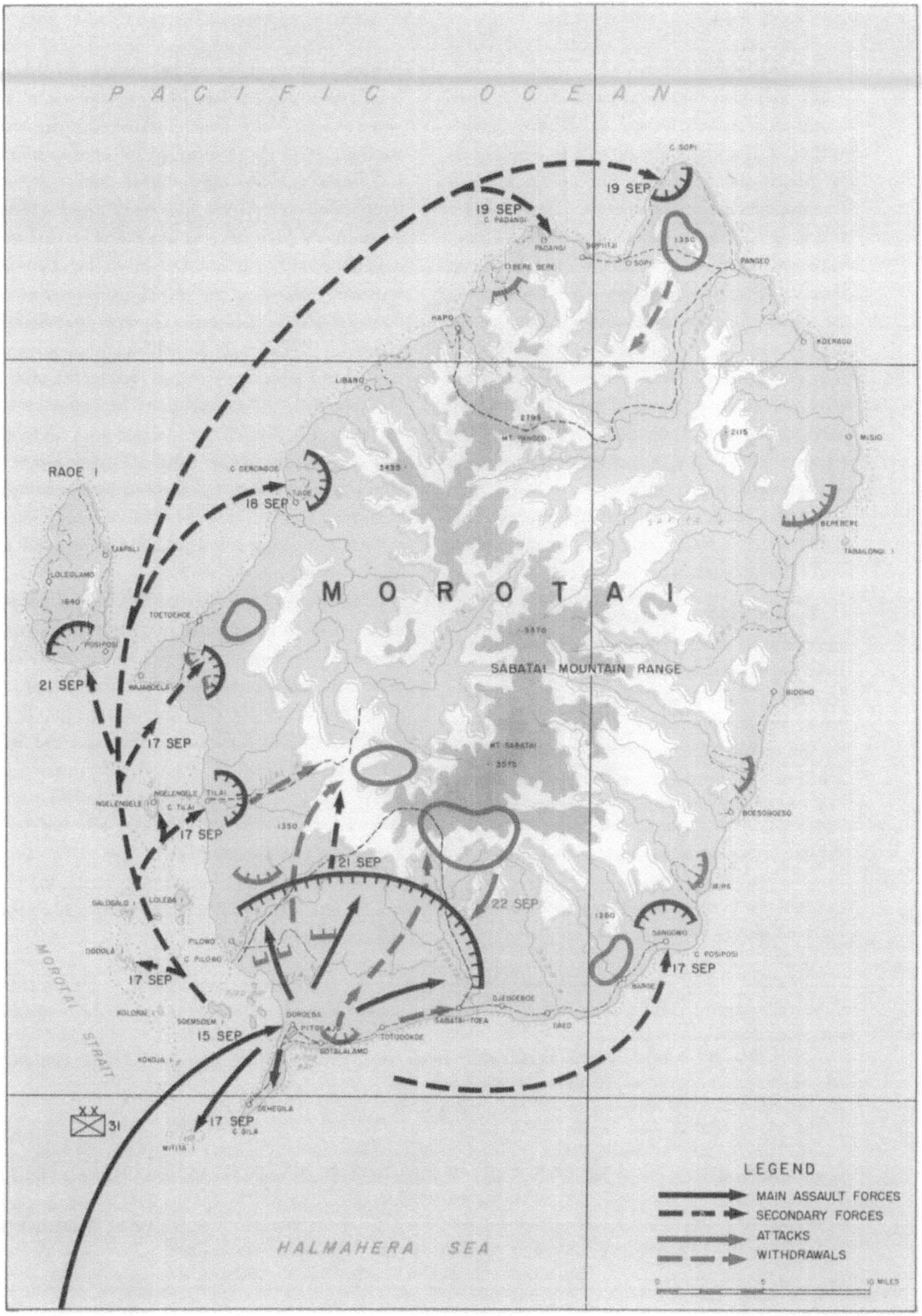

PLATE NO. 50

The Morotai Campaign, 15 September–4 October 1944

to the southern Celebes.[21]

Air and naval bombardment of shore positions on Halmahera, as well as Morotai, preceded the landings at Pitoe Bay on 15 September. An Alamo task force was composed of the 31st Division and 126th Regimental Combat Team of the 32nd Division plus supporting combat and service troops, directed by XI Corps. Seventh Amphibious Force convoys carrying the assault troops were protected by escort carriers and land-based fighters as they moved from Aitape and Toem. General MacArthur went with his troops and viewed the bombardment of Galela on Halmahera from aboard the cruiser *Nashville* and later went ashore on one of the beachheads.

A description of the landing appeared that day in the communique of the Southwest Pacific Headquarters:

We have landed in the Halmaheras three hundred miles beyond New Guinea. Our ground forces under cover of naval and air bombardment seized beachheads on the island of Morotai, the most northern of the Halmahera Group. The point of landing was unexpected, the enemy having apparently anticipated it in the lower parts of the islands where he had accumulated very strong forces in heavily defended positions. In by-passing these forces and landing farther to the north he was taken by surprise and his initial resistance is being rapidly overcome. Our ground losses up to the present time have been very light and we have had no naval or air losses.[22]

The main objectives on Morotai were reached by 16 September, and ground activities were limited to mopping up and extending the beachhead perimeter. (Plate No. 50) Air warning systems were set up, beach defenses coordinated, antiaircraft weapons brought into position and PT boats put on night patrol duty. By 21 September the Allies had established a firm position on Morotai and work was started immediately on facilities for pushing the offensive forward. Viewing the rapid development of Morotai with satisfaction, General MacArthur stated: "We shall shortly have an air and light naval base here within 300 miles of the Philippines."[23] An airstrip at Pitoe was opened operationally for fighter aircraft on 4 October and on that date the Morotai operation was officially terminated.[24] Mopping up operations, however, were carried out for a long period against a stubborn enemy who continued to land troops by barge traffic.

Strategically the Halmahera—Philippines line had been penetrated and the enemy's conquests to the south imperiled by an imminent threat of Allied envelopment. The rolling up of the remainder of this line would cut off in the Netherlands East Indies the Japanese Sixteenth and Nineteenth Armies, a force estimated at nearly 200,000 men, and would sever essential supplies of oil and other war materials from the Japanese mainland.

General MacArthur commented on the enemy and the future trend of the war in the following words:

We now dominate the Moluccas. I rejoice that it has been done with so little loss. Our campaign

21 COM3rdFLT Report 0085 to COMINCH, 25 Jan 45, G-3, GHQ, SWPA *Journal* (S). AAF, *Intelligence Summary* Nos. 235–238, 19–30 Aug 44, G-2, GHQ Admin (S).

22 GHQ, SWPA, Communique No. 891, 15 Sep 44.

23 GHQ, SWPA, Press Release, 15 Sep 44.

24 The capture of Morotai was the final task of the Alamo Force which had directed the operations of American combat units from the spring of 1943. The Alamo Force was officially dissolved on 25 September 1944 and General MacArthur ordered the Sixth Army to take over the initial Philippines campaign which previously had been assigned to the task force. No actual change was involved, since the Alamo Force consisted of units assigned or attached to the Sixth Army, and Headquarters Sixth Army also functioned as Headquarters Alamo Force.

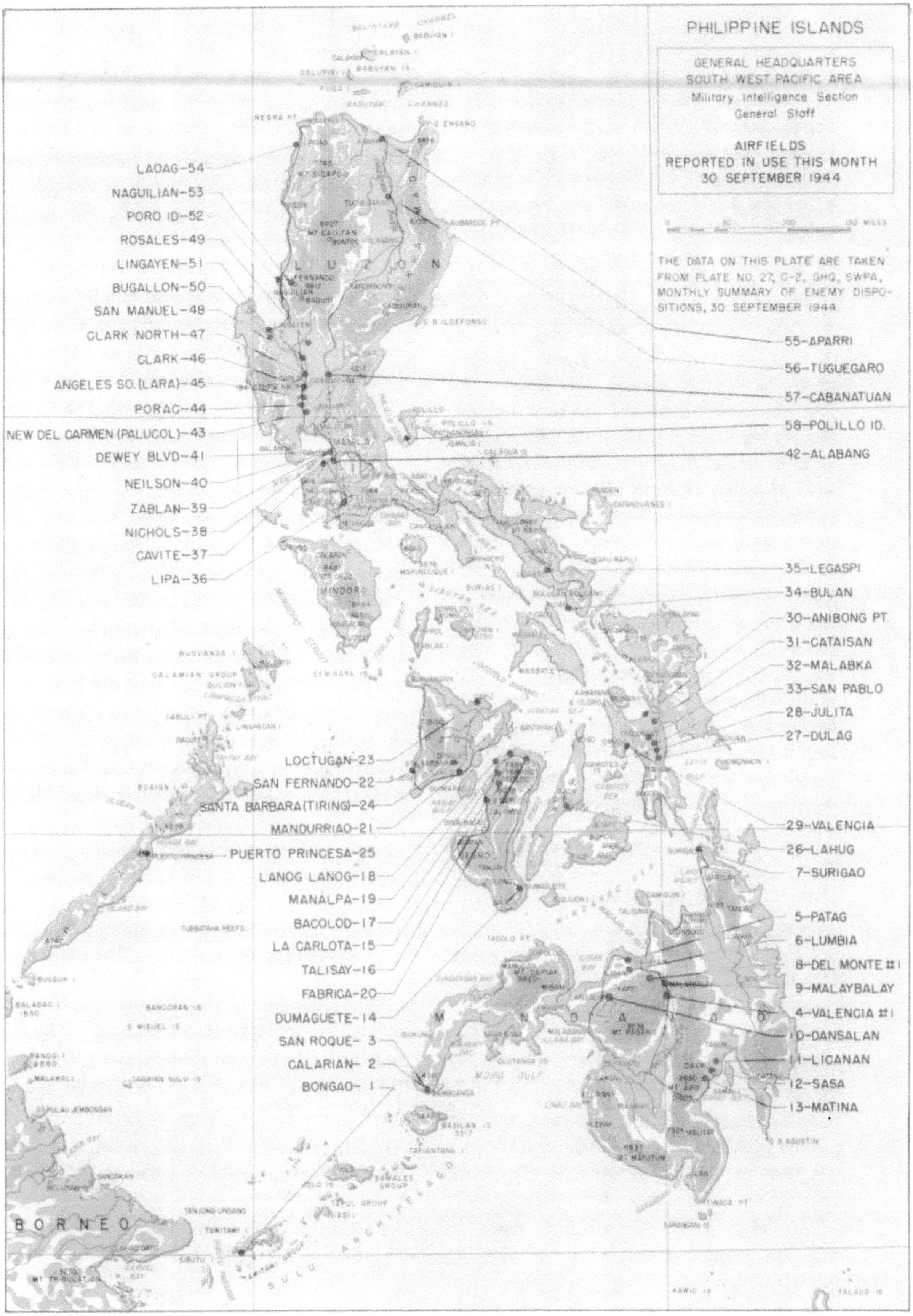

PLATE NO. 51

Enemy Airfields Reported in Use, September 1944

is entering upon its decisive stage. Japanese ground troops still fight with the greatest tenacity. The military quality of the rank and file remains of the highest. Their officer corps, however, deteriorates as you go up the scale. It is fundamentally based upon a caste and feudal system and does not represent strict professional merit. Therein lies Japan's weakness. Her sons are strong of limb and stout of heart but weak in leadership. Gripped inexorably by a military hierarchy, that hierarchy is now failing the nation. It has had neither the imagination nor the foresighted ability to organize Japanese resources for a total war. Defeat now stares Japan in the face. Its barbaric codes have dominated Japanese character and culture for centuries and have practiced a type of national savagery at strange variance with many basic impulses of the Japanese people. Its successful domination has been based largely on the people's belief in its infallibility. When public opinion realizes that its generals and admirals have failed in the field of actual combat and campaign, the revulsion produced in Japanese thought will be terrific. Therein lies a basis for ultimate hope that the Japanese citizen will cease his almost idolatrous worship of the military and readjust his thoughts along more rational lines. No sophistry can disguise the fact from him that the military had failed him in this his greatest hour of need. That failure may mark the beginning of a new and ultimately happier era for him; his hour of decision is close at hand.[25]

Palau Islands

Timed to coincide with the landings on Morotai, assaults against Peleliu and Angaur in the Palau Islands were carried out by forces of the Central Pacific. Preceded by two days of extensive minesweeping and bombardment, the 1st Marine Division went ashore on Peleliu on 15 September. The beach defenses, although heavily studded with mines, were quickly overrun, and by evening of the next day Peleliu airfield, the primary objective of the operation, had been taken. Additional progress, however, proved difficult. The Allies had to contend not only with a well organized and stubborn defense by the enemy but also had to surmount the obstacles offered by a series of limestone ridges, transformed into a natural fortress over a period of years. The 81st Division under Maj. Gen. Paul J. Mueller, which successfully invaded Angaur on 17 September, despatched a regimental combat team to assist in the capture of Peleliu and several smaller islands to the north. After heavy and prolonged fighting, Peleliu was finally subdued by the end of September.

Sweeping around Babelthuap, which was defended by a garrison of roughly one and one-half divisions, the Allies invaded Kossol Roads, a large body of reef-enclosed water about 70 miles north of Peleliu. Kossol Pass and its roadstead were cleared and used as an anchorage for seaplanes and light naval vessels. On 23 September, Ulithi Atoll, 35 miles to the northeast, was occupied by elements of the 81st Division, and became the main forward naval base of operations against the Philippines. With the termination of the Morotai and Palau operations, the threat of a possible enemy flank attack was removed and the Allied forces could now proceed freely with the impending assault on the Philippines.

Preparing for the Leyte Invasion

The operation to take Leyte was a most ambitious and difficult undertaking. The objective area was located over 500 miles from Allied fighter bases on Morotai and Palau, beyond the effective range of fighter cover. It was at the same time in the center of a Japanese network of airfields covering the Philip-

25 GHQ, SWPA, Press Release, 15 Sep 44. These views forecast the ultimate character of an enlightened and benevolent occupation of Japan and the administrative and reform methods of SCAP in the post-war period.

pines. (Plate No. 51) The islands would doubtless be defended to the limit of the enemy's capabilities, probably even at the risk of losing his heretofore husbanded navy, since a successful Allied landing on Leyte would presage the eventual reoccupation of the entire Philippine area. The Japanese could reinforce their positions by bringing in troops and supplies from their rear lines whereas, without air bases in the vicinity, the Allied forces would have to rely on naval aircraft to prevent enemy supply and reinforcement convoys from reaching the area. Again, as at Hollandia, SWPA forces would advance beyond the range of their own land-based fighter cover and put themselves under the temporary protection of carrier planes for the assault phase. The success of the operation, even after a landing was secured, would depend on the ability of Allied naval forces to keep the enemy from building up a preponderance of strength on Leyte and adjoining Samar, and to prevent enemy naval craft from attacking shipping in the beachhead area.

Careful study preceded the selection of the landing beaches and the direction of the inland thrust on Leyte. A special report of the Allied Geographical Section detailing airfields, landing beaches, and roads was furnished to the planning staff months ahead of the actual invasion.[26] Terrain studies covering all areas of the Philippines which were earmarked for invasion following Leyte gave necessary details on which to base the overall strategy. (Plate No. 52) Even the assault troops were supplied with pocket-size handbooks containing essential information, maps, and photographs on local topography, locations of objectives, and conditions to be encountered. The Allied forces were well informed.

On the basis of these intelligence reports, confirmed by last minute aerial photographic reconnaissance, the northeastern coastal plain of Leyte was selected as most suitable for the assault.[27] Seizure of the eighteen-mile stretch between Dulag and San Jose would permit the early capture of the important Tacloban Airfield and make possible the occupation and use of the airfield system under development at Dulag. It would permit domination of vital San Juanico Strait, and place the invading forces within striking distance of Panaon Strait to the south. Intelligence reports indicated that the beach area would not be heavily defended, although some fortifications were being prepared along the inland road net.[28]

Logistic Support

An entirely new organization called the Army Service Command was set up for logistic support of the assault force and for base development and construction on Leyte. It was made a subordinate echelon of the United

26 Allied Geographical Section, G-2, GHQ, SWPA, *Special Report No. 55, Airfields, Landing Beaches and Roads, Samar, Leyte, and Dinagat Group*, 10 July 44.

27 In the section of his report on the Leyte operation dealing with "Conclusions, Comments and Recommendations," Commander Seventh Fleet made the following statement with reference to intelligence on the landing areas: "Intelligence reports on beach areas were considered generally satisfactory. The Allied Geographical Section, SWPA, had completed Terrain Study #84 and Special Report #55, covering the landing area. The general landing beaches were selected from such reports, and photographs taken by carrier planes on 14 September confirmed the reports. Swamps, lagoons, and other similar obstacles required the leaving of small gaps between some of the beaches, but no material modification of the plans was necessary." Commander Central Philippines Attack Force (COM7thFLT), *Report of Operation for the Capture of Leyte Island including Action Report of Engagement in Surigao Strait and off Samar Island on 25 October 1944 (King Two Operation)*, 31 Jan 45, A16–3, Serial: 00302–C.

28 G-2, GHQ, SWPA, *Monthly Summary of Enemy Dispositions*, 30 Sep 44.

EXHIBIT 2A

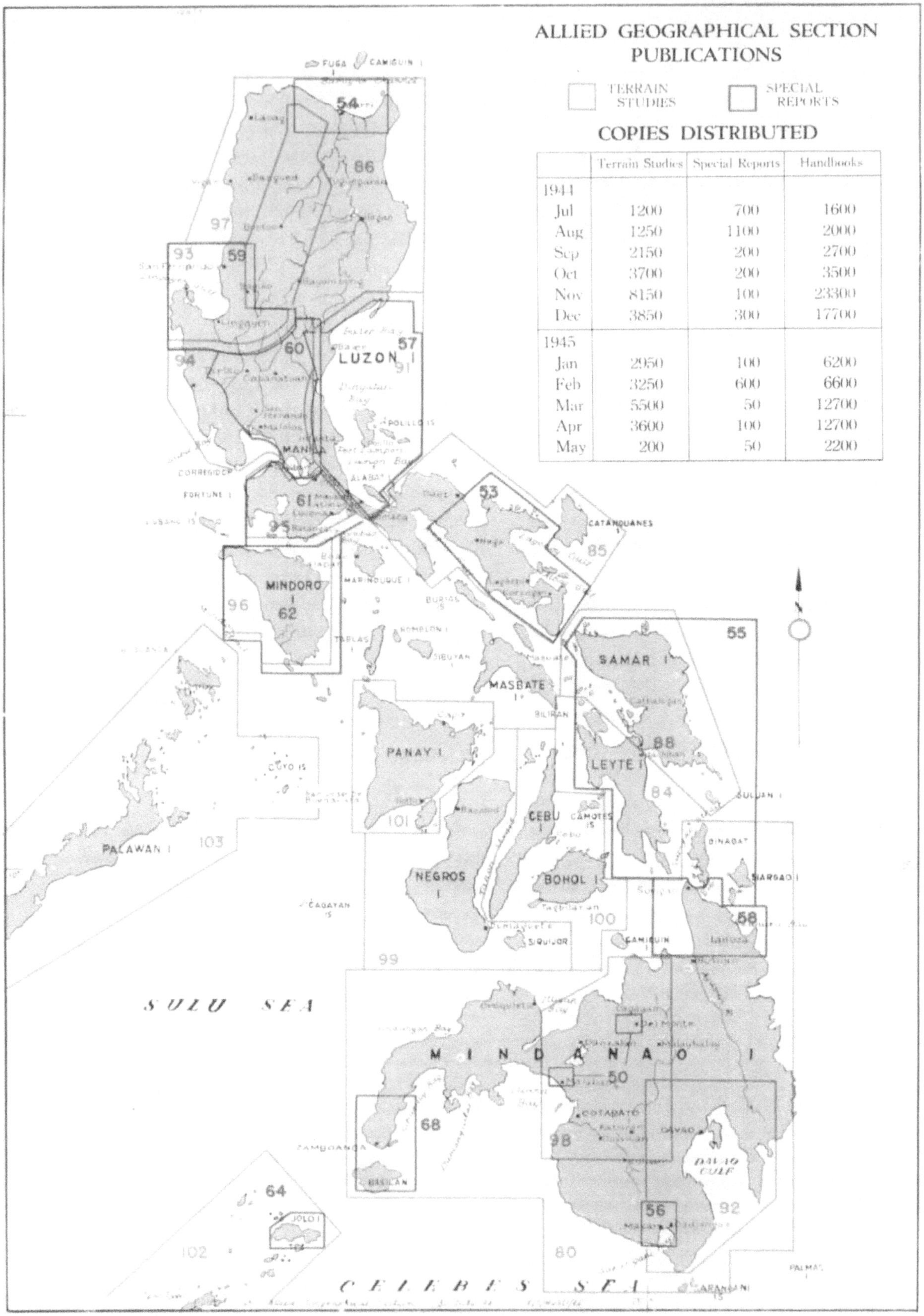

	Terrain Studies	Special Reports	Handbooks
1944			
Jul	1200	700	1600
Aug	1250	1100	2000
Sep	2150	200	2700
Oct	3700	200	3500
Nov	8150	100	23300
Dec	3850	300	17700
1945			
Jan	2950	100	6200
Feb	3250	600	6600
Mar	5500	50	12700
Apr	3600	100	12700
May	200	50	2200

PLATE NO. 52

Allied Geographical Section Publications

States Army Services of Supply, the organization which had furnished supplies to Allied troops in their drive from Australia to the Halmaheras. Maj. Gen. Hugh J. Casey, formerly General MacArthur's Chief Engineer, was given command of this new unit and directed to prepare and execute plans for the establishment and development of new bases in the Philippines in accordance with Services of Supply directives. Since the immediate problem was to develop an air base and transshipment port at Leyte, General Casey was instructed to plan and fulfill the combat logistical requirements for the operations of the Sixth Army after the Leyte landing was accomplished. To carry out this task as effectively as possible, the Army Service Command was put under direct control of General Krueger, meanwhile maintaining close liaison with Services of Supply from which it drew materiel and personnel.[29]

Organization of Forces

As soon as the Joint Chiefs of Staff approved the advanced date of 20 October for the Leyte attack, the machinery for the organization of the vast task force was set in motion. (Plate No. 53) Under the original December invasion plan, it was intended that SWPA would furnish the entire troop complement necessary to carry out the assault. Australian forces were to take over combat responsibility in New Guinea and New Britain and free United States units for the Philippine action. The relief of the U. S. 37th and 40th Divisions in these areas, scheduled originally for October and November, could not be accomplished in time to meet the new target date. Consequently, the Central Pacific Area made the XXIV Corps available to General MacArthur as a substitute for these divisions.

Southwest Pacific Area ground combat forces for the Leyte operation consisted of the X Corps, the major components of which were the 1st Cavalry and the 24th Infantry Divisions, both veteran units. This Corps was loaded on ships of Admiral Barbey's Seventh Amphibious Force, Task Force 78 of the Seventh Fleet, the same force which had so meritoriously conducted all previous major assault landings in the Southwest Pacific. Admiral Barbey's command was reinforced for the Leyte attack by shipping from Central Pacific sources. This augmented group, known as the Northern Attack Force, had the primary mission of landing the X Corps on beaches from Palo to Dulag on the eastern coast of Leyte. Part of the convoy carrying the X Corps Headquarters and Corps troops, the 24th Division, and the 6th Ranger Battalion, departed from Hollandia on 13 October. It rendezvoused on the 15th with the remainder of Admiral Barbey's fleet units which had carried the 1st Cavalry Division and reinforcing units from Manus Island.[30]

The XXIV Corps, less the 77th Infantry Division on Guam, had been combat loaded at Hawaii to be carried to Yap and Woleai by Vice Adm. Theodore S. Wilkinson's Third Amphibious Force, Task Force 33 of the Third Fleet. For the Leyte operation, Admiral Wilkinson's command was redesignated Task Force 79, the Southern Attack Force, and was transferred to the Seventh Fleet. The mission of the Southern Attack Force was to land and to establish the XXIV Corps, including the 7th and 96th Divisions, on beaches between San Jose and Dulag on the eastern coast of Leyte.

29 CG ASCOM Report, *Historical Record of the Army Service Command, 23 July–26 December 1944,* G-3, GHQ, SWPA *Journal* (S).

30 CG X Corps Report, *A History of the X Corps in the Leyte—Samar—Philippine Island Operation,* 15 Feb 45, G-3, GHQ, SWPA *Journal* (S).

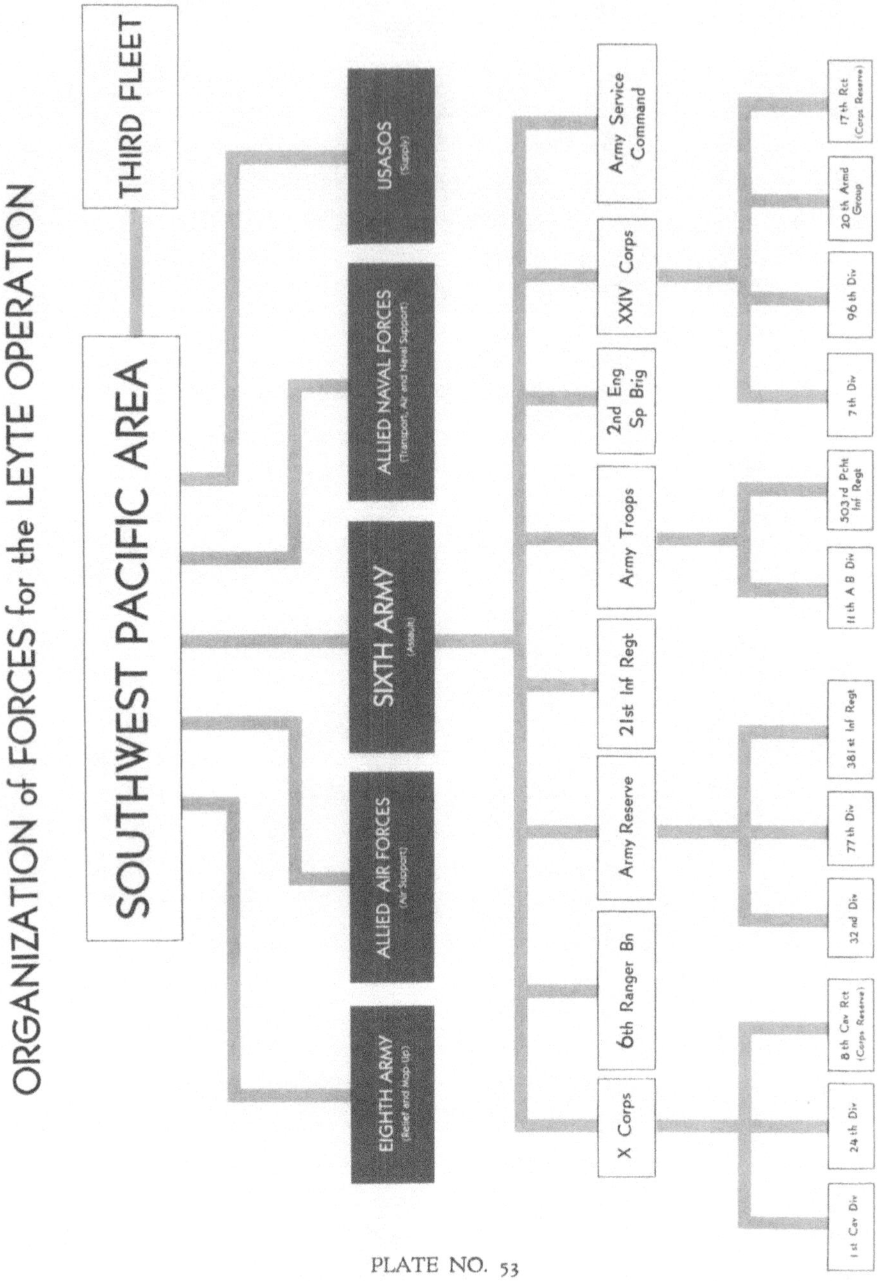

PLATE NO. 53

Organization of Forces for the Leyte Operation

The convoy, still loaded as for the Yap operation, left Hawaii on 15 September, resupplied at Eniwetok Atoll in the Marshall Islands, and then proceeded to the Admiralties where some changes were made in the loading dispositions. The LST group left Manus Island on 11 October and the remainder of the convoy, consisting of the faster transports, departed on the 14th.[31]

Both X Corps and XXIV Corps were under the Sixth Army, commanded by General Krueger who had led the American ground combat echelons of the Southwest Pacific from New Britain to Morotai with unvarying sucess.[32] Task Forces 78 and 79 were controlled by Admiral Kinkaid, Commander of the Allied Naval Forces in the Southwest Pacific Area and of the United States Seventh Fleet. In addition to responsibility for these two task forces, Admiral Kinkaid also had the over-all command of Task Force 77, the Central Attack Force. The battleships and cruisers of Task Force 77 were under Rear Adm. J. B. Oldendorf while the escort carriers were led by Rear Adm. T. L. Sprague. Many of these warships had seen long service and were borrowed for the occasion from units of the Central Pacific.[33]

Additional Fleet and Air Support

The size and scope of the Leyte assault demanded the full utilization of all available resources in the Pacific Theater. Accordingly, Admiral Halsey's powerful Third Fleet was called upon to assist in the operation by providing tactical and strategic support and by protecting the forces involved in the amphibious landings. The Admiral's flag covered four great striking groups, each consisting of fast carriers, cruisers, destroyers and the newest American battleships. These ships were to cruise as far as the Ryukyus and Mindanao, striking heavy blows against Japanese installations, planes, and surface craft whenever the opportunity arose in order to weaken the enemy's expected counteractions. Since the Leyte landing beaches were beyond the range of land-based fighter protection, carriers of the Third Fleet were scheduled to be back in position before 20 October to support the assault phases of the operation and to cover the ground troops. Additional air support of

31 COM3rdPHIBFOR (CTF79) Report to COM7thFLT, Serial 00323, 13 Nov 44. (S).

32 The major units operating under the Sixth Army in the Leyte assault were as follows:

Unit	*Commander*	*Phase*	*Staging Point*	*Landing Point*
Sixth Army	*Gen. Walter Krueger*			
6th Ranger Bn	*Lt. Col. H.A. Mucci*	*Assault*	*Finschhafen*	*Leyte Gulf Islands*
21st RCT	*Lt. Col. F.R. Weber*	*Assault*	*Finschhafen*	*Panoan Island*
2nd ESB	*Brig. Gen. W.F. Heavey*	*Follow-up*	*Hollandia, Biak, Manus*	*Leyte*
32nd Div	*Maj. Gen. W.H. Gill*	*Follow-up*	*Hollandia, Morotai*	
77th Div	*Maj. Gen. A.D. Bruce*	*Reserve*	*Guam*	
381st RCT	*Col. M.E. Halloran*	*Assault*	*Manus*	
X Corps	*Maj. Gen. Franklin C. Sibert*			
1st Cav Div	*Maj. Gen. V.S. Mudge*	*Assault*	*Manus*	*Leyte*
24th Div	*Maj. Gen. F.A. Irving*	*Assault*	*Hollandia*	*Leyte*
XXIV Corps	*Maj. Gen. John R. Hodge*			
7th Div	*Maj. Gen. A.V. Arnold*	*Assault*	*Oahu*	*Leyte*
96th Div	*Maj. Gen. J.L. Bradley*	*Assault*	*Manus*	*Leyte*
20th Armd Gp	*Lt. Col. W.A. Jensen*	*Assault*	*Hawaii, Manus*	
11th A/B Div	*Maj. Gen. J.M. Swing*	*Follow-up*	*Oro Bay*	
503rd Prcht RCT	*Col. G.M. Jones*	*Follow-up*	*Noemfoor*	

33 CG X Corps Report, *A History of the X Corps in the Leyte—Samar—Philippine Island Operation*, 15 Feb 45, G-3, GHQ, SWPA *Journal* (S).

the landing and convoy cover was to be provided by escort carriers of the Seventh Fleet.[34]

This naval air strength was coordinated with General Kenney's Allied Air Force which was assigned a secondary role—the protection of the rear and the southern flank until airfields could be developed on Leyte. With the completion of new bases, General MacArthur's air umbrella would move forward once more to shelter his troops, but in the meantime the Fifth and Thirteenth Air Forces were to hit targets on Mindanao and in the Netherlands East Indies from their bases in New Guinea and Morotai. Working in close conjunction with the Thirteenth Air Force, the Royal Australian Air Force Command of the Allied Air Force, under Air Vice-Marshal William D. Bostock, was to protect the rear and assist in the preliminary neutralization of Japanese installations in the Netherlands East Indies. To further the offensive effort, search planes of the Allied Air Force were assigned to cover the eastern areas of the Indies, most of the Philippine Archipelago, and parts of the South China Sea in strikes against enemy air bases.[35]

Operations Instructions

The operations instructions detailing the responsibilities and missions of the forces involved were published on 21 September. They are characteristic of the GHQ orders that moved complex armies along an advance of thousands of miles.

GENERAL HEADQUARTERS
SOUTHWEST PACIFIC AREA

A. P. O. 500
21 September 1944.

OPERATIONS INSTRUCTIONS
NUMBER 70

1. a. See current Intelligence Summaries and Annex No 3—Intelligence.

b. Allied forces occupy the line: Marianas—Ulithi—Palau—Morotai and control the approaches to the southern and eastern Philippines.

c. The Third Fleet, Admiral W. H. Halsey commanding, covers and supports the Leyte Gulf—Surigao Strait Operation by:

(1) Containing or destroying the Japanese Fleet.

(2) Destruction of hostile air and shipping in the Formosa, Luzon, Visayas and Mindanao areas during the period A-9 through A-3 and from A Day through A+30 as necessary to maintain their continued neutralization.

(3) Destruction of ground defenses and installations and shipping in the objective and adjacent enemy supporting areas from A-2 until the escort carriers assume the mission of direct support.

(4) Providing direct support of the landing and subsequent operations by fast carrier aircraft as required.

d. Coordination of operations of Third Fleet and Southwest Pacific Naval and Air Forces will be published later.

e. I Time (Zone-9) or Z Time will be used during the operation.

---

34 GHQ, SWPA, Opn Instr No. 70, 21 Sep 44.

35 Ibid. Comdr AAF Letter No. 322 to CINCSWPA, 8 Oct 44, subj: AAF Plan for Support of King II, 385 Plan 21, G-3, GHQ, Admin. (S).

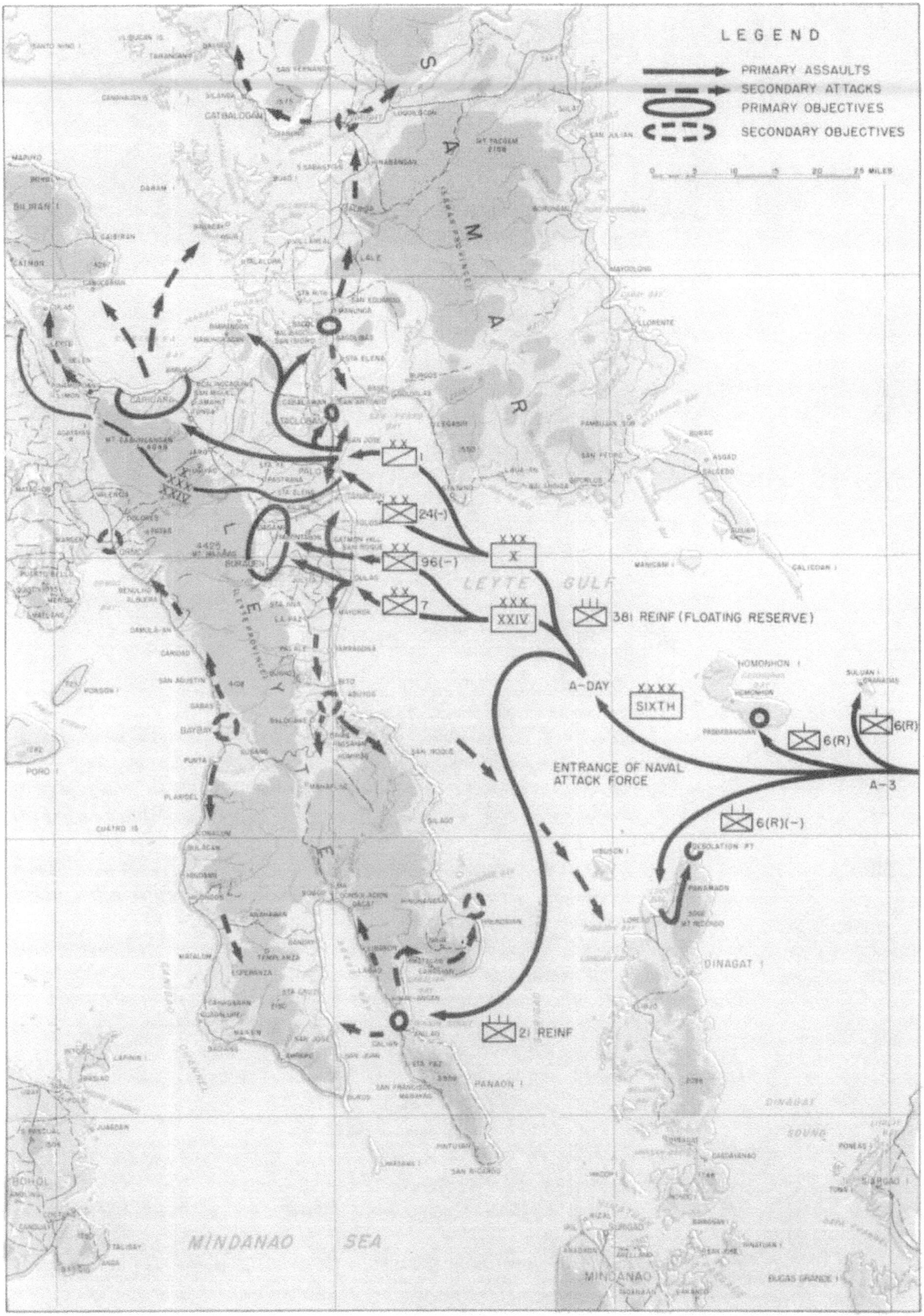

PLATE NO. 54

Plan of the Leyte Operation

2. *a.* *Forces of the Southwest Pacific, covered and supported by the Third Fleet, will continue the offensive to reoccupy the Philippines by seizing and occupying objectives in the Leyte and western Samar areas, and will establish therein naval, air and logistic facilities for the support of subsequent operations.*

b. *Target Date of A Day:* 20 *October 1944.*

c. *Forces.*

(1) *Sixth US Army—Lt General Walter Krueger, US Army.*

As constituted, less elements assigned by subsequent orders to Eighth US Army.

Sixth U. S. Army Reserve:

77th US Infantry Division—Guam.

6th US Infantry Division—Cape Sansapor, Dutch New Guinea.

Forces allocated for the operation as designated in Annex No 1. Tentative Troop List for the Operation. The exact composition of the landing force as designated by Commanding General Sixth US Army.

(2) *First Australian Army—Lt General V. A. H. Sturdee, CBE*

As constituted.

(3) *Eighth US Army—Lt General Robert L. Eichelberger, US Army.*

As later specified.

(4) *Allied Naval Forces—Vice Admiral T. C. Kinkaid, US Navy.*

As reinforced.

(5) *Allied Air Forces—Lt General George C. Kenney US Army.*

As constituted.

(6) *USASOS—Maj General J. L. Frink, US Army.*

As constituted.

3. *a.* *The Sixth US Army, supported by the Allied Naval and Air Forces, will:* (Plate No. 54)

(1) *By overwater operations seize and occupy:*

(*a*) *Objectives in the Tacloban and Dulag areas in Leyte and such adjacent areas as are required to initiate and insure uninterrupted naval and air operation therefrom.*

(*b*) *Objectives in the Homonhon and Dinagat Islands and such adjacent areas prior to the main assault in Leyte as will insure uninterrupted access for amphibious shipping into Leyte Gulf.*

(*c*) *Objectives in the Panaon Strait area that will permit passage of naval forces through the Panaon Strait for operations in the Camotes Sea. This objective will be secured simultaneously with* (1) (*a*) *above.*

(2) *Establish control of San Juanico Straits in order to permit passage of naval forces through the San Juanco Straits for operations in the Samar Sea.*

(3) *In subsequent operations, establish control over the remainder of Leyte Island; occupy and consolidate the western portion of southern Samar to include the Taft-Wright Highway and seize objectives that will permit opening of Surigao Straits for naval operations.*

(4) *Prepare to conduct such operations as may be later directed by this headquarters to:*

(*a*) *Complete the consolidation of Samar.*

(*b*) *Destroy or contain hostile garrisons in the Visayas.*

(5) *Occupy and defend sites for radar and air warning installations as arranged with the Commanders Allied Naval and Air Forces.*

(6) *Assume control of and direct the operations of Filipino Forces of the 9th Military District (Leyte—Samar).*

(7) *Establish facilities for minor naval operations at the earliest practicable date in the Leyte—*

Samar area as arranged with the Commander Allied Naval Forces and initiate the establishment of naval, air and logistic facilities for the support of subsequent operations to reoccupy the Philippines as directed in Annex No 4, Logistics, and Annex No 6, Engineer, and as later directed by this Headquarters.

(8) *Establish air facilities in the Leyte area with objectives as follows:*

(*a*) *First Objective:* *Immediately following the assault and by A+5 for:*	*1 fighter gp (P-38)* *1 fighter gp (P-40)* *1 night fighter sq*
(*b*) *Second Objective:* *Additional by A+15.*	*1 tactical reconnaissance sq* *1 photo sq* *1 medium bomb gp plus 1 sq P. O. A.* *3 PBY sqs (tender-based)* *1 VMR sq (Marine)*
(*c*) *Third Objective:* *Additional by A+30.*	*2 light bomb gps (A-20)* *1 air-sea rescue sq* *1 tactical reconnaissance sq* *1 fighter gp (P-38)*
(*d*) *Fourth Objective:* *Additional by A+45.*	*1 fighter gp (P-47)* *1 PB4Y sq (Air Ech)* *2 heavy bomb gps* *1 LAB sq*
(*e*) *Fifth Objective:* *Additional by A+60.*	*1 photo sq (F-5)* *1 PB4Y sq (Air Ech)* *2 troop carrier gps* *1 combat mapping sq (Air Ech)*

(*f*) *As later designated.*

The Commanding General Eighth US Army, supported by the Allied Naval and Air Forces will:

(1) *Relieve the Sixth Army of missions in New Guinea, the Admiralties, New Britain, and the Morotai area as later directed by this headquarters.*

(2) *Prepare to relieve the Sixth US Army in the Visayan area as later directed by this Headquarters.*

(3) *Assist the Commanding General Sixth US Army by training, staging and mounting units of the Sixth US Army in the Eighth US Army area of responsibility as arranged with the Commanding General Sixth US Army.*

c. *The First Australian Army, supported by the Allied Naval and Air Forces, will continue:*

(1) *The defense of naval and air installations within assigned areas of combat responsibility.*

(2) *The neutralization of Japanese forces within assigned areas, seizing every opportunity for the destruction of hostile forces.*

d. *The Commander Allied Naval Forces, while continuing present missions, will:*

(1) *Transport and establish landing forces ashore in Leyte Gulf—Surigao Strait areas as arranged with the Commanding General Sixth US Army.*

(2) *Support the operation by:*

(*a*) *Providing air protection for convoys and naval task forces and direct air support for the landing and subsequent operations, supplemented as arranged with the Commander Third Fleet and the Commander Allied Air Forces.*

(*b*) *Arranging direct air support and cover with carrier aircraft for minesweeping and preliminary landings in the Leyte Gulf area during the period A-2 to the time escort carriers assume the mission of direct support on A Day.*

(*c*) *Transporting supporting troops and their supplies as required to the Leyte Gulf—Surigao Strait area in naval assault shipping.*

(*d*) *Denying Japanese reinforcement of the Leyte area from the Samar, western Visayas and northeastern Mindanao areas.*

(*e*) *Clearing the Surigao Strait area of hostile naval forces and shipping and sweeping the Surigao Strait to open it for naval operations and shipping in the Camotes Sea and adjacent waters, in conjunction with operations of the Sixth US Army.*

(*3*) *Provide submarine offensive reconnaissance along probable routes of hostile naval forces and of waterborne reinforcements and supplies.*

(*4*) *Provide lifeguard services as required.*

(*5*) *Transfer to the Commander Allied Air Forces the mission of direct air support when land-based fighters and light bombers are established in the Leyte area, at a time as arranged with the Commander Allied Air Forces.*

(*6*) *Escort and protect shipping on the lines of communication into the Leyte and Samar areas.*

(*7*) *Establish in Visayan waters, naval forces required to support current and future operations.*

e. *The Commander Allied Air Forces, while continuing present missions, will:*

(*1*) *Support the operation by:*

(*a*) *Providing aerial reconnaissance and photography as required.*

(*b*) *Neutralizing, in coordination with carrier and land-based aircraft of the Third Fleet, hostile naval and air forces in areas within range in the Philippine Archipelago, intensifying the neutralization in the western Visayas and Mindanao areas from D-9 to cover the movement of naval forces, the landing and subsequent operations. (Coordination of air operations of the Third Fleet and Southwest Pacific Air Forces will be published later).*

(*c*) *Providing protection of convoys and naval forces and direct support of the landing and subsequent operations within capabilities and as required by Commander Allied Naval Forces.*

(*d*) *Assuming the mission of direct support of the operations in the Leyte—Samar area at the earliest practicable date after the establishment of fighters and light bombers in the Leyte area, as arranged with the Commander Allied Naval Forces.*

(*2*) *Continuing the destruction of hostile naval and air forces and shipping in the Arafura and Celebes Sea areas and by initiating strikes on northeastern Borneo and the Sulu Archipelago at the earliest practicable date; denying use of naval facilities in the Sulu Archipelago to the Japanese and protecting the western flank of the operation.*

(*3*) *Destroying hostile installations and sources of war materials in Eastern Netherlands East Indies.*

(*4*) *Establishing and operating radar and air warning facilities as required in the Leyte—Samar area, as arranged with the Commanding General Sixth US Army.*

(*5*) *Establishing air forces in the Leyte area in the priority as listed in paragraph 3a (8).*

x. (*1*) *For the coordination of planning the Commander Allied Naval and Air Forces will cause their respective close support commanders to report to the Commanding General Sixth US Army, who is charged with the coordination of plans.*

(*2*) *A brief of the coordinated plan of operations will be furnished this headquarters by the Commanding General Sixth US Army by 5 October 1944.*

(*3*) *Commanders Allied Naval and Air Forces, Southwest Pacific Area, will submit to this headquarters by 1 October 1944 their respective plans for general support to be afforded by their forces during the period of operations.*

(4) *During the amphibious movement and landing, the Commander Naval Attack Forces is in command of the amphibious operations; his command continues until the landing forces are established ashore. Command of the forces ashore is then passed to the Landing Force Commanders. The exact time of transfer of command from the Commanders Landing Forces will be announced by radio. The controlling considerations for fixing the time when the landing forces are established ashore will be as agreed by the Commander Allied Naval Forces and the Commanding General Sixth US Army, and will be announced by them to this headquarters and appropriate subordinates.*

(5) *For coordination of land-based and naval aircraft in support of the operation, see Standing Operating Procedure Instructions Number 16/1, this headquarters dated 10 August 1944.*

(6) *To coordinate the attack of Third Fleet carrier aircraft, the Commander Allied Naval Forces, in concert with the Commander Allied Air Forces and Commanding General Sixth US Army, will furnish the Commander Third Fleet at the earliest practicable date the following:*

(*a*) *Schedule and tracks of echelons.*

(*b*) *Target maps of air and surface bombardment.*

(*c*) *Communication plans.*

(*d*) *Naval gunfire plans.*

(*e*) *Other plans and data necessary for the support of the operation by the fast carrier forces.*

(7) *Areas of responsibility for naval and air operations of the Third Fleet and Southwest Pacific Forces will be designated later.*

(8) *Instructions for long range reconnaissance, and bombing and attack restrictions will be issued in subsequent Operations Instructions.*

(9) *Annex No 1 indicates the tentative troops list for the operation, and Annex 2 indicates the troop movements for the concentration.*

4 *See Annex No 4—Logistics. (to be issued later)*

5\. *a.* *See Annex No 5—Communications.*

b. *Command Posts.*

Pacific Ocean Areas—Hawaii
Third Fleet—Afloat
General Headquarters, Southwest Pacific Area—Hollandia
    Rear Echelon—Brisbane
    Advanced Echelon—Leyte (date and hour of opening to be announced later)
Sixth US Army—Leyte (as announced by Commanding General Sixth US Army)
    Rear Echelon—Hollandia
First Australian Army—Lae
Eighth US Army—Hollandia
Allied Naval Forces—Hollandia
    Rear Echelon—Brisbane
Allied Air Forces—Hollandia
    Rear Echelon—Brisbane
United States Army Services of Supply—Hollandia
    Rear Echelon—Brisbane

By command of General MacArthur:

R. K. SUTHERLAND
Lieutenant General, U. S. Army,
Chief of Staff.

OFFICIAL:

/s/ S. J. Chamberlin
S. J. CHAMBERLIN
Major General, G. S. C.,
Asst. Chief of Staff, G-3

ANNEXES: *(Omitted)*

Logistical Difficulties

The date for the invasion of the central Philippines had been advanced to strike the Japanese while their defense preparations were still generally incomplete. The original December target date would have given the enemy two more months to strengthen his positions and redeploy his forces.[36] By seizing the opportunity to strike him at a disadvantage, however, the already considerable logistic problems inherent in operations conducted over great water distances were further complicated by the fact that the change in schedule came only thirty-four days prior to the new assault date.

The cancellation of the landings on Mindanao which were originally intended to provide land-based air cover for the operation, altered the whole system of logistical support. These initially scheduled landings were to have been supplied from New Guinea bases. The Leyte operation, however, was partially supported directly from ports on the west coast of the United States. Coordination of the loading of supplies and equipment in two such distant areas necessitated the utmost care and timing.

An added logistic difficulty was presented with the decision to employ the XXIV Corps in the Leyte operation. This corps was a Pacific Ocean Areas corps and, at the time, was actually aboard ship ready to sail for the operation against Yap which was cancelled. Shipping for the support of this Corps had been loaded to accord with the plans of the Pacific Ocean Areas, which did not fit well into the plans of the Southwest Pacific for the support of the Leyte operation. This shipping was in fact suited only for the specific support of the XXIV Corps, and it lacked the flexibility of standard loading, employed by the Southwest Pacific, which would have made it suitable to support any troops without regard to type or composition. Utilization of the XXIV Corps shipping during the critical phases of the operation was extremely difficult, and for a time its use was postponed in favor of the more flexible standard-loaded ships.

Japanese Forces in the Philippines

An examination of Japanese shipping routes and facilities for supply and reinforcement disclosed substantial transport traffic between Japan proper and the Philippines. While their shipping potential was considerably reduced from the level of the previous year, the Japanese were still in a position to give ready

36 General Utsunomiya, Assistant Chief of Staff, Fourteenth Area Army, declared: "The far-sightedness in launching the Leyte Campaign two months earlier than the original schedule was one of the principal factors in its success." Maj. Gen. Yoshiharu Tomochika, Chief of Staff, Thirty-fifth Army in the southern Philippines, said that the advance of the Leyte invasion date caught the Japanese with their "defensive fortifications incomplete." He continued with the statement: "This American invasion, two months in advance of our estimation, in my opinion, was one of the basic reasons for the rapid collapse of our defenses in the Philippines." Interrogation Files, G-2 Historical Section, GHQ, FEC.

supply and reinforcement to their forces in the Philippines area.[37]

According to best reports, Japanese strength in the Philippines on 30 September 1944 included approximately 224,000 ground troops in addition to air personnel, improvised labor units, and civilians.[38] Reinforcements, moreover, were constantly arriving to bolster their positions. During the month of October the estimate of enemy strength was increased by roughly 50,000 men, the calculated increment of fresh troops and reinforcements.[39] This increase reflected continued preparations by the Japanese to strengthen their Philippine bastion. Evidence indicated that they were doing everything possible to meet the coming Allied assault on the Philippines which they considered increasingly imminent.

Headquarters of the Japanese Southern Army was located at Manila and the Thirty-Fifth Army had its Headquarters at Cebu. Under their command were some nine or ten divisions of ground troops, plus auxiliary units situated chiefly in Luzon and Mindanao. (Plate No. 55) This distribution afforded the Allies an excellent opportunity to drive a wedge between the two islands and divide the enemy forces. Air and naval personnel were scattered at various positions in the archipelago with the largest concentrations located in Luzon. Naval forces maintained mine fields at all important water approaches and inland sea areas which were considered likely points for Allied thrusts.

The Filipinos in the Plan of Operations

The invasion of the Philippines would bring the long-awaited day of liberation to 17,000,000 Filipinos. While anxious for an early defeat of the Japanese forces throughout the Islands, General MacArthur insisted on a careful selection of attack objectives and minimum destruction to Filipino life and property. He realized that here was a great moral issue involving a friendly civilian population :

> *One of the purposes of the Philippine campaign is to liberate the Filipinos ; they will not understand liberation if accomplished by indiscriminate destruction of their homes, their possessions, their civilization and their own lives ; humanity and our moral standing throughout the Far East dictate that the destruction of lives and property in the Philippines be held to a minimum, compatible with the assurance of a successful military campaign ; indications are that in some localities the Japanese are evacuating cities, leaving Filipinos in residence, either failing to warn them or compelling them to stay ; aerial bombing causes the greatest destruction ; our objective in areas we are to occupy is to destroy totally hostile effort in order to insure our own success ; in other areas we neutralize, to weaken any hostile effort which may tend to increase resistance to our occupation objectives ; in the latter areas, our attack objectives are primarily airfields and shipping, not metropolitan areas or villages or barrios ; to the extent possible, we must preserve port facilities that we plan to use. The Commander Allied Air Forces will, and CINC-POA is requested, to issue general instructions in*

---

37 G-2, GHQ, SWPA, *Monthly Summary of Enemy Dispositions*, 30 Sep 44.

38 Ibid. Admiral Kinkaid commented on the Allied intelligence estimates as follows : " By virtue of the excellent work of our guerrilla forces in Leyte and our coastwatcher personnel, we knew within very close limits before we undertook the attack exactly what enemy elements were in occupation, their capacities, and capabilities. The estimates of enemy air and land strength and of the naval forces available to him were remarkably exact and were made possible only by the careful study and evaluation of reports received from a wide variety of sources over a considerable period of time." Commander Central Philippines Attack Force (COM7thFLT), *Report of Operation for the Capture of Leyte Island including Action Report of Engagement in Surigao Strait and off Samar Island on 25 October 1944 (King Two Operation)*, 31 Jan 45, A16-3, Serial : 00302-C.

39 G-2, GHQ, SWPA, *Monthly Summary of Enemy Dispositions*, 31 Oct 44.

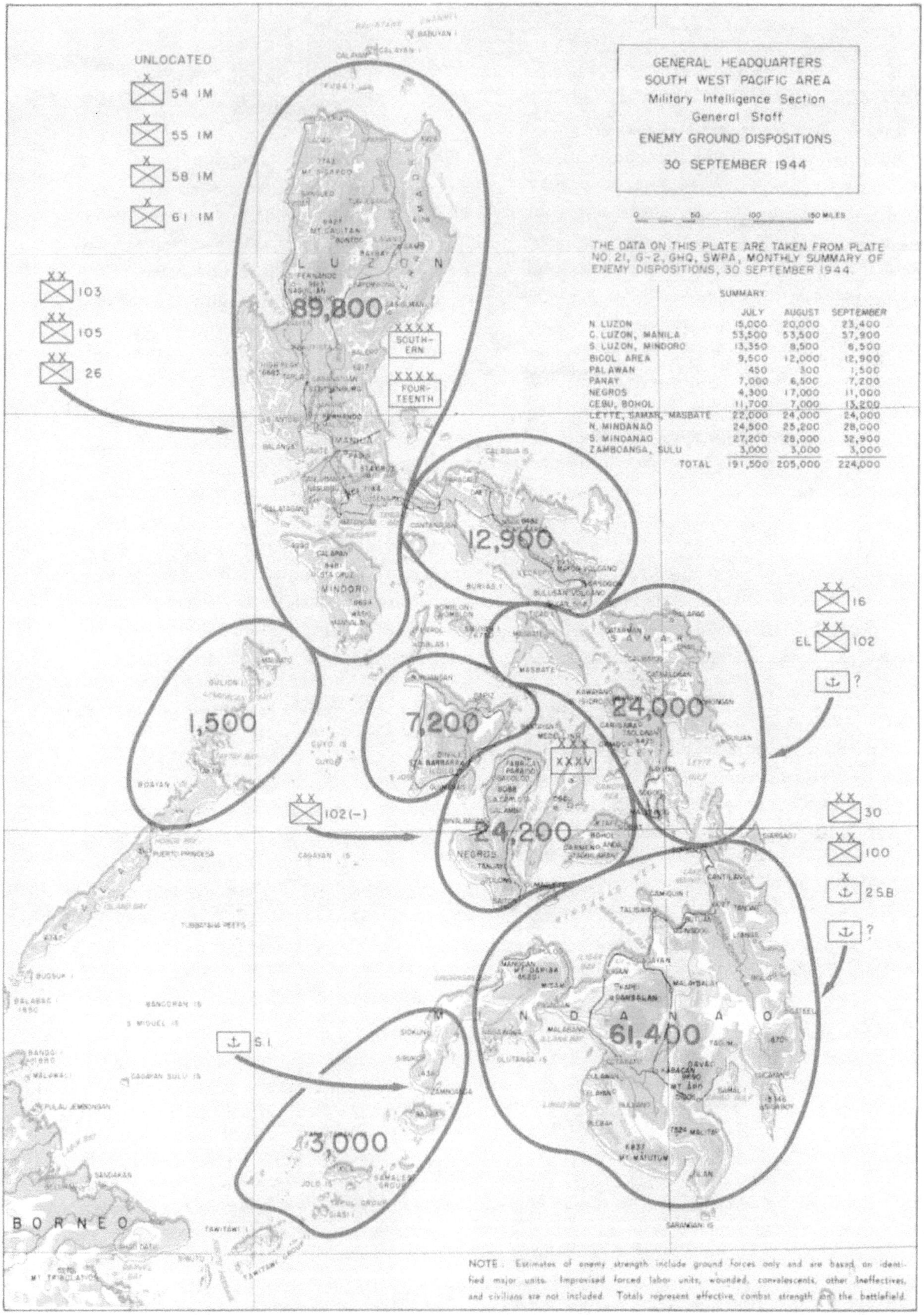

PLATE NO. 55

Enemy Ground Dispositions, 30 September 1944

consonance with the above objective of minimizing destruction of life and property of Filipinos. . . .[40]

Aside from the matter of physical destruction of Filipino life and property, General MacArthur was well aware of the importance of restoring to the Philippines a degree of freedom at least equal to that in existence before 1942. In a letter to Maj. Gen. John Hilldring, Director of the Civil Affairs Division of the War Department, General MacArthur stated:

> *It is essential . . . in any plans for the control of civil affairs that the measure of freedom and liberty given to the Filipino people be at least comparable to that enjoyed under the Commonwealth Government before Japanese occupation. It would be a matter of gravest concern if restrictions were imposed, whether by direct or by individual means, in excess of those existing before the war. If any impressions were created that the United States is curtailing rather than expanding liberties, the most unfortunate repercussions might be expected. The only restrictions which should be imposed are the minimum required by military necessity and these should be removed as quickly as possible. . . .*
>
> *I repeat, utmost care should be taken that an imperialistic attitude not be introduced into the situation under the guise of military operations and necessity. This would be entirely alien to the spirit of recent legislation passed by Congress and the announced purposes of the President of the United States.*[41]

Preliminary Air Bombardment

As the Leyte invasion date approached, the Allies increased the intensity of their air raids against important Japanese positions. Planes of the Far Eastern Air Force and the Third Fleet ranged over the western Pacific from Borneo to the Bonins and continued to pound the enemy's ground and harbor installations. The Seventh Air Force, from freshly built bases on Peleliu Island, searched the Palau area and raided the Bicol Provinces of Luzon. The Fourteenth Air Force, based in China, joined in the all-out assault with intermittent blows against Hong Kong and Formosa, in diversionary operations.

From 10–16 October, the U. S. Third Fleet launched its planes against Okinawa and Formosa in the greatest carrier strike of the war. The Japanese reacted violently to this intrusion into their inner perimeter and expended their planes on a lavish scale to counter the Allied attack. The resultant clash was one of the greatest air battles between carrier and land-based air power of the Pacific War. The Allied forces won a telling victory in the Formosan strike and inflicted great losses on the Japanese both in planes and in trained pilots.[42] Without a pause, successive attacks were then directed against Luzon and the Visayas as the Allied invasion forces put the finishing touches on their amphibious preparations.

In addition to the actual physical destruction inflicted on the Japanese in the Formosa attack, the Third Fleet's carrier raid had other important and far-reaching results. Formosa was an important arsenal for the supply and repair of the enemy air forces assigned to the protection of the Philippines, and the damage effected by the Allied strike seriously hampered the maintenance of subsequent air operations in Leyte and Luzon. Then, too, the approach of Admiral

---

40 CINCSWPA Radio, to Com AAF and CINCPOA, 2 Sep 44, SOPAC No. 517 C/S GHQ. (S).

41 CINCSWPA Ltr to Maj. Gen. John Hilldring, 2 Sep 44, WD No. 821, CS, GHQ (S).

42 The official communique of Japanese Imperial General Headquarters for 19 October 1944 publicly admitted the loss of 312 aircraft, in addition to an unspecified number of planes damaged, during the period 12–16 October. Allied Air Force estimates placed the figure at 396 Japanese planes of all types destroyed. In addition 35 ships of various sizes were reported as sunk and 74 ships as probably damaged. AAF *Intelligence Summary* No. 245, "Result of the Allied Air Attack on Formosa on 12–13 Oct 44," 15 Oct 44, G-3, GHQ, SWPA *Journal* (S).

Halsey's Third Fleet so close to Formosa led the Japanese High Command into a series of strategical errors.[43] The prospect of finishing off a major portion of the U. S. task force was too inviting to resist. As a result, the Japanese diverted 150 planes and the last of the trained pilot strength of Adm. Jisaburo Ozawa's Third Fleet to accomplish this foredoomed mission.[44] This attacking force was almost entirely destroyed, stripping the Japanese of any effective carrier plane power for use against the Allied landings in Leyte and influencing their defensive naval strategy in the Philippines. At this time, too, a naval task force under Adm. Kiyohide Shima was detached on the same futile task of destruction, additionally denuding Admiral Ozawa's fleet of cruiser and destroyer strength which it badly needed. This task force, instead of fulfilling its original mission as part of the Japanese Third Fleet's striking arm, was destined to play a lone and insignificant role in the forthcoming battle for Leyte Gulf. The subsequent enemy exaggerated damage reports caused the Japanese to reassess the probable date of the coming Allied invasion of the Philippines. Figuring that it was impossible for the Allies to replace their losses and repair the estimated damage to their fleet units before the end of November, the Japanese thought they had at least another month in which to strengthen their defensive measures in the Philippines.[45]

The "softening-up" operations were now finished and there remained only the final shore bombardment prior to the actual landing. Well over 1,000 enemy aircraft of all types had been destroyed by Allied air and naval forces during the preceding two weeks. Vast amounts of enemy shipping had been sunk or damaged. The enemy's most likely sources of reinforce-

43 The extent of Japanese exaggeration regarding damage inflicted upon the U. S. Fleet can be appreciated by an examination of their official reports at the time. A communique issued by Imperial General Headquarters on 19 October reads as follows: "Since 12 October our forces have been engaged in fierce attacks against enemy task groups in the seas east of Luzon and Formosa. We have routed these forces and destroyed over half their strength. Total results achieved are as follows: *Sunk:* 11 carriers, 2 battleships, 3 cruisers, 1 destroyer or cruiser. *Damaged:* 8 carriers, 2 battleships, 4 cruisers, 1 cruiser or destroyer and 13 ships of undetermined size. In addition at least 12 ships were left in flames."

44 Admiral Ozawa, Commander in Chief of the Japanese Third Fleet in the Battle for Leyte Gulf, stated, "... about 150 planes from the [Third Fleet] carriers were sent to Formosa; consequently our carrier strength was greatly reduced so the operation was changed to use land-based planes more frequently instead of carrier planes. My force of carrier planes became very much weakened; only 110 were left, so less than half remained." United States Strategic Bombing Survey, Naval Analysis Division, *Interrogation of Japanese Officials*, Vol I, p. 220.

45 Vice Adm. Shigeru Fukudome, Commander of the Second Air Fleet at the time of the Leyte landings, referring to the Japanese "success" in the Formosa raid, stated, "... while the results attained were probably not as great as reported at the time, I felt that considerable success had been attained and hence expected that some time would lapse before you [the Allies] would undertake the attack further south. However, your [Allied] thrust against the Philippines came much sooner than expected." USSBS, *Interrogations of Japanese Officials*, Vol II, p. 501. Lt. Col. Yorio Ishikawa, Staff Officer of the Fourteenth Area Army, also pointed out the effect of the magnified damage reports, saying, "It had been a prevalent conception among high commands that the United States was not thoroughly prepared to start the Philippines invasion when their landings actually took place on Leyte. The reason for this was that our navy reported crippling the enemy task forces in the air and sea battle off the shores of Formosa prior to the American landings on Leyte. The fact remained that our naval reports had been much exaggerated from the actual battle results and consequently we misjudged the potential move on the Philippines, thinking that the enemy was starting his operations without thorough preparations." General Tomochika stated, "We had estimated that approximately one half of the vaunted American air striking power had been crippled severely enough to prevent them from participating in the Philippines campaign." Interrogation Files, G-2 Historical Section, GHQ, FEC.

ment for planes and men—Luzon, Formosa, and the Ryukyus—had been severely pounded. Whether Japanese communications had been sufficiently crippled to prevent an effective reinforcement of the troops on Leyte would soon be revealed.

Guerrilla reports and other intelligence sources had given the Allies a good picture of what they would meet after their landing, but the enemy, on the other hand, had little precise information concerning Allied invasion plans. From the standpoint of strategy the Japanese calculated that the Allies would strike the Philippines soon after the Morotai and Palau landings. Tactically, however, they did not know precisely where, when, or in what strength the blow would fall. Estimates varied considerably both as to time and place. It was generally anticipated that the Allies would invade Mindanao first, then the central Philippines, and finally Luzon.[46]

The stage was set and the Allies were ready. In the Philippines, the people were eager to assist in the long-awaited battle for their liberation. After two and one half years of planning and fighting, General MacArthur could now fulfill his promise and reopen the gates of the Philippines to freedom.

---

46 The following statements are typical of the testimony given by Japanese Army commanders regarding the expected invasion of the Philippines: "There were various opinions concerning the anticipated date of the initial landings in the Philippines, both in the Thirty-Fifth Army and in the Fourteenth Area Army Command", said General Tomochika. "Some of the staff officers were of the opinion that the enemy would come by land via New Guinea to Mindanao Island utilizing the same type of invasion technique he had so successfully employed during the New Guinea campaign. They contended that the enemy by establishing air bases on the way would probably land on Mindanao by early November...." Maj. Kazuo Taguchi, Fourteenth Area Army Staff officer, added the following: "Based on the order of their expectancy, I had thought the first landings would take place as follows: (1) the Sarangani Bay area, South Mindanao, (2) the Davao area, South Mindanao, (3) the Leyte Gulf coast area. Lt. Gen. Iinuma, Chief of Staff of the Southern Army, had also estimated the first landing to take place as mentioned above. Since southern Mindanao was the closest of the Japanese held areas and within easy range of their bombers based on Halmahera, I felt strongly that the initial attempt would be made in the area." Interrogation Files, G-2 Historical Section, GHQ, FEC. The United States Strategic Bombing Survey synthesized the divergence of opinion within the Japanese Naval High Command on the Philippine invasion as follows: "... by August, and particularly after the Palau and Morotai landings on 15 September, it was agreed in the top naval commands that the next Allied move would be against the Philippines.... It seems, however, that the anticipated timing of the assault and its location in the Philippines remained matters of contention among the command and staff officers until the Leyte operation actually commenced. Apparently, most believed that the assault would come in November, although there was strong opinion that it would be in early October.... The original prevailing opinion that Mindanao was the most likely place for a Philippines invasion persisted, although Samar, Leyte, and Luzon were considered a likelihood for the next Allied move." USSBS, *Japanese Military and Naval Intelligence*, pp. 56–57.

CHAPTER VIII

THE LEYTE OPERATION

Strategic Value of Leyte

The Leyte operation was to be the crucial battle of the war in the Pacific. On its outcome would depend the fate of the Philippines and the future course of the war against Japan. Located in the heart of the archipelago, Leyte was the focal point where the Southwest Pacific forces of General MacArthur were to converge with the Central Pacific forces of Admiral Nimitz in a mighty assault to wrest the Philippines from the hands of the enemy. (Plate No. 56) With Leyte under General MacArthur's control, the other islands would be within effective striking distance of his ground and air forces. Leyte was to be the anvil against which he would hammer the Japanese into submission in the central Philippines, the springboard from which he would proceed to the conquest of Luzon for the final attack against Japan itself. Military necessity demanded that the Allies achieve a decisive victory on Leyte. General MacArthur and Admiral Nimitz were committed to employ the maximum resources at their command.

It was axiomatic, too, that the Japanese would fight to the bitter end to save the Philippines. The war situation had grown progressively worse for Japan and by the autumn of 1944 her position was both critical and desperate. No longer were the Allies held at the outer periphery of her defense system. They were now poised with their full power at the very threshold of her inner structure and if they should break through, the Homeland itself would stand dangerously exposed—an inviting target for the next invasion. With the initiative entirely in Allied hands, the war had reached that decisive stage where another important Japanese defeat would seal the ultimate fate of Japan's Empire, and destroy a centuries-old tradition of invincibility. The Philippines would probably offer the last chance for the Japanese Army to recover its lost prestige and gain a victory over General MacArthur's forces. To achieve this victory, the Japanese battle fleet would undoubtedly come forth in full strength to throw back all invasion efforts. If the Japanese were to realize the rapidly diminishing hope of stemming the surging American tide, the decisive battle would have to be in the Philippines. The alternative of failure would be inevitable defeat. The decision had to be pressed.

Clearing Leyte Gulf

The plan for the ground operations in the capture of Leyte comprised four main phases. Phase One covered minor preliminary landings to secure the small islands lying across the entrance to Leyte Gulf. Phase Two included the main amphibious assaults on Leyte from Dulag to Tacloban and called for the seizure of the airstrip, an advance through Leyte Valley, and the opening of San Juanico and Panaon Straits. Phase Three consisted of the necessary overland and shore-to-shore operations to complete the capture of Leyte and the

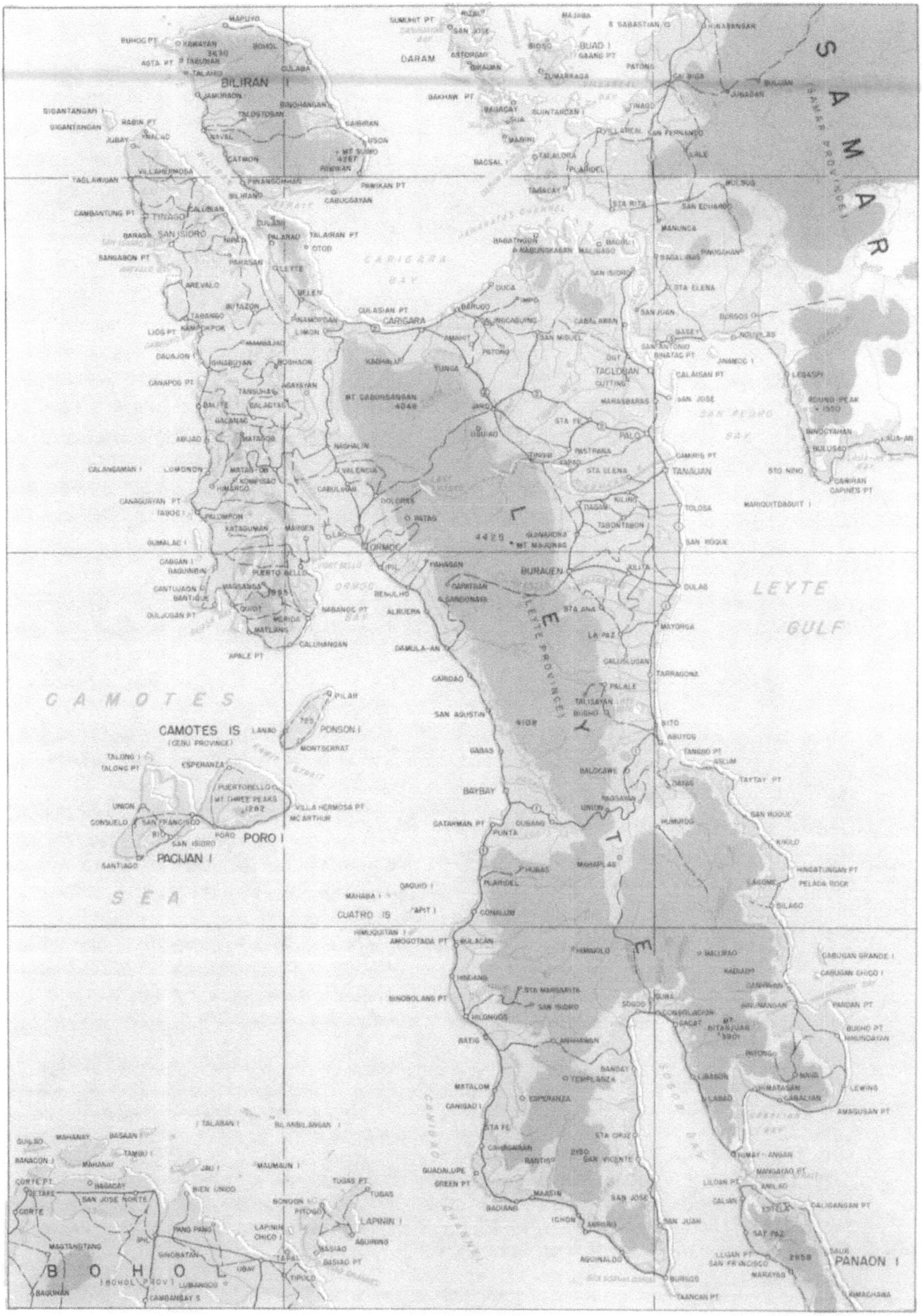

PLATE NO. 56

Leyte

seizure of southern Samar. Phase Four contemplated the occupation of the remainder of Samar and the further neutralization of enemy positions in the Visayas.

The first landings to mark the invasion of the Philippines were made on three small islands which guarded the eastern approaches to Leyte Gulf. Despite cyclonic storms and heavy seas, elements of the 6th Ranger Battalion, augmented by one company of the 21st Infantry, went ashore on Suluan and Dinagat Islands on 17 October 1944. (Plate No. 57) Heavy mists shrouded their approach and they were opposed only by the rough surf and battering winds. Homonhon Island was occupied the next day. All three islands were quickly cleared of small enemy garrisons and radio installations while mine sweepers and demolition teams co-operated to sweep the waters and the beaches of all obstacles potentially dangerous to the main operation.

The Return of MacArthur

On 20 October, the largest mass of naval assault craft and warships ever concentrated in the Pacific sailed boldly into Leyte Gulf. The landing beaches and tactically important rear areas had already been softened by a continuous two-day ship and plane bombardment.[1] After an additional morning barrage, the landing troops were ready to go ashore.

The main assault on the east coast of Leyte began at ten o'clock in the morning with landings along an 18-mile front between the two small villages of Dulag and San Jose. X Corps, comprising the 1st Cavalry and the 24th Divisions, covered the right flank of the landings to the north; XXIV Corps, consisting of the 7th and 96th Divisions, secured the left flank. Both shores of Panaon Strait at the southern tip of the island were seized by a single regimental combat team of the 21st Infantry which had gone ashore an hour prior to the main assault.

General MacArthur's stirring promise to return to the Philippines was fulfilled shortly after the main landings. In a drenching rain he strode ashore on the muddy beachhead near Palo, following close to the assault echelons, and heralded the coming hour of liberation. Speaking to millions of waiting Filipinos over a portable radio set, he declared:

This is the Voice of Freedom, General MacArthur speaking. People of the Philippines: I have returned. By the grace of Almighty God our forces stand again on Philippine soil—soil consecrated in the blood of our two peoples. We have come, dedicated and committed to the task of destroying every vestige of enemy control over your daily lives, and of restoring, upon a foundation of indestructible strength, the liberties of your people.

At my side is your President, Sergio Osmena, worthy successor of that great patriot, Manuel Quezon, with members of his cabinet. The seat of your government is now therefore firmly re-established on Philippine soil.

The hour of your redemption is here. Your patriots have demonstrated an unswerving and resolute devotion to the principles of freedom that challenges the best that is written on the pages of human history.

1 The severity of this bombardment accounted in great measure for the initial ease of the Allied landings. In addition to forcing the enemy from many of his beach entrenchments, it seriously disrupted his entire communication system. General Tomochika stated: "The positions of the 16th Division Artillery Regiment along the first line of defense were subjected to a severe pre-landing naval bombardment which resulted in the destruction of a great number of its field pieces. Regimental radio-telegraphic communications were disrupted by this bombardment, and direct communications with the 35th Army and with the 14th Area Army Headquarters in Manila were never re-established. Direct liaison between regiments and smaller units of the division could no longer be carried out effectively." 10th Information and Historical Service, HQ Eighth Army, *Staff Study of Operations of the Japanese 35th Army on Leyte*, Part I (R). Hereinafter cited as: Eighth Army, *Staff Study, Leyte*.

I now call upon your supreme effort that the enemy may know from the temper of an aroused and outraged people within that he has a force there to contend with no less violent than is the force committed from without.

Rally to me. Let the indomitable spirit of Bataan and Corregidor lead on. As the lines of battle roll forward to bring you within the zone of operations, rise and strike! For future generations of your sons and daughters, strike! In the name of your sacred dead, strike! Let no heart be faint. Let every arm be steeled. The guidance of Divine God points the way. Follow in His name to the Holy Grail of righteous victory![2]

Widening the Beachheads

Opposition at the landing beaches was negligible and first-day casualties resulted primarily from a few well-placed mortar and artillery pieces that remained unsilenced by the preliminary air and naval bombardment. In the San Jose sector, the X Corps advanced quickly and by midafternoon the 1st Cavalry Division, supported by tanks, had secured Tacloban Airfield, the most important early objective.[3]

At the same time, the 24th Division pressed forward, fighting its way inland in the area of the Palo—Tacloban Highway. (Plate No. 58) The objective in this sector was a height dominating the entire landing beach area near Palo labeled Hill 522. As this hill commanded both the highway system and the beaches, the Japanese had fortified it with trenches, caves, and cleverly concealed emplacements that would offer serious opposition to an attacking force. A slight but serious miscalculation on the part of the enemy, however, enabled the 24th Division to seize the hill with a minimum of loss. The severity of the naval, air, and artillery bombardment had forced the Japanese to leave their guns and take temporary refuge in the safer ground below. As soon as the shelling stopped, they began their move back up the slopes to reoccupy their commanding positions and open fire on the advancing Americans. They were too late. So swift was the progress of the invading forces, that troops of the 24th Division were already on the crest of the hill to meet and destroy the returning enemy.

In the Dulag sector, which was the responsibility of the XXIV Corps, the 96th Division advanced against stiffening resistance but by early afternoon it had carried its attack inland from 1,000 to 3,000 yards. On the left flank of the Corps, the 7th Division encountered little initial opposition after the landing and pushed steadily forward to secure Dulag airdrome on 21 October. The airfield, however, situated as it was in the flat, flood plain of the Marabang River, was not suitable for immediate use. Poor drainage, numerous small swamps, and the thick, sedimentary silt of the plain rendered the airstrips generally unusable until well toward the end of November.

In the X Corps sector, unloading was delayed by unfavorable terrain to the rear of the beach. Although sporadic enemy mortar and artillery fire hampered the fast discharge of LST cargo and landing of needed

2 General MacArthur's radio broadcast, 20 October 1944.

3 Although the Japanese had anticipated the landings at Dulag, they were not prepared for a direct assault on Tacloban and had even located their division headquarters there, thinking it would be well behind the battlefront. "We had misestimated the location of the initial enemy landings," said General Tomochika, "and consequently our defense in the area was very weak. We had estimated that there was a greater possibility of an enemy landing in the Dulag area since it was at the entrance to Leyte Gulf, instead of at Tacloban which was almost at the upper extreme end of the Gulf. The strategy employed by the enemy in landing at our weak spots can be attributed to the splendid intelligence system of the enemy, aided at times by the guerrilla agents who had infiltrated into our lines and had sent out vital information concerning our troop dispositions." Interrogation Files, G-2 Historical Section, GHQ, FEC.

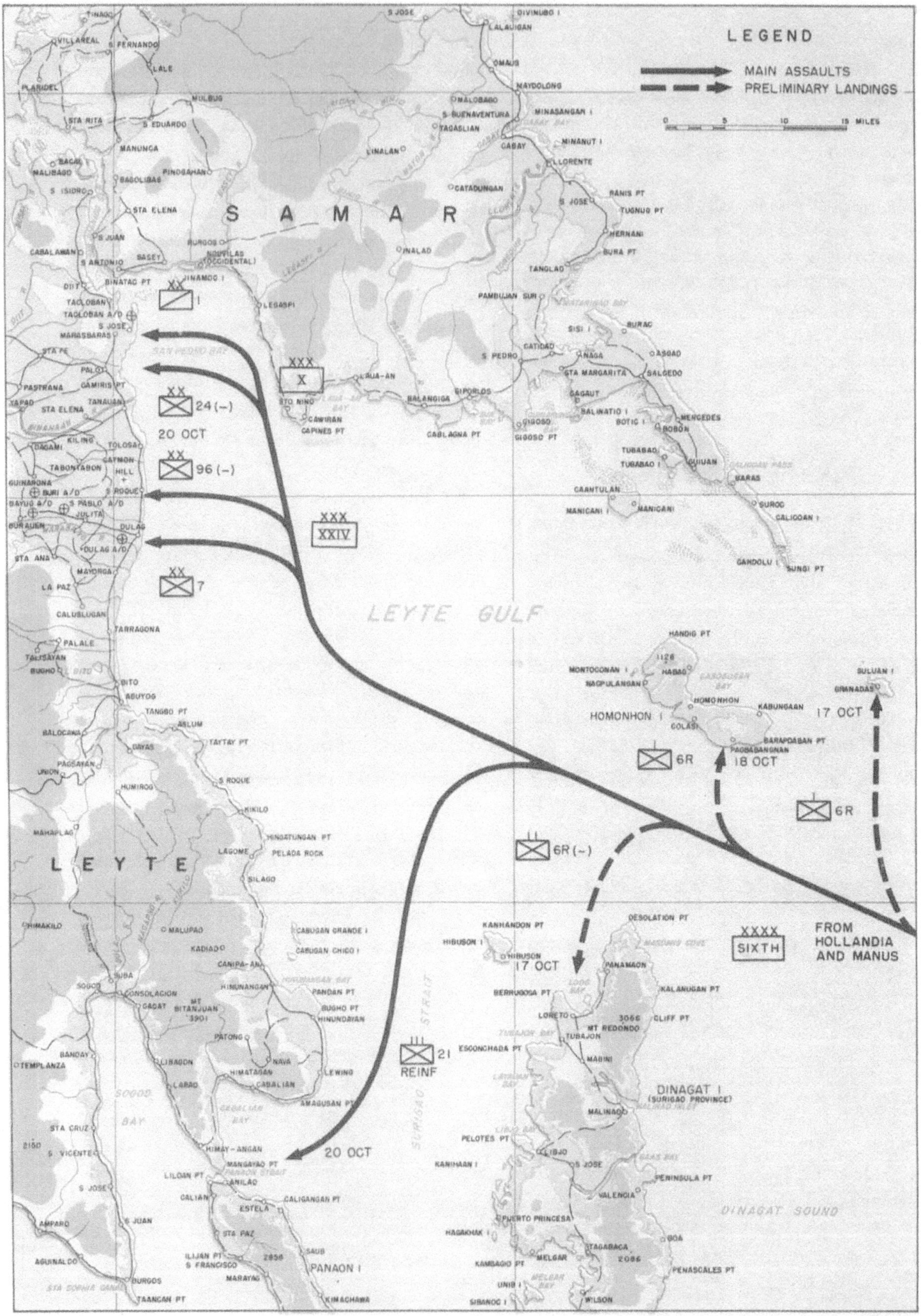

PLATE NO. 57

Sixth Army Landings on Leyte, 17–20 October 1944

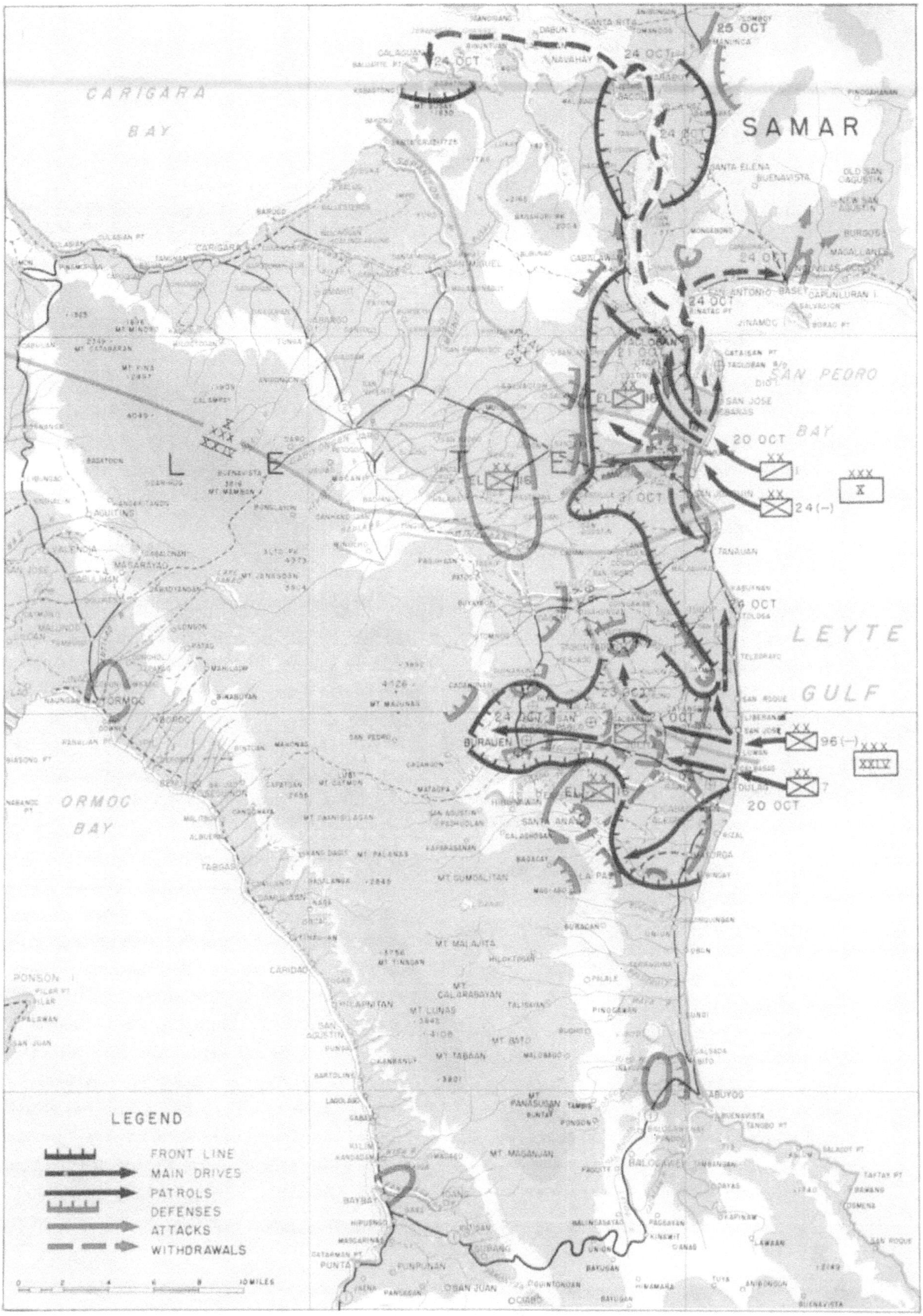

PLATE NO. 58

Leyte Assault, 20–25 October 1944

artillery, X Corps assault echelons completed unloading by the fifth day after the landing. They were ably assisted by the untiring efforts of the troops of the 2nd Engineer Special Brigade, veterans of almost every previous amphibious operation in the Southwest Pacific. Beach conditions in the XXIV Corps sector varied, but the presence of snipers and enemy mortar fire together with the lack of sufficient unloading personnel were temporary obstacles to the speedy disembarkation of men and materiel. Delay in unloading, however, did not prevent the establishment of beachheads and early advances inland by both corps.

General MacArthur, in personal command of the operation, summarized the landings and initial accomplishment of the invasion forces:

In a major amphibious operation we have seized the eastern coast of Leyte Island in the Philippines, 600 miles north of Morotai and 2,500 miles from Milne Bay from whence our offensive started nearly 16 months ago. This point of entry in the Visayas is midway between Luzon and Mindanao and at one stroke splits in two the Japanese forces in the Philippines. The enemy's anticipation of attack in Mindanao caused him to be caught unawares in Leyte and beachheads in the Tacloban area were secured with small casualties. The landing was preceded by heavy naval and air bombardments which were devastating in effect. Our ground troops are rapidly extending their positions and supplies and heavy equipment are already flowing ashore in great volume. The troops comprise elements of the 6th U.S. Army, to which are attached units from the Central Pacific, with supporting elements.

The naval forces consist of the 7th U. S. Fleet, the Australian Squadron, and supporting elements of the 3rd U. S. Fleet. Air support was given by naval carrier forces, the Far East Air Force, and the RAAF.

The enemy's forces of an estimated 225,000 include the 14th Army Group under command of Field Marshal Count Terauchi, of which seven divisions have already been identified: 16th, 26th, 30th, 100th, 102nd, 103rd, and 105th.[4]

The strategic result of capturing the Philippines will be decisive. The enemy's so-called Greater East Asia Co-Prosperity Sphere will be cut in two. His conquered empire to the south comprising the Dutch East Indies, and the British possessions of Borneo, Malaya and Burma will be severed from Japan proper. The great flow of transportation and supply upon which Japan's vital war industry depends will be cut as will the counter supply of his forces to the south. A half million men will be cut off without hope of support and with ultimate destruction at the leisure of the Allies a certainty. In broad strategical conception the defensive line of the Japanese which extends along the coast of Asia from the Japan Islands through Formosa, the Philippines, the East Indies, to Singapore and Burma will be pierced in the center permitting an envelopment to the south and to the north. Either flank will be vulnerable and can be rolled up at will.[5]

The assault continued after a rapid consolidation of the first day's objectives. Numerous enemy counterattacks were beaten off in all areas during the next few days as advancing forces reported increased resistance on every front. By the end of the third day, over 2,000 Japanese had been reported killed. On 24 October, elements of the X Corps began a drive up the Leyte side of San Juanico Strait, while farther south other units of the Corps pushed westward into Leyte Valley. At the same time, the XXIV Corps directed attacks northward and westward. The 96th Division moving inland from Dulag met heavy opposition from fortified positions on Catmon Hill, a terrain feature dominating the division's zone of action and giving protection to enemy mortars lobbing shells toward the assault shipping in Leyte Gulf. Catmon Hill was initially by-passsed, then neutralized by naval guns and field artillery, and finally cleared of the enemy by 31 October.

4 See Plate No. 55, "Enemy Ground Dispositions, 30 September 1944," Chapter VII.

5 GHQ, SWPA, Special Communique, 20 Oct 44.

Recognition of the effort by all arms was contained in General MacArthur's report of 25 October 1944:

Our ground forces have made extensive gains in all sectors. On the front of the XXIV Corps . . . the 7th Division penetrated the enemy's covering screen to seize San Pablo airdrome and fan out to the north toward Dagami. Elements of the 96th Division . . . have enveloped Catmon Hill and are approaching Tabontabon. In the northern sector, the X Corps has made substantial gains to the west of Palo and Tacloban and is pushing forward from the line of hills seized from the enemy which dominate the coast between Palo and Tacloban.

Carrier aircraft of the Seventh Fleet executed close support missions, attacking enemy ground installations, supply dumps and lines of communication. On October 22 and 23, our planes struck enemy airdromes in the western Visayas and northern Mindanao. Targets included Cebu, Bacolod, Alicante, Fabrica, Buluan and Del Monte. . . . Our successive raids from all sources on the enemy's air installations are neutralizing his attempts at staging in aircraft from Luzon and Borneo, and have greatly restricted the scale of his air counteroffensive. . . .[6]

Naval Threat to Leyte Gulf

While the Allies were fighting to improve their beachheads on Leyte, the Japanese were preparing to stake their whole remaining striking power on a gigantic gamble to maintain their positions in the Philippines.[7] The role of the Combined Fleet in this critical battle was vital. Brilliantly conceived and immense in scope, the Japanese plan was to deliver a crippling attack against the U. S. Navy and, with the strategic situation in their favor, to destroy in the same stroke General MacArthur's invasion at the beaches. (Plate No. 59) The Japanese were willing to risk the loss of their entire mobile fleet for the one opportunity of maneuvering their cruisers and battleships to within target range of the troop and supply transports in Leyte Gulf.

The decision to risk their most valuable military asset in an attempt to repel General MacArthur's invasion of Leyte emphasized again the importance which the Japanese attached to the Philippines. Admiral Soemu Toyoda, Commander in Chief of the Combined Fleet, was fully conscious of the hazards involved. "Since without the participation of our Combined Fleet", he stated, "there was no possibility of the land-based forces in the Philippines having any chance against your forces at all, it was decided to send the whole fleet, taking the gamble. If things went well we might obtain unexpectedly good results; but if the worst should happen, there was a chance that we would lose the entire fleet; but I felt that that chance had to be taken. . . . Should we lose in the Philippines operations, even though the fleet should be left, the shipping lane to the south would be completely cut off so that the fleet, if it should come back to Japanese waters, could not obtain its fuel

6 GHQ, SWPA, Communique No. 931, 25 Oct 44.

7 To protect their possessions against the Allied advance, the Japanese had developed an elaborate strategy of combined and interdependent Army, Navy, and Air Force actions known as Sho operations. These operations were intended to cover four critical areas of probable attack: Sho No. 1 applied to the Philippines; Sho No. 2, to Formosa, the Nansei Shoto (Ryukyu Islands) and southern Kyushu; Sho No. 3, to Shikoku, Honshu, and the Nampo Shoto (Southern Islands); Sho No. 4, to Hokkaido. The suddenness of the Leyte landings, together with the heavy losses in the battle off Formosa, caused several hurried modifications, although the main theme of the plan remained the same. Simultaneous operations would be undertaken by powerful naval and air forces against the Allied invasion units and landing points while, under cover of these movements, intensive attempts would be made to land troop reinforcements in the threatened area. Sho No. 1 was activated on 18 October as soon as the Japanese had ascertained that the Allied attack on Leyte was the main invasion effort.

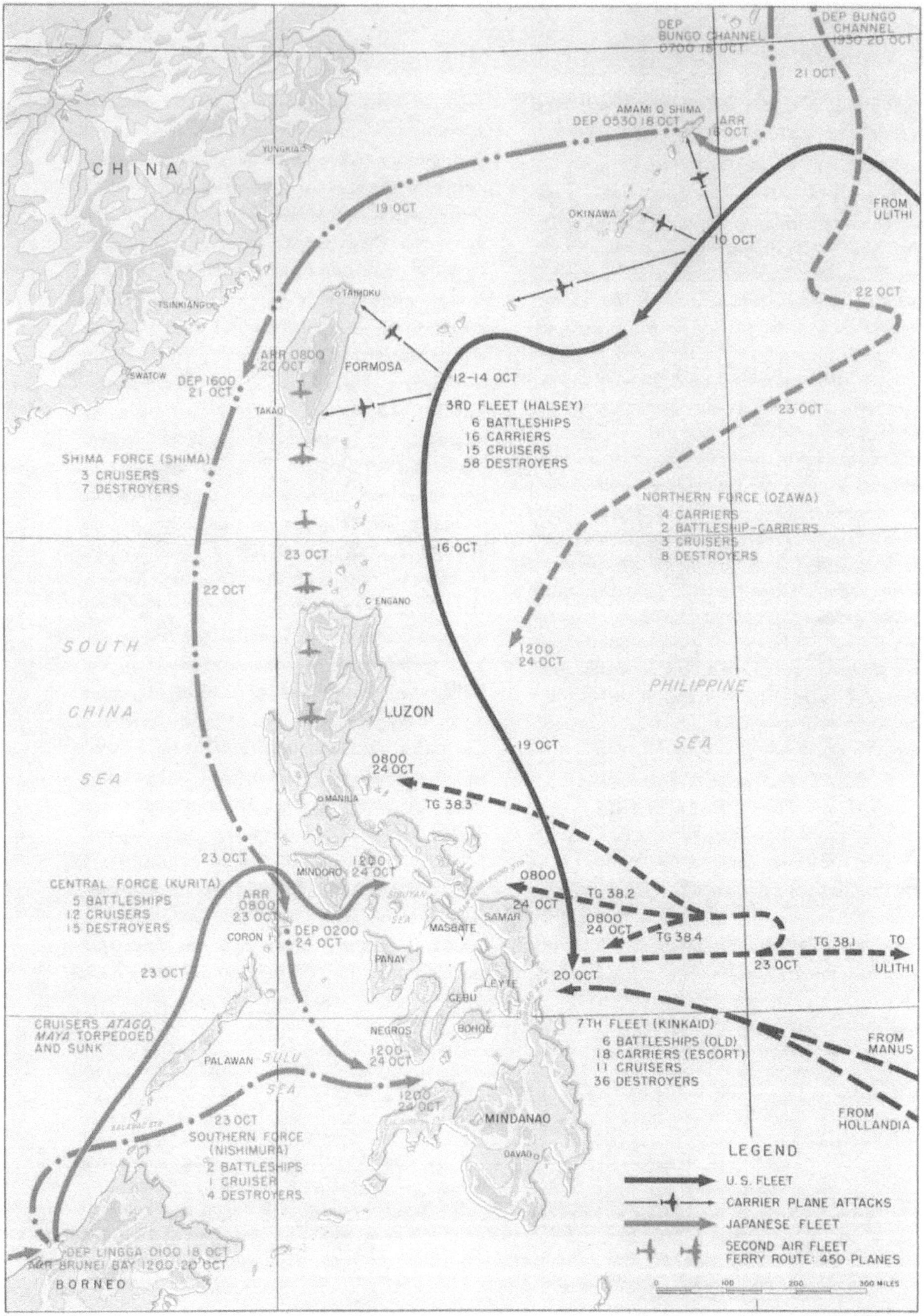

PLATE NO. 59

Approach of Naval Forces

supply. If it should remain in southern waters, it could not receive supplies of ammunition and arms. There would be no sense in saving the fleet at the expense of the loss of the Philippines."[8]

Based on similar strategic speculation, the Allied intelligence services assumed that the Japanese would employ the mass of their fleet to defend the Philippines.[9] As invasion day approached reconnaissance of every category was intensified to determine the location and movement of the Japanese Navy. It was expected that radio intercepts, aircraft and submarine sightings, and spot reports of A. I. B. agents and guerrilla units in the Philippines would provide the necessary clues on the operations and intent of the Japanese fleet well enough in advance for proper counter measures to be taken.

Long before the actual battle for Leyte Gulf, evidence began to accumulate on the dispositions of the Japanese Navy. As early as 16 October, it was recognized that the heaviest concentration of Japanese fleet units was in the Singapore—Brunei area, that most of the remaining ships were located in Japanese home waters and Formosa and that there were no important naval units in the Philippines.[10] The logical conclusion was that if the Japanese used their fleet against General MacArthur's landing forces in the Philippines they would probably converge from these two areas with the stronger elements coming up from the south.

Indications of Japanese Intentions

The Japanese tactical plan was also diagnosed in part during the first week in October. The division of the enemy fleet between the Singa-

8 USSBS, *The Campaigns of the Pacific War*, p. 281.

9 In a report of 10 January 1945, Admiral Kinkaid commented on the background for the assumption that the Japanese surface fleet would come out in force in case the Philippines were attacked: "The reconstruction of the Japanese fleet, after the sweeping Blue [U. S.] air victory over Truk on February 17th, into a strong task force organization termed the First Mobile Fleet; the capture of the 'Z' Operations Orders issued by the Combined Fleet Staff on March 8th, detailing the circumstances in which that fleet would be used to counter the Blue offensives; the fact that the striking force was brought out by the Japanese in mid-June to engage our forces in defense of the Marianas—all these factors, and many others combined to provide the background of subsequent intelligence bearing more directly on probable Japanese reactions to Blue landings in the Philippines. By September it was clear that the striking force . . . had been readied for use in the immediate future. The 7th Fleet Intelligence Officer, in a Staff memo on 24 September, estimated that the 2 diversion attack forces comprising the tactical organization of the First Mobile Fleet would be utilized for the defense of the Philippines, with a probable strength of 4–5 BB, 11 CA, 2 CL, 22 DD, plus 2 XCV-BB, 2 CV, 4 CVL." COM7thFLT Report Serial No. 000107 to CINCSWPA, 10 Jan 45.

10 A G-2 report of the time contained the following latest available information on the location of major enemy fleet units:

	Empire Formosa	*Singapore Brunei*	*Philippines*	*NE Area*
BB	1 u/s	6		
CA	1	12		2
CL	5	4		1
CV	5			
CVL	5			
CVE	2	2		
XCV	2			
DD	22	16	4	3
SS	54	2	8	1

G-2, GHQ, SWPA, *Special Intelligence Bulletin*, No. 526, 15/16 Oct 44, Appendix I.

pore—Brunei Bay area and Japan Proper led to the assumption that the enemy naval air strength would probably "cooperate in the East China Sea—Empire Area" and that the "whole remaining surface power" would "operate in the Sulu and South China Seas west of the Philippines." It was significantly recognized, too, that one of the immediate dangers to General MacArthur's amphibious landings in the Philippines lay in a possible sortie by strong Japanese surface craft from Brunei Bay via Surigao Strait.[11]

As D-day approached, the movements of the Japanese Fleet were kept under the closest surveillance. It was known on 14 October, for example, that a naval force, consisting of Vice Adm. Kiyohide Shima's minor units which were to participate belatedly in the Battle for Leyte Gulf, was to sortie from the Inland Sea at 2400 and that it was to pass through Bungo Channel early the next morning.[12] As confirmation of this movement, a U.S. submarine sighted Admiral Shima's force just off Bungo Channel on a southeasterly course at 0800 on 15 October.[13] Although Admiral Shima's cruiser and destroyer group was en route to the Formosa area at this time and did not receive definite orders to join the converging forces against Leyte Gulf until 1645 on 23 October, this was the first sighting of any of the units which were to participate in the crucial battle. Admiral Shima's force was sighted again on 16 October by Allied aircraft about 115 miles east of Okinawa,[14] and on 18 October by a U.S. submarine approximately 100 miles north of Okinawa.[15] The next day Japanese naval supply units were ordered to Bako for the replenishment of Admiral Shima's Force thus suggesting a possible sortie by his group.[16] It was subsequently sighted on 21[17] and 22 October the latter time about 150 miles northwest of Luzon en route to the Philippines.[18]

It was also known by 16 October that a force of major units under Vice Adm. Takeo Kurita in the Singapore area was preparing

11 Naval estimate, "Japanese Fleet Possibilities—King-Two Operation," 4 Oct 44, G-3, GHQ, SWPA *Journal*, 4 Nov 44. The following significant estimate of the possibility of Japanese fleet intervention in the Philippines appears in this report: "The surface danger to the King-Two Amphibious Forces lies in the First Diversion Attack Force, [under Adm. Kurita] . . . (unless it is previously damaged by air.) It is a typical Tokyo Express based at Brunei Bay, at present out of range of land-based and seaborne air but within striking distance of the landing force area. If the last half of the distance is negotiated during darkness its approach can only be detected by a submarine or airborne surface-search radar. Islands along the approach will cause protective shadows on other radars. That force already has received a directive to attack the Invasion Force at night. . . . If our bombardment forces retire to seaward at night Surigao Strait becomes an open back door. It must be assumed therefore that the Japanese Second Diversion Attack Force [Adm. Shima's Force] (battleships) will attempt to create diversions which are calculated to draw off our slow battleships, and our cruisers and destroyers, from screening positions in the vicinity of Surigao Strait. Conclusions are that an enemy night striking force is poised to attack the King-Two expedition through Surigao Strait, which should be forced and secured, as soon as control of the air has been obtained, at latest A-1 Day, and that surface forces superior to the enemy's First Diversion Attack Force should be maintained within intercepting distance of that force regardless of larger enemy fleet movements."

12 G-2, GHQ, SWPA, *Special Intelligence Bulletin*, No. 528, 17/18 Oct 44.

13 Ibid.

14 C. O. I. C. Situation Report No. 290, 17 Oct 44.

15 G-2, GHQ, SWPA, *Special Intelligence Bulletin*, No. 529, 18/19 Oct 44.

16 Op. cit. No. 531, 20/21 Oct 44.

17 C. O. I. C. Situation Report No. 295, 22 Oct 44.

18 G-2, GHQ, SWPA, *Special Intelligence Bulletin*, No. 534, 23/24 Oct 44.

for a possible further sortie.[19] Two days later it was recognized that this force might penetrate the Central Philippines to attack the United States landing forces. Although it was considered doubtful that "such a hazardous operation would be ordered," it was believed that it would "be detected at an early stage" and that United States supporting forces "with their air strength" would be able "to cope with the maximum estimated naval powers" which the Japanese could bring to bear.[20] On 20 October it was further reported that a Japanese "10,000 ton tanker, a repair ship and a coast defense vessel" were en route to Coron Bay in the Calamian Islands for the replenishment of Admiral Kurita's force.[21] Although this was the only indication up to that time of a "possible movement of major enemy fleet units to the southwest Philippines," it was a sign that could not be ignored.[22]

Vice Adm. Jisaburo Ozawa's carrier force was not actually sighted until 1640 on 24 October, but it was correctly estimated two days before that he was at sea as of 20 October[23] and "somewhere in the Formosa—P. I. Sea area on 22 October."[24] Significant too was the estimate of the same date that he was "moving probably toward the P. I. area in connection with counter measures to be taken against the Allied forces in the P. I. invasion."[25] Thus, with the landings on Leyte scarcely underway, a good general picture of Japanese fleet dispositions and probable intentions had been pieced together.

Approach of Enemy Naval Forces

Definite indications that the enemy was approaching the Philippines in force came from sightings of powerful enemy naval units by U. S. picket submarines previously assigned to search the waters off Borneo and Palawan. At 0200 on 22 October three large unidentified warships were reported in Palawan Passage on a northeast course with a speed of 21 knots.[26] No further sightings were reported that day in spite of the fact that extensive and prolonged air searches were made. At 0200 on 23 October, however, the submarine *Darter* sighted "three probable battleships" on approximately the same course. Within two hours the *Darter* estimated that this task force was composed of at least 9 ships with many radars. At 0430 the submarine *Bream* sighted more units and launched a torpedo attack obtaining two hits on one cruiser.[27]

Early the same morning the *Darter* reported sinking an *Atago* class cruiser and scoring four hits on another.[28] At 0700 the *Dace*, which had been assigned to cover the western approaches of Balabac Strait estimated that it had probably sunk a *Kongo* class battleship in a force of 11 ships which included battleships,

---

19 Op. cit. No. 528, 17/18 Oct 44.

20 Ibid.

21 Op. cit. No. 531, 20/21 Oct 44.

22 Ibid.

23 Op. cit. No. 533, 22/23 Oct 44.

24 Op. cit. No. 535, 24/25 Oct 44.

25 Ibid.

26 COM7thFLT Report to COMINCH US Fleet: *Report of Operation for the Capture of Leyte Island including Action Report of Engagements in Surigao Strait and off Samar Island on 25 October 1944 (King Two Operation)*, p. 37. Hereinafter cited as COM7thFLT Report.

27 Ibid. p. 13.

28 Ibid.

cruisers, destroyers, and a carrier.[29] Subsequent investigation determined that the *Darter* and *Dace* had sunk the heavy cruisers *Atago* and *Maya* and damaged the heavy cruiser *Takao*.[30] Before sunrise on 23 October the first blows of the fateful Battle for Leyte Gulf had been struck, with U. S. submarines performing yeoman service in their running fight with the enemy's large surface units. Not only did the submarines strike the first damaging blows against Admiral Kurita's ships but they alerted all United States forces to the approach of the main Japanese striking force two full days in advance of the final battle.

In the meantime the other Japanese naval forces which were to participate in the Battle for Leyte Gulf had not been sighted in Philippine waters. That Admiral Shima was nearing Coron Bay, however, was clear by 23 October, for he had requested that Japanese supply units be sent there early the morning of 23 October to fuel his force.[31] It was a highly probable assumption, too, that Admiral Ozawa's task force was prowling the waters of the Formosa —Philippine Sea area, although he had not yet been located. Later events were to prove this assumption correct. The fourth unit of the Japanese naval forces converging on the Philippines, Vice Adm. Shoji Nishimura's Southern Force, had also not been sighted by midnight of 23 October.

On 24 October, however, the various pieces of the Japanese naval puzzle were gradually pieced together. Early that morning continuous submarine sightings were made of the task force which had been moving through Palawan Passage. By 0030 it was estimated to consist of 15 to 20 ships including three probable battleships.[32] A few hours later the submarines *Angler* and *Guitarro* reported this force heading south through Mindoro Strait.[33] At 0810 U. S. carrier planes operating at a maximum range made a preliminary attack on the Japanese task force. Within the hour carrier search planes also reported that the enemy naval units were "moving northeast into Tablas Strait."[34] This force, designated as the Central Force, originally consisted of five fast battleships including the mysterious and dangerous 64,000 ton monsters the *Yamato* and *Musashi*, 12

29 Ibid.

30 Ibid.

31 G-2, GHQ, SWPA, *Special Intelligence Bulletin*, No. 534, 23/24 Oct 44. That the movement of enemy naval combat and supply units to Coron Bay provided an important clue to Japanese intentions is evident from Admiral Kinkaid's radio message of 23 October: "I view the approach of enemy combatant ships and tankers toward Coron Bay as the first phase of the build-up. Also magnified are express runs toward Leyte. I feel it highly probable that the tanker group which arrived at Coron Bay between 0200 and 0300/I on 23 Oct came for the purpose of refueling a major task force of the Japanese Fleet which has been assembling for several days in southern Formosa. Submarine reports indicate 3 probable battleships approaching from the south in position to arrive Coron Bay tonight, Monday. Another group of 11 enemy ships with many radars showing could reach that Bay at about the same time. There are signs of a concentration of a large number of enemy aircraft in the Luzon area. It is extremely important that early preparations for these enemy operations be discussed. Comdrs 5 and 13 AF are requested to continue their thorough reconnaissance of Coron Bay and its approaches, and to strike day and night if practicable. COM3rdFLT is requested to strike Coron Bay at the earliest practicable time, and to extend his search as far as it is possible to the west and northwest. The primary objective is enemy combatant ships and aircraft. It is possible that enemy carriers will protect their surface forces and will strike from west of Palawan. TG 77.4 is hitting the western Visayas twice daily." COM7thFLT Radio 230142/Z to CINCSWPA et al, 23 Oct 44, G-3, GHQ, SWPA *Journal*, 24 Oct 44.

32 COM7thFLT Report, p. 37.

33 Ibid.

34 Ibid.

accompanying cruisers, and 15 destroyers.[35]

At 0905 of 24 October aircraft sighted a second enemy fleet off Cagayan Islands well into the Sulu Sea on a course which indicated an intent to enter the Mindanao Sea. Aerial attacks were launched almost immediately with both bomb and rocket hits claimed on several of the Japanese units.[36] This Southern Force, under Admiral Nishimura, although not as large as the previously located Central Force, showed enough strength in battleships and cruisers to identify it as a major task formation.[37] It had left Brunei Bay on 22 October and at about noon of the 23rd passed through Balabac Strait and into the Sulu Sea. In the meantime Admiral Shima had departed from Coron Bay at 0400 on 24 October and sailed southward. After rounding the Cagayan Islands, his small force of 3 cruisers and 4 destroyers turned eastward where it was sighted west of Negros about noon. Admiral Shima followed Admiral Nishimura's Southern Force some 40 to 50 miles astern through the Mindanao Sea toward Leyte Gulf.

Estimate of the Enemy Plan

With the sighting of the Southern Force, the form of the Japanese plan began to take shape. The Central Force, under Admiral Kurita, would pass through the Sibuyan Sea, emerge from San Bernardino Strait and then set a course southward for Leyte Gulf. Meanwhile, the Southern Force, under Admiral Nishimura, would pace its approach via the Mindanao Sea and Surigao Strait so that both fleets, in a co-ordinated pincer movement, would converge simultaneously on the Allied flanks and attack the vulnerable "soft shipping" busy unloading in Leyte Gulf.[38] (Plate No. 60)

From the number and types of enemy warships moving through Philippine waters it was apparent to General MacArthur's staff that the Japanese were launched upon a full-scale operation to destroy or at least seriously upset the invasion. That they would not employ their carriers in such an ambitious plan was unthinkable and yet the Japanese carriers had not been sighted. It could only be concluded that the

---

35 Admiral Kurita's formidable force was consistently underestimated in the early sightings. It was composed of the battleships, *Yamato*, *Musashi*, *Nagato*, *Kongo*, and *Haruna*, 10 heavy cruisers, the *Atago*, *Maya*, *Takao*, *Chokai*, *Myoko*, *Haguro*, *Kumano*, *Suzuya*, *Tone*, and *Chikuma*, 2 light cruisers, the *Noshiro* and *Yahagi*, and 15 destroyers.

36 COM7thFLT Report, p. 37.

37 The Southern Force at this time comprised 2 battleships, the *Yamashiro*, and *Fuso*, the heavy cruiser *Mogami*, and 4 destroyers, the *Michishio*, *Asagumo*, *Yamagumo*, and *Shigure*.

38 In General Headquarters, Southwest Pacific Area, General MacArthur's G-2 Section assessed the enemy capabilities at this time as follows: "A late report . . . indicates a Japanese Task Force . . . in a position . . . 75 miles north by west of the northwest tip of Zamboanga Province [in the Sulu Sea]. . . . It is apparently heading for the Mindanao Sea, presumably either to bring reinforcements to the Leyte area or possibly to make a surface attack through Surigao Strait against the southern flank of our supply line.

"A second late report states that around noon two forces . . . on a southeast course in the Sibuyan Sea were also sighted. It is considered possible that this force may proceed through San Bernardino Strait in an attempt to attack our Leyte position from the north.

"It is considered significant that no carriers have been reported in either of these sightings, thus leaving the enemy the capability of bringing his carriers down the east coast of Luzon to support the possible attacks of his converging surface forces in the Leyte area. However, it is considered more probable that the carriers will be deployed in the vicinity of Mindoro, in order that their planes may strike across the Central Philippines, possibly staging through shore bases in the Visayas. Enemy air reaction since the landing on Leyte has been very weak. . . . This lack of activity gives rise to the enemy capability of launching a relatively strong air attack against our naval units, possibly in support of a naval engagement. . . ." G-2, GHQ, SWPA, *Daily Summary* No. 945, 23/24 Oct 44.

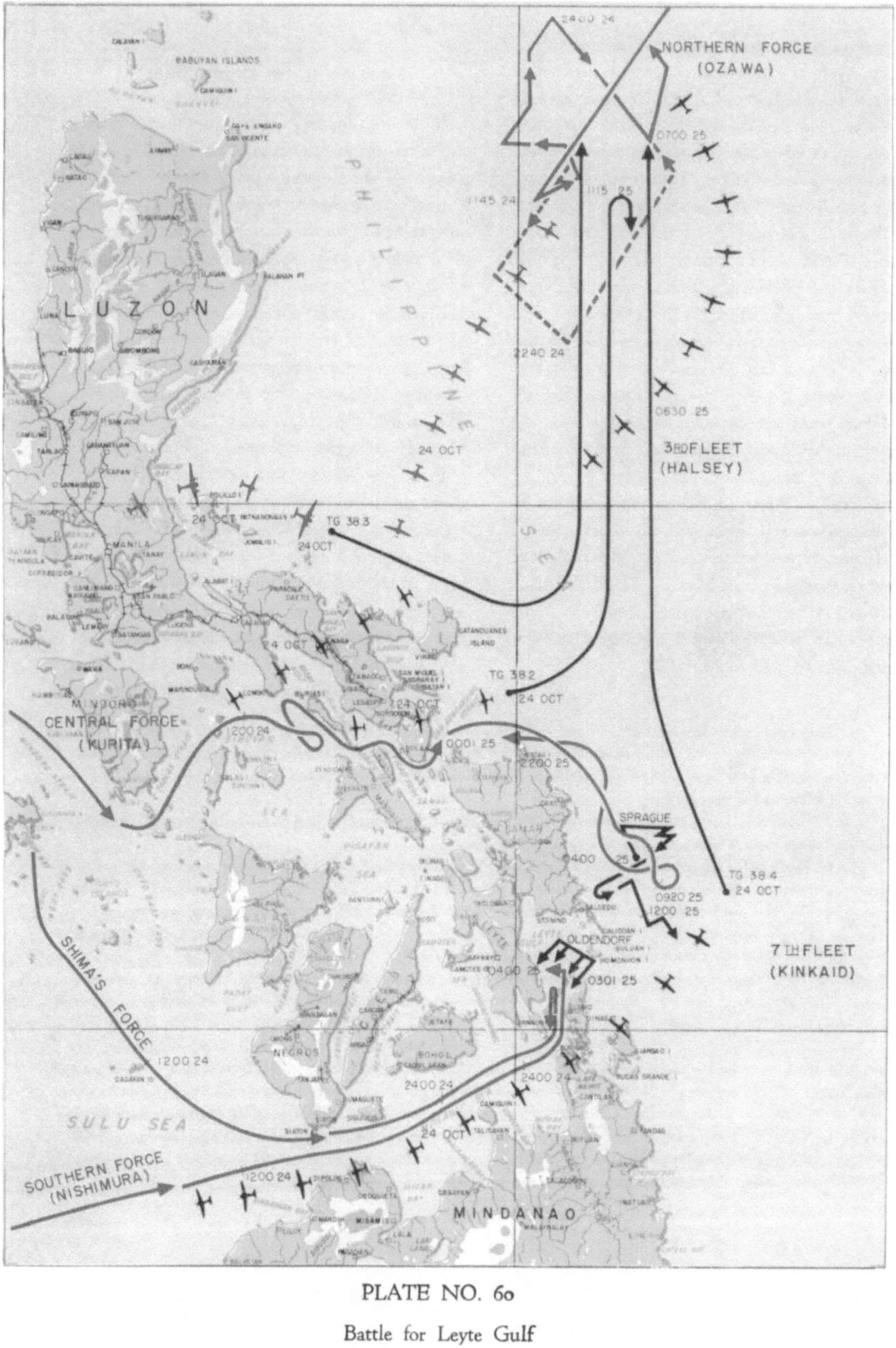

PLATE NO. 60

Battle for Leyte Gulf

missing carriers constituted still another force not yet located, probably Admiral Ozawa's force previously estimated to be somewhere in the Formosa—Philippine Sea area. This opinion was strengthened on the morning of the 24th when some 200 enemy planes launched a series of attacks concentrated mainly upon the carriers of the Third Fleet.[39] It was noticed that a great proportion of the attacking fighter planes consisted of carrier types. This fact, together with the direction of their approach, led Admiral Halsey to intensify his search to the north and east for some sign of enemy carriers.

In the crucible of the coming battle, General MacArthur's position was unique. The sweep of his forces along New Guinea had been consistently directed so that each operation would have available the full protection of his own land-based air force. Every advance had been predicated on the basic plan of procuring air bases 200 to 300 miles apart from which successive tactical advances could be made under a guaranteed air umbrella. By invading Leyte two months in advance of the original schedule, however, General MacArthur was compelled to send his troops into a landing area without his own controlled air coverage. The U. S. Navy therefore had a double responsibility. Initially, it had to assist the landing itself by its usual tasks of shore bombardment, convoy escort, and plane cover. In addition, it had the important mission of giving General MacArthur further air support for a period of time beyond the landing date until he could develop local airfields and stage his own air units forward from the south.

At Leyte General MacArthur was completely dependent on forces not under his control to protect his landing operation. Should the naval covering forces allow either of the powerful advancing Japanese thrusts to penetrate into Leyte Gulf, the whole Philippine invasion would be placed in the gravest jeopardy. It was imperative, therefore, that every approach to the Gulf be adequately guarded at all times and that an enemy debouchment via Surigao and San Bernardino Straits be blocked with adequate Allied naval strength.[40]

The naval forces protecting the Leyte invasion were disposed in two main bodies. The Seventh Fleet, under Admiral Kinkaid, protected the southern and western entrances to Leyte Gulf while the stronger Third Fleet, under Admiral Halsey operated off Samar to cover San Bernardino Strait and approaches from the north and east. Admiral Halsey's immediate superior was Admiral Nimitz in Hawaii; Admiral Kinkaid was responsible to General MacArthur. While it was realized that such a division of command entailed certain disadvantages, it was theoretically assumed that frequent consultation and co-operative liaison would overcome any difficulties in the way of proper co-ordination. The coming battle was to demonstrate the dangers involved in the lack of a unified command and the misunderstandings that can ensue during major operations in which the commander ultimately responsible does not have full control over all forces in the operation.

As Admiral Kurita's Central Force threaded its way through the narrow, reef-filled waters of

39 The light carrier *Princeton* was damaged so severely in these attacks that it had to be abandoned and sunk.

40 In pointing out the highly precarious position of the U.S. forces on Leyte, General MacArthur's G-2 stated: "There is no question that the Leyte position is vulnerable to the extent of U. S. Fleet availability; it is essential that its presence is continuous in adjacent waters, east of the Philippines, to interpose continuously between our land operations and possible enemy fleet intervention, which is essentially the primary mission of our Fleet." G-2, GHQ, SWPA, *Daily Summary* No. 944, 22/23 Oct 44.

the Sibuyan Sea during the morning and afternoon of the 24th, it was kept under repeated and severe attack by planes of the U. S. Third Fleet. The *Musashi*, one of the newest and largest of Japan's battleships which mounted 18-inch guns, was sunk; its sister ship, the mighty *Yamato*, was hit; a heavy cruiser was crippled; other cruisers and destroyers were damaged. The increasing force of these aerial blows together with the torpedoes of Seventh Fleet submarines caused Admiral Kurita, who was operating without air cover, to reverse his course for a time in order to take stock of the situation and reform his forces. This temporary withdrawal executed at 1530 was later reported by overly optimistic Allied pilots as a possible retirement of the Japanese Central Force. Admiral Kurita had no intention of abandoning his mission, however, and at 1714 he headed once more for San Bernardino Strait. Shortly thereafter he received a message from Admiral Toyoda in Tokyo: "All forces will dash to the attack, trusting in divine assistance."

Battle of Surigao Strait

In the meantime, Admiral Nishimura's Southern Force sailed doggedly on into the Mindanao Sea and passed Bohol despite the fact that it had been sighted and attacked off Cagayan Island.[41] Amply forewarned by sightings, Admiral Kinkaid had dispatched almost the whole of the Seventh Fleet's gunnery and torpedo force under Rear Adm. Jesse B. Oldendorf, to intercept and destroy the approaching Japanese warships.[42] Admiral Oldendorf took full advantage of both the geography of the battle area and his foreknowledge of the enemy's route of advance. PT squadrons were deployed at the entrance to Surigao Strait at a place where the Japanese would have to reform in column to negotiate the narrow passage. Behind the torpedo boats, covering the northern part of the strait, were posted the destroyer squadrons, cruisers, and battleships to form the horizontal bar to a "T" of vast fire power which the enemy would be forced to approach vertically as he moved forward. (Plate No. 61)

The ambush worked perfectly. As Admiral Nishimura's forces pushed forward through the smooth waters of Surigao Strait in the early hours of 25 October the PT boats, waiting in the darkness near Panaon Island, launched their torpedoes. Results were not clear but some damage was inflicted.[43] Three separate, co-ordinated attacks by U. S. Destroyer Squadrons 54, 24 and 56 followed in rapid succession as torpedo after torpedo found its mark. Destroyer Squadron 54 operating near Cabugan Grande Island and Hibuson Island launched its spread of torpedoes about 0300. A few minutes later, at 0327, Destroyer Squadron 24 fired its swift and deadly missiles from 700 yards off the port bow of the enemy formation. The third and final torpedo attack was made by Destroyer Squadron 56. At 0337 Admiral Oldendorf issued the order: "Launch attack—get the big boys."[44] Admiral Nishimura's force was not only confused and slowed by the tempest of torpedoes, it was mortally crippled. Virtually every unit in the formation was either sunk or badly damaged.

Despite this severe punishment, Admiral

41 The battleship *Fuso* suffered minor damage in this attack which took place at 0800 on 24 October.

42 Admiral Oldendorf's fleet comprised 6 battleships, 8 cruisers, 26 destroyers, and several squadrons of torpedo boats. The battleships included some veterans of the Pearl Harbor disaster. The more modern battleships were with Admiral Halsey.

43 COM7thFLT Report p. 18.

44 Ibid. pp. 19–20.

Nishimura's force steamed blindly ahead to its final doom in the death trap at the northern exit to Surigao Strait. There in full battle formation were Admiral Oldendorf's cruisers and battleships eagerly awaiting the ill-fated Japanese task force. At 0354 the battleships *West Virginia*, *Tennessee*, *California*, and *Maryland* opened fire in quick succession with range about 21,000 yards. The *Mississippi* followed with a single salvo at 0411.[45] Within a matter of minutes the withering hail of steel from Admiral Oldendorf's main battle line virtually completed the annihilation of Admiral Nishimura's force. The Japanese admiral himself went down with his flagship the *Yamashiro* which rapidly disintegrated after numerous torpedo hits. Only one lone destroyer, the *Shigure*, managed to survive the incredible carnage.

In the meantime Admiral Shima's cruiser and destroyer group, following some forty miles astern of Admiral Nishimura's Southern Force, entered Surigao Strait about 0300. The Southern Force had already suffered heavy losses inflicted by the U. S. Seventh Fleet and, from his flagship the *Nachi*, Admiral Shima could see the smoke and flash of gunfire in the distance. After brief blinker contact with the single surviving destroyer of the Southern Force and a careless collision with one of its crippled cruisers, the *Mogami*, which was maneuvering with broken steering gear, Admiral Shima retired southeastward into the Mindanao Sea and headed for Coron Bay. He did not escape unscathed, however, for he was attacked en route by Allied planes and lost one of his cruisers which had been torpedoed earlier in the action

With the destruction of the Japanese Southern Force and the retirement of Admiral Shima's units, the Battle of Surigao Strait had ended. The Seventh Fleet had performed its mission with precision and effectiveness. The southern entrance to Leyte Gulf had been closed successfully and General MacArthur no longer had to fear a Japanese naval threat from the south.

Third Fleet Goes North

While Admiral Kinkaid was preparing to meet the Japanese in Surigao Strait, Admiral Halsey was still trying to locate the suspected Japanese carriers In the late afternoon of the 24th, scout planes of the Third Fleet finally reported a large enemy task force, including several carriers, off the northeastern coast of Luzon about 300 miles from San Bernardino Strait.[46] As Admiral Halsey plotted the position and strength of the newly sighted enemy carrier force, he continued to received exaggerated reports of mounting damage inflicted by his attacking planes on the Japanese Central Force as it sailed through the Sibuyan Sea, toward San Bernardino Strait. As later events proved, the pilots' reports were inaccurate as to the status of both enemy forces. The firepower of the Northern Force was initially overestimated as was the damage inflicted upon Admiral Kurita's Central Force. This faulty information was primarily responsible for several important subsequent decisions.

The sighting of the carrier force presented a picture of three enemy fleets converging on the invasion area—the Northern Force moving southward in the Philippine Sea, the Central Force moving southeastward to San Bernardino Strait, and the Southern Force moving eastward toward the Mindanao Sea and Surigao Strait. Admiral Halsey felt that Admiral Kinkaid's Seventh Fleet had ample strength with which to meet the advancing Southern Force in Suri-

45 Ibid. p. 21.

46 COM7thFLT Report p. 14.

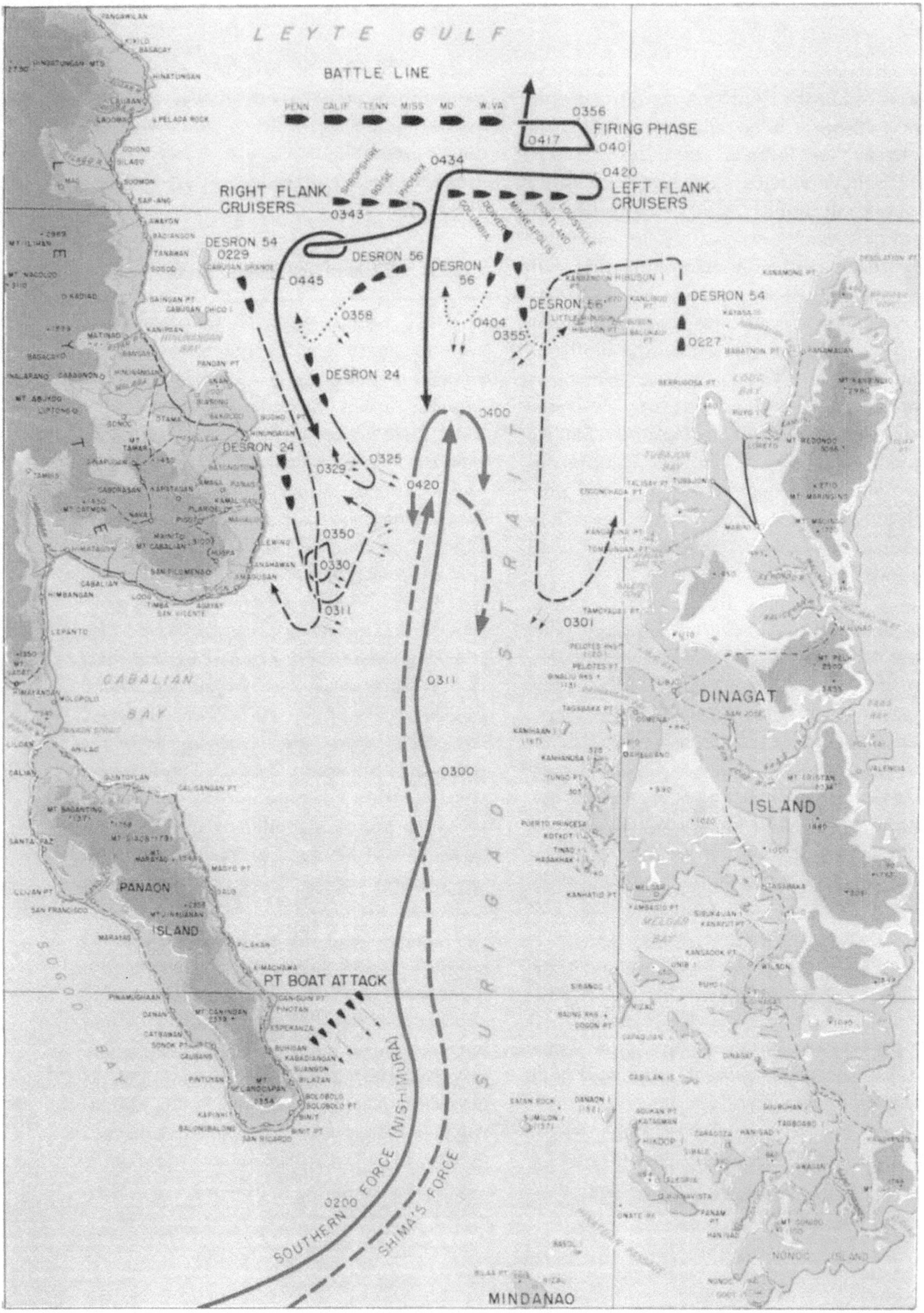

PLATE NO. 61

Battle of Surigao Strait

gao Strait. He judged from his aviators' optimistic reports that the Central Force had been greatly damaged, had perhaps retired, and in all likelihood had been removed as a serious menace. He reasoned that, in any event, its strength had been reduced to the point where any threat it presented would be satisfactorily overcome by Admiral Kinkaid's Seventh Fleet. He estimated that the Northern Force comprised the whole carrier strength of the Japanese and therefore constituted the most potent danger to be met.[47] Influenced by these factors, Admiral Halsey decided to move his entire force northward as a unit and intercept the Japanese carriers.

During the afternoon of 24 October, however, Admiral Halsey had organized Task Force 34 as a strong surface force comprising the battleships *Iowa*, *New Jersey*, *Washington*, and *Alabama*, 2 heavy cruisers, 3 light cruisers, and 14 destroyers.[48] According to Admiral Halsey, the dispatch forming Task Force 34 was not an executive dispatch but a tentative battle order and was so marked.[49] As everyone was greatly concerned about the Japanese Central Force, including Admiral Halsey who knew its position and formidable fire power, it was assumed that Task Force 34 would engage Admiral Kurita's force which was heading east through the Sibuyan Sea. In the early evening Admiral Halsey informed Admiral Kinkaid and others of the position of the Japanese Central Force and added that he was "proceeding north with three groups to attack the enemy carrier forces at dawn."[50] Accordingly, on the evening of the 24th, he withdrew the battleships, carriers, and supporting ships of the Third Fleet from San Bernardino Strait. It was a crucial decision and no end of confusion and uncertainty followed his action. As the fast battleships of the Third Fleet had been detached from the carrier groups and organized as Task Force 34, it was assumed that Task Force 34 was still guarding San Bernardino Strait. Although no definite statement had been made to this effect by Admiral Halsey,[51] Admiral Kinkaid thought that the big battleships were standing by awaiting the Japanese Central Force and that Admiral Halsey was going after the Japanese Northern Force with carrier units. Actually, however, Admiral Halsey took his three complete task groups on his run to the north and left San Bernardino Strait open.[52]

---

47 The Japanese Northern Force actually consisted of 2 converted battleship-carriers, the *Ise* and *Hyuga*, the large carrier *Zuikaku*, the 3 light carriers *Chitose*, *Chiyoda*, and the *Zuiho*, the 3 light cruisers *Oyodo*, *Tama*, and *Isuzu*, and 8 destroyers.

48 William F. Halsey and J. Bryan III, *Admiral Halsey's Story*, (New York, 1947) p. 214.

49 Ibid.

50 COM7thFLT Report p. 25.

51 Ibid.

52 Admiral Halsey explained the three alternatives before him at the time as follows: "1. I could guard San Bernardino with my whole fleet and wait for the Northern Force to strike me. Rejected. It yielded to the enemy the double initiative of his carriers and his fields on Luzon and would allow him to use them unmolested. 2. I could guard San Bernardino with TF 34 while I struck the Northern Force with my carriers. Rejected. The enemy's potential surface and air strength forbade half-measures; if his shore-based planes joined his carrier planes, together they might inflict far more damage on my half-fleets separately than they could inflict on the fleet intact. 3. I could leave San Bernardino unguarded and strike the Northern Force with my whole fleet. Accepted. It preserved my fleet's integrity, it left the initiative with me, and it promised the greatest possibility of surprise. Even if the Central Force meanwhile penetrated San Bernardino and headed for Leyte Gulf, it could hope only to harry the landing operation It could not consolidate any advantage, because no transports accompanied it and no supply ships. It could merely hit and run." Halsey and Bryan, op. cit., pp. 216–217.

Free passage through San Bernardino Strait was a pivotal point of the enemy's strategy in his daring scheme to strike the Allies in Leyte Gulf. Admiral Halsey did not know that Japan's carriers were to be deliberately sacrificed in a bold gamble to keep the Philippines from falling to the Allies. It was afterward revealed that the Northern Force under Admiral Ozawa, almost completely destitute of planes and pilots, had only one mission—to serve as a decoy and lure the most powerful units of the U. S. Fleet away from the Leyte area.[53] It was expected that the Northern Force would probably be destroyed, but it was hoped that this desperate device would enable the Japanese Central Force to pass unmolested through San Bernardino Strait and move southward into Leyte Gulf, timing its approach to coincide with the arrival of the Southern Force from Surigao Strait.

To accomplish his mission, Admiral Ozawa continually sent out radio messages in an effort to advertise his position to the U. S. Fleet. An undetected fault in his transmission system, however, prevented the Third Fleet from intercepting these signals. Equally important to later operations, communication difficulties and divided responsibility also prevented an adequate exchange of information between the enemy's Northern and Central Forces.

Battle off Samar

As the U. S. Third Fleet sought contact with the carriers of the Northern Force during the evening of the 24th, Admiral Kurita in a remarkable feat of high-speed navigation, skillfully led his warships through the darkness southeastward into the treacherous passes of the Sibuyan Sea. Since he had received no message from Admiral Ozawa off Luzon, he was actually unaware that the main body of the U. S. Fleet had been drawn away from the mouth of San Bernardino Strait. Nevertheless, Admiral Kurita in compliance with repeated orders from the Combined Fleet, continued on his original task, setting his course for Leyte Gulf. By midnight he was in the waters of San Bernardino Strait and at approximately 0035 of 25 October he debouched into the Philippine Sea. About 0530, as he was coming down the coast of Samar, Admiral Kurita received word of the loss of Admiral Nishimura's two battleships and of the damage to the cruiser *Mogami*. Although he was unaware of the fact, his was now the only Japanese force within striking distance of Leyte. The task of destroying the U. S. invasion units rested solely in his hands.

At dawn on 25 October, three groups of escort carriers under Admiral Thomas L. Sprague were disposed east of Samar and Leyte Gulf. The Northern CVE Group with 6 carriers, 3 destroyers, and 4 destroyer escorts was under the command of Rear Adm. Clifton A. F. Sprague.[54] It was on a northerly course about fifty miles east of Samar and approximately half way up the east coast of the island, directly in the path of Admiral Kurita's oncoming force.

53 "The Northern Force carriers had left the Inland Sea with a total of 108 planes of various types on board. By 24 October, Admiral Ozawa had lost 66 planes which failed to return from reconnaissance missions and attacks directed against the Third Fleet, leaving him with 42 planes on his carrier decks at the time he was sighted by Admiral Halsey's scout planes. After Admiral Ozawa felt that he was definitely contacted by the U. S. forces, he dispatched his entire operational bomber complement of 15 planes to airfields in Luzon leaving himself with 10 inoperable bombers and 17 fighters to meet the advancing Third Fleet on 25 October." Statement by Captain Ohmae, former Senior Staff Officer of Admiral Ozawa's First Mobile Fleet, 23 Sept 48. Interrogation Files, G-2 Historical Section, GHQ, FEC.

54 The escort carriers were the *Fanshaw Bay*, *St. Lo*, *White Plains*, *Kalinin Bay*, *Kitkun Bay*, and *Gambier Bay*; the destroyers were the *Hoel*, *Heermann*, and *Johnston*; the destroyer escorts were the *Dennis*, *J. C. Butler*, *Raymond*, and *S. B. Roberts*.

Some twenty or thirty miles east southeast was the Middle CVE Group with the same composition under Rear Adm. Felix B. Stump.[55] Farther to the south off northern Mindanao was the Southern CVE Group consisting of 4 carriers, 3 destroyers, and 4 destroyer escorts.[56] This group was under the direct supervision of the over-all escort carrier commander, Admiral Thomas Sprague.

Through a series of fatal misunderstandings directly attributable to divided command, ambiguous messages, and poor communication between the U. S. Third and Seventh Fleets, neither Admiral Kinkaid at Leyte Gulf, nor Admiral Sprague off Samar, realized that the exit from San Bernardino Strait had been left unguarded. They had no reason to expect such a situation ; previous commitments were firm. During the night of the 24th, however, Admiral Kinkaid became uneasy concerning the actual situation at San Bernardino Strait and decided to check on the position of Task Force 34. At 0412 on the 25th Admiral Kinkaid sent an urgent priority dispatch (241912) telling Admiral Halsey of the results in Surigao Strait and asking him the vital question : "Is TF 34 guarding San Bernardino Strait."[57] The reply to Admiral Kinkaid's dispatch was not sent until 0704, by which time the first salvos of Admiral Kurita's battleships were reverberating across the waters off Samar. Admiral Halsey's answer said: "Your 241912 negative. Task Force 34 is with carrier group now engaging enemy carrier force."[58]

It was a dramatic situation fraught with disaster. The forthcoming battle between the U.S. Seventh Fleet's slow and vulnerable "jeep" carriers and the Japanese Central Force of greatly superior speed and fire power gave every promise of a completely unequal struggle. The light carriers of Admiral Sprague were no match for the giant battleships and heavy cruisers of the Japanese Central Force. Should the enemy gain entrance to Leyte Gulf, his powerful naval guns could pulverize any of the eggshell transports still present in the area and destroy vitally needed supplies on the beachhead. The thousands of U. S. troops already ashore would be isolated and pinned down helplessly between enemy fire from ground and sea. Then, too, the schedule for supply reinforcement would not only be completely upset, but the success of the invasion itself would be placed in grave jeopardy. The battleships and cruisers of the Seventh Fleet were over 100 miles away in Surigao Strait with their stock of armor-piercing ammunition virtually exhausted by the pre-landing shore bombardment and the decisive early morning battle with the Japanese Southern Force. Admiral Halsey's Third Fleet was almost 300 miles away still in hot pursuit of the Northern Force and could not possibly return in time to halt the progress of Admiral Kurita.

On shore, General MacArthur had been powerless to do more than point out the absolute need for constant carrier air cover over the Leyte beaches. Under the divided command set-up, he had no effective control over the Third Fleet. Having advanced beyond the range of his own land-based aircraft, he was completely dependent upon carrier planes for

55 The escort carriers comprised the *Natoma Bay*, *Manila Bay*, *Marcus Island*, *Kadashan Bay*, *Savo Island*, and *Ommaney Bay*; the destroyers consisted of the *Haggard*, *Franks*, and *Hailey*; and the destroyer escorts were the *R. W. Suesens*, *Abercrombie*, *Leroy Wilson*, and *W. C. Wann*.

56 This group had the following escort carriers: *Sangamon*, *Suwannee*, *Santee*, and *Petrof Bay*; the destroyers were the *McCord*, *Trathen*, and *Hazelwood* and the destroyer escorts, the *R. S. Bull*, *R. M. Rowell*, *Eversole*, and *Coolbaugh*.

57 COM7thFLT Report, p. 25.

58 Ibid.

protection—a fact which he had emphatically made clear both before and during the planning for the invasion. On 21 October he had reiterated his conception of Admiral Halsey's mission saying:

> *The basic plan for this operation in which for the first time I have moved beyond my own land-based air cover was predicated upon full support by the Third Fleet; such cover is being expedited by every possible measure, but until accomplished our mass of shipping is subject to enemy air and surface raiding during this critical period; . . . consider your mission to cover this operation is essential and paramount. . . .*[59]

Now General MacArthur could do nothing but consolidate his troops, tighten his lines, and await the outcome of the impending naval battle.

Although Admiral Kurita's Central Force had suffered considerable damage in its course across the Sibuyan Sea, it was still a formidable fleet when it broke through San Bernardino Strait and headed for Leyte Gulf. The main force consisted of the mighty *Yamato* and three other fast battleships, the *Kongo*, *Haruna*, and *Nagato*. These were supported by the six heavy cruisers *Chikuma*, *Haguro*, *Chokai*, *Tone*, *Kumano*, and *Suzuya*. The two light cruisers *Yahagi* and *Noshiro* and eleven destroyers completed Admiral Kurita's fast and powerful task force. In quest of big game, its guns and shell hoists were loaded down with armor-piercing projectiles as it steamed southward along the coast of Samar.

First indication that the enemy might be approaching came at 0637 when the escort carrier *Fanshaw Bay* intercepted Japanese conversation on the interfighter director net. This was followed closely by sighting of antiaircraft bursts to the northwest about twenty miles distant. The final shock came at 0647 when a pilot aboard a plane from the *Kadashan Bay* frantically announced that a large enemy surface force of battleships, cruisers, and destroyers was closing in about twenty miles to the northwest at a speed of 30 knots. The frightful possibility had materialized; the Japanese Central Force had broken through San Bernardino Strait.

Admiral Kurita had located the U. S. escort carrier force at 0644 and a few minutes later at 0658 he gave the order to join battle. The *Yamato's* 18-inch guns fired first with the cruisers following as soon as they came within range. Never before had U. S. warships been subjected to such heavy caliber fire. With the world's biggest guns blazing away, Admiral Kurita pressed the attack at full speed. (Plate No. 62)

As soon as the enemy was sighted the Northern CVE Group changed course to 090°, due east, and began launching all available aircraft. Scarcely had the planes been sent aloft when large caliber shells began falling among the units of the formation. Admiral Kurita was closing rapidly, straddling the U.S. escort carriers with dye-marked salvos that landed with uncomfortable accuracy and bracketed their targets with red, yellow, green, and blue splashes. The situation was most critical and from this point on it was only a question of how to save the greatly outnumbered and outgunned U. S. ships from almost certain destruction.

At 0701 the commander of the Northern

59 CINCSWPA Radio to COM3rdFLT et al, 21 Oct 44, SOPAC No. 538, C/S, GHQ, (S). General MacArthur's message was in reply to a request from Admiral Halsey regarding withdrawal of Third Fleet units from the covering of the Leyte operations. Admiral Halsey's request read: "My present operations in strategic position to meet threat of enemy fleet forces are somewhat restricted by necessity of covering your transports and other overseas movements; request early advice regarding withdrawal of such units to safe position, which will permit me to execute orderly rearming program for my groups and allow further offensive operations." COM3rdFLT Radio to CINCSWPA et al, 21 Oct 44, SOPAC No. 537, C/S GHQ, (TS).

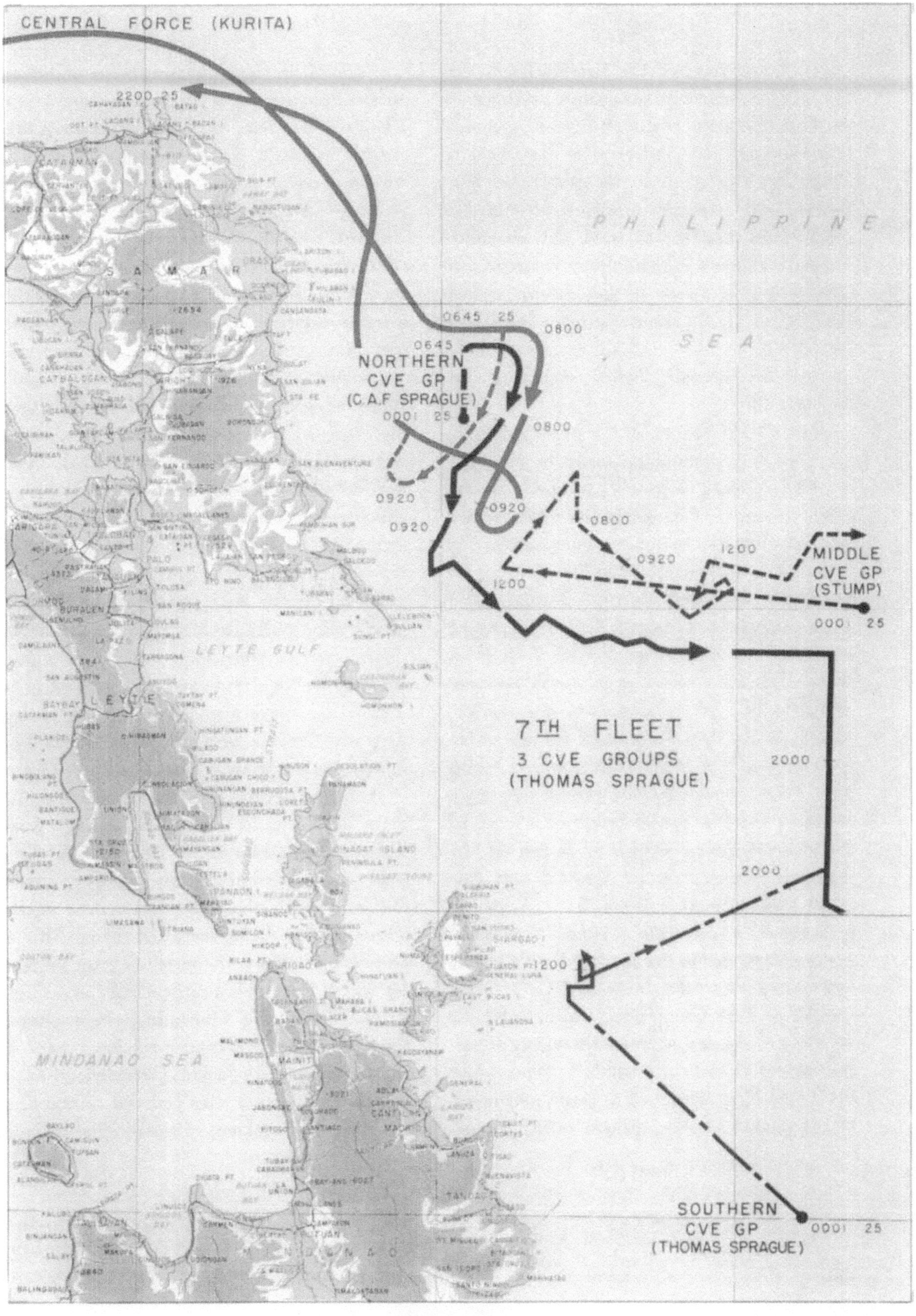

PLATE NO. 62

Battle off Samar

CVE Group reported the alarming news that he was in contact with the enemy and urgently needed assistance. He then ordered a general retirement to the southwest in the hope of obtaining support from the heavy forces in Leyte Gulf. Admiral Kinkaid received the request for support at 0724 and promptly ordered Admiral Oldendorf "to prepare to rendezvous his forces at the eastern end of Leyte Gulf"; the escort carrier planes were sent the same order, and a dispatch was transmitted to Admiral Halsey requesting immediate aid.[60]

Admiral Kurita pushed the attack vigorously. His ships gradually closed and continued to send a steady avalanche of shells at the escort carriers. Huge geysers of water erupted around Admiral Sprague's units as the big Japanese batteries began to find the range. As the battle developed, Admiral Kurita, intent on encircling the retiring U. S. ships, deployed several of his cruisers on the left flank from where they delivered the most troublesome fire of the battle. Several destroyers were also deployed on the right flank while the rest of his force pressed the attack from astern. Along with these surface attacks Japanese air units based in the Philippines launched a series of Kamikaze strikes against the U. S. carriers. In his great distress Admiral Kinkaid sent Admiral Halsey another dispatch: "Urgently need fast BB's Leyte Gulf at once."[61] Admiral Halsey responded to the Seventh Fleet's appeal by ordering Vice Adm. John S. McCain, Commander of Task Group 38.1, which had been refueling to the east, to attack the enemy at San Bernardino as soon as possible.[62] Meanwhile the Third Fleet continued to steam northward in hot pursuit of the Japanese carrier group.

Admiral Sprague's escort carriers used every trick of sea fighting to escape the heavy fire of the Japanese Fleet and to inflict losses on the enemy. Evasive tactics were ordered; thick smoke screens were laid down; temporary refuge was sought in a providential rain squall. In a desperate effort to stem the enemy advance, the destroyers and destroyer escorts fought back furiously. Interposing themselves between the carriers and their adversary, they boldly closed the range and fired their five-inch guns and torpedoes at cruiser and battleship targets. Both the *Johnston* and the *Heermann* challenged the *Kongo*. The escort carriers also engaged with gunfire whatever units of the enemy came within range. At the same time their planes attacked the enemy continuously and succeeded in putting several of his cruiser units out of action. They were greatly handicapped, however, by the damage inflicted on the carriers. The pilots, seeing their carrier decks ripped open or their ships sunk, made forced landings on the already overcrowded airstrip at Tacloban. Some also landed on the Dulag strip while others were compelled at the last minute to ditch their planes in Leyte Gulf.

As the battle progressed the situation became more desperate for the U. S. units. With disaster staring him in the face, Admiral Kinkaid sent Admiral Halsey another urgent dispatch which the latter received at 0900: "Our CVE's being attacked by 4 BB's 8 cruisers plus others. Request Lee cover Leyte at top speed. Request fast carriers make immediate strike."[63] By this time ammunition aboard the escort carriers was running low; some of the destroyers had expended their torpedoes and the torpedo planes were reduced to the dire expedient of making dummy runs on the

60 COM7thFLT Report p. 27.

61 Halsey and Bryan, op. cit., p. 219.

62 Ibid.

63 Op. cit. p. 220. Vice Adm. Willis A. Lee was commander of the battleships under Admiral Halsey.

enemy ships. Both the screening force and the carriers sustained considerable damage. The *Gambier Bay* was hit hard and sank at 0911 with an enemy cruiser pumping shells into her at a range of less than 2,000 yards. The destroyer *Johnston*, which had been under continuous heavy fire, was fatally struck and had to be abandoned. She rolled over and sank at 1010. The destroyer *Hoel* and the destroyer escort *Roberts* were also sunk. Early damage was also inflicted on the escort carriers *Santee*, *Suwannee*, and *Sangamon*. At 0922 Admiral Halsey had received a dispatch sent by Admiral Kinkaid at 0725: "Under attack by cruisers and battleships . . . request immediate air strikes. Also request support by heavy ships. My OBB's low in ammunition."[64]

Meanwhile, off Samar, victory lay within Admiral Kurita's grasp. After almost two and one-half hours of continuous battle the flanking enemy units began closing in, firing salvo after salvo at the escort carriers dodging desperately to avoid more damage. By 0920 the heavy cruiser *Tone* was within 10,000 yards of her targets and on the starboard flank the Japanese 10th Destroyer Squadron pressed the attack with torpedoes. The situation had become virtually hopeless for Admiral Sprague's task group, and few expected to come out of the ordeal afloat. Once more Admiral Kinkaid sent an insistent plea to Admiral Halsey, this time in the clear: "Where is Lee? Send Lee."[65] Almost simultaneously Admiral Halsey received an urgent dispatch from Admiral Nimitz in Pearl Harbor: "The whole world wants to know where is Task Force 34."[66]

Admiral Kurita Breaks Off Engagement

Then, as if in answer to fervent prayer, the unexpected happened. Admiral Kurita broke off the engagement.[67] His units had sustained no little damage,[68] and like everyone else he was unaware of the true battle situation. He

64 Ibid.

65 Ibid.

66 Ibid.

67 There has been much speculation and controversy regarding Admiral Kurita's crucial decision to break off the battle at this point and return to the north rather than continue on to Leyte Gulf. Perhaps the most concise summary of the reasons behind this move is given by Rear Adm. Tomiji Koyanagi, Admiral Kurita's chief of staff. "One reason", Admiral Koyanagi stated, "was that the Second Diversion Attack Force [Admiral Shima] reported the almost complete destruction of Admiral Nishimura's force. . . . The second reason was that Nishimura's force meant to go into the Bay in the morning and our force was to have entered at 0600 to coordinate but the approach of our force was greatly delayed. After the battle off Samar we finally decided that the cooperation of the two forces would not be effected at all, and if we ever decided to go into the Bay, the interval would be too much. The third reason, we intercepted a telephone message sent by your carrier to get reinforcements. We also intercepted an answer to the telephone call to the effect that it would be two hours before reinforcements of planes arrived; that is to say by the time we entered this Bay. We thought that planes would have come out and that warships would have come out and that the transports would have dispersed enough so that only a part would have been inside; they would escape from the danger zone. We figured by that time that the transports might have heard of the battle and started already in the morning. The fourth reason, we intercepted again a telephone message sent by carrier to the planes to the effect that all planes should go to Leyte shore strips; and also we thought reinforcement of the American carriers might come along and so your force would be very big, carriers and land-based planes, and it would not be advisable to go into this danger zone and be a target for attack by shore and reinforced carrier planes in narrow waters. The fifth reason, we surmised that your American Task Force might come down from the north; so after we failed to destroy this first American force, we thought that if we came back north now we might encounter another American Task Force, but we were very regretful that we failed to destroy your first American force. The sixth reason, if we continued the battle here at Leyte, it would consume more fuel; that was another reason for cruising north instead of staying around. We had no tankers anywhere around." USSBS, *Interrogation of Japanese Officials*, Vol. I, pp. 151–152.

68 As a result of the Samar Battle the Japanese lost the three heavy cruisers *Chokai*, *Suzuya*, and *Chikuma* and one destroyer. The U. S. Navy lost the escort carriers *Saint Lo* and *Gambier Bay*, the destroyers, *Hoel* and *Johnston*, and the destroyer escort *S. B. Roberts*. The escort carriers *Sangamon*, *Suwannee*, *Santee*, *Fanshaw Bay*, *White Plains*, and *Kalinin Bay* were also damaged as was the destroyer *Heermann* and the destroyer escorts *Dennis* and *R. M. Rowell*.

therefore ordered his forces to cease firing and reassemble to the north. For the U.S. carrier forces this retirement by the enemy meant a remarkable and completely unexpected escape. Admiral Sprague in summing up the results of the battle shortly thereafter stated:

. . . the failure of the enemy main body and encircling light forces to completely wipe out all vessels of this Task Unit can be attributed to our successful smoke screen, our torpedo counter-attack, continuous harassment of the enemy by bomb, torpedo, and strafing air attacks, timely maneuvers, and the definite partiality of Almighty God.[69]

The continuous and urgent dispatches from Admiral Kinkaid and the cryptic message from Admiral Nimitz finally led Admiral Halsey to change course and direct the bulk of his fleet southward toward Leyte. He accordingly directed Task Force 34 under Admiral Lee and Task Group 38.2 under Rear Adm. Gerald F. Bogan to proceed south toward San Bernardino Strait. At exactly 1115, when he was expecting the pagoda masts of Admiral Ozawa's force to appear over the horizon at any minute he made the crucial decision to return to Leyte Gulf. He expected to arrive early the next morning.

The other units of the Third Fleet, Admiral Mitscher's Task Force 38 with Task Group 38.3 under Admiral Sherman and Task Group 38.4 under Admiral Davidson were to continue against the enemy carrier force. Throughout the afternoon of 25 October Third Fleet carrier planes struck Admiral Ozawa's force again and again. When the battle was over the Japanese had lost the famous carrier *Zuikaku* and the three light carriers *Chitose*, *Chiyoda*, and *Zuiho*. One light cruiser and two destroyers were also sunk, but the two hermaphrodite battleship-carriers *Ise* and *Hyuga* escaped with minor damage. United States forces suffered no surface losses in the engagement.

After regrouping his forces and evaluating the situation, Admiral Kurita decided to make one last attempt against Leyte Gulf. At 1120 he ordered his ships to change course toward the target area to the southwest. He was en route approximately one hour, however, and only 45 miles from his objective when he finally decided to give up the attempt. At 1236 he ordered his ships about. As he retraced his course to the northwest his force came under attack by planes from the Seventh and Third Fleets. Although he had sustained considerable damage, Admiral Kurita managed to limp through San Bernardino Strait at 2130 on 25 October. Of his original force of 32 ships he escaped with 4 battleships, 4 cruisers and 7 destroyers. In the meantime Admiral Halsey came racing down from the north with his big battleships. They were not to fire a shot, however, for when they arrived it was too late. Admiral Kurita had escaped.

In an evaluation of the battle, Admiral Sprague's humble recognition of divine intervention must be read in conjunction with Admiral Halsey's radio dispatched on the day of impending disaster:

To prevent any misunderstanding concerning recent Third Fleet operations, I inform you as follows: It became vital on 23 October to obtain information of Japanese plans, so 3 carrier groups were moved to the Philippine coast off Polillo, San Bernardino and Surigao to search as far west as possible; on 24 October Third Fleet air struck at Japanese forces moving east through the Sibuyan and Sulu Seas; apparently the enemy planned a co-ordinated attack, but their objective was not ascertained and no carrier force located until afternoon of 24 October; merely

69 Rear Adm. C. A. F. Sprague Report to COMINCH, *Action Against the Japanese Main Body off Samar*, 29 Oct 44, G-3, GHQ, SWPA *Journal* (S).

to guard San Bernardino Strait while the enemy was co-ordinating his surface and carrier air force would be time wasted, so 3 carrier groups were concentrated and moved north for a surprise dawn attack on the carrier fleet; estimated the enemy force in the Sibuyan Sea too damaged to threaten Com7thFlt, a deduction proved correct by events of 25 October off Surigao; the enemy carrier force caught off guard, offering no air opposition over the target or against our force; evidently their air groups were land-based, arriving too late to fight; I had projected surface strike units ahead of our carriers in order to coordinate surface and air attacks against the enemy; Com7thFlt sent urgent appeal for assistance just when my overwhelming force was within 45 miles of the crippled enemy; no alternative but to head south in response to call, although I was convinced his opponents were badly damaged from our attack of 24 October, a conviction later justified by events off Leyte. . . .[70]

The command decisions during the Battle for Leyte Gulf require a realistic appraisal. The "events off Leyte" were the free entry of Admiral Kurita's Force into the Leyte area to a point where the complete destruction of the U. S. transports and escort carriers was an immediate and frightful possibility. Except for the superb and sacrificial intervention of a covering force under Admiral Sprague, the brilliant operation of Admiral Oldendorf in Surigao Strait, and the fortunate decision of Admiral Kurita, the results of the battle would have been different.

During the course of these critical naval operations, the Japanese air forces had intensified their attacks both on sea and shore. The Seventh Fleet was busily occupied in a struggle for survival, employing its air strength to fight off hordes of hostile planes attacking in co-ordination with enemy fleet units. Its problem was further complicated by the appearance of Kamikaze pilots whose novel suicide tactics initially caused considerable havoc.

On land, General MacArthur's troops without adequate air cover were exposed to continuous and damaging raids by enemy planes. Bombing and strafing from tree-top levels interfered with the unloading of supplies and hampered the progress of the advance into Leyte. In addition, under cover of their naval operations the Japanese had succeeded in landing 2,000 troop reinforcements from Cagayan at Ormoc on 25 October.[71]

Defeat of the Japanese Navy

The Japanese admirals had played for high stakes and lost. Their plan missed success, however, by only a slim margin. Had the Central Force adhered to its mission and proceeded into Leyte Gulf, the American invasion would in all probability have experienced a setback of incalculable proportions. The enemy's heavy guns would have experienced little trouble in pounding the remaining transports and landing craft. Shore positions and troop installations could have been bombarded almost at leisure and Admiral Kurita could then have continued on toward Surigao Strait. At this stage of the game, the weakened escort carriers and the ammunition-depleted battleships of the Seventh Fleet could have offered only relatively minor opposition. The intervention of the Third Fleet was too far removed to cause immediate tactical concern.

Within the narrow limits of a lost opportunity, the Japanese Navy, however, had suffered a crushing and fatal defeat.[72] The enemy fleet as an integral unit was no longer to be reckoned as a major factor in future operations, and his

70 COM3rdFLT Radio 251317/Z to CINCSWPA et al, 25 Oct 44, SOPAC No. 544, C/S GHQ (S).

71 Japanese Second Demobilization Bureau Report, *Philippine Area Naval Operations*, Part II.

72 The total losses inflicted on the Japanese Navy during the operations described were 3 battleships, 4 carriers, 9 cruisers, and 9 destroyers.

carrier force especially was now known to be impotent. The greatest sea battle in history left the remaining units of the Japanese Navy increasingly vulnerable to future Allied naval and air strikes. Against the background of this decisive naval victory, the strong wedge of General MacArthur's ground forces was driven solidly into the vulnerable Japanese flank on Leyte. If he could establish his forces securely in the central Philippines, the Japanese would be powerless to prevent him from overrunning the rest of the archipelago and bisecting the Japanese Empire.

In recognition of the overwhelming naval victory, General MacArthur sent the following message of appreciation to Admiral Nimitz :

> *At this time I wish to express to you and to all elements of your fine command my deep appreciation of the splendid service they have rendered in the recent Leyte operations. Their record needs no amplification from me but I cannot refrain from expressing the admiration everyone here feels for their magnificent conduct. All of your elements—ground, naval and air—have alike covered themselves with glory. We could not have gone along without them. To you my special thanks for your sympathetic and understanding cooperation.*[73]

Japanese Reaction to the Invasion

The decisive defeat of the enemy fleet in the waters of the Philippines meant the failure of only one phase of Japan's threefold plan for disrupting the Allied invasion of Leyte. Despite the loss of the naval battle, Japanese efforts in the air and on the ground were intensified rather than diminished. By bringing in plane reinforcements from Formosa and Kyushu to numerous existing bases in Luzon, they were able to maintain a continuous aerial offensive against the Allied transports and fleet units in Leyte Gulf throughout the landing period. Vigorous enemy air assaults began in the afternoon of 24 October with numerous low-level attacks upon Allied beachhead installations. During the next few days, a sustained series of heavy raids by 150–250 planes was directed against escort carrier forces and other surface units. Headquarters of the Sixth Army and General MacArthur's command post in Tacloban did not escape attack. Direct hits were scored within a block of General MacArthur's billet and the Headquarters Company was struck on Thanksgiving Day.

These raids marked the first appearance of the suicidal Kamikaze attack pilot, whose startling debut caused considerable consternation to Allied naval commanders and inflicted widespread destruction on U. S. fleet units in the Leyte area. On 29 October a suicide plane crashed into the Third Fleet carrier *Intrepid*. The next day, the carriers *Franklin* and *Belleau Wood* were struck, resulting in severe damage and loss of personnel. On 1 November the Seventh Fleet lost one destroyer to Kamikaze planes and suffered serious hits on five others. Four days later, the *Lexington's* signal bridge was blasted by a suicide plane. The necessity of dealing with the dangerous threat of these Kamikaze attacks forced the carriers to commit their planes to their own protection at the expense of furnishing support to the Leyte ground forces.

Allied efforts to counter these suicide attacks and to cover the advancing ground troops were hampered by a delay in construction of suitable airfields on Leyte. Heavy monsoon rains and difficult terrain had disrupted the schedule for development of the vital American air base at Tacloban. The Japanese naturally realized the importance of Tacloban and attacked the airstrip continuously, causing severe destruction to its closely parked planes and incomplete installations and seriously impeding its development.

73 CINCSWPA Radio 291311/Z to CINCPOA, 29 Oct 44, C/S, GHQ, SOPAC No. 545. (S).

One of these raids was particularly successful, destroying twenty-seven planes on the ground. The reckless Japanese pilots sometimes followed American flight formations into the landing areas. The explosion of ammunition dumps or oil storage tanks became an almost nightly occurrence. Except for the vicious air bombardment of Corregidor, at the outbreak of the war, never before in the Pacific had the Japanese blanketed an Allied position with such powerful, sustained, and effective air action.[74]

The land phase of the enemy's co-ordinated attack plan was also comparatively unaffected by the outcome of the naval battle. Taking full advantage of the temporary insecurity of the American air position, the Japanese poured a steady stream of reinforcements into Leyte from Luzon and the neighboring islands. The 2,000 troops which had been landed at Ormoc on 26 October were followed by other convoys of fresh forces. Additional units of the Japanese 30th and 102nd Divisions were sent from the Visayas and Mindanao and landed during the closing days of October. A first-class veteran outfit, the 1st Division, en route from Manchuria to Luzon, was diverted to Leyte and after disembarking successfully on 1 November was immediately committed to the defense of the Ormoc Corridor. At the same time elements of the 26th Division arrived to swell the total of Japanese troops on the island. These reinforcements made it clear that the enemy intended to hold Leyte at all costs and expected to fight for the Philippines in earnest.

The decision to make a major stand on Leyte was a modification of Japan's original plan to defend the Philippines. Prior to the Allied invasion it had been decided that the main ground battle would be fought on Luzon and that only delaying tactics would be employed in the southern Philippines and in the Leyte area.[75] This plan, it was hoped, would give the

74 The precarious air situation in the Leyte Gulf area was fully reported by the U. S. Navy at the time. "At present our control of the air is not satisfactory," said Admiral Kinkaid in his radio of 27 October. "Enemy air has been attacking in force for the past 72 hours, disrupting our unloading, refueling, and rearming. The activation of our dromes is being delayed as a result of enemy raids. Heavy combat air patrol is required and strikes against enemy airfields needed until enemy air strength is materially reduced. The support of at least 1 and preferably 2 fast carrier groups is required until we have control of the area. Remaining CVE's must refuel on 28 October, and can supply no planes. Army air squadrons are moving into the fields today, but difficult field conditions make doubtful effective operation." COMTF77 Radio 270221/Z to COM3rdFLT, CINCSWPA et al, 27 Oct 44, G-3, GHQ, SWPA *Journal*, 28 Oct 44 (S). On 26 October Admiral Halsey had reported as follows: "Two carrier groups will furnish support at Leyte on 27 Oct; after 17 days of fighting our fast carrier force is virtually out of bombs, torpedoes, and provisions and pilots are exhausted; I am unable to provide any extended direct air support. When will your shore-based air take over air defense at the objective?" COM3rdFLT Radio 261235 to CINCSWPA et al, 26 Oct 44, 385 Plan 21, G-3, GHQ, Adm (TS).

75 "Up until October 1944," stated General Utsunomiya, "strategy in the Philippines had been defensive in the Visayas and Mindanao areas and offensive in Luzon. . . . The strategy . . . did not directly change after the arrival of General Yamashita. The forces, according to General Yamashita's plan were still to delay the American units as long as possible to allow preparations for the meeting and annihilation of the enemy landing on Luzon." Lt. Col. Yorio Ishikawa stated: "Prior to the American landings on Leyte, the Fourteenth Area Army's plan of defense in the Philippines was to conduct a decisive battle on Luzon through the all-out employment of air, sea, and ground forces. . . . When the American invasion of Leyte was commenced, the Fourteenth Area Army was ordered to carry out a decisive battle on Leyte in spite of the fact that the Commanding General Yamashita, Chief of the General Staff Muto, and the majority of the staff members favored a decisive battle on Luzon." Interrogation Files, G-2 Historical Section, GHQ, FEC.

Japanese time to strengthen their forces on Luzon and thereby put them in a position to repel or at least greatly delay General MacArthur's conquest of the entire Philippines.

Several factors, however, influenced the Japanese High Command to alter the initial plan and to order that a decisive stand be made on Leyte. The mistakenly optimistic reports on the losses of the U. S. Third Fleet in the battle off Formosa and the consequent supposed weakness of the U.S. carrier fleet led to a belief that the Japanese land-based air forces could win control of the skies over the Leyte beachhead. It was also thought, in view of the fact that the U. S. landings came so soon after the Morotai operation and the Formosa battle, that General MacArthur might be committing his forces without full preparation or adequate protection. In addition, it was feared that should the Allies gain air bases on Leyte, any defense of Luzon would be made immeasurably more difficult. Therefore Imperial General Headquarters and Field Marshal Terauchi, in overall command of the Japanese Southern Army, felt that Leyte provided an excellent opportunity for a concerted effort by their combined air, naval, and ground forces to defeat and destroy the American invasion troops.

General Yamashita, in direct command of the Japanese ground forces in the Philippines, was opposed to this alteration of the original plan and was reluctant to pour his troops into Leyte for hasty deployment on an unprepared front at the expense of his Luzon defenses.[76] Large-scale U.S. carrier raids against the Philippines on the 17th, 18th, and 19th of October had convinced him that the U. S. naval forces were still extremely powerful and were evidently well prepared to cover any Allied invasion. Despite his objections, however, General Yamashita was ordered to dispatch reinforcements immediately and "annihilate the enemy invading Leyte." In accordance with these direct orders, he scrapped his original plan which called for delaying tactics on Leyte and commanded that "the Japanese will fight the decisive battle of the Philippines on Leyte." Despite numerous and damaging Allied attacks on the Leyte convoys, he had succeeded by early November in moving at least 25,000 additional troops and service elements into western Leyte to augment the original defense force.

The Advance Inland

While the Japanese were disembarking their reinforcements and maneuvering into position for a counterattack, General MacArthur's troops continued to drive inland and along the coast in a two-pronged attack and envelopment.[77] (Plate No. 63) Except for isolated strong points, such as Catmon Hill behind the XXIV Corps beachhead, only minor opposition was encountered, although enemy resistance was now more determined. By the end of October the XXIV Corps had secured a rough square bounded by Tanauan and Dulag on the coast and Dagami and Burauen, about 10 miles inland, at the western edge of Leyte Valley. Elements of the 7th Division moved south along the coast on Highway No. 1 toward Abuyog, eastern

76 Interrogation of General Yamashita after his capture on Luzon disclosed that he "insisted that sending reinforcements to Leyte was not his idea; that he was against committing additional strength and that the reinforcements were sent on orders from higher headquarters." G-3, Sixth Army, *Combat Notes*, Vol. X, p. 20 (R).

77 The guerrilla information net in the Philippines, backed by numerous radio stations, furnished valuable data in this critical period. The late General Roxas, first President of the Philippine Republic, helped to establish contacts in Manila in the highest circles. Through these channels important plans and major decisions of the Japanese High Command became known in fragmentary form. Messages were usually relayed to General MacArthur's Headquarters via Mindoro where an important radio outlet was maintained by SWPA secret agents.

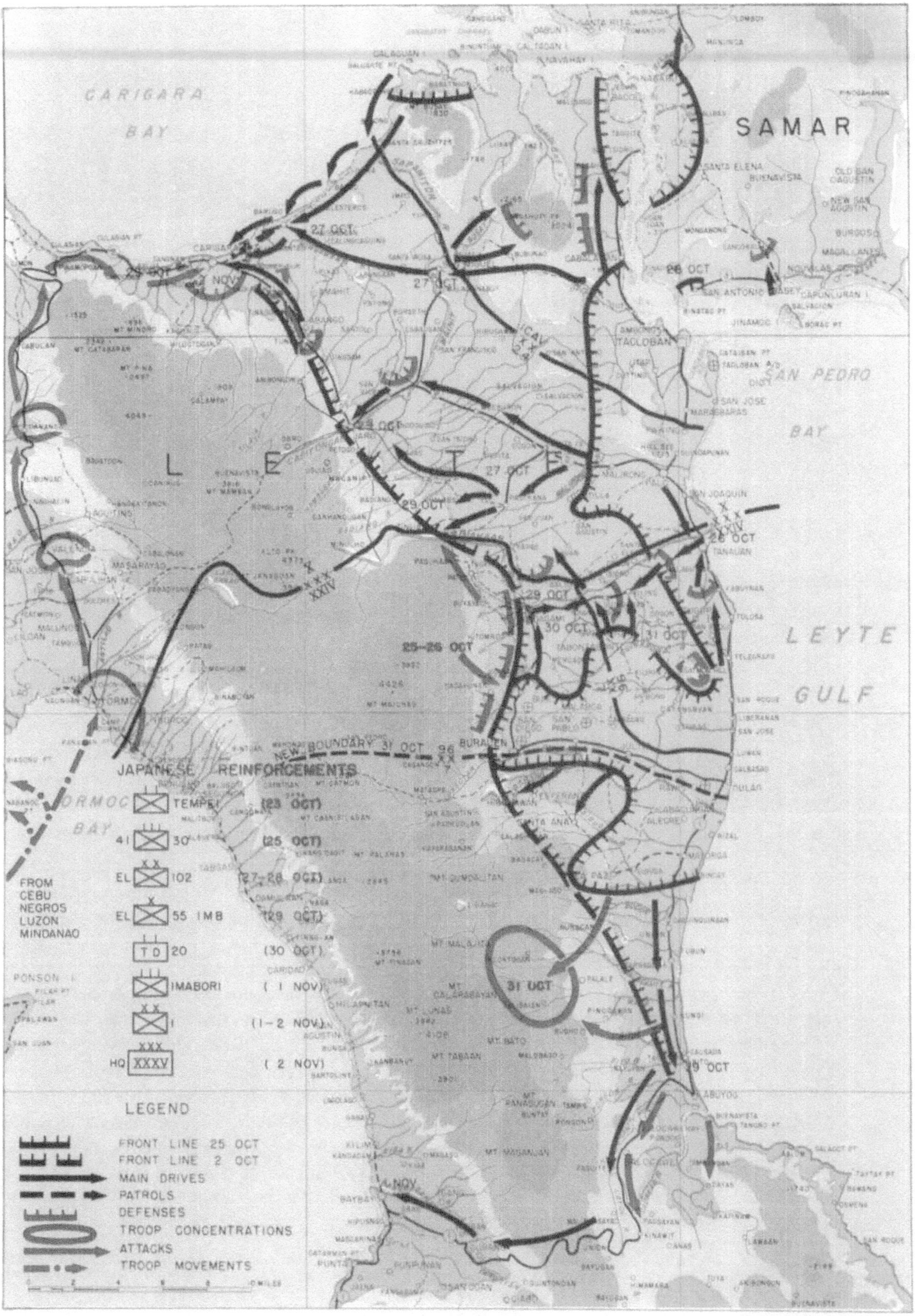

PLATE NO. 63

Leyte Attack Continues, 25 October—2 November 1944

terminus of a fairly good road running completely across Leyte from the Gulf to Baybay on the Camotes Sea. Other elements of this division, together with units from the 96th, established and maintained contact near Dagami. Stiffening enemy resistance was met west of the road from Dagami to Burauen, about eight miles to the south. The 96th Division established and maintained contact with the X Corps to the north.

In the zone of the X Corps the 24th Division, after securing Hill 522, moved west through Leyte Valley along Highway No. 2. Jaro, approximately 15 miles inland from Palo, was reached on 29 October in the face of stubborn resistance by the enemy intrenched in the hills and ridges along the valley. The 1st Cavalry Division meanwhile cleared both sides of San Juanico Strait, and by shore-to-shore movements instituted extensive patrolling on southwestern Samar. Similar amphibious operations took some of the cavalrymen around the northeastern tip of Leyte past Babatngon to Barugo on Carigara Bay.[78] Other units of the division, driving northwest from Tacloban into rugged mountainous terrain along the Diit River, moved through the northern reaches of the Leyte Valley to San Miguel, southeast of Carigara.

At this time several counterattacks in strength and increasing opposition in general indicated that the Japanese intended to make a determined stand at Carigara, a key terminus of the valley roads leading from Leyte's east coast. The X Corps, therefore, consolidated its units in preparation for a co-ordinated attack on Carigara by both the 1st Cavalry and 24th Divisions, moving along the shores of Carigara Bay and up Highway No. 2. On 2 November the combined attack was launched and the X Corps units moved forward only to discover that the town had been abandoned by the enemy. It became apparent that under cover of a successful delaying action, the Japanese had withdrawn to stronger positions in the high ground behind Carigara in an effort to spar for time until their reinforcements could be brought up from Ormoc.[79]

Resistance Stiffens

As the American forces pressed forward from the east coast of Leyte, the Japanese continued to move large reinforcements into Ormoc. There were indications late in November that the Japanese were sending troops of the 55th Independent Mixed Brigade from faraway Jolo in the Sulu Archipelago[80] in addition to the re-

78 The strategic consequences of this move were described by Maj. Chuji Kaneko, staff officer of the 102nd Division: "Our naval intelligence had estimated that the channel [between Leyte and Samar] could not be penetrated by boats, except small native boats. Elements of an American Division, I believe it was the 24th Division, made their way through the straits in small boats and barges from Tacloban and emerged onto Carigara Bay, where they effected landings on Barugo and Carigara. This surprise maneuver caught us off guard and upset the Thirty-fifth Army's plans to recapture Tacloban, since our units advancing on Tacloban by way of Carigara had now to contend with these troops that had now established positions to the south of Carigara." Interrogation Files, G-2 Historical Section, GHQ, FEC. This maneuver of the 1st Cavalry Division also prevented the intended landing of the enemy 68th Brigade on north Leyte for a pincer drive on the X Corps forces. Eighth Army, *Staff Study, Leyte.*

79 The Japanese were unable to retake Carigara. In the words of General Tomochika, "the loss of Carigara was a stunning blow to Japanese defense plans, especially to the 16th Division, whose main force was in the Dagami area, because Carigara was a key center of supply and communication to the entire Leyte Valley. Furthermore Carigara was important as a port for direct supplies by sea from Luzon. . . ." Eighth Army, *Staff Study, Leyte.*

80 G-2, GHQ, SWPA, *Daily Summary*, No. 977, 28/29 Nov 44.

mainder of the 1st and 26th Divisions which arrived from Luzon between 9 and 11 November.[81] Allied planes were not idle during these Japanese convoy movements. The entire enemy cargo of artillery and equipment which was sent in a subsequent shipment was destroyed in a savage Allied aerial attack and all five ships of the convoy including its destroyer escort were sunk. To circumvent the persistent attacks upon their convoys, the Japanese set up a barge shuttle system between Cebu and Ormoc to carry in troops from the Visayas under cover of darkness.[82] The enemy also turned Leyte's seasonal storms and unceasing rains to advantage by operating his convoys at full speed while the weather prevented U.S. planes from leaving their carriers.

The heavy and continuous tropical rainfall became a serious impediment to the scheduled progress of operations. Important roads were turned into rivers of mud, slowing and at times halting the movement of supplies to the advancing forces. Communication lines were constantly disrupted and contact between units was difficult to maintain. The poor drainage and swampy soil continued to delay the conditioning of the vital airstrips, leaving ground troops without adequate land-based air support for combat operations.

Again the key to the situation in the initial phase of the operation, as General MacArthur had anticipated, was air power. Not only was it important to provide adequate air cover for his ground operations but the immediate reduction of the enemy's defensive air strength was imperative. Realizing that the general strategic situation had changed, General MacArthur requested that the carriers of the Third Fleet continue to be used to give his forces additional air cover.

> *It is now evident that the enemy has decided to make a decisive stand in western Leyte in order to delay our further advance in the Philippines thereby gaining time for preparation for the defense of Luzon. The Leyte phase must be pressed with the utmost vigor to completion. It is therefore essential that the arrival of convoys now enroute be on schedule. The hazard to convoys must be accepted as a legitimate risk of war. In this general connection I am convinced that the support of fast carriers is likewise essential to current operations until Army air is established in Leyte in sufficient strength to hit all enemy air bases in the Philippines in full neutralizing force. I am requesting Admiral Nimitz to utilize the Third Fleet accordingly.*[83]

The dangerous shortage of airpower and the rapidly increasing opposition from a greatly strengthened enemy made it necessary for General MacArthur to bring additional forces into Leyte. Toward the middle of November the 32nd Division and the 112th Cavalry Regimental Combat Team, both veterans of many previous campaigns in the Southwest Pacific, were sent into action in the zone of X Corps.

---

81 The Japanese planned to use these two newly arrived divisions in a drive northwestward to secure the Carigara-Jaro sector. The general plan for the defense of Leyte was divided into three phases. Phase One: The 16th Division, reinforced, was to hold the line running through Burauen, Dagami, and northward in order to contain the United States forces on the east coast. Phase Two: By the middle of November, the bulk of the incoming reinforcements was to concentrate behind the 16th Division in the sector from Carigara to Jaro. The 1st Division was expected to occupy the area southeast of Carigara and later be joined by the 102nd Division and the 68th Brigade. The 26th Division was to advance from Ormoc, traverse the mountains and occupy the sector south of the 1st Division below Jaro. The 16th Division was to be further strengthened by the arrival of the main force of the 30th Division. Phase Three: A large-scale attack was to be launched on Leyte together with diversionary attacks against the rear lines from Samar. Interrogation of Col. Junkichi Okabayashi, Chief of Staff, Japanese 1st Division. Eighth Army, *Staff Study, Leyte.*

82 G-2, GHQ, SWPA, *Daily Summary*, No. 972, 19/20 Nov 44.

83 CINCSWPA Radio 130248/Z to COM7thFLT et al, 13 Nov 44, ANF No. 115, C/S GHQ (S).

Approach to the Ormoc Corridor

When it became apparent that the Japanese were committing the bulk of their forces in an offensive northward along the Ormoc Corridor, General Krueger decided to hold the initiative by constant pressure on all fronts. His strategy contemplated launching a strong assault southward from the direction of Pinamopoan while simultaneously pushing forward with units of both the X and XXIV Corps in order to penetrate into the Ormoc Valley. As soon as practicable, another arm of his forces would commence a drive northward along the coast from Baybay. If practicable, the vulnerable west flank of the Japanese would be attacked by an amphibious landing at Ormoc.

After occupying Carigara on 2 November, elements of the X Corps advanced westward to Capoocan and Pinamopoan, at the northern end of the Ormoc Corridor. (Plate No. 64) The 24th Division began an advance south from Pinamopoan down Highway No. 2 toward Valencia and Ormoc.[84] In the Limon area, however, the Japanese put up a stubborn fight from strong positions on the high ground of Breakneck Ridge and the progress of the division was bitterly contested. It was not until 14 November that the ridge defenses were overcome and the enemy forces cleared from the area. At this time, the exhausted elements of the 24th Division were relieved by fresh troops of the newly arrived 32nd Division which took up the fight for control of the Ormoc Corridor.

In an effort to approach the eastern flank of the Corridor, the 1st Cavalry Division, with units of the 24th attached, began a concerted drive southwest from the Jaro—Carigara road in the Mt. Badian sector. The mountainous terrain and torrential rains, combined with well-planned and fierce Japanese counterattacks, made progress exceedingly difficult. Even after the 112th Cavalry was added to the division on 14 November, the advance through the wild, hilly country against an unyielding enemy remained slow and arduous, and continued into December.

In the XXIV Corps sector the 7th Division reached Baybay, on the Camotes Sea, early in November.[85] Other elements of the 7th Division probed enemy positions in the mountains west of Dagami and Burauen. By the 25th of November, the bulk of the 7th Division had displaced to the west coast of Leyte. The 96th Division, taking up the attack in the central sector, moved westward in a series of strong actions, and then was halted by deeply intrenched enemy forces. At this point the Japanese were trying to hold in order to prepare a counterattack and all available remnants of the Japanese 16th Division were placed in

84 The Japanese countermeasures were described by General Tomochika: "The 1st Division was sent along the road north from Ormoc and then eastward along Carigara Bay to attack the American forces in the Carigara area. At the same time the 26th Division was sent over the mountains to attack in the direction of Jaro. The battle to retake Carigara never took place because, before the 1st Division could reach Carigara, American forces had landed near Pinamopoan and engaged the advance elements of the 1st Division. The 1st Division was compelled to divert its attention to repelling this landing force.... The best that the 1st Division could do was to set up battle lines along the mountains behind Pinamopoan." Eighth Army, *Staff Study, Leyte.*

85 Faulty intelligence caused the Japanese to commit a serious mistake in this action. "According to the information received from the 16th Division Headquarters, the important cross-island road from Baybay to Abuyog had been rendered impassable through the demolition of bridges and by bomb craters; therefore, no provisions were made to defend this road nor was it considered as a route of advance for Japanese units. The information proved to be erroneous and the mistake, made when the 35th Army failed to verify this information, resulted in great losses when the American 7th Division advanced westward along this road and flanked the 35th Army positions." Interrogation of General Tomochika. Eighth Army, *Staff Study, Leyte.*

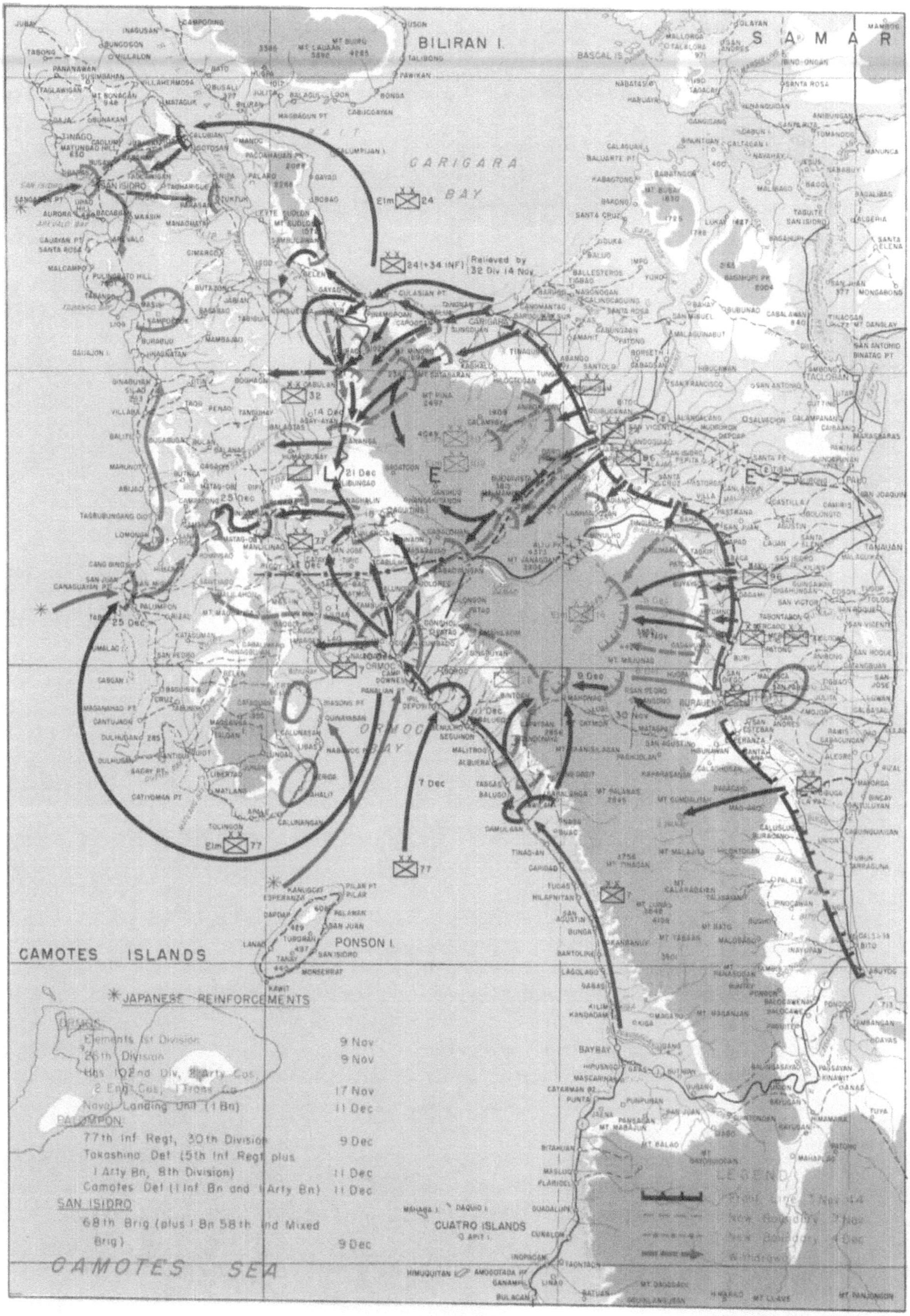

PLATE NO. 64

Ormoc—Carigara Corridor, 3 November—25 December 1944

positions well chosen to guard the southeastern approaches to the Ormoc Corridor. One regiment of the U. S. 96th Division moved northwestward from Dagami to outflank enemy positions and to secure the mountain passes leading westward in the Alto Peak area.

On 18 November, the 11th Airborne Division arrived on Leyte and was committed to action in the XXIV Corps area on the left flank of the 96th Division. Its mission was to control the passes into Leyte Valley and to secure the western exit from the mountains into the coastal corridor running along Ormoc Bay thus protecting the flank of the 7th Division. The paratroopers had to change practically overnight from an airborne to a mountain division as they moved across the rough terrain west of Burauen over a twisting, ill-defined road that wound toward Albuera on Leyte's west coast.

It was intended that the 11th Airborne Division ultimately link up with the 7th Division advancing north from Baybay. By 15 November, elements of the 7th Division had pushed their way to Damulaan about 13 miles north of Baybay. Here, it seemed, the Japanese planned to hold the Palanas River line just above Damulaan while they attempted to counterattack eastward toward Burauen. The 7th Division, therefore, consolidated its forces at Damulaan and prepared for a major drive northward in an effort to break through the enemy's river defenses. The concentration of the 7th Division on the west coast of Leyte compelled the Japanese to divert large forces from Ormoc and move them southward toward Damulaan. This not only helped disrupt the enemy's plan to recoup his losses by a drive eastward across the mountains but also served to weaken his garrison in the Ormoc area.

As the American ground attack pushed steadily forward, Army and Navy planes continued to give invaluable support to the Leyte operation. The Far Eastern Air Force interdicted potential reaction from the southwest by continuous and damaging raids against enemy bases on Borneo and in the Moluccas. Despite Kamikaze attacks and the handicap of adverse weather, carrier planes reduced the threat against U. S. shipping in Leyte Gulf by hitting at the source of the enemy's air counterattacks. Japanese airfields on Luzon were attacked repeatedly between the 5th and 19th of November. Whenever weather permitted, planes from the Third and Seventh Fleets did their utmost to harass enemy reinforcement shipping and hammered constantly at his convoys en route to Ormoc. After a final successful strike against enemy installations on Luzon on 25 November, the exhausted Third Fleet carrier force left the Philippines area for a merited and much-needed rest.[86] The protracted use of the carrier forces in the support of the Leyte operation necessitated extensive logistic replenishment, repair work, the replacement of casualties, and the recuperation of weary flight personnel.

The Japanese Attack Burauen

Heavy and unremitting pressure on every front kept the enemy off balance and prevented him from mounting any well-coordinated counteroffensive. Unable to mass his forces, he was compelled to commit in piecemeal fashion the reserves and reinforcements being brought into Leyte at such great sacrifice to his other battlefronts. The Japanese were determined to hold Leyte, however, and continued to invest heavily in its defense. By the end of November, despite large convoy losses and severe combat attrition, there were 10,000 more enemy troops on Leyte than there had been at the end of October.

86 CINCPOA Radio 292349/Z to CINCSWPA et al, 29 Nov 44, Navy No. 576, C/S GHQ (S).

In a desperate effort to gain the inititative, the Japanese attempted several ambitious airborne actions to disrupt U. S. airdrome installations prior to an attempted ground offensive. Their first efforts miscarried completely. On 26 November, enemy transport planes landed in the frothing surf on the east coast of Leyte in the sector of the XXIV Corps. Although some survivors managed to reach shore armed with demolition materials, they failed to accomplish any effective results. Another enemy transport plane attempted to land at Buri airfield but was quickly destroyed by antiaircraft fire.

Undeterred by these reverses, the Japanese maintained a vigorous defense of all Ormoc Corridor approaches, strengthened their positions, and regrouped their forces. On 6 December they undertook yet another airborne assault. This time the Tacloban and Dulag airfields were to be neutralized while the strips at Burauen were to be occupied and seized by airborne landings. At the same time, units of the Japanese 16th and 26th Divisions were to launch a drive to reach Burauen, penetrate the airfield defenses, and oust the Americans from control of the airdrome.

Preceded by bombing attacks and a smokescreen, about 250 hostile paratroops descended at twilight on the San Pablo and Buri airstrips. To coincide with these landings, enemy ground forces had begun a drive north and south of Burauen. Although the Japanese paratroops were initially successful, the ground forces in the vicinity had been so decimated and disorganized that their offensive efforts in the main were negligible and easily repelled. Scattered remnants of the enemy, however, managed to infiltrate onto the San Pablo and Bayug airstrips while a small concentration of parachutists held out on Buri.

United States forces took quick countermeasures to remove this threat to their positions. A strong assault by troops of the 11th Airborne Division was launched to regain control of the airdrome. By 11 December, the last remnants of the enemy had been eliminated and the Burauen area was declared secure. The Japanese had failed in their main objective but their short-lived assault did cause considerable damage and upset to some extent the supply mechanism of the Sixth Army areas.

Preparation for the Ormoc Landing

The slow development of airfields on Leyte, the resultant inadequacy of air support, and the urgent need for the rehabilitation of essential naval units caused a postponement in the next scheduled assault on Mindoro. Originally, the date for the Mindoro operation had been set for 5 December, but on 1 December General MacArthur directed that the operation be delayed for ten days.[87]

The deferment of the Mindoro landings had an immediate and important influence on the Leyte campaign. General Krueger had long recognized the decisive advantage to be gained by severing the enemy's supply line at Ormoc. Heretofore the lack of fighter cover to insure the safety of convoys entering the Camotes Sea and the fact that supporting naval forces did not have sufficient landing craft and resupply shipping to maintain amphibious operations prevented the mounting of any sizeable seaborne assault. The postponement of the Mindoro invasion released shipping previously earmarked

87 The postponement of the Mindoro operation came as a great relief to the Third Fleet, as explained by Admiral Halsey: "... the strain of the Kamikaze attacks, on top of our long stretch of combat, made an adequate rest period obligatory at once. MacArthur's next move, the invasion of Mindoro, was scheduled for December 5. We hated to request a postponement, but there was no help for it. Almost as soon as his obliging reply was decoded, we turned our prows toward Ulithi." *Admiral Halsey's Story*, p. 234.

for that operation and made available enough assault craft and naval support to carry out a landing in the Ormoc area. Moreover, the outlook for more effective land-based air support improved when five Marine fighter squadrons began operating from Leyte airfields on 4 December. General Krueger felt that the time was opportune for a strike on the west coast of Leyte and began immediate preparations to bring the necessary forces into position.

On 23 November the 77th Division, diverted en route from Guam to Guadalcanal, had landed on Leyte in the rear areas of the XXIV Corps in order to stage for the forthcoming assault on Mindoro. General Krueger decided to use this division in a major effort to force a rapid conclusion to the Leyte campaign. Accordingly he issued orders for the 77th Division to land in the Ormoc area in conjunction with a coordinated drive by units of the X and XXIV Corps.

The United States forces on Leyte commenced attacks in all sectors. On 5 December, the 7th Division at Damulaan drove northward and succeeded in breaching the enemy defenses at the Palanas River. At the same time the 11th Airborne Division on the east, and the X Corps on the north intensified their attacks on the Ormoc Corridor. In an effort to meet the heavy pressure developing simultaneously from three directions the Japanese hurriedly shifted all available forces to the threatened fronts. The combined land offensive by both X Corps and XXIV Corps units effectively diverted the bulk of the enemy reserves from the Ormoc sector and paved the way for an amphibious assault on Ormoc itself. The time had arrived when the enemy's "back door" to Leyte could be slammed shut.

Ormoc Captured

The assault on Ormoc marked an historic date. On 7 December, the third anniversary of Pearl Harbor, the 77th Division sailed around the southern tip of Leyte in a shore-to-shore movement from the east coast and landed near Deposito about three miles south of Ormoc.[88] The Japanese reacted immediately to the threatened sealing off of their main supply route and employed their remaining planes in the Visayas in an intensive effort to cripple the U. S. convoy. In spite of this aerial opposition which succeeded in inflicting serious damage on the ships and destroyers in the convoy, the division unloaded rapidly and established a firm beachhead.[89] The strategic implications of the Ormoc landing were described by General MacArthur in his communique of the day:

> *By this maneuver we have seized the center of the Yamashita Line from the rear and have split the enemy's forces in two, isolating those in the valley to the north from those along the coast to the south.*

88 "The Ormoc landing came as a complete surprise because the strait between Bohol and the southern coast of Leyte had been mined and Japanese Army Headquarters considered that not even LSTs could navigate through these waters. Furthermore, it was thought that the Bohol Straits were controlled by the Japanese navy and that the Americans would not attempt a maneuver so daring as a movement of troop transports into the landlocked waters. Because of these considerations, the defense of the Ormoc area had been limited to the 35th Army Port Unit (Mitsui), reserve elements of the 26th Infantry Division, and the right wing detachment of the 35th Army. These poorly equipped forces resisted stubbornly but were soon driven from the beaches. The Mitsui Port Unit was forced into the hills south of Ormoc. A counterattack was launched by the Imabori detachment on the night of the 8th, only to be repulsed. By the 11th, Ormoc had been completely overrun." Interrogation of General Tomochika. Eighth Army, *Staff Study, Leyte.*

89 The U. S. destroyers *Ward* and *Mahan* were so severely hit in this enemy attack that they had to be abandoned and sunk.

Both segments are now caught between our columns which are pressing in from all fronts. . . .[90]

The 77th Division moved northward against gradually stiffening opposition and advanced to the town of Ipil where heavy enemy defenses necessitated a temporary halt. In the meantime, the 7th Division had begun a drive from the Palanas River to pinch off the Japanese force caught along the coast, south of Deposito. This force, squeezed between the 7th Division on the south and the 77th Division on the north, had no choice but to retreat to the east and join other remnants of the battered units intrenched in the mountainous regions of central Leyte. Part of the 7th Division thereupon commenced a drive to link forces with the 11th Airborne Division still moving steadily westward from Burauen against well-fortified enemy positions.

The 77th Division resumed its advance from Ipil and, hammering down fierce enemy resistance, entered the town of Ormoc on 10 December. The next day the American divisions along the coast established contact at Ipil. This tactical junction was covered in a communique of 12 December:

The southern segment of the Yamashita Line has now been eliminated and the junction of the 7th and 77th Divisions has been effected. The enemy forces which were trapped between the two have been destroyed. Great quantities of supplies of all kinds were captured.[91] *The fighting in Ormoc itself before its fall was of the most desperate character, the entire defending garrison being destroyed. . . .*[92]

As soon as General Yamashita in Manila was notified of the American landings near Ormoc, he abandoned all plans for continuing the Burauen offensive and diverted additional troops to the west coast. Determined to hold Leyte at all costs, he ordered the Japanese 68th Brigade, whose original mission to support the Burauen attack was thwarted by the Ormoc landing, to land at Palompon on 7 December and approach Ormoc from the northwest. By this time General MacArthur's air arm had strengthened considerably, however, and the Japanese convoy was so battered en route by Fifth Air Force planes that the reinforcing brigade was compelled to make an emergency landing at San Isdro on the northwestern tip of Leyte where it found itself without artillery and completely cut off from its headquarters. Even in the face of these losses and a rapidly deteriorating battle situation, the enemy continued to send additional troops to Leyte. On 11 December another enemy convoy moving stubbornly into Palompon was attacked by U.S. planes and sustained heavy damage. On the same day enemy naval troops, ignorant of the American landing, disembarked at Ormoc itself—only a short distance from the occupying 77th Division. These final reinforcements, however, were destined to have little influence on the final outcome of the Leyte operation.

Leyte Secured

After consolidating in new positions above Ormoc, the 77th Division initiated an attack in full strength along and to the west of Highway No. 2. Its objective this time was Valencia,

90 GHQ, SWPA, Communique No. 975, 8 Dec 44.

91 The plight of the Japanese which resulted from the seizure of their supplies is told by Colonel Okabayashi: "On or about the third of December 1945, Japanese ships landing at Ormoc brought about a half month's supply of food and ammunition for the Japanese troops on Leyte. The American forces who landed near Ormoc on 7 December captured almost all of these supplies. Consequently, all Japanese troops on Leyte after the 1st of December were on a starvation diet and were compelled to live off the land gathering whatever they could find such as coconuts, bananas, papayas, native potatoes, a variety of grasses, bamboo shoots, and the heart fibres of coconut tree trunks. A powder of meal, prepared from native potatoes was used for making a crude bread." Eighth Army, *Staff Study, Leyte.*

92 GHQ, SWPA, Communique No. 979, 12 Dec 44.

about 10 miles to the north and a key position in the Ormoc Corridor. Although the Japanese fought bitterly for their last major stronghold on Leyte, Valencia and its adjoining airfield were in American hands by 18 December. The 77th Division requested authority to continue its drive northward and achieve a junction with the X Corps before the enemy could regroup his scattered forces. Permission was speedily granted and the division took up its advance along the highway.

In the meantime, the X Corps, operating in the north, had no easy time pushing down the Ormoc Corridor to meet the XXIV Corps units. The supply of forward troops had become a major problem. The 32nd Division, in the Limon area, and artillery along Carigara Bay were supplied by amphibious craft running between Carigara and Pinamopoan. To the southeast the 1st Cavalry Division was forced to supply itself by hand-carry over rain-soaked trails. Casualties, which could not be evacuated by Cub plane because of the lack of suitable landing terrain, were at first evacuated by native bearers and later by tractor.

From Limon, which was captured on 24 November, the 32nd Division pressed slowly southward against a fiercely defending foe, measuring many days' progress in mere hundreds of yards. The 1st Cavalry Division forces in the regions between the Ormoc Corridor and Leyte Valley continued to pick their way through the mountains until they were in a position from which they could debouch onto the main highway. On 19 December cavalry units captured Lonoy on the highway and cut off the enemy's supply line to his forces opposing the 32nd Division. Turning south, the cavalrymen then moved on to Cananga where contact was made with elements of the 77th Division.

The enemy had fought valiantly to hold his Leyte lifeline by maintaining control of the Ormoc Corridor but he found it impossible to cope with General Krueger's three-way offensive. Japanese forces were now chopped into isolated segments, either struggling in small pockets along Highway No. 2 or retreating into the comparative security of the mountains. Organized enemy resistance in Leyte's most fiercely contested region was finally brought to an end.

On 25 December, the last connecting road between the enemy's chief remaining port of entry for reinforcements and his troops inland was severed when 77th Division units, in an amphibious maneuver, landed near Palompon and seized the town. San Isdro Bay was likewise neutralized by a landing of a portion of the 24th Division at Calubian. The 24th Division troops bolstered the efforts of the guerrilla forces in that area and prevented further effective use of San Isdro harbor by the Japanese.

Samar, too, had been effectually denied to the enemy. Units of the 1st Cavalry Division had advanced against stiff resistance to the strategic Taft—Wright Highway which ran across central Samar. On 8 December the cavalrymen captured Wright, the western terminus of the highway, and then turned eastward to make contact with friendly guerrilla forces advancing from Taft on the opposite coast.[93] The tactical objective on Samar—control of the Taft-Wright Highway—had been achieved.

General MacArthur now controlled all major supply and communication routes of the enemy and on 25 December he declared Leyte secure. (Plate No. 65) The next day he asserted in his communique:

> *The Leyte—Samar campaign can now be regarded as closed except for minor mopping-up. . . . General Yamashita has sustained perhaps the greatest defeat in*

93 HQ Sixth Army, *Report of the Leyte Operation*, pp. 77–79.

the military annals of the Japanese Army.[94]

Leyte in Retrospect

The battle for Leyte demonstrated to a greater degree than ever before in the Pacific war the coordinated striking power of the American land, air, and naval team. Despite heroic efforts and great courage, despite the use of unprecedented numbers of ships, planes, and troops collected from all parts of their fast-shrinking empire, the Japanese were unable to stand before the overwhelming drive of American combined arms.

The Japanese suffered irreparable defeat in the Leyte campaign. Every phase of their ambitious Sho plan for the defense of the Philippines met with complete and irretrievable failure. Driven to near panic by the threat of losing the Philippines, Japan's wartime leaders in Tokyo had risked their remaining offensive power in one last effort to save their empire. Whatever chance the Japanese had to regain the initiative in the Pacific was lost on Leyte.

In the largest and most widespread sea battle in history, Japan's naval power had been smashed. The losses sustained by the Japanese fleet are among the most staggering on record. After Leyte, Allied warships could sail with impunity anywhere in the Pacific. Japanese air power as an organized force also was destroyed in the Philippines.[95] The employment of planes in suicide tactics not only reflected an extreme shortage of aircraft and trained personnel but also revealed the desperate plight of the minds that manipulated Japan's war strategy. The Kamikaze was the last fierce sputter of a burned out and dying air force.

On the ground General MacArthur's troops met the enemy for the first time en masse, with armies and corps pitted against each other in comparatively free maneuver. The Japanese, determined to retrieve their positions at any cost, dispatched troops from China and Manchuria and from all corners of the Philippines to the Leyte battlefront.[96] Paralleling their performance in New Guinea and in the Solomons, the Japanese yielded nothing without a bitter fight and it took daring and skillful maneuver on the part of the American soldier to pry them out of their well-prepared positions

94 GHQ, SWPA, Communique No. 993, 26 Dec 44.

95 The testimony of Admiral Fukudome, Commander, Second Air Fleet, July 1944 to 15 January 1945, tells the story of the defeat of the Japanese Naval Air Force in the Philippines: "Myself, together with my colleagues on the spot, felt that victory at Leyte was absolutely indispensable; and those in General Headquarters were of the same opinion. So there was agreement that every possible plane, as well as all possible Army forces, should be sent to the Philippines. I believe that up to the middle of December, the total air strength of between 600 and 700 planes was maintained. After that, however, replacement could not be continued to keep up that level. The losses increased as time passed and, from the middle of December, replacement could not keep pace with our losses; and by the early part of January, I had lost practically all of my planes, my air force had been practically wiped out. Replacements were not getting through owing to operations of your air force." USSBS, *Interrogation of Japanese Officials,* Vol II, p. 502.

96 The following is a brief historical resume of the enemy divisions which were engaged in the fighting on Leyte. The Japanese 1st Division from the Tokyo area was shipped directly to Leyte from Manchuria in November 1944 and was decimated in the northern Ormoc Corridor. The 16th Division from central Japan had participated in China operations and in the battle for Bataan; it remained in the Philippines and was the main force employed in the early phases of the Leyte fighting. The 26th Division had previously been stationed in North China and then on Luzon; it suffered heavy losses in convoy en route to Leyte and was destroyed in the final phases of the defense of the southern sector. The 102nd Division, activated in the central Visayas, was sent to Leyte in late October, and was defeated in the northern Ormoc Corridor. The 68th Brigade came to Manila from Formosa in December 1944 and was trapped on the west coast in the Palompon—San Isdro area. Elements of the 8th Division were committed in the Valencia sector where they were destroyed or driven into the hills.

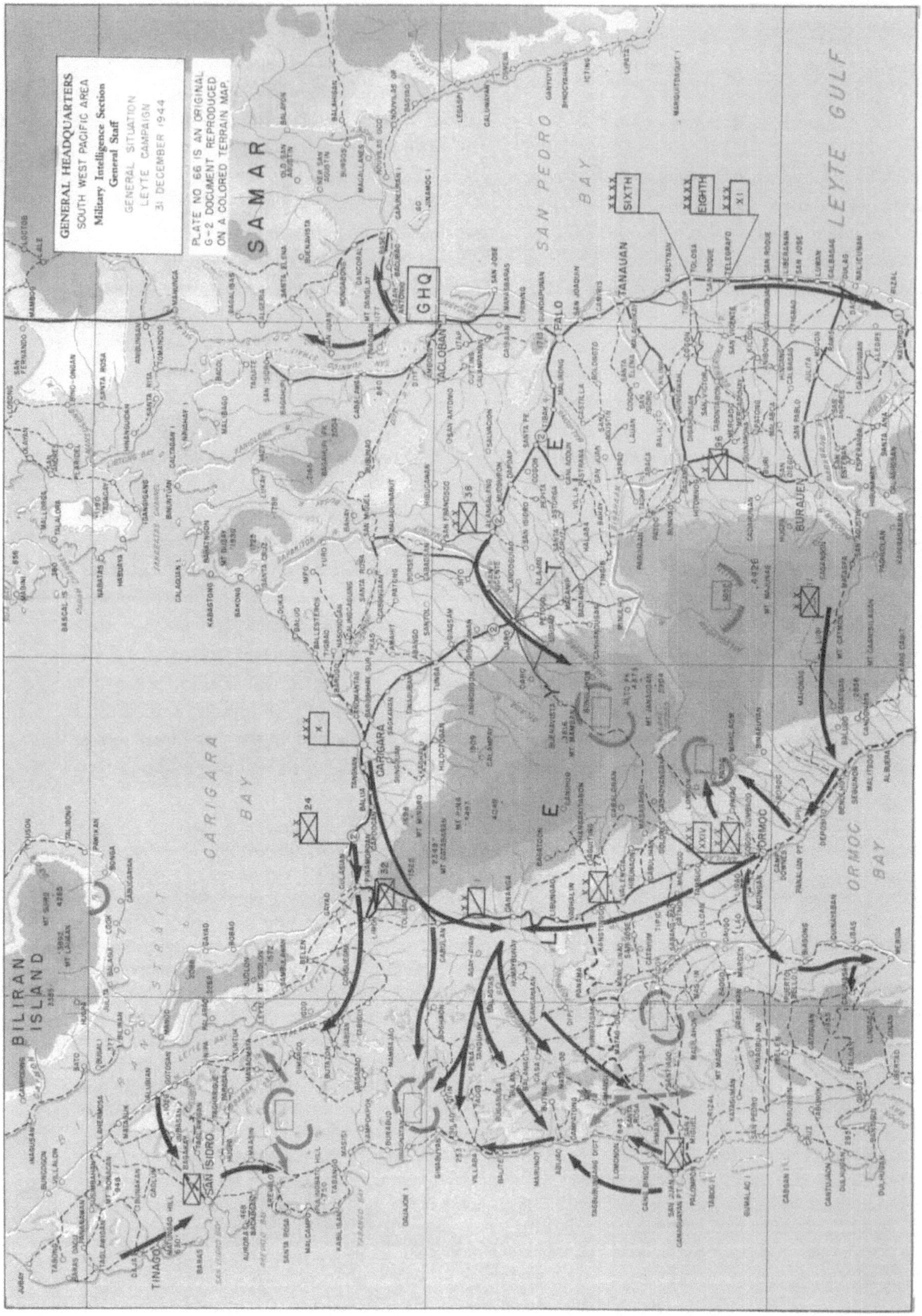

PLATE NO. 65

General Situation, Leyte Campaign, 31 December 1944

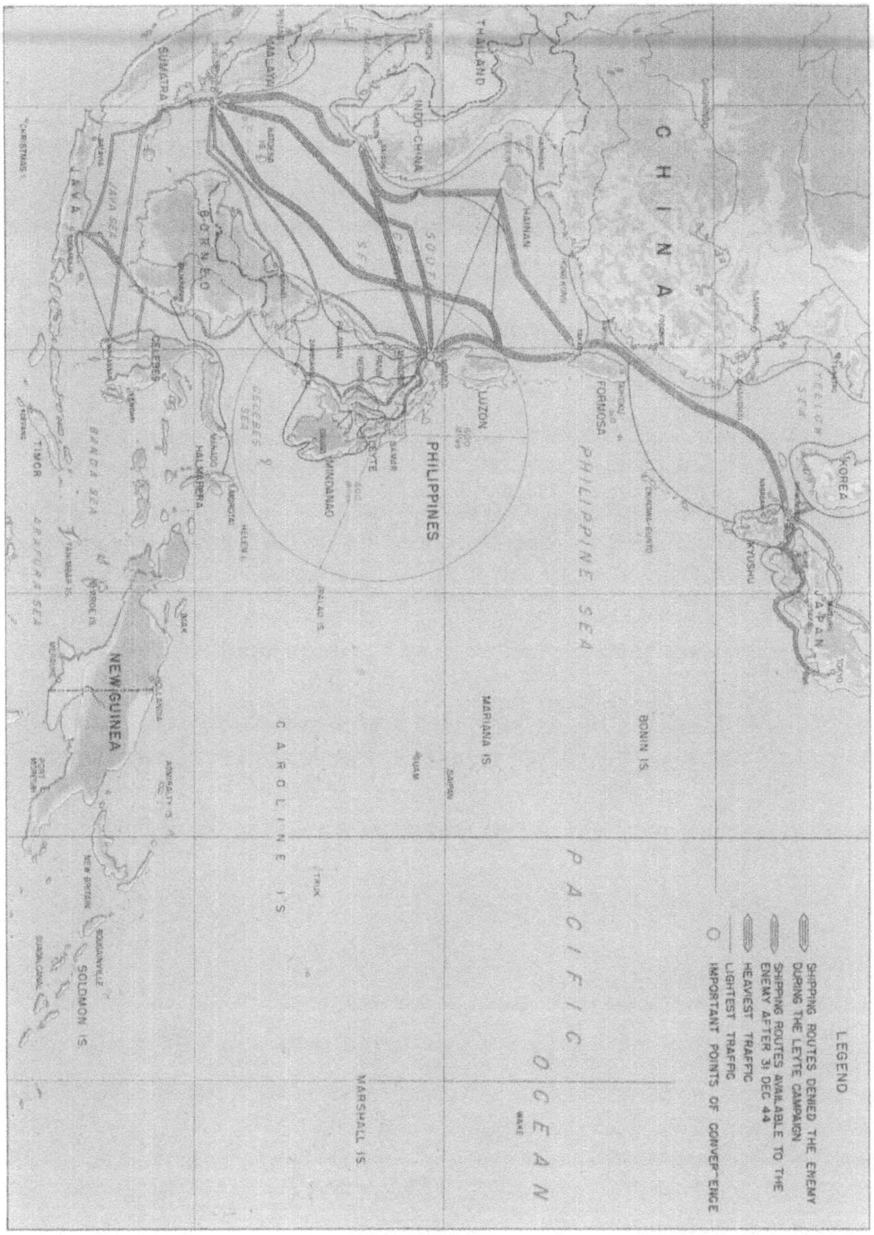

PLATE NO. 66

Enemy Shipping Routes Destroyed during the Leyte Operation

or break up their desperate counterattacks. All enemy efforts, however, proved in vain. Their movements were anticipated, their divisions decimated, and their troops forced to flee, starving and demoralized, into the hills and mountains of inner Leyte. General MacArthur's victory on Leyte was decisive and final.

Once more General MacArthur had caught the enemy off balance and had gained the decisive advantage of tactical surprise. The Japanese had anticipated an invasion of the Philippines but they did not know when or where the blow would fall. They were especially unprepared on Leyte. The speed with which the great convoy was unloaded and the rapid drive inland surprised and confused the Japanese.[97] In spite of having been consistently outmaneuvered in New Guinea, the Japanese apparently had learned little about the strength and versatility of the American ground, air, and naval machine. The Japanese failure on Leyte was also due in great measure to deficiencies in their procurement of intelligence and, by the same token, Allied success owed much to a superior intelligence organization.[98]

The enemy's piecemeal efforts to reinforce Leyte compromised his Luzon defense plan and drained the southern Philippines of combat strength. The cost to the Japanese in their reinforcement attempts was enormous. American air power crippled Japanese convoys and, while some troops were able to make shore, the bulk of combat materiel and supplies was often lost. Strong reinforcements did arrive at Leyte, but so great were the losses involved that General Yamashita's staff eventually accepted the view that an effective defense of the Philippines or even of Luzon alone had become impossible. The Japanese resolution to fight the decisive battle in the Visayas had come too late.

The battle for Leyte brought other valuable lessons into sharp relief. Although past operations in the Pacific had demonstrated what could be accomplished with controlled air power, Leyte showed how lack of aerial supremacy, at a critical juncture, could change the whole complexion of an operation. The definite need for the establishment of forward air bases in the early stages of a campaign was revealed more clearly in Leyte than in any previous operation. In the European Theater, bombers could operate from the same fields over long periods of time in view of relatively short flight distance, but Southwest Pacific air units seldom stayed in one locality more than two months before they had to establish new fields 300 to 500 miles away. Consequently,

97 The swift inland drive of the U. S. forces caught the Japanese unawares: "We had thought," said General Tomochika, "that the Allies, after landing, would try to secure their beachheads before proceeding with their advance; however, the enemy rather caught us off guard with their tactics of immediately starting their penetration inland even before their newly won beachheads had been secured." Interrogation Files, G-2 Historical Section, GHQ, FEC.

98 In a staff study dated 20 September 1944, Japanese Imperial General Headquarters stated: "The enemy has detailed information of our troop distribution in the Philippines." ATIS, G-2, GHQ, SWPA, *Current Translations*, No. 148, 6 Feb 45. Concerning intelligence in operations General Tomochika declared: "American intelligence was so far superior that a comparison is useless. It seemed to me as if we were fighting our battles blindfolded, while the enemy seemed to have ten times the intelligence we possessed." Major Kazuo Taguchi, Fourteenth Area Army staff officer, acknowledged that the "lack of an effective intelligence system and the ultimate failure to surmise correctly the mobility and speed of the American forces and the time and place of their actual landing attempts" was one of three "major Japanese mistakes." Said Colonel Kobayashi: "We had no inside information on American pre-invasion preparations. From ship movements and radio message traffic count we could estimate the probable site of landings. By past tactics and strategy, we were able to estimate the force, the method of landing and the time to some extent. We had no other intelligence." Interrogation Files, G-2 Historical Section, GHQ, FEC.

Pacific operations demanded mobile and flexible tactical air combat teams that could transfer their base of operations at a moment's notice to keep pace with the rapid forward strides of General MacArthur's ground forces.

The weakness in offensive air power in the early phases of the Leyte operation was direcly attributable to adverse weather and difficult terrain. A combination of continuous typhoons, heavy rains, poor drainage, and thick silt formations made it impossible to condition forward airfields as rapidly as the growing scope of the campaign demanded. The Japanese could therefore carry out their initial reinforcement program with comparative success. As a result, United States ground forces met increasing numbers of enemy troops and the successful conclusion of the battle was prolonged beyond schedule.

The Leyte operation emphasized the inherent risks of divided command. In the naval action, two key American commanders were independent of each other, one under General MacArthur, and the other under Admiral Nimitz. Both operated in the same waters and in the same battle. Divided command affected the Allies primarily in the sea fight for Leyte Gulf, but it also ran like a cancer throughout the operations of the Japanese armed forces in the Philippines.[99]

The Japanese defeat on Leyte brought American forces almost to the China Sea, placing them in a position to sever the sea routes connecting Japan with her Empire to the south. This menace to the safety of her sea lanes drastically reduced the volume of traffic between Japan and her conquered territories. (Plate No. 66) With access to her resources in the Netherlands East Indies and Malaya in danger of being completely cut off, Japan's hopes for successfully continuing the conflict were rapidly diminishing.

General MacArthur had kept his faith. His return to the Philippines was a dramatic and singular fulfillment of historical destiny. A year before the assault against Leyte, his forces had been deep in the tangled jungles and swamps of New Guinea, almost 1,500 miles from the Philippines. Now he had led them into the very heart of the islands—into a position to become masters of the archipelago.[100] On the third anniversary of the Japanese attack against the Philippines General MacArthur's forces were organizing the final blows which would lead to the capture of Manila and the surrender of General Yamashita's troops on Luzon. The dark shadow of defeat was edging ever faster across the face of the rising sun of Japan. The hour of total eclipse was not far off.

99 A summary of the cumbersome Japanese command situation in the Philippines is contained in a report by the U. S. Sixth Army: "General Yamashita's immediate superior was Field Marshal Terauchi, Commander of the Southern Army, which had its headquarters at Saigon and comprised the Malaya, Burma, French Indo-China. N.E.I., New Guinea, Solomons, Admiralty Islands, and Philippines area. Although General Yamashita came into direct command of all ground forces and of air and naval troops on Leyte and Luzon at the time of the U. S. landings on each of these islands, all tactical air units in the Philippines remained under Terauchi's direct command, and General Yamashita had no control over their employment. At the same time, all naval fleet units were controlled directly from Tokyo. This command set-up permitted no co-ordination of the three services and was a constant source of discord throughout the campaigns. Thus, for example, Yamashita knew nothing of the intents of the Japanese Fleet off Leyte until five days before the historic 24–25 October actions." Sixth Army, *Combat Notes*, Vol X, p. 19.

100 General MacArthur's long and brilliant record in the service of his country won him promotion to the newly created rank of General of the Army on 18 December 1944.

CHAPTER IX

THE MINDORO AND LUZON OPERATIONS

Plans for the Northern Philippines

The curtain was raised on another stage of the Philippine theater while the concluding act of the Leyte operation was still in progress. This time the scene was set in Mindoro—the last steppingstone to the island of Luzon. The occupation of Mindoro as an immediate prelude to the main assault on Luzon was outlined in the final "Musketeer" plan of the Southwest Pacific Area.

The sudden advancement of the date for the Leyte invasion had necessitated certain revisions in General MacArthur's original scheme of maneuver for the recapture of the Philippine Islands. To include the latest developments in the fast-changing picture, a third "Musketeer" plan was prepared on 26 September, covering operations to take place as soon as the seizure of Leyte was successfully accomplished.[1]

Under "Musketeer III," the first operation scheduled to follow the initial entry into the Philippines was the seizure of southwest Mindoro, contemplated for 5 December 1944. (Plate No. 67) The full-scale invasion of Luzon with landings along Lingayen Gulf was projected for 20 December, when it was expected that the airfields on Mindoro would be ready for use. Contingent provision was also made to support the Lingayen landing, if necessary, by an assault at Dingalan Bay on the east-central coast of Luzon. To provide for a fluctuating tactical situation, a preliminary attack at Aparri, in northern Luzon, was also tentatively outlined. This last operation, however, would take place only if carrier-borne aircraft could not insure uninterrupted transit of naval assault shipping around northern Luzon should such an alternate route be chosen. Concurrently with these operations, all other available Southwest Pacific forces were to undertake the consolidation of the Visayas, Mindanao, Palawan, and the Sulu Archipelago. It was planned that the Eighth Army would conduct these mop-up campaigns with such assistance as could be obtained from guerrilla units organized throughout the central and southern Philippines.[2]

Approval of "Musketeer III" by the Joint Chiefs of Staff was dependent upon a final decision on the next major operation to be undertaken after Leyte. General MacArthur's plans called for an immediate invasion of Luzon as soon as success on Leyte was assured. Other ideas were being studied for by-passing Luzon and shifting the main effort of the Allies farther north.[3] For some time, a direct attack on Formosa had been advocated to eliminate that island as an enemy assembly area and advance air transfer point. Allied occupation of Formosa would enable the utilization of its

1 GHQ, SWPA, *Musketeer III, Basic Outline Plan for Revised Philippine Operations*, 26 Sep 44. Hereinafter cited as: Musketeer III.

2 Ibid.

3 JCS (thru Marshall) Radio No. W-50007 to CINCSWPA and CINCPOA, 17 Jun 44, C/S GHQ, WD No. 759 (S).

airfields for heavy bombers to operate in conjunction with the newly acquired bases in the Marianas; at the same time the island would provide a staging area close to the heart of Japan. In addition, it was pointed out, Formosa could be used by the Allies to assist operations in China and prevent the consolidation of the Chinese coastal regions by the Japanese.[4]

General MacArthur did not believe that any operation should be undertaken against Formosa until the northern Philippines were firmly in Allied hands. Without Luzon, an invasion of Formosa would be logistically precarious in view of its great distance from the nearest available Allied bases, necessitating the movement of forces over extremely long and attenuated lines of supply.[5] On the other hand, he felt that the Central Plain—Manila area on Luzon could be cleared of organized enemy units by February 1945 if the Lingayen landings were made on schedule, permitting operations north of the Philippines to be carried out earlier and with less risk than would otherwise be feasible since it was generally agreed that an assault against Formosa could not be prepared before February. Then, too, General MacArthur estimated that, with the obvious tactical and strategic advantages inherent in the establishment of extensive air and supply bases on Luzon, an invasion of Formosa would become unnecessary.

Another major consideration in the choice of targets was the general shortage of service troops in the Pacific Theater. An invasion of Formosa would undoubtedly necessitate a large-scale transfer of service personnel from the Southwest Pacific Area. Already critically short for operations in progress, General MacArthur was unable to commit any service units from his command to an assault on Formosa. In this respect, he pointed out that the native population was an important factor in planning future operations. In Luzon, the Allies could depend upon thousands of loyal Filipinos to augment service troops by working as stevedores, engineers, carpenters, drivers, and in other fields of skilled and semi-skilled labor; the guerrillas, too, would undoubtedly play a large part in assisting a Luzon assault. In any event, General MacArthur contended, an invasion of Luzon as the next step in the Allied advance would be the most logical development in the strategy of Pacific operations since Formosa would not be a secure and satisfactory base so long as Luzon remained in enemy hands.

On 3 October 1944, after careful examination of all the complex factors involved, the Joint Chiefs of Staff made their decision. In accordance with the general outline of SWPA's "Musketeer III" plan, General MacArthur was instructed to occupy Luzon with a target date of 20 December 1944. (Plate No. 68) He was also to establish bases on northern Luzon to support further Allied advances, including an assault by the Central Pacific forces against the Nansei Shoto (Ryukyu Islands), an operation set tentatively for 1 March 1945.[6] Directives for possible operations against Formosa were withheld, pending further developments in the Philippine Campaign.

Choosing the Route to Luzon

As the Leyte operation progressed and the enemy's Kamikaze attacks became more damaging, some questions again arose as to the advisability of a direct approach from Leyte to Luzon. Naval commanders feared that an Allied convoy passing through the narrow,

4 C/S WD Radio No. 55718 to CINCSWPA, 24 Jun 44, C/S GHQ, WD No. 764 (S).

5 CINCSWPA Radio No. CX-13891 to C/S WD, 18 Jun 44, C/S GHQ, WD No. 761 (S).

6 JCS Radio No. WARX 40782 to MacArthur and Nimitz, 3 Oct 44, C/S GHQ, WD No. 852 (S).

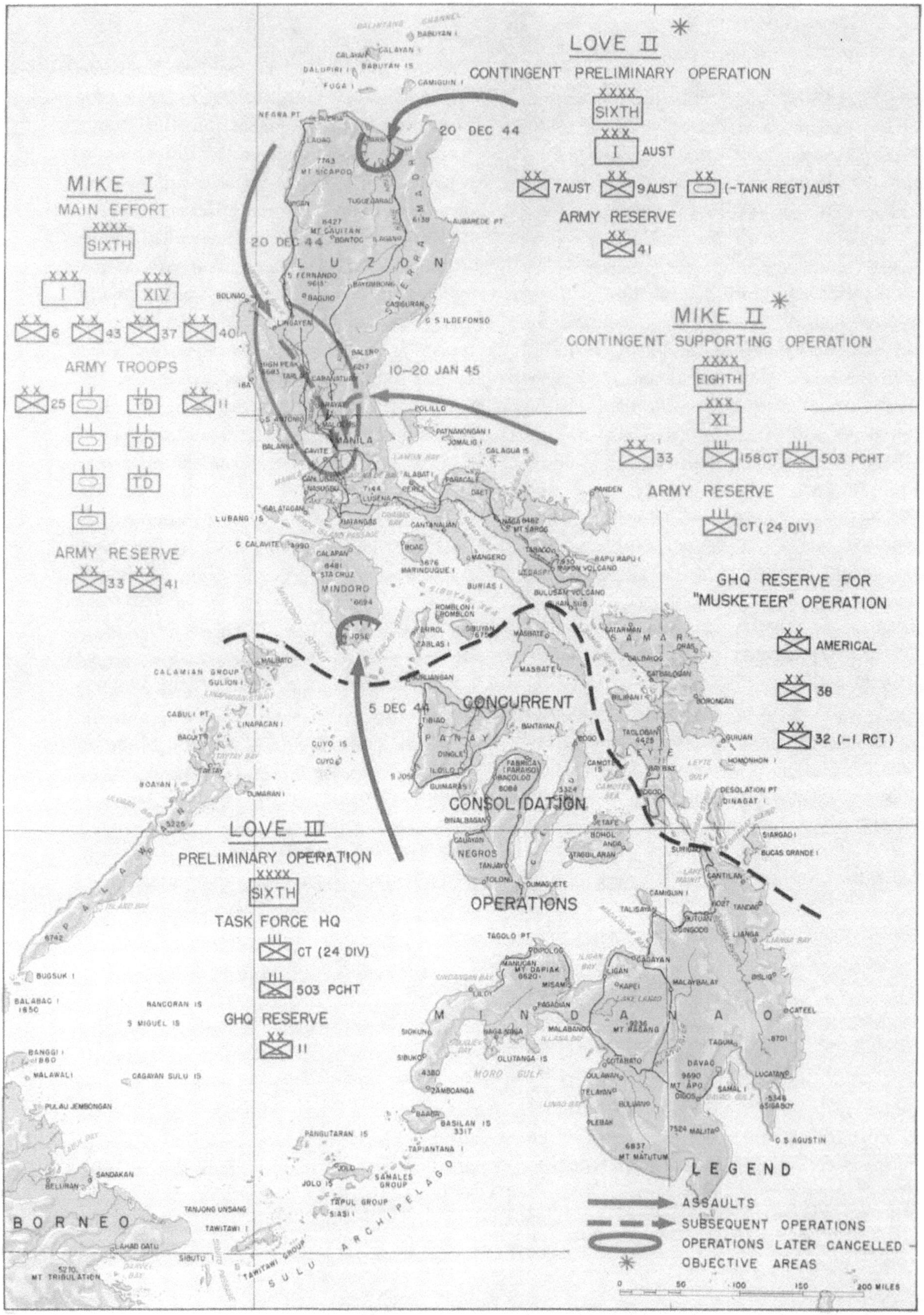

PLATE NO 67

"Musketeer III" Plan: Love and Mike Operations

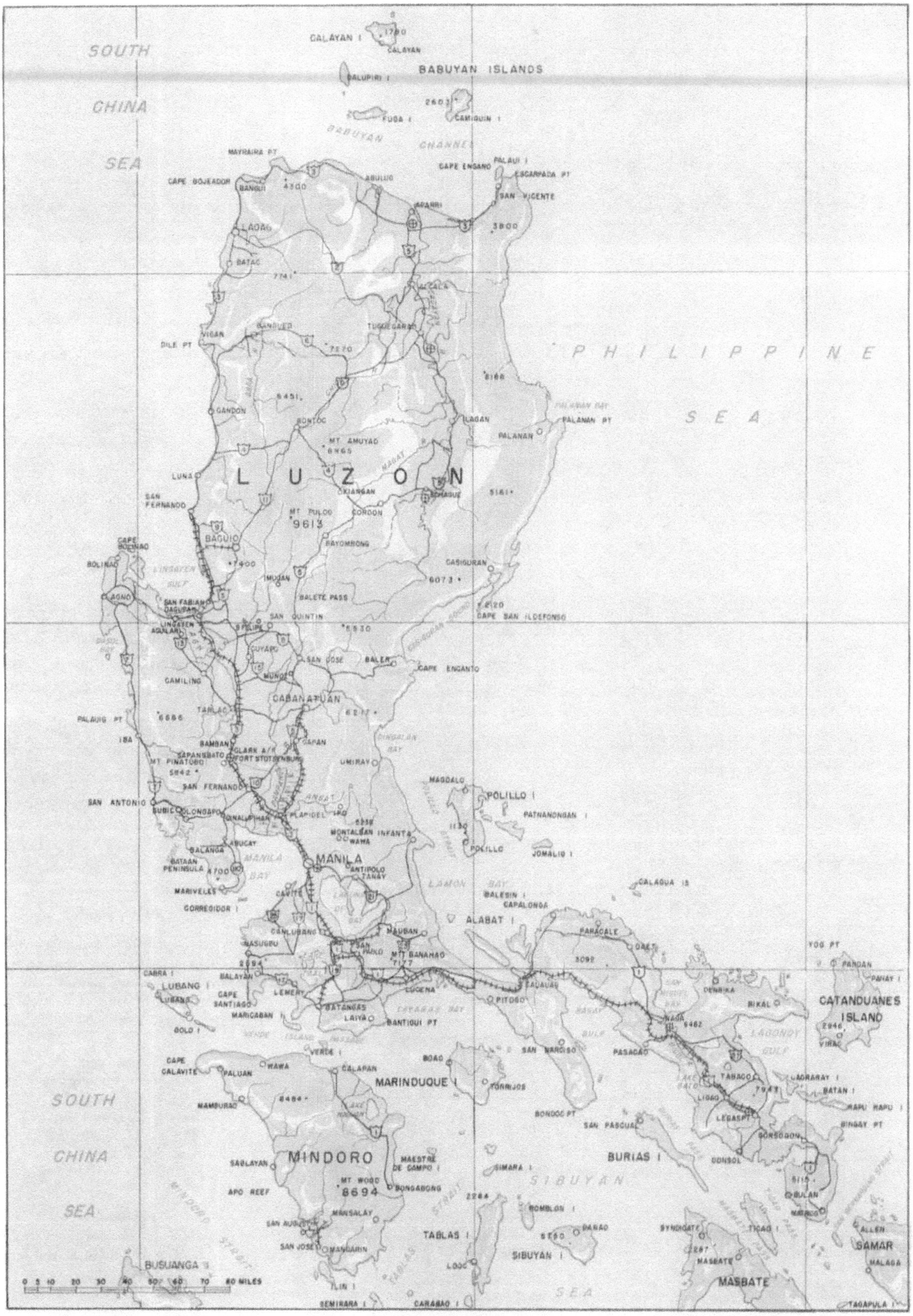

PLATE NO. 68
Luzon

restricted waters of the central Visayas would be subject to heavy air attack from the many enemy airfields in the surrounding islands. It was proposed that the alternate plan for the prior seizure of positions on Aparri be carried out to insure fighter cover for resupply ships which could then be routed around northern Luzon where more open seas permitted greater protection and safer passage.[7]

General MacArthur and his staff were in favor, however, of moving the Seventh Fleet and the necessary amphibious forces to Luzon along the southern route from Leyte via Mindoro. Although the danger of enemy air attacks would have to be faced during the initial invasion of Mindoro and the subsequent supply of forces located there, it was felt that once the new airfields were established, the Allies would have a much shorter route to Lingayen Gulf with greater air protection and less exposure to the vicissitudes of typhoon weather than the suggested route around northern Luzon. In addition to these considerations, there was the added factor that an invasion of Aparri would delay the all-important Lingayen operation by at least a month.[8]

After a series of conferences between members of General MacArthur's and Admiral Nimitz's staffs early in November, it was generally agreed that the major amphibious forces would approach Lingayen Gulf through the Visayan waters while the powerful ships and carrier planes of the Third Fleet operated north of Luzon in cover and support missions; it was also decided that the Aparri operation need not be undertaken.[9]

Final Plans for the Mindoro Landing

Even before the Sixth Army had left Hollandia for the Leyte assault, General MacArthur had issued instructions to General Krueger to prepare tentative field orders for the Mindoro invasion. Many of the details for the projected landings were worked out aboard ship, while the convoy was en route to the Leyte beaches. To conduct the ground operations on Mindoro, General Krueger constituted the Western Visayan Task Force, under the command of Brig. Gen. William C. Dunckel.[10] The initial plan contemplated a landing on 5 December in a combined airborne and amphibious maneuver which would secure the San Jose area near the southwest coast in order to make

---

7 Walter Karig, Russell L. Harris, and Frank A. Manson, *Battle Report*, Vol V, "Victory in the Pacific" (New York, 1949), p. 156.

8 CINCSWPA Radio No. C-51429 to C/S WD, 2 Nov 44, GHQ, WD No. 873 (S).

9 CINCSWPA Radio No. C-51706 to C/S WD, 7 Nov 44, G-3, GHQ Admin 385 Plan 21 (TS).

10 The major units participating in the Mindoro assault were as follows:

Ground: *Western Visayan Task Force (under Sixth Army)—Brig. Gen. William C. Dunckel*

Unit	Commander	Phase	Staging Point	Landing Point
19th RCT, 24th Div	*Lt. Col. G. H. Chapman*	*Assault*	*Leyte*	*San Jose*
503rd Prcht RCT	*Col. G. M. Jones*	*Assault*	*Leyte*	*San Jose*

Air: *Fifth Air Force—Maj. Gen. Ennis C. Whitehead*

Unit	Commander	Unit	Commander
310th Bomb Wing	*Col. J. A. Wilson*	*49th Fighter Gp*	*Lt. Col. G. A. Walker*
17th Bomb Gp	*Lt. Col. H. S. Ellmore*	*418th Night Fighter Sq*	*Maj. C. C. Smith, Jr.*
3rd Bomb Gp	*Col. J. P. Henebry*	*17th Tac Rcn Sq*	*Maj. J. C. Wise*
90th Bomb Gp	*Lt. Col. W. H. Banks*	*318th Troop Carrier Sq*	*Maj. C. G. Carter, Jr.*
85th Fighter Wing	*Col. P. H. Greasley*	*25th Photo Sq*	*Maj. H. A. Curran*
58th Fighter Gp	*Col. G. G. Atkinson*		

Naval: *Seventh Fleet—Adm. T. C. Kinkaid*

Unit	Commander	Unit	Commander
Task Group 78.3	*R. Adm. A. D. Struble*	*Task Group 77.3*	*Adm. D. E. Barbey*
Task Grou 77.12	*R. Adm. T. D. Ruddock (CVE)*	*Task Group 70.1.4*	*Lt. Comdr. N. B. Davis, Jr.*

immediate use of its airstrips for the support of the Luzon operations[11] and to counter the many Japanese airfields located on Luzon. (Plate No. 69)

The delay in the development of Leyte airdromes and the continuing need for air support of the Leyte ground forces caused several changes in this original Mindoro plan. Fighters and bombers occupied all available space on the Leyte airstrips, and there was no room for the large transports which would have to carry the paratroops to Mindoro. Consequently, the airborne phase was cancelled and, instead, arrangements were made to transport the parachute regiment by water in LCI's.[12]

The status of airfield construction on Leyte also made it questionable whether land-based planes would be in position by 5 December to cover the dangerous interim period between the landings and the activation of new strips on Mindoro. The peculiarity of weather conditions in the Philippines further complicated the problem. Fifth Air Force planes based on Leyte were handicapped by the seasonal bad weather common to the eastern Visayas, while flying conditions prevailing over enemy airfields in the Sulu Sea area were usually excellent. Hostile aircraft flying in from the west would thus be permitted to attack Allied convoys at Mindoro with little fear of counteraction by grounded Allied planes.

After thoroughly considering the airfield construction problem, various naval recommendations, and the relatively low level of supplies on Leyte, General MacArthur decided to postpone the Mindoro operation for ten days to 15 December.[13] The interlude was used to good advantage. While the Leyte airstrips were being developed and repaired, the Allied landings at Ormoc sped the Leyte operation to a successful culmination. In addition, the orders for the Mindoro assault were amended to include the seizure of the Calapan area on the northern coast and the occupation of Marinduque Island, to the northeast, to help foster an illusion that the next major Allied strike would be against the Batangas—Bicol region of Luzon.

Occupation of Mindoro

On 12 December 1944, the Western Visayan Task Force, escorted by warships of the Seventh Fleet, sailed from the Dulag area of Leyte and set a course for Mindoro via Surigao Strait. The voyage through the restricted waters of the Philippines was not without incident. The Japanese were quick to spot the convoy and, although most of their efforts were relatively minor, one strike was particularly damaging. In the early afternoon of 13 December, a single-engine Kamikaze plane, swooping in from the direction of Siquijor Island, skimmed swiftly across the glassy waters of the Mindanao Sea and crashed headlong into the super-structure of the *Nashville*, the flagship of the Task Force.[14] Losses in the resultant explosions were heavy, with estimated casualties of 131 killed and 158 wounded. General Dunckel and several of his staff were hit by flying fragments and the *Nashville* itself was put out of action and forced to return to Leyte.[15]

As the amphibious forces sailed through the Sulu Sea, powerful groups of the Third Fleet threw a neutralizing blanket of carrier planes over Luzon from the east. Continuous air

11 HQ Sixth Army Report to TAG, 22 Apr 45, G-3, GHQ, SWPA *Journal* (S).

12 Ibid.

13 CINCSWPA Radio No. 301358/z to CINCPOA, COM3rdFLT, 30 Nov 44, C/S GHQ, Navy No. 577 (S).

14 CTG 78.3 Action Report, Serial 00308 to COMINCH, *Report of the Mindoro Operation*, 7 Mar 45, G-3, GHQ, SWPA *Journal* (S).

15 COMPHIBGP 9 Report, Serial 00308, to COMINCH, 7 Mar 45, File No. A16–3 (S).

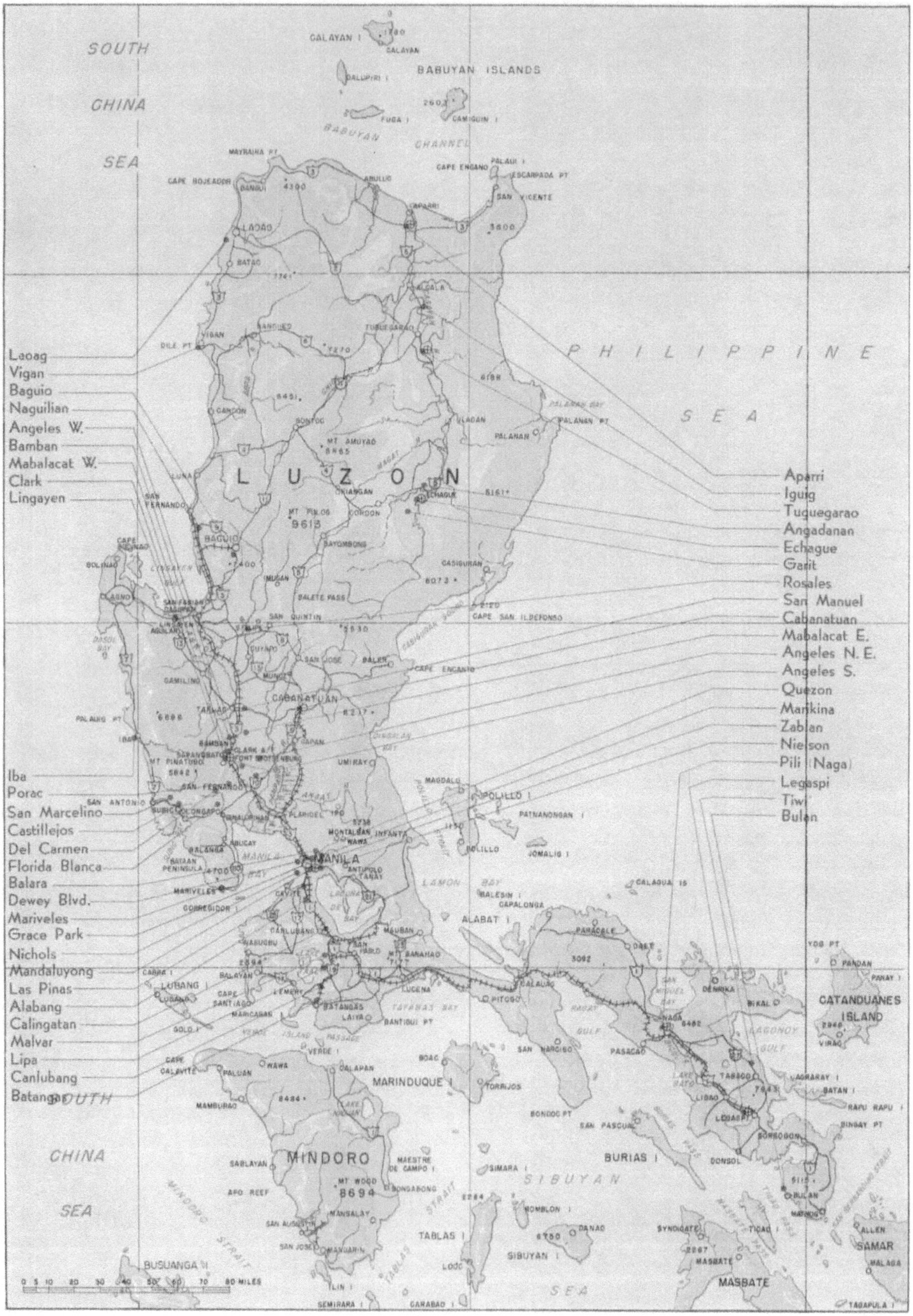

PLATE NO. 69
Japanese Operational Airfields on Luzon, 31 December 1944

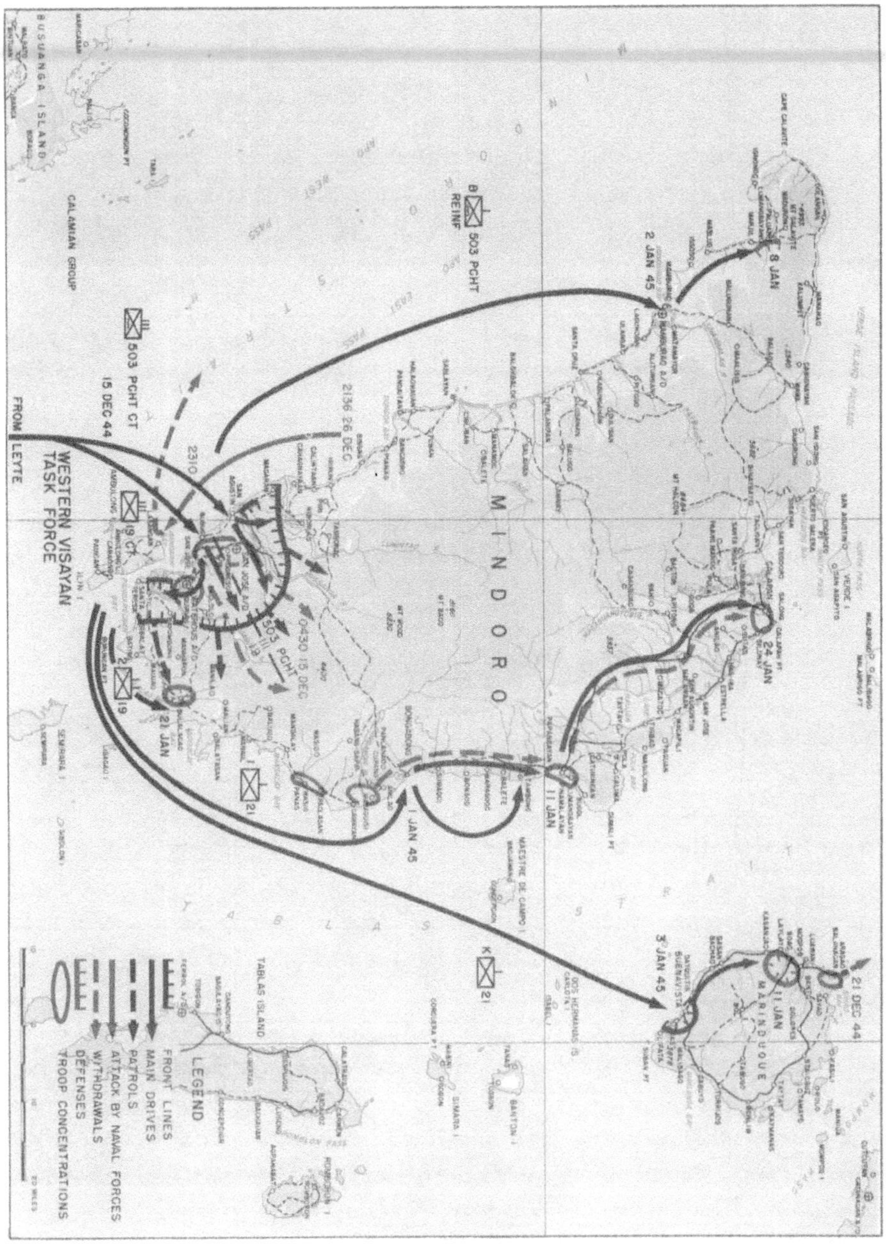

PLATE NO. 70
Mindoro—Marinduque

patrols over enemy airfields prevented any sizeable sorties by Japanese aircraft. Control of the air corridors over Luzon was held until the evening of the 16th when a raging typhoon struck with devastating effect and forced the Third Fleet to postpone its air operation.[16]

Although the direction and timing of General MacArthur's next move after Leyte was, in general, correctly estimated by the Japanese, an actual landing in force on Mindoro was not considered probable. In early November, the staff of the Japanese Fourteenth Area Army judged that the U.S. Sixth Army would achieve tactical control over Leyte by the first part of December. General MacArthur's next assault, it was assumed, would be made somewhere in the western Visayas, probably on Panay or Negros.[17]

This conclusion was based upon the supposition that any amphibious force of the size needed to invade an island as large as Luzon would require air bases farther advanced and better situated with regard to weather than Leyte was able to provide. Air sites in the western Visayas, especially on Panay, were already developed and far superior to those on Mindoro. From the point of view of the Japanese, who were without bulldozers and pierced steel mat runways, the terrain of Mindoro was generally unfavorable for airbase construction and, except for the small airfield at San Jose, there seemed little to tempt an American invasion. It was presumed, therefore, that one of the islands in the western Visayas would be General MacArthur's next target and that Mindoro would be by-passed on his route of advance to Luzon.[18]

So unexpected was an American assault on Mindoro prior to a main invasion of Luzon that even when the Task Force was sighted in the Mindanao Sea on 13 December, the Japanese estimated its destination as Negros or Panay. During the 14th, enemy patrol planes searched the beach waters of these two islands, fully expecting to find American landing operations already under way. Even when the convoy had progressed beyond the western Visayas, there were still some elements in the Japanese High Command, especially in the naval and air forces, who held that a landing directly on Luzon was in the offing.[19] Only when the American convoy was finally anchored off San Jose did the Japanese discover that Mindoro was the ultimate objective of the amphibious assault forces.

On 15 December, after sailing past Negros and Panay with further interference from the enemy limited to weak raiding missions, the assault ships of the Western Visayan Task Force drew close to the beaches of southwest Mindoro. Following a short naval bombardment, troops of the 19th RCT and the 503rd Parachute Regiment went ashore without opposition and pushed rapidly inland. (Plate No. 70) The landing phase of the invasion was accomplished without the loss of a single Allied soldier, although several suicide planes managed to penetrate the transports' air cover, sinking two LST's and damaging a destroyer. In order that the convoy be exposed to enemy

16 COM3rdFLT Report, Serial 0085, 25 Jan 45, File No. A16-3/(16) (S).

17 Statement by Maj. Eizo Hori, Intelligence Staff Officer, Fourteenth Area Army, Interrogation Files, G-2 Historical Section, GHQ, FEC.

18 Statement by Col. Misoo Matsumae, Staff Officer, Fourth Air Army, Interrogation Files, G-2 Historical Section, GHQ, FEC.

19 The staff of the Japanese Southwest Area Fleet estimated that the Mindoro-bound convoy would pass through Mindoro or Tablas Strait for a landing at either Tayabas Bay or Batangas Province on Luzon. Japanese Second Demobilization Bureau Report, *Philippine Area Naval Operations*, Part III, pp. 8–9, G-2 Historical Section, GHQ, FEC. See also Colonel Matsumae, op. cit.

air attack for as short a time as possible, some 1200 combat troops of the 77th Division had accompanied the assault forces, to be utilized at the beachhead for the sole purpose of unloading the necessary equipment. The innovation, improvised to overcome the shortage of service troops, was highly successful and all materiel was transferred to the beaches before nightfall. Their work completed, the 77th Division troops were then returned to Leyte with the emptied ships.[20]

By noon of the invasion day, the town of San Jose had been occupied and work was begun on its airstrips. Australian construction troops assisted in the conditioning of the airfields, two of which were already in use by 23 December. Mindoro was but lightly defended by the Japanese, and Sixth Army operations in the objective area consisted mainly of patrolling and light skirmishing.

The main efforts of the Japanese to interfere with the Mindoro operation were confined primarily to actions by their air and naval forces. Allied resupply convoys ran a gantlet of determined and damaging Kamikaze attacks as they traversed the dangerous waters en route to the Mindoro beachhead. For a time the destruction wrought by these suicide assaults created a serious problem, especially with regard to aviation gasoline and air force technical supplies. Troops ashore were also subjected to sporadic and fatiguing night raids by hostile aircraft.[21]

During the evening of 26 December, a Japanese naval task force, consisting of 2 cruisers, 3 destroyers, and 3 destroyer-escorts, attempted an ambitious bombardment of Allied shore installations, but the enemy plan fell far short of its purpose. In its approach to the beachhead, the Japanese task force lost the destroyer *Kiyoshimo* to Allied air attack, and the damage suffered by the other ships limited the accuracy and power of its gunnery. The total effect of the bombardment was negligible and the damage was easily repaired. Several attempted torpedo strikes against ships at anchor were similarly abortive. This was the last time that major surface units of the Japanese Fleet interfered with Allied shore operations in the Philippines.

The primary purpose of the seizure of Mindoro was to establish airfields from which land-based aircraft could bomb selected targets on Luzon and at the same time protect the assault and resupply shipping en route to Lingayen Gulf. Supplementing this aim was an extensive deception plan to obfuscate Japan's military leaders as they tried to anticipate General MacArthur's next target after Mindoro. This plan was twofold in intent. From a broad strategic viewpoint, it attempted by means of extensive naval operations in nearby waters to direct the enemy's attention to a possible Allied threat against Formosa and southern Japan. Tactically, it aimed to undertake such measures as would lead the enemy to believe that the main thrust of any Allied offensive on Luzon would be launched against western

20 According to Lt. Gen. Akira Muto, Chief of Staff, Fourteenth Area Army, the American landing at Mindoro thwarted Japanese plans for a counterattack in force at Carigara in northern Leyte. "We began to plan and prepare in earnest for the counterlanding at Carigara Bay," General Muto stated in describing the early December staff discussions of the Fourteenth Area Army. "On 15 December, the enemy ships stopped at San Jose on Mindoro Island, where forces were landed and the construction of a base was begun. As San Jose is 250 kilometers from Manila, all communications between Manila and the southern Philippines were cut off. This sudden turn in the situation put the counterlanding operation at Carigara Bay out of the question and made the supplying of munitions to the Visaya district impossible. Thus ended the Leyte campaign." Memoirs of Lt. Gen. Akira Muto, Interrogation Files, G-2 Historical Section, GHQ, SWPA. Hereinafter cited as: Memoirs of General Muto.

21 According to Japanese records, 53 special-attack (Kamikaze) planes supported by 160 fighters, attacked American vessels approaching Mindoro between 16 December and 5 January. Japanese First Demobilization Bureau Report, *Philippines Air Operations Record*, Phase II, Chart 17, G-2 Historical Section, GHQ, FEC.

Batangas or the Bicol Provinces. Ground operations in furtherance of this plan were directed by the Eighth Army which took over control of Mindoro from the Sixth Army on New Year's Day, 1945.

As the first step in the tactical deception effort, one company of the 21st Infantry of the 24th Division moved on Bongabong along Mindoro's east coast on 1 January. Other troops of the same regiment then advanced by shore-to-shore movement to Calapan, the main town on northeastern Mindoro, while enemy-held villages on the northwestern side were also cleared. In all of these actions substantial assistance was rendered by organized guerrilla forces. Occupation of Marinduque Island, situated close to southern Luzon's Bicol Peninsula, was the next operation undertaken. On 3 January, a small force of the 21st Infantry landed unopposed at Buenavista, on the island's southwestern shore, and consolidated positions for the establishment of radar installations.

Concurrently with these ground operations, additional steps were taken to conceal the Lingayen invasion plan from the Japanese. While United States bombers struck carefully selected targets on Luzon, other aircraft flew photographic and reconnaissance missions over the Batangas—Tayabas region and transport planes made dummy drops over the same area to simulate an airborne invasion. At the same time, Seventh Fleet motor torpedo boats patrolled the southern and southwestern coasts of Luzon as far north as Manila Bay from new bases on Mindoro, and mine sweepers cleared Balagan, Batangas, and Tayabas Bays. Landing ships and merchantmen also approached the beaches in these areas until they were fired upon by the enemy and then slipped away under cover of darkness. On instructions from GHQ, the guerrillas in lower Luzon intensified their activities and conducted ostentatious operations designed to divert Japanese attention to the south. Although it is not clear exactly to what extent these deception measures influenced the operational plans of the Japanese Fourteenth Area Army, it is certain that, in the period during and immediately after the Mindoro assault, the Japanese centered their attention on southwestern Luzon. To guard against the double threat of the Mindoro convoy either continuing on to Batangas or being the forerunner of a second strike destined for Luzon, additional forces were sent southward.

On 14 December, the Japanese 71st Infantry was dispatched to Batangas and the 39th Infantry, reinforced, to Bataan. There is little doubt that, until General Yamashita began the actual deployment of his troops in accordance with his Luzon defense plan of 19 December, he remained greatly concerned over the possibility of an imminent invasion of the southern area of the island.[22]

In addition to providing an advanced base

22 Maj. Gen. Haruo Konuma, Deputy Chief of Staff, Fourteenth Area Army, made the following statement regarding the Japanese estimate of the situation prior to the American landing on Luzon: "I recall that although the Area Army had originally thought that the main landing points of the American forces would be the Batangas and Lingayen areas, it became greatly concerned with the Batangas area about 18 December 1944. During the period immediately subsequent to my arrival at my post (about 18 December 1944), the aftereffects of the landing of the American forces on Mindoro were the factors that had influenced the estimate of the situation. Headquarters was under tension and was devoting its efforts to strengthening the defenses of the Batangas area and the vicinity of the southern strongholds. The estimate of the situation which I heard from the Area Army Chief of Staff was as follows: 'Although the main landing points of the American forces on Luzon cannot be definitely limited to Batangas and Lingayen, there are indications that they may make landings in the Batangas area earlier than we expect (the middle of January).' I think that about 26 or 27 December, it was thought that the main landing of the American forces would take place first in the Lingayen area.... In short, during the period immediately subsequent to the American landing on Mindoro the Area Army's estimate on the main landing points of American forces on Luzon was the Batangas area. Later, however, the Area Army began to pay attention to the Lingayen area as a possible landing point. I believe it was not until just prior to the sighting of the American convoy in the waters west of Bataan Peninsula that the Area Army was able to estimate definitely that the landings would be made in the Lingayen Area." Interrogation Files, G-2 Historical Section, GHQ, FEC.

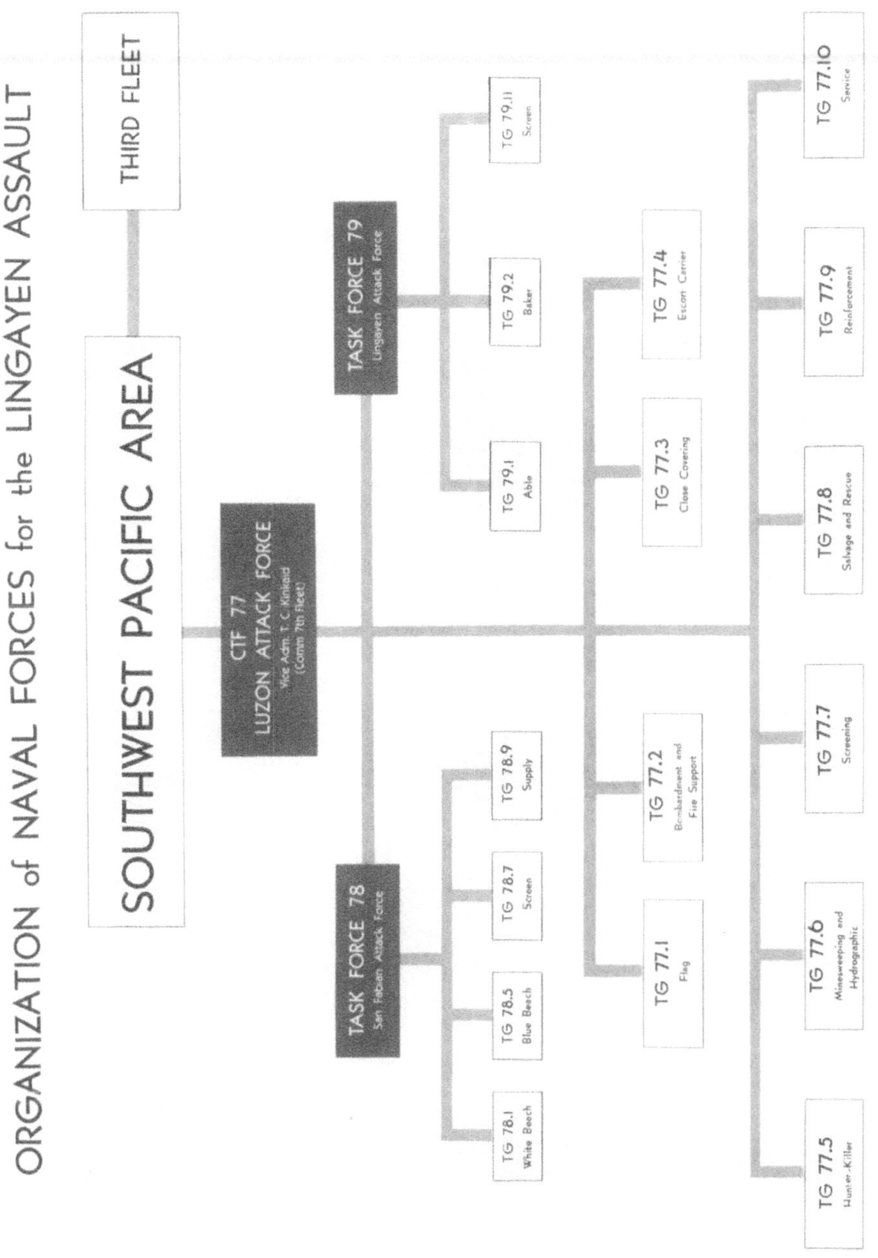

PLATE NO. 71
Organization of Naval Forces for the Lingayen Assault

for airfields and for deceptive operations, the invasion of Mindoro broke the enemy's final link with the Visayas; his communication lines with the southern and central Philippines were completely disrupted, and any further efforts to move his troops from Luzon to the Leyte area were now utterly out of the question.[23]

The enemy was also compelled to make major changes in his transport routes between Japan and the southern areas. With the loss of Mindoro, Japan halted shipments of additional reinforcements to the Philippine Islands. Manila Bay was abandoned as a main stopping point for convoys moving to and from the Netherlands East Indies region, and all remaining shipping lanes were pushed toward the China coast.[24]

Assault at Lingayen Gulf

General MacArthur's campaigns in the Southwest Pacific Area now neared their climax. The battle for Luzon was at hand. In many respects the scheme of maneuver and organization of forces for the invasion of Luzon paralleled the plan of the Leyte operation.[25] Naval forces were divided into three main sections with Admiral Kinkaid in over-all command. (Plate No. 71) The Bombardment and Fire Support Group and the Escort Carrier

23 Commenting on the transport and communication problem, Lt. Col. Yorio Ishikawa, Staff Officer of the Japanese Fourteenth Area Army, stated: "After the American landing on Mindoro, our supply line between Luzon and the Visayas was completely cut off. As for our communications, we had to depend entirely on wireless to contact our forces in Leyte, Cebu, Ormoc, etc. from Manila, after the United States had seized Mindoro and Marinduque Islands." Interrogation Files, G-2 Historical Section, G-2, GHQ, FEC.

24 G-2, GHQ, SWPA, *Monthly Summary of Enemy Dispositions*, 31 Dec 44.

25 The major units participating in the Lingayen assault were as follows:

Ground: *Sixth Army—General Walter Krueger*

Unit	*Commander*	*Phase*	*Staging Point*	*Landing Point*
I Corps	*Maj. Gen. Innis P. Swift*			
6th Div	*Maj. Gen. E. D. Patrick*	*Assault*	*Sansapor*	*San Fabian*
43rd Div	*Maj. Gen. L. F. Wing*	*Assault*	*Aitape*	*San Fabian*
32nd Div	*Maj. Gen. W. H. Gill*	*Follow-up*	*Leyte*	
33rd Div	*Maj. Gen. P. W. Clarkson*	*Follow-up*	*Finschhafen-Toem*	
XIV Corps	*Maj. Gen. Oscar W. Griswold*			
40th Div	*Maj. Gen. R. Brush*	*Assault*	*Gloucester*	*Lingayen*
37th Div	*Maj. Gen. R. S. Beightler*	*Assault*	*Bougainville*	*Dagupan*
1st Cav Div	*Maj. Gen. V. D. Mudge*	*Follow-up*	*Leyte*	
112th Cav RCT	*Brig. Gen. J. W. Cunningham*	*Follow-up*	*Leyte*	
6th Ranger Bn	*Lt. Col. H. A. Mucci*	*Reserve*	*Leyte*	
25th Div	*Maj. Gen. C. L. Mullins, Jr*	*Reserve*	*Noumea*	
13th Armd Gp	*Col. M. E. Jones*	*Reserve*	*Leyte, Oro Bay, Hollandia*	
158th RCT	*Brig. Gen. H. MacNider*	*Reserve*	*Noemfoor*	
4th ESB	*Brig. Gen. H. Hutchins, Jr.*	*Assault*	*San Fabian*	*Dagupan*
533rd EB&SR	*Col. W. S. Moore*	*Assault*	*Aitape*	*Lingayen*
543rd EB&SR	*Col. G. E. Galloway*	*Assault*	*Sansapor*	*San Fabian, Dagupan*

Air: *Fifth Air Force—Maj. Gen. Ennis C. Whitehead*

Unit	*Commander*	*Unit*	*Commander*
308th Bomb Wing	*Brig. Gen. D. W. Hutchison*	*35th Fighter Gp*	*Lt. Col. E. A. Doss*
86th Fighter Wing	*Col. J. O. Guthrie*	*71st Rcn Gp*	*Lt. Col. H. C. Thompson*
38th Bomb Gp	*Col. E M. Gavin*	*547th Night Fighter Sq*	*Lt. Col. W. C. Odell*
18th Fighter Gp	*Col. M. B. Adams*		

Naval: *Seventh Fleet (Supported by units of the Third Fleet)—Adm. T. C. Kinkaid*

Unit	*Commander*	*Unit*	*Commander*
Task Group 77	*Adm. T. C. Kinkaid*	*Task Group 79*	*V. Adm. T. S. Wilkinson*
Task Group 78	*Adm. D. E. Barbey*		

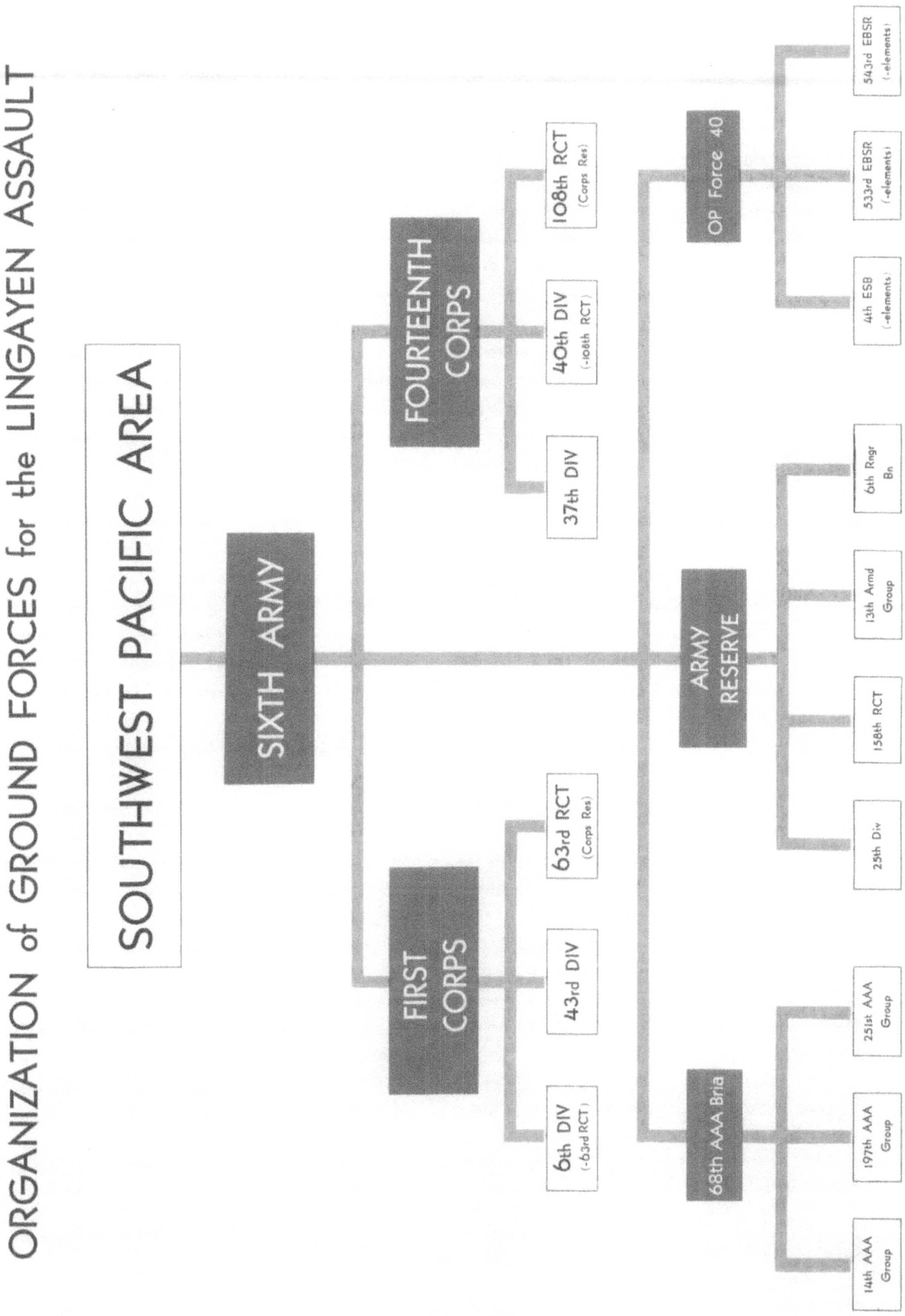

PLATE NO. 72
Organization of Ground Forces for the Lingayen Assault

Group were under Admiral Oldendorf and Admiral Berkey, respectively; the San Fabian Attack Force under Admiral Barbey; and the Lingayen Attack Force under Admiral Wilkinson.[26] The Third Fleet would again provide strategic support for the operation by neutralizing enemy airfields, and at the same time it would be prepared to give direct support whenever necessary. The Fifth and Thirteenth Air Forces' land-based planes were to protect the flanks and rear by overwater searches and by strikes against enemy installations in the southern Philippines and the Netherlands East Indies. In addition, army aircraft were to give protection to convoys moving through central Philippine waters and, as far as possible, lend air support to ground operations. As at Leyte, the Sixth Army, under General Krueger, was to land two corps abreast, this time I and XIV Corps, containing the 6th, 37th, 40th, and 43rd Divisions in the assault phase, and the 25th Division, 158th Regimental Combat Team, 13th Armored Group, and 6th Ranger Battalion in reserve.[27] (Plate No. 72)

The postponement of the Mindoro operation to 15 December caused a corresponding delay in the Lingayen Gulf landing, originally set for 20 December. A new target date, 9 January 1945, was chosen to give time for the reorganization of forces and for the establishment of airfields on Mindoro and also to take advantage of the favorable moon and tide conditions. The few extra days were used to replenish and repair fleet units, conduct landing rehearsals, and further co-ordinate plans for air and naval support.

The Third Fleet struck its first blows in support of the Lingayen operation on 3 January 1945 in the Formosa and Nansei Shoto regions. Bad weather hampered most of the carrier strikes, but moderate successes were achieved against the enemy's airfields and harbor shipping. On 6 January, the Third Fleet moved back into position to begin its neutralization strikes against Japanese installations on Luzon, preparatory to the imminent Sixth Army landings.

It was fairly certain that the Japanese would throw their remaining air power in the Philippines against General MacArthur's Luzon invasion forces, and it was also anticipated that the major portion of their air effort would consist of the suicide technique so effectively introduced at Leyte. Both these expectations were soon realized. The passage of the advance echelons of the bombardment and fire support units of the task force from Leyte to Lingayen Gulf was marked by persistent and damaging Kamikaze attacks. (Plate No. 73) One day after Admiral Oldendorf sailed forth on his voyage to Luzon, Japanese planes began to plummet down from the sky into his ships. On 4 January, the *Ommaney Bay*, an escort carrier, was damaged so badly that it had to be sunk. The following afternoon, the *Louisville*, *Stafford*, *Manila Bay*, *Savo Island* and HMAS *Arunta* and *Australia* suffered hits or damaging near misses. On 6 January, off Lingayen Gulf, the fire support groups were again attacked by the " largest and most deadly group of suicide planes encountered during the operation."[28] At least sixteen vessels were struck during the course of the day resulting in extensive casualties both to ships and personnel. Admiral Oldendorf's flagship, *California*, received serious damage, as did the battleship *New*

---

26 Commander Luzon Attack Force Report, Serial 001200, to CINC USFLT, *Action Report—Luzon Attack Force, Lingayen Gulf—Musketeer Mike One Operation, 15 May 1945*. Hereinafter cited as: Luzon Attack Force Action Report.

27 G-3, GHQ, SWPA, *Monthly Summary of Operations*, Jan 45.

28 COMBATRON 1 (CTG 77.2) Report, Serial 0018, to COMINCH, via COM3rdPHIBFOR, COM7thFLT, CINCSWPA, *The Bombardment and Occupation of Lingayen Gulf*, 28 Jan 45, File No. FC1-1A16-3, (S).

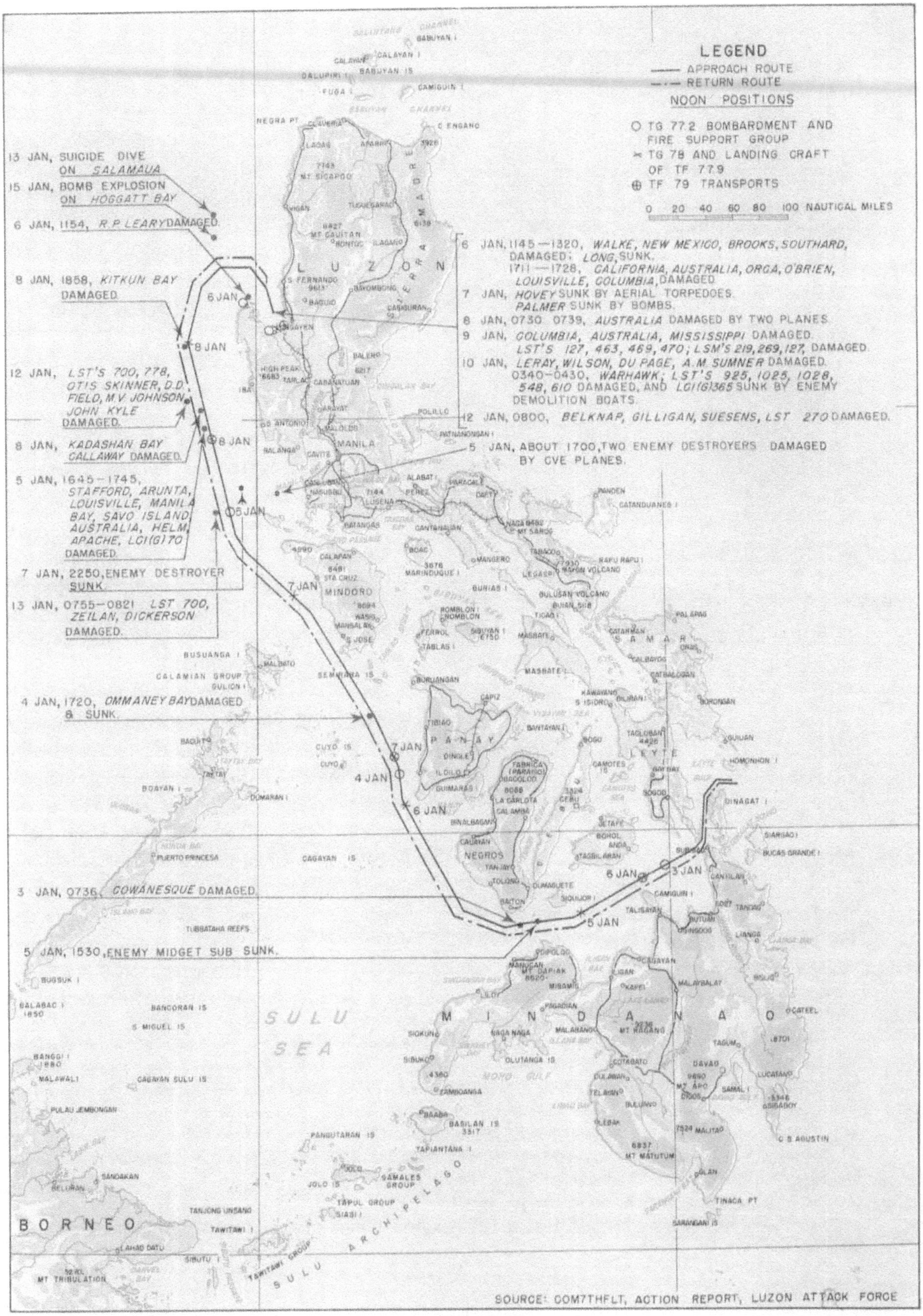

PLATE NO. 73

Approach of Forces to Lingayen

Mexico and the cruisers, *Columbia* and *Louisville*.[29]

So determined and damaging were the Japanese attacks that a bombardment of the beachhead areas was impossible that day. Bad weather had minimized the effectiveness of the air attacks by the Fifth Air Force and the Third Fleet, thrusting the brunt of the air defense mission on the escort carrier planes. An attempt had been made by the Third Fleet to maintain a continuous air patrol over all enemy airstrips during 6 January, but a solid overcast prevented the blanketing of the more northerly areas. Escort carriers of the Seventh Fleet did their best to protect the accompanying warships, but their efforts were met by a resourceful and skillful enemy who had improved his tactics greatly since the days of Leyte Gulf. The Japanese pilots were now relatively well trained and their deception measures excellent. They made full use of land masses, "window", and counterfeit identification devices to escape radar detection. In addition, continuing inclement weather, together with the relatively large area to be covered, prevented the Third Fleet's carrier planes from effectively covering the numerous Japanese-held airdromes on Luzon and grounded the army aircraft based on Mindoro.

As a result of the heavy air opposition by the Japanese, Admiral Kinkaid requested that the Third Fleet repeat and intensify its strikes against Luzon on 7 January. This action necessitated cancellation of a planned sortie against Formosa, but the new mission was completed satisfactorily despite adverse weather.[30] Enemy air attacks began to diminish sharply, indicating that his plane strength was nearing exhaustion; on the 8th, as the Third Fleet retired to refuel, only a handful of enemy aircraft sortied into the Lingayen Gulf area. The few remaining suicide bombers now appeared to be coming from the northwest. To eliminate this potential threat, General MacArthur asked that the China-based XX Bomber Command shift its main effort against Formosa from Keelung Harbor to the southern part of the island.[31] It was also decided that a carrier-borne aircaft attack on Formosa would be preferable to continued Third Fleet presence in the Luzon region. Admiral Halsey's Third Fleet thereupon moved to the north and west while the Seventh Fleet's escort carriers, aided by all available land-based planes of the Fifth Air Force, took up the task of direct support of the amphibious landings.

To insure adequate protection for the assault troops, General MacArthur requested Admiral Nimitz to retain in Luzon waters a reasonably large portion of the bombardment and fire support vessels temporarily assigned to the Seventh Fleet. These warships, on 7 and 8 January, had accomplished their assigned mission by thoroughly pulverizing the Lingayen beach areas with naval gunfire. On the 9th they stood by, ready to deliver call-fire. Initial intelligence from photographs and numerous guerrilla reports indicated that the shores of the

---

29 Luzon Attack Force Action Report, p. 15. Available Japanese documents contain only the number of suicide and escort planes used against the Lingayen invasion force. Attack missions by conventional aircraft are not included. According to these records, on 5 January, 20 suicide planes and 5 escort fighters were directed against the Allied Task Force; on 6 January, a total of 58 suicide planes and 17 escort fighters were employed in attacks on the ships in the general area of Lingayen Gulf. Japanese Second Demobilization Bureau Report, *Philippine Area Naval Operations*, Part II, Appended Chart 8. See also Japanese First Demobilization Bureau Report, *Philippine Air Operations Record,* Phase II, Appended Chart No. 17, G-2 Historical Section, GHQ, FEC.

30 Luzon Attack Force Action Report, pp. 15–16.

31 CINCSWPA Radio No. CX-56001 to GG Army Air Forces, 7 Jan 45, G-3, GHQ, Admin. 385 Plan 21 (TS).

Gulf had been fortified by the Japanese.[32] Enemy gun installations were found only on the eastern side of the Gulf, however, and a relatively small amount of counterbattery fire was required. Since the Japanese had withdrawn from the beach areas, an accurate assessment of the effectiveness and value of the naval gunfire was difficult to make. The work of mine sweepers, operating in close conjunction with the bombardment and fire support units, was highly successful. Few mines were found in the area and sweeping progressed well ahead of schedule, thus making the Gulf safe for shipping and amphibious craft.

The assault convoys, meanwhile, proceeded through the Visayas and up the west coast of Luzon with only sporadic interference from the enemy. One notable attack took place on 5 January when the *Boise*, with General MacArthur aboard, became the target of two torpedoes from a Japanese midget submarine. Fortunately, both torpedoes were avoided by prompt and skillful maneuvering and the submarine was rammed and depth-charged immediately by a covering destroyer.[33] In the early morning hours of 9 January all amphibious shipping arrived in Lingayen Gulf. Japanese suicide boats succeeded in sinking or damaging several of the LST's and LCI's but the landings proceeded smoothly.

The entry plan for Task Forces 78 and 79, transporting I and XIV Corps, respectively, was so well co-ordinated that the four attack groups, one for each division participating in the assault, were in position off their designated beaches almost simultaneously along a front of twelve miles. I Corps under Maj. Gen. Innis P. Swift, consisting of the 6th and 43rd Divisions, landed on the left flank near San Fabian. XIV Corps, under Maj. Gen. Oscar W. Griswold and made up of the 37th and 40th Divisions, went ashore on the right, in front of Lingayen and Dagupan towns. (Plate No. 74) The scheduled landing hour was 0930 and by 0940 all landing waves had hit the beach. Initial opposition on the beaches was limited to mortar fire from the hills in the San Fabian region which damaged some of the landing craft. By late afternoon, the four division commanders had assumed control ashore.

The Luzon landing operation was announced in a communique of 10 January:

> *Our forces have landed in Luzon. In a far-flung amphibious penetration our troops have seized four beachheads in Lingayen Gulf. . . . The decisive battle for the liberation of the Philippines and the control of the Southwest Pacific is at hand. . . .*
>
> *General MacArthur is in personal command at the front and landed with his assault troops. His ground forces of the Sixth Army are under General Krueger, his naval forces of the Seventh Fleet and Australian Squadron are under Admiral Kinkaid, and his air forces of the Far East Air Force are under General Kenney. The Third Fleet under Admiral Halsey is acting in coordinated support.*[34]

32 The guerrillas did excellent work in leading the escort carrier planes to enemy war materiel. The commander of the air support control units reported, "Guerrilla sketches of enemy installations in the interior prepared by Lt. Col. Volckmann's command were invaluable. By means of these sketches CSA (Commander Air Support Aircraft) was able to direct pilots to specific buildings within a town where arms and fuel were stored, thus sparing from needless destruction other parts of the town. The sketches were amazing in their accuracy and were put to good use throughout the operation." Comdr Air Support Control Units Report to COM7thFLT, *Report of Support Aircraft Operations, Lingayen Gulf Landings*, 20 Jan 45, G-3, GHQ, SWPA *Journal*. See also G-2, GHQ, FEC, General Intelligence Series Vol I, *The Guerrilla Resistance Movement in the Philippines* and Vol II, *Intelligence Activities in the Philippines during the Japanese Occupation*.

33 Luzon Attack Force Action Report, p. 12.

34 GHQ, SWPA, Communique No. 1008, 10 Jan 45.

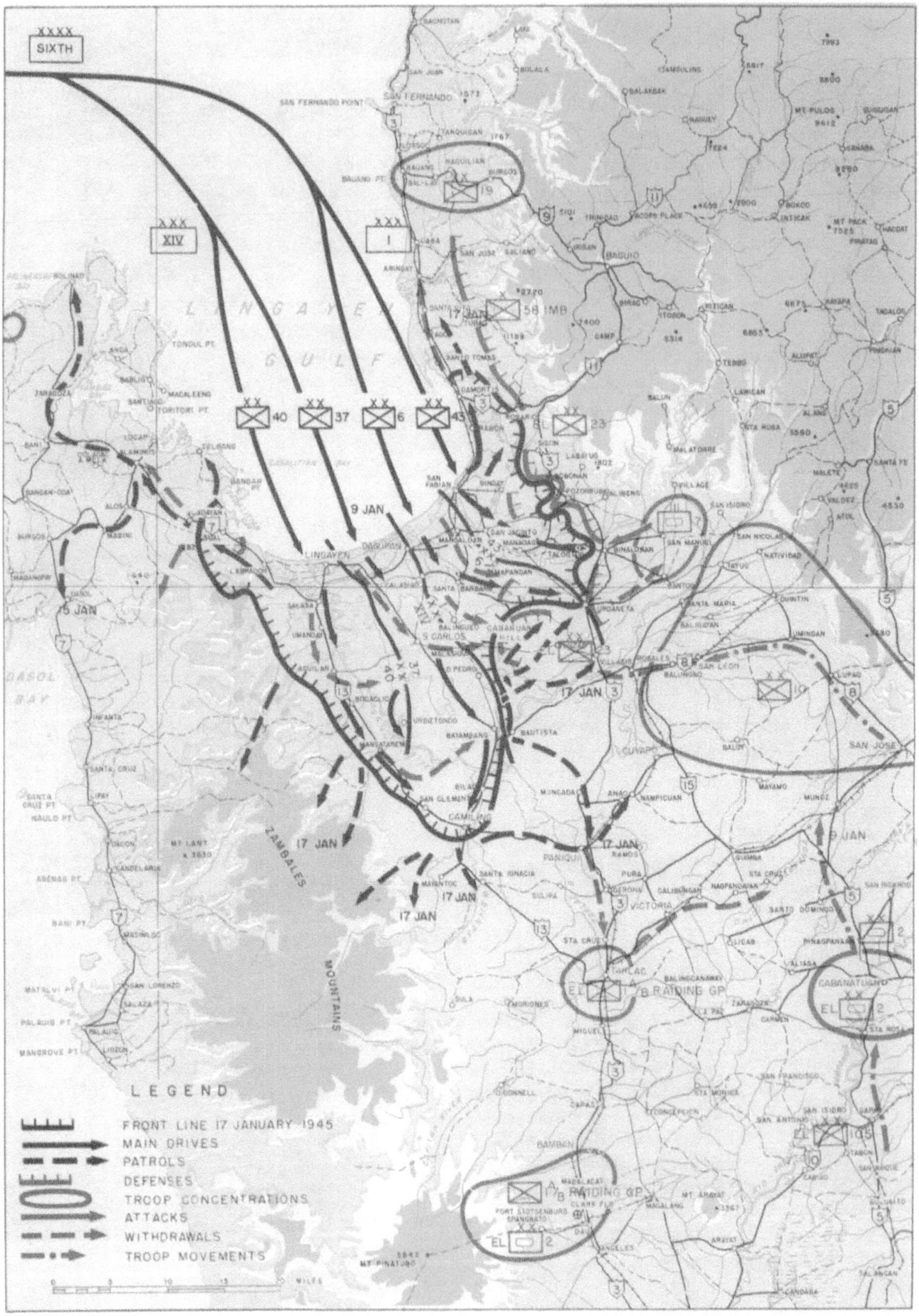

PLATE NO. 74
Sixth Army Landings, Lingayen Gulf, 9–17 January 1945

I Corps Advances Eastward

The strongest opposition to the initial inland progress of the Allied invasion force was met in the I Corps zone of action. As the five regimental combat teams[35] moved into the mountains north and east of the San Fabian—Damortis beachhead, they encountered a long arc of well-emplaced enemy positions covering all avenues of approach to Baguio and to the Cagayan Valley. This line of defense, built along a series of ridges beginning at Damortis, ran eastward and then southward through Rosario, Binalonan, and Urdaneta to anchor itself in the precipitous and rugged ground of the Cabaruan Hills. Again the Japanese had skillfully utilized the advantages of the terrain to protect the strategic passes and roadways with heavily fortified and mutually supporting cave and tunnel systems, fully supplied with all types of automatic weapons, mortars, and artillery. Most of these caves were self-sufficient strongholds containing enough materiel and food to withstand a protracted seige. Although artillery fire could be directed against positions located in the hillsides and ravines, it accomplished little toward neutralizing the powerful guns which were installed within the caves themselves.

By the morning of 10 January, troops of the 43rd Division were in positions from San Jacinto to Rabon. After containing and then by-passing a strong concentration of Japanese entrenched in the Binday area, the division, reinforced on 11 January by the 158th Regimental Combat Team, began an arduous, stubbornly-resisted advance along the Damortis—Rosario road. The Japanese had done a thorough job of demolition to impede the progress of the American forces. Engineers were called upon to perform herculean tasks in the face of accurate enemy artillery fire, clearing the roads, removing cleverly constructed obstacles, throwing temporary bridges across streams, and, in one case, even ripping up twenty miles of railroad track to convert the roadbed into a motor highway. Rabon fell on the 12th; Damortis was entered on the 13th. By 16 January, troops of the 43rd Division and the 158th RCT had taken positions north and south of the Damortis—Rosario road near the Apangat River and continued to press against strong enemy defenses along the Rosario—Pozorrubio—Binalonan line.

The 6th Division, on the right flank of the I Corps sector, was similarly slowed down in its move from the landing beaches. Strong enemy positions were met on the outskirts of the Cabaruan Hills, about twelve miles inland, southeast of the beachheads. The Japanese had established their principal defense point along a horseshoe-shaped promontory a short distance west of the Cabaruan barrio. All ridges to the west and northwest which covered the approaches to the main defenses were also well fortified and so cleverly camouflaged that even close-range scrutiny failed to reveal their location. In addition to the conventional pillboxes, dugouts, and interlocking trenches, ingeniously concealed sniper posts were strategically placed to protect the road nets to those positions.

On 11 January, the 25th Division moved into the I Corps zone between the 43rd and 6th Divisions and instituted a drive in the area north and east of Urdaneta. On the 18th, elements of the 25th Division attacked heavy enemy forces at Binalonan, a key town on the network of roads leading to Baguio and the Cagayan Valley. From Binalonan, the division pushed northeastward to assault San Manuel. The town was strongly defended by the main force of the Japanese 7th Tank Regiment (the

35 Three regimental combat teams of the 43rd Division, the 63rd RCT of the 6th Division, and the 158th RCT.

Shigemi Detachment) and during the initial phase of the attack the advance of the American troops was slowed by heavy casualties. Artillery pieces and mortars fired down from the hills on the northwest of the town while all approaches were covered by enfilading and cross-fire of well-placed automatic weapons and antitank guns. Despite these formidable obstacles, the division forces moved stubbornly and steadily ahead and on 28 January, after repulsing a fierce tank counterattack in which most of the enemy's armor was destroyed, San Manuel was occupied and secured.

Throughout the I Corps sector, the enemy plan of defense revealed itself as an attempt to command key terrain positions and control strategic points on the main highways. In many places, tanks and other armored equipment were discovered buried deep in the ground with their turrets protruding, for use as pillboxes in close support of infantry positions. Hard as such installations were to destroy, this unorthodox employment of tanks was not without benefit to the American forces, since the Japanese were forced to sacrifice many excellent opportunities for more logical and effective maneuver.[36] By mid-January, it was evident that the enemy was concentrated in strength in the sector guarding the roads to Baguio and to the Cagayan Valley and that any progress in that direction would necessitate intensive and sustained effort by the I Corps forces.

XIV Corps Advances Southward

On the right flank of the Allied invasion forces in the XIV Corps zone, the open terrain and paved highways of the Luzon Central Plain permitted a rapid advance southward on the road to Manila. General Krueger's strategy called for XIV Corps to drive southward through the Central Plain while I Corps protected the left flank by containing the enemy in the mountains to the northeast.[37]

The Central Plain, formed by alluvial deposits flowing from the high ranges bordering on the east and west, cuts a swath 30–50 miles wide through Luzon from Lingayen Gulf to Laguna de Bay—a distance of about 120 miles. A highly developed irrigation system, fed by numerous streams coursing down from the surrounding mountains, turns this vast region into an immensely fertile valley for the profitable cultivation of rice, sugar, and other staples. With its extensive plantations and well-developed transportation system of railroads and motor highways, the Central Plain constitutes the wealthiest and most important area in the Philippines.

XIV Corps' 37th and 40th Divisions, moving in parallel columns southward from Lingayen into the Central Plain, were virtually unopposed by the Japanese. The 40th Division moved along Highway No. 13 while the 37th Division, about eight miles on the left, moved down a second highway that runs from Lingayen to San Carlos and Bayambang. On 12 January,

---

36 In reply to his critics, Col. Shigeo Kawaii, Operations Staff Officer of the Japanese 2nd Tank Division on Luzon, said the following: "The employment of the tank division in the Philippines is generally considered a great blunder. The fact remains that the American forces had command of the air, preventing movement along the highways and cross-country movement in an area covered with rice paddies was impossible. Consequently, even though the tanks were organized for combat maneuvers, they were soon immobilized because of the lack of air cover and the destructive American aerial attacks which the tanks could not counter. They were, therefore, converted into armored, fixed defenses to be used by the infantry in key positions along the defense lines. This adaptation of the tanks was so successful that, in one instance, a line 60 kilometers long was held for a period of one month." 10th Information and Historical Service, HQ Eighth Army, *Staff Study of Japanese Operations on Luzon*, Part IV, p. 7 (R).

37 HQ Sixth Army, Field Order No. 34, 20 Nov 44, G-3, GHQ, SWPA *Journal*, 29 Nov 44.

the front line of XIV Corps ran from Bayambang on the east flank to Aguilar on the west. By 15 January, the Agno River had been crossed and still the Japanese had not been encountered in any sizeable force. Advance patrols of the 40th Division, instead of meeting enemy machine guns and mortars were greeted by joyous civilians who lined the roads and shouted encouragement.[38] In fact, the very rapidity of the advance, coupled with the conspicuous failure of the Japanese to employ their usual tactics of defending key road positions, created a suspicion that some trap was in the making based on the hope that the Allied forces would overextend their lines. XIV Corps, therefore, kept a wary eye to the east, since its left flank was somewhat exposed when I Corps to the northeast was blocked in its drive by the heavy enemy defenses in the Cabaruan Hills. To safeguard this open flank, XIV Corps turned temporarily westward to seize Paniqui and Anao before continuing in its southward movement. With both these positions in Allied hands by 19 January, the advance to Manila was resumed. (Plate No. 75)

On 21 January, the 40th Division occupied Tarlac without opposition while the 37th Division to the east held the line at Victoria. The advance of the 40th Division continued unabated until it reached the Bamban River. The Japanese had left a small garrison at this point to delay the capture of the airfield near Bamban and a brief but sharp skirmish ensued before the troops of the 40th Division were able to move ahead.

As the forward drive continued southward from Bamban toward Fort Stotsenburg, the American forces began to encounter increasingly strong pockets of resistance which indicated that they were approaching the first outposts of the Japanese defense line. On 24 January, XIV Corps regrouped its two divisions and prepared for a co-ordinated advance in full strength. By 26 January, just 17 days from the date of landing, XIV Corps had advanced 59 miles to reach forward positions extending 124 miles, with 77 miles held by the 40th Division and 47 miles controlled by the 37th Division.[39]

Enemy Plan of Defense

The lack of enemy opposition to the 40th Division's advance west of the Agno River was not a trap but part of General Yamashita's general plan of defense. Forced to dispatch the cream of his original combat forces to the bitterly contested operation on Leyte, he no longer saw any possibility of winning decisive victories on Luzon. At first he had hoped that if additional replacements were sent from Japan, he could strengthen his thinly spread lines sufficiently to allow an aggressive and full-scale battle against the expected invasion forces. By the end of November, however, the decimation of the Japanese troops on Leyte and the tardy arrival of reinforcements from Japan convinced him that his only course of action was one which embodied a delaying defense to postpone the inevitable destruction of his forces as long as possible and gain time for the other Japanese forces north of Luzon.[40]

38 HQ XIV Corps, *After-Action Report, M-1 Operation*, 29 Jul 45, p. 57, AG 314.7 W. Hereinafter cited as: XIV Corps, M-1 Operation Report.

39 HQ Sixth Army, *G-3 Daily Report*, 23 Jan 45.

40 Interrogation of Col. Shujiro Kobayashi, Senior Staff Officer, Fourteenth Area Army and later Tactical Staff Officer of the Japanese Forty-first Army, Interrogation Files, G-2 Historical Section, GHQ, FEC. Col. Ryoichiro Aoshima, Chief of Staff, Line of Communications, Fourteenth Area Army described the evolution of Japanese strategy on Luzon as follows: "Prior to the Leyte Campaign, Fourteenth Area Army had planned to make Luzon the decisive battleground, but from November on, plans were gradually changed from an aggressive counterattack to delaying tactics. Not, however, until the landing on Mindoro, on December 15th, 1944, were the actual preparations for delaying action gotten underway." 10th Information and Historical Service, HQ Eighth Army, *Staff Study of Japanese Operations on Luzon*, Part I, p. 11 (R).

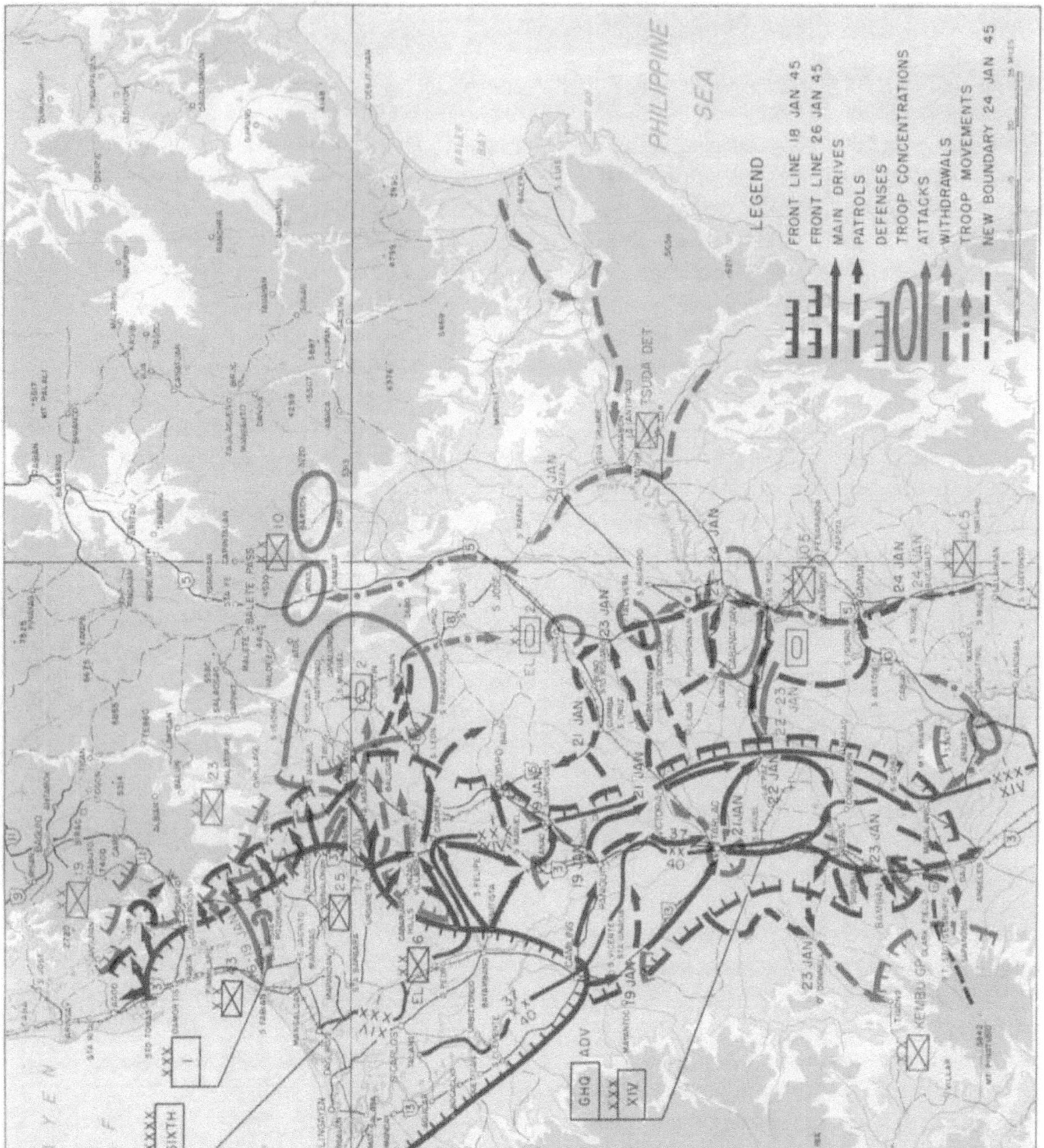

PLATE NO. 75
Drive to Manila, 18–26 January 1945

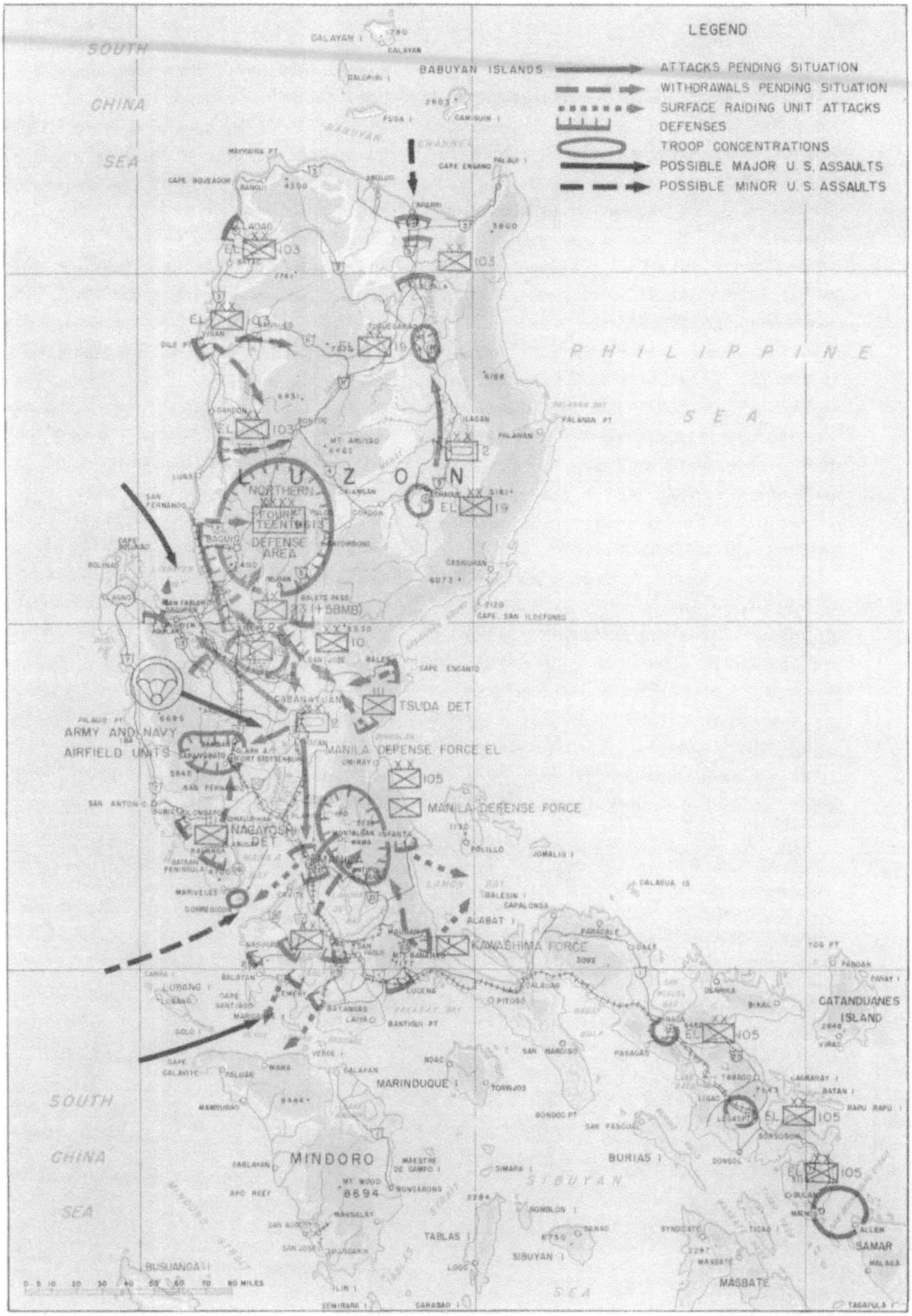

PLATE NO. 76
Japanese Plan for Defense of Luzon, 19 December 1944

On 19 December, General Yamashita issued a final plan under which the main strength of his forces was to be deployed to hold the northern Luzon area and to protect the main approaches to the fertile Cagayan Valley.[41] (Plate No. 76) In addition, certain key sectors, especially around the region west of Clark Field (later organized under the "Kembu Group") and in the mountainous and strategic region east of Manila (later under the command of the "Shimbu Group"), were also to be defended in strength. The forces at these points were ordered to "coordinate their operations with the objective of containing the main body of the American forces on Luzon and destroying its fighting strength, and at the same time, prepare for protracted resistance on an independent, self-sufficient, basis."[42]

Although General Yamashita was prepared to fight the landing forces in the eastern sector of Lingayen Gulf along roads which led to his headquarters at Baguio, he contemplated no more than a token effort in the sector west of the Agno River. He felt that lack of air and artillery support for his fuel-short tanks and their consequent lack of mobility, combined with a scarcity of combat troops, allowed no adequate defense of the wide plain which ran down from the Lingayen beaches southwest of the river.[43] Accordingly, he directed: "Against an enemy planning to land on the western peninsula of Lingayen Gulf, endeavor to reduce his fighting strength through raiding attacks using locally stationed units. . . ."[44]

Because the American landings on the eastern shores of Lingayen Gulf directly threatened his northern defense sector, General Yamashita made a few hurried modifications in the 19 December plan. Strengthening the area surrounding his headquarters at Baguio, he disposed his forces as follows: the 19th Division was placed in the San Fernando area; the 58th Independent Mixed Brigade, from Naguilian to Rosario; and the newly arrived 23rd Division, from Rosario to Urdaneta and the Cabaruan Hills. This last division had suffered shattering losses en route to Luzon as a result of heavy Allied attacks. Supporting the 23rd Division was the 7th Tank Regiment, reinforced.

Japanese efforts to transfer their troops into position according to the orders of the hastily drawn Luzon defense plan of 19 December were seriously disrupted by the constant attacks of American aircraft on their transportation

41 The extent of the Japanese dependence on the Cagayan Valley for food was emphasized by General Muto in his story of the Philippine Campaign. "I have already mentioned," he stated, "that the Japanese forces endeavored to requisition rice on Luzon because of the shortage of provisions. We knew that, even if central Luzon rice could be requisitioned as anticipated, the Army could not expect much after the amount required by the Filipinos had been subtracted, but we were counting very heavily on the Cagayan River valley rice; indeed, our lives depended on it. When General Yamashita had decided on the Luzon defense positon, the stationing of the Japanese main force in northern Luzon had of course been based upon the deployment of our troops at that time and the enemy situation estimate, but the availability of food in the Cagayan River valley was also an important factor." Memoirs of General Muto, p. 28.

42 HQ Fourteenth Area Army, *Outline of Operational Policy for Luzon Island*, 19 Dec 44; Appendices I and II, issued separately 20 Dec 44, G-2 Historical Section, GHQ, FEC.

43 General Muto held that the Leyte battle had revealed the impossibility of contending with American mechanized units in open maneuvers. "Ample demonstration had been given in the Leyte campaign," he declared, "of the manner in which American fire power and mechanized strength could maneuver and advance under aerial protection. To the question of what could be expected to result from a clash on an open plain and without air support between the Japanese Army and its superior adversary, the most steadfast believer in the power of spiritual factors had perforce to answer that the outlook was dark for Japan." Memoirs of General Muto, p. 12.

44 HQ Fourteenth Area Army, *Outline of Operational Policy for Luzon Island*, 19 Dec 44, G-2 Historical Section, GHQ, FEC.

systems. In addition, the ceaseless harassing tactics of guerrilla units played havoc with their communication lines. As a result, General MacArthur's invasion of Luzon on 9 January gave these forces little and, in some cases, no time for the proper preparation of their assigned defenses. Many units were caught in the process of reorganizing or before their transfer movement could be completed. Contact between field troops and headquarters was poor, and some elements were cut off entirely as soon as the American troops began to push inland from their beachheads.

XI Corps Cuts across Bataan

While XIV Corps battled its way into the Clark Field—Fort Stotsenburg area, General MacArthur moved another force to the shores of Luzon. On 29 January, XI Corps, commanded by Maj. Gen. Charles P. Hall and consisting of the 38th Division and the 34th RCT of the 24th Division, landed on the west coast of Zambales Province to push an additional wedge toward the Central Plain and Manila.[45] This landing was also intended to prevent the enemy from attempting to duplicate General MacArthur's successful strategy of three years before in withdrawing into Bataan Peninsula for a prolonged and stubborn defense. The assault force, carried by the Seventh Fleet from Leyte through the Sulu Sea and Mindoro Strait, went ashore in the San Felipe—San Antonio area and began its advance inland.[46] (Plate No. 77) Acting upon information furnished by guerrillas of the Zambales Military District, the pre-arranged naval bombardment had been cancelled although naval support remained to assist with call-fire if necessary.[47]

There was no opposition at the beaches and, after securing the airstrip at San Marcelino and occupying Grande Island at the entrance to Subic Bay, the main body of XI Corps pressed southeastward toward Castillejos and the town of Subic. Progress was unimpeded, and by the evening of the day of invasion both of these positions were in American hands. During the next morning, XI Corps moved into Olongapo, and by 5 February the vital passes to the Dinalupihan—Hermosa Road were denied to the enemy by flanking forces of the 38th Division and advance elements of the 40th Division. Any hostile movement to or from Bataan Peninsula was now blocked.

45 The major units participating in the Zambales landing were as follows:

Ground: *Eighth Army—General Robert L. Eichelberger*

Unit	Commander	Phase	Staging Point	Landing Point
XI Corps	*Maj. Gen. Charles P. Hall*			
38th Div	*Maj. Gen. H. L. C Jones*	*Assault*	*Leyte*	*San Narciso*
34th RCT	*Lt. Col. C. A. Daklen*	*Assault*	*Mindoro*	*San Antonio*

Air: *Thirteenth Air Force (supported by elements of the Fifth Air Force)—Maj. Gen.St. Clair Streett*

Unit	Commander	Unit	Commander
XIII Fighter Comd	*Brig. Gen. E. W. Barnes*	*348th Fighter Gp*	*Col. R. R. Rowland*
XIII Bomber Comd	*Col. C. A. Brandt*	*421st Night Fighter Sq*	*Capt. R. D. Kiick*
309th Bomb Wing	*Col. N. D. Sillin*	*5th Bomb Gp*	*Col. T. C. Musgrove*
86th Fighter Wing	*Col. J. O. Guthrie*	*307th Bomb Gp*	*Col. C. H. Rees*

Naval: *Seventh Fleet—Adm. T. C. Kinkaid*

Unit	Commander	Unit	Commander
Task Group 78.3	*R. Adm. A. D. Struble*	*Task Group 77.3*	*R. Adm. D. E. Barbey*
Task Group 77.4	*R. Adm. T. L. Sprague*		

46 The American landing along Zambales Province cut off the Japanese forces west of Clark Field from all contact with the rest of their units on Luzon. General Muto stated: "American elements landed at San Antonio on 30 January, putting the Kembu Group between two enemy forces and leading subsequently to its complete encirclement. Communication between the Kembu Group and General Yamashita was cut off in the middle of January, and we lost all track of it until the end of the war." Memoirs of General Muto, p. 25.

47 HQ XI Corps, *Historical Report, Luzon Campaign*, Advance Copy, Section II, p. 3.

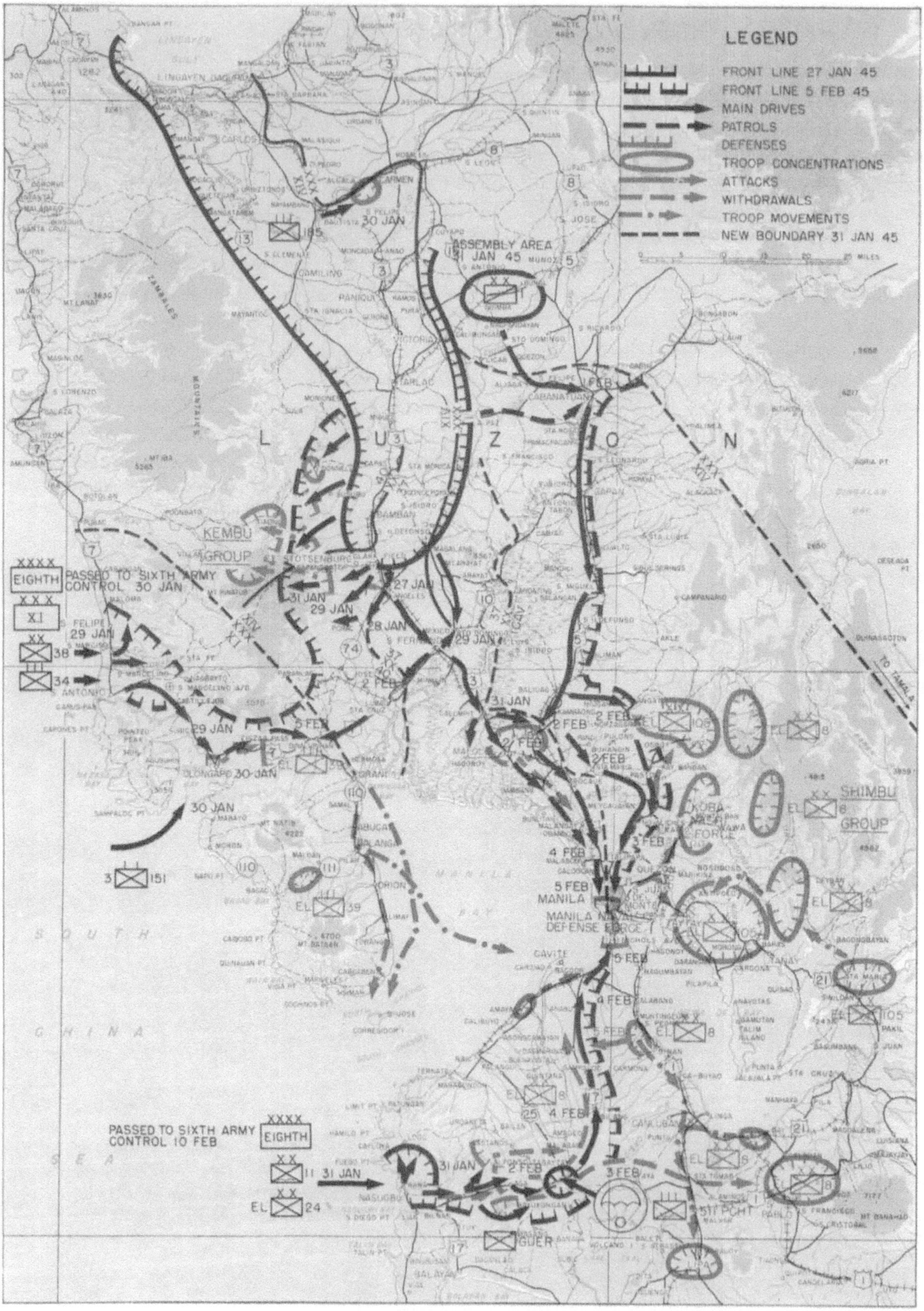

PLATE NO. 77

The Envelopment of Manila, 27 January—5 February 1945

XI Corps resumed its attack along Highway No. 7 but began to meet stubborn resistance as it approached well-entrenched Japanese forces and the rough terrain in the mountains of the Zigzag region. Zigzag Pass was strongly defended by mutually supporting, concealed caves and excellently camouflaged concrete emplacements. A deadly stream of machine gun and mortar fire poured out from these defenses, taking a heavy toll of the probing advance elements of the 34th RCT. After severe fighting and with the timely aid of several napalm bombing strikes by Fifth Air Force planes, Highway No. 7 was finally opened to traffic by 14 February. XI Corps continued to the south and east to secure Bataan Peninsula and also to assist in the opening of Manila Bay.[48]

Recapture of Clark Field an Fort Stotsenburg

While XI Corps to the southwest was advancing across Zambales Province and I Corps was still hammering away at the tenacious and deeply entrenched Japanese forces in the hills and mountain ranges northeast of the Central Plain, XIV Corps prepared to attack the Clark Field—Fort Stotsenburg defenses. It was apparent from the strength and disposition of the opposing forces that the advance of XIV Corps to Manila would be materially affected by the evident enemy determination to make any capture of the airfield area as costly as possible.[49]

The question arose whether to contain the defended region with the 40th Division and move on to Manila with the 37th, or whether to launch a combined assault with the full strength of both divisions. There were numerous indications that the Japanese occupied the hills west of the main highway in considerable depth. Enemy counterattacks in force against the 40th Division, if it were left unsupported, could conceivably cut eastward to cross the highway and interrupt the XIV Corps' line of supply to any forward units. The Corps commander, General Griswold, decided therefore to throw the entire weight of the Corps into a drive which would push the enemy forces westward into the hills and away from the highway.[50] With the immediate areas of Clark Field and Fort Stotsenburg secure, the 37th Division could then resume the advance to Manila; the 40th Division, meanwhile, would push southwestward to pinch off the enemy by a juncture with XI Corps advancing northeastward along Highway No. 7.

On the night of 27 January, while the 40th Division inched slowly across the rugged ground of the Bambam Hills on the west, the 37th Division launched a combined tank and infantry attack against Clark Field and Fort Stotsenburg from the east. The 37th Division moved steadily into the objective area against heavy automatic and artillery fire, thickly-sown tank mines, and numerous enemy counterattacks. By 29 January, despite this fierce resistance, both Clark Field and Fort Stotsenburg had been captured and occupied by XIV Corps troops. The next day, in response to an invitation from General Griswold, General

48 Ibid., Sections III and IV.

49 The Japanese had a total of approximately 30,000 troops of the "Kembu Group" defending the Clark Field—Fort Stotsenburg sector. This figure included 8,000 troops of three army detachments (Takayama, Eguchi, and Takaya Butai), 15,000 naval troops deployed in the hills to the northwest in the rear of the Clark Field defenses, and 7,000 miscellaneous air service troops who had been located at the various airfields. Japanese First Demobilization Bureau Report, *Philippines Operation Record*, Phase III, Supplement 2 to Vol III, "Kembu Group Operations in the Clark Sector," G-2 Historical Section, GHQ, FEC.

50 XIV Corps, M-1 Operation Report, p. 69.

Krueger officiated at the flag-raising ceremony. With Japanese artillery shells still falling on the western edge of the military reservation, the "Stars and Stripes" were flown once again over the post where, a little over three years before, General Wainwright had received the first startling news of the attack against Pearl Harbor.[51]

In their hasty retreat into the mountains, the Japanese were forced to abandon huge quantities of supplies to the advancing troops of the Sixth Army. A communique of 29 January reported:

Tremendous depots of enemy quartermaster, signal, ordnance, and air technical supplies captured in the Clark Field area include over two hundred new aircraft engines, many radio transmitters and receivers, great quantities of miscellaneous radio and telephone equipment, several months' stores of ammunition, food and equipment, and over forty artillery pieces of various calibers. . . .[52]

The 40th Division found it a difficult task to contain the enemy north and west of Fort Stotsenburg. The Japanese had chosen an ideal location from which to make their defense. The terrain in this region is characterized by steep ridges and isolated peaks interspersed with deep ravines and narrow, winding defiles which usually are the only paths of access to the main roads. Murderous machine gun and artillery fire from cave positions in the cliffs made any advance difficult and costly. At times the reduction of each cave individually was necessary before the American troops could move on to their next position.

By the beginning of February, the advance elements of the 40th Division, XIV Corps, and the 38th Division, XI Corps, were in contact at Dinalupihan on Highway No. 7. The 37th Division, after securing the Fort Stotsenburg area was reforming in preparation for the drive to Manila. To the north, I Corps was moving slowly eastward. The 43rd Division had captured Pozorrubio on 18 January and pushed on against stiff resistance through Pimmilapil and Concepcion to improve its position on the Damortis—Rosario road. The 25th Division, battling courageously against a stubborn enemy southwest of Asingan, progressed toward Baligayan. On the right flank of I Corps, the 6th Division had driven through the Cabaruan Hills to occupy Urdaneta, and advance forces of the division were pressing southwest toward Talevera and Munoz.

Final Drive to Manila

The final drive to Manila was initiated by General Krueger's Field Order No. 46 issued on 30 January 1945. This order fixed the boundary between XIV Corps and I Corps, putting the towns of Dagupan, San Carlos, Malasiqui, Carmen, and Victoria within the XIV Corps zone and the positions from Licab to Cabu and Tamala, in the I Corps zone. The line of demarcation on the southwest, between XIV Corps and XI Corps, ran from Purac on the west coast through Dinalupihan and Orani. The 1st Cavalry Division, which had landed on Luzon on 27 January, was attached to XIV Corps four days later with orders to approach Manila on the left flank of the 37th Division by initiating an advance across the Pampanga River to secure the line from Hagonoy through Sibul Springs and Cabanatuan.[53]

In the early morning hours of 1 February, the 1st Cavalry Division under Brig. Gen. Verne D. Mudge, moved out of its assembly area at Guimba and began an advance southward. The Pampanga River was quickly crossed,

51 Ibid., p. 72.

52 GHQ, SWPA, Communique No. 1027, 29 Jan 45.

53 XIV Corps, M-1 Operation Report, p. 79.

and by evening advance elements of the division had penetrated as far as Cabanatuan.[54] Here two squadrons of Brig. Gen. John H. Stadler's 1st Cavalry Brigade formed a "flying column" in conjunction with the 44th Tank Battalion and moved in swiftly to surprise the Cabanatuan defenders. The Japanese had placed a 3000-pound charge of dynamite to destroy the bridge across the Pampanga River and planned to detonate the explosives with mortar fire before the American troops could effect a crossing. Quick action by General Mudge in leading his men onto the bridge to throw the dynamite into the river even though the mortar shells had already begun to fall, foiled the enemy plan, preserved the bridge intact, and enabled the cavalrymen to proceed with little difficulty. Once the bridge was crossed, further progress was relatively unopposed as forward elements of the 1st Cavalry Division raced down Highway No. 5 to contact units of the 37th Division at Plaridel near the Angat River.

The movement of the 37th Division along Highway No. 3 was greatly impeded by the thorough demolition work of the retreating Japanese as a result of which almost every important bridge along the route of advance had been destroyed. On 2 February, elements of the division passed through Calumpit and Malolos while other units engaged an enemy road block just north of Plaridel close by the Angat River where first contact was made with the cavalry patrols. After a bitter all-night battle, Plaridel and the river crossing were secured. The 1st Cavalry Division meanwhile forded this river at a point to the east and then continued on rapidly toward Novaliches. By evening, XIV Corps, fulfilling its assigned task, held the line from Hagonoy through Sibul Springs to Cabanatuan.[55]

Battle in Manila

General Krueger's strategy for retaking Manila involved an assault on the capital from two directions. The 1st Cavalry and 37th Divisions of XIV Corps would launch the main attack from the north, while a smaller but nonetheless strong effort would be carried out on the south by combined airborne and infantry forces.[56] On 31 January, the Eighth Army landed troops of the 11th Airborne and 24th Divisions on Batangas Province to form the lower jaw of the Manila pincer movement.

The 11th Airborne Division, less the 511th Parachute Infantry but reinforced by two battalions of the 24th Division, landed by water to establish a virtually uncontested beachhead at Nasugbu, a little over 30 miles southwest of Cavite.[57] The suicide thrusts of small enemy

54 Two days earlier, during the evening of 30 January, a force of guerrillas, assisted and led by elements of the 6th Ranger Battalion, raided the enemy prison camp near Cabanatuan. Over 500 Allied prisoners of war were liberated in this successfully executed attack and the Japanese garrison was virtually annihilated. See Chapter X, p. 318.

55 XIV Corps, M-1 Operation Report, p. 82.

56 Ibid., p. 79.

57 The major units participating in the Nasugbu—Tagaytay assaults were as follows:

Ground: *Eighth Army—General Robert L. Eichelberger*

Unit	Commander	Phase	Staging Point	Landing Point
511th Prcht RCT	*Col. O. D. Haugen*	*Prcht Assault*	*Mindoro*	*Tagaytay*
11th A/B Div (–1 RCT)	*Maj. Gen. J. M. Swing*	*Assault*	*Leyte*	*Nasugbu*

Air: *Fifth Air Force—Maj. Gen. Ennis C. Whitehead*

Unit	Commander
54th Troop Carrier Wing	*Brig. Gen. P. H. Prentiss*

Naval: *Seventh Fleet—Adm. T. C. Kinkaid*

Unit	Commander
Task Group 78.2	*R. Adm. W. Fechteler*

torpedo boats attempting to attack the landing craft were promptly and easily crushed. General MacArthur described the tactical implications of this landing as follows:

> *This operation places the Eighth Army on the south side of Manila which is now the center of the converging columns of the Sixth and Eighth Armies. It largely seals off the possibility of the enemy troops south of Manila joining those to the north and definitely outflanks the enemy's defense lines in southern Luzon. . . .*[58]

On 3 February, the remaining parachute regiment dropped on Tagaytay Ridge on the north shore of Lake Taal. A juncture of the two forces was quickly effected and the combined paratroop units moved northward along Highway No. 17 toward Manila. Opposition was slight and the troops drove rapidly ahead until they reached the permanent concrete and steel emplacements which guarded Nichols Field. Here the division halted to consolidate until heavy artillery could be brought in from XIV Corps.[59]

Both the northern and southern outskirts of Manila were now reached, but the relative ease of movement which had hitherto characterized the American advance down the Central Plain was by no means an indication of the nature of subsequent progress. Although General Yamashita had begun to evacuate the bulk of his forces and supplies from the Manila area in mid-December while the Allies were carrying out their invasion of Mindoro, American intelligence estimates listed between 15,000 and 18,000 enemy troops still remaining intrenched within the city.[60] These troops, consisting mostly of naval base units, were intended not only to protect the removal of the rest of the vast quantities of supplies and equipment which had been stored in the capital, but also to deny the use of Manila's harbor facilities, airfields, and other important installations for as long a period as possible.[61] Although the defenses inside the city were generally improvised and adapted to meet a rapidly shifting tactical situation, the tenacity and bitterness of the resistance encountered made the American seizure of central Manila an exhaustive and prolonged effort.

As the forward elements of the 1st Cavalry Division advanced from the Angat River, word was brought that some 4,000 American and other Allied prisoners were interned at Santo Tomas University and that others might possibly be held at Malacanan Palace. The same "flying column" that had so effectively spearheaded the drive into Cabanatuan was immediately dispatched to free these internees lest they be harmed while the battle for Manila was in progress.[62] At dawn on 3 February, the cavalrymen left Santa Maria and sped swiftly southward. The bridge at Novaliches

58 GHQ, SWPA, Communique No. 1031, 2 Feb 45.

59 XIV Corps, M-1 Operation Report, pp. 91–94.

60 These forces were under the command of Rear Adm. Sanji Iwabuchi, Commander of the Japanese 31st Naval Special Base Force. According to Japanese records, naval troops in the Manila—Cavite area at the time of the Lingayen invasion numbered about 17,000 men. In addition, approximately 3,000–4,000 army troops were sent to the Manila area and placed under the control of Admiral Iwabuchi for ground operations. A considerable number of these troops succeeded in effecting a withdrawal to join forces with the "Shimbu Group" east of Manila. Japanese First Demobilization Bureau Report, *Philippines Operations Record,* Phase III, Supplement I to Vol III, "Shimbu Group Operations East of Manila," p. 21, G-2 Historical Section, GHQ, FEC. See also: Japanese Second Demobilization Bureau Report, *Operations by Manila Naval Defense Force,* May 47, Part I, pp. 49–64 and Part II, pp. 49–50, G-2 Historical Section, GHQ, FEC; and Statement of Comdr. Koichi Kayashima, Operations Staff Officer, 31st Naval Special Base Force, Interrogation Files, G-2 Historical Section, GHQ, FEC.

61 HQ Fourteenth Area Army, *Outline of Operational Policy for Luzon Island,* 19 Dec 44, G-2 Historical Section, GHQ, FEC.

62 XIV Corps, M-1 Operation Report, p. 83.

over the Tuliahan River was reached just in time to extinguish the still burning fuses of an enemy demolition charge. After the bridge was crossed, numerous small groups of Japanese in ambush along the road tried to delay the advancing troops with small arms fire.

When the cavalrymen entered the northern suburbs of Manila, the hangars and airfield equipment at Grace Park were already ablaze and little could be saved. The "flying column" proceeded down Rizal Avenue to Santo Tomas University, meanwhile diverting one troop of cavalry and a platoon of tanks to Malacanan Palace. Resistance on the University grounds was stiff but, with tank support, the Americans forced the main gates and wiped out the enemy troops in the area. All internees were liberated with the exception of 221 who were held as temporary hostages and released the following morning.[63] Malacanan Palace was also reached against sporadic rifle fire from across the Pasig River but only Filipino police guards and attendants were found to occupy the building.

The bulk of the remaining cavalry units was assembled in the vicinity of Novaliches by 3 February. Further progress was temporarily interrupted while repairs were made on the bridge which the enemy had finally managed to destroy after the "flying column" had passed. The advance was soon resumed, however, and by 5 February, the cavalry troops had begun to gather in the Grace Park area. (Plate No. 78)

After its brief contact with patrols of the 1st Cavalry Division at the Angat River, the 37th Division pushed along Highway No. 3 against constant automatic and mortar fire. The Japanese had blown the bridges at every stream crossing and progress was relatively slow. Malanday and Caloocan were occupied on 4 February, and Manila was entered on the same day. The division effected its own rescue mission when some of its units entered Bilibid prison and discovered 800 American prisoners of war who had been abandoned by their jailers. The brilliant record of the Sixth Army in the release of prisoners of war and internees on Luzon was described in a communique of 6 February:

> *The 37th Infantry Division in capturing Bilibid prison released more than 800 prisoners of war and about 500 civilian internees including women and children. With the 3,700 internees from Santo Tomas released by the 1st Cavalry Division, this brings the total rescued to approximately 5,000. About 4,000 were Americans and the rest British, Australian and other nationalities. Every facility of the armed forces is being devoted to the care and attention of those who have been rescued. . . .*[64]

As the 37th Division approached the Pasig River, it was met by a devastating enemy machine gun and rifle barrage. Incessant detonations and collapsing structures filled the air with deafening concussions. The entire sky was lighted with the roaring fires of conflagrant buildings and at times the mixture of smoke, heat, and dust became so overpowering that substantial progress through the city became an almost impossible task. Amid this holocaust and bedlam, elements of the division effected a crossing of the Pasig River near the Presidential Palace. The entire XIV Corps then began an envelopment from the east as troops of the 1st Cavalry and 37th Divisions pushed laboriously through the streets and avenues of the capital toward Manila Bay.

General MacArthur's victorious entry into Manila was made on 7 February. A group of officers and men which included General Griswold, General Mudge, General Chase, and part of the "flying column" which had so recently distinguished itself, met him at the city limits. General MacArthur congratulated everyone on

63 Ibid., p. 84.

64 GHQ, SWPA, Communique No. 1035, 6 Feb 45.

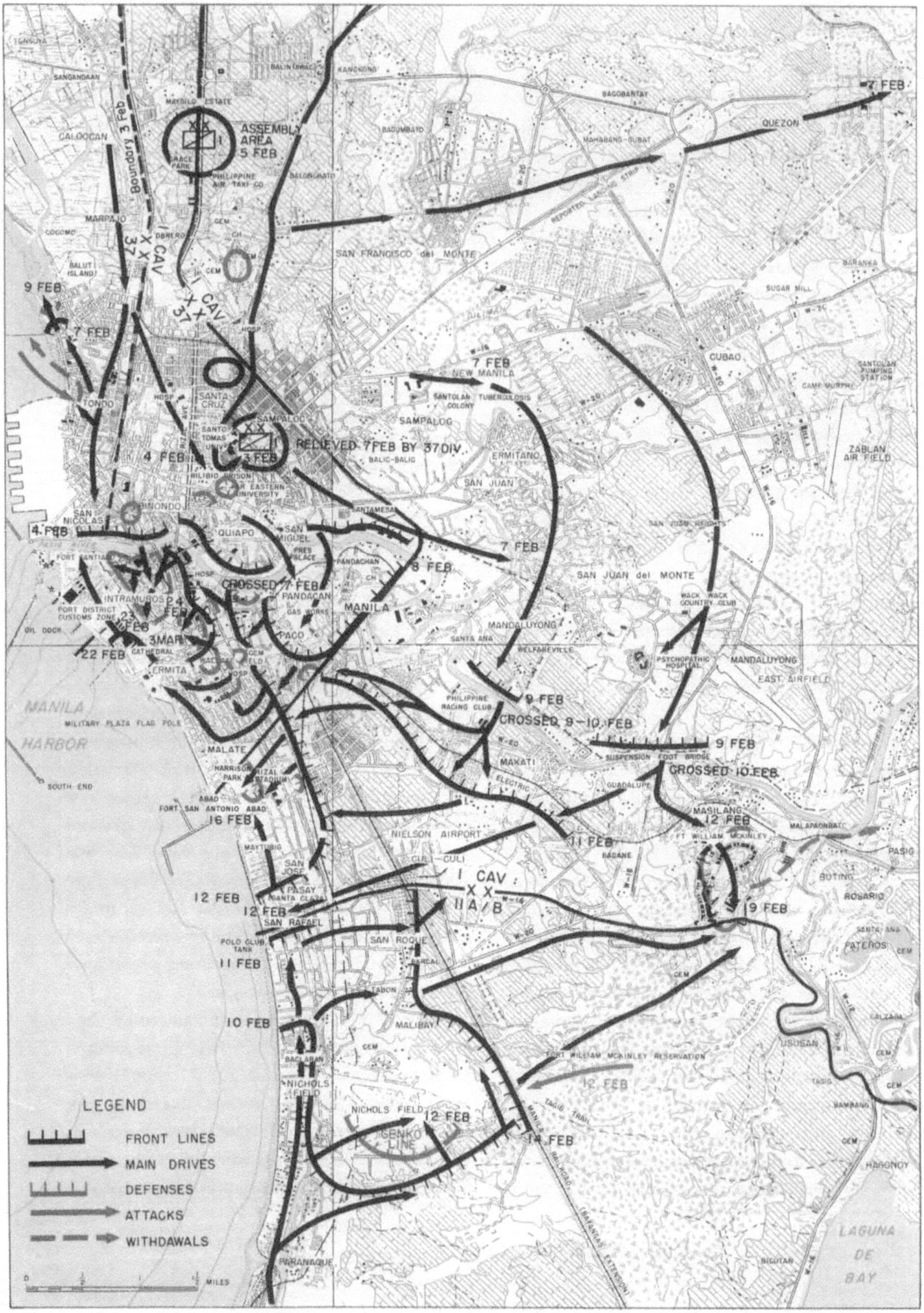

PLATE NO. 78
The Battle of Manila, 3 February—3 March 1945

a job well done and then drove through the war-torn Philippine capital amidst the acclaim of a grateful populace. Sniping and artillery fire continued in almost every section of the city as he visited the Malacanan Palace and the front-line troops engaging the enemy along the Pasig River.[65]

On 10 February, control of the 11th Airborne Division, drawn up south of Manila, passed from the Eighth Army to the Sixth Army.[66] On the same day, XIV Corps artillery poured a steady concentration from the north into the enemy concrete installations on Nichols Field, placing the shells with deadly accuracy in front of the forward paratroop positions. Under cover of this barrage the airborne division moved its tanks against the thick pill-boxes. General Swing's plan was to circle northward and move on the west flank of the Japanese defense line. By the end of the day, the paratroops had seized positions to within 1000 yards of the Polo Club—the main core of enemy resistance northwest of the airfield.

Thus, in the first week of February, General MacArthur had three divisions inside Manila: the 37th Division, attacking south across the Pasig River and on toward the Intramuros area; the 1st Cavalry Division, moving southwestward across San Juan Heights toward Neilson Field; and the 11th Airborne Division, pressing north and east across Nichols Field toward Fort McKinley. Despite this sizeable force, the occupation and clearing of Manila was an arduous task. The Japanese troops in the city fought bitterly, knowing that their chances of escape were small. Improvised positions were set up behind piles of fallen debris, barricaded windows, and sand-bagged doorways. Every vantage point was manned and fiercely defended with a solid curtain of machine gun and rifle fire.

The heaviest fighting took place in the sector assigned to the 37th Division. The Japanese in this area struck out viciously from every position, fighting from building to building and from room to room without surrender. It was not until 17 February that the division was able to launch its assault on the Intramuros, the venerable XVI century citadel in western Manila near the mouth of the Pasig River. Even by modern standards this ancient "Walled City" was a formidable fortress, ringed with a stone wall 15 feet high and widening from 8 to 20 feet at the top to 20 to 40 feet at the base. Four of the main gates were covered by mutually protecting redoubts backed by a heavily fortified concrete building.

Complicating the problem of breaching this massive bastion was the fact that many nonbelligerents, mostly women and children, were within the city. Because of these helplessly imprisoned civilians, all thought of pulverization of the Intramuros area by air bombardment had to be abandoned. A plea was broadcast to the Japanese entrenched within, either to surrender or at least to evacuate the civilian population and prevent unnecessary bloodshed. No answer was received. There was no choice but to order a time-consuming infantry assault to move in, after the way had been prepared by artillery and mortars.

The attack started with 105 mm and 155 mm howitzer shells blasting huge chunks out of the ancient walls. On the 19th, under cover of a heavy smokescreen, 37th Division troops began to pour through the breaches and over the rubble to meet the waiting Japanese. The enemy positions in the immediate vicinity of

65 During this inspection tour of Manila, General MacArthur was accompanied by General Willoughby and Brig. Gen. Bonner Fellers. At this time also, General MacArthur visited briefly the liberated internees at Bilibid Prison and Santo Tomas University.

66 XIV Corps, M-1 Operation Report, pp. 93–94.

the walls had been effectively destroyed by the terrific power of the preliminary bombardment, and the initial incursions of the American forces met with comparatively light losses. Resistance mounted swiftly, however, as the troops advanced. To add to the difficulty, movement became greatly impeded by the streams of refugees that swarmed out of the buildings and milled around the streets. Fire had to be withheld until these scattered masses of civilians could be removed from the battle zone. By 24 February, after a week of savage fighting characterized by numerous hand-to-hand engagements and room-to-room combat, the entire Intramuros was in Allied hands

On 28 February General MacArthur made the following address upon reestablishing the Commonwealth Government in the city of Manila:

> *More than three years have elapsed—years of bitterness, struggle and sacrifice—since I withdrew our forces and installations from this beautiful city that, open and undefended, its churches, monuments and cultural centers might, in accordance with the rules of warfare, be spared the violence of military ravage. The enemy would not have it so and much that I sought to preserve has been unnecessarily destroyed by his desperate action at bay but by these ashes he has wantonly fixed the future pattern of his own doom.*
>
> *Then, we were but a small force struggling to stem the advance of overwhelming hordes treacherously hurled against us, behind the mask of professed friendship and international good will. That struggle was not in vain! God has indeed blessed our arms! The girded and unleashed power of America supported by our Allies turned the tide of battle in the Pacific and resulted in an unbroken series of crushing defeats upon the enemy, culminating in the redemption of your soil and the liberation of your people. My country has kept the faith!*
>
> *Its soldiers come here as an army of free men dedicated, with your people, to the cause of human liberty and committed to the task of destroying those evil forces that have sought to suppress it by brutality of the sword. An army of freemen that has brought your people once again under democracy's banner, to rededicate their churches, long desecrated, to the Glory of God and public worship; to reopen their schools to liberal education; to till the soil and reap its harvest without fear of confiscation; to reestablish their industries that they may again enjoy the profit from the sweat of their own toil; and to restore the sanctity and happiness of their homes unafraid of violent intrusion.*
>
> *Thus to millions of your now liberated people comes the opportunity to pledge themselves—their hearts, their minds and their hands—to the task of building a new and stronger nation—a nation consecrated in the blood nobly shed that this day might be—a nation dedicated to making imperishable those sacred liberties for which we have fought and many have died.*
>
> *On behalf of my Government I now solemnly declare, Mr. President, the full powers and responsibilities under the Constitution restored to the Commonwealth whose seat is here reestablished as provided by law.*
>
> *Your country thus is again at liberty to pursue its destiny to an honored position in the family of free nations. Your capital city, cruelly punished though it be, has regained its rightful place—Citadel of Democracy in the East.*[67]

On 9 and 10 February, units of the 1st Cavalry Division crossed the Pasig River and fanned out in three directions—toward Manila Bay on the west, Neilson Field on the south, and Fort McKinley on the southeast. Elements of the 11th Airborne Division, meanwhile, swept the enemy from Nichols Field by the 13th and moved on to Neilson Field and Fort McKinley.[68] Rather than be caught between the two advancing forces, the Japanese aban-

67 Manila *Free Philippines*, February 28, 1945, p. 1.

68 On 23 February, a specially constituted task unit, composed of guerrillas and 11th Airborne Division troops, speared into the Japanese prison camp at Los Banos to add more than 2,100 Allied civilians to the long list of liberated prisoners. See Chapter X, p. 318.

doned Neilson Field without a struggle. During the last two weeks of February, the enemy defenses throughout the city were reduced to widely scattered and rapidly shrinking pockets of opposition. Rizal Stadium, Fort McKinley, and the buildings along the harbor district were occupied after short but intense fighting. On 3 March, the last organized enemy defenses in the Welfare Park area were overcome. The battle for Manila was over.

Bataan and Corregidor Retaken

After the capture of Zigzag Pass and the opening of Highway No. 7 on 14 February, XI Corps continued on its mission to occupy Bataan Peninsula and clear the entrance to Manila Bay. To strengthen the Corps, one regiment of the 6th Division, the 1st Infantry which had been relieved at Urdaneta by the 25th Division on 18 January, was attached to the 38th Division. For the recapture of Bataan, two forces were designated to carry out simultaneous operations along both sides of the peninsula. On 15 February, a "South Force" of the 151st RCT, 38th Division, sailed from Olongapo on Subic Bay and landed at Mariveles on Bataan's south coast to take up the advance northward along Highway 110. (Plate No. 79) Meanwhile an "East Force" of the 1st Infantry, 6th Division, moved overland from Dinalupihan on the northern end of this same highway and advanced down Bataan's east coast to establish contact with the "South Force". The 1st Infantry also had the mission of securing Highway No. 111 which runs laterally across Bataan from Pilar on the east to Bagac on the west.

The "East Force" seized Pilar without opposition and, after diverting two battalions westward along Highway No. 111, it continued south against scattered resistance to take Orion and Limay in quick succession. Progress of the "South Force" was delayed only by a group of fortified enemy pillboxes in the mountains immediately northeast of Mariveles. This obstacle was quickly eliminated, however, and the advance was resumed.

By 18 February, both forces had joined at Limay to occupy the entire length of highway along the east coast of Bataan. A co-ordinated attack by 38th Division forces, assisted by Fifth Air Force planes, dislodged the stubbornly resisting groups of Japanese infantry who were deeply entrenched near Bagac along Highway No. 111. By 21 February, the 38th Division had secured all objectives on the peninsula. Thus, within the short space of seven days, Bataan was once more in American hands.

Just as in the dark days of 1942 when the fall of Bataan had preceded the capitulation of Corregidor, so now, almost three years later with the roles of the opposing armies completely reversed, the recapture of the peninsula was a prelude to the recovery of the "Rock". Corregidor had been pounded steadily from the sea and air since the last week in January. In addition to the heavy naval shelling, the Fifth Air Force alone had dropped over 3,000 tons of bombs and napalm in some 2,000 sorties, shattering the island's outer fortifications and crumbling the exposed concrete installations into a mass of jagged rubble.

On 16 February, after a last powerful naval bombardment, XI Corps launched a co-ordinated airborne and seaborne invasion against the strategic rock fortress. (Plate No. 80) A battalion of the 503rd Parachute Infantry made the first assault, dropping on the western portion of the island; it was followed shortly afterward by a battalion of the 34th RCT which landed by water on San Jose beach slightly southwest of Malinta Hill. Another battalion of airborne troops was dropped later that afternoon but because of the extremely rough terrain in the landing area and the increasing casualties

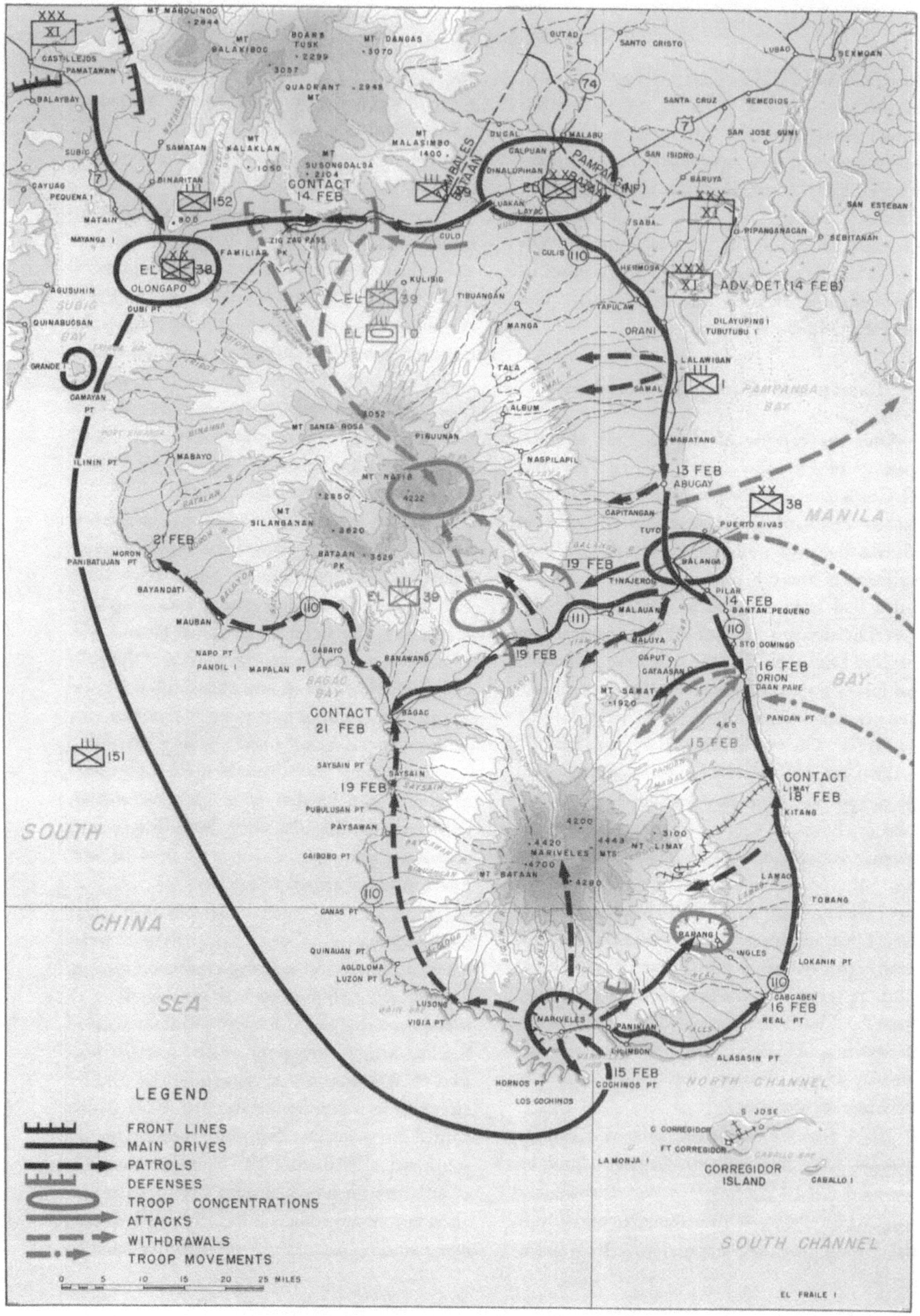

PLATE NO. 79

Bataan Retaken, 13–21 February 1945

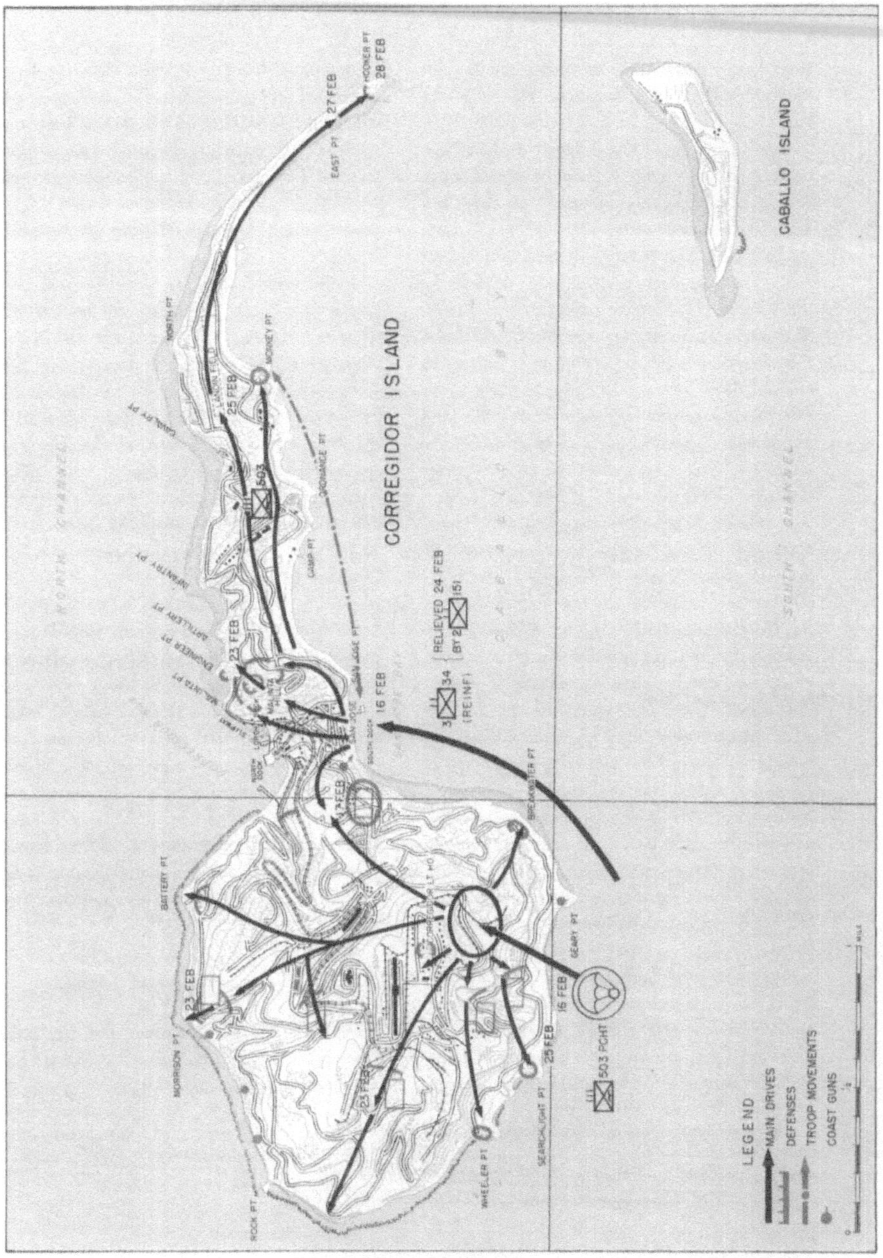

PLATE NO. 80

Recapture of Corregidor, 16–28 February 1945

from heavy enemy fire, the third battalion of paratroops came in by water on the following day.

The 503rd Parachute Infantry and the battalion of the 34th RCT quickly joined forces to eliminate the main system of cave and tunnel defenses running through the Malinta Hill district. The Japanese fought to the bitter end, defending each position with devastating fire. Rather than be dislodged by the irresistable onslaught of tanks, bazookas, and flame throwers, they blew up many of their tunnels in suicidal desperation. During the night of 23 February, a terrific explosion from the vast ammunition stores in Malinta Hill shook the entire island of Corregidor and sent its reverberations along the whole of Manila Harbor.

Although these self-sealing tactics did cause many casualities to American troops caught in the immediate vicinity of the blasts, the overall effect was to lighten the task of cleaning out each individual cave. By 27 February, the American forces had seized complete control of all commanding ground on the island, and on the next day Corregidor was pronounced secured. The twelve-day fight had been vicious and bloody. Virtually the entire Japanese force of 4,700 troops had been annihilated with only a handful taken as prisoners-- somber and undeniable attestation to the tenacity of the enemy's resistance. Another chapter had been added to the long story of bitter battles that charactrized the campaigns in the Southwest Pacific Area.

The capture of Corregidor by United States forces marked a historic milestone in the war against Japan. The " Rock ", after a hard and bitter three-year campaign which carried General MacArthur to Australia and back, was again in American hands. For the Commander-in-Chief this was more than a victory in arms over the enemy; it was the fulfillment of a personal crusade. In a stirring ceremony on 2 March, General MacArthur, in the presence of members of his pre-war staff,[69] gave orders to raise the colors once more over the tiny, battle-scarred island.[70]

In the meantime the area of operation of XI Corps had been expanded to include the enemy's " Kembu " defense sector west of Fort Stotsenburg. On 21 February, the Corps assumed the responsibility for the direction of the 40th Division which had been engaged in the lone and difficult task of clearing large concentrations of well-equipped and tough Japanese troops from their strongly prepared hill positions in the Zambales Mountains. On 26 February, the exhausted troops of the 40th Division were replaced by the 43rd Division which, in turn, was relieved of its mission in the I Corps zone by the newly arrived 33rd Division. The 43rd Division was ordered to launch a co-ordinated frontal and enveloping attack to eliminate the main centers of resistance in the Zambales area. Substantial advances were made on all fronts and, on 7 March, the 43rd Division was again withdrawn, this time for action in the region east of Manila, while the 38th Division moved up from Bataan to continue the attack in Zambales and complete the detailed mopping-up of both provinces.

The Assault East of Manila

Manila had fallen, Bataan and Zambales Provinces were occupied, and Corregidor had been recaptured, but the toughest fighting of

69 This staff included Generals Sutherland, Willoughby, R. K. Marshall, Marquat, and Casey. Even before the outbreak of the war these officers had been associated with General MacArthur, serving with the Military Advisor's Group, with the Philippine Department, or with USAFFE.

70 HQ XI Corps, *Historical Report, Luzon Campaign*, Section V, p. 12.

the Luzon campaign was still in the future. In consonance with his general plan of a protracted defense of the island, General Yamashita had grouped his forces in strategic positions to protect the approaches to the Cagayan valley and the mountain fastness of northern Luzon. There was still much action ahead for XIV Corps in the Batangas and Bicol regions but the "Yamashita Line," extending from Antipolo on the south through the hills around Montalban and Ipo Dams on to Cabanatuan and then into the northern mountains, was to prove the greatest obstacle to the battling forces of the Sixth Army.

The portion of the "Yamashita Line" that ran east of Manila from Antipolo through Montalban to Ipo Dam was known as the "Shimbu Line" and was defended by a heterogeneous collection of Japanese combat and service units numbering approximately 30,000 troops. These troops were distributed among three enemy forces protecting the main positions east and northeast of Manila: the Kawashima Force in the northern Ipo sector, the Kobayashi Force in the central Marikina sector, and the Noguchi Force in the Antipolo sector to the south. In addition there were about 13,000 miscellaneous troops scattered through the rear areas and some 9,000 troops which had withdrawn from Manila during the first stages of the battle for the city. These Manila escapees comprised mostly naval units and took little part in the fighting, filtering through the lines toward Infanta and then melting into the mountains along eastern Luzon.[71]

The "Shimbu Line" presented a triple threat to XIV Corps since, potentially, the enemy could launch a counterattack against the extended left flank of the Corps or, by controlling the vital dams, dominate or destroy the water supply to Manila or, with long range artillery, lob shells into the capital itself. The 6th Division, transferred from I Corps to XIV Corps on 17 February, pressed against the central and northern sectors of the enemy's line of defense while the 1st Cavalry Division and later the 43rd Division moved against the Antipolo region. (Plate No. 81)

On 5 March,[72] after the fall of Manila, XIV Corps ordered a combined assault by the 1st Cavalry and 6th Divisions to cut between the Wawa and Antipolo areas and effect an envelopment from the south of the enemy's strong points on Mt. Mataba and Mt. Pacawagan. On 8 March, after a heavy two-day plane bombardment of enemy positions north of Antipolo, XIV Corps began the offensive to the east. Resistance was fierce and progress slow, but by 14 March considerable advances had been made. It was on this date that Maj. Gen. Edwin D. Patrick, commander of the 6th Division, was fatally struck by a sudden burst of enemy machine gun fire which raked his advance observation position.[73]

The formidable cave defenses in the Antipolo—Wawa line were well described in a GHQ communique:

> *The enemy line of defense consists almost entirely of caves in the hill-sides closely spaced for interlocking fire. A cave normally consists of a ten foot shaft large enough for a man to climb up or down a rope ladder. At the foot of the shaft a small tunnel leads to a compartment some twenty by thirty feet. Four or five lateral tunnels lead from the compartment. Just before a lateral reaches the surface slope*

71 Japanese First Demobilization Bureau Report, *Philippines Operations Record,* Phase III, Supplement 1 to Vol III, "Shimbu Group Operations East of Manila," attached map No. 1, G-2 Historical Section, GHQ, FEC.

72 This date also marked the day that Lt. Gen. Walter Krueger received his promotion to General, crowning a brilliant career of outstanding achievement and service to his country.

73 HQ XIV Corps, M-1 Operation Report, p. 155.

PLATE NO. 81
Assault Eastward from Manila, 8 March—28 May 1945

of the hill it makes a sharp bend to prevent shelling from entering the tunnel. Against frontal assault methods the fire power developed from these miniature forts is deadly. Our method of attack consists in preliminary saturating bombardments by the air, artillery and all infantry weapons calculated to confine the enemy within the tunnels. When this has been accomplished a small demolition group from the infantry moves in with white phosphorous grenades, flame throwers and demolitions. Several hundred pounds of explosives are placed inside the entrance to a lateral and blown to close the opening. In this manner the four or five laterals are sealed off and the vertical shaft then blown from the top. Each cave with its laterals can hold about twenty-five men. When trapped in this way the enemy is suffocated to death and destroyed without taking any compensating toll from our troops. As an example, a Brigade of the 1st Cavalry Division in forty-eight hours in the attack now developing took 137 caves and blew 446 of the outlets to these caves with practically no loss themselves, while the enemy's loss is estimated at several thousand men. Similar methods have been used throughout the New Guinea and the Philippine campaigns. Progress of the attack is slowed by these methods but an enormous saving in life and materiel is affected.[74]

To speed the penetration of "the Shimbu" defenses, General Krueger once more regrouped his forces and ordered a change in the Corps' boundaries. The 43rd Division had already relinquished its task of clearing the Zambales Mountains to the 38th Division and was en route to relieve the 1st Cavalry Division in the Antipolo region. On 15 March, XI Corps replaced XIV Corps in the "Shimbu" sector, taking over control of the 6th Division and the 43rd Division less the 169th RCT. XIV Corps was, in turn, assigned the task of clearing Batangas and the Bicol Peninsula.

On 19 March, the 6th Division initiated a drive eastward to secure the Wawa Dam sector. The Japanese in this area were committed to a last-ditch defense and were well equipped with mortars, artillery, and even crudely constructed but nevertheless destructive rockets. The dominating enemy positions on Mt. Mataba and Mt. Pacawagan gave strong protection to Wawa Dam and made the advance of the 6th Division extremely difficult. It was not until the end of March that positions were reached from which direct assaults could be carried out against Mt. Mataba. The Japanese poured unceasing fire from the mountain ridges, and all attempts to outflank their positions met with failure. Enemy entrenchments which were almost inaccessible to ordinary artillery and mortar shelling were bombarded repeatedly from the air with heavy napalm strikes. Despite all efforts, however, the daily gains were small, and it was not until the end of April that the defenses on both mountains yielded to the stubborn pressure of combined infantry, artillery, and air assaults and the 6th Division troops could consolidate for the final drive on Wawa Dam itself. On 30 April, the 6th Division (less the 145th Infantry), tired and depleted as a result of almost continuous fighting since its landing at Lingayen on 9 January, exchanged places with troops of the 38th Division in the Zambales Mountains.[75] Other elements of the 38th Division had completed their task of clearing the remainder of Manila Bay with their capture of Caballo Island by 29 March and of Carabao Island and Fort Drum by 3 April, and 18 April, respectively. The 145th Infantry, which had left Manila on 18 April to relieve the 20th RCT and was comparatively fresh, was attached to the 38th Division to continue the offensive against the "Shimbu Line."

The 43rd Division, in the southern sector of the "Shimbu Line," meanwhile pressed its

---

74 GHQ, SWPA, Communique No. 1068, 10 Mar 45.

75 HQ Sixth Army, *Sixth Army on Luzon*, Section 3, "Cave Warfare in the Mountains," p. 50.

attack in the Antipolo region. A swift 25-mile sweep eastward by the division's mechanized forces broke through the Japanese defenses and resulted in the seizure of numerous fortified positions east of the Morong River valley. Exploiting this advantage, the division went on to capture Tanay on 19 March, and three days later the main fortifications along Mt. Tanauan had been eliminated. Other units of the division swung through Teresa to seize Mt. Caymayuman northeast of Antipolo.

By the end of March, the badly shattered forces of the enemy had been driven into the hills east of the Boso Boso River, and the 43rd Division had accomplished its main objective of driving a wedge between the southern and central sectors of the "Shimbu Line." The division then turned along the eastern shore of Laguna de Bay to take Mabitac, Siniloan, Famy, Pangil, Pakil, Paete, Longos, and San Juan in quick succession. For the remainder of April, it continued operations in this area, cleaning up by-passed pockets of enemy troops and making contact with elements of the 1st Cavalry Division, XIV Corps, which were moving north, near San Juan in Batangas. Thus by the end of April, the "Shimbu Line" east of Manila was broken in the south, penetrated in the center, and contained in the north.

Drive in the North

While XI and XIV Corps were pushing across the southern half of Luzon, I Corps was still battling furiously against the enemy in the north. The Japanese had committed the major strength of their forces to the protection at all costs of the main passes to their headquarters at Baguio and to their "rice bowl" in the Cagayan Valley. When the U. S. 33rd Division relieved the 43rd Division and the 158th RCT in the Rosario region on 13 February, it was directed to assault the Baguio area. Although the division immediately launched a strong three-pronged drive—from Bauang on the coast, from Pugo along the center, and from Rosario on the south—the operation proved to be bitter and long. (Plate No. 82)

Baguio, the pre-war summer capital of the Philippines, at this time served as a headquarters for both General Yamashita's forces and for the collaborationist government.[76] The Japanese protected the mountain roads leading to this city with their usual thoroughness, covering every strategic pass with excellent and mutually supporting fortifications. Daily progress of the 33rd Division through the tortuous hills and along the narrow ridges was slow and painstaking and, in many cases, the engineers had to carve their own roads over otherwise impassable terrain. Elements of the division, moving from the coast, were aided in their advance by guerrilla units contacted near Bauang.[77] The key town of Naguilian and its adjacent airstrip on the Bauang—Baguio road were seized by the end of March. Other units of the division slugged along the roads from Pugo and from Rosario, seizing commanding terrain positions against fierce opposition.

On 26 March, part of the 37th Division was sent to relieve the guerrilla force at San Fernando on the coast. On 19 April, the

76 The cabinet moved to Baguio on 22 December 1944; General Yamashita's headquarters, on 4 January 1945. Memoirs of General Muto, pp. 19, 22. Besides the puppet president of the Philippines, Jose P. Laurel, some of the other high officials included in the transfer to Baguio were Benigno S. Aquino, Speaker of the Congress; Jose Yulo, President of the Supreme Court; Manuel Roxas, Minister of the Planning Board; and Claro M. Recto, Foreign Minister. Statement of General Utsunomiya, 18 Jul 49, Interrogation Files, G-2 Historical Section, GHQ, FEC.

77 These guerrillas were under Col. Russel W. Volckmann, USAFIP, NL, who was the recognized leader of all guerrilla forces in the northern sector of Luzon. See G-2, GHQ, FEC, *Intelligence Series*, Vol I, "The Guerrilla Resistance Movement in the Philippines"; see also Chapter X, p. 320.

remainder of the division, fresh from a rest period following the fall of Manila, arrived at the Baguio front to assist the 33rd Division in its operations. The campaign against the enemy's stronghold at Baguio continued well into April. The Japanese, gradually being forced back into the hills, decided that the city was no longer tenable and began a withdrawal of their main forces into the chain of mountains enclosed by the triangle, Baguio—Bontoc—Bambang.[78] Overcoming the rear guard left to protect this retreat, the forces of the 33rd and 37th Divisions entered Baguio on 26 April. The 33rd Division spent the next month clearing out the remaining enemy positions in the neighboring mountains, while the 37th Division moved south to replace the 25th Division in the Balete Pass area.

The Japanese defense of the Cagayan Valley centered upon two strategic passes, Balete on Highway No. 5 and Salacsac along the Villa Verde trail. At the time that General MacArthur's forces were preparing to land at Lingayen Gulf, General Yamashita had ordered the 105th Division to move from the Manila area to reinforce the vital positions at the entrance to the Cagayan Valley. Large stock piles of ammunition were accumulated in the caves dug along the sharp ridges and heavily wooded mountain slopes. The enemy's utilization of every contour of the terrain for defensive purposes was, as usual, detailed and expert, and the lack of secondary roads added greatly to the strength of his positions.

At the beginning of February, I Corps had three divisions engaged in the task of breaking into the Cagayan Valley. The veteran 32nd Division had arrived in Luzon at the end of January to take up positions in the San Manuel—Asingan region. The 25th Division moved into the central sector of the I Corps line to continue its eastward advance toward the gates of the Cagayan Valley. The 6th Division, meanwhile, drove along the southern flank to Munoz, a strongly fortified town on the road to San Jose and Balete Pass.

The enemy had assembled the bulk of his remaining armor in the Munoz sector in a strong effort to keep the road into the north open for the transfer of troops and supplies. The main elements of the Japanese 105th Division, moving up from the south, were concentrated at Cabanatuan on 25 January en route along Highway No. 5 to the strategic passes of the Cagayan Valley. The enormous quantities of materiel which had been brought up from Manila and stored at San Jose had to be removed before the escape route into the northern redoubt could be cut off. For these reasons, General Yamashita had ordered the Japanese 2nd Armored Division to make a firm stand in defense of this critical area to delay the American forces as long as possible.[79]

The fight for Munoz was a hard one. Despite the fact that the town had been blasted by repeated air strikes, the Japanese troops were able to offer determined and effective resistance. Heavy and accurate fire from well-emplaced artillery and dug-in tanks, employed as pill-boxes, slowed down the advancing American troops. Finally after a week of heavy fighting in which Munoz was cut off from the rear, the Japanese attempted an unsuccessful retreat to the north and yielded the battered town to the Americans.[80] The 6th Division then pushed rapidly southeastward to seize

78 On 19 April, General Yamashita again moved his headquarters, this time from Baguio to Bambang, preparatory to ultimate transfer to Kiangan. Memoirs of General Muto, p. 36.

79 Japanese First Demobilization Bureau Report, *Philippines Operations Record*, Phase III, Vol III, pp. 83–84, G-2 Historical Section, GHQ, FEC.

80 Commenting on the Japanese attempt to escape from Munoz, Colonel Kawai said, "The enemy [6th Division] cut off the only retreat route and covered it with anti-tank guns. Thus, this force [at Munoz] was annihilated. Retreat through other areas was impossible because of the flooded rice fields." 10th Information and Historical Service, HQ Eighth Army, *Staff Study of Japanese Operations on Luzon*, Part IV, p. 22.

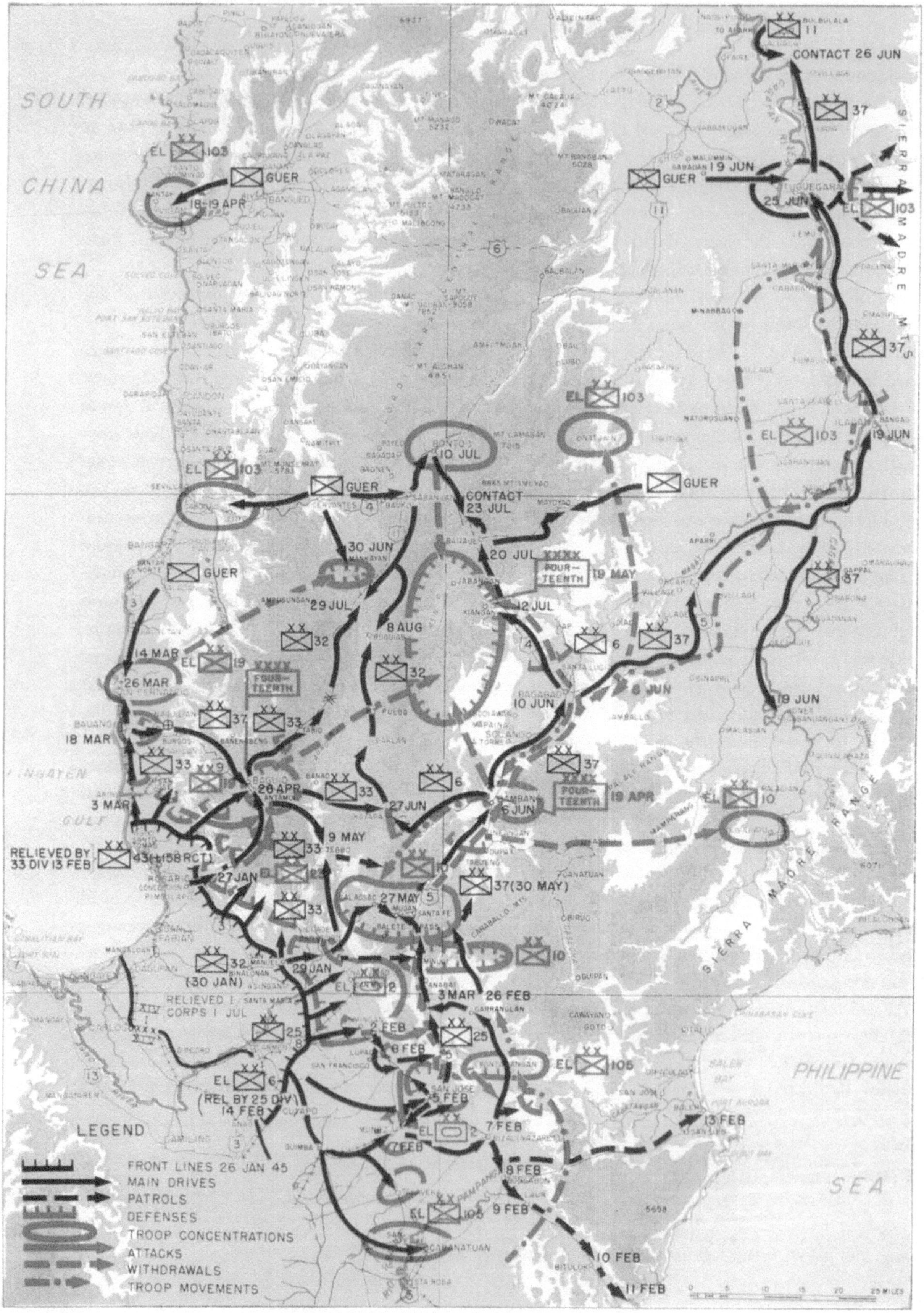

PLATE NO. 82
Operations in Northern Luzon, 26 January—8 August 1945

Rizal and Bongabon while strong patrols sliced across to Dingalan Bay on the east coast to cut Luzon in half.

On 14 February, the 25th Division took over the southeastern flank of I Corps while the 6th Division was diverted to the attack in the "Shimbu" region. The 25th and 32nd Divisions then prepared a two-pronged assault to open the last doors to the Cagayan Valley. The 32nd Division speared northeastward to begin a drive along the Villa Verde trail toward Salacsac while the 25th Division started to advance from San Jose to Balete along both sides of Highway No. 5.

At this time, the 32nd Division occupied the area near Santa Maria at the beginning of the Villa Verde trail—a narrow corridor that wound along the ravines and cliffs of the Caraballo Mountains past Salacsac and Imugan to join Highway No. 5 at Santa Fe. Following the unchanging pattern of yard-by-yard progress in northern Luzon, the struggle to penetrate the elaborate defense works of the Japanese in this area was intensive and prolonged. The extreme scarcity of water along the first half of the trail, added to the terrific heat and the dust from the roads, made the advance doubly difficult. Careless of life, the enemy fought tenaciously from cave to cave and from foxhole to foxhole, covering each move with protective fire. Heavy Japanese reinforcements were brought down from the north to bolster the defenses at Salacsac Pass and, although the 32nd Division had reached this strategic area by mid-April, it was not until 27 May that the pass could be secured.

Along Highway No. 5, the story of the 25th Division's advance was much the same. The campaign to seize Balete Pass was one of bitter fighting which continued all through March and April, and lasted well into May. There were few battles in the Pacific war which were waged along more formidable terrain. Gen. Joseph J. Stillwell, on a visit to the Balete front, commented: "This seems to be as tough as anything could be.... In Burma it was thick, almost impenetrable, jungle. Here there are cliffs that are almost impossible to scale as well as the worst sort of mountain terrain."[81] Heavy rainstorms during the latter part of April, flooding already impassable roads, taxed the ingenuity and endurance of the engineers to the utmost.[82] From the end of April to the middle of May, the division made maximum use of aircraft to assist the infantry advance, an average of eighty planes daily flying tactical support missions.

The enemy continued to contest every inch of ground, however, and it was not until 13 May that troops of the 25th Division finally broke through Balete Pass. After long weary weeks of flushing Japanese from their underground burrows, sealing the deep caves, building new roads, and improving old trails, the American forces had finally forced an entrance into the Cagayan Valley. Contact between the 25th and 32nd Divisions was made at Imugan and I Corps began to move on to Santa Fe and into the valley itself.

---

81 HQ 25th Division, *Tropic Lightning News*, No. 115, 12 Jun 45, p. 2.

82 General Muto expressed great admiration for the feats of American engineers in breaking through the strategic passes of the Cagayan Valley. "In our estimate, based on the past concept of tactics," he stated, "the terrain features of these areas provided impregnable fortification. However, the American forces started attacking in the beginning of February and kept it up incessantly. The superior enemy bombardment and shelling gradually obliterated the jungle area. Bulldozers accomplished the impossible. Tanks and artillery appeared in positions where we had thought they would never penetrate. Our front line troops destroyed bulldozers, tanks and artillery by valiant hand to hand fighting. However, the enemy advanced inch by inch, capturing this mountain, taking that hill." Memoirs of General Muto, pp. 36–37.

Ipo and Wawa Dams Captured

While I Corps endeavored to enter the fertile Cagayan Valley, XI Corps maintained its assaults against the central and northern sectors of the tough "Shimbu Line." After the 6th Division withdrew to western Luzon at the end of April, the 38th Division increased the pressure in the Wawa Dam area, advancing from Mt. Mataba and Mt. Pacawagan. At the beginning of May, the 43rd Division relinquished tactical responsibilities in the Laguna de Bay region to the 1st Cavalry Division and moved northward to take up positions for a concerted effort along the Ipo Dam front.

Ipo Dam normally supplied over 30 percent of Manila's fresh water supply and its early recapture from the Japanese was imperative to the capital's welfare. On 6 May, the 43rd Division initiated the attack to recover the dam. Its drive was aided by guerrilla forces on the other side of the Angat River which were bearing down on the enemy from the northwest. By the middle of May, the high ground dominating the dam on both the north and south sides had been seized and the enemy's positions to the west were being pinched off by the Allied forces advancing in a fast enveloping movement. To speed the drive, Fifth Air Force planes poured the heaviest concentration of napalm ever used in the southwest Pacific on the Japanese defenses surrounding the dam.[83] Aided by powerful searchlight illumination of the battle areas during the hours of darkness, the guerrilla units from the north and the 43rd Division from the south continued to advance with unremitting pressure. On 17 May, Ipo Dam was captured intact and the critically needed water supply was restored to Manila. The Japanese had set demolition charges along the dam but the swiftness of its capture had prevented the planned destruction.

South of Ipo, the 38th Division advanced down the slopes of Mt. Pacawagan to seize Mt. Binicayan on 7 April, and a final base was established for the projected assault of Wawa. Aided by flame-throwing tanks, bulldozers, and air support, the division advanced steadily while the retreating Japanese melted into the hills on both sides of the dam. On 28 May, Wawa Dam was secured, also undamaged, against comparatively light resistance. The last important stronghold of the "Shimbu Line" had been broken, and by the end of June activity in the XI Corps zone consisted of mopping up isolated enemy troops.[84] On 15 June, the Corps assumed control of all combat operations on Luzon south of Cabanatuan and, finally, on 1 July, it was relieved of all tactical responsibilities in the Philippines.[85]

Clearing Batangas and the Bicol Peninsula

After XI Corps had relieved XIV Corps in

83 On 17 May, over 70,000 gallons of napalm were dropped by 288 aircraft preceding infantry troops. G-3, GHQ, SWPA, *Monthly Summary of Operations*, May 45.

84 In describing the rapidly deteriorating condition of the remnants of the Japanese units which had withdrawn into the mountains of eastern Luzon, Colonel Kobayashi stated: "From the beginning of the retreat into the hills, the basic strategy had been to secure positions which could be easily defended by taking advantage of the contours of the land. Offense against the American invaders was to consist of night infiltration raids. Until the end of June we had been able to continue these operations, but after this time the troops were so weak from hunger that only those areas near positions could be guarded, and no infiltration attacks were carried out. From this time on, groups were organized to hunt for food. These often infiltrated American lines to steal supplies. They had not the power to attack any sizeable group of American or Filipino troops." Interrogation Files, G-2 Historical Section, GHQ, FEC.

85 G-3, GHQ, SWPA, *Monthly Summary of Operations*, Jun 45.

the " Shimbu " sector, the latter assumed the offensive against enemy remnants in southern Luzon. On 15 March, the 6th Division, with the 112 RCT attached, passed to the control of XI Corps, and the 37th Division was put in Sixth Army reserve and given the mission of patrolling Manila. XIV Corps now included the 1st Cavalry Division, which had moved south for a ten-day rest at Alabang along Laguna de Bay, and the 11th Airborne Division with the 158th RCT attached. The Corps held a line stretching from Laguna de Bay, southwest to Lake Taal, and thence to Batangas City which had been taken on 10 March.[86] (Plate No. 83)

On 19 March, the 1st Cavalry Division, on the north, and the 11th Airborne, on the south, were ordered to begin a double enveloping drive around Lake Taal to open the highway from Santo Tomas to Batangas City.[87] This maneuver was successfully accomplished by the end of the month. On 24 March, the 158th RCT was relieved in the southern Batangas area by the 11th Airborne Division and instructed to prepare for an amphibious landing at Legaspi on the southeast coast of Bicol Peninsula.

The 1st Cavalry Division continued its eastward drive and, on 6 April, contacted elements of XI Corps near San Juan on the eastern shore of Laguna de Bay. The 11th Airborne Division wiped out the remaining enemy pockets around Mt. Macalod and Mt. Malepunyo near Lipa, and then moved on eastward to cut all routes leading to the Bicol Peninsula. On 19 April, after both the cavalry and the airborne division troops reached Lamon Bay on Luzon's east coast, boundaries between XI and XIV Corps were changed to permit the 1st Cavalry Division to push northeast into the Santa Maria valley and to Infanta, a key coastal city. This drive along the coast was intended to turn the southeastern flank of the " Shimbu " defenses, prevent scattered enemy troops from reorganizing, and deny the Japanese the use of Infanta as a concentration or evacuation point. The tasks were completed with substantial assistance from guerrilla forces, and Infanta was seized with little opposition on 25 May. The remaining enemy forces withdrew in scattered groups into the rugged mountains to the north and northwest. Organized resistance in Batangas Province had ended on 3 May.[88]

XIV Corps was now free to proceed with the liberation of the Bicol Provinces. Great strides had already been made in this direction. On 1 April, the 158th Regimental Combat Team carried out its scheduled amphibious assault at Legaspi on the southeastern tip of Luzon. Opposition was negligible, except at one isolated pocket, since the Japanese had transferred the single division garrisoning the Bicols to the north during January. Although many prepared defenses were encountered, they were manned mainly by army base and naval service troops, together with a few remnants that had escaped from the Leyte—Samar battlegrounds. The 158th Regimental Combat Team had little trouble against such heterogeneous personnel, and on 2 May contact was established with the 1st Cavalry Division, which had been driving into the Bicols from the northwest. The division's advance had been delayed only by the poor condition of the roads and the necessity of an over-water movement across Ragay Gulf. Further combat patrolling uncovered small scattered enemy groups and abandoned defense positions. The Bicol region could also be considered secured. By 31 May, all areas in southern Luzon had been

86 Ibid., Mar 45.

87 XIV Corps, M-1 Operation Report, p. 174.

88 G-3, GHQ, SWPA, *Monthly Summary of Operations*, Apr and May 45.

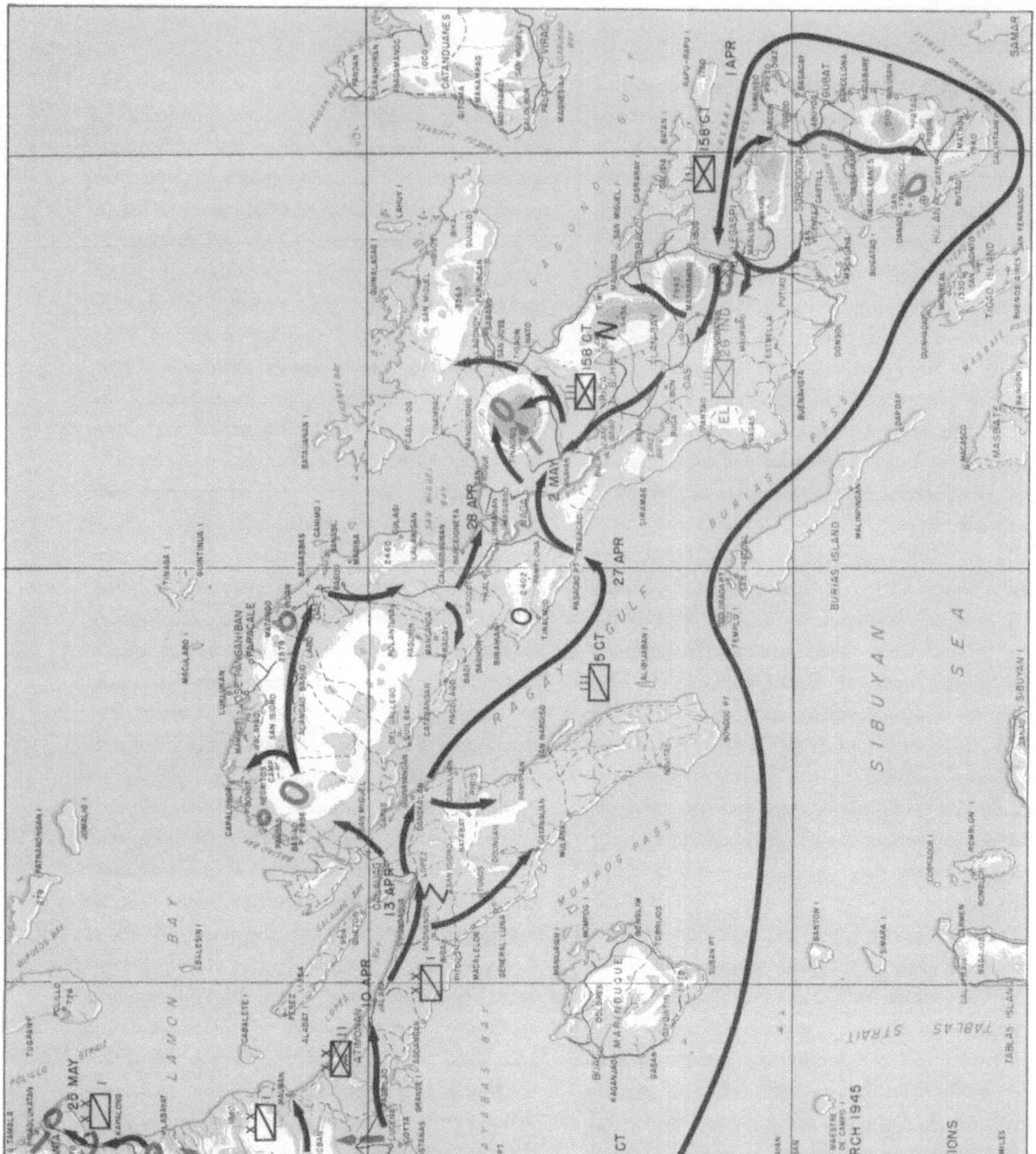

PLATE NO. 83

Operations in Southern Luzon, 15 March—25 May 1945

cleared of major enemy units, and on 1 June the Commanding General of XIV Corps declared that organized Japanese resistance in his sector was at an end. On 15 June, the Corps was relieved of tactical responsibility in southern Luzon and transferred to the north.[89]

Occupation of Kiangan and the Cagayan Valley

The battle for Luzon now entered its final phase. Operations in the Baguio and Cagayan Valley sectors were still under the control of I Corps. At the end of May, I Corps' 37th Division relieved the 25th Division in the Balete Pass area, while the 33rd Division expanded its responsibility to include that part of the Corps' left flank previously patrolled by the 37th Division. Troops of the 33rd Division then commenced a drive southeast from Baguio to Tebbo while other units of the same division pushed toward the same town from the south, after relieving elements of the 32nd Division in the Agno River Valley and in the mountains to the west. Both arms of the division's pincer movement established contact at Tebbo on 9 May, and patrolling was instituted into the hilly country to the east.

More spectacular progress was being made in the Cagayan Valley. The Japanese in this area had put forth their last massed effort in the furious battle to protect the vital Salacsac and Balete Passes, and now their weakened defenses crumbled rapidly before the advancing 37th Division. By 9 June, the town of Bagabag, about 40 miles north of Balete Pass on Highway No. 5, had fallen. At this town, Highway No. 4 enters the Cagayan Valley from the northwest and joins Highway No. 5. The capture o this important road junction cut off the Japanese escape route to the east, leaving their troops concentrated in a tight pocket centered on Kiangan.

Three drives were immediately instituted to reduce the Kiangan sector. The 6th Division, after moving north from central Luzon to follow the 37th Division into the Cagayan Valley, turned northwest at Bagabag and pushed along Highway No. 4 against heavy enemy opposition. By 30 June, the 6th Division's drive had carried to within five miles of Kiangan itself. Another column of this division moved from Bambang in the direction of Kayapa to outflank the Kiangan pocket from the southwest. The 33rd Division meanwhile continued a drive that the 37th Division had started in late May, northward and eastward from Baguio along Highway No. 11. One column of the 33rd Division made contact with the 6th Division near Kayapa on 27 June. A second movement established and maintained a road block on Highway No. 11 north of Tabio until relieved by elements of the 32nd Division on 30 June. The third drive on Kiangan came from Cervantes where guerrilla units were striving to move southeastward to scissor through the Kiangan pocket and meet the 6th Division. By the end of June this last drive had netted over ten miles and had reached Mankayan.

In the Cagayan Valley, the 37th Division continued northeastward from Bagabag, advancing as rapidly as the rate of logistic support would permit. On 18 June, the division crossed the Cagayan River south of Ilagan and next day secured the town. On the 21st, U. S. Rangers and guerrilla forces assaulted and captured Aparri. Two days later, elements of the 11th Airborne Division dropped by parachute and glider on Camalaniugan Airfield near Aparri and drove south in an effort to effect a junction with the 37th Division. Operating between these two forces, guerrilla units fought for Tuguegarao against determined

89 XIV Corps, M-1 Operation Report, p. 220.

Japanese attacks. The 37th Division reached Tuguegarao on the 25th of June and, pressing onward, met the 11th Airborne Division on the 26th.

This union marked the end of organized Japanese resistance in the Cagayan Valley. General MacArthur announced the end of this operation on 28 June:

> *Our northern and southern columns have joined forces, securing the entire length of the Cagayan Valley, heart of northern Luzon. This juncture climaxes a campaign which overran the 200 mile valley in twenty-eight days. Battered enemy remnants have been driven into the rugged mountain ranges to the east and west cut off from all sources of supply.*
>
> *Except for isolated operations this closes the major phases of the Northern Luzon Campaign, one of the most savage and bitterly fought in American history. No terrain has ever presented greater logistical difficulties and none has ever provided an adversary with more naturally impregnable strongholds. . . . Our troops comprised the I Corps and the north Luzon guerrillas, all of the Sixth Army, closely and most effectively supported by the Far Eastern Air Force and the Seventh Fleet. The entire island of Luzon, embracing 40,420 square miles and a population of 8,000,000 is now liberated. . . .*[90]

Only General Yamashita's last stronghold, the Kiangan pocket, held on with any semblance of organization.[91] The 37th Division, together with guerrilla units in the area, turned east to mop up enemy remnants in the Sierra Madre Mountains. The 6th Division, assisted by guerrilla units, undertook the task of crushing the enemy forces at Kiangan.

The liberation of Luzon brought the following message from Secretary of War Stimson to General MacArthur:

> *Your announcement that all Luzon has been liberated marks the achievement of a great military success. It has been brought about with a minimum of casualties. My congratulations go to you and to all officers and men in your command for this most skillful and heroic accomplishment.*
>
> *From my own service in the islands and my close association with their government, I have retained a high respect and warm friendship for the Philippine people. They have suffered cruelly under the Japanese occupation. I share their great rejoicing at the liberation of the main island of their commonwealth. Your great victory hastens the day when the last of the oppressors will have cleared Philippine soil.*[92]

At this time, a number of changes were being prepared in the area commands of the Philippines. Four divisions were assigned to XIV Corps—the 6th, 32nd, and 37th from I Corps and the 38th from XI Corps. On 15 June, XIV Corps was directed to move to northern Luzon to relieve I Corps Headquarters and on 1 July the transfer was effected. On that same date, the Sixth Army, exclusive of XIV Corps, was relieved of operational control of American and Filipino forces on Luzon and was started on a program of training and preparation for its next objective—Japan itself. Operations on Luzon then became the responsibility of the Eighth Army, which was engaged in clearing the Japanese from the southern and central Philippines.[93]

All drives instituted during the period of I Corps control continued unabated after XIV Corps took over. The 6th Division pushed

---

90 GHQ, SWPA, Communique No. 1178, 28 Jun 45.

91 General Yamashita still had under his command in northern Luzon at this time remnants of the 2nd Armored, 19th, 23rd, and 105th Divisions and the 58th Independent Mixed Brigade. These forces, heavily depleted by casualties, sickness, and ineffectives, were operating at less than half strength. Most of their equipment, artillery, and other weapons had been either lost or destroyed in battle and their rations were virtually exhausted. Japanese First Demobilization Bureau Report, *Philippines Operations Record*, Phase III, Vol III, "Operations on Luzon," p. 274, G-2 Historical Section, GHQ, FEC.

92 SECWAR Stimson Ltr to CINCSWPA, 3 Jul 45, G-2 Admin File.

93 G-3, GHQ, SWPA, *Monthly Summary of Operations*, Jun 45.

on against stubborn enemy resistance along Highway No. 4, and reached the town of Kiangan on 12 July.[94] Several guerrilla units were held in the northwest until the middle of July to reduce Japanese fortified positions in the Mankayan area. Other Filipino forces moved northeast from Cervantes to seize Sabangan and Bontoc by 10 July. From Bontoc these same forces pressed south along Highway No. 4 toward Banaue, a town which finally was taken by the 6th Division on 20 July. Three days later contact was established between the American and Filipino units, thus splitting the Kiangan pocket into two sectors. Numerous enemy strong points continued to maintain a determined defense in mountain positions west of Highway No. 4 until the month of August, but the Japanese forces in the eastern half of the Kiangan sector were again bisected when Philippine guerrillas struck from the northeast, seized Mayoyao, and met elements of the 6th Division moving east from Banaue.[95]

Concurrently with the operations of the 6th Division, the 32nd Division, after relieving the 33rd Division in the Tabio area, drove north along Highway No. 11. Philippine Army units pushed south along the same road from Sabangan and these two advancing forces met on 29 July, cutting the enemy's last north-south line of communication in the Kiangan sector. The same forces then pressed eastward from the highway to secure the upper Agno River valley. A battalion of the 32nd Division departed from Daklan, which had been captured by the 33rd Division late in June, and pressed northward along the same valley. At Buguias, on 8 August, contact was established with Filipino troops which had moved down the valley from the north. With the valley thus secured, a firm foothold was obtained on the western slopes of the mountains defended by the enemy. It was only a matter of time before the three-pronged drive formed by the 6th Division from the east, the 32nd Division from the south and west, and the Filipino forces from the north and west, would accomplish the destruction of the last organized remnants of General Yamashita's troops.

To the north and east of the Kiangan pocket, combat activity during July and August consisted mainly of patrols into the Sierra Madre Mountains, east of the Cagayan Valley. The enemy, by this time, was concerned more with problems of physical survival than with continuation of the battle. The critical condition of his troops was much in evidence. In the Sierra Madre Mountains the 37th Division found many Japanese dead from starvation. Farther south, the 38th Division and Philippine guerrillas eliminated the isolated and weakened troops of the once formidable "Shimbu" defense sector east of Manila.

By August all enemy defenses, except for those in the Kiangan sector, were broken, disorganized, and largely ineffective.[96] Lack of food, widespread malnutrition, and prolonged hardships in the field had rendered General Yamashita's remaining troops capable of little

94 Continuing his retreat, General Yamashita had moved his headquarters to Kiangan in mid-May. In June, he went farther into the mountains, establishing himself at Jabangan on the 23rd. As the American forces pursued northward, a new headquarters was established on 12 July at a location along the upper reaches of the Asin River. Memoirs of General Muto, pp. 40–41, 49–54.

95 G-3 GHQ, SWPA, *Monthly Summary of Operations*, Jul 45.

96 On 20 August, the Eighth Army turned over command of the Philippines to the Commanding General, Army Forces Western Pacific. This Headquarters, previously responsible for the logistic support of United States forces in the Southwest Pacific, undertook to bring about the individual surrender and disposition of all Japanese troops remaining in the Philippines.

more than nuisance efforts. To these scattered survivors was given a Hobson's choice of surrender or eventual starvation.[97]

97 According to General Muto, it was impossible to carry on any longer. Describing the conditions in Japanese Headquarters at the beginning of August, he stated: "The wireless facilities of the Area Army, which for a long period had made no contact with the Southern Area General Army in Saigon and the Imperial Headquarters, re-established communication on about 30 July. I felt that it was necessary to make a report on the general situation of the Area Army and I personally drafted a report concerning the disposition of troops, military resources, the enemy situation, the food situation, etc. In conclusion I stated that the Area Army estimated that its organized battle would cease in early September. General Yamashita, after making a few changes, authorized its dispatch. Thus, General Yamashita had decided that the final stage would be in early September." Memoirs of General Muto, p. 56.

CHAPTER X

GUERRILLA ACTIVITIES IN THE PHILIPPINES

The Philippine Resistance Movement

In the wake of military conquest by foreign invaders there have developed great and powerful resistance movements from within subjugated peoples. These rebellions against imposed authority have usually brought swift and ruthless suppression. For their own protection, invading armies have branded the patriot *franc-tireur* and guerrilla as a bandit and a criminal in an effort to alienate him from his people and crush his efforts to gain liberation for his country.

Fighters of the underground have usually received payment in the form of a one-way ticket to the gallows or the firing squad, very often by way of the torture chamber. World War II rings with the echoes of many rifle volleys directed against blindfolded rebel patriots standing stolidly against a stone wall or tied hurriedly to a convenient telegraph pole.

The foreign conqueror, however, is usually vulnerable in his attenuated lines of communication ; the *franc-tireur* is an elusive opponent. The spirit of free men thrives on oppression. In the European Theater, the French *Maquis* has become a valiant and symbolic figure, in his untiring struggle against the Nazis. Equally impressive in the Pacific is the rise of the Filipino against the Japanese invader. The Filipino guerrillas fought for the same principles as the European underground, against the same background of peril, ruthlessness, and hardship.[1]

After abortive efforts to draw the people of the Philippines into the "Greater East Asia Co-Prosperity Sphere" by propaganda, quislings, bribery, and subversion, the Japanese were forced to resort to wholesale arrests, punitive expeditions, and summary executions in an attempt to stem a steadily rising tide of opposition.[2] Repressive measures, however, only increased the determination of the Filipino patriot to resist.

The 7000-odd islands of the Philippines, sprawling along 1000 miles of ocean, made it impossible for the Japanese armies to garrison more than key towns in the populated and cultivated districts. The sparsely settled farmlands and the virtually inaccessible mountains of the interior were left relatively unoccupied by enemy troops. Since the Philippine Constabulary had been demobilized with the invasion, the mass of people in the outlying areas

1 "The saga of a people's refusal to bow before a foreign aggressor has been written many times in this war, especially throughout Europe. But in the Pacific, it fell to the Filipino to show the world that the love for freedom and democracy was not an exclusive quality of the Occident." Manila *Free Philippines*, May 19, 1945, Editorial, p. 2.

2 A good indication of the extent and severity of Japanese anti-guerrilla measures is evident in the captured records of court trials held in the Philippines. These records list the names of hundreds of Filipinos who were tried and summarily sentenced to death or to long prison terms on charges of "guerrilla" or "baneful action." ATIS, G-2, GHQ, SWPA, *Enemy Publications*, No. 398, "Trial Records of Filipino and Chinese Guerrillas and Civilians and Japanese Soldiers and Civilians," 22 Sep 45 (C).

was left without adequate police protection. Such a situation soon encouraged the rise of numerous marauding parties which roamed the countryside in search of easy loot and tribute to be taken from the defenseless farmers.[3]

As usual in such cases, the harassed people took matters into their own hands. The small vigilante groups that were formed to combat these raiders soon banded together for greater strength and mutual protection, and it did not take long to eradicate the lawless elements as an immediate threat to their homes. These vigilantes later combined forces with the unsurrendered soldiers of the United States and Philippine forces and provided the nucleus for the guerrilla resistance movement.[4] As the occupation became harsher and more onerous, opposition and active rebellion began to spread rapidly throughout the islands.

This spirit of resistance was vividly expressed by Tomas Confesor, pre-war governor of Iloilo Province on Panay, who headed the free civil government of that island during the Japanese occupation. In 1943, Confesor wrote a stinging letter of rebuke to a Filipino quisling who had appealed to him to surrender in return for a promise of personal safety:

There is a total war in which the issues between the warring parties are less concerned with territorial questions but more with forms of government, ways of life, and those that affect even the very thoughts, feelings and sentiments of every man. In other words, the question at stake with respect to the Philippines is not whether Japan or the United States should possess it but more fundamentally it is:—what system of government would stand here and what ways of life, system of social organizations and code for morals should govern our existence. . . .

You may not agree with me but the truth is that the present war is a blessing in disguise to our people and that the burden it imposes and the hardships it has brought upon us are a test of our character to determine the sincerity of our convictions and the integrity of our souls. In other words, this war has placed us in the crucible to assay the metal in our being. For as a people, we have been living during the last forty years under a regime of justice and liberty regulated only by universally accepted principles of constitutional governments. We have come to enjoy personal privileges and civil liberties without much struggle, without undergoing any pain to attain them. They were practically a gift from a generous and magnanimous people—the people of the United States of America. Now that Japan is attempting to destroy those liberties, should we not exert any effort to defend them? Should we not be willing to suffer for their defense? If our people are undergoing hardships now, we are doing it gladly, it is because we are willing to pay the price for those constitutional liberties and privileges. You cannot become wealthy by honest means without sweating heavily. You very well know that the principles of democracy and democratic institutions were brought to life through bloodshed and fire. If we sincerely believe in those principles and institutions, as we who are resisting Japan do, we should contribute to the utmost of our capacity

3 The Japanese also recognized the need for reconstituting the Philippine Constabulary to assist in maintaining order, as reported by Comdr. Charles Parsons in June 1943: "The kempei (military police) soldiers are being gradually replaced by Filipino policemen, and as the reorganized Philippine Constabulary soldiers are trained and sent to the various provincial units to take over the maintenance of law and order, treatment of the civil population may be expected to be even better." GHQ, SWPA, *G-2 Information Bulletin*, "Report on Conditions in the Philippine Islands," Jun 43.

4 The geographical subdivision of the Philippines and the deliberate formation in semi-independent corps were elements in a projected resistance movement; this was the issue between General Wainwright and General Homma in their surrender meeting on Bataan on 6 May 1942. General Wainwright initially claimed that his authority did not extend beyond Luzon whereas General Homma insisted on the surrender of all forces in the Philippines. The territorial organization for the mobilization of the Philippine divisions formed the nucleus and administrative rallying areas for the resistance movement in its organizational phase.

to the cost of its maintenance to save them from destruction and annihilation and such contribution should be in terms of painful sacrifices the same currency that other peoples paid for those principles.

You were a member of the Constitutional Convention that adopted the Constitution of the Philippine Commonwealth. You did not only subscribe to it but you became a Filipino citizen by virtue thereof. Now that the hour of test has come, how dare you advise the people, as you do now, to forsake that sacred document and accept anything for peace and tranquility? Should I harken to you, I would be conspiring with you and the Japanese military authorities to destroy the Constitution that you and I signed with all solemnity, and everything for which that Constitution stands. . . .

I firmly believe that it is not wise and statesmanly for our leaders, in this their darkest hour, to teach our people to avoid sufferings and hardships at the sacrifice of fundamental principles of government and the democratic way of life. On the contrary, it is their bounden duty and responsibility to inspire our people to willingly undergo any kind of difficulties and sacrifices for the sake of noble principles that they nourish deep in their hearts. . . .

America is at war with Japan not because she wants to keep the Philippines but to uphold and maintain the principles of democracy therein. In the speeches of Japanese military authorities, especially that of Gen. Homma, formerly commander in chief of the Japanese Imperial Forces in the Philippines, they condemned democracy and the principles of liberty under such a system of government. It is, therefore, evidently fallacious and insincere on your part to state that you are not pro-Japanese nor a pro-American but a pro-Filipino. What do you mean by being a pro-Filipino? What are the principles for which you stand as a pro-Filipino? What national objectives do you have in mind when you expressed the thought that you are a pro-Filipino and not a pro-Japanese nor a pro-American? What ideals do you propose to realize as a pro-Filipino? If you have any objectives and ideals at all, do you believe in realizing them more effectively under a totalitarian and absolute system of government than under a democracy?. . .

You are decidedly wrong when you told me that there is no ignominy in surrender. That may be true in the case of soldiers who were corralled by the enemy consisting of superior force with no way of escape whatsoever. For when they gave themselves up they did not repudiate any principles of good government and the philosophy of life which inspired them to fight heroically and valiantly—to use your own words. Should I surrender, however, and with me the people, by your own invitation and assurance of guarantee to my life, my family and those who follow me, I would be surrendering something more precious than life itself: "The principles of Democracy and the Justice of the Honor and Destiny of our **people.**"

I noted you emphasized in your letter only peace and the tranquility of our people. . . . You and your fellow puppets are trying to give them peace and tranquility by destroying their honor and dignity, without suffering or if there is any, the least possible. On the other hand, we endeavor to inspire them to face difficulties and undergo any sacrifice to uphold the noble principles of popular rule and constitutional government thereby holding up high and immaculate their honor and dignity at the same time. In other words, you are trying to drive our people to peace and tranquility on the road of **ignominy,** *to borrow your own language. Peace and tranquility are easy to achieve if you choose the easy way but in that case, however, you would be living beneath the dignity of a human being. You would be reducing our people as a result thereof to the status of a dumb animal like the good carabao which lives in peace and tranquility you are talking about—that of a carabao? Would this not be clearly ignominious?. . .*

You may have read, I am sure, the story of Lincoln who held firmly to the conviction that the secession of the Southern States from the Northern was wrong. . . . If Lincoln had revised his convictions and sacrificed them for the sake of peace and tranquility as you did, a fatal catastrophe would have befallen the people of America. With this lesson of history clearly before us, I prefer to follow Lincoln's example than yours and your fellow puppets. . . .

I hope I have made myself clear enough to make you understand my position. I will not surrender as long as I can stand on my feet. The people may suffer now and may suffer more during the next months. To use the words of St. Paul, the Apostle: "The sufferings of the present are not worthy to be compared with the glory to come that shall be revealed in us. . . .[5]

Confesor's feelings were echoed in the hearts of the majority of his countrymen.[6] Their struggle was bitter and sometimes marked by internal discord and petty bickering, but the Filipino guerrillas managed to keep the torch of freedom burning during the period from the fall of Corregidor to the liberation of Luzon.[7] During the course of the war every island produced its local forces to resist the enemy. (Plate No. 84)

Although the size and geography of the Philippines made it difficult for the Japanese to interfere with the continuted growth of the numerous, small guerrilla bands that had sprung up on the various islands, this very topography was a tremendous handicap to any effective unification of strength. Isolated from each other and from the outside world, the guerrillas at first dissipated their efforts in unco-ordinated raids against the enemy. These minor operations were generally fruitless and often did more harm than good since they brought swift and severe retaliatory measures by the Japanese. Even on the larger islands of Luzon, Mindanao, and Leyte, the terrain and poor communications caused a multiplicity of initially independent guerrilla commands to arise with intransigent leaders pursuing their own particular interests. A single driving force was badly needed to direct the guerrilla potential into channels which could produce maximum results. As soon as the facts concerning Filipino resistance became known in 1942, it was General MacArthur's purpose to provide this direction and to weld the scattered groups into unified and responsible forces through the designation and support of responsible local commanders.

Activities of the Allied Intelligence Bureau

The story of guerrilla activities in the Southwest Pacific Campaign can be divided into three phases, Phase One consisting of the initial exploration of the guerrilla movement by the Allied Intelligence Bureau under the operational control of G-2, Phase Two comprising its development under the Philippine Regional Section, and Phase Three composed of the merging of all guerrilla activities with the actual invasion of the Philippine Islands.[8]

The fall of Corregidor in May 1942 cut off virtually all communication with the Philippines. A single radio station operated in the

5 Letter to Fermin Caram, Japanese Puppet Governor of Iloilo Province, Panay, 20 Feb 43. Philippines Files, G-2 Historical Section, GHQ, FEC.

6 "This was why we kept fighting; this was why we continued to hope. We believed in Quezon and in the fighting spirit of the Philippines. We believed in MacArthur—and America." Carlos P. Romulo, *I Saw the Fall of the Philippines*, pp. 176–177.

7 In addressing the Philippine Congress on 9 July 1945, General MacArthur praised the spirit and resistance of the Filipino people in the following words: "Your combat record on Bataan and the magnificent spiritual and physical resistance of the great masses of your people to the enemy efforts at pacification has given to the world the true strength of your character and established your undisputed spiritual capacity for self-government under any standards. . . ." Manila *Free Philippines*, July 10, 1945, p. 3.

8 For a detailed history of the growth, problems, and activities of the various guerrilla groups on each island in the Philippines see G-2, GHQ, FEC, *Intelligence Series*, Vol I, "The Guerrilla Resistance Movement in the Philippines," hereinafter cited as: The Guerrilla Resistance Movement. See also Vol II, "Intelligence Activities in the Philippines during the Japanese Occupation," hereinafter cited as: Intelligence Activities in the Philippines. These were published as war records and have been distributed to U.S. Army service schools.

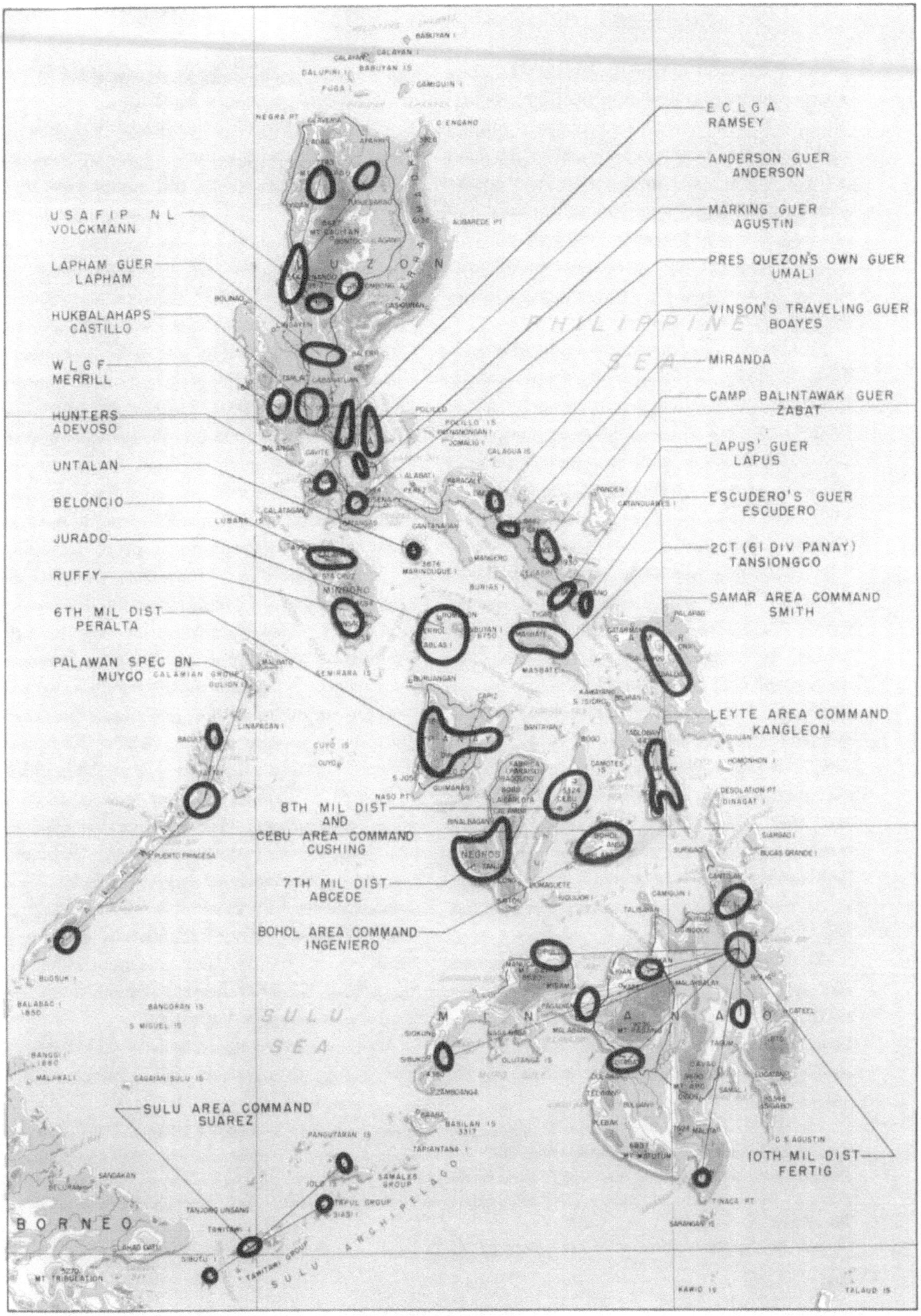

PLATE NO. 84
Major Guerrilla Forces in the Philippines, 1942–1945

Luzon province of Nueva Ecija by a Philippine Army Officer, Lt. Col. Guillermo Z. Nakar, Philippine Army, and several other Filipino and American escapees continued to provide a slender thread of information for several months. Messages from this station broke off in August 1942, however, and with the subsequent capture and execution of Colonel Nakar by the Japanese, radio contact with the Philippines was temporarily lost.

In October 1942 two unsurrendered officers, Capt. William L. Osborne and Capt. Damon J. Gause, who had made a hazardous journey from Luzon in a remarkable feat of navigation, arrived in Australia with the first reports of guerrilla activities on southern Luzon, Palawan, and Tawi Tawi. In December other escaped personnel brought in more detailed information concerning numerous guerrilla groups in operation on central Luzon, Leyte, Samar, Cebu, Negros, and Panay. During this same period, radio contact was re-established, and intercepted calls from guerrilla commanders on northern Luzon and Panay added to the picture.[9] By the beginning of 1943, it was clear that organizations to combat the Japanese were forming everywhere in the islands and that with proper exploitation valuable intelligence could be obtained locally for use in planning future operations. Steps to penetrate the Philippines by clandestine methods began in earnest.

At this early stage, General MacArthur was obviously limited in the amount of aid he could give the guerrilla organizations to help them through their embryonic stages. His own supplies were inadequate and the great distances involved made the problem of tangible assistance a formidable one.

The initial task of contacting the guerrillas and laying the groundwork for an extensive intelligence net in the Philippines was given to the Allied Intelligence Bureau under the operational control of the Theater G-2. A long-range program was developed, based on previous experience with AIB operations behind enemy lines in the Solomons—New Guinea area. In October 1942 the AIB established a special Philippine Sub-Section for the exclusive handling of operations to assist the guerrillas. Solutions were worked out for the best methods of dispatching supplies and funds to the Philippines; areas of responsibility were defined; and plans were made for communication channels to forward information to Australia.

The dispatch of a pioneering party to explore the prevailing situation in the islands and develop specific information on the military, political, and economic aspects of the Japanese-dominated Philippine Government, as well as on the attitudes of the guerrillas themselves, became a priority project. On 27 December 1942, a Filipino aviator, Capt. (later Maj.) Jesus A. Villamor, together with a party of five, left Australia on the submarine *Gudgeon* to organize an intelligence net, determine means of receiving emergency supplies, and obtain general information on Japanese activities. The party landed successfully on Negros and, on 27 January, just one month after his departure, Major Villamor established efficient radio contact with Australia.[10]

Meanwhile two agents sent by Col. Wendell W. Fertig, an unsurrendered American officer

9 The first radio contacts with the Philippine guerrillas were established by station KFS in San Francisco and were relayed to station KAZ in Darwin, Australia. In November 1942, KFS established contact with Maj. Ralph B. Praeger in northern Luzon and with Col. Macario Peralta in Panay. Shortly afterwards, KAZ made direct communication with station WZE in Panay. Cebu and Negros were contacted in early 1943. Intelligence Activities in the Philippines, pp. 12–13.

10 Intelligence Activities in the Philippines, pp. 10–12.

who had become a guerrilla leader on southern Mindanao, reached Australia by sailboat. Their reports indicated that this large and strategically placed island could be made into a major guerrilla base for further expansion to the north. Commander Charles Parsons, USNR, because of his wide and intimate knowledge of the Philippines, was selected to lead a secret fact-finding mission to Mindanao and carry in cipher materials and token supplies. This party arrived in Zamboanga, the westernmost province of Mindanao, on 5 March. After contacting Colonel Fertig, Commander Parsons presented General MacArthur's concept of guerrilla activities and then went on to visit the other islands in the archipelago.[11]

Following these two initial penetrations, additional parties were sent in as rapidly as strained transportation facilities permitted. Submarines carried supplies to Panay in April, and to Tawi Tawi and Mindanao in May. A number of concealed radio transmission stations were established in these islands and material support was given to the local guerrillas.[12]

This initial exploratory period also saw an outstanding episode of clandestine operations. On 16 June 1943, Major Emigdio Cruz, P.A., arrived in Australia from Washington on the first leg of a secret mission to Manila on instruction of Manuel Quezon, President of the Philippine Government-in-Exile. After conferring with General MacArthur and members of his staff, Major Cruz sailed aboard the submarine *Thresher* and landed on Negros on 9 July. From there he worked his way ingeniously across the intervening islands to Luzon, posing at various stages along the way as an itinerant trader, a vendor of fowl, and a vegetable peddler. Several times he narrowly missed discovery by the Japanese but, despite frequent arrests and searching interrogations, he finally arrived safely in Manila on 22 October.

Major Cruz' main mission in Manila was to contact General Manuel Roxas, a well known Filipino politician with an intimate knowledge of high-level Japanese activities in the Philippine Puppet Government, who was in constant communication with the various guerrilla leaders on the islands. After a series of conferences with General Roxas and personal contacts with the other government officials, Major Cruz had accumulated sufficient important data on the inner workings of the Philippine puppet regime to dictate his return to Australia.

On 8 November he left Luzon for Negros and by the end of February 1944, he had retraced his difficult course to complete a brilliant and extremely hazardous mission.[13] Besides the highly important intelligence brought back, Major Cruz' journey showed that, despite the great risks involved, the occupied islands of the Philippines could be traversed by a person with sufficient daring, judgment, and ability.

The information collected by these few penetration parties provided a good working basis for future plans. The guerrilla units could be classed into three main categories: (a) those built around a nucleus of unsurrendered United States and Philippine Army troops; (b) those of purely local origin, under the leadership of prominent civic personages or former Constabulary, which sprang up more or

11 Ibid., pp. 12–14.

12 Ibid., pp. 20–21.

13 In recounting his journey to Manila, Major Cruz quoted General MacArthur as saying in farewell, "Cruz, this is a very tough job. Personally, I believe you have no chance to get through. With your connection to President Quezon you have become very well known. I give you 10 percent chance to enter Manila, but honestly I believe you have no chance to get out." Intelligence Activities in the Philippines, Vol II, Documentary Appendices, Appendix XX, p. 1.

less spontaneously to combat the immediate threat of uncontrolled banditry; (c) those, like the Hukbalahaps, which were an outgrowth of pre-war semi-political organizations. There were also a few roving bands of the outlaw variety which were motivated more by the lucrative prospects of brigandage under cover of guerrilla warfare than by any consideration of patriotism.[14]

Although the majority of the guerrillas shared a common antipathy for the Japanese, they were often divided among themselves, separated into intractable rival factions engaged in a bitter struggle for power. There was no established demarcation of authority and no defined chain of command. All reports of returning AIB agents stressed the necessity of achieving greater co-operation and more unified control among the guerrilla organizations.

It was considered that the best way to meet this problem would be to reactivate the pre-war Philippine Military Districts. Based on population densities, these territorial entities had been used by the Philippine Army for administrative and mobilization purposes. This device had the advantage of being based on legal precedent and would probably be the most acceptable method of division to the majority of de facto guerrilla leaders.

In accordance with this concept, the first district commanders were appointed in February 1943. Colonel Fertig was given command of the 10th Military District on Mindanao and Lt. Col. Macario Peralta, of the 6th Military District on Panay. Since these officers already exercised considerable influence over adjacent islands, Colonel Fertig was also assigned responsibility or the 9th Military District, embracing Leyte and Samar, until a permanent commander could be selected; Colonel Peralta was similarly given temporary control over the 7th and 8th Districts of Negros and Cebu. (Plate No. 85)

Meanwhile the development of the Philippine communications net progressed steadily. Navy and Signal Corps departments co-operated closely with the Philippine Sub-Section of AIB in working out a co-ordinated program to meet immediate needs and at the same time provide for future expansion. Separate networks were mapped out for guerrillas and AIB parties, with additional provisions for a special naval coastwatching system to cover important strategic waterways.

June 1943 marked the end of the preliminary phase of SWPA's penetration into the Philippines—the exploration of the guerrilla potential. This initial period had seen effective liaison established with guerrilla groups on Mindanao, Negros, and Panay. Agents in Manila had also been contacted. In co-operation with the U.S. Navy, supplies and trained personnel had been transported by submarine to Tawi Tawi, Mindanao, Cebu, and Panay and put in the hands of the local leaders. The groundwork for a wide-spread intelligence net had been begun under Major Villamor with heartening results. Steps had been initiated to expand

---

14 In this class may also be included the Pulahanes and Ganaps, who were backed by the Japanese and remained under their influence. To counteract the effect of guerrilla activities, the Japanese tried to set up pro-Japanese organizations among the Filipinos. One such ineffectual attempt on Luzon is described by Colonel Kobayashi, G-3, Fourteenth Area Army, as follows: "Because the American-supported guerrillas were extremely active we tried to organize a similar group called the Ganap to guard the roads, especially north of Manila and the food-producing area around Laguna. They were also to harass and mis-direct American troops in operations after the American landings. The group was never fully organized, however, and never worked well. It had no effect on operations. While the Americans steadily received intelligence from their guerrillas, our group never gave us any information that we could use." Maj. Gen. Naotake Utsunomiya, Assistant Chief of Staff, Fourteenth Area Army, declared that he "felt that Ganap was not fully reliable." Interrogation Files, G-2 Historical Section, GHQ, FEC.

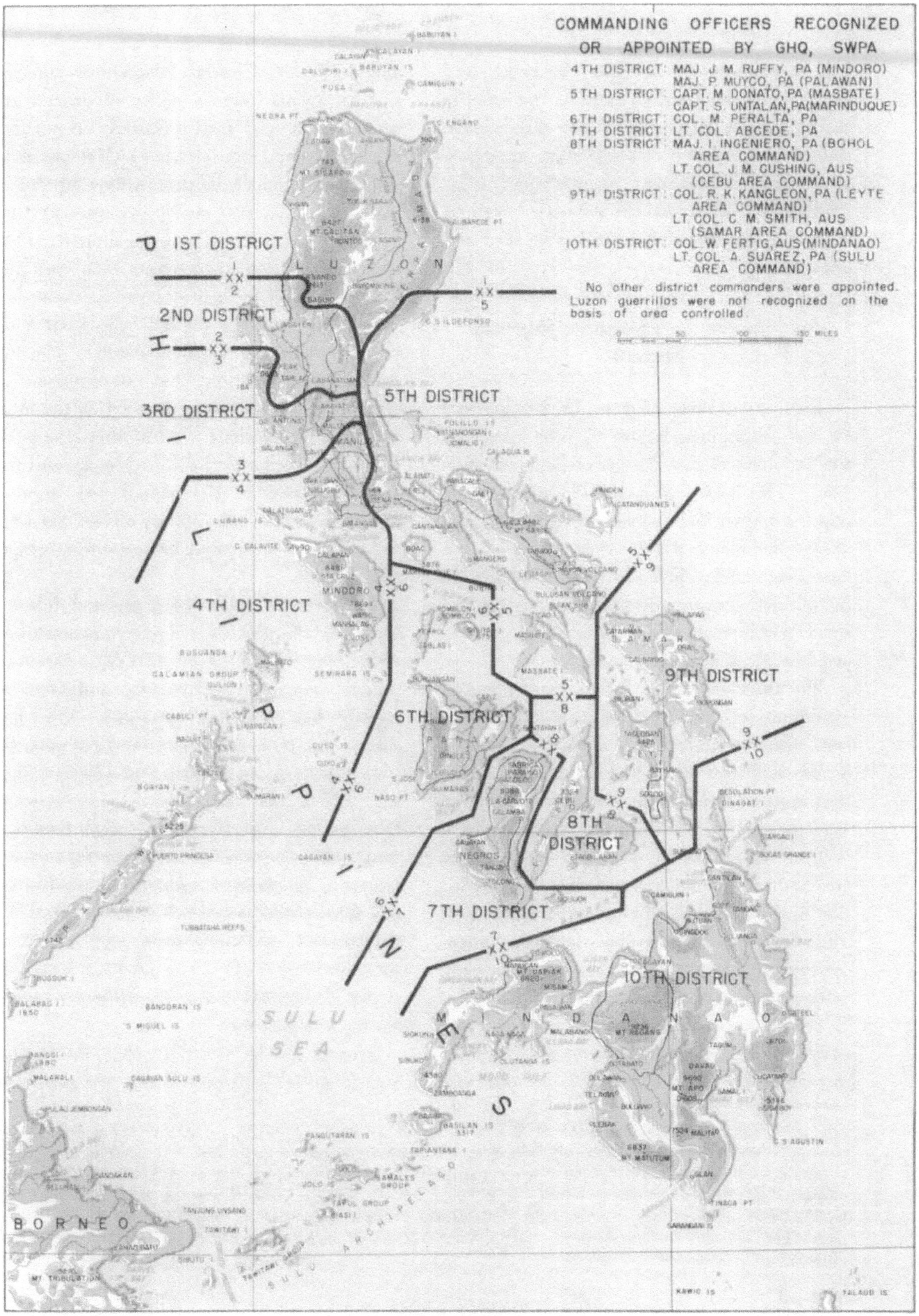

PLATE NO. 85

Military Districts, 1943–1945

the procurement of supplies, weapons, and capable personnel and to increase the number of intelligence parties sent into the islands. Radio stations had been established and developed until the operation of an efficient, comprehensive communications system was well on its way. A good and encouraging start had been made.

Activities of the Philippine Regional Section

The second phase of guerrilla development in the Philippines began in June 1943 with the activities of the Philippine Regional Section. This new section, formed from the original Philippine Sub-Section of AIB and given semi-autonomous status, was organized in late May, under Col. Courtney Whitney, to handle the increasing problems inherent in the rapid development of events and the growing availability of supply facilities.

The assistance and co-ordination of guerrilla operations was continued on an enlarged scale, and efforts were intensified to push on from the bases established on Mindanao and Panay into the islands to the north. Additional parties were prepared for the Visayas, and plans were laid for the penetration of Luzon via Mindoro and Samar. To aid this program, facilities for the transportation of supplies under the general direction of Commander Parsons were augmented by the acquisition of more cargo-carrying submarines from the U.S. Navy.

To guide the various guerrilla leaders in the prosecution of their operations and to make maximum use of their services in the war against Japan, General MacArthur directed that his agents follow a policy of general encouragement and careful instruction without direct command interference which might incur resentment. Guerrilla groups were advised to assist in maintaining civil order so that they might receive reciprocal popular support. They were also cautioned to refrain from open and aggressive warfare against Japanese troops lest they bring reprisals on the people out of all proportion to the results achieved. The collection, co-ordination, and transmission of useful intelligence were stressed as the most important, immediate contributions the guerrillas could make to the Allied cause until the actual invasion of the islands was begun.[15] Before that time, all military operations were to be limited to strategic harassment, sabotage, and ambush.

On Panay, despite strong personal differences, Tomas Confesor's civil government worked with Colonel Peralta's guerrilla organization to collect voluntary contributions and taxes to support the resistance movement. On Mindanao, the civil government and the guerrilla forces under Colonel Fertig were closely affiliated. With the authorization of President Quezon, the guerrillas issued their own currency and even carried on their own postal system. A planned agricultural production and distribution program was also mapped out to insure a maximum food supply. Such measures were indicative of Japanese weakness in the Philippines and of the Filipino's potential for independence.

During the latter half of 1943, the Philippine Regional Section sent two new parties to

15 This policy was emphasized in GHQ's instructions to its agent in Mindoro: "You should carefully refrain from any move which might be construed by local leaders as encouragement for more aggressive guerrilla action with a consequent bid to heavy enemy reinforcement and increased danger of detection of your position and activity. Your mission is one of secret intelligence, and while it is desired that you extend friendly co-operation to local guerrillas and loyal residents, your participation in their affairs to any greater extent could tend to compromise the success of that mission and should be carefully avoided." GHQ, SWPA, Radio to Maj. Phillips, No. 6, 13 Dec 43. Intelligence Activities in the Philippines, pp. 36–37.

pierce the line north of the Central Visayas. (Plate No. 86) In October, Maj. L.H. Phillips led a group of agents to the island of Mindoro, just south of Luzon. Major Phillips was able to establish a radio station and develop some contacts in Manila. Unfortunately, the encouraging start made on Mindoro was short-lived. In February 1944, just three months after Major Phillips' arrival, the Japanese managed to discover his hide-out in Mt. Calavite and, after a futile attempt to escape, Major Phillips was killed and his headquarters destroyed.[16] In July 1944, Comdr. George F. Rowe, USNR, succeeded Major Phillips as GHQ representative and was able to re-weld the severed radio link between Mindoro and Australia.

A party sent to Samar in November 1943, under Maj. Charles M. Smith, was more successful in its efforts. Major Smith set up a radio control station which was in contact with GHQ by 20 December, and established a firm base of operations for further advances. From his position on Samar, he dispatched a number of his men to Masbate, Cebu, and to south and central Luzon. The planting of agents in these various localities was to bring forth valuable intelligence information to aid in planning Philippine invasion operations.

By the end of 1943 a communications net had been established covering most of the southern Philippines. (Plate No. 87) This net formed a framework for later development and extension into the areas to the north during 1944.

Meanwhile, Commander Parsons had returned to Australia in the late summer of 1943 from his fruitful mission to Mindanao and its neighboring islands. After the information he carried back had been co-ordinated with the plans and activities of the Philippine Regional Section, Commander Parsons set out a second time, in October, to expand the contacts made on his previous trip. He again remained on Mindanao for several months, helping Colonel Fertig to consolidate his control and to increase the efficiency and value of his organization. In February 1944, Commander Parsons conducted still another supply run to Mindanao, Tawi Tawi, and Mindoro. His name became well known throughout the southern islands of the Philippines and his "life line" supply service was famous among the important guerrilla leaders.

The first half of the year 1944 saw a marked speed-up in the activities of the Philippine Regional Section. The number of submarine-borne parties was increased and the tonnage of transported materials considerably augmented. In January, supplies were landed on Panay and Negros; in February, on Tawi Tawi and Mindanao. The month of May was a particularly active one. A large party of specially trained agents was dispatched to Colonel Smith on Samar and another to Colonel Fertig on Mindanao. Additional quantities of supplies were brought into this latter island for distribution to the north. During May, too, the first agents were sent to the island of Palawan. In June, a party with complete equipment for transmission of weather information was sent to Negros.

By the middle of 1944, as plans for the invasion of the Philippines were fast ripening, the scope of the Philippine Regional Section's expansion in the archipelago had reached the point where direct participation by the various staff sections of General MacArthur's Headquarters became desirable. The vital early steps of arranging contacts, ascertaining actual conditions and the problems facing the guerrilla movement, sending in supplies and equip-

16 Ibid., pp. 38–39.

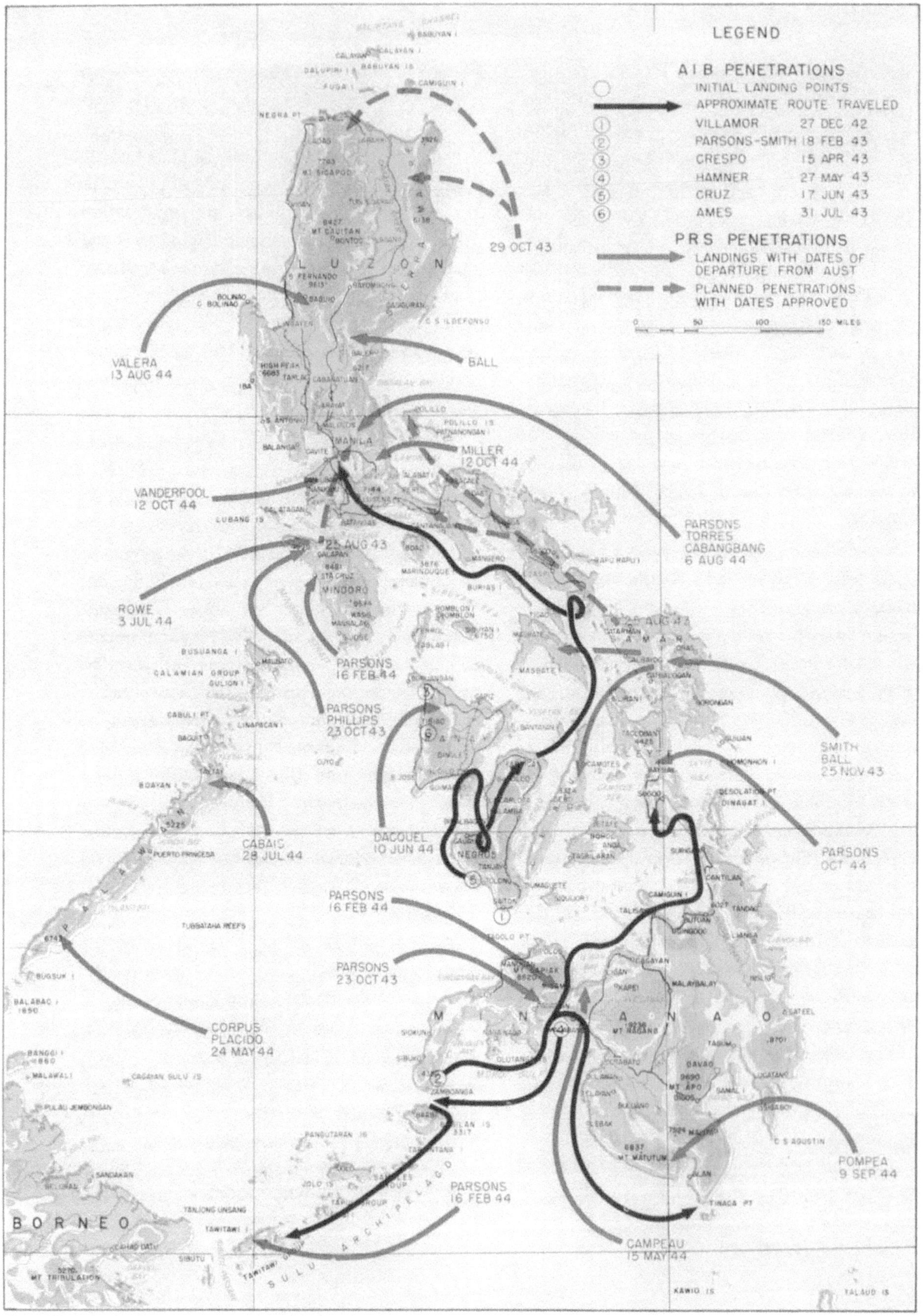

PLATE NO. 86

AIB and PRS Penetrations of the Philippines, 1943–1944

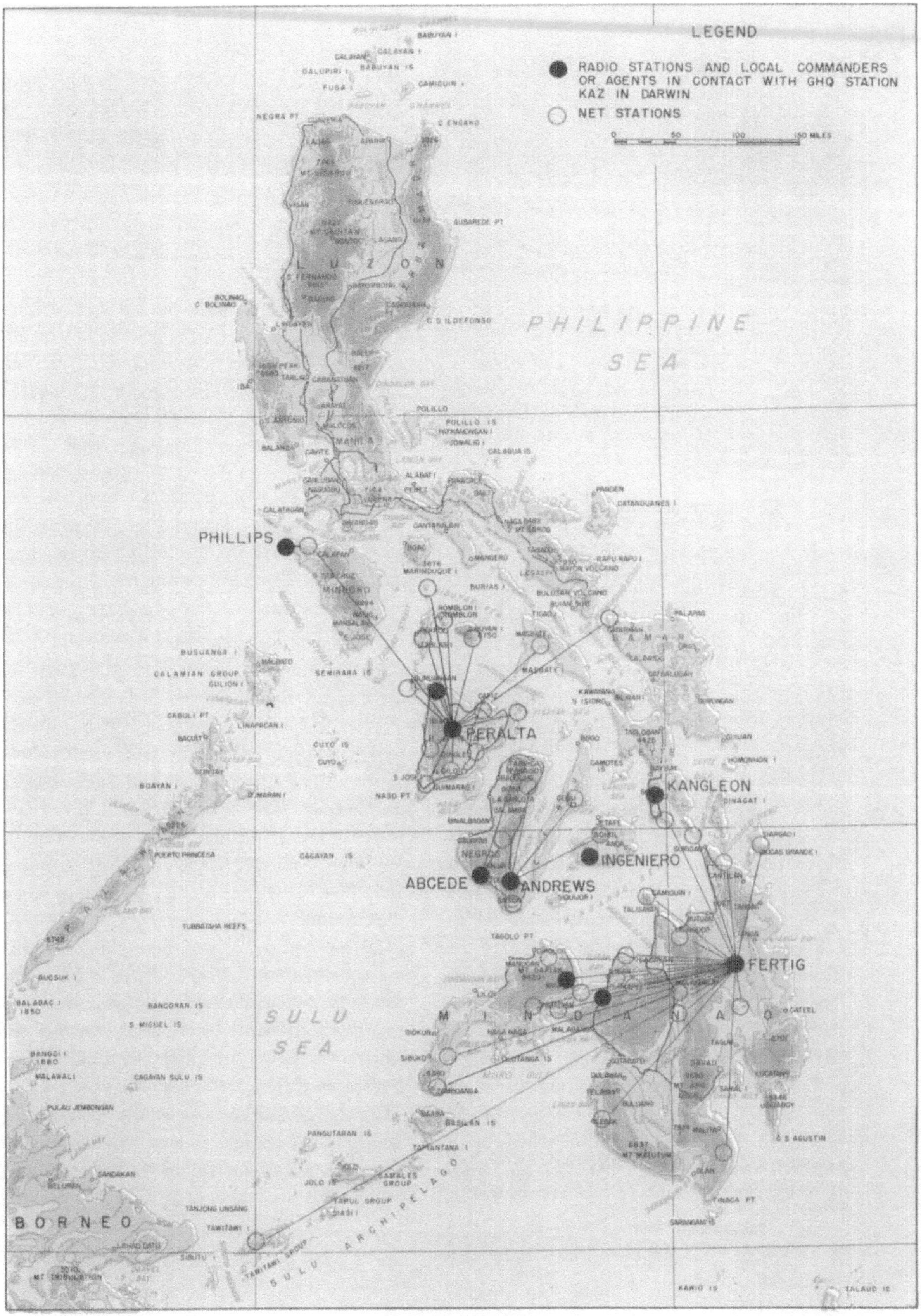

PLATE NO. 87

Philippine Islands Communications, 15 December 1943

ment, establishing an intelligence net, and organizing military commands within the areas of guerrilla activity had been largely completed. (Plate No. 88) Guerrilla activities had to be thoroughly co-ordinated with operational plans of the Southwest Pacific Area for the coming assault. In June, therefore, the functions of the Philippine Regional Section were decentralized, and the further direction of the guerrilla movement was apportioned among the General Staff sections so that the optimum result in each phase of activity could best be achieved. A nucleus of the Philippine Regional Section continued as a co-ordinating and advisory agency.[17]

The Guerrillas on Mindanao

With all operations between SWPA and the Philippine guerrillas now channeled directly into the invasion planning of specific GHQ staff sections, the third phase of development was opened. During this stage, the guerrillas emerged from their hideouts to take their places in battle beside the advancing American divisions.

Because of its large size, its rugged terrain, and its location farthest from the center of Japanese occupation in the Philippines, Mindanao was particularly adaptable to the easy formation of guerrilla groups. Japanese troops held only a few main cities along its 1,400-mile coastline and paid little attention to the interior of the island. It enjoyed comparative freedom from Japanese surveillance and pressure and was consequently the scene of early development of guerrilla organizations.

The growth of the guerrilla movement on Mindanao was in general prototypic of the movements in the rest of the Philippine Islands. On Mindanao, however, the movement matured earlier and with less hostile interference. With the complete absence of Japanese inland patrols, small guerrilla bands quickly made their appearance all through the interior. The many mountains, limited road nets, and primitive communication facilities at first kept these groups isolated from each other, and a certain mistrust and jealousy on the part of the guerrilla leaders prevented any initial attempts at consolidation. In addition, the vast expanse of the island, with almost three weeks required to journey from east to west, increased the obstacles in the way of operational co-ordination.

As time passed, however, the bond of common purpose and the advantages apparent in unification induced the various leaders to seek some means of co-operation. The smaller groups soon blended into larger ones and finally Colonel Fertig emerged as the generally accepted commander of the Mindanao guerrillas. Colonel Fertig was a former American mining engineer who had fought on Bataan and then, upon its surrender, escaped to Mindanao to serve with General Sharp. When Mindanao, in turn, fell to the Japanese, Colonel Fertig took a group of officers and men into the hills to form the nucleus of a responsible resistance movement. By perseverance and diplomacy Colonel Fertig gradually won the respect of the other guerrilla leaders, and by October 1942 he had built up a fairly cohesive guerrilla organization.[18]

In November, Colonel Fertig decided that the time was ripe to notify General Headquarters of the potentialities of his organization and to request assistance. He dispatched his two emissaries, Capt. J. A. Hamner and Captain Smith, on their trip to Australia which resulted in the subsequent contact by Commander Parsons. With authority over the 10th Military District as conferred by GHQ at the time

17 Ibid., p. 58.

18 The Guerrilla Resistance Movement, pp. 83–84.

of Commander Parsons' first trip and, with the approval of the Philippine Government-in-Exile, Colonel Fertig attempted to establish a smoothly functioning civil government to parallel his military organization. Former Philippine officials were appointed as provincial governors and to other civic posts. By early 1943, conditions on Mindanao had become so stable that President Quezon authorized the creation of the Mindanao Emergency Currency Board to issue its own monetary notes for use as a medium of exchange among guerrilla forces.

Before the swift-moving events of the war made it advisable for General MacArthur to make his first strike in the Philippines directly at Leyte, it had been planned to retake the islands by an initial invasion of Mindanao. This fact, together with a convenient geographical position which eased the problem of transportation by submarine, constituted the main reason why the Mindanao guerrillas were the first to be supplied extensively. It was a sound strategical investment.

With the assistance of SWPA, the Mindanao guerrilla organization eventually became the largest and best equipped in the Philippine Islands. By January 1945, Colonel Fertig's command included a force of about 38,000 men.[19] (Plate No. 89) His radio and intelligence network consisted of some seventy transmitter stations and an excellent and extensive coast-watcher system. GHQ was furnished with a constant stream of information which, within its limits of accuracy, helped considerably in the planning of operations against the Japanese in the Philippines. The guerrillas had also prepared airfields at Dipolog, Labo, Lala, and Barobo.[20]

When General MacArthur was ready to retake the Philippine Islands, the guerrillas on Mindanao were in a position to contribute substantially to military operations.[21] With the American invasion of the southern Philippines in early 1945, they began to strike openly against the Japanese forces occupying the island. They seized the airfield at Dipolog, held it until elements of the 21st Infantry landed, and later helped them defend it against strong Japanese counterattacks while a squadron of American fighters used the field as a base for operations to the south. When the American forces chased the Japanese from Zamboanga City, guerrillas set up strong positions behind the retreating enemy troops to form a wall against any further escape into the mountains. On 12 April, five days before

19 This command comprised a tentative organization of six divisions and the "Maranao Militia Force"; the Maranao Force was composed of the proud and temperamental Mohammedan Moros who maintained a separate group within the Mindanao guerrillas. Less than half of these forces were armed, however, and because of their poor training and outmoded weapons their fighting potential could not be judged by their numerical strength. Ibid., p. 85.

In treating the subject of guerrilla forces and organization it must be borne in mind that there was an understandable tendency on the part of their leaders to use the rather ambitious military nomenclatures of corps and divisions, based on the pre-war District mobilization pattern.

20 Ibid.

21 The tactics employed by the guerrillas in their warfare against the Japanese on Mindanao were described in an intelligence report covering the period 1–30 April 1944 issued by Headquarters, Japanese Fourteenth Area Army: "(1) The bandits occupy and utilize key points of communications, firing on and making surprise attacks against our military traffic. They flee whenever we attack. (2) They construct obstacles on the roads and destroy bridges and, when we are engaged in clearing the way or in repair work, they execute surprise attacks. (3) By cutting wires, kidnapping people, burning homes, and other actions calculated to disturb the peace, they draw out our forces; they execute aggressive attacks on a considerable scale. (4) The enemy draws us out by using small units and then carries out an enveloping attack with his main force. When our forces outnumber theirs, the enemy, particularly the Moros, lies in wait in jungle areas for our return and attacks fiercely." ATIS, G-2, GHQ, SWPA, *Enemy Publications*, No. 359, Part I, 28 Apr 45, p. 7.

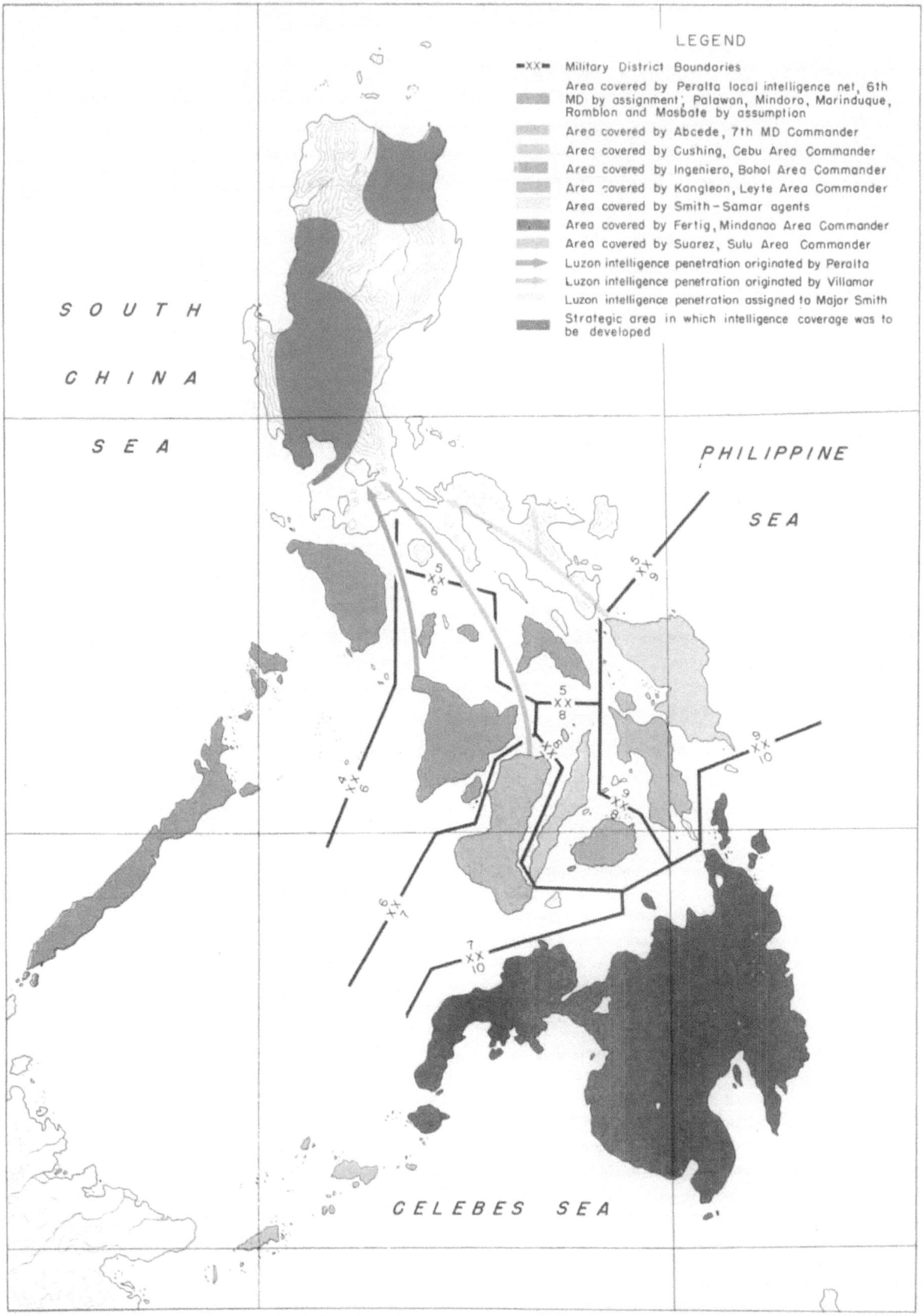

PLATE NO. 88

General Philippine Intelligence Coverage, 1943–1944

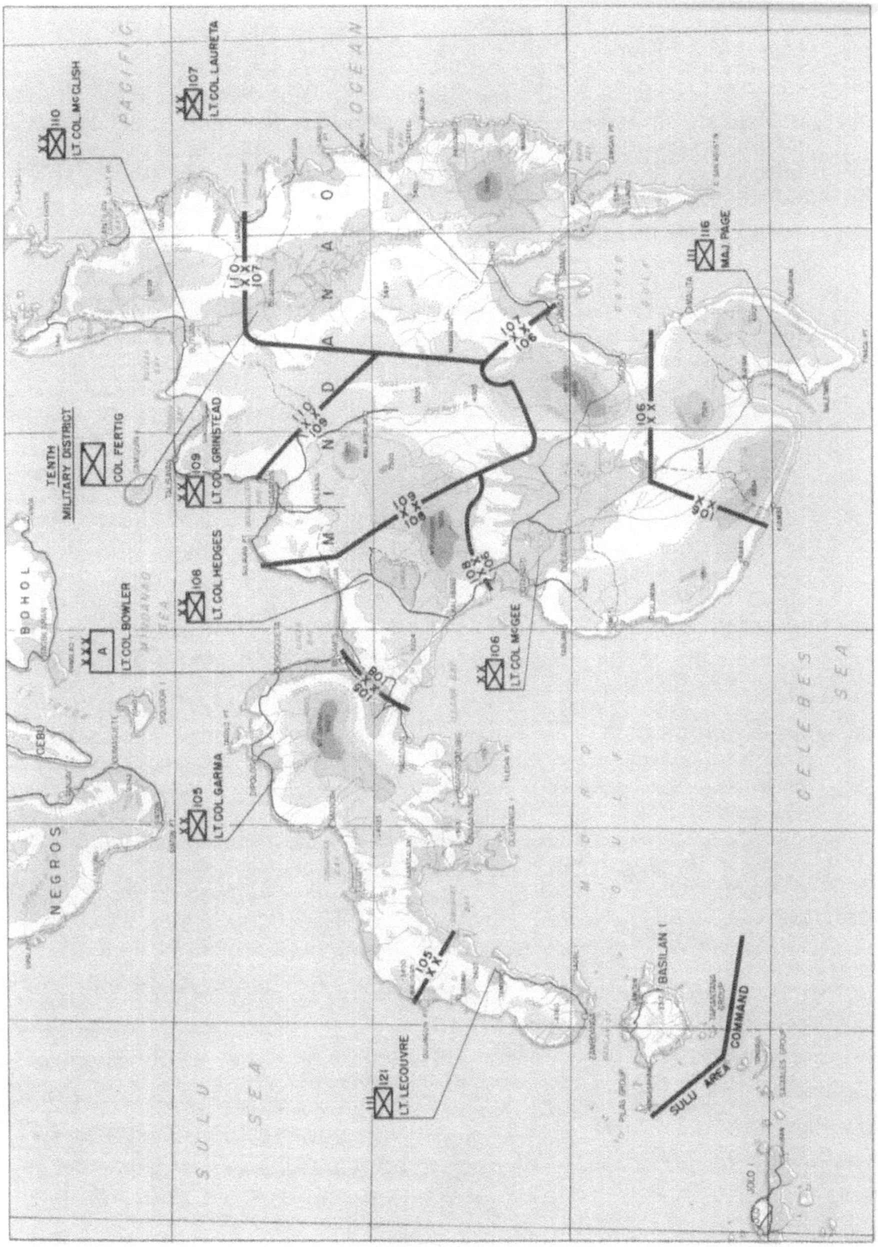

PLATE NO. 89

Mindanao Guerrilla Organization, 31 January 1945

the first Eighth Army landings along Illana Bay on Mindanao's west coast, Colonel Fertig notified General Eichelberger that the initial objective of Malabang and its airfield already had been captured by the guerrillas. Acting on this information, the American forces made their assault further down the coast at Parang, for a drive on the enemy-held town of Cotabato. On 10 May, when elements of the U.S. 40th Division landed near Bugo on northern Mindanao's Macajalar Bay, they found that the guerrillas had cleared the Japanese from the beaches and were ready to assist in the advance to the important town of Cagayan. Aiding the drive of the U.S. 24th Division, Colonel Fertig's forces guarded Highway No. 1 from Kabakan to the Tanculan River so that the Americans could race across the island without fear of an unguarded flank. Guerrilla troops also seized the Tagum River area on north Davao Gulf, as well as Talikub Island in the Gulf itself.[22]

The Guerrillas on Negros, Cebu, and Bohol

Slightly to the northwest of Mindanao lie the three islands of Negros, Cebu, and Bohol. The growth of guerrilla organizations on these three islands followed a pattern very similar to that of Mindanao. (Plate No. 90) After the usual birth pains of interfaction rivalries and petty jealousies, the islands gradually evolved their own de facto commands which eventually were unified under a few main leaders. Unlike Mindanao, where the guerrilla movement developed primarily under the leadership of American officers, these commands were mainly Filipino-organized and Filipino-led.

On Negros, the chief character to arise as commander of the central and northern portions of the island was Capt. Salvador Abcede, P.A. Captain Abcede originally was sent to Negros in November 1942 by Colonel Peralta of Panay. His initial efforts to spread Peralta's control throughout the entire island met with considerable initial opposition from other guerrilla chiefs, particularly from Lt. Col. Gabriel Gador, P.A., in southern Negros.[23]

Major Villamor's dramatic arrival in January 1943 as the forward representative of General Headquarters marked the first genuine progress toward real unification of the Negros guerrillas. His presence as an advisor accredited by SWPA was welcomed by the majority of the guerrilla leaders and, after his appointment as temporary commander of the 7th Military District in May, a general accord was worked out among the various dissident groups. Major Villamor organized the 7th Military District Headquarters, appointed a civil administrator, Henry Roy Bell, and secured authority for the establishment of a free civil government under Alfredo Montelibano, pre-war governor of Negros Occidental.

Upon his return to Australia in October 1943, Major Villamor nominated Captain Abcede as the man best qualified to assume permanent command of the 7th Military District. When this nomination was finally approved and made official in March 1944, Captain Abcede worked aggressively to improve the efficiency of his command. The remainder of the guerrilla units, including Colonel Gador's group in southern Negros, was absorbed into his organization and an amicable and satisfactory relationship was achieved with the free civil government. Captain Abcede gave particular attention to the development of his intelligence network and succeeded in providing a wide coverage of enemy activities. By December 1944, the strength of Colonel

22 *Report of the Commanding General, Eighth Army, on the Mindanao Operation, Victor V*, pp. 3, 17, 23, 35, 43, 55.

23 The Guerrilla Resistance Movement, pp. 69–72.

Abcede's organization on Negros numbered approximately 13,000 men.[24]

East of Negros is the long, narrow island of Cebu, the most densely populated of all the Philippine Islands. The story of the guerrilla movement on Cebu is primarily that of two men—Harry Fenton and Lt. Col. James H. Cushing, both American-born. With the Japanese capture of Cebu City in the central part of the island, a large number of USAFFE forces escaped to the hills, taking along a sizeable quantity of arms and supplies. Numerous guerrilla units were soon formed in the enemy-unoccupied regions to the north and south and these gradually combined into larger groups. By the middle of 1942, the guerrilla organization on Cebu was split into two bodies ; one under Fenton, in the north, and the other under Colonel Cushing, in the south.

Despite marked differences in personalities and methods of operation, these two leaders quickly recognized the advisability of uniting their resources for co-ordinated action against the Japanese. A joint command was established which put administration under Fenton and combat activities under Colonel Cushing. A single staff served for both factions and areas of control were delegated to subordinate leaders.[25]

This arrangement functioned satisfactorily until mid-1943 when critical food shortages and rapidly dwindling supplies, coupled with intensive countermeasures by the Japanese, seriously disrupted the Cebu guerrilla organization. To aggravate the situation, both leaders fell seriously ill and all activities were temporarily curtailed. During this period, dormant animosities between the two factions were again aroused ; disagreements arose between Fenton and his associates. Instituting a reign of terror and persecution, Fenton engaged in a series of reckless and injudicious actions which alienated many of his officers. On 15 September he was tried and executed and his command was reorganized.

Colonel Cushing meanwhile recovered his health and, in the face of persistent Japanese anti-guerrilla campaigns, began to rebuild his weakened groups for further operations. In January 1944 he was designated by GHQ as commander of the 8th Military District and shortly afterward his organization was sent supplies and radio equipment from Australia. Colonel Cushing broadened and improved intelligence coverage on Cebu and his guerrillas throughout the island worked with increased efficiency. By April, GHQ was receiving a gratifying volume of information on Japanese movements and military operations.[26]

On Bohol, the oval-shaped, coral island bordering Cebu on the southeast, several guerrilla groups developed, with Maj. Ismael Ingeniero emerging as the leader of the specially created Bohol Area Command.[27] Internal friction among the guerrilla groups had to a degree alienated the civil populace and when the Japanese landed in force on the island in June 1944 the guerrilla organizations collapsed. Following the Japanese partial withdrawal in

---

24 Ibid., p. 77. Capt. Abcede was promoted to the rank of Lieutenant Colonel on 1 October 1944.

25 Ibid., pp. 35–37. Many of the puppet officials installed by the Japanese on Cebu worked secretly with the Cushing—Fenton guerrillas. A noteworthy example of such an official was Hilario Abellana, pre-war Governor of Cebu, who escaped from the Japanese in July 1943 and joined the Cebu guerrillas. He proved invaluable to the guerrillas by giving them aid in carrying on civil government functions, raising funds, and looking after the interests of the free civilian communities.

26 Ibid., pp. 37–38.

27 Ibid., pp. 80–81.

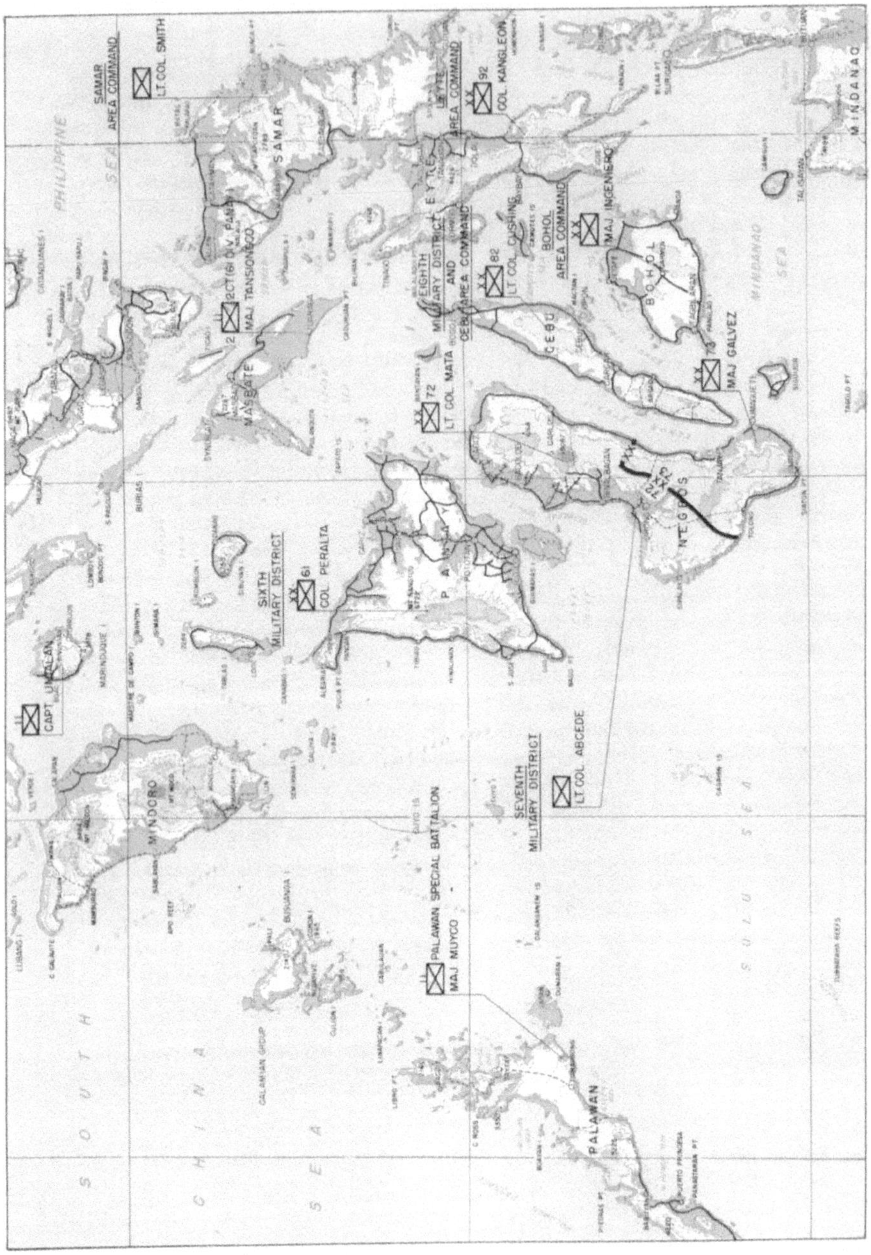

PLATE NO. 90

Central Philippines Guerrilla Organization, October—November 1944

July, Major Ingeniero's organization was largely reestablished but little useful information was extracted from Bohol before the major invasion of the Philippines.

Despite deficiencies in the resistance organizations on the islands of Negros, Cebu, and Bohol, the guerrilla forces played a significant part in the liberation of their territories when the Eighth Army invaded the southwest Visayas.

Colonel Abcede's units on Negros had done valuable preliminary work to assist the invasion troops. The guerrillas held the Japanese to a line stretching from Bacolod, the capital, on the west to San Carlos on the east. Most important towns south of Bacolod were under guerrilla control. After the landing of the U. S. 40th Division, the guerrillas, familiar with the jungle terrain, served efficiently as scouts and guides in helping to rout the Japanese from hidden retreats and successfully executed numerous combat missions assigned by the division.[28]

Colonel Cushing's guerrillas on Cebu played havoc with Japanese patrols and movements prior to the arrival of the Americal Division. Enemy lines were disrupted and the task of the invading troops was made considerably easier.[29] The guerrillas had also developed an airstrip and had control of all but a few areas in east and northern Cebu. After the landing, Colonel Cushing's guerrillas joined the combat patrols of the Americal Division in trailing enemy remnants which had fled to the mountainous interior.[30]

Most of Bohol Island was free of Japanese and under surveillance of Major Ingeniero's guerrillas before the coming of the American forces on 11 April. The landing parties were met by the news that no enemy forces were in the area.[31]

The Guerrillas on Panay and Adjacent Islands

The resistance movement on Panay was unique. It developed rapidly; there was a minimum of discord; and a dynamic leader emerged at an early time. The guerrilla structure on Panay was built around a core of refugee troops of the Philippine 61st Division who had taken to the hills immediately after the surrender orders were published. Scarcely ten weeks after the Japanese invasion, Colonel Peralta, former G-3 of the division and a man of strong and driving character, assumed undisputed control of the main guerrilla groups. The early emergence of a generally accepted leader and the availability of a relatively large amount of salvaged supplies and equipment gave a powerful impetus to the formation of a smoothly working guerrilla command. In addition, the efforts of Panay's intrepid governor, Tomas Confesor, whose free civil government was left comparatively unmolested by the light Japanese garrison, strengthened the framework of the Panay organization and bolstered the morale of the people.[32]

Colonel Peralta made rapid progress. By November 1942 he had reactivated the Philip-

28 Manila *Free Philippines*, June 7, 1945, p. 4.

29 Maj. Kohei Takahashi, who served as staff officer with both the Japanese Thirty-fifth Army and the Fourteenth Area Army, made the following comment on useful intelligence furnished by the Cebu guerrillas to the Allied forces: "When the U.S. task force raided Cebu City in September 1944, it hit only the buildings occupied by the Japanese Army. No other places were hit. This was far from an accident. It was due to knowledge acquired through the guerrillas." Interrogation Files, G-2 Historical Section, GHQ, FEC.

30 *Report of the Commanding General, Eighth Army, on the Panay—Negros and Cebu Operations, Victor I and II*, p. 59. HQ Eighth Army, *Operational Monograph on the Cebu—Bohol, Negros Oriental Operation, Victor II*, pp. 13–14.

31 Ibid., pp. 14–15.

32 The Guerrilla Resistance Movement, pp. 46–51.

pine 61st Division, initiated an intensive training program, and established first radio contact with Australia. He also began to extend his influence to the adjacent islands in the Visayas and even to Mindoro and Palawan. In February 1943, GHQ appointed Colonel Peralta as de facto commander of the 6th Military District which included Panay, the Romblon Islands, and Guimaras Island. While this appointment solidified Colonel Peralta's control over his own territory, it had the effect of cancelling any official authority in other regions where he had aspired to establish his influence.

In spite of this limitation, Colonel Peralta's activities in adjacent areas continued to flourish spontaneously. The small guerrilla bands on Masbate, Marinduque, Mindoro, and Palawan, having no outstanding leaders of their own, remained under the domination of the 6th Military District. Colonel Peralta soon developed one of the most extensive and efficient intelligence systems in the Philippines. He had radio contacts and courier service with the principal guerrilla chiefs in the Visayas and Mindanao and his agents were in operation as far as Luzon. Voluminous intelligence reports flowed in a steady stream from the north and the east via Masbate and Tablas Island into Panay Headquarters where they were collated and relayed to SWPA.[33]

When the U. S. 40th Division went ashore on Panay in March 1945, Colonel Peralta's forces made a large contribution toward eliminating the Japanese. Even before the landings, his guerrillas had cleared the enemy from the outlying districts and had won possession of nine airstrips in the northern and southern parts of the island. To aid the advance of the American troops, all important bridges were repaired, roads were serviced, and key junctions were kept under control.[34]

After the 40th Division forces had moved inland from the beaches, the guerrillas were used as guides and patrols. Guerrilla troops joined in the liberation of the capital city of Iloilo late in March and in the subsequent attack on the strong Japanese garrison at San Jose.

In the neighboring islands of Mindoro, Masbate, and Palawan, guerrilla units, though not as strong or as well integrated as those on Panay, were also helpful. On Mindoro, the Japanese fugitives in the interior were hunted down in the mountains and through the jungles; on Masbate, the guerrillas conducted their own amphibious assault and occupied the capital town; on Palawan, guerrilla groups confined the Japanese to the area of Puerto Princesa and joined in the elimination of scattered enemy pockets. With the assistance of the various guerrilla units of Colonels Peralta, Fertig, and Abcede, the invasion tasks of the Eighth Army forces in Mindanao and the western Visayas were immeasurably simplified and greatly shortened.

The Guerrillas on Leyte and Samar

Until the Spring of 1943, a dozen different guerrilla leaders contested bitterly for authority on Leyte. Although most of these men shared a desire to work against the Japanese, any thought of unification was subordinated to their individual interests. There was apparently no leader unselfish enough to put aside his personal motives for the common good or strong enough to enforce obedience from the others.

On a visit to Leyte in April 1943, Commander Parsons persuaded Col. Rupert K. Kangleon, former commander of the Philippine

33 Ibid.

34 *Report of the Commanding General, Eighth Army, on the Panay—Negros and Cebu Operations, Victor I and II*, pp. 18–45; GHQ, SWPA, Communique No. 1080, 22 Mar 45.

81st Infantry, to attempt a consolidation of the dissident factions on the island under the guidance of SWPA. By a judicious mixture of force and diplomacy and by the strength of his own prestige, Colonel Kangleon eventually succeeded in winning the allegiance of the principal guerrilla groups to begin a reorganization of the pre-war Philippine 92nd Division.[35] In October 1943 he was appointed by GHQ to head the Leyte Area Command, and by Fall of the following year, Leyte boasted a well-trained guerrilla force of some 3,200 troops.[36]

On Samar, as on Leyte, numerous irreconcilable groups contended for supremacy after the removal of the Philippine Government and the dissolution of the Philippine Constabulary. Since the island was of little value strategically, there were few Japanese troops to fear, and conditions fostered the unhampered existence of a multitide of guerrilla bands. Samar, however, did not possess a man of sufficient caliber to harmonize the various prevailing differences, and as a result the island remained without any centralized authority until October 1944, the month of General MacArthur's invasion of the Philippines.

The two largest groups on Samar were commanded by Col. Pedro V. Merritt, P. A., who was established in the north, and by Manuel Valley, an escapee from Bataan, who led an organization in the south. An attempt in September 1943 by Colonel Kangleon's emissary, Lt. Col. Juan Causing, to unite these two leaders was unavailing, and although much good work was done independently by the guerrilla units on Samar, very little was contributed to aid the plans of General MacArthur's Headquarters until after the assault on Leyte.[37]

Samar's main value lay in its use as a base of operations by GHQ's representative, Colonel Smith, whose agents working on Luzon and in the Bicols relayed accumulated information on the Japanese to Australia. Although Colonel Smith did not take an active part in guerrilla affairs, his advice was often sought and his suggestions generally heeded; he gradually won the confidence of both Colonel Merritt and Manuel Valley, and in September 1944 the two principal guerrilla groups agreed to accept him as their co-ordinator. In early October, GHQ appointed Colonel Smith as commander of the Samar Area. Colonel Smith was in the process of reorganizing the Samar units when the American forces landed on the island.[38]

General MacArthur's invasion of Leyte on 20 October 1944 sounded the signal for the Phillippine guerrillas to throw off the cloak of concealment and come forth in open warfare against the Japanese. Shortly before the assault forces were due to sail for their objective, General MacArthur issued the following alert to the Visayan guerrilla commanders:

> *The campaign of reoccupation has commenced. Although your zone is not at present within the immediate zone of operations, it is desired that your forces be committed to limited offensive action with the specific mission of harassing the movement of the enemy within your area and as far as possible contain him in his present positions. Intelligence coverage must be intensified in order that I be fully and promptly advised of all major changes in enemy disposition*

35 The Guerrilla Resistance Movement, pp. 1–3.

36 A captured Fourteenth Area Army intelligence report concerning the guerrillas on Leyte contained the following statement: "The bandit groups in every area, particularly the Kangleon bandit group (Leyte), have blind faith in the return of the American forces.... They are preparing for an uprising simultaneously with the return of the American forces and are laying plans to link up with them." ATIS, G-2, GHQ, SWPA, *Enemy Publications*, No. 359, Part I, 28 Apr 45, p. 4.

37 The Guerrilla Resistance Movement, pp. 5–8.

38 Ibid.

or movement.[39]

It was on Leyte that the Filipino guerrilla and the American soldier first joined forces in battle. With the initial Sixth Army landings on the beaches at Tacloban and Dulag, Colonel Kangleon's units went into action. They dynamited key bridges to block Japanese displacement toward the target area; they harassed enemy patrols; and they sabotaged supply and ammunition depots. Information on enemy troop movements and dispositions sent from guerrilla outposts to Colonel Kangleon's Headquarters was dispatched immediately to Sixth Army.

The guerrillas also performed valuable service in maintaining public order and in keeping the roads and highways free of congestion. After the American beachheads were established, the Leyte guerrilla groups were attached directly to the Sixth Army corps and divisions to assist in scouting, intelligence, and combat operations.

On neighboring Samar, a regiment of the 1st Cavalry Division, which landed on 23 October, was aided extensively in its mission by the guerrilla units on the island. The main objective of seizing and controlling the strategic Taft—Wright Highway was achieved by a dual drive of cavalry and guerrilla forces. While the 8th Cavalry battled to capture Wright at the western terminus of the Highway, the guerrillas fought the Japanese from Taft on the east. A junction of the two forces in December cleared the enemy from the heart of Samar and prevented his reinforcement of Leyte from the northeast.[40]

The Guerrillas on Luzon

In contrast to the rest of the Philippine Islands, which in general were lightly garrisoned, the main island of Luzon was heavily occupied by Japanese military forces. Thorough policing and frequent, intensive clean-up campaigns prevented any effective unification of the numerous guerrilla groups which sprang into existence after the surrender of the USAFFE units.

One of the earliest organizations developed on Luzon was headed by Col. Claude Thorp who, in January 1942, worked his way from Bataan through the Japanese lines to establish a headquarters in the Zambales Mountains. From this retreat, Colonel Thorp attempted to centralize operations in the various regions of the island including northern Luzon and the Bicol Peninsula. Though he made substantial progress in this direction, his efforts were brought to an untimely end. In October 1942 Colonel Thorp and several of his staff were trapped in a Japanese raid and subsequently executed. After Colonel Thorp's death, a multiplicity of independent guerrilla commands began to develop throughout the provinces of Luzon.[41]

In the southern half of the island, three units were particularly outstanding in their growth and operations. (Plate No. 91) These were the forces of Maj. Bernard L. Anderson

39 Intelligence Activities in the Philippines, p. 92. The Japanese were much concerned with the plans of the guerrillas to rise up against them simultaneously with the American invasion of the Philippines. In an intelligence report issued by Fourteenth Area Army Headquarters, covering the period 1–30 April 1944, the following information was given: "The bandit groups have become more and more aggressive.... they appear to be planning an uprising simultaneously with the return of the American forces, and in the meantime are conserving their forces and building up their military strength. They are directing most of their efforts toward reconnaissance of our forces, principally of our defense formations, and toward fifth column destruction of our rear lines of communications units." ATIS, G-2, GHQ, SWPA, *Enemy Publications*, No. 359, Part I, 26 Apr 45, p. 3.

40 HQ Sixth Army, *Report of the Leyte Operation*, p. 79.

41 The Guerrilla Resistance Movement, p. 9.

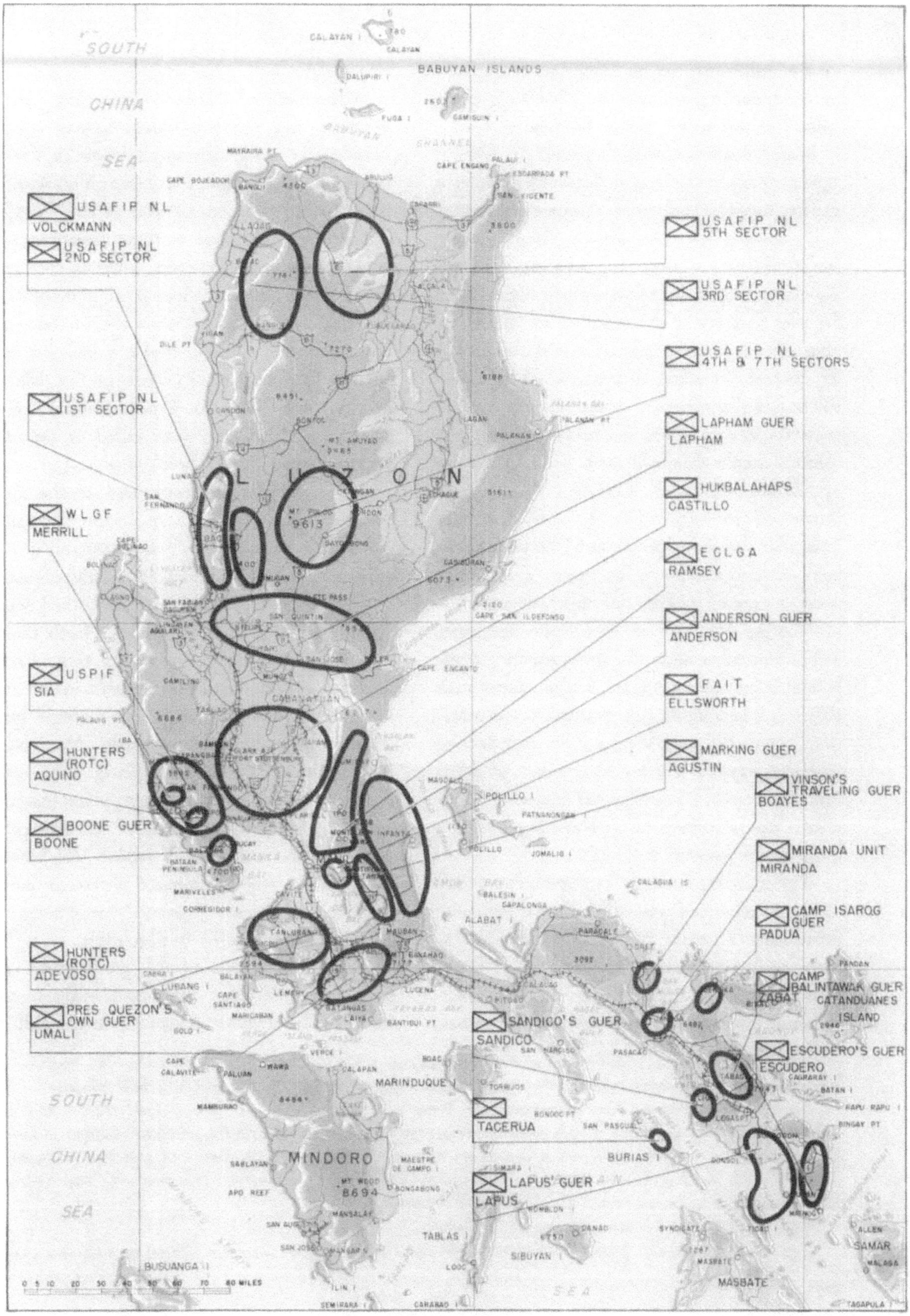

PLATE NO. 91
Guerrilla Forces on Luzon, October—November 1944

in the eastern region, of Maj. Robert Lapham in the central region, and the "Marking Guerrillas" in the sector east of Manila.[42]

Major Anderson was contacted by SWPA agents in mid-1944 and in September began to receive supplies and radio equipment brought in by the ever-busy submarines. Major Anderson's efforts to achieve co-ordination, for intelligence purposes, were enormously complicated by the concurrent existence of so many independently active organizations in the area to be covered. In addition to the Marking Guerrillas and the forces of Major Lapham, there were the Hukbalahaps in Pampanga, the East-Central Luzon Guerrilla Area (ECLGA) units of Colonel Edwin P. Ramsey in east-central Luzon, the Hunters in Cavite, the Fil-American Irregular Troops in Rizal, and President Quezon's Own Guerrillas in Batangas. These units were of varying quality and effectiveness.

Making the best of a difficult situation, Major Anderson succeeded in forwarding much valuable information from Luzon directly to SWPA Headquarters in Australia. In addition, he distributed some of the supplies he received, especially radio equipment, to other units in southern Luzon in an endeavor to increase the efficiency of the intelligence and propaganda network.[43]

The guerrilla situation in the northern half of Luzon remained generally obscure until well into 1944. Distance, difficulty of communications, and the extensive countermeasures of the Japanese hampered any effective SWPA penetration of the upper provinces either for liaison or supply.

After Colonel Nakar's execution by the Japanese and the subsequent loss of contact between his headquarters and Australia, a series of successors attempted to carry on his work in the northern mountains and in the Cagayan Valley. The Japanese in these areas were particularly watchful, however, and, as each new leader arose, he was tracked down and eliminated. In a heroic and desperate effort to continue the movement, Colonel Nakar's intrepid lieutenants, Lt. Col. Arthur Noble, Lt. Col. Martin Moses, Maj. Ralph B. Praeger, and Lt. Col. Manuel P. Enriquez, were killed or captured by the enemy before the close of 1943.[44]

In early 1944 the command of the main guerrilla forces in northern Luzon ultimately fell to Maj. Russell W. Volckmann, an unsurrendered American officer. Major Volckmann designated his organization as the United States Army Forces in the Philippines, North Luzon (USAFIP, NL) and set about the task of revising the whole guerrilla movement in his area in order to weld all groups into a single force responsible to a central authority. He divided his command into sectors, giving each sector commander full power to consolidate and control the fragmentary outfits in his particular area. At the same time, Major Volckmann built up a strong intelligence system to funnel all information on Japanese movements in northern Luzon to his headquarters.

Major Volckmann's forces grew rapidly and by the end of 1944 numbered some 10,000 men.[45] The greatest drawback to the full re-

42 Ibid., pp. 10–24. The commander of this organization was a natural leader, Col. Marcos V. Agustin. A fearless fighter, he was aided by the brilliant Yay Panlillo, Filipino woman journalist, whom he later married.

43 Colonel Kobayashi, G-3, Fourteenth Area Army, paid tribute to Major (later Colonel) Anderson in the following words: "About the middle of October 1944, we attempted to extirpate the guerrilla movement in Luzon. Unfortunately for us, however, Colonel Anderson was too good a leader and the American guerrillas continued to function. . . . In recalling the final stage of the war in the Philippines . . . I remember how famous Colonel Anderson became among us." Interrogation Files, G-2 Historical Section, GHQ, FEC.

44 The Guerrilla Resistance Movement, pp. 39–42.

45 Ibid., pp. 43–44. Major Volckmann was promoted to the rank of Lieutenant Colonel on 9 October 1944.

alization of his efforts, however, was the lack of radio contact with General Headquarters, SWPA. Finally, in September 1944, he succeeded in putting a makeshift radio into operation and, through this lone channel of communication, messages began to be sent and received. It was then that the guerrillas in northern Luzon first learned of General MacArthur's imminent return to the Philippines. Agents and equipment sent from Australia in November and December helped to co-ordinate Colonel Volckmann's operations with the American invasion plans.

When General MacArthur landed at Lingayen Gulf, the Japanese were caught in the midst of a general redeployment of their forces throughout Luzon. Seizing the advantages of the moment, the guerrillas broke out in full force. Roads were torn up, bridges destroyed, mountain passes blocked, and rail and motor facilities sabotaged at every turn to interfere seriously with Japanese troop and supply movements.

Shortly after the American landings, Colonel Anderson was requested to form a Filipino battalion to be attached to General Krueger's Sixth Army forces. Colonel Anderson responded by taking the best personnel at his disposal to form the first "Anderson Battalion." This unit performed efficiently and valiantly throughout central and eastern Luzon and built a battle record of 3,000 Japanese killed and 1,000 captured.

Major Lapham's guerrillas in central Luzon played a prominent part in effecting the dramatic rescue of over 500 Allied internees from the ill-famed Cabanatuan prison camp. (Plate No. 92) The first in a series of bold liberations of Allied prisoners from enemy hands, this daring raid was carried out 25 miles behind Japanese lines by a mixed force of 286 guerrillas, 121 troops of the 6th Ranger Infantry, and 13 Alamo Scouts. The guerrillas acted as the eyes of the raiding force to guide it through the brush and as its ears to be on the alert for any surprise flanking movement by the enemy. They constructed roadblocks at the northeast and southwest approaches to the stockade to hold up hostile reinforcements and also arranged for food caches so that the liberated prisoners could be fed at convenient points along the return route.

The attack was launched on the night of 30 January 1945. Within thirty minutes the entire Japanese garrison had been wiped out and the last prisoner removed from the prison area. The Rangers' return was covered by a guerrilla delaying action which successfully fought off approximately 800 enemy reinforcements sent to assault the strategically placed roadblocks. Meanwhile the litter patients from the camp were transported by guerrilla-organized carabao cart train to Sibul Springs, whence they were evacuated to a hospital at Guimba.[46]

On 23 February, in another equally brilliant and even more extensive liberation of Allied internees, Luzon guerrillas helped troops of the 11th Airborne Division to release more than 2,100 prisoners from the Los Banos prison camp in Laguna Province on the shores of Laguna de Bay.

For several nights prior to the attack, guerrilla units infiltrated through the Japanese lines to gather in the area of Los Banos. On the morning of the 23rd, one element of the 11th Airborne crossed Laguna de Bay in amphibious craft while another element took off by plane for a spectacular parachute drop. All forces converged in a swift and co-ordinated attack which caught the Japanese guarding the camp in the middle of their morning calisthenics. The entire garrison was annihilated with prac-

46 G-3, HQ Sixth Army, *Combat Notes*, No. 5, 21 Mar 45, pp. 1–10.

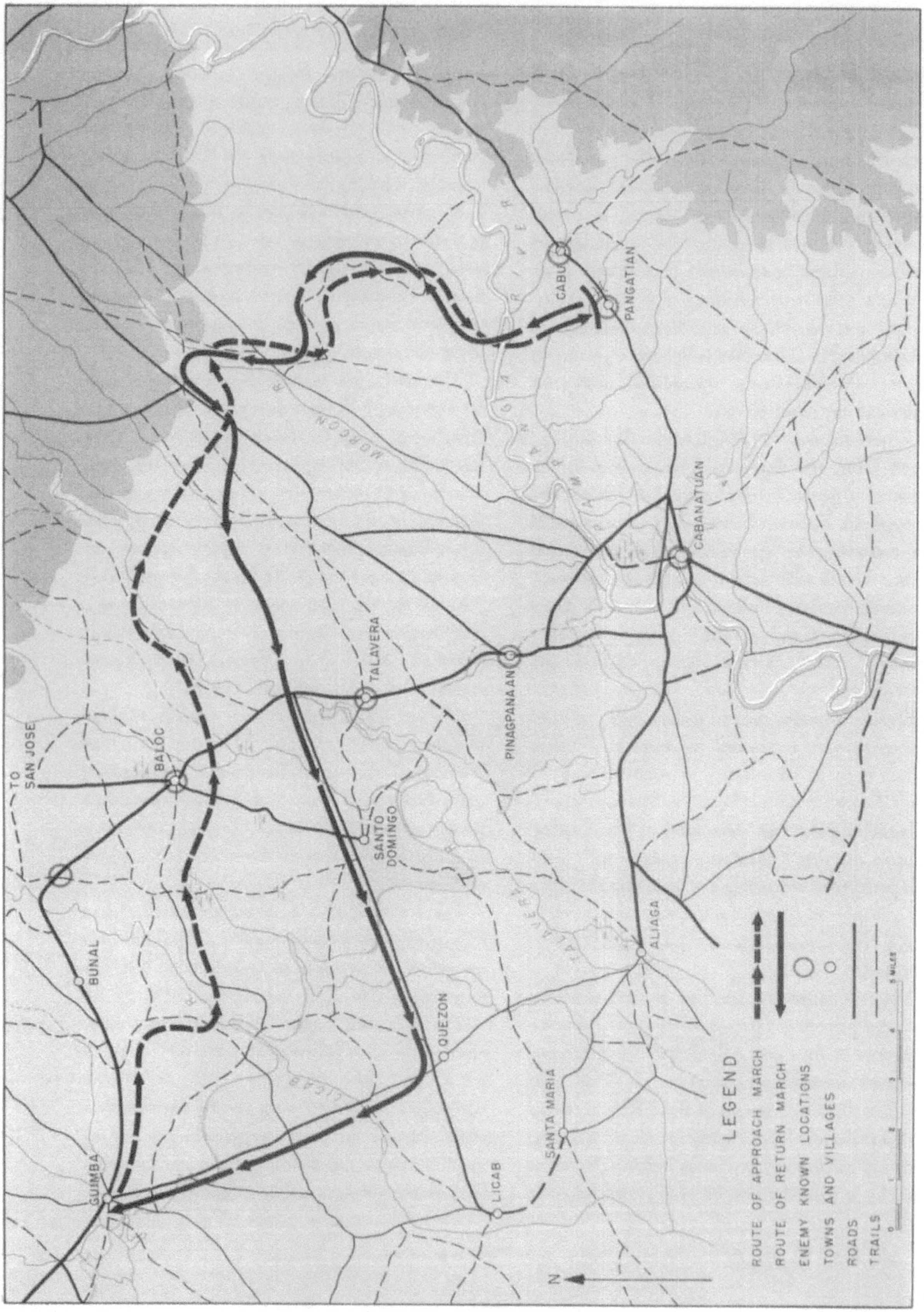

PLATE NO. 92
Cabanatuan Prison Raid, 30–31 January 1945

tically no loss to the Allies, and the Los Banos prisoners were evacuated across the Bay.[47]

The Marking Guerrillas, led by Col. Marcos V. Agustin, carried out extensive combat operations in the mountains northeast of Manila. After a month's hard training and fighting with the troops of the U. S. 43rd Division, Colonel Agustin's force, numbering some 3,000 men, was assigned a part in the powerful assault on Ipo Dam, the largest of the three dams supplying Manila. While two prongs of the 43rd Division converged on the dam from the south and west, the guerrilla force formed a third prong that came down from the northwest. The Japanese were routed from their defense positions, and the dam was captured intact. After successfully completing their assignment in the Ipo sector, the Marking Guerrillas pursued the fleeing remnants of the enemy into the hills and later aided considerably in other missions which, in the words of the 43rd Division Commander, Maj. Gen. Leonard F. Wing, "otherwise would have required costly and protracted action by American forces.[48]

In the northern half of Luzon, Colonel Volckmann's units fought effectively with I Corps against General Yamashita's beleaguered forces in the high mountains around Kiangan and in the plains of the Cagayan Valley. Supported by planes of the Fifth Air Force, these guerrillas were able to clear the enemy entirely from Ilocos Norte Province.[49] They captured San Fernando on the eastern shore of Lingayen Gulf and took part in the drives on Baguio and the succeeding operational bases used by the Japanese in their retreat into the northern hills.[50] Guerrilla destruction of the bridges on the Bagabag—Bontoc road reduced the Japanese to a single carabao trail for transport of supplies. Cervantes, on the way to Bassang Pass, was taken by guerrillas and, in a hard-fought battle, they captured the Pass itself to break into a part of the Japanese defense perimeter.

In the bitter fight for Balete Pass in the Caraballo Mountains, guerrilla infiltration of enemy lines paved the way for the final assaults on this key approach to the Cagayan Valley.[51] In the Valley itself, Colonel Volckmann's forces seized the important town of Tuguegarao and occupied its adjoining airfield.

So widespread and effective had been the assistance rendered by the Filipino guerrillas in the liberation of their country that General Krueger, upon withdrawing the Sixth Army from combat on Luzon, said in acknowledgment:

> *The gallant Filipino forces, despite tremendous difficulty and with very limited means at their disposal, rendered invaluable support to our operations. Their accomplishments are worthy of high praise.*

By the first half of July 1945, when the Eighth Army assumed combat responsibility for Luzon, the Japanese had been driven deep into the mountains, their main power broken and their destruction or surrender inevitable. A

---

47 GHQ, SWPA, Special Communique No. 1054, 24 Feb 45; HQ 32 Div, "Daily Summary of Enemy Information" No. 73, 26 Apr 45.

48 Manila *Free Philippines*, June 2, 1945, p. 3.

49 GHQ, SWPA, Communique No. 1063, 5 Mar 45.

50 GHQ, SWPA, Communique No. 1079, 21 Mar 45.

51 General Muto explained this penetration in the following words: "Filipino guerrilla units penetrated the jungle and appeared at our rear. The initial force of some 15 men increased to 100 the following day, and by the third day the group expanded into a complete unit augmented by American troops, who then occupied key positions behind our front lines." Unpublished memoirs of Lt. Gen. Akira Muto: *Hito Sakusen no Shinso—The Truth of the Philippines Campaign*, 15 Jun 47, p. 52, G-2, GHQ, FEC.

large part of the painstaking task of mopping up these dismembered but dangerous forces was performed by the various guerrilla groups whose elusive fighting tactics were particularly well suited for jungle and mountain warfare against isolated enemy troops.

General MacArthur's Tribute to the Philippine Guerrillas

The enormous volume of valuable military information sent by the various guerrilla units in the Philippines to General Headquarters constituted a contribution fully as important as their direct combat participation. The extent and degree of intelligence coverage are evident in the complex radio communication system developed under the noses of the Japanese during the days of their occupation. The entire archipelago from north to south and from east to west was literally dotted with guerrilla transmitting and receiving stations. (Plate No. 93)

Perhaps the best recapitulation of the rise of the guerrilla movement in the Philippines and its gradually growing part in the liberation of the Filipino people from the domination of the Japanese was given by General MacArthur shortly after his memorable return to Leyte when he said:

As our forces of liberation roll forward the splendid aid we are receiving from guerrilla units throughout the immediate objective area and adjacent islands causes me at this time to pay public tribute to those great patriots both Filipino and American who had led and supported the resistance movement in the Philippines since the dark days of 1942. These inadequately armed patriots have fought the enemy for more than two years. Most are Filipinos but among these are a number of Americans who never surrendered, who escaped from prison camps, or who were sent in to carry out specific missions.

Following the disaster which, in the face of overwhelming superior enemy power, overtook our gallant forces, a deep and impenetrable silence engulfed the Philippines. Through that silence no news concerning the fate of the Filipino people reached the outside world until broken by a weak signal from a radio set on the Island of Panay which was picked up, in the late fall of that same fateful year, by listening posts of the War Department and flashed to my Headquarters. That signal, weak and short as it was, lifted the curtain of silence and uncertainty and disclosed the start of a human drama with few parallels in military history.

In it I recognized the spontaneous movement of the Filipino people to resist the shackles with which the enemy sought to bind them both physically and spiritually. I saw a people in one of the most tragic hours of human history, bereft of all reason for hope and without material support, endeavoring, despite the stern realities confronting them, to hold aloft the flaming torch of liberty. I gave this movement all spiritual and material support that my limited resources would permit.

Through the understanding assistance of our Navy I was able to send in by submarine, in driblets at first, arms, ammunition and medical supplies. News of the first such shipment spread rapidly throughout the Philippines to electrify the people into full returning consciousness that Americans had neither abandoned them nor forgotten them.

Since then, as resources increased, I was enabled, after formalizing the guerrilla forces by their recognition and incorporation as units of our Army, to send vitally needed supplies in ever increasing quantities through Philippine coastal contacts by four submarines finally committed exclusively to that purpose.

I would that at this time I might name the gallant heroes of this epic in Philippine-American history, but considerations of security for the individuals, their families and the cause require that I limit myself to a generalization of their work and a statement of their brilliant achievements.

Of the latter I need but point out that for the purposes of this campaign we are materially aided by strong, battle tested forces in nearly every Philippine community, alerted to strike violent blows against the enemy's rear as our lines of battle move forward and that now are providing countless large areas adjacent to military objectives into which our airmen may drop with assurance of immediate rescue and protection. We are aided by the militant loyalty of a whole people —a people who have rallied as one behind the standards of those stalwart patriots who, reduced to wretched material conditions yet sustained by an unconquerable spirit, have formed an invincible center to a resolute over-all resistance.

We are aided by the fact that for many months our plans of campaign have benefited from the hazardous labor of a vast network of agents numbering into the hundreds of thousands providing precise, accurate and detailed information on major enemy moves and installations throughout the Philippine Archipelago. We are aided by the fact that through a vast network of radio positions extending into every center of enemy activity and concentration throughout the islands, I have been kept in immediate and constant communication with such widespread sources of information. We are aided by the fact that on every major island of the Philippines there are one or more completely equipped and staffed weather observatories which flash to my Headquarters full weather data morning, afternoon and night of every day and which in turn provides the basis for reliable weather forecasts to facilitate and secure the implementation of our operational plans. Widely disseminated to our forces throughout the Pacific and in China the information from this weather system has materially aided our military operations over a large section of the world's surface.

We are aided by an air warning system affording visual observation of the air over nearly every square foot of Philippine soil established for the purpose of flashing immediate warning of enemy aircraft movement through that same vast network of radio communications. We are aided by provision of all inland waterways and coastal areas of complete observation over enemy naval movement to give immediate target information to our submarines on patrol in or near Philippine waters. This information has contributed to the sinking of enemy shipping of enormous tonnage, and through such same facilities was flashed the warning to our naval forces of the enemy naval concentration off the western Philippines during the Marianas operation.

Finally we are aided by the close interior vigilance that has secured for our military use countless enemy documents of great value, among which were the secret defensive plans and instructions of the Commander-in-Chief of the combined Japanese areas and complete information on the strength and dispositions of enemy fleet and naval air units. That same Commander-in-Chief of the Combined Japanese Fleets was a prisoner of one of our guerrilla units prior to his death from injuries sustained in an air crash.

All of these vital aids to our military operations, and there are many more still unmentioned, are responsive to the indomitable courage of the military and civil leaders whom I shall in future name and their loyal followers both Filipino and American; to gallant Filipinos, residents of the United States, who have volunteered to infiltrate into the islands in succor of their countrymen and Americans who have infiltrated with them; and finally to the militant loyalty and unconquerable spirit of the masses of the Filipino people.

As Commander-in-Chief of the forces of liberation I publicly acknowledge and pay tribute to the great spiritual power that has made possible these notable and glorious achievements—achievements which find few counterparts in military history. Those great patriots, Filipino and American, both living and dead, upon whose valiant shoulders has rested the leadership and responsibility for the indomitable movement in the past critical period, shall, when their identities can be known, find a lasting place on the scroll of heroes of both nations—heroes who have selflessly and defiantly subordinated all to the cause of human liberty. Their names and their deeds shall ever be enshrined in the hearts of our two peoples in whose darkest hours they have waged relentless war against the forces of evil that sought, through ruthless brutality, the enslavement of the Filipino people.

To those great patriots to whom I now pay public tribute I say stand to your battle stations and relax not your vigilance until our forces shall have swept forward to relieve you.[52]

52 GHQ, SWPA, Special Release, 25 Oct 44.

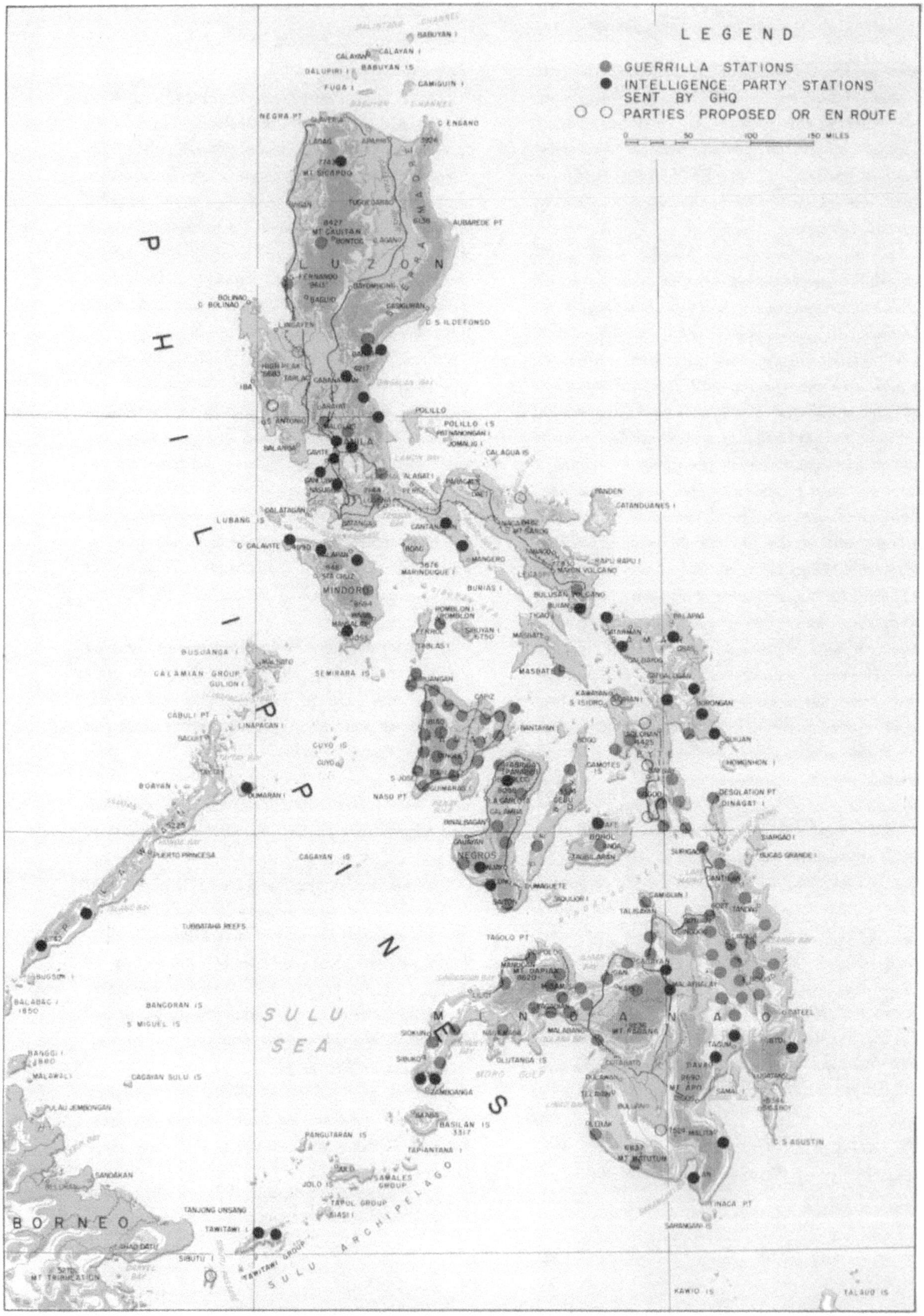

PLATE NO. 93

The Philippine General Radio Net Developed during the Japanese Occupation, 9 October 1944

CHAPTER XI

OPERATIONS OF THE EIGHTH ARMY IN THE SOUTHERN PHILIPPINES

Southward Advance

Operations to free the islands of the southern Philippines followed in swift succession upon the capture of Leyte and the successful invasion of Luzon. To assure complete domination of the shipping lanes of the southern and central Philippines and unquestioned control of the entire archipelago, General MacArthur ordered the immediate seizure of the strategically situated islands along the water route from San Bernardino Strait through the Sibuyan Sea to Verde Island Passage. The task of defeating the Japanese in the land areas of the Visayas and Mindanao was scheduled for the last stage of the Philippines Campaign. During this final phase when General MacArthur's right flank was secured, a limited operation for the recapture of Borneo was to be undertaken as the first step toward freeing the Netherlands East Indies.

Plans for the operations in the southern Philippines and in the Netherlands East Indies had been completely changed after the Leyte operation was substituted for the previously scheduled invasion of Mindanao. With Leyte as the entering wedge into the Philippines, the advance into the Visayas and Mindanao had to be replanned. "Montclair III" was the name given to the new plan, under date of 25 February 1945.[1]

"Montclair III" was divided into two separate but co-ordinated operations—*Victor* for the southern Philippines and *Oboe* for the Netherlands Indies. The Victor operations were to be conducted entirely by United States troops of the Eighth Army, while Australian forces were to assume responsibility for the seizure of Borneo, Java, and adjacent islands. Timing was dependent upon the progress of the fighting in Luzon since the plan involved the commitment of available troops and means as soon as requirements of the Luzon operation had been met.[2] The use of guerrillas in the central and southern Philippines was expected to permit rapid progress without necessitating great naval or aerial support.[3]

The operations for the central and southern Philippines were to be carried out in rapid succession. The landings on Palawan were to take place on 28 February 1945, at Zamboanga on 10 March, on Panay five days later, on Cebu fifteen days later, and on central Mindanao on 12 April.[4] (Plate No. 94) Landings on Borneo under the *Oboe* operations

1 G-3, GHQ, SWPA, *Montclair, Basic Outline Plan for Reoccupation of the Western Visayas—Mindanao—Borneo—NEI, Edition No. 3*, 25 Feb 45. Hereinafter cited as: *Montclair III*.

2 Ibid.

3 CINCSWPA Radio CA-50178 to AC/S, G-3, GHQ, SWPA, 30 Jan 45, G-3, Admin 381/13 (TS).

4 These operations were designated as follows: Palawan, *Victor III*; Zamboanga, *Victor IV*; Panay and northern Negros, *Victor I*; Cebu, southern Negros, and Bohol, *Victor II*; central and eastern Mindanao, *Victor V*. The date of the last-mentioned operation was changed to 17 April by Amendment No. 2, 6 Mar 45, to *Montclair III*.

were scheduled for May and June.

Clearing the Visayan Passages

A preliminary to the *Victor* operations was initiated during the early part of February as the Sixth and Eighth Armies jointly undertook to clear a sea route through the central Philippines. Although United States forces had by this time established bases on Leyte, Samar, Mindoro, and Luzon, the Japanese still controlled most of the southern islands. Their positions along the waterways of the central and southern Philippines could menace Visayan shipping from Leyte and the Philippine Sea. The shortest and most important route thus threatened by enemy forces was from San Bernardino Strait to Verde Island Passage around southern Luzon. Control of these waters by American forces would not only secure a vital shipping lane but would also deny the Japanese an important channel for escape or reinforcement. The chain of islands stretching along this route for a distance of 325 miles was accordingly selected as the next operational target.[5]

On 5 February General MacArthur directed the Eighth Army to " institute operations at the earliest practicable date to clear the northern coast of Samar and the islands in the Cape Verde Passage with the objective of securing the southern exits to San Bernardino Strait and Verde Island Passage."[6] The Sixth Army was assigned the northern shores of this seaway including all of Luzon and the small island of Maricaban at the mouth of Batangas Bay.

Two simultaneous landings were made on 19 February on the northwest corner of Samar and on a small island just off shore. This action initiated a series of amphibious operations conducted by troops under Eighth Army command to capture the islands obstructing San Bernardino Strait, the Sibuyan Sea, and Verde Island Passage. (Plate No. 95) The numerous landings were in every instance successful, many of them assisted in large part by guerrilla units, especially in the mop-up phases. The relief of the 108th Infantry on Masbate by Filipino troops on 4 May marked the completion of the Eighth Army's mission in the waters of the northern Visayas. The sea route from the Lubang Islands in Verde Island Passage to the Balicuatro Islands in San Bernardino Strait was now cleared. Although scattered enemy remnants roamed the islands for some time thereafter, General MacArthur had control of the Visayan sea lanes and their principal entrances and exits; his supply route was secure.[7]

Operations on Palawan

While amphibious landings were in progress on the enemy-held islands within the Visayan passages, attention was focused on the operations designed to destroy Japanese dominance in the Visayas and the southern Philippines. The Eighth Army, early in February 1945, was assigned the task of liberating the islands bordering the Sulu Sea—Palawan region to the north and west, the Sulu Islands to the south, and Zamboanga Peninsula and Basilan Island to the east.[8]

---

5 HQ Eighth Army, *Operational Monograph on the Palawan Operation, Victor III*, p. 4. Hereinafter cited as: Eighth Army, *Palawan Operation. Report of the Commanding General, Eighth Army on the Leyte—Samar Operation*, pp. 20–21. Hereinafter cited as: *Leyte—Samar Operation*.

6 *Leyte—Samar Operation*, pp. 20–21.

7 *Leyte—Samar Operation*, pp. 23–32.

8 GHQ, SWPA, Operations Instructions No. 89, 6 Feb 45; *Report of the Commanding General, Eighth Army on the Palawan and Zamboanga Operations*, p. 5. Hereinafter cited as: *Palawan—Zamboanga Operations*.

The task of recapturing these areas was divided into two operations. The first, *Victor III*, called for the invasion of Puerto Princesa, Palawan, on 28 February; it would be followed shortly thereafter by *Victor VI*, the seizure of Zamboanga, Basilan Island, and the Sulu Archipelago. Possession of these land areas would ensure control of two vital sea lanes—the north-south route between the Netherlands East Indies and Japan and the east-west route from the central and southern Philippines to Singapore. Co-ordinated large-scale operations would then be conducted against the Japanese on Panay and on Negros Occidental, with Cebu, Negros Oriental, and Bohol as the targets to follow.[9]

The island of Palawan was of strategic importance because it dominated the passages between the southern Philippines and the South China Sea. It possessed several air strips, moreover, from which Allied planes could establish complete control of the three straits of Mindoro, Linapaoan, and Balabac. The capture of Palawan would aid American forces in severing the Japanese supply lines between the East Indies and the Japanese Homeland. In addition, Palawan could serve as a supporting point for further Allied operations against Mindanao and the other islands in the southern Philippines.[10]

General Eichelberger designated the 41st Division, under Maj. Gen. Jens A. Doe, for the tasks of liberating Palawan, Zamboanga, and the Sulu Archipelago. The 186th Infantry Regimental Combat Team was assigned to the Palawan invasion, while the 162nd and 163rd Infantry were scheduled for the Zamboanga and Sulu operations.

Responsibility for the amphibious phase of *Victor III* was given to Rear Adm. William M. Fechteler, commanding Naval Task Force 78.2, named the Palawan Attack Group. The Thirteenth Air Force and the First Marine Aircraft Wing were designated as the assault air forces, with additional support being furnished by Maj. Gen. Ennis C. Whitehead's Fifth Air Force.[11]

In preparation for the attack against Palawan, the 41st Division arrived at Mindoro from Biak on 9 February. Final plans were completed and a separate task force headquarters was formed, with the 186th Regimental Combat Team (reinforced) as a nucleus. This task force, designated the Palawan Force, was

9 Eighth Army, *Palawan Operation*, p. 5; *Palawan—Zamboanga Operations*, p. 4.

10 Eighth Army, *Palawan Operation*, p. 5; *Palawan—Zamboanga Operations*, p. 4.

11 HQ Eighth Army, Field Order No. 20, 17 Feb 45, G-3, GHQ, SWPA *Journal*, 18 Feb 45 (S); Eighth Army, *Palawan Operation*, p. 31; *Palawan—Zamboanga Operations*, pp. 5–6. The major units participating in the Puerto Princesa landing were as follows:

Ground: *Eighth Army—General Robert L. Eichelberger*

Unit	*Commander*	*Phase*	*Staging Point*	*Landing Point*
186th RCT, 41st Div	*Col. O. P. Newman*	*Assault*	*Mindoro*	*Puerto Princesa*

Air: *Thirteenth Air Force (supported by elements of the Fifth Air Force)—Brig. Gen. P. B. Wurtsmith*

Unit	*Commander*	*Unit*	*Commander*
XIII Fighter Comd	*Brig. Gen. E. W. Barnes*	*43rd Bomb Gp*	*Col. J. T. Pettus, Jr.*
347th Fighter Gp	*Col. L. F. Dusard, Jr.*	*419th Night Fighter Sq*	*Maj. J. A. Shulmistras*
42nd Bomb Gp	*Maj. H. C. Harvey*	*403rd Troop Carrier Sq*	*Col. H. J. Sands, Jr.*
85th Fighter Wing	*Col. P. H. Greasley*		

Naval: *Seventh Fleet—Adm. T. C. Kinkaid*

Unit	*Commander*
Palawan Attack Gp (Task Force 78.2)	*Adm. W. M. Fechteler*

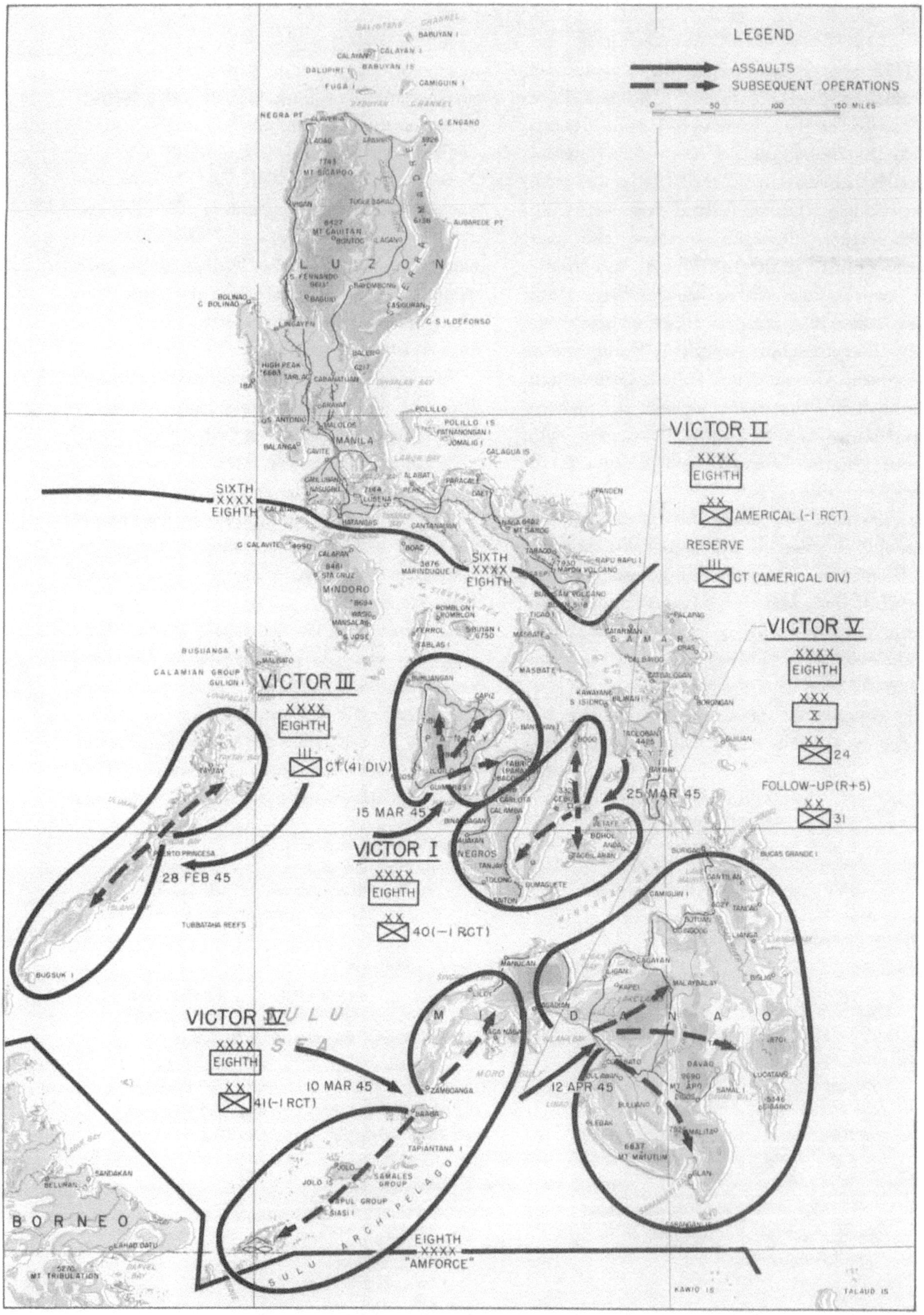

PLATE NO. 94

Plan of "Montclair III": Victor Operations, 25 February 1945

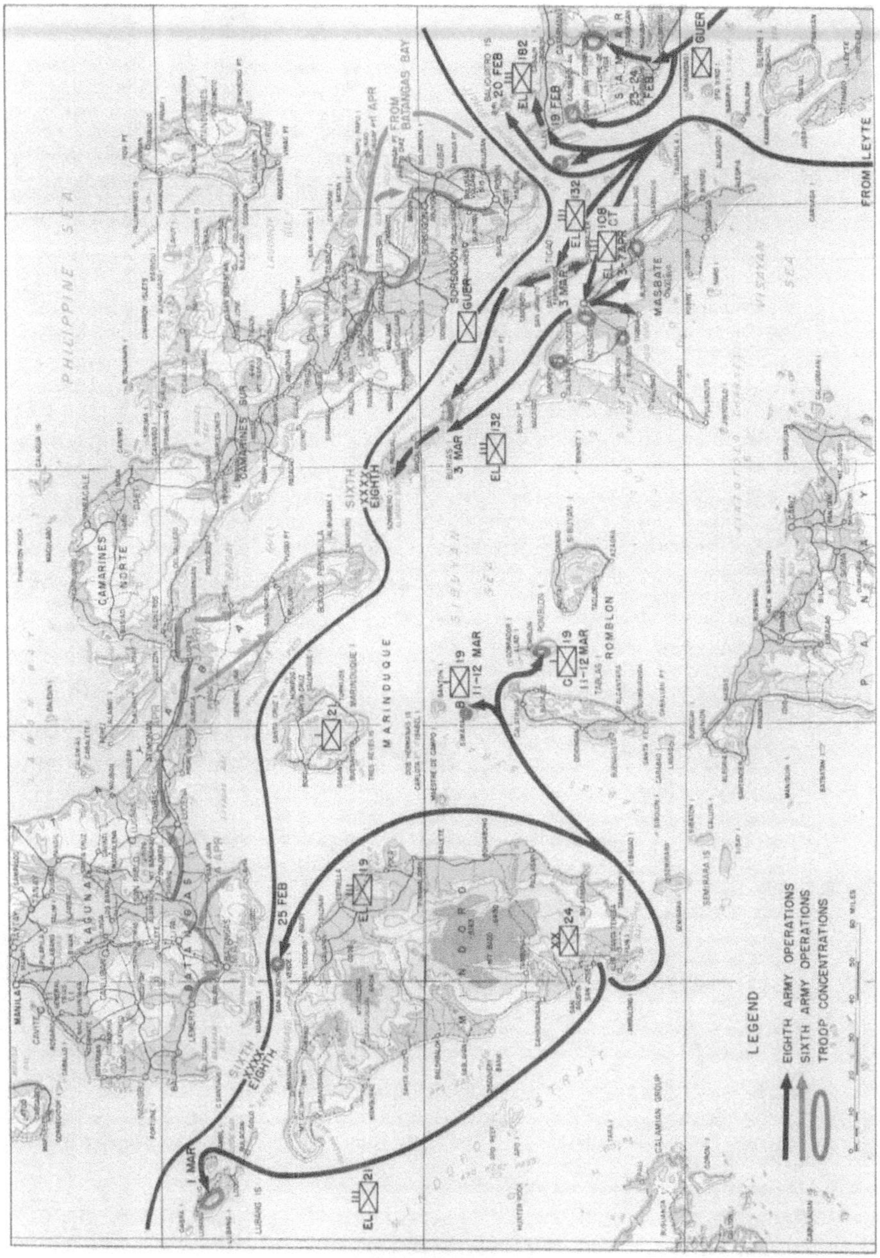

PLATE NO. 95

Clearing of the Visayan Passages, 19 February—8 May 1945

activated on 17 February under the command of Brig. Gen. Harold H. Haney.[12]

Palawan Force departed from Mindoro on 26 February and arrived off the southwest shore of Puerto Princesa early on the morning of the 28th. (Plate No. 96) The landing beaches had already been under air bombardment for forty-eight hours, but an additional intensive concentration of naval gunfire and rocket barrage was laid down just before H-hour. The assault troops encountered no enemy opposition, and a beachhead was quickly established.[13]

The Palawan landings and their strategic significance were covered in the communique of 2 March:

We have landed on Palawan and seized control of this fifth largest and most westerly island of the Philippine Archipelago. Engrossed in operations elsewhere the enemy again failed to diagnose our plans and properly prepare his defense. Elements of the 41st Division of the Eighth Army, coordinated with naval and air support, landed near Puerto Princesa, the capital and major city of the island, and pushed swiftly inland to seize the town and two nearby airfields. The enemy fled to the hills and our losses were very light. The air bases of Palawan command the western end of the southern water passage through the Philippines by way of the Sulu Sea and interdict the north and south channel of the South China Sea, the enemy's main water transportation line to Indo China, Thailand, Malaya, Singapore, Burma, and the East Indies. It places us within 500 miles of the enemy bases of Brunei Bay and Tarakan in North Borneo. It will therefore not only help to insure the safe passage of our own sea transport but render hazardous that of the enemy. The progressive securing of the Philippines as a base thus tends to cut the enemy in two and condemn all his conquests to the south to recapture.[14]

Operations ashore on Palawan were concentrated primarily on clearing the Puerto Princesa area.[15] On all sides there was evidence of hasty Japanese withdrawal, and it was not until 2 March that the retreating forces were contacted.[16] From that time on scattered Japanese units on Palawan were hunted down with the aid of organized guerrillas. On 9 March reconnaissance troops went ashore on Dumaran Island, northeast of Palawan, and reported that the enemy had evacuated the island. By 20 March the airstrips near Puerto Princesa, which had once been an important link in the Japanese ferry and transport route to the Netherlands East Indies, had been rebuilt into all-weather fields and put into operation. The liberation of Palawan was completed by the end of March.

In mid-April patrols of the 18th Infantry cleared Busuanga Island to the northeast of Palawan and Balabac and Pandanan Islands to the south. With this closing of the western exit to the Visayan Sea, the fate of the remaining enemy forces on Panay, Negros,

12 Eighth Army, *Palawan Operation*, pp. 32–33.

13 Ibid., pp. 61–62; COMPHIBGP 8 (Admiral Fechteler) Report to COMINCH, Serial 0015, 25 Mar 45, G–3, GHQ, SWPA *Journal*, 1 Mar 45 (S).

14 GHQ, SWPA, Communique No. 1060, 2 Mar 45.

15 "The Puerto Princesa area," declared Major General Yoshiharu Tomochika, Thirty-fifth Army Chief of Staff, "was considered important . . . since it offered the enemy a possible staging base for air operations against our supply lines from our homeland to the southwest Pacific area." Major Chuji Kaneko, a staff officer of the 102nd Division, stated: "It was considered most important to secure the air and sea route between . . . Borneo and our homeland via Manila." Interrogation Files, G-2 Historical Section, GHQ, FEC.

16 "In case of an enemy landing," asserted Major Kaneko, "our troops were to resist as long as possible and to protect . . . the Puerto Princesa airfield. . . . After the Americans had landed on Luzon, an order was handed down from Thirty-fifth Army to the units on Palawan to engage in guerrilla warfare and to resist from the mountains instead of Puerto Princesa." Interrogation Files, G-2 Historical Section, GHQ, FEC.

Cebu, and Bohol was similarly sealed; the Japanese were cut off on all sides. As new naval bases were constructed on Palawan, motor torpedo boats of the Seventh Fleet began to patrol the east and west coasts of the island, later extended their activities to adjacent areas, and finally reached as far south as British North Borneo.[17] Aircraft meanwhile carried out searches over the South China Sea and bombed Japanese installations from Borneo to Indo China. The ground troops, too, were expanding their scope of operations, as the 41st Division turned its attention to the western tip of Mindanao—to Zamboanga and the Sulu Archipelago.

The Zamboanga Landing

Basilan Strait, off the southwest tip of Mindanao, constitutes one of the two main approaches to the Asiatic mainland from the southwest Pacific; it is dominated by the city of Zamboanga and the peninsula of the same name. In American hands, Zamboanga would further constrict the enemy-held positions in the Philippines and virtually complete their encirclement.

To the south of Basilan Strait lies the Sulu Archipelago, a long chain of islands connecting Mindanao and Borneo. The islands lie directly astride the shortest water routes from the China Sea to the Celebes, the Moluccas, and New Guinea. Occupation of this area coupled with control of Zamboanga would completely isolate the Netherlands East Indies from the Philippines and provide valuable bases for future offensives against the Japanese in Borneo.[18] *Victor IV*, the assault against Zamboanga and the Sulu Archipelago, was consequently selected as the next operation for the Eighth Army. (Plate No. 97)

Planning for the operation against Zamboanga and Sulu had been undertaken concurrently with the planning for the attack on Palawan, and their seizure was directed in the same field order.[19] Again the 41st Division was designated as the assault force, with 10 March 1945 set as the date for the landing.[20]

---

17 S-3, Palawan Force, Report to TAG, "Narrative History of the Palawan (V-3) Operation," G-3, GHQ, SWPA *Journal*, 30 Jun 45. G-3, GHQ, SWPA, *Monthly Summary of Operations*, May 45.

18 *Palawan—Zamboanga Operations*, p. 37; HQ Eighth Army, *Operational Monograph on the Zamboanga—Sulu Archipelago Operation*, pp. 5–6. Hereinafter cited as: Eighth Army, *Zamboanga—Sulu Operation*.

19 HQ Eighth Army, Field Order No. 20, 17 Feb 45, G-3, GHQ, SWPA *Journal*, 18 Feb 45.

20 The major units participating in the Zamboanga assault were as follows:

Ground: *Eighth Army—General Robert L. Eichelberger*

Unit	Commander	Phase	Staging Point	Landing Point
41st Div (–186 RCT)	*Brig. Gen. J. A. Doe*	*Assault*	*Mindoro*	*San Mateo, Zamboanga Peninsula*

Air: *Far East Air Forces (supported by elements of the Seventh Air Force and certain Marine Air Units)—General George C. Kenney*

Unit	Commander	Unit	Commander
Fifth Air Force	*Maj. Gen. E. C. Whitehead*	*22nd Bomb Gp*	*Lt. Col. L. T. Nicholson*
Thirteenth Air Force	*Brig. Gen. P. B. Wurtsmith*	*43rd Bomb Gp*	*Col. J. T. Pettus, Jr.*
XIII Fighter Comd	*Brig. Gen. E. W. Barnes*	*42nd Bomb Gp*	*Col. C. C. Kegelman*
XIII Bomber Comd	*Col. C. A. Brandt*	*307th Bomb Gp*	*Col. C. H. Rees*
85th Fighter Wing	*Col. P. H. Greasley*	*Marine Air Gp 12*	*Lt. Col. Verne McCall*
494th Bomb Gp	*Col. L. B. Kelley*	*Fleet Air Wg No 10*	
5th Bomb Gp	*Maj. A. W. James*	*1st Marine Aircraft Wg*	

Naval: *Seventh Fleet—Adm. T. C. Kinkaid*

Unit	Commander
Palawan Attack Gp (Task Force 78.3)	*Adm. F. B. Royal*

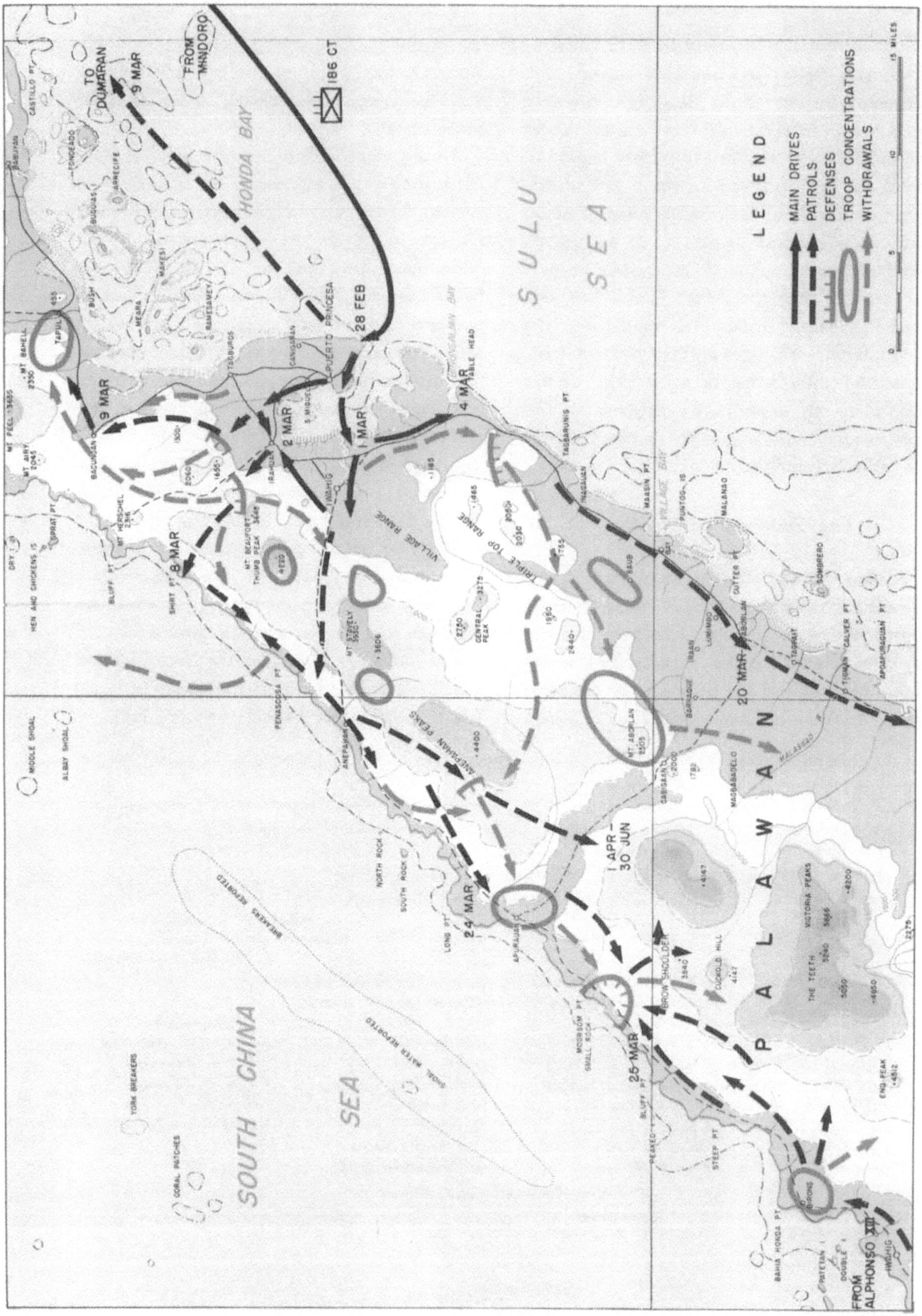

PLATE NO. 96

Operations on Palawan, 28 February—30 June 1945

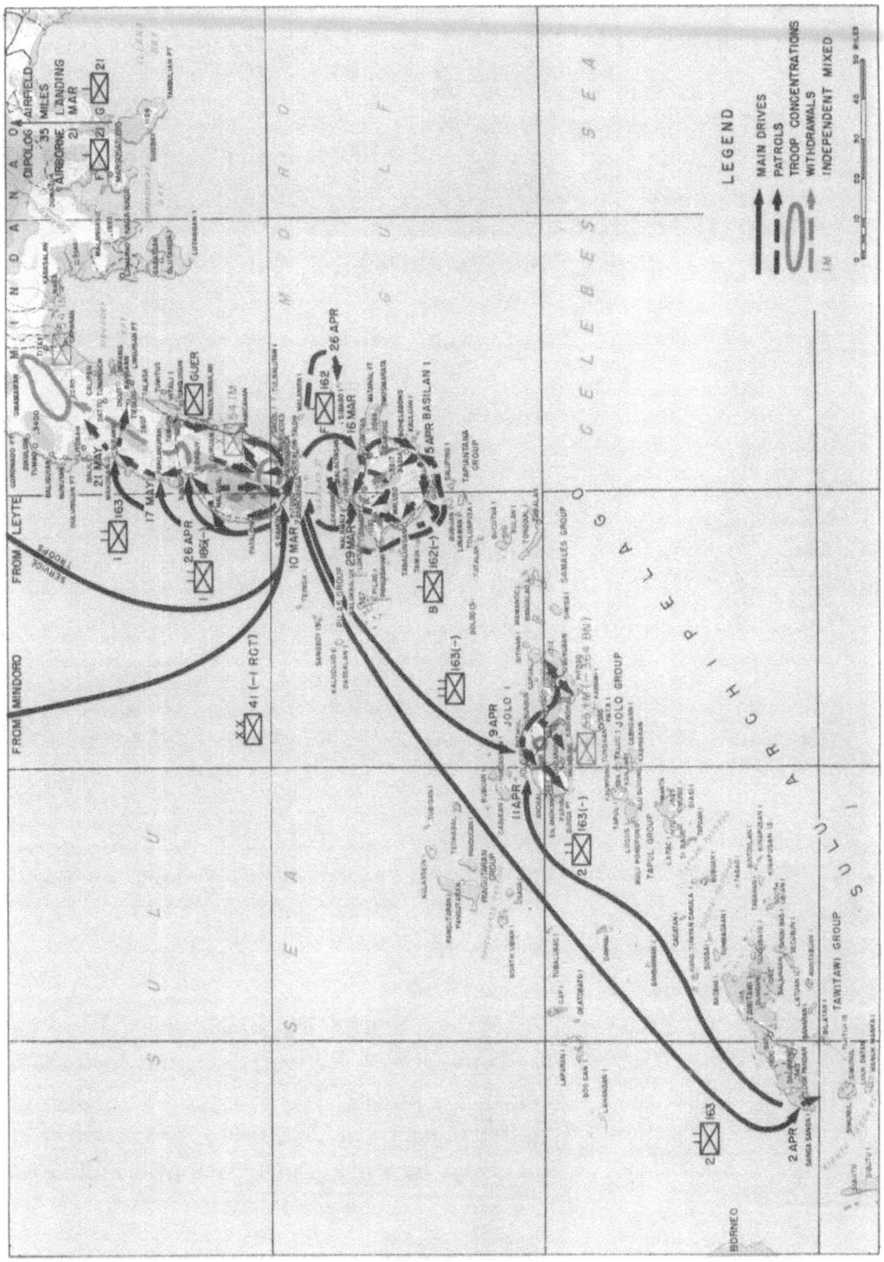

PLATE NO. 97

Seizure of Zamboanga and the Sulu Archipelago, 10 March—20 June 1945

These operations were to be followed by an assault on Mindanao proper, the principal strategic objective in the southern Philippine Archipelago.[21]

Despite this planned sequence, the first invasion of Mindanao by American troops actually preceded the scheduled invasion of Zamboanga. In order to take advantage of certain guerrilla actions, an airborne landing was made between 8–10 March on the north coast near Dipolog by two companies of the 24th Division. Their mission was to reinforce and secure the guerrilla-held Dipolog Airfield in order to provide an air base from which planes of the Far East Air Forces could operate in support of the Zamboanga invasion. Aided in their defense of the airfield by guerrilla forces, the American troops remained until nearly the end of March, permitting a squadron of fighters based on the airfield to share in the intensive preliminary aerial bombardment of Zamboanga. The garrison companies were evacuated by air on 27 March in the face of a strong advancing enemy force.[22]

Loading for the Zamboanga operation was completed at Mindoro on 7 March, while six LST's were loaded with special equipment at Leyte. The Zamboanga Attack Group, under Rear Adm. Forrest B. Royal, departed from Mindoro Strait early the following day, encountering no hostile naval or air opposition. Rendezvousing with the elements from Leyte on the afternoon of the 9th for the passage through the Sulu Sea, the task force proceeded to Basilan Strait and stood off Zamboanga early on the morning of 10 March.

Following an intense air and naval bombardment, reinforced units of the 41st Division, less the 186th Regimental Combat Team which was still mopping up on Palawan, poured ashore against light opposition at San Mateo, just west of Zamboanga City. The Thirteenth Air Force, which had been carrying out pre-invasion bombardment of the objective area since 1 March, continued the aerial pounding throughout the operation.[23]

General MacArthur was able to report on 12 March:

> *We have landed near Zamboanga on the southwesternmost tip of Mindanao. . . . The bulk of the enemy garrison, caught off guard, has fled to the hills in disorder. We now control the entire length of the western shores of the Philippine Islands from the northwestern tip of Luzon to the southwestern tip of Mindanao, a distance of approximately 800 miles. The blockade of the South China Sea and the consequent cutting off of the Japanese conquests to the south is intensifying.*[24]

Consolidation of the beachhead following the landing proceeded rapidly against sporadic enemy resistance.[25] The Japanese defensive positions were superior to any previously encountered by the Eighth Army in the Philip-

21 Eighth Army, *Zamboanga—Sulu Operation*, p. 5.

22 Ibid., pp. 31–32.

23 HQ 41st Div. Report to TAG, "G-3 Report on the V-4 Operation, 10 Mar–20 Jun 1945" G-3, GHQ, SWPA *Journal*, 20 Jun 45 (S). Hereinafter cited as: 41st Div. *V-4 Report*.

24 GHQ, SWPA, Communique No. 1070, 12 Mar 45.

25 Although the American invasion of Zamboanga obviously was not totally unexpected, subsequent interrogations of surviving Japanese revealed that the landings caught them unprepared. Maj. Yasura Hanada, Chief of Staff and only surviving staff member of the 54th Independent Mixed Brigade, Thirty-fifth Army stated: "We expected that when the Americans landed on Luzon they would proceed to the Homeland. We did not expect Zamboanga to be attacked. However, in February, there were indications of a landing, and by then it was too late to do much about it. That was where our error lay." 10th Information and Historical Service, HQ Eighth Army, *Staff Study of the Japanese Activities in the Zamboanga (Victor IV) Operation*. Hereinafter cited as: Eighth Army, *Staff Study, Zamboanga*.

pines, but they were frequently unoccupied. Many strong points had been abandoned by the Japanese who had become jittery and thoroughly demoralized by the heavy air and naval bombardment.

The initial objective, which included Wolfe Airstrip, was secured before noon. The following day San Roque Airfield and Zamboanga City were taken. The Japanese troops had withdrawn to the heights north of the city and the only defense of the landing area came from sporadic mortar and artillery fire directed from well-constructed gun emplacements in the hills outside Zamboanga. The roads to and within the city, however, were heavily mined, planted with booby traps and flanked by concrete pillboxes protected by wire entanglements.[26]

The beachhead was rapidly expanded along the east and west coasts of Zamboanga Peninsula, while to the north of the city, where the enemy had set up a strong defense area protecting artillery emplacements, progress was more difficult.[27] Behind the enemy on still higher ground, guerrillas had stretched a series of strong points across the peninsula, forming a barrier to their further retreat. The continuous pressure of the advancing 41st Division tightened the jaws of a vise which closed relentlessly on the Japanese forces. Only a small part of the enemy managed to escape through the guerrilla cordon.

Grim and stubborn Japanese resistance slowed the advance into the mountains. With the arrival of the 186th Regimental Combat Team in the line late in March, however, the drive gained fresh vigor. By early April organized resistance had ceased and enemy opposition had degenerated into weak forays by scattered, unco-ordinated elements which were later sealed off and decimated by a series of American landings along the western coast of the peninsula.[28]

Clearing the Sulu Archipelago

Having established firm control on Zamboanga, the 41st Division could now begin the second phase of Operation *Victor IV*—the clearing of the Sulu Archipelago to the south. First action against the numerous islands comprising the Archipelago took place on 16 March when elements of the 162nd Regimental Combat Team landed unopposed near Lamitan on the northeast coast of Basilan Island. Two days later, the same troops occupied Malamaui Island, northwest of Basilan.[29]

Thrusting deep into the Sulu Archipelago, a battalion of the 163rd Regimental Combat Team on 2 April by-passed Jolo Island, a major enemy stronghold, and landed unopposed on Sanga Sanga Island in the Tawi Twai Group. Objectives were quickly secured and later the same day a beachhead was established on nearby Bongao Island by the same battalion. Hostile action was limited to occasional sniper fire, and all organized Japanese resistance was broken by 6 April, when guerrillas were brought in to garrison Bongao. Guerrilla forces eliminated the few Japanese on Tawi Tawi Island.[30]

With the fall of Tawi Tawi the Japanese

26 *Palawan—Zamboanga Operations*, pp. 49–51; 41st Div. *V-4 Report*.

27 In commenting on the Japanese tactics of withdrawing to the hills rather than putting up a stiffer resistance on the beaches of Zamboanga, Major Hanada said: "It was indicated from the Battle of Leyte that such tactics would have resulted in a great loss of troops to us. Rather than take such a risk and lose so much, it seemed wiser to remain in the hills and take a delayed action strategy." Eighth Army, *Staff Study, Zamboanga*.

28 G-3, GHQ, SWPA, *Monthly Summary of Operations*, Mar and Apr 45.

29 *Palawan—Zamboanga Operation*, pp. 56–58; 41st Div. *V-4 Report*.

30 Eighth Army, *Zamboanga—Sulu Operation*, pp. 53–54; G-3, GHQ, SWPA, *Monthly Summary of Operations*, Apr 45.

last opposition dominating their shipping routes to the oil-rich east coast of Borneo. General MacArthur emphasized the strategic significance of the seizure of this small group of islands in his communique of 5 April:

The seizure of Tawi Tawi secured the last link in the chain of blockading airfields which the recapture of the Philippines had made possible. Along the entire coast from the northern end of Luzon to the southern tip of the Sulu Archipelago, a distance of more than a thousand miles, our airfields flank the waters of the South China Sea between the coast of Asia and the Philippines, while our fields stretching for two thousand miles from New Guinea to the Philippines cover the waters necessary for communication from Japan to Java, Borneo, Celebes, the Moluccas and all the other islands in the Southwest Pacific. The width of the South China Sea averages only a few hundred miles so that its expanse as well as the coast line of the Asiatic mainland is easily covered by our bombers in addition to our submarines. This cuts off enemy sea traffic to the conquered possessions to the south and severs the so called Empire life line to the East Indies. The effectiveness of this blockade results from its great depth permitting multiplicity of successive fields to bear upon the target. The enemy can easily under favorable weather conditions slip through waters covered only from one set of bases but when he runs into successive belts flanked by fields extending for thousands of miles he finds it impossible to pierce the blockade. He may well penetrate the thinness of a line blockade but it is difficult indeed to run the gauntlet of successive bases which cover the length of the way for thousands of miles. The campaign of the Southwest Pacific Area along the coast of New Guinea through the Halmaheras and throughout the length of the Philippines had this end in view as one of its primary strategic objectives. Already our blockade has sunk many hundreds of thousands of tons of enemy shipping and with the acquisition of the Tawi Tawi base is now in complete operation. For many weeks it has been difficult for Japan to ship oil, rubber, or other essential military commodities from the Southwest Pacific to the homeland, or military supplies from Japan to garrisons in those waters, and as the air pressure of our blockade reaches its peak, such traffic will become practically impossible. The entire Dutch East Indies are now isolated not only for exploitation but from enemy reinforcement and supply. It is the culmination of one phase of the campaign which started from Milne Bay late in June 1943 and progressing through Kiriwina, Salamaua, Lae, Finschhafen, Arawe, Gloucester, Saidor, Hollandia, Sarmi, Biak, Noemfoor, Sansapor, Leyte, Mindoro, Luzon, Palawan and Zamboanga, has now reached its far end at Tawi Tawi.[31]

Jolo, central island of the Sulu Archipelago, represented the only point remaining under Japanese control. American reconnaissance patrols had landed on the island on 2 April, and the main landing followed seven days later when the 163rd Regimental Combat Team attacked in force. Guided by the advance scouts, the assault troops hit the beach unopposed about five miles from the town of Jolo.[32]

By 10 April, Jolo, together with Zettel Airfield, was seized without difficulty. The heaviest opposition was met at Mount Daho, a heavily fortified hill five miles to the southeast of the town, where approximately 400 Japanese had firmly intrenched themselves. This position required four days of shelling by artillery and thirty-six dive bombing strikes before it could be reduced. By 22 April, however, the

31 GHQ, SWPA, Communique No. 1094, 5 Apr 45.

32 The original plans of the 55th Independent Mixed Brigade for the defense of Jolo Island called for an all-out defense at the shoreline, with vigorous counterattacks, to be followed by a retreat to prepared positions inland, if necessary, for a final stand. This plan, however, was altered early in March when the incidence of malaria and jungle ulcers had cut the strength to a point where it was considered impracticable to have a double line of defense. The constant attrition from guerrilla attacks and the fear of a guerrilla attack from the rear to aid the American landing were other factors in the change of plans. Interrogation of Maj. Tokichi Tenmyo, CO, 365th Battalion, 55th IMB, Eighth Army, *Staff Study, Zamboanga.*

back of the Japanese resistance was broken. The enemy troops which managed to escape wandered in small groups throughout the island and soon fell easy prey to pursuing Moro guerrilla bands. Sibago, an inlet just to the north of Basilan, was taken on 26 April and by 10 May the Japanese forces on Jolo had been destroyed.[33]

After all major positions in the Sulu Archipelago were secured, Seventh Fleet motor torpedo boats, based on Tawi Tawi, patrolled the entire length of the island chain. In addition, airstrips along the archipelago were repaired and enlarged for the use of Allied planes. On Zamboanga Peninsula, active patrolling continued until 20 June, when *Victor IV* operation was officially declared closed.[34] During the two months following the close of the operation, however, over 200 Japanese troops were killed and more than 500 prisoners were captured.

Planning for Operations in the Southern Visayas

The Eighth Army next turned its attention to the reconquest of Panay, Negros, Cebu, and Bohol. These four islands constitute a strategic geographical unit commonly called the Southern Visayas. Within this smaller theater, however, the topography required the division of operations into two areas and, consequently, the plan for their liberation developed into two campaigns by separate task forces. The areas of these campaigns were separated by the steep mountain range that runs roughly along the north-south axis of Negros Island. To the west and northwest of this range lie Panay and Negros Occidental; to the east, Negros Oriental, Cebu, and Bohol. The mission of liberating Panay and Negros Occidental (*Victor I* operation) fell to the 40th Division with the 503rd Parachute Regimental Combat Team attached. (Plate No. 98) The capture of Negros Oriental, Cebu, and Bohol (*Victor II* operation) was assigned to the Americal Division.

At the time these operations were launched in March and April of 1945 the Southern Visayas as a group were isolated from the rest of the Philippine Archipelago. American forces controlled most of the surrounding larger islands. To the north, Mindoro and the most important areas of Luzon were in American hands ; to the south, the Zamboanga Peninsula and the Sulu Archipelago had been invaded ; to the west, Palawan had been captured ; and to the east, Leyte and Samar were under the control of American forces. With all escape routes cut off by Allied naval and air superiority, the Japanese forces in the Southern Visayas were caught in a trap from which there was no escape.[35]

The Seizure of Panay

Panay was the first of the Southern Visayas to fall before the fast-moving onslaught of the Eighth Army. The 40th Division (less the 108th Regimental Combat Team) was withdrawn from operations on Luzon in the middle of March and placed under the control of the Eighth Army. This division, commanded by Maj. Gen. Rapp Brush, formed the nucleus of the assault task force, while the amphibious

---

33 Eighth Army, *Zamboanga—Sulu Operation*, pp. 55–61 ; G-3, GHQ, SWPA, *Monthly Summary of Operations*, Apr 45.

34 CO 2nd Bn, 163rd Inf Report to TAG, "Sulu White, V-4 Operations," G-3, GHQ, SWPA *Journal*, 10 Apr 45 (S).

35 HQ Eighth Army, *Operational Monograph on the Panay—Negros Occidental Operation*, pp. 5–6. Hereinafter cited as : Eighth Army, *Panay—Negros Occidental Operation*. *Report of the Commanding General, Eighth Army on the Panay—Negros and Cebu Operations*, p. 1. Hereinafter cited as : *Panay—Negros—Cebu Operations*.

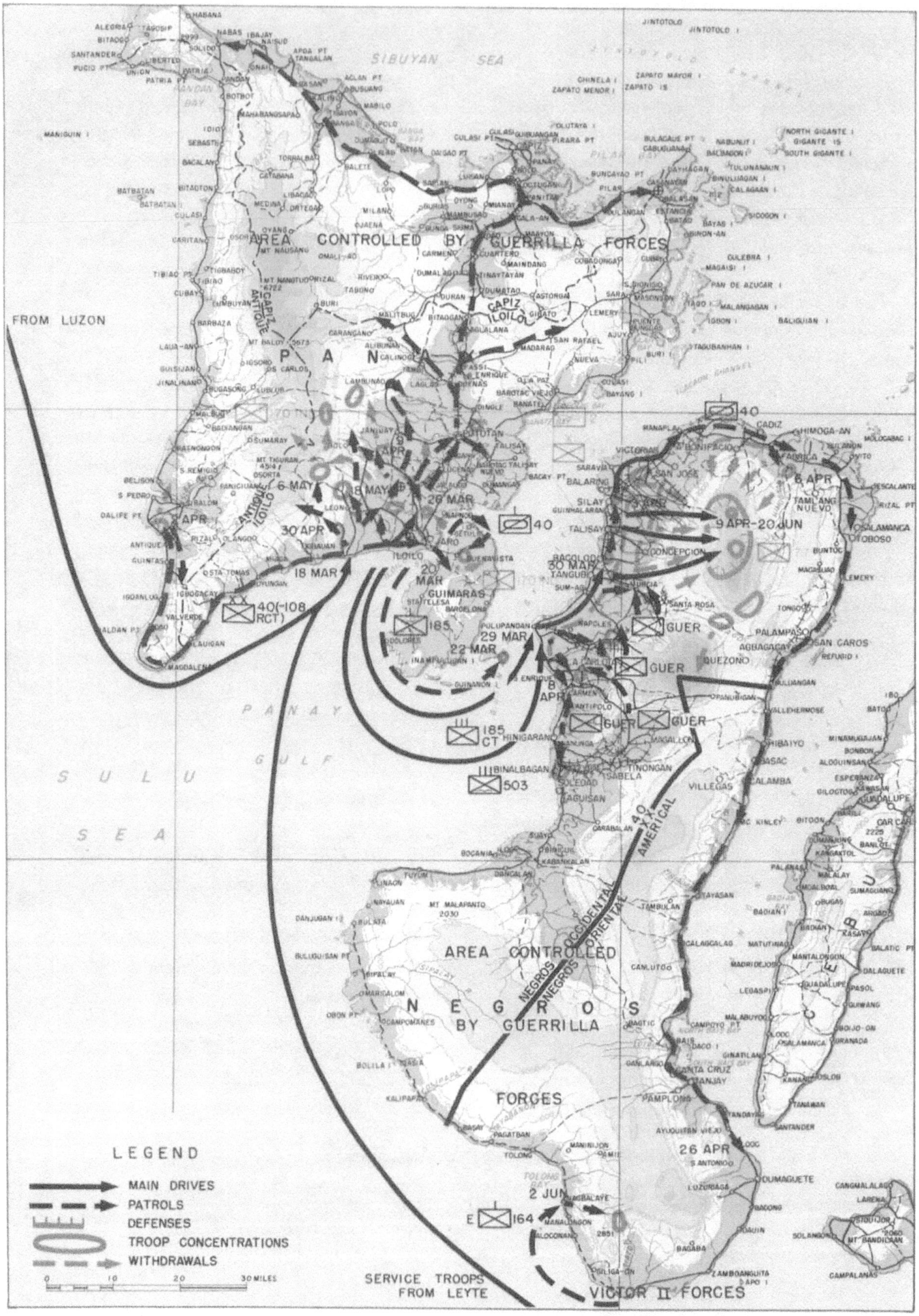

PLATE NO. 98

Operations in Panay and Negros Occidental, 18 March—20 June 1945 phase of the operation came under Rear Adm. Arthur D. Struble, Commander of the Panay Attack Group.[36]

As in previous amphibious landings, the objective area was softened up by aerial pounding before the invasion by ground forces. Starting on 1 March, nearly three weeks before the assault forces were to land, Marine and Thirteenth Air Force fighters made almost daily aerial attacks on Panay, aiming at enemy installations and troop concentrations. Meanwhile, Allied naval forces harassed the enemy's overwater communications in Guimaras and Iloilo Straits.[37]

The task force departed from Luzon on 15 March and arrived off Panay three days later. After destroyer bombardment of the landing beaches, the assault waves landed near Tigbauan. As they waded ashore the lead troops were greeted by men of Colonel Peralta's guerrilla forces.

By evening of the first day, troops of the 185th Regimental Combat Team had advanced approximately ten miles along the coast to Arevalo in the direction of Iloilo. Only light opposition was encountered, most of the enemy having retreated north along the Sibalon River. By evening the second day, Mandurriao Airfield had been secured and the 185th Regimental Combat Team stood poised for an attack on Iloilo, the largest city on Panay.[38]

The attack was launched on the morning of 20 March and Iloilo was taken almost immediately. The Japanese garrison had beat a hasty retreat the previous night, after destroying approximately 70 per cent of the city.[39] Before the landing, Colonel Peralta's guerrillas had virtually cleared the interior and coastal sectors of the island and had bottled up most of the Japanese garrison within the Iloilo area. As

---

36 The major forces participating in the Panay assault were as follows:

Ground: *Eighth Army—General Robert L. Eichelberger*

Unit	*Commander*	*Phase*	*Staging Point*	*Landing Point*
40th Div (–108 RCT)	*Maj. Gen. R. Brush*	*Assault*	*Lingayen*	*Tigbauan*

Air: *Thirteenth Air Forces (supported by elements of the Fifth and Seventh Air Forces and certain Marine Air Groups)—Brig. Gen. P. B. Wurtsmith*

Unit	*Commander*	*Unit*	*Commander*
XIII Bomber Comd	*Col. C. A. Brandt*	*347th Fighter Gp*	*Col. L. F. Dusard, Jr.*
XIII Fighter Comd	*Brig. Gen. E. W. Barnes*	*85th Fighter Wing*	*Col. P. H. Greasley*
307th Bomb Gp	*Col. C. H. Rees*	*403rd Troop Carrier Gp*	*Col. H. J. Sands Jr.*
42nd Bomb Gp	*Maj. H. C. Harvey*	*Marine Air Gp, Zamboanga*	*Col. C. Jerome*
5th Bomb Gp	*Lt. Col. A. W. James*		

Naval: *Seventh Fleet—Adm. T. C. Kinkaid*

Unit	*Commander*	*Unit*	*Commander*
Task Gp 78	*Adm. D. E. Barbey*	*Task Gp 78.3.2*	*Capt. R. W. Cavenaugh*
Task Gp 78.3	*R. Adm. A. D. Struble*		

37 Eighth Army, *Panay—Negros Occidental Operation*, pp. 31–32.

38 CG 40th Div. Report to TAG, "Victor One & Victor One Able Operations, 40th Inf. Div." G-3, GHQ, SWPA *Journal*, 1 Jul 45 (S). Hereinafter cited as: 40th Div. *Victor I Report*. Eighth Army, *Panay—Negros Occidental Operation*, pp. 37–42.

39 The Japanese retreat was explained by Capt. Sadoyoshi Ishikawa, Staff Officer of the 77th Brigade, 102nd Division: "Lacking adequate supplies of ammunition and equipment, the Japanese units were broken up and scattered soon after the American assault landing. Believing that direct combat would prove disastrous, an all-out clash with the Americans was avoided and the fighting limited to defensive tactics. This was particularly true in the mountain areas where the final stand was made." 10th Information and Historical Service, HQ Eighth Army, *Staff Study of Japanese Operations on Panay Island.*

the fleeing enemy units hurried north from Iloilo, to escape the American forces, they were met by guerrillas in defensive positions north of Jaro. The guerrilla strength was insufficient to stop the Japanese completely, however, and the enemy troops broke through the lines and continued their retreat.

Pursued by American and Filipino forces, the Japanese were systematically ferreted out of their retreats and eliminated. Santa Barbara Airfield was secured and by 22 March all organized enemy resistance on Panay had been crushed. Overwater and overland patrols by guerrillas, assisted by elements of the 40th Division, effectively reduced the isolated pockets along the north and west coasts. A small enemy garrison on Inampulugan Island was wiped out, while a landing party to nearby Guimaras Island found the Japanese had retired to the hills. The scattered enemy groups were gradually hunted down by guerrillas and American mop-up patrols.[40]

Meanwhile, the 40th Division prepared to move across Guimaras Strait to Negros Island. It was estimated that the relatively strong enemy garrison on Negros Occidental, consisting of the remnants of two divisions, would be able to prevent a rapid completion of the operation by the initial landing force of two regimental combat teams. Rather than withdraw units from other operations and cause a postponement of their scheduled dates of execution, however, it was decided to proceed according to the original plan. After the first landings, the 503rd Parachute Regimental Combat Team would be made available to augment the forces of the 40th Division.[41]

The Attack on Negros Occidental

On 29 March the 40th Division, less one regimental combat team, crossed Guimaras Strait from Iloilo and landed at Pulupandon on Negros Occidental. There was no preliminary naval bombardment and the landing was made without opposition. The landing force pushed rapidly inland and the strategic Bago River Bridge was secured after a brief skirmish. Lack of initial resistance was largely due to the assistance of the strong guerrilla units of Colonel Abcede, which were instrumental in confining the enemy principally to the north and northwest coasts of the island.

Troops of the 185th Regimental Combat Team moved northward, crossed the Magsungay River in the face of intense enemy fire and began their attack on Bacolod.[42] After destroying a portion of the city, the Japanese garrison had withdrawn to the north and east, leaving a small delaying force capable of only limited action. Bacolod and its airdrome were secured by 30 March, and the outskirts of Talisay were reached the following day. By this time it had become apparent that the enemy did not intend to defend the coastal areas but would make a strong stand in the rugged high ground of the north-central part of the island.[43]

The Japanese fought tenaciously at Talisay

40 G-3, GHQ, SWPA, *Monthly Summary of Operations*, Mar 45; GHQ, SWPA, Communique No. 1080, 22 Mar 45.

41 AC/S G-3 Memo to C/S GHQ, SWPA, 17 Mar 45, G-3, GHQ Admin 381/36-P (TS).

42 Lt. Col. Shigekatsu Aritomi, Staff Officer of the 77th Infantry Brigade, Japanese 102nd Division, revealed that it had been planned to destroy all bridges on the route to Bacolod, but the rapidity with which the American forces secured them, plus the fact that the detonators of the explosive charges were mechanically defective, prevented their destruction. Continuing, Colonel Aritomi said: "Since our supplies were cut off, our policy was to obstruct the Americans as long as possible and to destroy the airfields so that they would be useless to the Americans. We also planned to destroy all the bridges but failed in this." 10th Information and Historical Service, HQ Eighth Army, *Staff Study of Japanese Operations on Negros Island.*

43 *Panay—Negros—Cebu Operations*, pp. 27–35.

but the town and airfield fell on 2 April. Silay, a small barrio to the north, was taken the next day. Meanwhile, the 160th Regimental Combat Team had moved into positions at Bacolod vacated by the 185th Regimental Combat Team in its drive northward. The primary objectives in Negros Occidental had been secured; the 40th Division controlled the most important section of the west coast, extending from Silay to Pulupandan, while the area south of Pulupandan was mostly in guerrilla hands.

The division's next attack was directed inland against the Japanese prepared defenses in the mountain ranges of north-central Negros. This eastward advance was rapid and opposed only by delaying actions of a minor nature. By 4 April, the only remaining enemy defense areas in Negros Oriental were the pockets in the central mountain range and the fortified town of Dumaguete.

On 8 April, the 503rd Parachute Regimental Combat Team made an overwater movement from Mindoro to assist the 40th Division. Taking charge of activity on the left flank, it began a concerted attack against the enemy main defense line the next day. Japanese resistance to the 40th Division's advance increased with time, bearing out intelligence predictions that the thrust would encounter the enemy's main defenses.[44] Following the first day's operations in the division's assault, it became evident that progress would be slow and tedious for the Japanese were well-entrenched in rugged terrain.

In spite of this vigorous opposition, the attacking forces advanced steadily as the Japanese fell back from strong point to strong point. During the latter part of April the enemy defense deteriorated more quickly; however, it was not until mid-May that operations in the hills were considered to have passed into the mopping-up stage. Even then, with close support from bombers of the Thirteenth Air Force and guerrilla reinforcements, the 503rd Parachute Regimental Combat Team made but slow progress in the hill mass south of Fabrica. Isolated though they were, the Japanese in that area continued to resist until the cessation of hostilities on 15 August.

Reconnaissance elements of the division meanwhile swung around the north coast of Negros and advanced down the east coast almost to Dumaguete. Here they were met by a combat team of the Americal Division which had landed on 26 April as a part of the *Victor II* operation. In this eastern area no organized enemy resistance was found and, by the end of May, pursuit of these fleeing remnants was taken over by Filipino guerrillas. In the mountains southwest of Dumaguete, United States forces encountered strong Japanese defenses which were not overcome until the second week in June. On 9 June the 503rd Parachute Regimental Combat Team received orders to relieve the troops engaged in Negros Oriental and garrison the island. It thereupon assumed the responsibility for all further operations on Negros.[45]

Liberating Cebu

The veteran Americal Division, while still engaged on Leyte, was given the triple task of liberating Cebu, Negros Oriental, and Bohol. Pre-invasion plans called for an intense five-day air bombardment of Japanese garrison installations and principal coast defenses prior

44 G-2, GHQ, SWPA, *Monthly Summary of Enemy Dispositions,* Mar 45.

45 Op. cit., Apr 45; 40th Div, *Victor I Report;* CO 503rd PRCT, Report to TAG, "Historical Report for the Operation V-1, 7 April–20 June 45, (503rd RCT)" G-3, GHQ, SWPA *Journal,* 20 Jun 45 (S); CO 164th Inf. Report to TAG, "Operations Report, Cebu, Negros, 164th Inf." G-3, GHQ, SWPA *Journal,* 30 Jun 45 (S).

to the scheduled landing near the city of Cebu.[46]

Early on 26 March, following a devastating naval pounding, the Americal Division (less the 164th Regimental Combat Team) under Maj. Gen. William H. Arnold, swept ashore near Talisay, five miles southwest of Cebu City.[47] (Plate No. 99) Although enemy reaction was limited to sporadic small arms and mortar fire, the assault waves ran into one of the most elaborate and effective beach defenses yet encountered in the Philippines. The first troops to hit the beaches were halted abruptly by a system of buried land mines which knocked out most of the landing craft.

The 182nd Regimental Combat Team, which landed nearest Talisay, rapidly breached the mine and barrier defenses. The 132nd Regimental Combat Team, however, was able to continue the advance inland only after considerable delay in clearing paths through the mine fields. The Japanese defensive installations were not confined to the beach area, although it was found that the many well-prepared positions in Talisay had been abandoned as a result of the pre-landing barrage. In addition to planting numerous mines and booby traps, the enemy had constructed many pillboxes and barricades commanding all road junctions and the main highway leading into Cebu City. These defensive installations were unmanned, however, and only intermittent sniper fire opposed the advance. With the assistance of guerrillas, Buhisan Reservoir was seized and the city of Cebu was entered the day following the landing. After almost completely destroying the city the main Japanese forces had withdrawn to positions in the hills overlooking Cebu and the Lahug Airfield. Here the enemy concentrated his forces, pouring down heavy automatic fire from numerous pillboxes.[48]

Lahug Airfield was attacked and quickly seized on 28 March and nearby Mactan Island was taken without opposition the same day. Fanatical resistance was encountered, however, on the high ground above Lahug Airfield. It became necessary to bring up tanks to support the infantrymen in the face of heavy automatic fire from mutually-supporting caves and camou-

46 *Panay—Negros—Cebu Operations*, p. 56; HQ Eighth Army, *Operational Monograph on the Cebu—Bohol—Negros Oriental Operation*, pp. 1 and 19. Hereinafter cited as: Eighth Army, *Cebu—Bohol—Negros Oriental Operation.*

47 The major forces participating in operations against Cebu, Bohol, and Negros Oriental were as follows:

Ground: *Eighth Army—General Robert L. Eichelberger*

Unit	Commander	Phase	Staging Point	Landing Point
Americal Div (–164 RCT)	*Maj. Gen. W. H. Arnold*	*Assault*	*Leyte*	*Talisay Beach*
164th RCT	*Col. W. J. Mahoney*	*Follow-up*	*Leyte*	*Talisay Beach*

Air: *Thirteenth Air Force (supported by elements of the Fifth Air Force and certain Marine Air Groups)—Maj. Gen. P. B. Wurtsmith*

Unit	Commander	Unit	Commander
XIII Fighter Comd	*Brig. Gen. E. W. Barnes*	*Marine Air Gp, Zamboanga*	*Col. C. Jerome*
XIII Bomber Comd	*Col. C. A. Brandt*	*5th Bomb Gp*	*Lt. Col. I. J. Haviland*
42nd Bomb Gp	*Maj. H. C. Harvey*	*307th Bomb Gp*	*Col. C. H. Rees*
403rd Troop Carrier Gp	*Col. H. J. Sands, Jr.*		

Naval: *Seventh Fleet—Adm. T. C. Kinkaid*

Unit	Commander	Unit	Commander
Task Force 78.3.3	*Adm. D. E. Barbey*	*Task Force 78.2.8*	*Adm. D. E. Barbey*
Task Force 74.3	*Adm. D. E. Barbey*		

48 *Panay—Negros—Cebu Operations*, pp. 61–64; Eighth Army, *Cebu—Bohol—Negros Oriental Operation*, pp. 41–46; G-3, GHQ, SWPA, *Monthly Summary of Operations*, Mar 45.

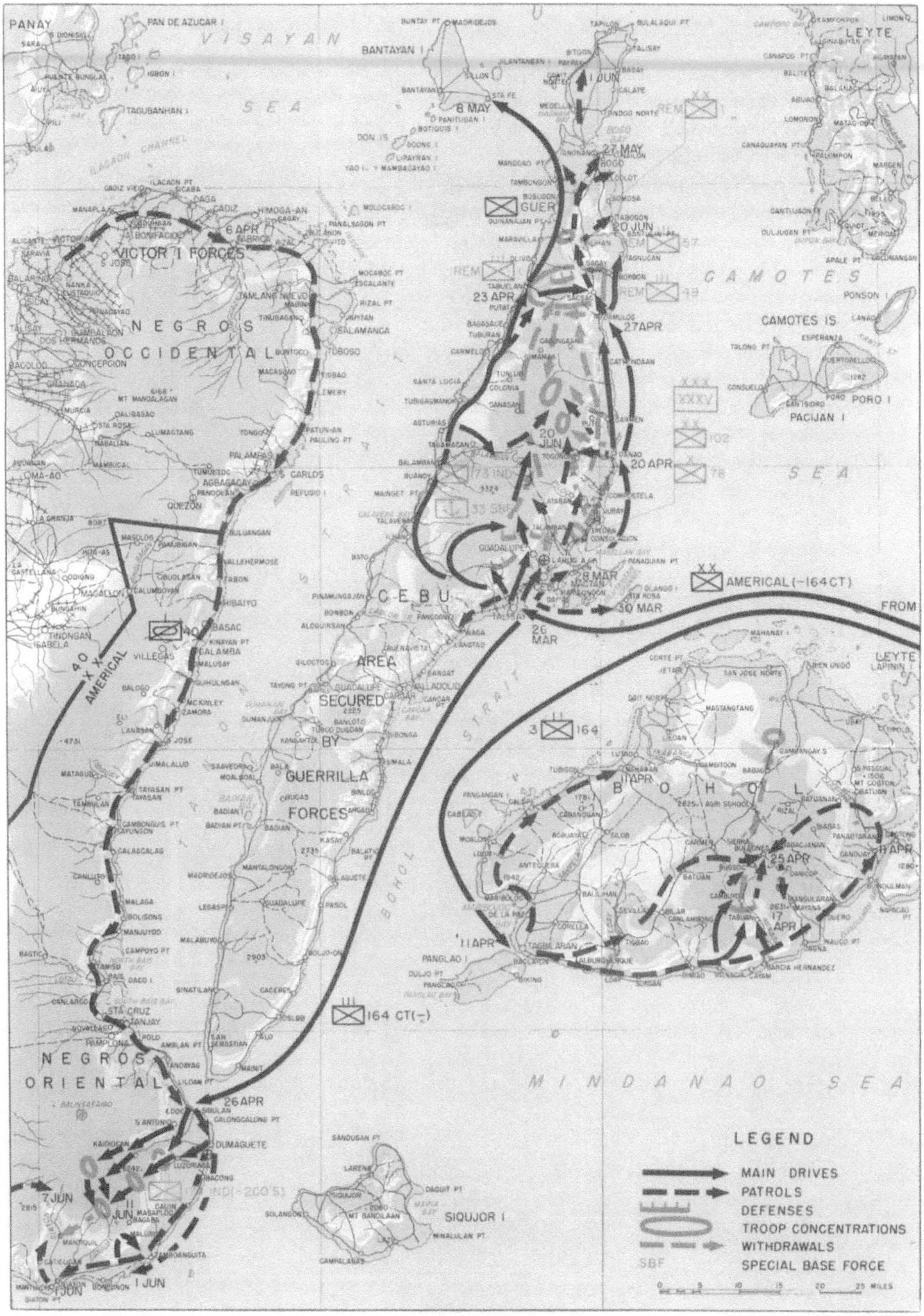

PLATE NO. 99

Landings on Cebu, Bohol, and Negros Oriental, 26 March—20 June 1945 flaged pillboxes. These installations, together with demolition traps, formed an elaborate defense system which was overcome only with the greatest difficulty.[49]

Although the enemy was unable to launch a full-scale counteroffensive, he continually resorted to harassing tactics and local attacks. Infiltration parties frequently penetrated the American defenses and in one case succeeded in destroying an ammunition dump.

In view of the excellent key positions occupied by the Japanese on all principal terrain features, Americal Division forces at first were unable to dislodge them. The 164th Regimental Combat Team, less the 3rd Battalion which was sent to Bohol, was committed on 9 April to bolster the units already employed. Converging in a wide envelopment from the south around the enemy's right flank, the fresh forces joined in a concerted attack on 12 April which soon forced the Japanese to abandon their positions. The principal Japanese defenses in the Cebu City area were overrun by 20 April and the enemy withdrew northward, hoping to arrange for evacuation to Negros. With this action the main fighting on Cebu was finished and the bulk of the enemy forces throughout the Visayas had been overcome. General MacArthur's communique of the following day described the final fighting on Cebu and stressed the significance of the Visayan operations:

On Cebu our troops in a wide enveloping movement which caught the enemy unawares attacked his positions from the west flank and rear and completely defeated him. His losses were very heavy, 5,000 dead being left on the field, with the remnants scattered into the hills to be hunted down by our guerrilla forces. This virtually concludes the Visayan campaign.

Our forces now control the entire Visayan group, comprising an area of more than 33,000 square miles and liberating a population of more than 6,400,000. In addition to the main islands of Leyte, Samar, Mindoro, Panay, Palawan, Cebu, Negros, Bohol and Masbate, Japanese garrisons on a score of lesser islands including Bantayan, Biliran, Burias, Guimaras, Homonhon, Lubang, Poro, Marinduque, Siquijor, Tablas, Tabao and the Sibuyan group have been eliminated. This sweeps clear the center of the Philippines and leaves the only remaining enemy organized resistance in Mindanao on the south and upper Luzon on the north. From the Visayas will come many of the foodstuffs and raw materials for rehabilitation and reconstruction of the Philippines and for prosecution of the war against Japan itself, including hemp, copra, metals, lumber and cement. They already provide important air, land and naval bases for attacks against vital areas in the Netherlands East Indies and Southeastern Asia, and for our increasingly tight blockade of enemy shipping lanes down the coast of China, Indo-China, Thailand and Malaya.

Our losses in this campaign were extraordinarily light due largely to the enemy's continued inability to diagnose our point of attack and to understand our local tactics of combat. His ground forces throughout the Philippine campaign in numerical strength have equalled and in some cases exceeded our own. Yet he has seemed bewildered and confused and although one of the most tenacious of fighters has permitted himself to be constantly surprised, divided and destroyed in detail without being able to inflict more than a minor fraction of compensating losses.

Mopping-up operations on Cebu continued with movements up the coast as the Americal Division blocked Japanese attempts to reorganize or to evacuate the island. Control of the coastal areas forced the enemy to retreat to the hilly interior where guerrillas and American combat patrols continued to hunt down and destroy scattered pockets of Japanese troops. By 20 June the enemy on Cebu

---

49 *Panay—Negros—Cebu Operations*, pp. 64–67; Eighth Army, *Cebu—Bohol—Negros Oriental Operation*, pp. 46–61, G-3, GHQ, SWPA, *Monthly Summary of Operations*, Mar and Apr 45.

50 GHQ, SWPA, Communique No. 1110, 21 Apr 45.

was no longer capable of offering effective resistance. Mopping-up action continued well into August, however, as Americal Division troops criss-crossed the northern part of the island, eliminating small groups of Japanese.[51]

Final Operations in the Visayas—Bohol and Negros Oriental

While the 132nd and 182nd Regimental Combat Teams of the Americal Division were engaged in operations on Cebu, the division received orders to invade Bohol and Panglao Islands and destroy all hostile forces thereon. The 164th Regimental Combat Team, which was en route to Cebu from Leyte, was ordered to divert a battalion to accomplish the mission. Since guerrilla reports indicated that the enemy garrison in the coastal areas had withdrawn to the hills, little or no opposition to the landing was expected.

The landing was made on 11 April at Tagbiliran on the southwestern tip of Bohol Island. Not until the 15th was any contact made with the enemy. A co-ordinated attack was then launched on 17 April against the Japanese in the hilly interior. Despite desperate resistance and attempted counterattacks, the objectives were seized. Some of the enemy forces retreated northward, where subsequent pursuit destroyed the bulk of their strength. By the end of the month, only ineffective delaying action was encountered, and the battalion forces were withdrawn by 7 May. Final reduction of the enemy stragglers was left to the Filipino guerrillas, assisted by a detachment of 50 Americans.[52]

The final phase of the *Victor II* operation began on 26 April when troops of the Americal Division waded ashore north of Dumaguete on Negros Oriental. The landing by the 164th Regimental Combat Team (less one battalion) was completely unopposed. Pushing rapidly south, unhindered except by a few scattered mine fields, the assault forces soon secured Dumaguete and its airfield. The Japanese garrison there had withdrawn after burning part of the town.

In the mountains to the west, however, stiffening enemy resistance developed. Prior to the arrival of the American invasion force, the Japanese had taken advantage of the commanding terrain to create a formidable series of cave and pillbox defenses from which they could be ejected only with great difficulty. Meanwhile, the enemy had an unobstructed view of American movements and could put up a tenacious and effective resistance to delay the progress of the attacking forces.[53]

Reducing the well-entrenched Japanese forces became a slow and tedious process for the two battalions of the 164th RCT. Following their usual custom, the Japanese withdrew from one strong point to another as their positions became precarious. Heavy artillery fire followed by direct infantry assaults in the face of suicidal opposition became the pattern of combat. The American tactics of sustained assaults upon position after position, however, proved effective. The Japanese were forced farther and farther to the south and relentless attacks finally broke organized opposition on 12 June. The remaining scattered enemy groups were left to be hunted down by guerrilla patrols and the 164th RCT was withdrawn.[54]

---

51 G-3, GHQ, SWPA, *Monthly Summary of Operations*, Apr–Aug 45; *Panay—Negros—Cebu Operations*, pp. 67–70.

52 *Cebu—Bohol—Negros Oriental Operations*, pp. 89–94.

53 Ibid. pp. 100–113; *Panay—Negros—Cebu Operations*, pp. 71–72.

54 *Cebu—Bohol—Negros Oriental Operation*, pp. 113–114.

Operation *Victor II* was officially declared closed on 20 June 1945. American control by that date extended over all the major Visayan Islands. The Japanese who still remained in the central Philippines faced the alternatives of surrender or eventual destruction. Only Mindanao remained to be secured.[55]

The Enemy Situation in Mindanao

Mindanao had been the initial target area of the Allied forces moving northwestward up the New Guinea coast toward the Philippines but, with the change in plans which substituted Leyte as the point of invasion, the attack on Mindanao was postponed for almost six months. During this time the Japanese defense plans and troop dispositions were radically changed. Some units were sent to the Leyte battleground and others were shifted to meet new possible directions of attack. By the end of January 1945 the largest Japanese force was being assembled in central Mindanao along the Sayre Highway, running from Macajalar Bay in the north to a junction with Route No. 1 from Davao Gulf in the southeast. Other activity was apparent on the western and southern approaches to the city of Davao and at the southern end of Zamboanga Peninsula.[56] Late in February, the Japanese plan appeared to be to hold the Cotabato—Parang coast of Illana Bay, defend the Davao and Zamboanga areas, and maintain a mobile reserve in central Mindanao. For the rest, Mindanao's lengthy coast lines remained lightly held and guerrillas were in control of large portions of the island.[57]

X Corps Invades Mindanao

Victor V operation against central Mindanao was carried out shortly after the end of organized Japanese resistance on the Zamboanga Peninsula and in the Sulu Archipelago.[58] (Plate No. 100) The assault landing by General Sibert's X Corps was scheduled to take place at Malabang on the eastern edge of Moro Gulf on 17 April. Five days before the landing, however, the guerrilla commander on Mindanao notified General Eichelberger that

55 G-3, GHQ, SWPA, *Monthly Summary of Operations*, Jun 45.

56 G-2, GHQ, SWPA, *Monthly Summary of Enemy Dispositions*, 31 Jan 45.

57 Op. cit. 28 Feb 45.

58 The major forces that participated in the assault against central Mindanao were as follows:

Ground: *Eighth Army—General Robert L. Eichelberger*

Unit	*Commander*	*Phase*	*Staging Point*	*Landing Point*
X Corps	*Maj. Gen. Franklin C. Sibert*			
24th Div	*Maj. Gen. F. A. Irving*	*Assault*	*Mindoro*	*Malabang and Parang*
31st Div	*Maj. Gen. C. A. Morten*	*Follow-up*	*Morotai and Sansapor –22 Apr 45*	
108th RCT	*Lt. Col. M. T. Strata*	*Follow-up*	*Leyte*	*Macajalar Bay*

Air: *Thirteenth Air Force (supported by certain Marine Air Units)—Maj. Gen. P. B. Wurtsmith*

Unit	*Commander*	*Unit*	*Commander*
XIII Bomber Comd	*Col. C. A. Brandt*	*Marine Air Gp 24*	
XIII Fighter Comd	*Brig. Gen. E. W. Barnes*	*Marine Air Gp, Zamboanga*	
5th Bomb Gp	*Lt. Col. I. J. Haviland*	*307th Bomb Gp*	*Col. C. H. Rees*
42nd Bomb Gp	*Maj. H. C. Harvey*		

Naval: *Seventh Fleet—Adm. T. C. Kinkaid*

Unit	*Commander*
Task Force 78.2	*R. Adm. A. G. Noble*

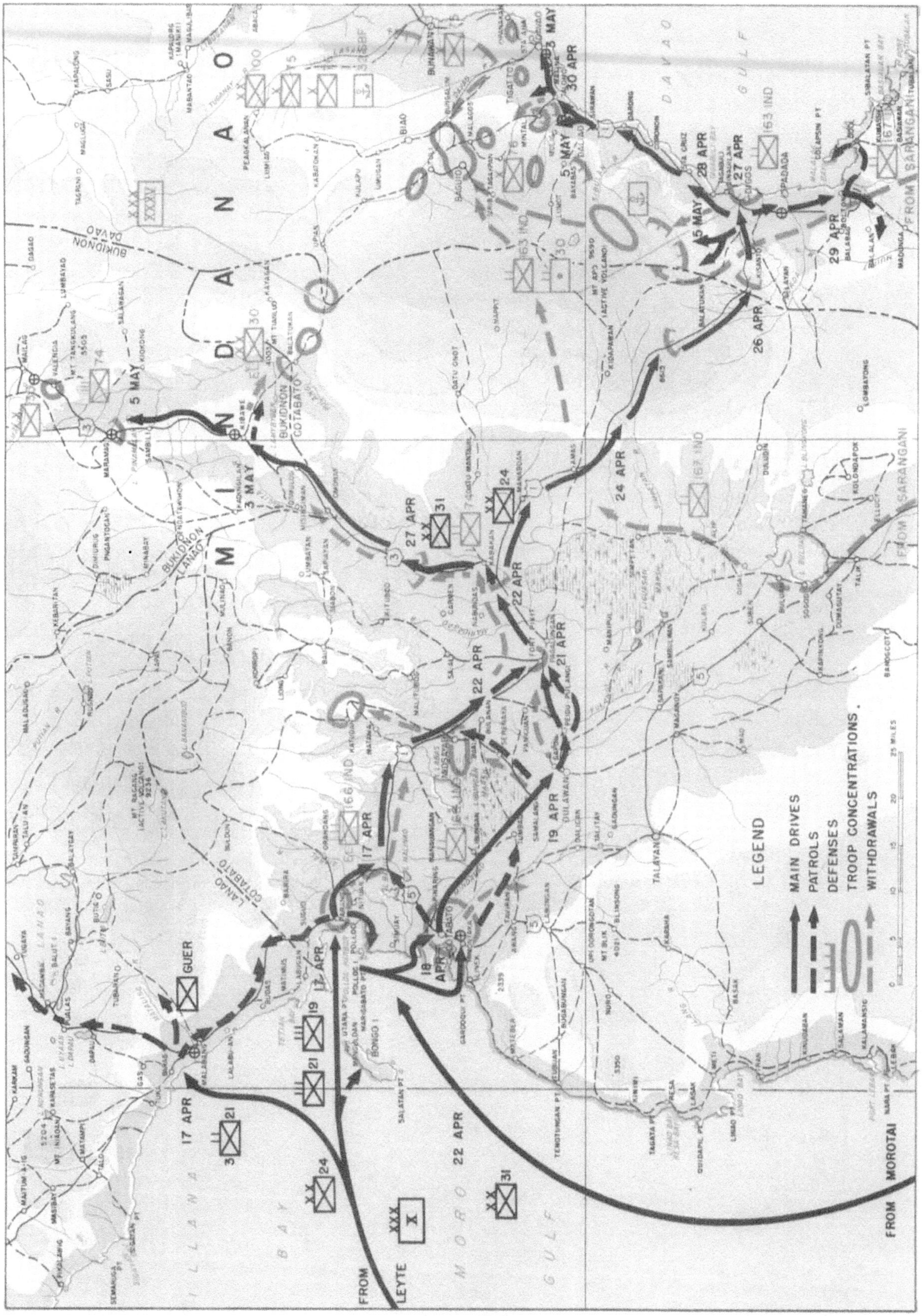

PLATE NO. 100
The Assault against Central Mindanao, 17 April—5 May 1945

the initial objectives—the town and airfield at Malabang—were already in guerrilla hands. He further stated that the Japanese had withdrawn inland, leaving the coast in that area undefended.[59] Receipt of this information caused a change in plans; the main landing was shifted farther south to Parang while only a reduced force was ordered into Malabang.

Staging at Mindoro, western Mindanao, and Leyte, the invasion force, comprising initially the 24th Division reinforced, rendezvoused off Zamboanga on 16 April. Since no enemy opposition was expected in the Malabang area, there was no preliminary bombardment and a battalion of the 21st Regimental Combat Team waded ashore unopposed early the next morning. Malabang was quickly occupied and the airfield immediately put to use. At the same time, other elements landed on nearby Bongo Island, east of Polloc Harbor, and found it unoccupied.[60]

Meanwhile, Parang and Cotabato had been subjected to an intense pre-invasion shelling by cruisers and destroyers of the Mindanao Attack Group under Rear Adm. Albert G. Noble. Several bombing and strafing missions by the supporting Thirteenth Air Force, followed by a climactic rocket barrage, completed the pre-assault bombardment, which evoked no enemy naval or air reaction. Assault elements of the 19th Regimental Combat Team stormed ashore near Parang, again only to find that the enemy had fled.

They drove rapidly inland to secure Parang and the high ground overlooking Polloc Harbor. They then turned southeast along Highway No. 1, toward the vital junction where Highway No. 5 branches off in the direction of Cotabato and Sarangani Bay. Few enemy stragglers were encountered, and the advance was retarded only by the destroyed bridges along the road. By the next morning, the road junction had been secured and the main force headed eastward toward Kabakan in the central part of the island.[61]

On the morning of 18 April, elements of the 21st Infantry set out from Parang in a shore-to-shore movement to the mouth of the Mindanao River with the mission of seizing Cotabato and Tamontaka. Their

59 Lt. Gen. Ryosaku Morozumi, Commanding General, Japanese 30th Division estimated the general situation prior to the landings as follows: "I had anticipated American landings on Mindanao to be in the order of Davao, Cagayan, and Cotabato respectively. Approximately ten days prior to the American landings at Cotabato, the guerrilla units in the area carried out violent activities and surprise attacks on our airfields, etc.; furthermore, air attacks by American aircraft became very intense and more or less gave us a clue to the fact that American landings would be carried out in the very near future. However, we still did not believe that they were going to land at Cotabato, and even when an American task force moved into Cotabato Bay [Polloc Harbor], we had figured that they were merely trying to fake a landing there when they were actually headed towards points further north. . . . Just after the task force moved into Cotabato Bay, troops started to land on shore near Parang and it was at this time that we first learned that they were really going to make landings in that area." Interrogation Files, G-2 Historical Section, GHQ, FEC.

60 *Report of the Commanding General, Eighth Army on the Mindanao Operation*, p. 23. Hereinafter cited as: *Mindanao Operation.* HQ Eighth Army, *Operational Monograph on the Mindanao Operation*, pp. 33, 52–54. Hereinafter cited as: Eighth Army, *Mindanao Monograph.*

61 The following comments on the American invasion of Mindanao are based on interrogations of General Morozumi: "On the 18th of April, following the [American] landing at Parang defense plans were put into effect. The American assault proved so forceful, however, that the 166th Battalion, the only combat unit in the area, experienced a serious setback. Realizing the futility of holding at this point, the Division Commander decided to withdraw the Battalion to Palma [4 kilometers north of Kibawe] so as to protect the rear of the Division. This plan, however, never materialized because of the swiftness of the American advance and the difficult terrain did not allow the necessary time. . . ." 10th Information and Historical Service, HQ Eighth Army, *Staff Study of Japanese Operations on Mindanao Island*, pp. 7–8. Hereinafter cited as: Eighth Army, *Staff Study, Mindanao.*

landings followed heavy air and naval bombardment and were unopposed. After the seizure of the town and the airfield at Cotabato, part of the same force proceeded up the Mindanao River. By this time, the initial mission of the 24th Division—the seizure of the Malabang—Parang—Cotabato area—had been completed, four days ahead of the originally scheduled date.

With the beach area secured, a two-pronged drive across the island to Davao Gulf was begun. A battalion of the 21st Regimental Combat Team started up the Mindanao and Pulangi Rivers by gunboat and landing craft while the 19th Regimental Combat Team began an overland movement along Highway No. 1. The objective was Kabakan, at the junction of the east-west Highway No. 1 and the north-south Sayre Highway (Highway No. 3). Control of the Kabakan area would mean control of the island's road network. It would also prevent the Japanese from concentrating their forces and opposing the advance inland; in addition, it would isolate enemy pockets of resistance and establish a secure supply route into the central area. Finally, it would open the way to Digos and Davao Gulf.[62]

The 34th Regimental Combat Team, which had remained in floating reserve during the first two days following the landing, meanwhile disembarked and moved up to assist the 21st RCT. Since progress up the Mindanao River proved to be more rapid than along Highway No. 1, a battalion of the 34th RCT was assigned to capture Fort Pikit in a double drive, by land and by water, from the town of Paidu Pulangi which had already been reached in the advance up the river. Aided by Navy gunboats the river force landed safely near Fort Pikit joining the rest of the battalion and moving into the old American fort by the afternoon of the 21st. The enemy had again retreated just ahead of the advancing Americans and by the following day, elements of the 34th RCT secured the critical junction of Highway No. 1 and Sayre Highway, just south of Kabakan. The town of Kabakan itself was entered the next morning, after a brief clash with Japanese patrols.

Control of this area placed the X Corps in position to strike north along the Sayre Highway to Malabalay and Macajalar Bay and also southeast to Davao Gulf. By the same token, the main line of communications between the two principle Japanese defense sectors was severed. The 24th Division could now proceed down Highway No. 1 to seize Digos and split Mindanao in half. The 31st Division, which had arrived on Mindanao five days after the 24th Division, was available to proceed north up the Sayre Highway to block the enemy's retreat from the east coast and secure the main north-south supply route. The final drive against the Japanese on Mindanao was thus divided into two separate and distinct actions along the two highways leading out of Kabakan.[63]

Kabakan to Davao

On 24 April the 24th Division, which had advanced 60 miles in seven days, launched its thrust along Highway No. 1 toward Davao Gulf. Although the retreating Japanese offered resistance in the form of light small-arms fire, destroyed bridges, and mines emplaced along the route of approach, the ad-

62 CTG 78.2 COMPHIBGP 8, Comdr Victor V Attack Group (Adm. A. G. Noble) Report to COMINCH, Serial 00145, 22 May 45; CG X Corps Report to TAG, "History of X Corps on Mindanao, 17 Apr 45–30 June 45."

63 Ibid. Eighth Army, *Mindanao Monograph*, pp. 60–74; G-3, GHQ, SWPA, *Monthly Summary of Operations*, Apr 45.

vance was nowhere appreciably retarded.[64] As the leading elements approached Digos, Japanese disorganization was reflected in the large quantities of abandoned equipment and supplies found along the way.

The drive on Digos proceeded rapidly. On 27 April, advance elements of the division crossed the Digos River and entered the town. After a sharp fight enemy positions were soon overrun. Davao Gulf was reached and Padada Airdrome secured on the 28th, making the drive to Davao Gulf one of the most rapid infantry advances of the Pacific War. Approximately 110 miles had been covered in slightly more than ten days after the initial landing at Parang and the front door to Davao was now open.

While mopping up continued around Digos, the 24th Division turned north along Highway No. 1 toward Davao. Resistance was encountered from enemy positions en route but the advance continued. After driving the Japanese from the high ground commanding the approaches to Davao, the city was entered on 3 May. The next day the mines and booby-traps were cleared and the remaining enemy troops were killed or captured. Seizure of Davao, the last major city held by the Japanese in the Philippines, came as a climax to a drive that had covered 145 miles in 15 days. Still remaining, however, was the task of finding and destroying the large enemy force that had withdrawn to the hills to the north and northwest.[65]

Clearing the Sayre Highway

The 31st Division, commanded by Maj. Gen. Clarence A. Martin, had meanwhile swung into high gear for its drive northward along the Sayre Highway. Starting on 27 April, the division directed its initial thrust toward the town of Kibawe, at which point the Talomo Trail, running northwest over the mountains from Davao Gulf, joins the Sayre Highway. Although slowed considerably by destroyed bridges, the division met only scattered delaying action and was able to capture Kibawe and its airstrip by 3 May.[66] With this mission accomplished, the lines of retreat utilized by the enemy to evacuate his troops from the Davao Gulf area to central Mindanao were blocked and his forces in the Davao sector were effectively isolated.

The 31st Division continued to advance along the Sayre Highway from Kibawe concurrently with the 24th Division's drive on the defense above Davao. (Plate No. 101) The enemy fought tenaciously to permit movement of his remaining combat units into the Malaybalay area. By 10 May, however, elements of the division had reached Maramag, secured its airstrip, and continued the drive to the north. On the 15th, the Japanese made a desperate banzai attack in the vicinity of the airstrip, but it was successfully repulsed. This defeat ended

64 The following statements concerning the Japanese delaying tactics during this retreat to Digos are based on an account of Gen. Jiro Harada, Commanding General, Japanese 10th Division: "To delay the rapid advance, the division commander ordered the 163rd Independent Infantry Battalion to destroy the bridges on the Digos—Kabakan road. The completion of this mission was reported to General Jiro Harada, but this action did not delay the Americans. They had quickly surmounted these obstacles, either repairing or rebuilding the bridges, and were continuing the headlong advance. In a last desperate attempt to hold the enemy long enough to permit the Japanese garrison in Davao to effect an orderly withdrawal the division commander directed the Digos troops to make a stand along the Digos—Davao Road. This attempt to delay also proved futile." Eighth Army, *Staff Study, Mindanao*, p. 14.

65 G-3, GHQ, SWPA, *Monthly Summary of Operations*, Apr, May 45.

66 The following comment is based on an account by Lt. Gen. Morozumi: "The destruction of bridges coupled with our harassing actions along the [Sayre] highway did not delay the [American] advance as much as was anticipated. The Americans' ability to keep up with our rapid withdrawal caused considerable amazement and consternation." Eighth Army, *Staff Study, Mindanao*, p. 8.

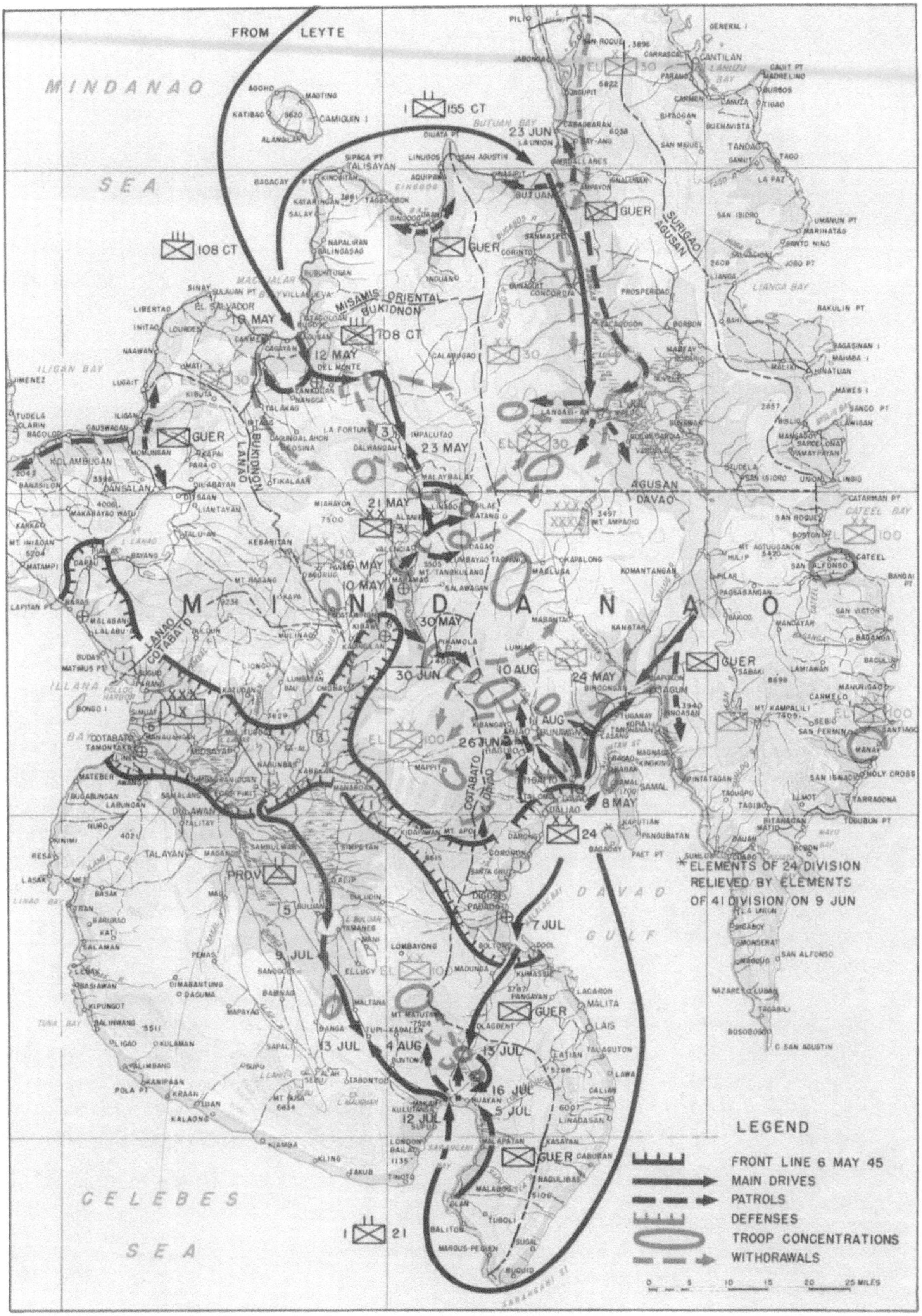

PLATE NO. 101

Final Operations on Mindanao, 6 May—11 August 1945

the last major offensive action by the Japanese along the Sayre Highway and from then on torrential rains became the main deterrent to the advance. Valencia fell on 16 May and Malaybalay, capital of Bukidnon Province, was seized on the 21st.[67]

The Landing at Macajalar Bay

Late in April, it became apparent that the enemy was planning to consolidate for a final stand in the hills northwest of Davao. General Eichelberger therefore decided to land a regimental combat team at the rear of the enemy in the Macajalar Bay area of northern Mindanao. This force would then drive down the Sayre Highway to meet the 31st Division advancing from the south.

The 108th Regimental Combat Team of the 40th Division from Leyte was selected for the operation. It landed unopposed, near Bugo on 10 May. The Japanese had withdrawn from the area and the beach was already in the hands of guerrillas who had also captured the nearby town of Cagayan. The entire northeast coast was reported cleared of the enemy.

The drive toward Del Monte Airdrome was only lightly opposed, and the airfield was secured on 12 May. The following day, stiffening opposition was encountered at the entrance to a canyon below Del Monte where the enemy had set up strong, well-placed defenses. Reduction of this strong point was completed by 18 May, however, and the remaining Japanese troops were overcome within the next few days. The meeting between the 108th Regimental Combat Team and the 155th Regimental Combat Team of the 31st Division took place just outside Impalutao on 23 May. The juncture of the two forces marked the end of Japanese resistance along the Sayre Highway.[68]

Reduction of Enemy Forces in East-Central Mindanao

With the Sayre Highway opened and the Davao area liberated, the X Corps' campaign was strategically ended. The fighting, however, was not yet over. There remained the task of ousting the Japanese from the mountain areas to which they had fled. As usual, the enemy held fanatically to his hill positions, especially those east of the Sayre Highway near Malaybalay. In that area, elements of both the 108th RCT and the 31st Division continued to send out strong patrols to the east and west throughout May and June.

Late in June a battalion of the 31st Division moved by water to the shores of Butuan Bay and advanced southward up the Agusan River toward Waloe. The objective of this maneuver was to deny the valley to remnants of the Japanese forces east of the Sayre Highway and to destroy any of the enemy who might be moving south through the Agusan River Valley from the Surigao Peninsula to join the groups in that area. By 1 July, the battalion had reached Waloe and had overcome all organized resistance. The remaining enemy groups fled to the hills. Operations against these pockets in the area between Malaybalay and Waloe were still under way on 15 August, the date of the Japanese acceptance of the terms of the Potsdam Declaration.[69]

In the Davao area the Japanese clung stubbornly to their positions near the western terminus of the Talomo Trail. The battle for the ground on the flank of the 24th Division's

67 G-3, GHQ, SWPA, *Monthly Summary of Operations*, May 45; Eighth Army, *Mindanao Monograph*, pp. 231–260.

68 Ibid., pp. 260–263.

69 *Mindanao Operation*, pp. 70–74; Eighth Army, *Mindanao Monograph*, pp. 263–269, 274–276.

route to Davao raged throughout May and into June. Elements of the 41st Division moved from Zamboanga to enter the struggle and on 9 June relieved battle-weary units of the 24th Division. Not until the middle of June, however, did the enemy's defenses along the lower reaches of the Talomo Trail begin to disintegrate. From that time on, the fighting developed into the usual, protracted mopping-up operations. The advance along the trail continued on to Kibangay, which was captured on 26 June. Beyond this point the road became impassable and showed no signs of enemy fortifications. At the other end of the trail, elements of the 31st Division instituted a drive southeastward from Kibawe. Entrenched Japanese slowed the progress of the advancing forces, but by 30 May, they secured a bridgehead across the Pulangi River. From there the drive continued slowly to the village of Pinamola which was taken on 30 June. During the ensuing weeks the drives from both ends of the Talomo Trail encountered stubborn resistance which continued until the end of hostilities.[70]

Final Operations on Mindanao

On 30 June, when the enemy had been driven out of both Kibangay and Pinamola on the Talomo Trail and had been forced to flee into the mountains east of Malaybalay, the Mindanao Operation was officially declared closed. Action after that date was confined mainly to patrols, but an additional landing operation proved necessary to complete the destruction of enemy units on the island. Sarangani Bay in southern Mindanao, which had been considered in the early planning phase of the Philippine Campaign as the initial landing spot for the return of General MacArthur to the Philippines, remained as the last point to be taken. Planning for the reduction of this area began early in July, with the 24th Division designated as the landing force. The scheme of maneuver called for a three-pronged offensive, two forces proceeding by overland routes, while a third made an amphibious landing on the north shore of Sarangani Bay. By this strategy the enemy was to be hemmed in by concurrent offensives from the south, the northeast, and the northwest.

An Expeditionary Battalion of Filipino troops was landed at Malalag Bay on 7 July and began its march overland toward Buayan the following morning. It encountered no organized resistance and arrived at Buayan on the 16th. The amphibious landing to the west of Buayan by a battalion combat team of the 24th Division was made on 12 July. Opposition was lacking since the area had already been secured by guerrilla forces. Meanwhile, a battalion of provisional antiaircraft personnel serving as infantry had departed from Fort Pikit by landing craft to Lake Buluan and had proceeded overland toward Tupi where, on 13 July, it met advance elements of the battalion striking northwestward from its landing beach of the previous day. Vigorous and extensive patrolling was carried out but, except for occasional skirmishes, there was little resistance. The enemy retired northward into the mountains where it required several weeks to rout him out of his positions. By 11 August, however, the mop-up of the Sarangani Bay area was considered terminated and the operation declared at an end.[71]

End of the Philippine Campaign

With the completion of the Mindanao Operation the mission of the Eighth Army in the

70 *Mindanao Operation,* pp. 62–68; Eighth Army, *Mindanao Monograph,* pp. 181–209, 269–274.

71 Ibid., pp. 290–301. *Mindanao Operation,* pp. 75–78.

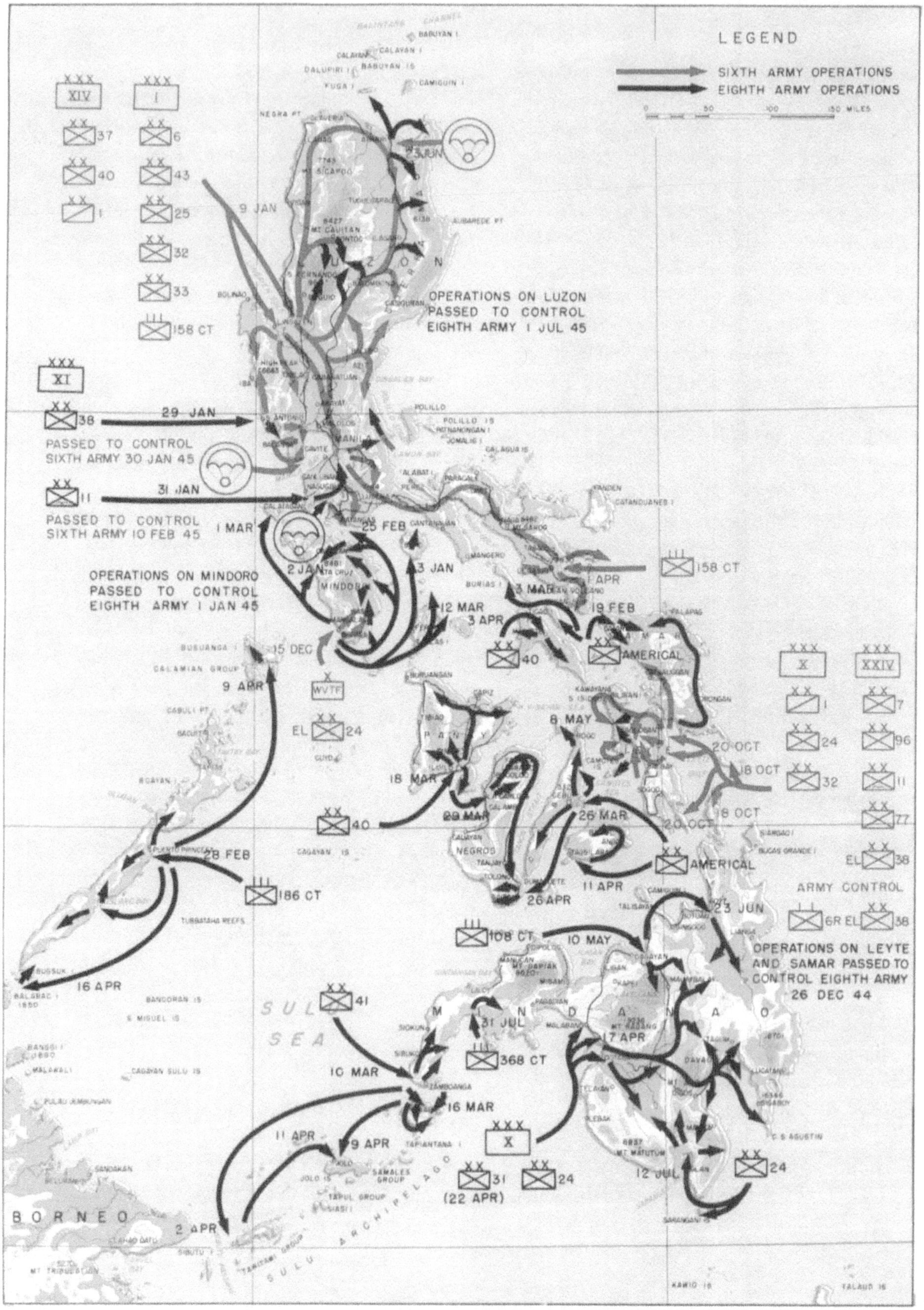

PLATE NO. 102

The Philippine Campaign, October 1944—July 1945

southern Philippines was ended. General Eichelberger's forces in a series of simultaneous and rapid offensives over a large area of diverse terrain conditions had destroyed the enemy and left his decimated troops scattered in disorganized bands throughout the hills and jungles of the archipelago. All that remained was to accept the surrender of the helpless Japanese remnants. Now that the arduous task of the Eighth Army in the Philippines was finished, a new job lay before it—a part in the occupation of the Japanese Home Islands.

In a special communique to the press from his headquarters in Manila, General MacArthur had announced the official end of the Philippine Campaign in the following words:

The entire Philippine Islands are now liberated and the Philippine Campaign can be regarded as virtually closed. [Plate No. 102] Some minor isolated action of a guerrilla nature in the practically uninhabited mountain ranges may occasionally persist but this great land mass of 115,600 square miles with a population of 17,000,000 is now freed of the invader.

The enemy during the operations employed twenty-three divisions, all of which were practically annihilated. Our forces comprised seventeen divisions. This was one of the rare instances when in a long campaign a ground force superior in numbers was entirely destroyed by a numerically inferior opponent.

The Japanese ground forces comprised the following divisions or equivalents: 1st, 8th, 10th, 16th, 19th, 23rd, 26th, 30th, 100th, 102d, 103d, 105th, 2d Armored, 2d Airborne Brigade (reinforced to divisional strength), the 54th, 55th, 58th, 61st and 68th Independent Mixed Brigades (reinforced to divisional strength), three divisional units known as the Kobayashi, Suzuki and Shimbu commands, organized from twenty-eight independent battalions, three naval divisions comprising a Provisional Naval Command of corps strength, under Admirals Iwabuchi and Shiroya, and a large number of base and service elements. The total strength approximated 450,000 men. [Chart]

Naval and air forces shared equally with the ground troops in accomplishing the success of the campaign. Naval battles reduced the Japanese Navy to practical impotence and the air losses running into many thousands have seriously crippled his air potential. Working in complete unison the three services inflicted the greatest disaster ever sustained by Japanese arms.

The objects of the campaign were as follows:

1. To penetrate and pierce the enemy's center so as to divide him into north and south, his homeland to the north, his captured Pacific possessions to the south. Each half could then be enveloped and attacked in turn.

2. The acquisition of a great land, sea and air base for future operations both to the north and to the south comparable to the British Islands in its use as a base for allied operations from the west against Germany.

3. The establishment of a great strangulating air and sea blockade between Japan and the conquered possessions in the Pacific to the south so as to prevent raw materials from being sent to the north and supply or reinforcement to the south.

4. The liberation of the Philippines with the consequent collapse of the enemy's imperial concept of a Greater East Asia Co-Prosperity Sphere and the reintroduction of democracy in the Far East.

5. The liberation of our captured officers and men and of internees held in the Philippines.

6. A crippling blow to the Japanese Army, Navy, and Air force.

All of these purposes were accomplished.[72]

In contrast to the New Guinea Campaign which had required two long years of uphill battle, the Philippine Campaign had been won in nine months of rapid operations. During this short period American combined arms had routed the strong and hard-fighting Japanese Fourteenth Area Army from its prepared defenses and broken its power as a military force. The issue had been decided

72 GHQ, SWPA, Communique No. 1185, 5 Jul 45.

JAPANESE GROUND FORCES IN THE PHILIPPINES
18 October 1944 to the End of the War

1. ARMY GROUND FORCES

A. Present on 18 October 1944:

Southern Army GHQ	3,204
14th Area Army	
Headquarters	9,544
8th Division	18,496
26th Division	13,726
103rd Division	16,491
105th Division	16,101
2nd Tank Division	8,948
58th Independent Mixed Brigade	10,761
61st Independent Mixed Brigade	4,855
L of C and Service Personnel	55,760
Other Units	35,043
35th Army	
Headquarters	842
16th Division	20,311
30th Division	17,747
100th Division	14,960
102nd Division	15,518
54th Independent Mixed Brigade	5,210
55th Independent Mixed Brigade	3,900
L of C and Service Personnel	11,037
Other Units	5,574
TOTAL	288,028

B. Arriving Subsequent to 18 October 1944:

1st Division	13,058
10th Division	14,602
19th Division	9,380
33rd Division	13,503
68th Independent Mixed Brigade	6,707
Locally Organized Units	20,836
Other Units	26,011
TOTAL	104,097

2. NAVAL FORCES ASHORE (Exclusive of Air Force Personnel)

A. Present on 18 October 1944:

Northern Philippines	27,683
Central Philippines	5,570
Southern Philippines	16,761
TOTAL	50,014

B. Arriving Subsequent to 18 October 1944:

Northern Philippines	16,884
Central Philippines	1,306
Southern Philippines	19,624
TOTAL	37,814

3. AIR FORCE BASE AND SERVICE PERSONNEL

A. Present on 18 October 1944:

Manila Army Air Depot	11,104
Other Units	38,745
TOTAL	49,849

B. (No later reinforcements received)

GRAND TOTAL	529,802

SOURCE: Japanese First and Second Demobilization Bureaus Reports, "Strength of Japanese Forces in the Philippines, 1944–1945."

early in the campaign when the best of Japan's army, navy, and air force had been crushed at Leyte.[73] The invasion of Luzon a few months later climaxed the Philippine Campaign and brought final and utter defeat to General Yamashita's divisions. The Eighth Army's widespread series of operations covering the Visayas, Palawan, the Sulu Archipelago, and Mindanao destroyed the remainder of Japan's once powerful Thirty-fifth Army and completed the liberation of the Philippine Archipelago. By the summer of 1945 every major island and every important strategic area was freed from the control of the enemy. General MacArthur had redeemed his promise and the people of the United States their pledge to drive out the Japanese invader. The Philippines were again free.

Victory in the Philippines had been achieved by the joint efforts af all forces involved. The intensity of the fighting had varied from area to area according to terrain conditions and the state of the Japanese defenses but in all operations the smooth co-ordination of United States land, sea, and air forces had been the keynote of the American success. The reaction of Japanese Fourteenth Area Army staff officers to the part played by the component American forces was a tribute to this strategy of three-dimensional warfare. Maj. Gen. Haruo Konuma, Deputy Chief of Staff, Fourteenth Area Army, commented on the combined use of the three branches of the United States forces in the following words: "The U. S. Navy and American air power together made important contributions to American victory in the Philippines by their protection of landing convoys. Once ashore, moreover, the landing forces were given excellent protection by the supporting naval and air elements. The Air Force and the Navy contributed greatly to the success of the ground forces during and following the initial assaults, and then continued to keep open the routes to supply the land forces.... The basic reason for American victory in the Philippines was America's ability to concentrate and maintain the necessary men and materials in the front line while at the same time cutting the Japanese lines of communication. The well co-ordinated action of their land, sea, and air forces was also a substantial factor in achieving victory."[74]

Speaking of the U. S. Navy, Col. Shujiro Kobayashi, Tactical Staff Officer said: "Through the action of submarines and carrier-based Grummans, Japanese supply was effectively cut off in the Philippines. The naval big guns also blasted beachheads before invasions, forcing our troops into the hills. Indeed the American Navy seemed everywhere at once and powerful."[75]

Maj. Masaaki Kawase, Supply Staff Officer of the Fourteenth Area Army, indicated how American air power had disrupted Japanese operations during the Philippine Campaign when he stated: "I believe that American

---

73 Even the average Japanese soldier in the field realized that the fall of Leyte was the beginning of the end in the Philippines, as shown by the following statement of Col. Shujiro Kobayashi, Tactical Staff Officer, Fourteenth Area Army: "Until almost the end of the Leyte campaign, morale was indeed high as the men and officers fully expected that the remnants of the invasion fleet would be annihilated in Leyte. After this campaign resulted in complete failure for the Japanese and it appeared to be only a prelude to the invasion of Luzon, morale while still high was that rather of men doomed eventually to extinction but willing to take as great a toll as possible before the end. By the time the Luzon assault was under way, morale had fallen very low, due to the fact that there no longer seemed any hope of communication with Japan. Isolated and starving, the troops had lost almost all will to continue." Interrogation Files, G-2 Historical Section, GHQ, FEC.

74 Interrogation Files, G-2 Historical Section, GHQ, FEC.

75 Ibid.

airpower had a great deal to do with their victory in the Philippines and it was the decisive factor in their success. The American pre-invasion air attacks on the Philippines had a marked effect in neutralizing our defense preparations. Damage caused by their frequent air attacks on airfield installations, aircraft, ships, etc. was tremendous. When our convoy ships were struck at Manila Bay, we lost 200,000 tons of war materials. At the time of the American invasion of Leyte, our air force was literally helpless and failed to conduct effective air operations against enemy convoy ships. Supremacy in the air and on the sea was quickly attained by the enemy in the Visayan area, thus making it impossible for us to reinforce and supply the Thirty-fifth Army in Leyte at an opportune time to meet the invading forces. In Luzon we had been always overwhelmed by the enemy air attacks and our action was limited during the day. Above all, our supply lines were completely cut off and with food and materiel shortages we could not display but a fraction of the strength deployed on Luzon."[76]

A third Staff Officer of the Japanese Fourteenth Area Army, Lt. Col. Yorio Ishikawa, had the following comments to make with regard to the American ground forces' conduct of the campaign : "The United States forces with their superior equipment and techniques made surprise moves by overcoming terrain and inclement weather believed humanly impossible. For instance, the United States forces made landings on Leyte when a severe typhoon prevented our patrol planes from operating. In another instance, United States forces advanced through the difficult terrain near Baguio by constructing roads which we did not believe they could do."[77] To these comments, Colonel Kobayashi added : "Generally, American leadership and planning were superb. To mention the outstandingly good points, locations for operations were well chosen, and landings were expertly timed. In addition, the Americans had succeeded in shutting off our supply lines from Formosa and French Indo-China. Through the American-supported Filipino guerrillas, our supply lines within the Philippines were cut. The guerrillas also transmitted intelligence which led to effective bombing of our positions and supply dumps."[78] In over-all appraisal of General MacArthur's conduct of the Philippine Campaign, Lt. Gen. Shuichi Miyazaki, Chief of Operations Bureau, Imperial General Headquarters, stated : "Strategic plans, strategic preparations, operational decisions—these were splendid. I came to the decision in Tokyo that the combined use of air, ground and naval forces and in general all war plans involving the co-operation of these three together were especially notable for their success."[79]

The Philippine Campaign proved the advantages of General MacArthur's adherence to a master plan which, though broad in scope, was fluid in execution and sufficiently flexible to cover operations from Port Moresby to Manila. His central plan permitted rapid intermediate strokes along its stupendous axis, as in the cases of the Admiralties operation, Hollandia, and even Leyte itself. The Allies had carefully selected definitive and progressive objectives and concentrated their power in each co-ordinated assault. In comparison, Japanese planning was extemporized and spas-

76 Ibid.
77 Ibid.
78 Ibid.
79 Ibid.

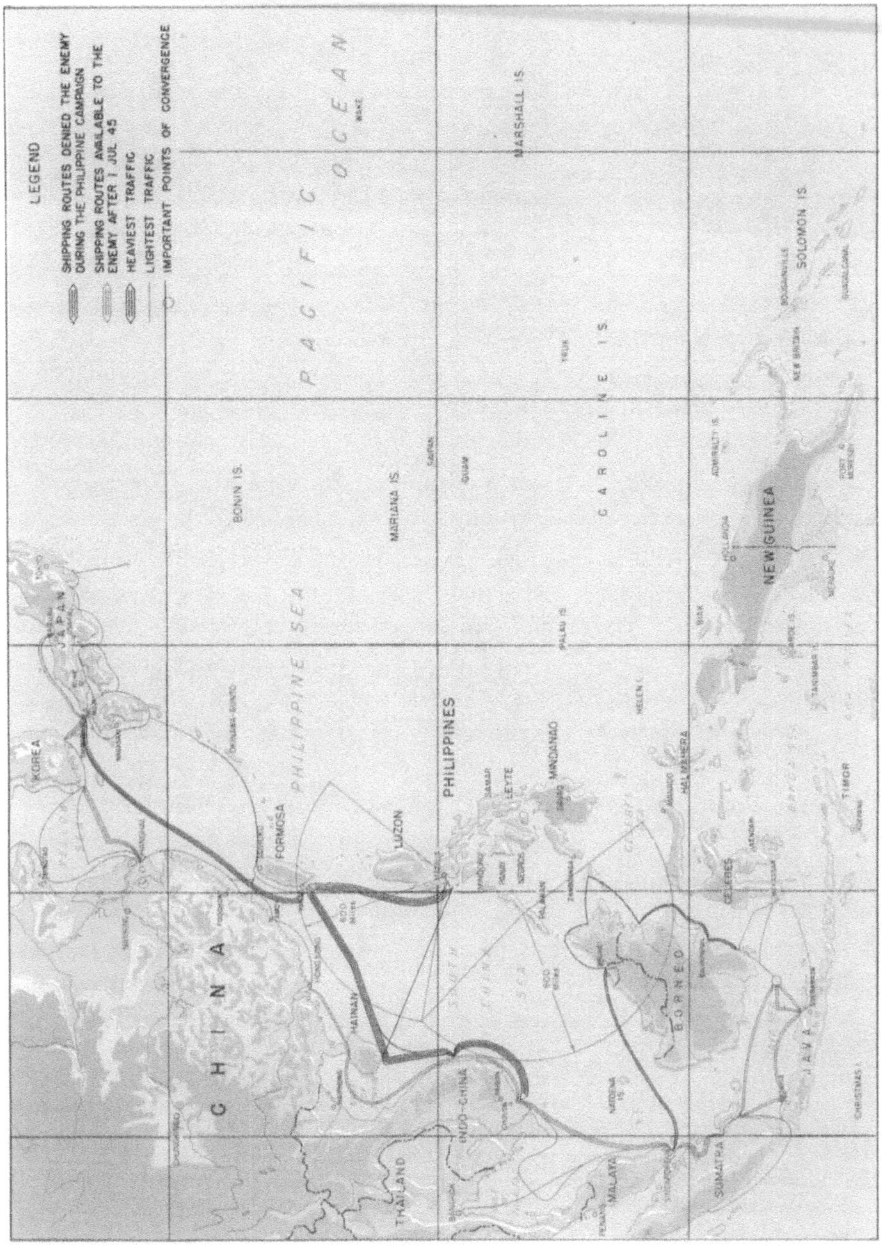

PLATE NO. 103

Enemy Shipping Routes Destroyed during the Philippine Campaign modic, and their leadership often vacillating.[80] Their last-minute decision to alter all previously prepared plans and force a decision on Leyte instead of Luzon, for example, exemplified the makeshift character of Japan's military strategy in the later stages of the war.

On the other hand, the long planning and brilliant strategy of General MacArthur and his staff which led to the liberation of the Philippines were well rewarded by the great destruction and enormous losses inflicted on the enemy on Leyte, Luzon, and the other islands of the central and southern Philippines. The choice of this archipelago as the decisive battleground of the Pacific War had ensured American forces the advantages of a friendly native population and familiar terrain. The careful consideration of such factors was characteristic of Allied planning in the Southwest Pacific Area and one of the chief reasons for the unbroken succession of victories achieved.

Defeat in the Philippines confirmed the worst fears of the Japanese leaders. The Japanese empire of conquest was cut in two and the volume of sea traffic between Japan and the southern regions was reduced to a mere trickle. (Plate No. 103) The Netherland East Indies and Malaya—the Eldorado to the south which had lured Japan's leaders into war in 1941, and which for four years had supplied much of her resources—were now lost. The words of Adm. Mitsumasa Yonai, Navy Minister in the Koiso and Suzuki Cabinets, July 1944—August 1945, were unmistakably clear: "When you took the Philippines, that was the end of resources, in cutting off the southern supplies."[81]

With access to the southern regions cut off and with the ever-growing wall of the Allied sea and air blockade virtually impenetrable, Japan's time was fast running out. Her stock piles of raw materials to feed her hungry war machine were almost exhausted. Japan's security in space was a thing of the past. The great ally of geography which had separated the Japanese Homeland from the Allied forces and had proved such a tangible obstacle to the Allied prosecution of the war in the Pacific was now practically overcome. Distance was no longer a protector of Japan's factories or population and the approach of imminent peril could no longer be concealed from her people. The Philippines would soon become a powerful new base for the final blows to end the war. Once the cornerstone of Japan's military conquests southward and a strong bulwark in her pattern of defense, the Philippines now presented a close and dangerous threat to the Japanese nation itself.

---

80 General Miyazaki, Chief of Operations Bureau, Imperial General Headquarters, analyzed the reasons for the Japanese defeat in the Philippines as follows: "In the final analysis I believe that success and failure were accounted for by the absolute disparity between the Japanese and American fighting power and the qualitative value of the defense against attack. . . . Historically speaking, the greatest errors that the Japanese Army committed were probably the following: (1) The Japanese Army collapsed at the very front in operations in the southern regions, but then lacked interest and force to make thorough and serious preparations for the American attacks which were due to come. They failed to utilize their precious time. (2) In the Philippines the headquarters of General Terauchi and of General Yamashita, or rather his successor, were set up parallel, and then when the islands were about to be invaded by the American forces General Terauchi's headquarters were moved to Saigon. This invited any number of inconveniences and deficiencies as far as the division of responsibility of the two headquarters and the shift in command of the subordinate units were concerned. (3) As soon as the invasion of Leyte was known, the defense plans for all the Philippines were changed. Ground units were taken from Luzon and committed to battle in Leyte. As soon as the battle in Leyte had proved to be a failure, the defense preparations in Luzon also automatically failed. Just before the American troops were about to land on Luzon, the general defense plans for Luzon were all completely changed. Disposition of units was entirely changed but execution of movement was almost impossible due to the fact that air superiority was in the hands of the American air force. For this reason, there was not enough time before the American landings to prepare defenses." Interrogation Files, G-2 Historical Section, GHQ, FEC.

81 USSBS, *Interrogation of Japanese Officials*, Vol II, p. 330.

CHAPTER XII

FINAL SWPA OPERATIONS AND ORGANIZATION OF AFPAC

Changing Concepts

With his conquest of the Philippines, General MacArthur had thrust a solid wedge deep into the heart of Japan's war-acquired empire. His battling forces in the Pacific had advanced more than 3,000 miles through enemy-controlled sea and land areas without a single defeat.

Until mid-summer of 1944, the strategy of the Joint Chiefs of Staff in the war against Japan was designed to accomplish two general objectives. On the one hand, forces of the Southwest Pacific and the Pacific Ocean Areas would sever the enemy's lines of communications with the Netherlands East Indies and China, interdict the waters of Japan's Home Islands, and secure advance positions from which the invasion of the Japanese mainland could be launched should such a step "prove to be necessary." The other and equally important objective was to obtain bases from which the long-range bombers of the Twentieth Air Force could profitably strike important targets within the inner defense zone of the enemy's island possessions.[1]

In July 1944, the unbroken series of victories of the Allied forces and the increasing potential of the United States made it necessary to redefine and restate these over-all objectives in terms of the current situation. At that time, it was generally felt that, although it was possible to defeat Japan by sustained aerial bombing and the destruction of her sea and air forces, such methods would involve an unacceptable delay.

In view of the marked deterioration of Japanese resistance, the increasing superiority of Allied ground, air, and sea power, and the prospect of augmenting Allied strength in the Pacific after the surrender of Germany, it was concluded that the quickest and most effective way to win the Pacific War would be to seize the industrial heart of Japan by amphibious invasion. The actual landings, of course, were to be preceded by air and sea blockades, by intensive aerial bombardment, and by continued assaults against the remaining elements of the Japanese Navy and Air Forces.[2] This concept became the guiding principle for the final year of the war.

During the early months of 1945, Japan's economic and military power was crumbling rapidly under the combined weight of the Allied offensives. The Japanese still had many divisions in the Southwest Pacific and Pacific Ocean Areas but the effectiveness of these forces was neutralized by blockade, isolation, and fast-dwindling supplies.

In the Southwest Pacific Area, eight Japanese armies had been either defeated or rendered powerless to conduct more than delaying actions. In the New Guinea—Solomons region, the Japanese Second, Seventeenth, and Eighteenth Armies had been crushed. In the Philippines, the Japanese Fourteenth Area and Thirty-fifth Armies were being rapidly annihi-

1 JCS Memo for CCS, 11 Jul 44, CCS 417/3, G-3, GHQ Ex JCS and CCS Papers 4 (TS).

2 Ibid.

lated by the advancing forces of Generals Krueger and Eichelberger. In the Borneo—Celebes area, the Japanese Sixteenth, Nineteenth, and Thirty-seventh Armies were helplessly cut off and constituted no threat to General MacArthur's drive toward Japan.

In the Pacific Ocean Areas, the penetration of Japan's outer defense was completed with the seizure and occupation of Saipan, Guam, and Tinian in the Marianas. Operations had been launched in February to reduce the island of Iwo Jima and establish it as a base for fighter planes to support the Marianas-based B-29 bombers. With the capture of this important island in the Bonins a month later, the United States gained possession of a strategic military position only 750 miles from Tokyo and pushed its forces far into Japan's inner line of island fortifications. They would be pushed even farther with the projected assaults on Okinawa and the Ryukyus, the next operations scheduled for the Pacific Ocean Areas.

On the continent of Asia, the picture was no less grim for Japan. In Burma, Allied forces of the Southeast Asia Command, under Lord Admiral Louis Mountbatten, had the Japanese in full retreat by the end of January. The enemy's supply lines to the Asiatic coast had been cut by General MacArthur's successive landings in the Philippines and U. S. Navy operations in the China Seas. With the capture of the key port city of Rangoon on 3 May, the Burma campaign was virtually completed. Some enemy units were able to retreat eastward into Thailand and eastern Burma, but thousands of Japanese troops remained penned in isolated pockets without chance of escape.

In China, although the initiative for the most part still rested with the Japanese, their offensive power had fallen off considerably. Newly trained and re-equipped Chinese forces had begun to counterattack vigorously with encouraging results. In addition, the Japanese were becoming increasingly concerned about the safety of their Home Islands and were carrying out a general withdrawal of their troops from southern and central China in an effort to strengthen their positions along the eastern Asiatic coast.

These millions of enemy troops in Asia and in the islands of the Pacific could never contract their lines to keep pace with the ever-narrowing arena of conflict. They were unable to conduct an orderly retreat in classic fashion to fall back on inner perimeters with forces intact for a last defense of Japan's main islands. It was a situation unique in modern warfare. Never had such large numbers of troops been so effectively outmaneuvered, separated from each other, and left tactically impotent to take an active part in the final battle for their Homeland. (Plate No. 104)

From his strategic position in Manila, General MacArthur began planning for the final invasion of Japan itself in accordance with the directives of the Joint Chiefs of Staff. In the absence of any indications to the contrary, it appeared that the command structure which had existed throughout the war in the Pacific would continue unchanged. General MacArthur believed, however, that the Pacific Command required considerable revision if ultimate victory were not to be needlessly jeopardized.

In December 1944, he expressed his views in a special radio to the War Department and stressed the necessity for reorganization. In his opinion, the command had to be greatly simplified and the completely arbitrary area boundaries, which originally had been conceived for defensive purposes, had to be abolished altogether.[3] As matters stood, the U.S. Army ground, air, and service forces were unevenly

3 CINCSWPA Radio No. C-55018 to WARCOS, 17 Dec 44, C/S GHQ, WD No. 890 (TS).

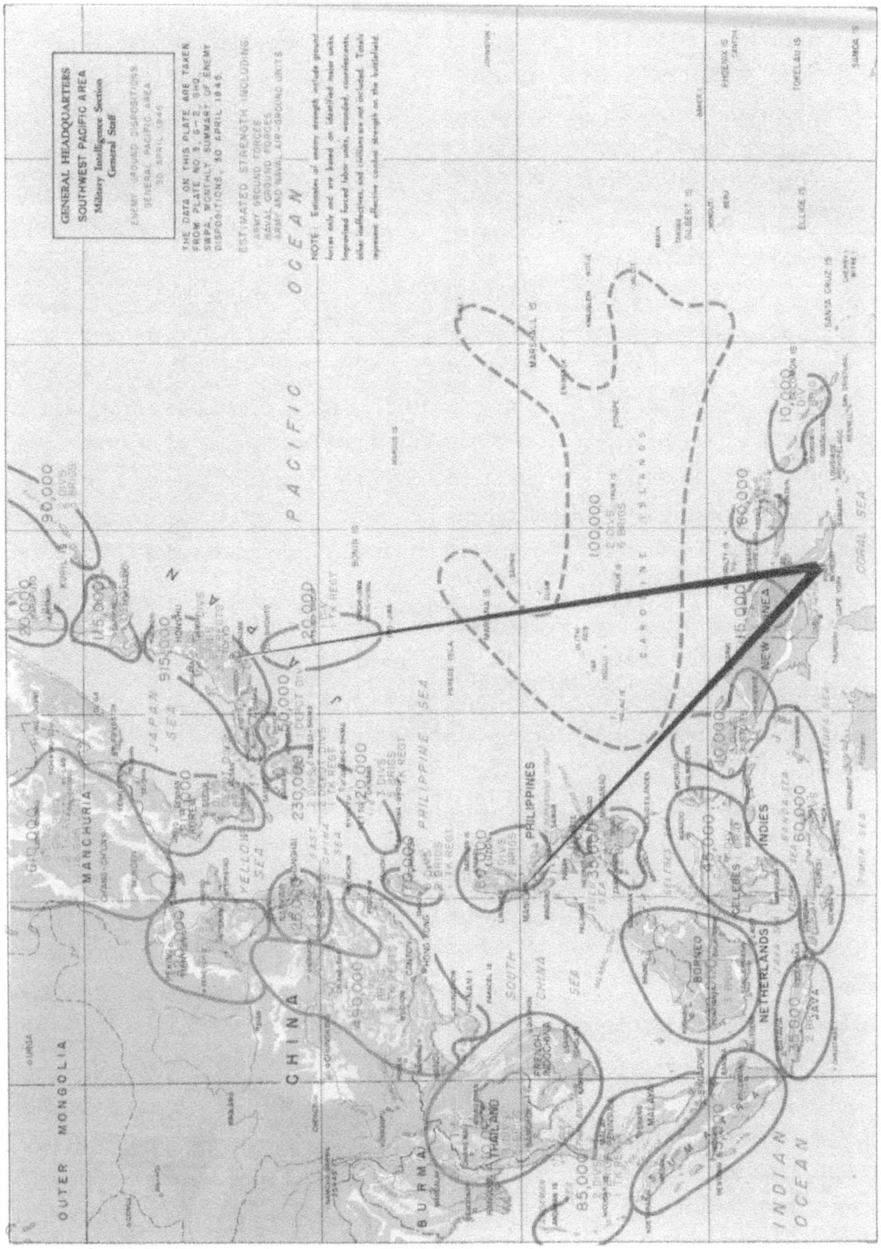

PLATE NO. 104

Enemy Ground Dispositions, General Pacific Area, 30 April 1945 and, in some cases, inequitably distributed between the Southwest Pacific Area and the Pacific Ocean Areas. Divided under two separate and independent commands, these forces could not be shifted with the speed and facility demanded by a rapidly changing battle situation nor could they be employed with maximum efficiency.

Until the campaign to liberate the Philippines, the individual operations in the Pacific had been conducted by relatively small forces, the size of a division or a corps. In the Philippines, however, an integrated army was employed on Leyte and two armies were required on Luzon. The invasion of Japan Proper would demand the use of an army group. General MacArthur proposed that a single commander should be responsible for the operational co-ordination of such large bodies of troops to permit their deployment to the best advantage.[4]

General MacArthur felt that the change in command structure should apply only to United States forces. In early January 1945, he recommended that army and navy Allied responsibilities continue as in the past, but that elements of the Australian Fleet should come under the operational control of the U. S. Navy.[5] This would involve no difficulty since the respective elements of the two fleets had previously operated in close harmony.

Australian ground and air units had been co-ordinated with the United States forces for nearly three years. "Both from political and military points of view," General MacArthur declared, "it is considered inadvisable to effect a reorganization." He felt that the guiding principle to be followed was to retain under a single command all army forces which were engaged in a campaign in a theater against one enemy force. "Any deviation from this," he maintained, "merely weakens the potential, prolongs the war, and increases the cost in blood."[6]

He believed that the greatest efficiency could be secured by placing all naval forces under a naval commander and all army forces under an army commander, with the Joint Chiefs of Staff exercising over-all direction and control. Only in this way could there be attained that complete flexibility and efficient employment of forces essential to victory. Unity of command for specific operations would be achieved by the creation of joint task forces. The task force commander in each instance would be chosen from the service having the paramount interest. General MacArthur also proposed that the same principle be applied to rear area troops and installations, with each service having its own supply and service organizations.

In general, such a determination of command would place the great land masses, such as Hawaii, New Guinea, and the Philippines, under the Army, and the outlying posts, such as Guam, Kwajalein, and Manus Islands, under the Navy. "This proposed organization," General MacArthur concluded, "will give true unity of command in the Pacific, as it permits the employment of all available resources against the selected objective."[7]

The Directives of 3 April 1945

On 3 April 1945, the Joint Chiefs of Staff issued a directive which reorganized the command structure in the Pacific. Under this directive, General MacArthur was designated Commander in Chief, United States Army Forces in the Pacific (AFPAC) in addition

4 Ibid.

5 CINCSWPA Radio No. C-55848 to WARCOS, 3 Jan 45, C/S GHQ, WD No. 904 (TS).

6 Ibid.

7 Ibid.

to Commander in Chief, Southwest Pacific Area and placed in administrative control of all army resources in the Pacific with the exception of the Twentieth (Strategic) Air Force, the Alaskan Command, and the army forces in the Southeast Pacific Area. All naval resources in the Pacific, except those in the Southeast Pacific Area, were placed under the administrative control of Admiral Nimitz.[8]

This reorganization, in essence, meant that the Joint Chiefs of Staff would exercise strategic jurisdiction over the whole Pacific Theater, assigning missions and fixing command responsibility for specific major campaigns. General MacArthur normally would be responsible for the conduct of all land operations and Admiral Nimitz, for sea operations. Each would have under his control the entire resources of his own services and each was authorized to establish joint task forces or appoint commanders to co-ordinate the conduct of operations for which he had been made responsible.

Thus, General MacArthur's proposal for two co-operating commanders in chief for the Pacific, one responsible for army forces and one responsible for navy forces, had in general been accepted. Actually, the Twentieth Air Force constituted a third distinct command, since it would continue to bomb Japan in accordance with directives from the Joint Chiefs of Staff to General Henry H. Arnold, Commander of all Army Air Forces.[9]

In a separate directive issued simultaneously with the reorganization order, General MacArthur was instructed specifically (1) to complete the occupation of Luzon and conduct such additional operations in the Philippines as would directly contribute to the defeat of Japan and the liberation of the Filipinos; (2) to make plans for occupying North Borneo, using Australian combat and service troops; (3) to plan and prepare for the campaign against Japan Proper, co-operating with Admiral Nimitz in the naval and amphibious phases of the invasion.[10]

Pacific Theater Command Reorganization

With the issuance of the directives of 3 April by the Joint Chiefs of Staff, General MacArthur began a gradual regrouping of his forces in such a way that the necessary changes would be accomplished with a minimum disruption to operations already in progress.

Effective 6 April, the War Department announced the establishment of the United States Army Forces in the Pacific (AFPAC) with General MacArthur as Commander in Chief. Eleven days later, Lt. Gen. Robert C. Richardson, Jr., commanding the United States Army Forces in the Pacific Ocean Areas, was ordered to report his command to AFPAC for administration. Meanwhile, the Seventh Fleet was transferred from General MacArthur's control to the administrative command of Admiral Nimitz.[11]

Representatives of General MacArthur and Admiral Nimitz met at Guam early in April 1945 and agreed in general on the following principles to govern the reorganization of com-

8 JCS Radio No. WX-62773 to CINCAFPAC, CINCPOA, and CG 20AF, 4 Apr 45, C/S GHQ, WD No. 956 (TS).

9 Ibid.

10 JCS Radio No. WX-62774 to CINCAFPAC, CINCPOA, and CG 20AF, 4 Apr 45, C/S GHQ, WD No. 957 (TS).

11 Ibid. WARCOS Radio No. WX-63939 to Distribution List, 6 Apr 45, C/S GHQ, WD No. 959 (S); GHQ, AFPAC General Order No. 1, 6 Apr 45, G-3, GHQ Admin 300-4 (S); COMINCH Radio No. 062116/Z to COM7thFLT, 8 Apr 45, C/S GHQ, Navy 612 (TS).

mand in the Pacific:

1. General MacArthur and Admiral Nimitz would immediately assume administrative command of their respective services.

2. Each would release operational control of all forces of the other service, except those considered essential to the functioning, development, or defense of their respective geographical areas or to the success of previously scheduled operations. Resources to be released by each commander would include depots and supply systems of the other service.

3. Each commander would assume as rapidly as possible full supply responsibility for the forces of his own service.

4. The existing army and navy responsibilities within the Pacific Ocean Areas for the joint support of positions in the Marshalls, the Carolinas, the Marianas, and the Ryukyus would continue in effect until modified by mutual agreement.

5. Both commanders would establish as quickly as possible their respective command organizations necessary for the planning and conduct of the phases pertaining to the invasion of Japan.[12]

The administrative reorganization was quickly completed. The transfer of operational control was the more important phase of the reorganization but the Guam agreement did not specify any definite schedule for the actual shifting of forces. Both army and navy resources in the Pacific were still split between the areas as both services went ahead with plans and staff studies for future operations.

Army Air Force Reorganization

During this same period, army air forces in the Pacific were also regrouped for the final phase of the war. Before 3 April, there were three army air commands in the Pacific, each with a separate mission. The Far East Air Force, consisting of the Fifth and Thirteenth Air Forces, provided air support for operations in the Southwest Pacific Area. The Air Service Command, together with units of the Army Air Forces, performed the vital functions of maintaining adequate bases, keeping up a continual flow of parts and supplies, and providing rapid maintenance service. The Twentieth Air Force was an independent Pacific command, operating under the direct control of General Arnold. The super-bombers of the Twentieth Air Force carried out long-range and photographic missions in strategic support of operations in the Southwest and Central Pacific Areas.[13]

General MacArthur had recommended to the War Department on 14 May 1945 that the Pacific Theater be divided between the Far East Air Force (to include the Fifth, Seventh, and Thirteenth Air Forces) and the Twentieth Air Force (to consist of the long-range bombers in the Marianas and the Seventh Fighter Command on Iwo Jima). He also recommended that the Army Air Forces of the Pacific Ocean Areas be abolished and their personnel transferred to the Far East and Twentieth Air Forces.[14]

On 2 June, the Joint Chiefs of Staff announced a regrouping and reassignment of the various air forces in the Pacific.[15] Headquarters of the Twentieth Air Force was scheduled to be transferred from Washington to Guam on 1 July and to be simultaneously redesignated the U. S. Army Strategic Air Force, under General Carl Spaatz. The XXI Bomber Command based in the Marianas took the title of the Twentieth Air Force and was placed

12 CINCPOA Radio No. 160250/Z to CINCSWPA, 16 Apr 45, C/S GHQ, Navy No. 613 (TS); CINCAFPAC Radio No. CX-14288 to CINCPOA, 18 Apr 45, C/S GHQ, Navy No. 615 (TS).

13 CG AAF Ltr to CINCSWPA, 4 Oct 44, C/S GHQ, WD No. 853 (TS).

14 CINCAFPAC Radio No. CX-16199 to WARCOS, 14 May 45, C/S GHQ S, WD No. 1016 (TS).

15 JCS Radio No. WX 104631 to CINCAFPAC and CINCPAC, 2 Jun 45, C/S GHQ, WD No. 1042 (TS).

under Lt. Gen. Nathan F. Twining. The XX Bomber Command, scheduled for deployment in the Ryukyus, was redesignated the Eighth Air Force under the command of Lt. Gen. James H. Doolittle. Both the Eighth and the newly regrouped Twentieth Air Forces were incorporated into the command of the Strategic Air Force, giving General Spaatz, when he assumed command on 16 July, control of the mightiest fleet of super-bombers ever assembled.[16]

In July the Seventh Air Force, with headquarters in the Ryukyus, was transferred to General Kenney's Far East Air Force. This completed the concentration of the Army's main air power in the Pacific under two major commands, the Far East Air Force and the Strategic Air Force. The Fourteenth Air Force continued to operate in China as part of the Eastern Air Command under Lt. Gen. George E. Stratemeyer.

Plans for the Borneo Campaign

As plans for the invasion of Japan developed, General MacArthur made every effort to complete the liberation of the Philippines and carry out the projected assaults in the Netherlands East Indies as rapidly as possible. It was important to free his command for the planned operations against Japan, for it was logical to assume that his well-organized veteran staff and forces would constitute the nucleus of the great land and air team assigned to defeat the Japanese in their home islands.

The liberation of key areas in the Netherlands East Indies was a task which General MacArthur considered a United States obligation.[17] In a radio to the War Department on 28 February, he outlined three main reasons for the necessity of a campaign in the Netherlands East Indies. First, he pointed out, the United States was obligated under the international agreement establishing the Southwest Pacific Area to undertake such a campaign. To overlook the Netherlands East Indies after freeing United States and Australian territories and restoring the former governments therein would represent a failure on the part of the United States to keep faith. Secondly, the re-establishment of the Netherlands East Indies government in Batavia would enhance the prestige of the United States in the Far East. With the occupation of Batavia, the Southwest Pacific Area would accomplish its mission except for the ever-necessary mop-up operations. General MacArthur felt that when the time came for such mopping-up the SWPA command should be dissolved and the responsibility for the conduct of civil affairs and for the further consolidation of territory should be turned over to the British and Dutch governments. Such measures would fulfill every obligation of the United States and, in addition, free General MacArthur from all future operational commitments outside of the areas necessary for the actual launching of the assault on Japan. A third reason was that an attack against the Netherlands East Indies sector would provide a remedy for the current comparative inactivity of the Australian troops. Operations in the New Guinea area did not require the full capabilities of the Australian forces and, until the final decisions were made regarding the employment of forces for the invasion of Japan, the Borneo sector furnished ideal action targets.[18]

A specific proposal to undertake the Netherlands East Indies operations had been made in the summer of 1944, when the British sug-

16 WARCOS Radio No. WX 30438 to CINCAFPAC, 12 Jul 45, C/S GHQ, WD No. 1073 (TS).

17 CINCSWPA Radio No. CA-50688 to WARCOS, 26 Feb 45, C/S GHQ, WD No. 937 (TS).

18 Ibid.

gested that a separate Commonwealth task force under a British commander be established. This plan contemplated the removal of Australia, the Indies, and Borneo from General MacArthur's command once the forces of the Southwest Pacific Area had been established in the Philippines. The British forces involved would then operate independently of General MacArthur.

The proposal was made originally to the Australian Prime Minister, John Curtin, at a Dominion conference in London.[19] Mr. Curtin, however, recommended that the plan be rejected because of the existing Allied and Australian arrangements relating to command in the Southwest Pacific Area. He emphasized that Australian naval, land, and air units were included in the forces available to implement Allied strategy in the Pacific and were assigned to the Commander in Chief of the Southwest Pacific Area. He felt that there was danger of grave misunderstandings with the United States if the Australians were taken from General MacArthur's command and placed under an Empire commander. There was a tradition of successful association and collaboration between the Australian Government and General MacArthur's Headquarters, and the Australian Prime Minister was opposed to any change, except upon recommendation of the Combined Chiefs of Staff with the consent of the Australian Government.

General MacArthur was in full accord with the suggestion of the Joint Chiefs of Staff to accept all the British and Dominion forces that could be made available to him for use in the Pacific. He was completely opposed, however, to any proposition which might restrict the manner in which he would employ these forces. No steps should be taken which would impose an unjustifiable limitation upon his conduct of operations or unduly complicate the existing command channels. In a radio to the War Department on 27 August 1944, General MacArthur said: "To attempt to segregate such a force into an entirely self-contained command consisting of ground, naval, and air components would not only introduce a clash of command authority but would require the complete reorganization of the present set-up, wherein the Australian, New Zealand, and American forces are amalgamated along service lines and coordinated under my immediate control. . . ."[20] He also stressed that in other than exceptional circumstances, British or Dominion forces would serve under their respective commanders and would be given independent missions, but he re-emphasized that the co-ordination of all ground, naval, and air forces should remain under his control.[21]

To conduct the campaigns in the Netherlands East Indies, General MacArthur decided to use the Australian I Corps. The 6th, 7th, and 9th Divisions would be moved into the Hollandia—Morotai area to be staged for assaults along the coasts of Borneo.

Logistic factors made it impossible to execute these plans without some modifications. Difficulties were encountered in obtaining the necessary shipping to support the scheduled operations. In reply to a request for forty-eight cargo vessels, General MacArthur was advised that no additional tonnage was available. There were, however, ten trans-Pacific troop ships on hand which could be employed.[22] These ships, together with a change

19 CINCSWPA Radio No. C-16679 to WARCOS, 27 Aug 44, C/S GHQ, WD No. 813 (TS).

20 Ibid.

21 C/S GHQ, SWPA Radio No. C-17683 to C/S WD, 13 Sep 44, C/S GHQ, WD No. 836 (TS).

22 CINCSWPA Radio No. C-57530 to WARCOS, 3 Feb 45, G-3, GHQ Admin 381 (TS); WARCOS Radio to CINCSWPA, 7 Feb 45 (TS).

in the program for moving surplus supplies, service units, and bases from Australia, New Guinea, and other points in the South Pacific to the Philippines, provided a partial solution to the problem.

To cut shipping requirements still further, the original plan for the movement of the Australian troops was also changed. Only one division and a proportionate share of corps troops would be moved in heavy shipping; the remainder would be transported in amphibious craft. An initial schedule was established which called for the occupation of Tarakan Island, off the northeast coast of Borneo, on 29 April 1945 to provide land-based air support for subsequent operations against Balikpapan. The assault against Balikpapan, farther south on the same coast, would be launched on 22 May, while the move into the Bandjermasin area of southeastern Borneo was to take place on 1 June.

Air power based on Bandjermasin would, in turn, support the attack against Soerabaja, which was set for 30 June. If British carriers were available for support at this time, Soerabaja would be by-passed and the amphibious forces would strike directly at Batavia. The campaign would be concluded with one task force driving through Java and then through the remaining islands of the Indies within the Southwest Pacific Area, while a second task force moved up to occupy British North Borneo.[23]

Additional changes were required when it was decided not to relieve the Australian 6th Division from its task of reducing Japanese remnants in New Guinea. Lt. Gen. Sir Leslie Morshead, commanding Australian I Corps, was ordered to carry out the Tarakan and Balikpapan operations using only the 7th and 9th Divisions. It was also decided that priority should be given to the capture of the Brunei Bay area in British North Borneo, because this region would provide an excellent advance base for the British Pacific Fleet and would also serve as a source of raw rubber, a critical commodity.[24] New operations instructions were issued incorporating these latest modifications. The Tarakan assault was moved back to 1 May; Brunei Bay was scheduled for attack on 10 June and the Balikpapan operations were set back five weeks to 1 July.

Final Southwest Pacific Operations: Borneo

Approximately 600 miles wide and 800 miles long, Borneo is the largest and yet one of the most sparsely populated of the islands in the Netherlands East Indies. The topography of the interior is hilly and mountainous and the entire island is covered with large tropical rain forests and swamps. Mechanized movement of any sort is virtually impossible except in small areas located mainly along the coasts.

During the early years of the war, Borneo was vitally important to the Japanese, both from an economic and a military standpoint. The enemy's selection of the northern oil fields as invasion points in 1942 indicated the emphasis which was placed on Borneo's petroleum supply. It was estimated that during the years of 1943 and 1944, this rich island supplied 40 per cent of Japan's fuel oil and 25-30 per cent

---

23 G-3, GHQ, SWPA, *Montclair III*, 2nd Amendment, 6 Mar 45 (TS); CINCSWPA Radio No. CA-50688 to WARCOS, 26 Feb 45, C/S GHQ, WD No. 937 (TS).

24 JCS Radio No. W-66056 to CINCSWPA, 11 Apr 45, C/S GHQ, WD No. 973 (S); CINCSWPA Radio No. CA-51543 to C/S WD, 12 Apr 45, G-3, GHQ Admin 381/43–P (TS); JCS Radio No. W-70717 to CINCAFPAC, 21 Apr 45, C/S GHQ, WD No. 984 (TS).

of her crude and heavy oils.[25] Borneo's principal oil fields were located in the region of Samarinda on the east coast, on the island of Tarakan, and between Miri and Seria at the northern boundary of Sarawak on the west coast.

In addition to its resources, the island occupied a strategic position on Japan's southern shipping lanes. Borneo's coasts were bordered by the two most important north-south sea routes off southeast Asia, the South China Sea to the west and Makassar Strait to the east. The Java Sea to the south provided east-west access to key ports throughout the Indies and to Malaya, Burma, and Thailand. Bases and harbors on the island's coasts served as staging, fueling, and transfer points for ships and planes moving along the lifelines of Japan's empire.[26]

During the first months of 1945, the Japanese regrouped and shifted their forces to prepare the island for defense against an anticipated Allied attack. The firm and extensive occupation of Borneo by the Japanese made the procurement of intelligence an extremely difficult task and great care had to be exercised in sending Allied intelligence agents into the area. Despite the many handicaps, however, much valuable information on enemy strength and dispositions had been collected from agents of the Allied Intelligence Bureau and from other G-2 sources. During February, March, and April 1945, the estimates of enemy forces in the Borneo area remained fairly static, listing an aggregate of from 25 000 to 30,000 troops of the Japanese Thirty-seventh Army and miscellaneous naval units.[27] (Plate No. 105)

Beginning their largest amphibious operation of the Pacific War, Australian troops sailed from Morotai on 27 April to launch the opening attack against Tarakan Island, off the northeast coast of Borneo. The assault against this island was the initial step in the *Oboe* operations of the over-all "Montclair" plan. By assaulting the important oil port of Tarakan, the Australians would strike their first blow to regain the rich colonial empire which had been seized by the Japanese in the dark days of January 1942. At that time, the oil storage tanks on Tarakan had been fired by the Dutch in a desperate effort to prevent their advantageous use by the invaders. Now again those same tanks, having been rebuilt by the Japanese, vanished in smoke and flame, this time under the bombardment of American and Australian-manned Liberators.

Backed by aerial assault and the devastating fire power of a naval task force composed of United States and Australian cruisers and destroyers, the Tarakan Attack Group under Rear Adm. Forrest B. Royal, arrived unopposed off Tarakan on the morning of 1 May. The reinforced 26th Infantry Brigade of the Australian 9th Division and a battalion of the Royal Netherlands Indies Army assailed the elaborate system of Japanese defenses at Lingkas, southeast of the Tarakan airdrome. Additional support was provided by artillery fire from Sadau Island, approximately six miles up the coast, which had been previously occupied by Australian commandos. Three days had been spent prior to the assault in sweep-

25 Joint Army-Navy Intelligence Study of Borneo (JANIS 150) Chapter 1, Brief, 15 May 45, pp. 1-4, 1-9, 1-29, 1-35.

26 Ibid.

27 G-2, GHQ, SWPA, *Monthly Summary of Enemy Dispositions*, 30 Apr 45. As of 31 December 1946 repatriation figures from the Japanese Foreign Office showed that 15,607 army personnel and 5,766 navy personnel had been returned to Japan, in addition to 3,321 civilians, while 865 persons were still awaiting repatriation. The Japanese sustained 5,693 casualties during the fighting on Borneo and lost 536 as prisoners of war captured by the Allies. The total for army and navy personnel coincides closely with the G-2 estimates before the start of the Borneo Campaign.

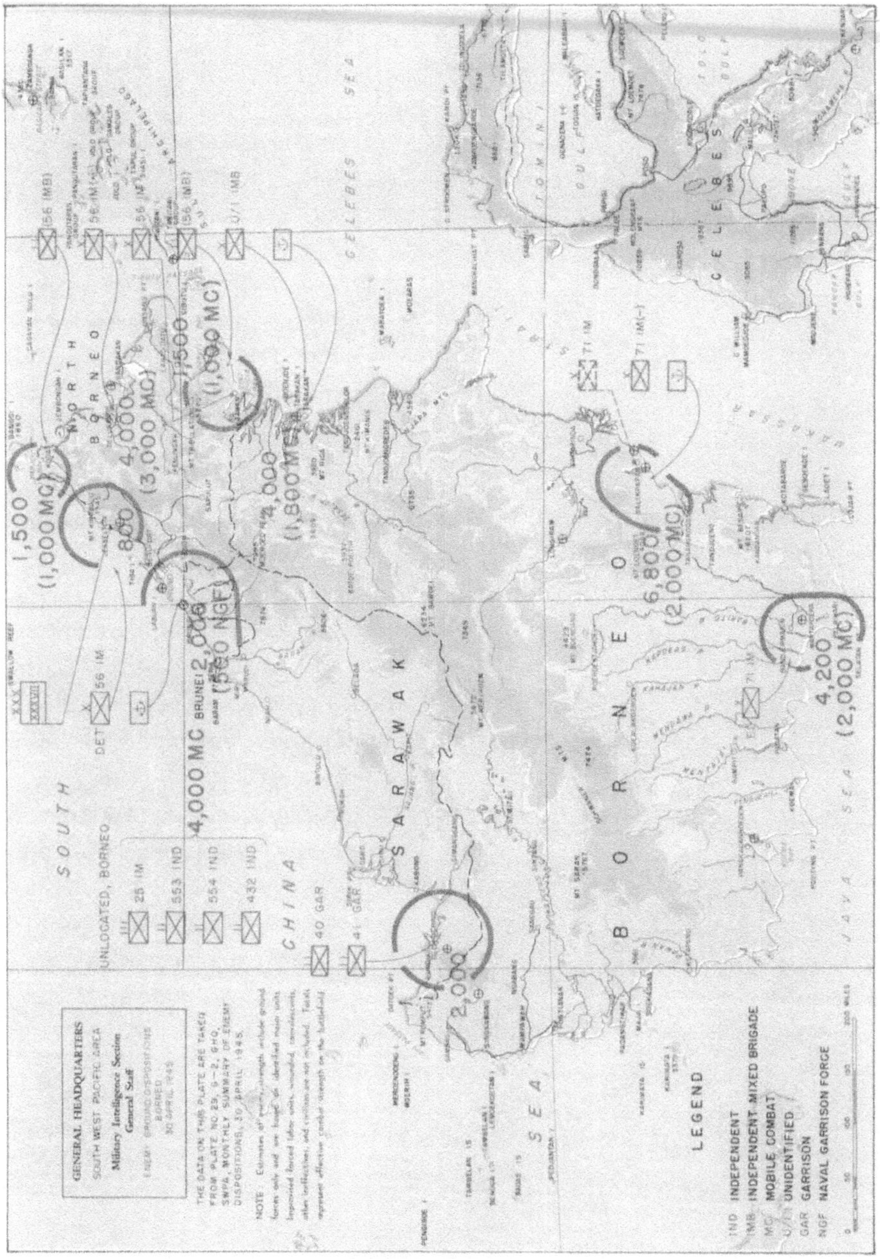

PLATE NO. 105
Enemy Dispositions on Borneo, 30 April 1945

ing the waters of mines, under the protective cover of naval guns. Steel rails, embedded in mud near the shore, and underwater barbed wire entanglements formed an additional obstacle which was extremely hazardous to move. Depth charges, land mines buried on the shore, oil pipes emplaced to spurt flame and smoke—all helped to form an elaborate and destructive beach defense.[28]

The communique of 3 May detailed the importance of the Tarakan move:

Australian ground forces from one of its most famous divisions, veterans of New Guinea and the Middle East, have landed on the key island of Tarakan, off the eastern coast of Borneo. Following intense air attacks by Royal Australian and Far East Air Forces and a four-day naval bombardment by units of the United States Seventh Fleet and the Royal Australian Navy, our troops in amphibious tanks and fast landing craft swept ashore near Lingkas, two miles east of Tarakan airfield. A beachhead was quickly established before the enemy garrison could offer effective opposition and our troops are advancing toward the airfield and town. There has been no enemy air or naval reaction.

Earlier denied the fruits of his rich Borneo oil and rubber conquests by our air and submarine blockade, his actual possession is now directly challenged. This operation virtually severs the enemy's holdings in the south. His forces in the eastern portion of the Netherlands East Indies, including the Celebes, Moluccas, Lesser Sundas, and other island outposts, are effectively isolated.

The establishment of this base will complete our chain of airfields extending from Luzon in the north to Darwin in the south, and enable our bombers to strike at will the enemy's forces anywhere in the Southwest Pacific Theater and constantly interdict his lines of supply and communication. Enemy shipping in these waters will be hunted down and destroyed as it has been already in the China Sea.[29]

Although the strong enemy defense measures delayed the landing, they failed to repel it. The Australians went ashore at several places along the beach near Lingkas as the Japanese withdrew to prepared inland positions. (Plate No. 106) Aided by precision bombing of the Liberators and strong tank support, the Australian troops drove toward the island airfield, but as they approached Tarakan Town they began to meet increasingly fierce resistance. Heavy machine gun and mortar fire from a labyrinth of interconnecting tunnels and pillboxes opposed the advance. Fierce hand-to-hand fighting broke out along the ridges and in the jungle. Fanatic suicide charges were made by the Japanese defenders as they counterattacked with savage determination.[30]

Despite this strong enemy reaction, the Aus-

28 Adv HQ, ALF, SWPA, "Australian Military Forces Report on Operation 'Oboe One,' Tarakan, Borneo," pp. 18–19; COMPHIBGP 6 Report to COMINCH, Serial 0188, (05), 5 May 45, G-3, GHQ, SWPA *Journal*, 3 May 45 (C); Hugh Buggy, *Pacific Victory, A Short History of Australia's Part in the War Against Japan*, (Australian Ministry for Information), pp. 284–285.

29 GHQ, SWPA, Communique No. 1122, 3 May 45.

30 G-3, GHQ, SWPA, *Monthly Summary of Operations*, May 45; Buggy, op. cit., pp. 285–288. Interrogation of Captain Saichiro Matsumoto, former commanding officer of the 455th Independent Infantry Battalion, Thirty-seventh Army, revealed that, until the arrival of the Japanese 455th Battalion in December 1944, military preparations on Tarakan were neglected and totally inadequate to defend the island. After the battalion landed, the construction of fortifications was begun in earnest and, by the time of the Allied invasion, gun emplacements and other beach defenses were almost completed. Most of these fortifications were destroyed, however, by the intense naval and aerial bombardment. Because of the lack of shipping, no supplies or reinforcements were forthcoming and resistance had to be continued with what was on hand. Fighting strength and morale deteriorated with the unfavorable progress of the battle and the non-arrival of reinforcements and supplies. By mid-June, the remaining troops had disbanded into small groups to carry on guerrilla warfare but actually these remnants were primarily concerned with physical survival and escape from the island. About half of the deaths incurred resulted from disease and malnutrition. Interrogation Files, G-2 Historical Section, GHQ, FEC.

tralians, by envelopment from the southeast and northwest, secured the Tarakan airstrip by 6 May. The Japanese were driven into the hills east of the airfield where they fell back on well-prepared ground to continue a dogged resistance. Tanks blasted the Japanese from their holes and trenches while repeated infantry attacks reduced their defenses to small individual pockets.

Supported by planes, tanks, artillery, and flame-throwers, the Australians pushed onward and soon captured the heavily damaged Djoeata and Sesanip oil fields. Cape Djoeata and nearby Boenjoe Island were cleared and, by the end of the month, the remaining Japanese had retreated into the central mountains of the island where they stubbornly resisted all attempts to dislodge them. Attack after attack was made against the numerous pillboxes and honeycombs of tunnels along the crests and ridges but, despite constant shell-fire and bombing attacks, the Japanese clung desperately to their positions. In early June, these positions were subjected to low-level air attacks in which incendiary belly tanks were dropped to burn out the remaining enemy troops. By 21 June, all organized Japanese resistance on Tarakan had ceased.[31]

Assault along the West Coast

Five weeks after the Tarakan invasion, the Australians launched their second blow against Borneo. Furthering General MacArthur's plan to segment the Japanese forces in the Netherlands East Indies, the Brunei Attack Group, under Admiral Royal, left Morotai on 4 June to secure the great naval base at Brunei Bay on the western coast of British North Borneo. Arriving off the objective area six days later, the powerful Allied task force poured a heavy concentration of shells and bombs on the enemy's beach defenses. There was complete integration of United States and Australian naval forces, Australian ground forces, and United States and Australian air forces as the 20th and 24th Brigades of the Australian 9th Division swept ashore simultaneously on Labuan and Muara Islands and at Brooketon, just north of Brunei.[32]

The communique of 12 June described the strategic value of the objective:

> *The Brunei area is rich in oil, rubber, coal, lumber, iron and other resources and in the 500 square miles of its sheltered bay numberless ships of any size can ride at anchor. The establishment of air and naval facilities at Brunei Bay, combined with those in the Philippines, will complete a chain of mutually supporting strategic bases 1500 miles long, from which naval and air forces are within continuous effective range of the Asiatic coast from Singapore to Shanghai. Operations from the Philippine bases have already practically destroyed enemy shipping in the South China Sea aud we shall now be able to interdict his overland lines of communication and escape routes in Indo China and Malaya.*
>
> *At Brunei and Tarakan our columns stand not at the gates but at the geographical center of the enemy-occupied Celebes, Bali, Java, Sumatra, Malaya, and Indo-China. His garrisons there can now only await piecemeal destruction at will.*
>
> *With his conquest in the East Indies cut off and isolated from the Empire, the rich resources rendered valueless, his naval and air arms in the Indies impotent, his ground troops immobilized and unable to obtain reinforcements or supplies, the enemy invader has definitely lost the war of strategy in the Southwest Pacific.*[33]

General MacArthur personally supervised

31 G-3, GHQ, SWPA, *Monthly Summary of Operations*, May, Jun, 45; Buggy, op. cit., pp. 288–289.

32 Buggy, op. cit., pp. 289–290; CTG 78.1 (COMPHIBGP 6, R. Adm. Royal) Confidential Action Report 0322 to COMINCH, 19 Jun 45, G-3, GHQ, SWPA *Journal*, 17 Jun 45.

33 GHQ, SWPA, Communique No. 1162, 12 Jun 45.

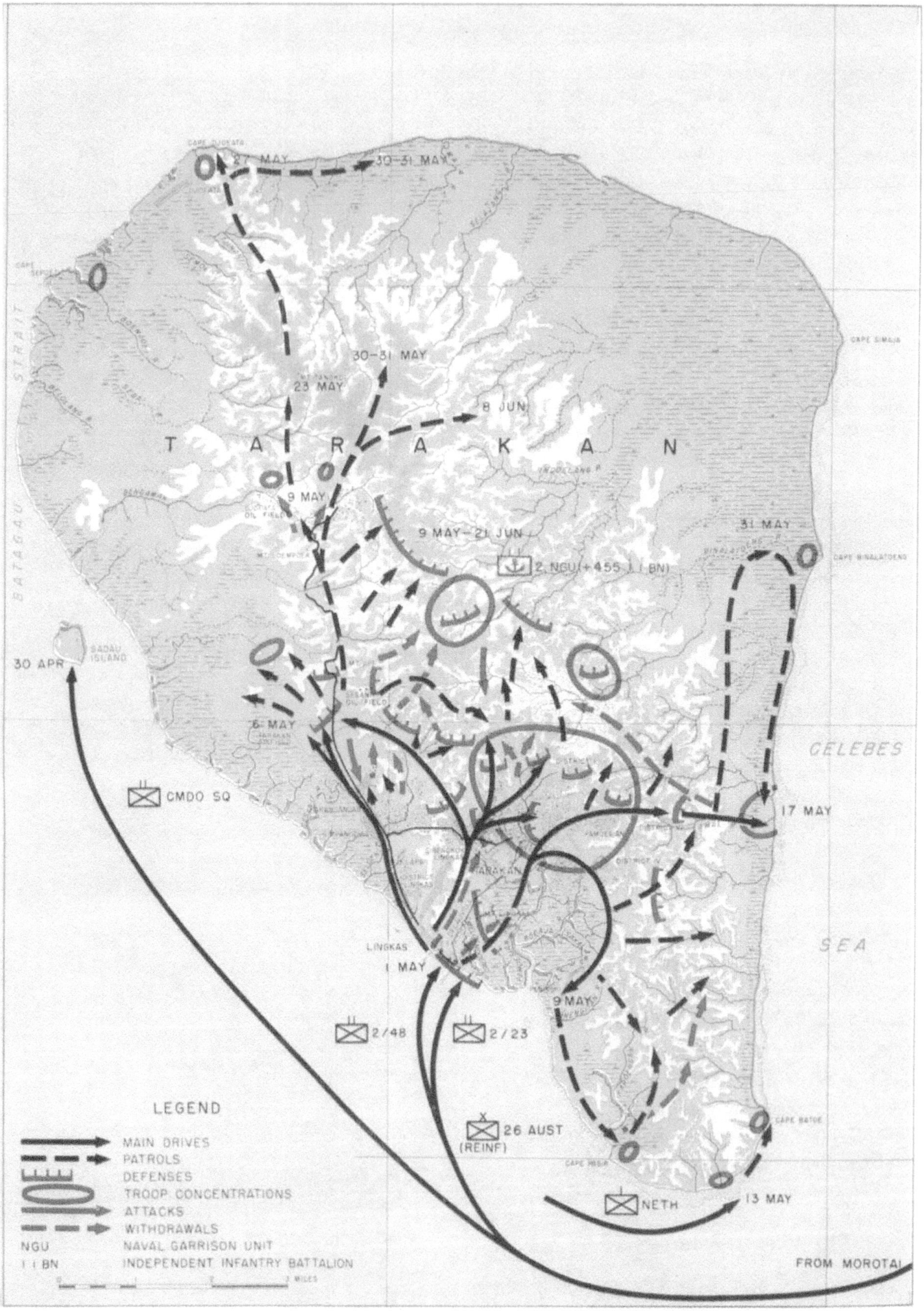

PLATE NO. 106
Tarakan Operation, 1 May—21 June 1945

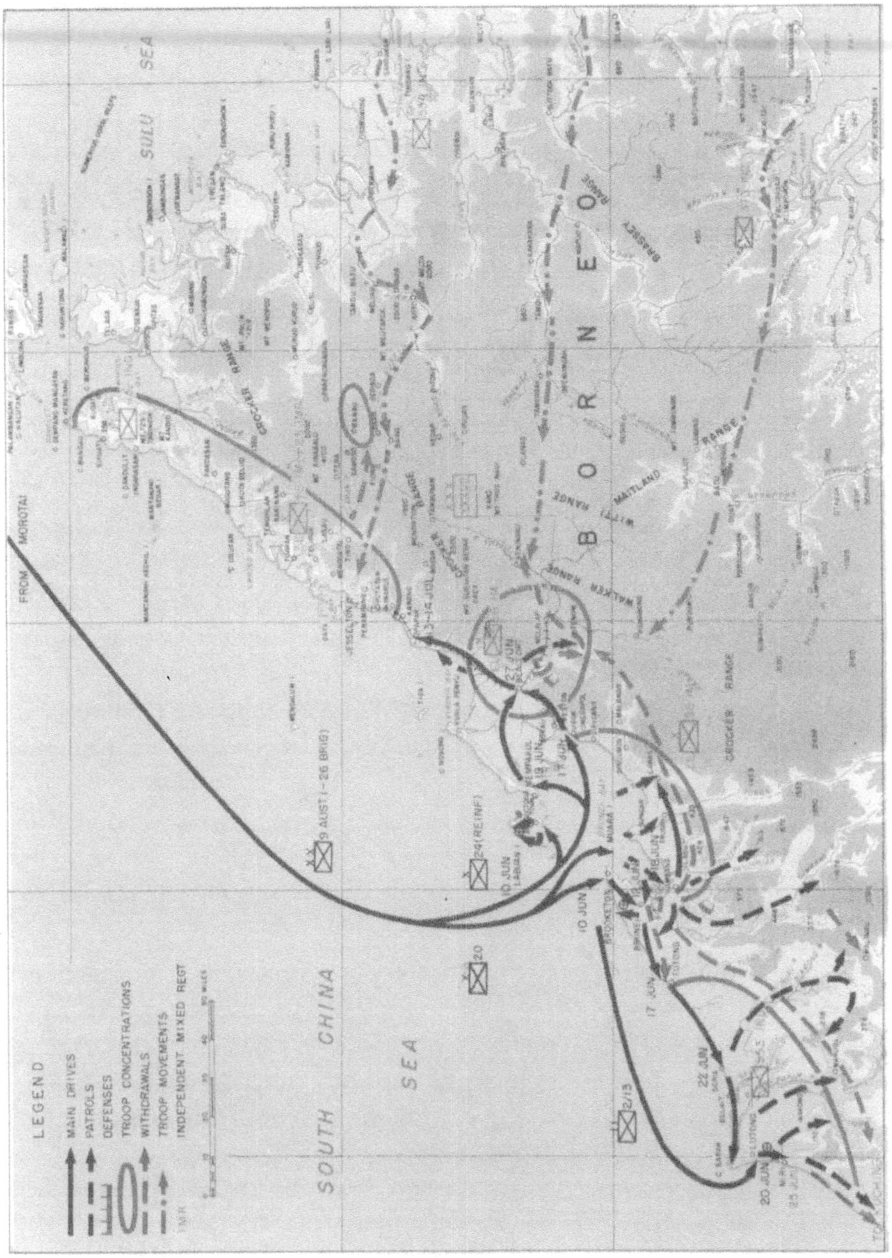

PLATE NO. 107
Brunei Bay Operations, 10 June—14 July 1945

the successful landings on Labuan Island and, together with General Morshead, Australian I Corps Commander, and General Kenney, Commander of Allied Air Forces, went ashore with the assault waves. "Rarely is such a great strategic prize obtained at such low cost," General MacArthur commented when he observed the success of the initial landings.[34]

Labuan (Victoria) Town was virtually destroyed by a naval bombardment three times greater than that which had been visited upon Tarakan. Japanese beach defenses lay shattered and only a few scattered parties of dazed enemy troops attempted to offer any resistance. The main opposition to the inland movement came in the form of glutinous mud which impeded infantry and vehicles alike. By nightfall, however, the Australians held the ruined town and the bomb-devastated airstrip. By 15 June, all organized resistance had been terminated on Labuan Island.[35]

On Brunei Peninsula, the advance southward proceeded rapidly. By 12 June, Australian troops had secured Brunei Airfield. (Plate No. 107) The town of Brunei fell the next day, after a swift march along the Cape and an amphibious movement across Brunei Bay. The Japanese had beaten a hasty retreat and only feeble counteraction met the Australian forces.

Accompanying the Australian troops into Brunei were members of the British North Borneo administration unit who were charged with restoring civil administration in the liberated areas of British Borneo and administering to the general welfare of the native population of Labuan and Brunei Peninsula. Positions in and around Brunei were consolidated and the drive continued southwest along the coast. Tutong, an oil refinery center located 35 miles south of Brunei Bay, was captured on 17 June. Meanwhile, a new and easy landing was effected by 9th Division troops at the port of Weston on the eastern shore of Brunei Bay.[36]

On 19 June, in the wake of an artillery bombardment from Labuan Island, 9th Division units in a shore-to-shore move landed at Mempakul, on the northeast shore of Brunei Bay. This unopposed landing gave the Australians complete control of the shores bordering both entrances into the bay. At the same time, the elements which had landed at Weston were pressing toward Beaufort, the strategic center on the single-track railway to Jesselton.[37]

Still another unopposed amphibious landing was made on 20 June at Lutong, the refinery center for the Seria and Miri oil fields, eighty miles down the coast from Brunei Bay. The town fell without a fight. Oil storage tanks

---

34 Buggy, op. cit., p. 290. The Commander in Chief of the Southern Army, Field Marshal Terauchi, had estimated that Allied forces would land on the strategic west coast of Borneo sometime after March 1945. He therefore ordered part of the main Japanese dispositions around Tawau on the east coast to shift to the west coast around Brunei. Because Japanese sea communications had by this time been cut off, the move to the Brunei area was made over the difficult mountainous terrain of central Borneo, a distance of more than 300 miles. During this move, many losses were incurred because of the difficult terrain, bad weather, sickness, and lack of provisions. Those troops which succeeded in reaching the vicinity of Brunei had insufficient weapons and supplies and their physical strength was decreased by malaria and fatigue. Their defense preparations, therefore, could not be completed prior to the Allied landing. Japanese First Demobilization Bureau Report, *Borneo Operations Record*, Vol II, G-2 Historical Section, GHQ, FEC.

35 Buggy, op. cit., p. 290; G-3, GHQ, SWPA, *Monthly Summary of Operations*, Jun 45.

36 G-2, GHQ, SWPA, *Daily Summary*, No. 1170, 17/18 Jun 45; G-3, GHQ, SWPA, *Monthly Summary of Operations*, Jun 45; GHQ, SWPA, Communique No. 1169, 19 Jun 45.

37 G-2, GHQ, SWPA, *Daily Summary*, No. 1172, 19/20 Jun 45; GHQ, SWPA, Communique No. 1171, 21 Jun 45.

and refineries were captured intact, along with stores of abandoned equipment.[38]

The uninterrupted advance along the west coast continued toward Seria, 26 miles southwest of Tutong. The Australians reached Seria on 22 June and three days later took Miri, one of the oldest oil fields in Borneo.[39] At Miri, 300 oil wells with a peacetime annual production of 1,318,000 barrels had been set afire by the Japanese and the 600,000-barrel storage tanks had been destroyed. Nevertheless, the capture of the northwest Borneo oil fields had been virtually completed, and Dutch oil experts were soon at work exploiting the vast liquid wealth still underground.[40]

Beaufort, an important railroad junction approximately 60 miles northeast of Brunei Bay, was captured on 27 June with little trouble and by the end of the month Australian forces were pushing along the Jesselton railway. The Australians now dominated 5,000 square miles of northwest Borneo and their lines stretched along the coast for 135 miles. Except for patrolling and mopping up, the North Borneo Campaign was over.[41]

The Brunei Bay operation had proceeded smoothly both in timing and execution. Nowhere on the mainland had the Japanese put up a concerted defense. The important oil fields of northwestern Borneo, strategic Brunei Bay, and the terminal of the northern narrow-gauge railway had been yielded by the enemy with only minor skirmishing. The Japanese on the west coast had shown little of the tenacity displayed on Tarakan, preferring instead to withdraw whenever possible. Rather than face the power of the Allied forces, many enemy troops had retreated from vital strategic areas to the mountainous interior, moving southward along the western coast of Kuching and into the area northeast of Brunei Bay.[42]

Assault along the East Coast

The third and most shattering blow in the reconquest of Borneo fell upon Balikpapan. Located roughly midway along Makassar Strait on Borneo's east coast, the Balikpapan area was one of the world's richest oil centers. When the Japanese attacked this great oil port at the beginning of 1942, the Dutch fired the storage tanks and blew up all installations in an attempt to prevent the exploitation of the oil fields by the enemy. Despite these efforts, the capture of Balikpapan had been a major and lucrative acquisition for the Japanese, whose well-prepared technicians and laborers had restored production facilities in record time.

On 26 June, in the largest amphibious operation under General MacArthur's command since the landings at Lingayen Gulf, the ships of the Balikpapan Attack Group, under Admiral Noble, sailed out of Morotai. Three days later, the invasion force stood off the shores of Balikpapan to begin the most intense pre-landing naval barrage ever put down in the Southwest Pacific Area.[43] More than 45,000 rounds of five and six-inch shells were fired by

---

38 G-2, GHQ, SWPA, *Daily Summary* No. 1173, 20/21 Jun 45; GHQ, SWPA, Communique No. 1172, 22 Jun 45.

39 G-3, GHQ, SWPA, *Monthly Summary of Operations*, Jun 45; G-2, GHQ, SWPA, *Daily Summary* No. 1175, 22/23 Jun 45; GHQ, SWPA, Communique No. 1174, 24 Jun 45.

40 G-3, GHQ, SWPA, *Monthly Summary of Operations* Jun 45; Buggy, op. cit. pp. 291–292.

41 G-3, GHQ, SWPA, *Monthly Summary of Operations*, Jun 45; G-2, GHQ, SWPA, *Daily Summary* No. 1181, 28/29 Jun 45; GHQ, SWPA, Communique No. 1180, 30 Jun 45; Buggy, op. cit., p. 292.

42 G-2, GHQ, SWPA, *Daily Summary* No. 1175, 22/23 Jun 45; Buggy, op. cit., pp. 292–293.

43 Buggy, op. cit., p. 293; CTG 78.2 Report to COMINCH, Serial 0235 "Action Report for the Balikpapan Operation, 15 Jun—6 Jul 45," 14 Aug 45, G-3, GHQ, SWPA *Journal*, 6 Jul 45 (C).

cruisers and destroyers of the United States Seventh Fleet and Australian and Dutch Fleet elements. This terrific shelling had been preceded by twenty days of bombing by the Royal Australian Air Force and the U. S. Fifth and Thirteenth Air Forces. Averaging 200 tons of bombs a day, the Liberators hammered the oil port relentlessly and effectively neutralized all enemy airfields within range of Balikpapan. Fifteen days prior to the landings, mine sweepers were at work clearing the surrounding waters of the thousands of mines that had been laid successively by the Dutch, the Japanese, and the Allies as the course of the war changed with the years.[44]

On 1 July, the 18th and 21st Brigades of the Australian 7th Division charged ashore in the region of Klandasan. (Plate No. 108) The 25th Brigade was held as a floating reserve to be employed according to operational developments.

The communique for 2 July announced the strategic implications on the landings:

Australian ground forces have made a third major landing on the vast island of Borneo. . . .

Swiftly following our seizure of Brunei Bay on the northwestern coast and Tarakan on the northeastern, the enemy's key Borneo defenses are now isolated or crushed, and his confused and disorganized forces are incapable of effective strategic action. The speed, surprise, and shock of these three operations have secured domination of Borneo and driven a wedge south splitting the East Indies.

Strategic Makassar Strait, gateway to the Flores and Java Seas, is now controlled by our surface craft as well as by air and submarine. Development of already existing air facilities at Balikpapan will enable our aircraft of all categories to disrupt and smash enemy communications on land and sea from Timor to Sumatra. The whole extent of Java and the important ports of Soerabaja and Batavia are now within easy flight range and subject to interdiction. Our shipping can now sail with land-based air cover to any point in the Southwest Pacific.

It is fitting that the Australian 7th Division which in July three years ago met and later turned back the tide of invasion of Australia on the historic Kokoda Trail should this same month secure what was once perhaps the most lucrative strategic target in our East Indies Sector and virtually complete our tactical control of the entire Southwest Pacific.[45]

Landing at Klandasan Beach, two miles from Balikpapan, the Australians moved rapidly inland and within six hours had established a beachhead three miles in length. Aircraft and naval gunfire formed a barrage ahead of the advancing troops as the Japanese withdrew from the beach areas. Oil tanks became flaming infernos when supporting warships hurled shells into the Japanese defenses and into the town and the extensive oil storage areas.[46] General MacArthur, landing on the beach with the last wave of assault troops, declared as he strode ashore, "I think today we settled the score of the Makassar Strait affair of three and a half years ago".[47]

Meeting only sporadic opposition, the Australians seized Mount Malang and the high ground overlooking the town. The advance continued eastward along the coast toward the

44 G-3, GHQ, SWPA, *Monthly Summary of Operations*, Jul 45.

45 GHQ, SWPA, Communique No. 1182, 2 Jul 45.

46 Buggy, op. cit. pp. 293–294; G-3, GHQ, SWPA, *Monthly Summary of Operations*, Jul 45. Maj. Misuaki Yamada, former Commanding Officer of the 454th Independent Infantry Battalion, Thirty-seventh Army, was highly impressed with Allied naval and air power in the invasion of Balikpapan. He stated that the Japanese were never able effectively to combat this overwhelming superiority. Interrogation Files, G-2 Historical Section, GHQ, FEC.

47 Manila *Free Philippines*, July 2, 1945, p. 1. In his statement, General MacArthur referred to the 22–24 January 1942 sortie of four U. S. destroyers through the waters of Makassar Strait in a vain attempt to block the approach of a huge Japanese invasion convoy.

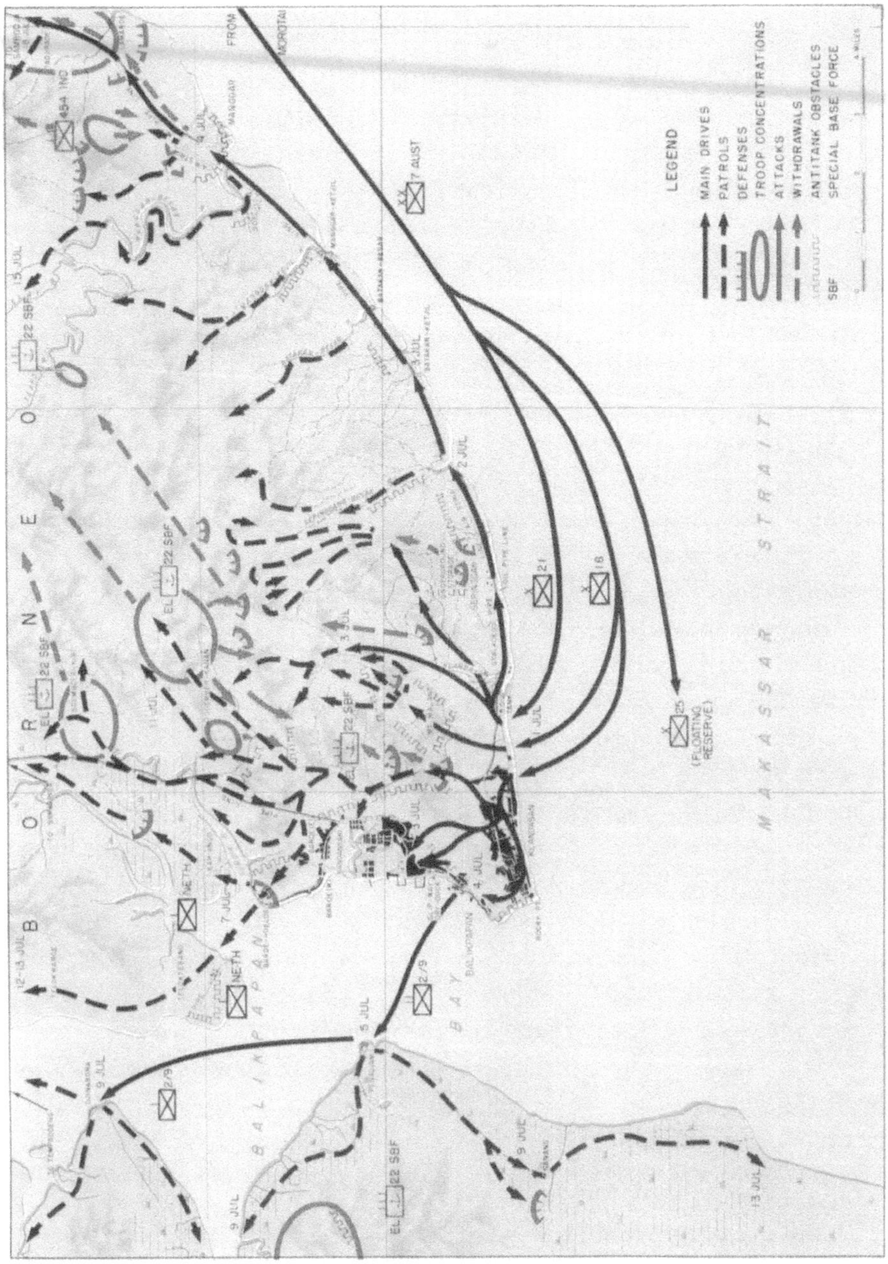

PLATE NO. 108
Balikpapan Operation, 1–18 July 1945

Japanese-built Sepinggan Airfield, by-passing several small enemy pockets along the way. Elaborate defenses, including large pillboxes, were abandoned by the Japanese as they fled before the fast-moving onslaught of the Australian 7th Division.

Resistance stiffened on 2 July, but a determined drive up the coastal highway gained the town of Sepinggan and its 5,000-foot airstrip. The next day, the center of Balikpapan was captured and by the 4th the entire Klandasan Peninsula was secured. Other elements of the 7th Division advanced steadily inland and enveloped the Pandansari oil refinery area.[48]

Thrusting along the coast, the Australians reached the Manggar airstrip, 13 miles east of Balikpapan, on 4 July.[49] Ridges and gullies along the way were criss-crossed with Japanese trenches and tunnels while the hills were dotted with log pillboxes and bunkers, but flame-throwing tanks and field guns drove the Japanese from their positions. Destroyers and rocket ships moved inshore to deliver fire in support of the advance as heavy fighting developed west of Manggar.

A series of overwater movements and shore landings was carried out to tighten the hold on strategic Balikpapan Bay, while enemy cave positions farther inland were methodically eliminated. Backed by naval gunfire, Australian troops on 5 July landed unopposed at Penadjam on the western shore of the bay. Netherlands East Indies forces made two amphibious landings on the northern shore of Balikpapan Bay on 7 July.[50] Two days later, 7th Division troops made another overwater movement to Djinabora, 4 miles north of Penadjam, and established complete control over the Balikpapan Bay shore area.[51]

By 11 July, the Australians had forced the enemy from the Pandansari area and had driven a wedge inland northeast of Balikpapan along the Balikpapan—Samarinda road. In the nearby hills, enemy cave positions were gradually eliminated. Although stubborn delaying action was encountered along the Balikpapan—Samarinda highway, the resistance met did not match the caliber of previous campaigns against the Japanese.[52]

Northeast of Manggar, elements of the 7th Division moved along the road toward Sambodja without enemy contact. Other Australian units drove inland, expanding their control over the entire Balikpapan area. On 18 July, the Australians secured Sambodja, a huge oil field 28 miles northeast of Balikpapan, against steadily weakening resistance by the bewildered and demoralized enemy. Further action in the Balikpapan area consisted of little more than mop-up skirmishes with scattered enemy remnants who continued to retreat into the jungles.[53]

By the end of July, scarcely three months after the fighting had begun, every objective

48 Buggy, op. cit., pp. 294–295; G-3, GHQ, SWPA, *Monthly Summary of Operations*, Jul 45; G-2, GHQ, SWPA, *Daily Summary*, No. 1186, 3/4 Jul 45.

49 G-2, GHQ, SWPA, *Daily Summary* No. 1187, 4/5 Jul 45.

50 G-2, GHQ, SWPA, *Daily Summary* No. 1190, 7/8 Jul 45; GHQ, SWPA, Communique No. 1190, 10 Jul 45.

51 G-2, GHQ, SWPA, *Daily Summary* No. 1192, 9/10 Jul 45; GHQ, SWPA, Communique No. 1191, 11 Jul 45.

52 G-2, GHQ, SWPA, *Daily Summary* No. 1196, 13/14 Jul 45; Buggy, op. cit. p. 295; GHQ, SWPA, Communique No. 1192, 12 Jul 45; G-3, GHQ, SWPA, *Monthly Summary of Operations*, Jul 45. Major Yamada reported that the severe losses of supplies forced the Japanese virtually to live off the land. During the battle there was no hope of reinforcement or resupply, and the result was a serious loss of morale among the Japanese forces. Resistance became ineffective and the Japanese were reduced to fighting merely a delaying type of warfare with no hope of final victory. Interrogation of Maj. Misuaki Yamada previously cited.

53 G-2, GHQ, SWPA, *Daily Summary* No. 1201, 18/19 Jul 45; GHQ, SWPA, Communique No. 1200, 20 Jul 45; Buggy, op. cit., pp. 295–296; G-3, GHQ, SWPA, *Monthly Summary of Operations*, Jul 45.

of the Borneo Campaign had been attained. (Plate No. 109) The swiftly moving 7th and 9th Divisions had thoroughly crushed the enemy. The Japanese defenders were completely defeated and their beaten, sickly remnants were driven into the wooded hills of the interior to live off the land. The campaign had netted two great naval bases—Brunei Bay and Balikpapan, seven important airfields, the rich Seria—Miri oil wells, the refineries at Lutong, and huge stores of Japanese equipment.[54]

Enemy garrisons remaining in the Celebes, Bali, Java, Sumatra, Malaya, and Indo-China areas were further isolated from their empire with no future but surrender or eventual destruction at the hands of Allied mop-up forces. Just before the Japanese capitulation, the total enemy casualties in the Borneo operations were given at 5,693 dead and 536 prisoners. Allied casualties, in sharp contrast, were 436 killed, 3 missing, and 1,460 wounded.[55]

Final Actions in New Guinea, New Britain, and Bougainville

General MacArthur's successful by-passing tactics along the New Guinea coast, followed by his invasion of the Philippines, had left thousands of Japanese contained in the various islands of the Southwest Pacific. At the end of 1944, there were over 110,000 enemy troops scattered in the Solomons, New Britain, New Ireland, and in Eastern and Western New Guinea.[56]

The urgent need for the veteran soldiers of the Sixth Army to invade Leyte left to the Australians the task of finishing operations in these by-passed zones of enemy-occupied territory. General MacArthur proposed to continue the neutralization of the pocketed enemy forces but the tactical methods for accomplishing this mission were left entirely to the discretion of the Australian commanders.[57] Australian First Army troops replaced United States forces in November and December 1944 and took over full responsibility for operations in New Guinea, New Britain, and Bougainville.[58]

The fighting in all three of these areas followed the same general and unvarying pattern. (Plate No. 110) Small enemy garrisons, cut off from supplies or reinforcements and clinging desperately to their defensive positions had to be painstakingly and methodically eliminated. Activity in the Wewak—Aitape area during the first two months after the Australians assumed responsibility consisted mainly of energetic patrol actions against harassing parties of Japanese. Later, as the Australians began their drive from Aitape, some enemy forces retreated across the rugged Toricelli mountain range while other units withdrew to Wewak. By late December, troops of the Australian 6th Division had advanced 34 miles along the coast toward Wewak and had pushed 40 miles inland beyond the Toricelli Mountains.[59]

In mid-March 1945, a coastal drive east-

---

54 Buggy, op. cit., pp. 293–296; G-3, GHQ, SWPA, *Monthly Summary of Operations*, Jul 45.

55 GHQ, SWPA, Communique No. 1225, 14 Aug. 45.

56 Japanese First Demobilization Bureau, Special Report No. 39, "Japanese Army Strength in New Britain, New Ireland, Bougainville, and New Guinea in 1945," 30 Jul 49, G-2 Historical Section, GHQ, FEC. See also G-2, GHQ, SWPA, *Monthly Summary of Enemy Dispositions*, Dec 44.

57 CINCAFPAC Ltr to J. B. Chifley, Acting Prime Minister of Australia, 20 May 45, C/S GHQ, Aust 920.

58 GHQ, SWPA, Opn Instr No. 65, 30 Aug 44, and No. 78, 5 Nov 44, G-3 Planning File 370.2 (S).

59 Buggy, op. cit., pp. 276–278.

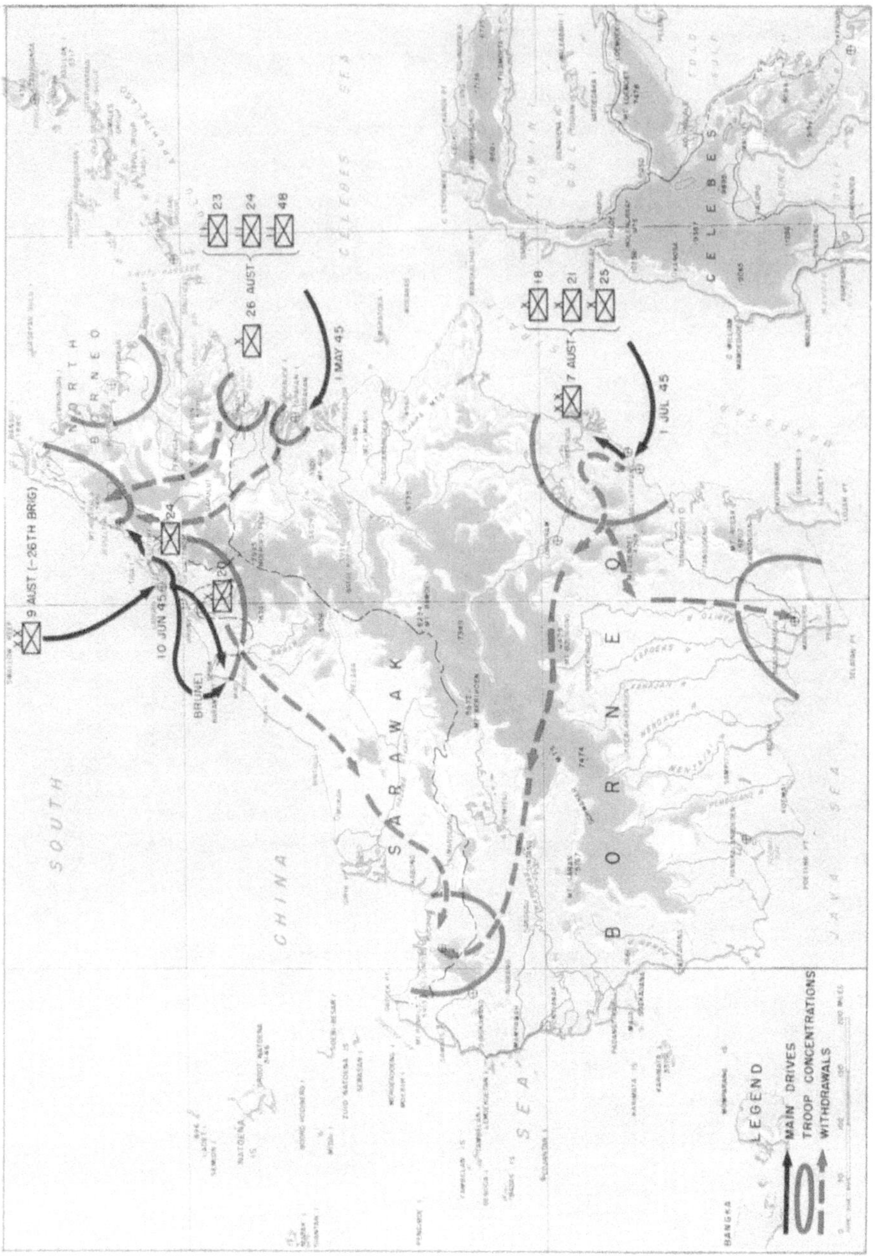

PLATE NO. 109
The Borneo Operations, May—July 1945

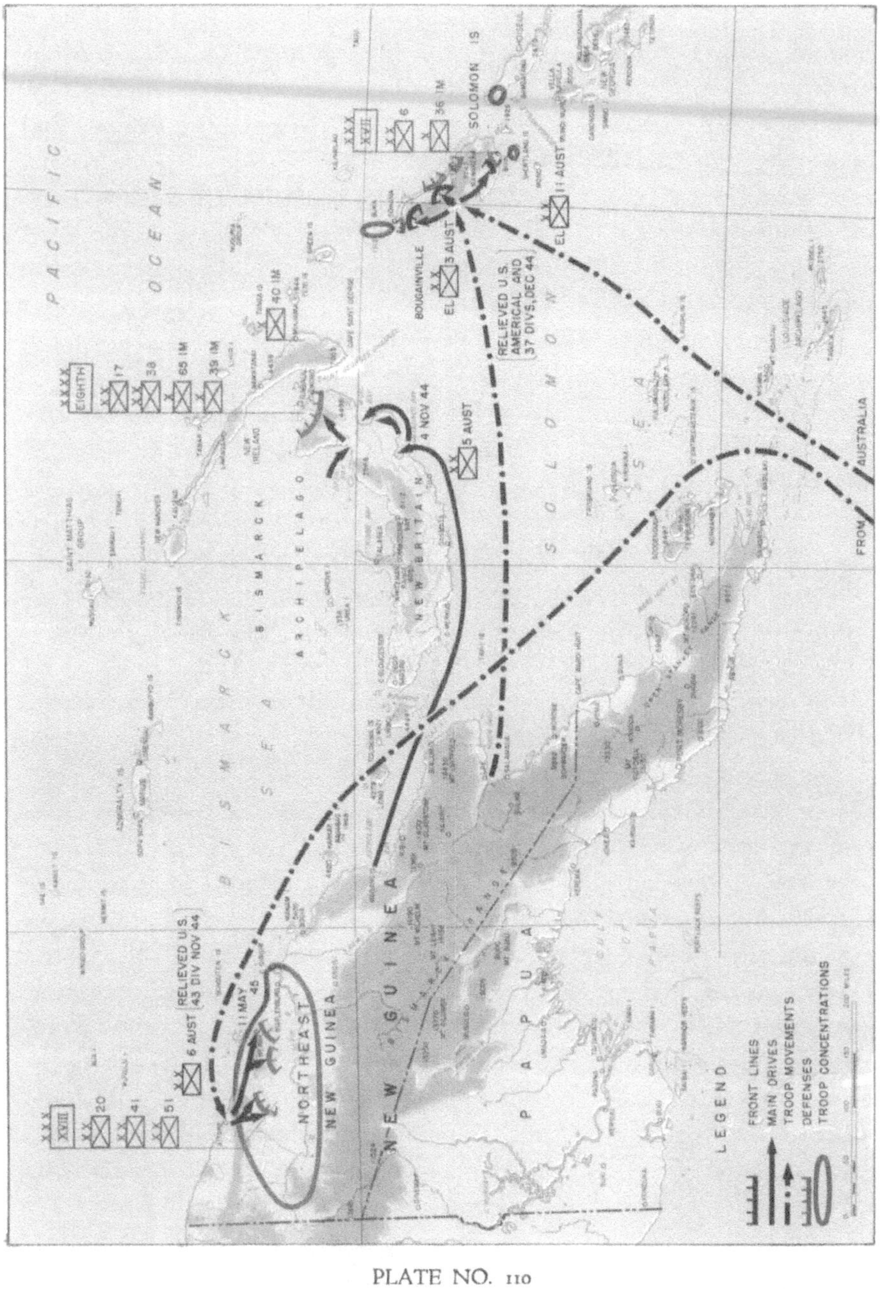

PLATE NO. 110
Mop-Up Operations in Eastern New Guinea, New Britain, and Bougainville

ward resulted in the capture of the But and Dagua airfields.[60] A month later the coastwise advance had reached a point within twenty miles of Wewak. Meanwhile, the inland offensive encountered determined resistance as the Australians approached the heavily defended Maprik area.[61]

A general attack against Wewak was launched on 10 May. Supported by tanks, artillery, and air and naval bombardment, the Australians drove westward against the Japanese Eighteenth Army's last main positions on the shore of Eastern New Guinea. An amphibious landing on the east coast of Cape Moem cut the enemy's coastal escape route and menaced Wewak from the east. After fierce fighting, the Wewak Peninsula was wrested from the Japanese on 11 May. The whole Wewak coastal area was then cleared when the eastward and westward drives joined on 23 May. Except for a pocket remaining at Maffin Bay, the New Guinea coastline was free of enemy resistance as far west as Geelvink Bay.[62]

New Britain was returned to Australian control early in November 1944. By-passing positions seized earlier in the war, troops of the Australian 5th Division landed first at Jacquinot Bay on the south coast of the island to place themselves only 100 miles from the Japanese stronghold at Rabaul. Later they made another landing across the island at Wide Bay which cut this distance in half. The Royal Australian Navy and Air Force supported the operations.[63]

Drives along the north and south coasts forced the Japanese into the mouth of the narrow Gazelle Peninsula and bitter fighting ousted them from their strong positions in the Open Bay area by April 1945. Although the Australians carried out vigorous patrolling and fought occasional skirmishes with the enemy, their activity was generally limited after May. The remaining Japanese forces were effectively confined within the limits of the Gazelle Peninsula.

A campaign to destroy the Japanese Eighth Area Army's heavily garrisoned bastion at Rabaul was made unnecessary by Japan's capitulation and surrender. Such an operation would undoubtedly have proved bitter and costly. Post-war investigation disclosed that the high ground around Rabaul was honeycombed with strong underground defenses heavily stocked with ammunition and explosives.

Most of Bougainville and all of Buka Island to the north were still under the control of the Japanese Seventeenth Army when the Australians took over the defensive perimeter at Empress Augusta Bay in November 1944. After relieving United States forces, Australian First Army troops expanded the perimeter considerably and began a gradual elimination of enemy units on the island. They moved along the western coast and across the island toward Numa Numa, the main enemy base on the east side of the island. Stubborn resistance was encountered in all sectors, particularly in the central areas where the Japanese lines of communications to the north were seriously threatened.

In their drive northward toward Buka Passage, the Australians were faced by a well-entrenched enemy at Tsimba Ridge in the north-west part of the island. In February, heavy artil-

60 G-2, GHQ, SWPA, *Daily Summary* No. 1089, 20/21 Mar 45; GHQ, SWPA, Communique No. 1080, 22 Mar 45, and Communique No. 1083, 25 Mar 45.

61 G-2, GHQ, SWPA, *Daily Summary* No. 1112, 20/21 Apr 45.

62 G-2, GHQ, SWPA, *Daily Summary* No. 1133, 11/12 May 45; G-3, GHQ, SWPA, *Monthly Summary of Operations*, May 45; GHQ, SWPA, Communique No. 1132, 13 May 45, and Communique No. 1133, 14 May 45.

63 Buggy, op. cit., p. 283; ONI, USN, *Office of Naval Intelligence Weekly*, Vol. IV, No. 3, 17 Jan 45, p. 205. Hereinafter cited as: O. N. I. Weekly.

lery barrages and fierce bombing and strafing attacks by Australian and New Zealand airmen were instituted to dislodge the Japanese from their positions. During the same month, the Motupena Peninsula in the southwestern part of the island was completely occupied against light opposition.[64]

To speed the northward advance, Australian units made two landings in March on the Soraken Peninsula in the northwestern sector. A drive across the peninsula to the east coast compressed the Japanese into the northern tip of the island and effectively severed their lines of communication.[65] The Soraken Peninsula was cleared of the enemy by May as the Australians moved northward toward Buka Passage. The advance to the east coast in the central part of the island reached Numa Numa. In the south, meanwhile, the thrust toward Buin progressed slowly against bitter opposition.[66]

During June, successive enemy pockets of resistance were steadily eliminated. To the north, active patrolling was carried out as small bands of raiding Japanese were annihilated. In the central and southern sectors the Australians increased pressure on the strong enemy positions emplaced along the Mibo River. The situation remained generally static thereafter with the defenders being gradually compressed to the south and east along the coastal sectors and north to Buka Island until the surrender.[67]

The final mop-up of the northeast New Guinea, New Britain, and Bougainville areas which were spread over 1,000 miles of land and water had been a tedious task. At the end of July 1945, General MacArthur's Headquarters in Manila announced that a total of 12,385 Japanese had been killed on these islands since the first of the year.[68]

Interim Plans

Pending termination of the European conflict, for which the amphibious invasion of Japan waited, consideration was given to other possible operations which would require relatively small resources and would not interfere with preparations for the main effort. A large-scale offensive by United States forces on the Asiatic continent was deemed impractical because of difficult terrain, inadequate communications, and strong Japanese ground force opposition. The China coast, however, had objectives suitable for limited operations subsequent to the conquest of the Ryukyus. Seizure of positions below Shanghai would tighten the blockade of Japan from the south while occupation of areas north of Shanghai would cut Japanese lines of communication to Korea and Manchuria across the Sea of Japan and the Yellow Sea. The Shantung Peninsula, the Shanghai area, the Ningpo Peninsula, and the Korean Archipelago, all within range of Tokyo, offered favorable bases for the intensification of aerial bombardment.[69]

Certain operations in the North Pacific could be undertaken simultaneously. Existing schedules called for the defeat of Japan without the assistance of the Soviet Union. The possible use of United States and Russian heavy bombers and long-range fighters from

64 O. N. I. Weekly, Vol. IV, No. 7, 14 Feb 45, p. 529.

65 Buggy, op. cit., p. 283; O. N. I. Weekly, Vol. IV, No. 13, 28 Mar 45, p. 1026 and No. 14, 4 Apr 45, p. 1114.

66 O. N. I. Weekly, Vol. IV, No. 18, 2 May 45, pp. 1472–73, and No. 20, 16 May 45, p. 1605.

67 G-3, GHQ, SWPA, *Monthly Summary of Operations*, Jul 45.

68 GHQ, SWPA, Communique No. 1208, 28 Jul 45.

69 Joint Staff Planners (JSP) Report to JCS, 31 Dec 44, JCS 1215, G-3, GHQ Ex JCS and CCS Papers 6 (TS); COMGENCHINA Radio No. CFBX 32476 to CINCSWPA and CINCPOA, 6 Feb 45, G-3, GHQ Admin 381/20-P (TS).

bases in Siberia and the Maritime Provinces was strategically desirable but not considered essential at any time. During late 1944, tentative plans were made for securing a water route through the Sea of Okhotsk to Russian ports once the Russian entry into the war against Japan became imminent. These plans were dropped, however, because necessary resources were not available and also because such a maneuver might have precipitated the premature entry of the Soviet Union into the Pacific War.[70]

The Soviet Union was not disposed to enter the war until after the defeat of Germany. The most advantageous time, from the Russian point of view, would be after United States forces made their initial lodgment in Kyushu, drawing Japanese troops from Manchuria. Conversely, the most favorable time from the American standpoint would be three or four months after the surrender of Germany and about three months prior to the invasion of Kyushu. This correlated timing would have ensured that the Soviet Union had sufficient strength to eliminate the possibility of a successful Japanese counterattack which might disrupt the Russian advance or necessitate aid from the United States. At the same time, it would prevent the displacement of hostile troops from Manchuria, Korea, and North China to Japan's Home Islands.[71]

Toward the end of 1944 and in early 1945 the question of Russian intervention in the Pacific appeared occasionally in international discussions. The political, economic, and military effect of such intervention seemed to have become a vital factor in the hitherto secret understandings. From the viewpoint of GHQ, AFPAC, any intervention during 1945 was not required. The substance of Japan had been gutted; the best of its army and navy had been defeated; the Japanese Homeland was at the mercy of air raids and invasion. Although General MacArthur in 1941 had urged Russian participation to draw the Japanese away from the South Pacific and Southeast Asia, by 1945 such intervention had become superfluous.

In February 1945, the combined Chiefs of Staff favored an invasion of Kyushu—Honshu in late 1945 or early 1946, following the defeat of Germany. The date of Germany's capitulation was estimated, at the earliest, as 1 July 1945 and, at the latest, as 31 December 1945. It was also estimated that Japan would be defeated eighteen months after Germany.[72]

Additional positions to further the blockade and air bombardment of Japan would be seized following the Okinawa operation and prior to the invasion of Kyushu. The air bombardment of Japan would then be intensified, thereby further reducing Japan's major military forces and creating a situation favorable to the direct invasion of the industrial heart of Japan via the Tokyo Plain.

General MacArthur considered that an invasion of Kyushu was undoubtedly the most advantageous operation to undertake in the year 1945. He was convinced that any movement or allocation of resources that did not directly contribute toward this goal should be eliminated. He felt that the full power of the combined resources—ground, naval, and air—in the Pacific was sufficient to initiate such an operation by November 1945, regardless of the status of redeployment from Europe and without consideration of Russia's entry or non-entry into the Pacific war.

The real crux of the problem was the supply

70 JSP Report to JCS, 23 Nov 44, JCS 1176, G-3, GHQ Ex JCS and CCS Papers 5 (TS); CCS Report to the President and the Prime Minister, 9 Feb 45, CCS 776/3, G-3, GHQ Ex JCS and CCS Papers (TS).

71 JSP Report to JCS, 23 Nov 44, JCS 1176, G-3, GHQ Ex JCS and CCS Papers 5 (TS).

72 Ibid.

situation but this, General MacArthur thought, could be solved if, as sole commander, he were given a high degree of operational authority to reorganize all army supply agencies in the Pacific. To provide the required allotment of service troops for the invasion of Kyushu there would have to be a ruthless thinning of rear areas and a comprehensive pooling of army and navy resources. It would be necessary for the Navy to assist in moving forward service troops, equipment, and supplies from New Guinea and the South Pacific. The War Department would have to allocate additional shipping for the amphibious movement and for the direct resupply to Kyushu.[73]

Command Changes for "Olympic"

The issuance of a Joint Chiefs of Staff directive on 25 May clarified the command organization for the operations against the main island of Japan and set the date for the "Olympic" invasion of Kyushu at 1 November 1945.[74] This target date allowed about five months for the preparation of the ground and amphibious forces and for the completion of the necessary logistic arrangements.

The first step taken was a reorganization of the army command. General MacArthur assumed the responsibility for the Japanese campaign as Commander in Chief of the newly constituted United States Army Forces in the Pacific rather than as Commander in Chief of the Southwest Pacific Area, the capacity in which he had directed all previous operations. The duties and responsibilities of the two positions were quite different. As Southwest Pacific Area Commander, General MacArthur had exercised operational but not administrative control over ground, air, and naval forces of the United States, Australia, and the Netherlands East Indies. Simultaneously, as Commanding General of the United States Army Forces in the Far East, he was administrative commander of the Sixth and Eighth Armies, the Far East Air Force, and the United States Army Services of Supply, all of which comprised the major American elements in the Southwest Pacific.

As Commander in Chief, United States Army Forces in the Pacific, General MacArthur's responsibilities were expanded to include both operational and administrative control over all United States Forces in the Pacific except the Twentieth Air Force and certain troops in Alaska and the Southeast Pacific Area. (Plate No. 111) The need for a separate administrative headquarters was ended. The organization of the United States Army Forces in the Far East was discontinued except as a nominal agency to permit the approval of certain financial expenditures in the Philippines as required by law. The United States Army Forces in the Western Pacific (AFWESPAC), a command subordinate to AFPAC, was created at Manila on 1 June 1945, under the command of Lt. Gen. Wilhelm D. Styer. This command was designated to take over the functions of the United States Army Forces Service of Supply and some of the functions of the deactivated USAFFE. AFWESPAC would control all American Army forces within the Southwest Pacific Area except major combat commands. It would also be responsible for the logistical support of operations, except for air corps technical supplies.

A similar organization, the United States Army Forces in the Middle Pacific (AFMIDPAC) was established on 1 July under Lt. Gen. Robert C. Richardson, Jr. This command was formed to take over the forces and instal-

73 CINCAFPAC Radio No. C-14442 to WARCOS, 20 Apr 45, C/S GHQ, WD No. 982 (TS).

74 The events leading up to this directive are discussed in Chapter XIII.

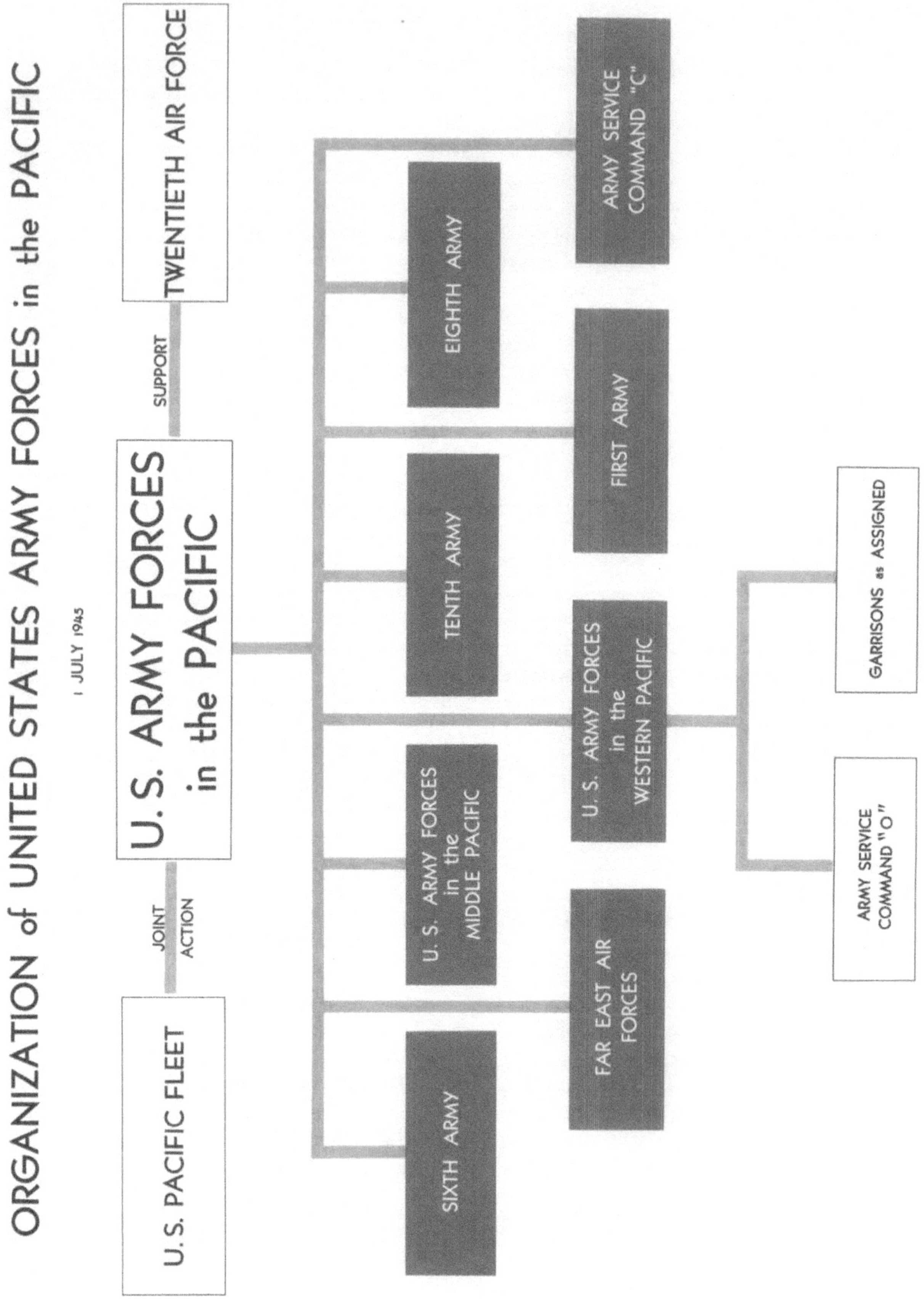

PLATE NO. 111
Organization of United States Army Forces in the Pacific

lations of the United States Army Forces of the Pacific Ocean Areas and the Hawaiian Department as they were released from the operational control of Admiral Nimitz.[75]

Okinawa, strategically the most important island of the Ryukyus, had been invaded by the Tenth Army's XXIV Corps and the Marine III Amphibious Corps on 1 April 1945. After some of the heaviest fighting of the Pacific War in which severe losses in men and ships were suffered by both sides, the campaign was declared officially closed on 21 June, except for mopping-up disorganized enemy remnants.

With the capture of Okinawa, the Allies had acquired new airfields from which almost any type of plane could operate against targets in Japan's Home Islands. Okinawa also provided several excellent anchorages within 350 miles of southern Kyushu. Japan had lost the last outer fortress protecting her communication lines to Korea, the Chinese mainland, and to the Indo-China and Singapore areas. Formosa was cut off and left helpless against Allied air and sea attack.

On 19 July, Admiral Nimitz was directed to transfer to General MacArthur by 1 August control of the American-held areas in the Ryukyus and all army forces there, including the Tenth Army. At that time, General MacArthur would assume responsibility for the defense of these areas and for the logistical support of the Strategic Air Forces based there. Admiral Nimitz would retain command of naval forces, installations, and bases.[76]

By these reorganizations, General MacArthur was finally given operational as well as administrative control of all army resources in the Pacific, with the exception of General Richardson's AFMIDPAC and certain other garrison and service troops in the islands of the Pacific Ocean Areas. On 28 July, General MacArthur attempted to secure operational responsibility of these forces also, so that all army forces in the Pacific would be under one commander. He again recommended that the area boundaries in the Pacific be abolished because they had long ceased to serve any useful purpose, were patently artificial, and complicated the proper strategic and tactical handling of the United States forces in that theater of operations. He pointed out that the existing demarcations prevented unification of command within each of the services and thus were contrary to the operational principles of the theater.[77] General Richardson, however, continued to function under dual control until the end of the war, with responsibilities to the commanders in chief of both services.

Division of the Southwest Pacific Area

The final disposition of the Southwest Pacific Area presented another problem that had to be resolved before General MacArthur could concentrate upon his responsibilities as Commander in Chief of the United States Army Forces in the Pacific. He had recommended on 25 February 1945 that the Southwest Pacific Area be dissolved with the completion of the Australian operations in the Netherlands East Indies. The consolidation of liberated territories and the conduct of civil affairs could then be transferred to British and

75 CINCAFPAC Radio No. CA-51477 to WARCOS, 10 Apr 45, C/S GHQ S, WD No. 971 (TS); CINCAFPAC Radio No. C-16063 to WARCOS, 12 May 45, C/S GHQ S, WD No. 1014 (TS); WARCOS Radio No. W-87663 to CINCAFPAC, 26 May 45, C/S GHQ S, WD No. 1030 (TS); GHQ, AFPAC, General Order No. 13, 20 Jun 45, G-3, GHQ Admin 300–4 (S).

76 JCS Radio No. V-73BG152 to CINCAFPAC and CINCPAC, 19 Jul 45, C/S GHQ, WD No. 1076 (TS).

77 CINCAFPAC Radio No. C-29008 to WARCOS, 28 Jul 45, C/S GHQ, WD No. 1091 (TS).

Dutch authorities to permit the concentration of American resources for the invasion of Japan.[78]

On 27 June, General MacArthur again recommended that the areas south of the Philippines be removed from United States control, turned over to the British, and handled by them in co-ordination with the Dutch. As an initial step, all ground, naval, and air forces other than United States forces would be released and transferred to the commanders designated. Thereafter, ports, airfields, and bases would be released progressively as their use by United States forces was terminated.[79]

Negotiations with the British for this purpose were opened in April. The British at first were reluctant to put an additional burden upon the Southeast Asia Command until after 1 January 1946. It was expected that Admiral Mountbatten by that time would have completed the recapture of Singapore. The British were also uncertain of the extent to which they could absorb the great numbers of men and ships and the vast quantities of materiel which would be transferred when the withdrawal of United States control in this area was effected.[80]

The Americans believed, however, that Australian and Dutch units could garrison the Netherlands East Indies without affecting Admiral Mountbatten's planned operations. The final decision was made at the Potsdam Conference. The Southeast Asia Command was enlarged to include Borneo and the Celebes and Admiral Mountbatten was directed to assume control as soon as practicable after 15 August. The Southwest Pacific Area was continued as an Allied command under General MacArthur but its operations were limited to minor rear area activities.[81]

As the date for the transfer of responsibilities approached, General MacArthur dispatched messages to the Governments of Australia, New Zealand, and the Netherlands in tribute to the Allied troops who had fought under him in the arduous campaigns of the Southwest Pacific. To the Australian soldiers, sailors, and airmen, he wrote:

Since the 18th of April 1942, it has been my honor to command you in one of the most bitter struggles of recorded military history—a struggle against not only a fanatical enemy under the stimulus of early victory, but the no less serious odds of seeming impenetrable barriers of nature—a struggle which saw our cause at its lowest ebb as the enemy hordes plunged forward with almost irresistible force to the very threshold of your homeland.

There you took your stand and with your Allies turned the enemy advance on the Owen Stanleys and at Milne Bay in the fall of 1942, thus denying him access to Australia and otherwise shifting the tide of battle in our favor. Thereafter, at Gona, Wau, Salamaua, Lae, Finschhafen, the Huon Peninsula, Madang, Alexishafen, Wewak, Tarakan, Brunei Bay, and Balikpapan, your irresistible and remorseless attack continued.

Your airmen ranged the once enemy-controlled skies and secured complete mastery over all who dared accept your challenge—your sailors boldly engaged the enemy wherever and whenever in contact in con-

78 CINCSWPA Radio No. CA-50688 to WARCOS, 26 Feb 45, C/S GHQ, WD No. 937 (TS).

79 CINCAFPAC Radio No. C-21640 to WARCOS, 27 Jun 45, C/S GHQ, WD No. 1065 (TS); WARCOS Radio No. V-93BG165 to CINCAFPAC, 19 Jul 45, C/S GHQ, WD No. 1077 (TS).

80 Ibid.

81 The actual transfer of command of the areas took place on 2 September 1945, the date on which the surrender was formally signed by the Japanese. At the same time, Allied Land Forces, Allied Air Forces, and Allied Naval Forces, the three operational commands of the Southwest Pacific Area, were inactivated. WARCOS Radio No. W-89740 to CINCAFPAC, 30 May 45, C/S GHQ S, WD No. 1031 (TS); CINCAFPAC Radio No. C-17403 to WARCOS, 30 May 45, C/S GHQ S, WD No. 1033 (TS); JCS Radio No. V-169BG248 to CINCAFPAC, 22 Jul 45, G-3, GHQ Ex JCS and CCS Papers (TS).

temptuous disregard of odds and with no thought but to close in battle so long as your ships remained afloat.

These, your glorious accomplishments, filled me with pride as your commander, honored for all time your flag, your people and your race, and contributed immeasuably to the advancement of the sacred cause for which we fought.

I Shall shortly relinquish this command which, throughout its tenure, you have so loyally and so gallantly supported. I shall do so with a full heart of admiration for your accomplishments and of a deep affection born of our long comradeship in arms. To you of all ranks, I bid farewell.[82]

Demobilization, Redeployment, and Replacements

When the United States Eighth Army assumed the task of cleaning out remnants of the Japanese defenders in the Philippines, General Krueger's Sixth Army was released to prepare for the invasion of Kyushu. During a two-year period, the Sixth Army had conducted 12 major operations, involving 21 separate amphibious landings, and had advanced a cumulative distance of approximately 3,000 miles. Many of its units were understrength as a result of battle losses and the readjustment of high-point personnel. To remedy these shortages, veteran troops from the European Theater were to be obtained as replacements.[83]

Before the war ended in Europe, plans had been made for a partial demobilization of the Army immediately following the defeat of Germany. On 16 April 1945, General Marshall urged all theater commanders to plan for the movement of the maximum number of troops likely to be eligible for discharge to the United States immediately upon the termination of the war in Europe.[84] After the German capitulation, it was estimated that 2,000,000 men would be discharged during the ensuing year. Approximately 300,000 of these would be drawn from the Pacific Theater.[85]

Under the initial adjusted service rating score of 85, the Sixth Army, which was being prepared for the invasion of Kyushu, would lose 23,000 enlisted men. In addition, there were 20,712 officers in the theater with 85 points or more who were eligible for discharge, even though there was already an acute shortage of officers for combat units. To correct this situation, plans were made to ship 10,000 selected enlisted men and 4,500 officers directly from Europe to the Pacific Theater in September.[86]

General MacArthur and his staff were opposed to a further lowering of points prior to the invasion of Kyushu, as this would release veteran non-commissioned officers and specialists and would endanger projected operations. There was not sufficient time to replace personnel of invasion units who had critical scores slightly below 85, nor time to train the replacements if they were furnished. Recognizing the urgency of the situation, the War Department agreed to retain the 85 score temporarily to meet operational requirements in the Pacific.

The Southwest Pacific Area had been handicapped continuously by a shortage of personnel during most of the war. Only after Germany's defeat could sufficient forces and shipping be accumulated to make the final effort against Japan. When General MacArthur became Commander in Chief, AFPAC, on 6 April

82 CINCAFPAC Radio No. 210555 to Prime Minister Chifley of Australia, with enclosure, "Personal Message of Farewell to Australian Forces," 21 Aug 45, C/S GHQ S, Aust 952.

83 G-3, GHQ, SWPA, *Monthly Summary of Operations*, Jun 45.

84 Marshall Radio No. WX-68469 to MacArthur, Eisenhower, etc. 16 Apr 45, G-3, GHQ Admin 370 (S).

85 CINCAFPAC Radio No. C-29256 to WARCOS, 29 Jul 45, G-3, GHQ Admin 370 (TS).

86 Ibid. CINCAFPAC Radio No. C-23346 to WARCOS, 5 Jul 45, G-3, GHQ Admin 370 (TS). WARCOS Radio No. W-42164 to CINCAFPAC, 2 Aug 45, G-3, GHQ Admin 370 (TS).

1945, the combined forces under his new command consisted of 1 airborne division, 1 cavalry division, 19 infantry divisions, and approximately 53 air groups. Proposed redeployment would build this strength to 5 armored divisions, 1 airborne division, 1 cavalry division, 29 infantry divisions, and 125 air groups (plus 17 squadrons). The total forces of the Pacific area would be increased from approximately 1,400,000 army troops as of 30 June 1945, to 2,439,400 as of 31 December 1945.[87]

Securing qualified replacements was a task that had required careful planning and close co-ordination with the War Department. The situation had become particularly difficult for the Southwest Pacific Area in February 1945. Many units were understrength and there were cases where combat casualties and battle fatigue had whittled rifle companies down to only thirty men. The effective combat strength of the theater had been so reduced that General MacArthur had even considered inactivating a division and using its personnel as replacements.

The situation was relieved considerably in the spring of 1945. The reorganization of the Army command in the Pacific and the approaching end of the war in Europe permitted an increased flow of replacements. In May, the understrength was reduced to 4,971 by the arrival of 46,420 new men. At the end of the following month, the theater had gained 23,029 overstrength.

In the first part of August 1945, 115,000 troops passed through the Panama Canal on their way from Europe to Pacific assignments. The advance echelon of the First Army's Headquarters had arrived in the Philippines on 7 August. The VII and XVIII Corps were scheduled to arrive by 15 October and six infantry divisions were slated to arrive in the Philippines during September, October, and November.[88] The tremendous forces and supplies necessary to execute the greatest amphibious invasion in history were being moved rapidly into position.

87 Memo to G-3, GHQ, AFPAC, 5 May 45, summarizing JCS 521/12, 29 Mar 45, JCS 521/15, 11 Apr 45, and JCS 521/16, undated, JCS, GHQ, and CCS No. 3 (TS).

88 WARCOS Radio No. W-18922 to CINCAFPAC, 20 Jun 45, G-3, GHQ Admin 370 (TS); Teleconference TT-3844 (Manila—Washington) 16 Aug 45, G-3, GHQ Admin 384-1/12-2-P (TS).

CHAPTER XIII

"DOWNFALL"—THE PLAN FOR THE INVASION OF JAPAN

Evolution of "Downfall"

While the war was being fought in the Philippines and Okinawa, plans were ripening rapidly for the largest amphibious operation in the history of warfare. "Downfall," the grand plan for the invasion of Japan, contemplated a gargantuan blow against the islands of Kyushu and Honshu, using the entire available combined resources of the army, navy, and air forces.

The plans for "Downfall" were first developed early in 1945 by the Combined Chiefs of Staff at the Argonaut Conference held on the tiny island of Malta in the Mediterranean. On 9 February, just a few days before the historic Three-Power meeting at Yalta, President Roosevelt and Prime Minister Churchill were informed of the conclusions reached at Argonaut. At that time, the strategic concept of future operations in the Pacific embodied the defeat of Japan within eighteen months after Germany's surrender and included the following series of proposed objectives:

a. *Following the Okinawa operation, to seize additional positions to intensify the blockade and air bombardment of Japan in order to create a situation favorable to:*

b. *An assault on Kyushu for the purpose of further reducing Japanese capabilities by containing and destroying major enemy forces and further intensifying the blockade and air bombardment in order to establish a tactical condition favorable to:*

c. *The decisive invasion of the industrial heart of Japan through the Tokyo Plain.*[1]

On 29 March, the U.S. Joint Chiefs of Staff, working on the assumptions that the war in Europe would be over by 1 July 1945 and that the forthcoming Okinawa operation would be concluded by mid-August of 1945, set a tentative schedule for the invasion of Japan. The invasion plan was assigned the cover name "Downfall" and consisted of two main operations: "Olympic," the preliminary assault on the southern island of Kyushu, which was slated for 1 December 1945, and "Coronet," the subsequent landing on Honshu, which was scheduled for 1 March 1946. (Plate No. 112) It was proposed that forces already in the Pacific be used to the fullest extent possible in planning for the assault and follow-up phases of "Olympic." Reserve and follow-up divisions for "Coronet" would be obtained by redeployment, either directly or via the United States, of troops and equipment from the European Theater.[2]

On 3 April 1945, the Joint Chiefs of Staff issued a directive in which General MacArthur was instructed to complete the necessary operations in Luzon and the rest of the Philippines, prepare for the occupation of North Borneo, and "make plans and preparations for the campaign in Japan." The amphibious and aerial phases of the projected Homeland invasion

1 CCS Report to President Roosevelt and Prime Minister Churchill, 9 Feb 45, CCS 776/3, JCS and CCS Papers, G-3 Ex (TS).

2 JCS Planners Report, 29 Mar 45, JCS 521/13, JCS and CCS GHQ (TS).

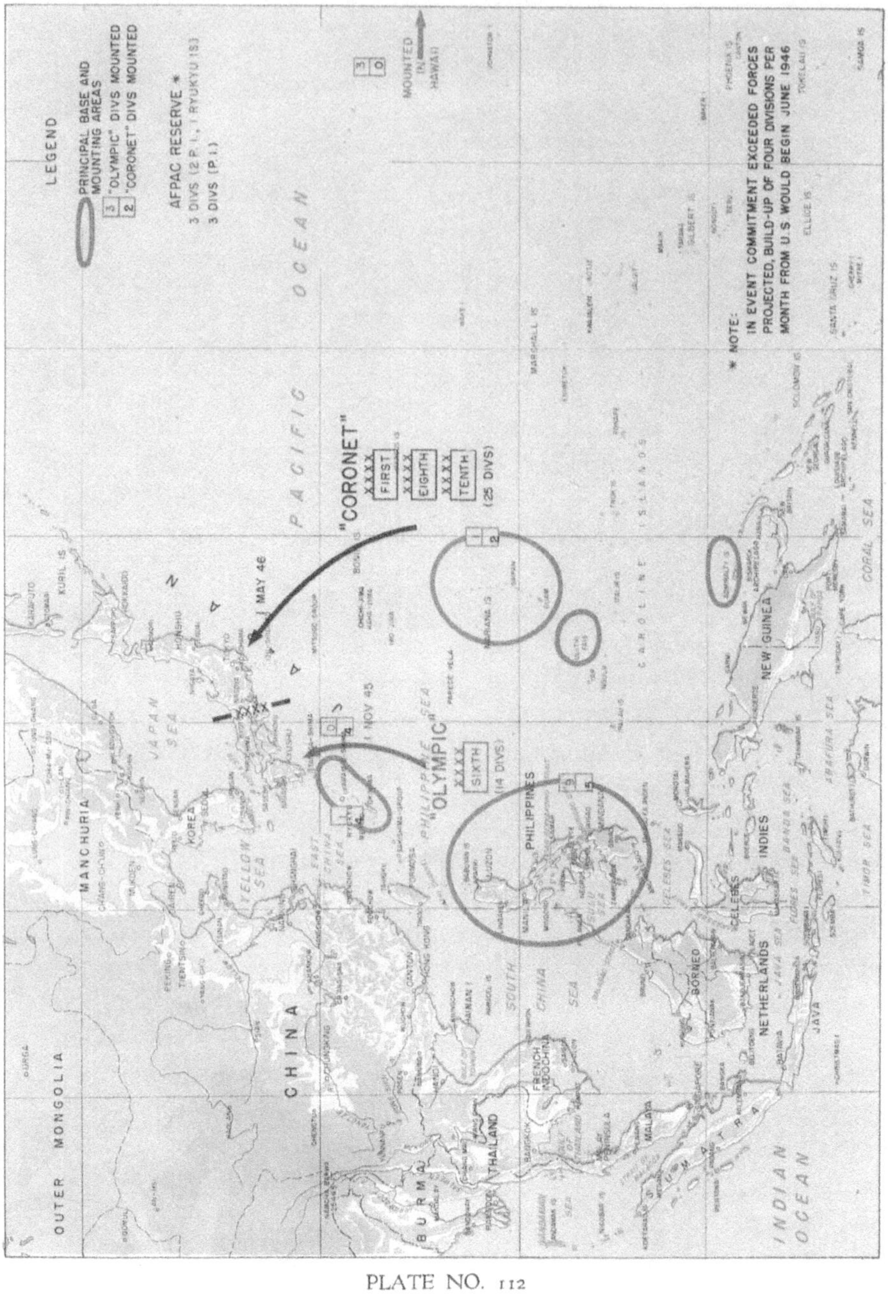

PLATE NO. 112

"Downfall" Plan for the Invasion of Japan, 28 May 1945

were to be formulated in co-operation with Admiral Nimitz and General Arnold.[3]

Strategies under Consideration

As was to be expected in the consideration of an assault of such magnitude, there were numerous and varying opinions as to the best strategy to follow in developing the final plans. These different opinions flowed into two main channels of thought. On the one hand, it was felt that much more preparation was needed than would be possible under the tentative target dates of 1 December and 1 March for the Kyushu and the Honshu operations. To reduce the risks inherent in any assault on the well fortified and strongly garrisoned islands of Japan, a preliminary and far-reaching campaign of air-sea blockade and bombardment was advocated. To implement such a concept, prior operations along the China coast at Chosan and Shantung, in Korea, and in the Tsushima Strait area were envisaged. Such a strategy, it was held, though admittedly more prolonged than direct assault, would minimize the number of casualties, further reduce hostile air potential, and cut off reinforcements from Asia to Japan. Furthermore, it was conceivable that such a program could force Japan's surrender without the necessity of a major landing on the Home Islands.

The other school of thought believed in driving directly to the heart of Japan as soon as the necessary forces could be mounted from the Philippines and land-based planes could be established in the Ryukyus. The proponents of this strategy contended that Japanese air and sea power was already a relatively minor factor and that by the end of 1945 it would be weakened sufficiently to permit a successful invasion. Any sizeable reinforcement to Japan from China or Manchuria could be effectively interdicted by the powerful ships and planes of the U. S. Fleet and by air strikes from Okinawa. In addition, the enemy's shipping potential had declined to a level that did not permit material strengthening of forces in Japan by transport from the Asiatic coast. By December, the combined efforts of B-29's and carrier planes would have devastated large areas in Japan to soften the landing sectors and prohibit the rapid maneuver of major Japanese forces. From a broad perspective, it was argued, immediate invasion was considered to be "the quickest way to assure the end of the war."[4]

In reply to a query from General Marshall requesting his opinion on the problem, General MacArthur pictured the future strategy in the Western Pacific as presenting three principal courses of action. First, the Allies could encircle Japan by further Allied expansion to the westward, at the same time deploying maximum air power preparatory to attacks on either Kyushu or Honshu in succession, or on Honshu only. A second course would be to isolate Japan completely by seizing bases to the west and endeavoring to bomb her into submission without actually landing in force on the Homeland beaches. The third course open was to attack Kyushu directly and install air forces to cover a decisive assault against the principal island of Honshu.[5]

General MacArthur went on to analyze the relative merits of each of the choices offered. The first course, he felt, provided greater air power and a high degree of pre-assault neutralization, in addition to achieving the eventual isolation of Japan, but it had the great disad-

---

3 JCS Radio No. WX-62774 to CINCAFPAC, CINCPAC, and CG 20AF, 4 Apr 45, C/S GHQ S, WD No. 957 (TS). See Chapter XII.

4 C/S WD Radio No. WAR-67098 to CINCSWPA, 12 Apr 45, C/S GHQ S, WD No. 977 (TS).

5 CINCAFPAC Radio No. C-14442 to WARCOS, 20 Apr 45, C/S GHQ, WD No. 982 (TS).

vantage of deploying the bulk of available resources off the main axis of advance. Such a course would fail to increase short-range air coverage of vital portions of the Japanese islands and yet, by spreading Allied strength in the Pacific over a wide area, would prevent an attack on the Japanese mainland until after redeployment from Europe could be accomplished. General MacArthur also expressed the fear that United States forces would become progressively more involved in the China area, perhaps necessitating a further postponement of the Honshu operation.

The second course of action—reliance upon bombing alone—General MacArthur considered to be capable of accomplishment with a minimum loss of life but at the risk of prolonging the war indefinitely. In addition, he stated:

[It] would fail to utilize our resources for amphibious offensive movement; assumes success of air power alone to conquer a people in spite of its demonstrated failure in Europe, where Germany was subject to more intensive bombardment than can be brought to bear against Japan, and where all the available resources in ground troops of the United States, the United Kingdom, and Russia had to be committed in order to force a decision.[6]

The third course, involving assaults directly against Kyushu and Honshu, had several strong arguments in its favor and was analyzed by General MacArthur as follows:

[It] would attain neutralization by establishing air power at the closest practicable distance from the final objective in the Japanese islands; would permit application of full power of our combined resources, ground, naval, and air, on the decisive objective; would deliver an attack against an area which probably will be more lightly defended this year than next; would continue the offensive methods which have proved so successful in Pacific campaigns; would place maximum pressure of our combined forces upon the enemy, which might well force his surrender earlier than anticipated; and would place us in the best favorable position for delivery of the decisive assault early in 1946. Our attack would have to be launched with a lesser degree of neutralization and with a shorter period of time for preparation.[7]

Examination of the advantages and disadvantages of the various courses of possible action, General MacArthur concluded, indicated that the last of these outlined strategies was clearly the preferable plan. Unless there was insufficient supporting air power available or the resources in the Pacific could not be gathered in time to initiate the assault by 1945, he advocated the immediate adoption of this last course.[8]

General MacArthur had little doubt that sufficient power to overcome Japanese resistance could be massed in the Fall of 1945 for an invasion of Kyushu. (Plate No. 113) He stated his reasons as follows:

I am of the opinion that the ground, naval, air, and logistic resources in the Pacific are adequate to carry out Course III. The Japanese Fleet has been reduced to practical impotency. The Japanese Air Force has been reduced to a line of action which involves unco-ordinated, suicidal attacks against our forces, employing all types of planes, including trainers. Its attrition is heavy and its power for sustained action is diminishing rapidly. Those conditions will be accentuated after the establishment of of our air forces in the Ryukyus. With the increase in the tempo of very long range attacks, the enemy's ability to provide replacement planes will diminish and the Japanese potentiality will decline at an increasing rate. It is believed that the development of air bases in the Ryukyus will, in conjunction with carrier-based planes, give us sufficient air power to support landings on Kyushu and that the establishment

6 Ibid.

7 Ibid.

8 Ibid.

of our air forces there will ensure complete air supremacy over Honshu. Logistic considerations present the most difficult problem.[9]

From the standpoint of weather, which was a determinative factor in any plan for operations again Japan, General MacArthur thought that the Kyushu assault should be made in November rather than December and suggested that the tentative date for "Olympic," previously adopted by the Joint Chiefs of Staff, be moved forward one month.

Admiral Nimitz, although suggesting several modifications in the general strategy, was in full agreement with General MacArthur's opinion that the invasion of Kyushu should be effected at the earliest date. He concurred in the selection of 1 November as the day on which to launch the invasion.[10] In a series of conferences at Manila in mid-May, General MacArthur's and Admiral Nimitz's planning staffs formulated the broad principles to be incorporated into the plan for "Downfall" and submitted their conclusions to Washington.[11]

On 25 May, the Joint Chiefs of Staff issued the directive for the "Olympic" operation against Kyushu, setting the target date for 1 November 1945. Under this directive, General MacArthur was given the primary responsibility for the conduct of the entire operation including control, in case of exigencies, of the actual amphibious assault through the appropriate naval commander. In addition, he was directed to make plans and preparations for the continuance of the campaign in Japan and to co-operate with Admiral Nimitz in the formulation of its amphibious phases. Admiral Nimitz, in turn, was charged with the responsibility of carrying out the naval and amphibious phases of "Olympic." The land campaign and its requirements were to be given first priority in all subsequent preparations for the Kyushu assault. The Commanding General of the Twentieth Air Force was directed to co-operate in the planning and execution of "Olympic" and also to assist the campaign in Honshu.[12]

Concept of "Downfall"

The concept of "Downfall" visualized attainment of Japan's surrender by two successive operations: the first, to advance Allied land-based air forces into southern Kyushu in order to develop air support for the second—a "knockout blow to the enemy's heart in the Tokyo area."[13] These operations would be expanded and continued until all organized resistance in the Japanese Home Islands could be brought to an end.

When the plans for "Downfall" were drawn in April 1945, certain suppositions regarding both Allied and enemy capabilities were made. It was assumed that the entire resources of the Pacific would be at the disposal of General MacArthur and Admiral Nimitz for the conduct of the operation; that the flow of redeployed United States Army forces to the Pacific Theater would be maintained after the surrender of Germany; and that required base establishments, staging facilities, and heavy cargo shipping would be available for the mounting and continued support of the forces

9 Ibid. General MacArthur's strategy was predicated primarily on the basis of available American resources. The consideration of possible aid from the Soviet Union was not included in these estimates.

10 CINCPOA Radio No. 280235/Z to CINCAFPAC, 28 Apr 45, C/S GHQ, Navy 625 (TS).

11 Report on CINCAFPAC—CINCPAC Conference at Manila, P.I., 16 May 45, C/S GHQ, Navy 633 (TS).

12 JCS Directive 1331/3 to CINCAFPAC, CINCPAC, and CG 20AF, 25 May 45, JCS and CCS Papers 6, G-3 Ex (TS).

13 GHQ, AFPAC, *"Downfall" Strategic Plan for Operations in the Japanese Archipelago*, 28 May 45 (TS). Hereinafter cited as: *Downfall.*

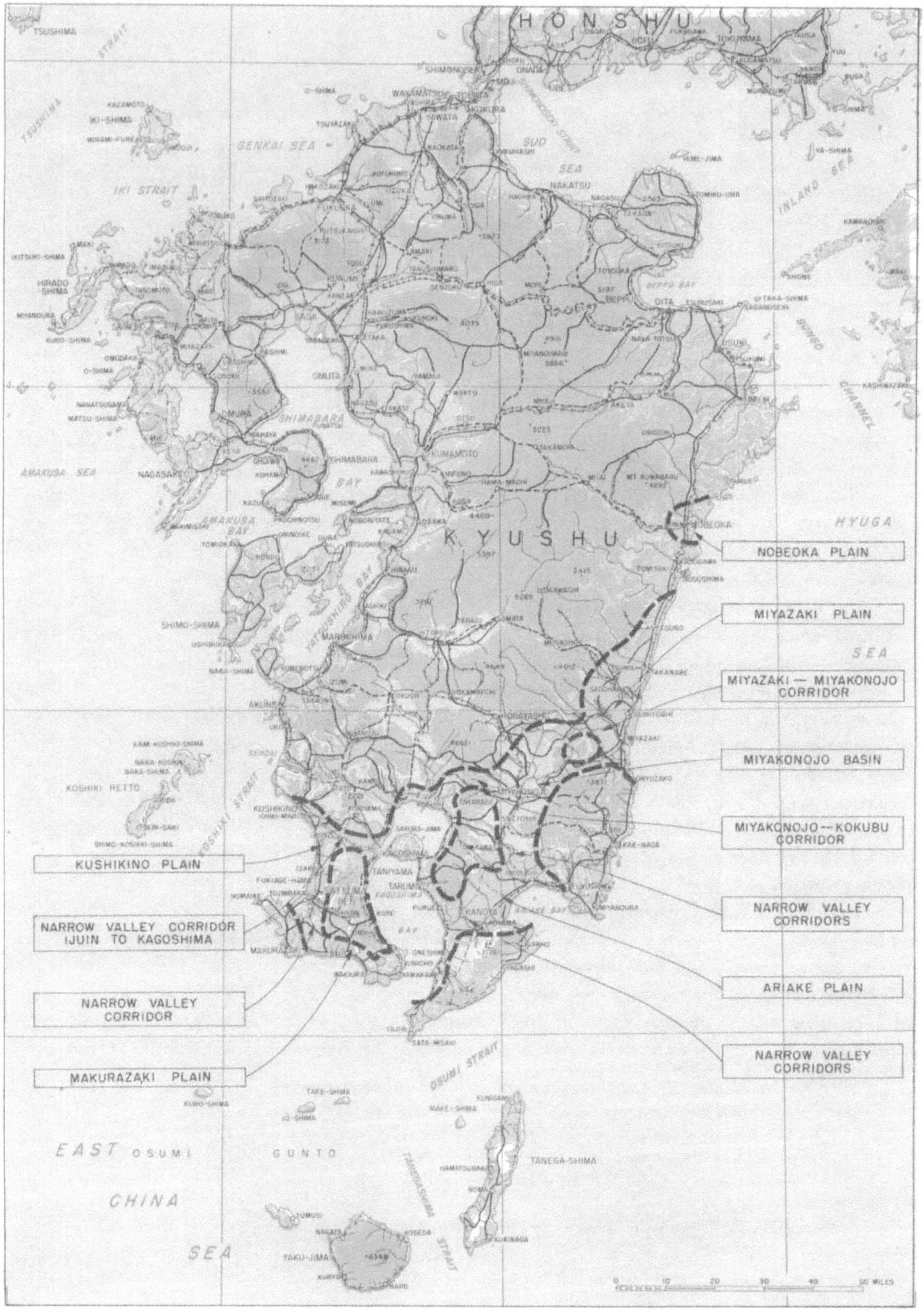

PLATE NO. 113

Kyushu

committed.

It was also assumed that, at the initiation of the assault, United States forces would be established on the Bonins—Ryukyus line. It was estimated that by then land-based planes would have attained air superiority over southern Kyushu while the U. S. Pacific Fleet, operating from forward naval bases in the Philippines, Ryukyus, and Marianas, would dominate the waters east of Japan and the southern part of the East China Sea.

For the purpose of planning, it was presumed that the Japanese would continue the war to the limit of their capabilities, bending every effort to defend the Homeland with their full strength. Such a defense would include not only opposition by their organized military forces but also a fierce and active resistance by the entire population. The power of the Japanese Navy was discounted as constituting only a minor threat. It was estimated that the bulk of the enemy fleet remnants would be withdrawn to the Yellow Sea or to the western side of the Japan Sea. Submarines and suicide boats would undoubtedly oppose the landings with reckless counterattacks and mines would probably be sown in large numbers. The most effective opposition to the Kyushu assault was expected to come from massed Kamikaze attacks by the remaining strength of the Japanese air forces. It was hoped, however, that continued pounding of enemy airfields and factories would minimize these suicide strikes to the point where Allied losses would not be excessive.[14]

Japanese Plans and Estimates

By early 1945, Japan's overseas military forces were precariously close to disaster. Confronted by decisive defeats on every battlefield and by a rapidly decreasing war potential on the home front, the Japanese Government and Imperial General Headquarters took serious steps to bolster their last remaining defense lines. Plans for the final struggle in the Homeland provided for the utilization of all available resources, both of men and material, within a wide resistance sphere encompassing the key areas in Japan, Manchuria, and China.[15]

In January 1945, Imperial General Headquarters estimated that the basic strategy of the Allies included the following broad concepts: (1) isolation of Japan from the continental and southern resources areas; (2) destruction of her vital industries; (3) elimination of Japan's air, naval, and land forces as threats to an amphibious invasion; and (4) extension of the effective range of U. S. fighter planes to the heart of the Japanese Homeland.[16]

To carry out these objectives and to prepare for a full-scale invasion it would be necessary, according to the Japanese reasoning, for the United States to strengthen its existing bases and acquire others still closer to Japan. For this reason, Iwo Jima and the Ryukyu Islands were regarded as almost certain targets for future assault.[17]

At the beginning of 1945, Japan was far from ready to meet a full-fledged Allied invasion. Plans to defend the Home Islands were not considered until after the fall of Saipan in

---

14 Ibid.

15 Report No. 9 to the Supreme Council for the Direction of the War, "Outline of Emergency Procedure for the Decisive Battle," 25 Jan 45, G-2 Historical Section, GHQ, FEC.

16 Statement by Col. Takushiro Hattori, Chief of the Operations Section, Imperial General Headquarters, Army Section, 10 Jun 49, Interrogation Files, G-2 Historical Section, GHQ, FEC.

17 Report No. 10 to the Supreme Council for the Direction of the War, "Estimate of the World Situation," 15 Feb 45, G-2 Historical Section, GHQ, FEC.

July 1944 and even then the outlined preparations were restricted to naval and air force actions with little provision made for ground combat. Even these halfway measures were soon vitiated by the tremendous plane and ship losses incurred in the Philippines Campaign. In the four main islands of Japan Proper only 11 first-line divisions (including 1 armored division) and 3 brigades were available for ground defense until new or reserve units could be mobilized on a large scale. Strategic coastal areas were partially fortified by the construction of heavy artillery positions but no clear-cut Homeland battle strategy had yet been formed.[18]

By mid-January 1945, it had become apparent to the Japanese High Command that a new and comprehensive defense policy was imperative. On 20 January, Imperial General Headquarters issued an outline of policy for future military preparations.

After first stating that the final decisive battle would be waged in Japan Proper, the outline called for an immediate fortification in depth of Japan's defense perimeter. Included in this perimeter were Iwo Jima, Formosa, Okinawa, the Shanghai region, and the south coast of Korea. Resistance in the Philippines was to be maintained as long as possible to delay the advance of the United States forces toward these key positions. While this outer perimeter was being assaulted, the forces in the Homeland would bend every effort to complete preparations for the decisive battle by early Fall of 1945.

On the basis of this general policy directive, Imperial General Headquarters hastened to construct the Homeland defenses. By the end of July 1945, the ground forces in Japan had been increased to a basic strength of 30 line-combat divisions, 24 coastal-combat divisions, and 23 independent mixed brigades, 2 armored divisions, 7 tank brigades, and 3 infantry brigades.[19]

Defense preparations in Kyushu and in the Tokyo—Yokohama and Nagoya districts of Honshu were particularly emphasized. The various area armies were called upon to protect all points strategically important for tactical defense, for production, and for communications and to complete their preparations to crush any American invasion.[20]

By April 1945, the course of events made it increasingly probable that Japan would be invaded in late summer or autumn. All key points in the Philippines had fallen; Iwo Jima was lost; American forces had gained a firm foothold on Okinawa; Russia had started to strengthen her forces in Siberia; and Germany

---

18 Japanese First Demobilization Bureau Report, *Homeland Operations Record*, Vol. I, Nov 46, G-2 Historical Section, GHQ, FEC.

19 Japanese First Demobilization Bureau Report, *Homeland Operations Record*, Vol. IV, Oct 46, pp. 37–38, Map No. 2; Vol. V, Oct 46, pp. 4–5, 17–21, 28–30, 32, 38, G–2 Historical Section, GHQ, FEC. The Japanese coastal-combat divisions were organized specifically for beach defense and compared favorably in strength with the regular line-combat divisions. Although the coastal-combat divisions lacked the heavy artillery weapons and mobility of the regular line divisions, they excelled the latter in infantry fire power, being well equipped with rifles, machine guns, 75mm and 80mm mortars, and mountain and anti-tank guns. The training of the coastal-combat divisions was not as thorough as that of the regular divisions but it was adequate to carry out the assigned mission of all-out beach defense. The Japanese had also established civilian National Volunteer units under the Volunteer Service Act of 23 June 1945. These civilian units, organized in every prefecture, were expected to aid in rear area maintenance and repair of transportation, communication, supply, and production facilities. First Demobilization Bureau Report No. 1026, 31 Mar 46, submitted in answer to questionnaire by Japanese Liaison Section, GHQ, SCAP; Statement of Maj. Tosaku Hirano, Staff Officer, Second General Army, 21 Oct 49, Interrogation Files, G-2 Historical Section, GHQ, FEC.

20 General Defense Command, "Homeland Defense Army Operations Order A No. 2," 9 Feb 45, G-2 Historical Section, GHQ, FEC.

was on the verge of final defeat. Victory in Europe would provide an additional reservoir of troops and supplies for Allied operations in the Pacific. These climactic developments spurred Imperial General Headquarters to put its paper policies into concrete form and accelerate its preparations for combating a full-scale amphibious invasion.[21] On 8 April, a plan for the final campaign on the Japanese main islands was completed and issued. This plan, designated "Ketsu Operation," embodied an extensive program to utilize Japan's remaining potential in the forthcoming last-ditch battle for the Homeland.[22]

Stressing the necessity for gearing every aspect of national life to the war effort, the "Ketsu Operation" plan provided for the concentration and rapid reinforcement of forces in the critical battle areas of Kyushu and of the Kanto Plain embracing the Tokyo—Yokohama region. Should Kyushu be the initial invasion point, four divisions would be diverted immediately from the Nagoya (Tokai) district, the Kobe—Osaka sector, western Honshu, and Shikoku to reinforce the strength already deployed on the Kyushu battlefronts. Replacements for these divisions would be provided in the centrally located Kobe—Osaka area by the diversion of three or four additional divisions from the Kanto Plain and northern Honshu. On the other hand, should the Tokyo—Yokohama area of the Kanto Plain be invaded first, eight reinforcing divisions would be moved to Nagano and Matsumoto Prefectures north of Nagoya—three from the northern Honshu district, three from the Kobe—Osaka sector, and two from the Kyushu district. If the situation permitted, two divisions from the Nagoya district would also be placed in the Tokyo—Yokohama area. Two divisions advancing from Hokkaido to northern Honshu and five divisions coming from Kyushu and Shikoku to the Kobe—Osaka area would be available for reinforcement.[23]

Considerable difficulty, however, was anticipated in carrying out this plan. Assuming that the United States would destroy land and water communications, the Japanese made provision for these troop movements to be made on foot, with railroads and ships to be used wherever possible. It was estimated that, if Honshu should be assaulted first, sixty-five days would be required to shift the necessary divisions from Kyushu to the Nagoya district, with ten additional days required to arrive at the battlefield.[24]

In April 1945, the accelerated preparations for the defense of the Homeland brought about a reorganization of the Japanese army command. The problems raised by the increasing numbers of troops in Japan, the slow-up of defense preparations caused by the lack of manpower and materials, aggravated transportation difficulties, together with the need for co-ordination of operational and administrative functions, made it necessary to simplify the command system.

Accordingly, Imperial General Headquarters divided Japan into two general army districts—the eastern and western—with a general army headquarters in each. (Plate No. 114) Eastern Japan was put under the First General Army of Marshal Gen Sugiyama, while western Japan came under the Second General Army of

21 Japanese First Demobilization Bureau Report, *Imperial General Headquarters Army High Command Record*, Nov 46, G-2 Historical Section, GHQ, FEC.

22 Imperial General Headquarters, Army Directive No. 2438, Separate Table, "Outline of Preparations for the Ketsu Operation," 8 Apr 45, G-2 Historical Section, GHQ, FEC.

23 Ibid., Separate Table, "Outline of Plans for Concentration in the Ketsu Operation."

24 Ibid.

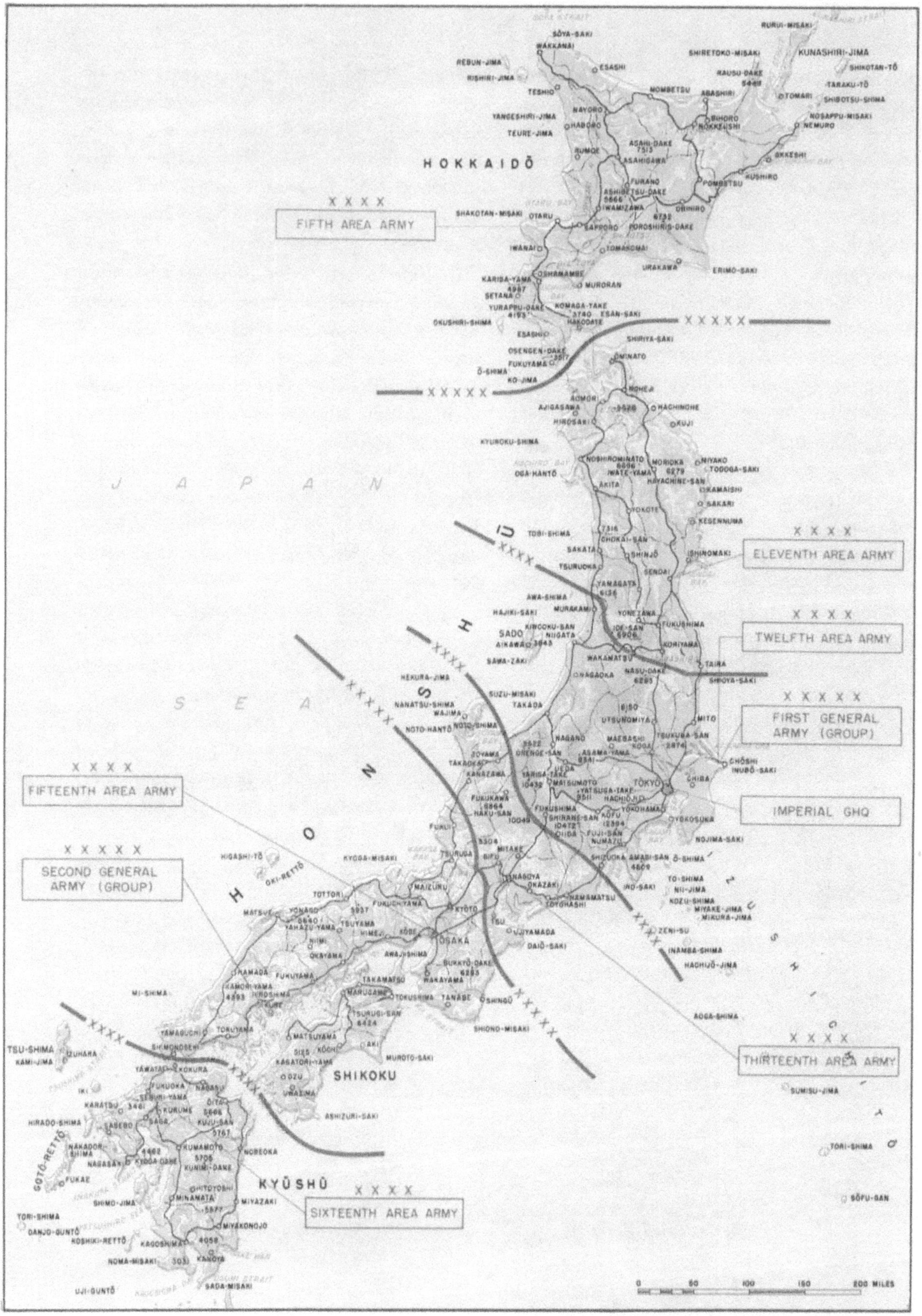

PLATE NO. 114

Disposition of Japanese Army Ground Forces in the Homeland, April 1945

Marshal Shunroku Hata. A new General Air Army Command was established for unified control of army air operations in Japan and Korea with headquarters for the First General Army and for the General Air Army in Tokyo, while Second General Army Headquarters was set up in Hiroshima.[25]

Under the command of the First General Army were the Eleventh (northern Honshu), Twelfth (Tokyo—Yokohama), and Thirteenth (Tokai) Area Armies. The Second General Army commanded the Fifteenth (Kobe—Osaka, western Honshu, Shikoku) Area and Sixteenth (Kyushu) Area Armies. The General Air Army controlled the First Air Army (eastern Japan), the 51st, 52nd, and 53rd Air Divisions, and the 30th Fighter Group. The Sixth Air Army (western Japan) remained under the Combined Fleet.[26]

Almost immediately after the stepped-up home defense program was instituted, American air raids became increasingly severe. Transportation and production facilities were greatly damaged and defense preparations were seriously handicapped. It was almost impossible for the Japanese to put up any effective defense against these air attacks. The loss of the Marianas and the Ogasawara chain, left Japan without bases from which to carry out patrolling activities and, consequently, the Japanese air forces had little or no warning of impending raids. In addition, the small number of planes being produced had to be carefully husbanded for the forthcoming great battle in the Homeland and could not be committed to immediate air defense.

Simultaneously with the growing intensity of these air raids, Japan suffered a long series of reverses between April and June 1945. The loss of the Philippines, Germany's capitulation, the fall of Rangoon, and the American capture of Okinawa—all came as major disasters to Japan. United States activities in the Pacific, meanwhile, gave clear indication that a new offensive would be forthcoming against Japan Proper in the near future.[27]

Although the Japanese hastened their preparations against the threat of impending invasion, they were faced with a multitude of serious difficulties. Transportation and communications facilities were disrupted, the people were gradually learning of their country's grave plight, and the entire economic situation was deteriorating rapidly. In May 1945, Homeland defense measures, particularly air and naval, were far behind schedule, while the overall civilian defense program was still in a disorganized state.[28]

Despite these obstacles, certain air and naval measures had been accomplished. By the summer of 1945, approximately 8,000 suicide or special-attack planes had been produced by converting army and navy fighters, bombers, trainers, and reconnaissance planes. Plans included 2,500 more to be made available by the end of September. Training of pilots was stressed, primarily to develop quickly the ability necessary to fly these craft on suicide missions. Naval preparations also included intensified production of special-attack boats and midget submarines.[29]

With the final loss of Okinawa in late June

---

25 Japanese First Demobilization Bureau Report, *Imperial General Headquarters, Army High Command Record*, G-2 Historical Section, GHQ, FEC.

26 Ibid.

27 Imperial General Headquarters Army Section, "Situation Estimate for the Latter Half of 1945 and Spring of 1946," 1 Jul 45, G-2 Historical Section, GHQ, FEC.

28 Japanese First Demobilization Bureau Report, *Homeland Operations Record*, Vol. III, G-2 Historical Section, GHQ, FEC.

29 Japanese First Demobilization Bureau Report, *Homeland Air Operations Record*, Dec 46, G-2 Historical Section, GHQ, FEC. Japanese Second Demobilization Bureau Report, *Preparations for Counterattacks against Enemy Landings on the Homeland*, Supplement No. 4, May 47, G-2 Historical Section, GHQ, FEC.

and the large-scale air raids over their main cities, it became increasingly evident that Japan was in desperate straits—and that the time was fast approaching when the war would be waged in their own islands.

Before any extensive defense measures could be put into effect, the Japanese High Command had to formulate a concrete estimate of Allied intentions. Only by an accurate assessment of the timing and strategy projected in United States plans and by a correct disposition of their own strength, could Japan's military leaders hope to upset or repel an invasion of the mainland.

Opinions within Imperial General Headquaters differed on the question of impending Allied operations. The various views fell into two main categories: one maintaining that the United States would initiate a long-range program of intensifed blockade and strategic air bombardment to destroy completely Japan's combat potential, while the other considered that the war would be brought to a decisive stage by an immediate amphibious invasion of Japan Proper. Although these two possibilities were injected into all discussion on strategy, they were not given equal prominence. The majority of Japan's military planners adhered consistently to the latter view—that an Allied invasion in force would be launched as soon as the necessary men and ships could be massed.

In order to carry out definite defense measures, a decision as to which course of action to prepare for became necessary. By 1 July 1945, Imperial General Headquarters adopted the official position that the United States would seek a quick end to the war by an all-out ground force invasion coupled with intensified sea and air operations. It was assumed that new forward bases for air and naval action would be seized in the northern Ryukyus, the Izu Islands, and possibly Quelpart Island. After these preliminary moves, the Japanese expected strong amphibious assaults against the southern part of Japan Proper. Tanega Island, Osumi Peninsula, and other strategic areas in south Kyushu and along the southern coast of Shikoku were named as the most likely targets to be occupied by Allied forces. The possibility of a diversionary feint at Hokkaido to cover the main landings was also taken into account.[30]

In general, the time of the southern Japan operations was placed in the Fall of 1945 and the date of the decisive Kanto Plain operation, in the Spring of 1946. The date of invasion, the Japanese thought, would depend to a large degree upon the number of troops and the amount of shipping the United States considered necessary for large-scale successful landings. It was calculated that by the Fall of 1945, the United States would be able to mount a total of thirty divisions for amphibious operations against the Homeland and that a cumulative total of fifty divisions could be massed by the Spring of 1946.

The general conclusion drawn in the Japanese estimate of Allied capabilities in July 1945 was that the United States was mustering enormous and overwhelming military strength for use against Japan and that the great battle would be joined between early Fall of 1945 and Spring of 1946.[31]

Geography and Road Net of Kyushu

Kyushu, the southernmost of Japan's four main islands, extends about 200 miles from

30 Imperial General Headquarters Army Section, "Situation Estimate for the Latter Half of 1945 and Spring of 1946," Jul 45, and Supplement No. 1, G-2 Historical Section, GHQ, FEC; also statements by Lt. Gen. Shuichi Miyazaki, 20 May 49, and Rear Adm. Sadatoshi Tomioka, 20 Aug 49, Interrogation Files, G-2 Historical Section, GHQ, FEC.

31 Ibid.

north to south and has a general width of about 80 to 120 miles from east to west. More than three-quarters of its territory consists of mountainous terrain, with a few plainlands scattered along the coasts.

The road net running across the island was limited and in poor condition. (Plate No. 115) The *Kokudo* or national highway was a euphemism for a single-loop, two-lane gravel road, badly torn by the heavy military traffic in constant flow over its surface. The highway was built along the island's coast, running on the west down to Kushikino, cutting across to Kagoshima, along the north bay to Miyakonojo, thence north and east to Miyazaki and finally up the east coast. The inland prefectural roads were, for the most part, one and a half lanes wide interspersed with frequent "passing" locations and suitable for light transport only. The remaining roads were narrow, primitive, one-way dirt tracks virtually impassable in wet weather.

The main railroad line paralleled generally the route of the highway and consisted of a single-track system with numerous bridges and tunnels that could be easily and quickly blocked when necessary. In the cultivated lowlands, the roads were built on fills rising four to five feet above the surrounding ground. By-passing and detouring would prove extremely difficult. In the mountains, both the roads and railroads were channeled through many cuts, any of which could be sealed against hostile passage. The entire transportation network was subject to complete disruption to prevent movement by attacker and defender alike.[32]

General Plan of "Olympic"

The American plan for the invasion of Kyushu contemplated the seizure of only the southern part of the island below a line drawn from Tsuno on the east coast to Sendai on the west. The 3,000 square miles thus marked off were deemed sufficient to provide the air bases necessary for short-range support of the final operations planned against the industrial centers of Honshu.[33]

In this southern portion selected for invasion under "Olympic" plan, there were four lowland areas suitable for the development of major airfields. One of these extended from Kagoshima, on the western shore of the bay bearing the same name, through a narrow corridor to the Kushikino plain on the East China Sea. A second ran northward from Shibushi on Ariake Bay, through a twisting valley to Miyakonojo; the third started at Kanoya, east of Kagoshima Bay and followed along the coast of Ariake Bay. The fourth and largest was located north of Miyazaki on the east coast.[34]

To acquire these valuable areas in the shortest possible time, the initial assaults were to be directed toward securing Kagoshima and Ariake Bays as ports of entry. The inland advance would then be extended as far as the Tsuno—Sendai line to block mountain defiles and prevent enemy reinforcements from the north.[35]

The southern Kyushu landings were to take place on 1 November 1945 under cover of one of the heaviest neutralization bombardments by naval and air forces ever carried out in the Pacific. (Plate No. 116) From bases in the Marianas and the Ryukyus, the Twentieth Air

32 Allied Geographical Section, G-2, GHQ, SWPA, Terrain Study No. 130, *S.E. Kyushu*, 31 May 45.

33 GHQ, AFPAC, *Staff Study "Olympic," Operations in Southern Kyushu*, 28 May 45. Hereinafter cited as; *Olympic*. This plan was later redesignated "Majestic" but is usually referred to by the former name.

34 AGS, G-2, GHQ, SWPA, Terrain Study No. 130, *S.E. Kyushu*, 31 May 45.

35 *Downfall*.

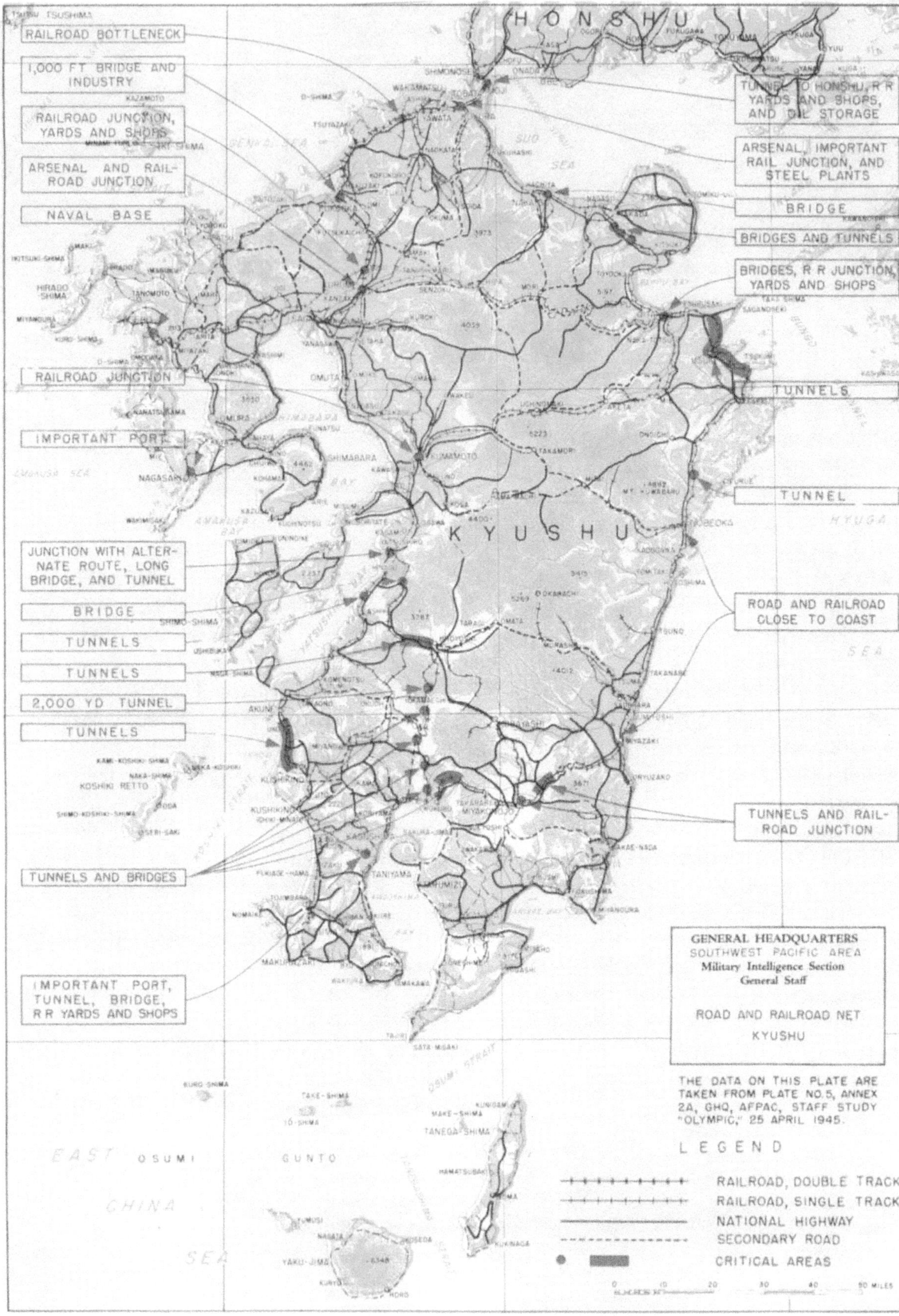

PLATE NO. 115

Road and Railroad Net, Kyushu

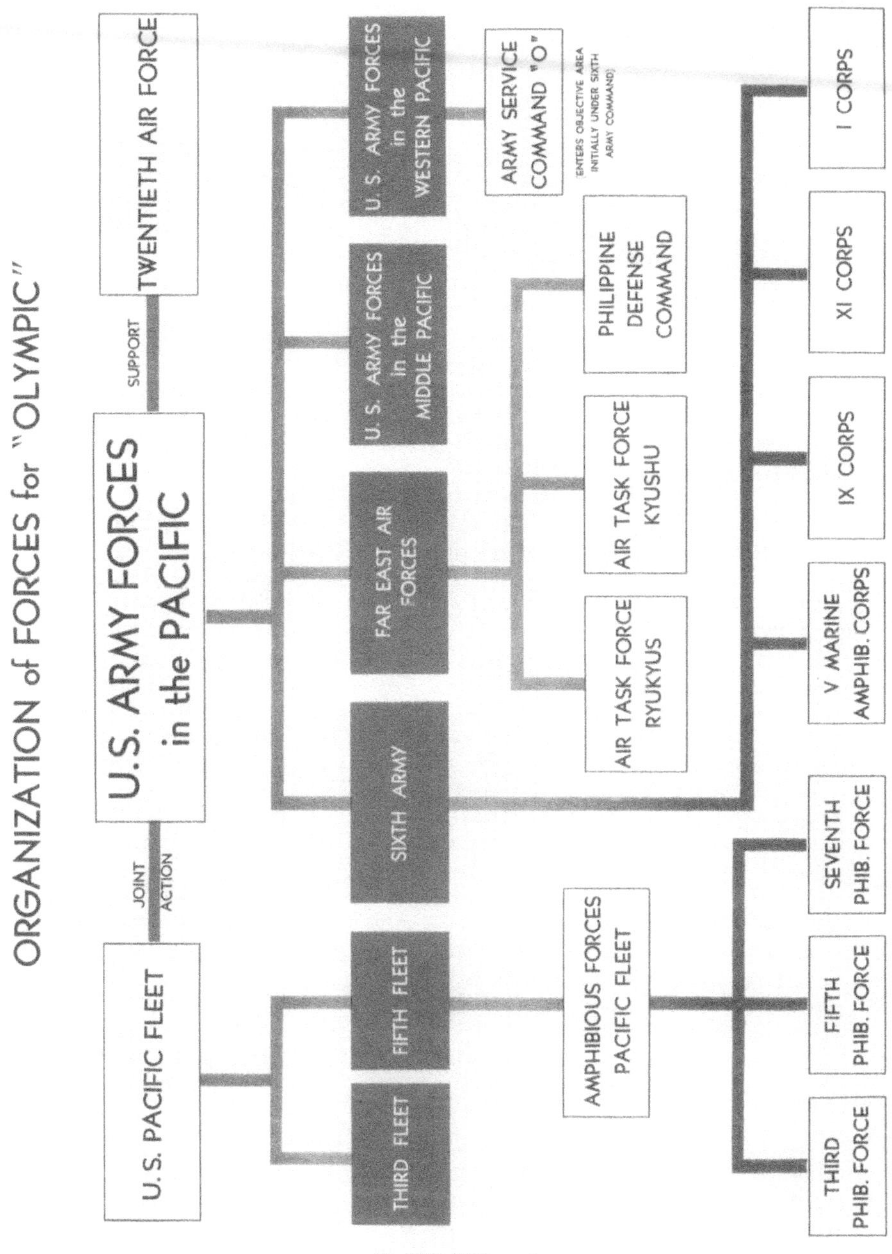

PLATE NO. 116
Organization of Forces for "Olympic"

Force would attempt to seal the source of Japan's industrial potential by striking ruthlessly at her factories and transportation system. The steady pounding by the huge B-29 bombers, increasing in tempo and fury with each passing day, was expected to reduce materially Japan's ability to maintain her large military organization intact or to distribute effectively her remaining power.[36]

Simultaneously, carrier task forces would carry out repeated raids against vital coastal areas, attacking enemy naval and air forces, disrupting communications on shore and at sea, and aiding the long-range bombers in attacks against strategic targets. The Far East Air Force, based in the Ryukyus, would concentrate on selected objectives calculated to destroy Japan's air arms both in the Homeland and in nearby North China and Korea. By intercepting shipping and shattering communications, the Far East Air Force would complete the isolation of southern Kyushu and prepare it for amphibious assault.[37]

It was intended that these air raids be intensified with the approach of the target date to culminate in an all-out effort from X-10 to X Day. These last ten days before the landing phase would see the massed bombing power of all available planes, both land and carrier-based, directed in a mighty assault to reduce the enemy's defenses, destroy his remaining air forces, isolate the objective area, and cover the preliminary mine sweeping and naval bombardment operations. The fortifications within the selected landing areas would be smothered under tons of explosives, while naval vessels and engineer units moved in to eliminate underwater mines and barriers. Supported by such overwhelming and concentrated power, it was expected that the amphibious forces would be able to stage the asssault landings without unreasonable losses.

To conduct the preliminary operations of clearing the routes of approach to the landing beaches, seizing emergency anchorages, and initiating the shore bombardment, an Advanced Attack Force would be launched from the Philippines to arrive off Kyushu on X-4. Strong air cover would be furnished this advance force.[38]

The Main Attack Force would proceed from the four key Pacific bases—Hawaii, the Marianas, the Philippines, and the Ryukyus—and, protected by the U. S. Pacific Fleet and a solid air umbrella, would converge on southern Kyushu by X Day for a three-pronged landing in the Miyazaki area on the east, at Ariake Bay on the south, and at Kushikino, on the west.[39] (Plate No. 117)

Two main deception efforts were planned, one strategic and the other tactical. To encourage the belief that Allied attention was focused primarily on the Asiatic coast and the Chusan Archipelago, the China Theater Command and the Southeast Asia Command would conduct diversionary ground movements in China and in the Malay Peninsula. The tactical diversionary effort would be made by a feinting blow against the island of Shikoku. A floating reserve of two divisions, a part of the Main Attack Force, would appear off eastern Shikoku from X-2 to X Day and then, after its deceptive mission was performed, would proceed to the Ryukyus to be available for reinforcement call.

After the beachheads were safely established, reserve and service troops would be brought forward, land-based aviation installed progressively, and logistic facilites developed. Military

36 *Olympic.*

37 Ibid.

38 Ibid.

39 Ibid.

Goverment would be instituted as soon as practicable after the objective areas were consolidated.[40]

Employment of "Olympic" Ground Forces

Since the Joint Chiefs of Staff directive of 3 April 1945 was based on the premise that forces from Europe would not be available in time for "Olympic," General MacArthur's plan for the employment of his ground forces involved primarily those combat troops already within the theater. The assault on Kyushu was assigned to the veteran Sixth Army under General Krueger. (Plate No. 118)

The 40th Division, reinforced, was named as the advance attack force and given the mission of seizing positions in the Koshiki Island group located off Kyushu's southwest coast opposite Sendai. The division was expected to establish emergency naval and seaplane bases in these islands and clear the sea routes to the coastal invasion area of Kushikino. The 40th Division was also given the mission of making preliminary landings in the four islands of Tanega, Make, Take, and Io off the southern tip of Kyusyu. These latter islands would be occupied to safeguard the subsequent passage of friendly shipping through the strategic Osumi Strait.[41]

On X Day, three corps would effect simultaneous assault landings in the three objective areas. V Marine Amphibious Corps (3rd, 4th, and 5th Marine Divisions) would go ashore in the vicinity of Kushikino, drive eastward to secure the western shore of Kagoshima Bay, and then turn north to block the movement of enemy reinforcement from upper Kyushu. XI Corps (1st Cavalry, Americal, and 43rd Divisions) would land at Ariake Bay, capture Kanoya, advance to the eastern shore of Kagoshima Bay, and then move northwestward to Miyakonojo. I Corps (25th, 33rd, and 41st Divisions) would make its assault at Miyazaki on the east coast, move southwestward to occupy Miyakonojo and clear the northern shore of Kagoshima Bay to protect the northeast flank. XI Corps (81st and 98th Divisions), initially in the Sixth Army Reserve afloat, was selected to carry out the diversionary threat off the island of Shikoku while the other three assault forces moved on the actual landing beaches. Should these four corps prove insufficient to accomplish the tasks assigned, a build-up from the elements earmarked for "Coronet" would be instituted at the rate of three divisions per month beginning about X+30. The "Coronet" operation would be adjusted accordingly.

If he deemed it advisable, General MacArthur would have the reserve elements conduct an amphibious assault near Wakiura on the southern coast as soon as adequate naval support could be assured. The primary aim of all operations was to secure areas suitable for the immediate construction of air bases and to clear Kagoshima Bay for use as a port and naval base. As soon as these objectives were accomplished, General Krueger would consolidate the area south of the Tsuno—Sendai line and prepare to stage four divisions and other forces to aid in the execution of "Coronet."[42]

Logistic Arrangements

Under "Olympic" plan, General MacArthur was charged with the responsibility for the logistic support of all army forces engage in the assault operation, including the Marine Corps units under his command. Admiral Nimitz

40 Ibid.
41 Ibid.
42 Ibid.

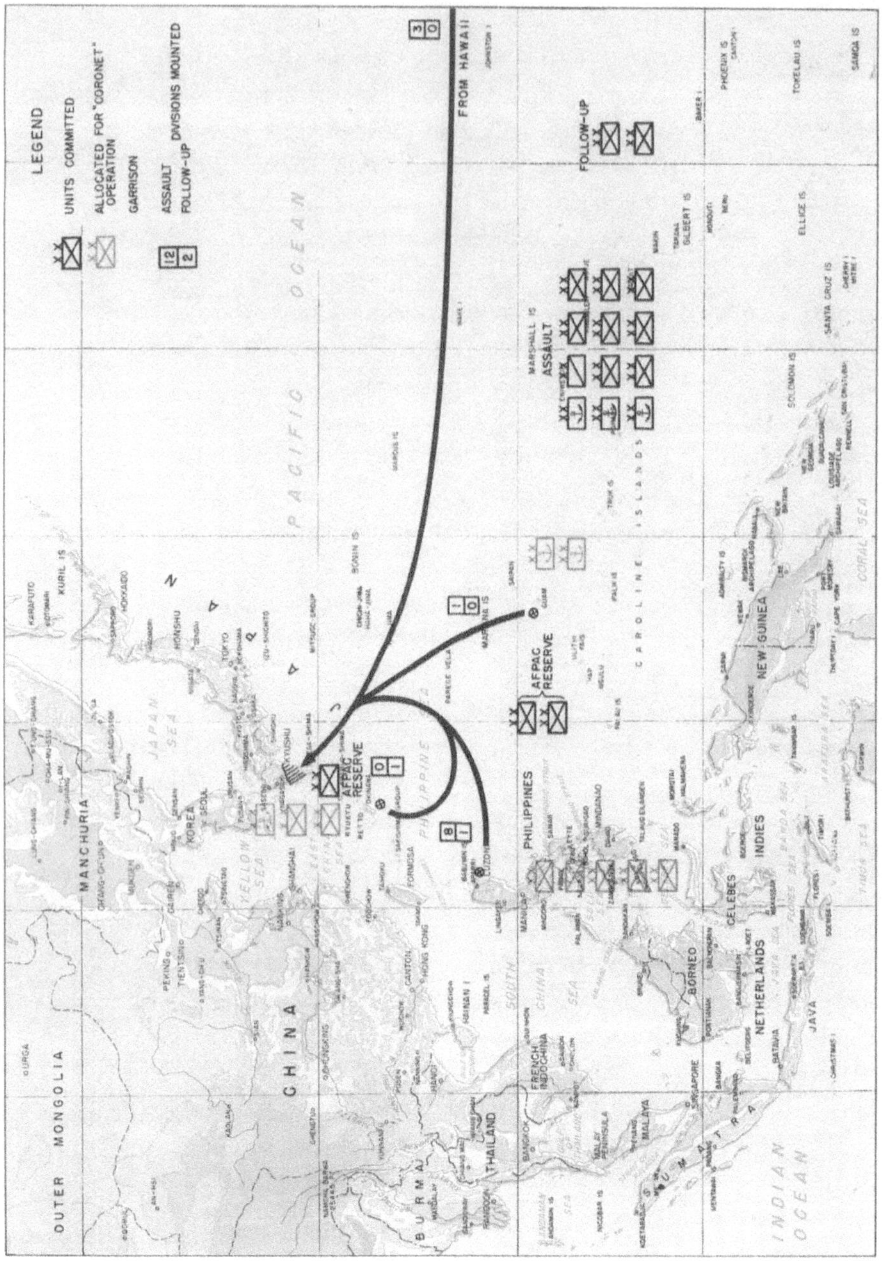

PLATE NO. 117

Staging of Forces for "Olympic"

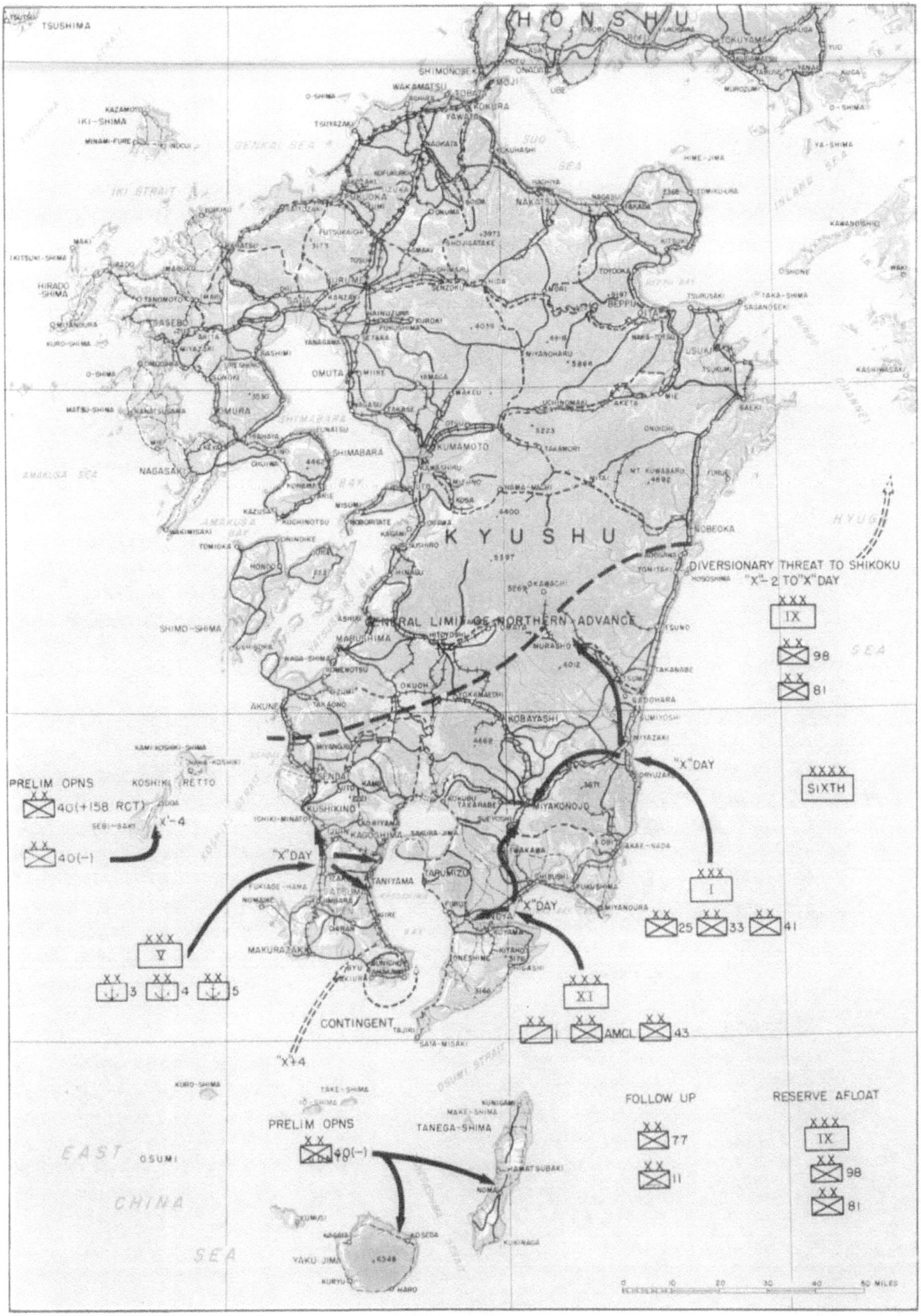

PLATE NO. 118

"Olympic," the Invasion of Kyushu

had similar responsibility for maintaining all naval forces and, in addition, was to furnish the organizational equipment and to mount the supplies necessary for the Marine corps passing to Army control. The Twentieth Air Force would handle its own logistics.[43]

Manila was designated as the base to provide the initial supply of troops being staged in the Western Pacific, and Hawaii was selected to perform the same function for troops sent out from the Middle Pacific. Re-supply and the bulk of construction materials would come directly from the United States, augmented as necessary from Pacific bases. AFWESPAC was made responsible for organizing Army Service Command "Olympic" to provide logistic support within the Kyushu combat zone.

The Philippines, Ryukyus, Marianas, and Hawaii were the four Pacific bases chosen to stage, equip, and mount the invasion forces. Naval shipping would carry the assault and reinforcing elements forward from the mounting areas and transport the heavy equipment and stores. Following the successful completion of the assault phase, the Kyushu ports of Kagoshima and Shibushi would be developed as soon as possible to support further penetration inland and to the north. It was also planned to make the maximum use of available local resources and civilian labor to help speed the progress of the Allied advance.[44]

The Enemy Situation

A cardinal problem confronting General MacArthur before every operation was the question of enemy dispositions and capabilities in the projected invasion area. The following extracts from the intelligence estimate prepared for "Olympic" provide an excellent illustration of the comprehensive preparation and staff work that preceded a major military campaign. By the end of July, General MacArthur had a fairly complete picture of what to expect when his forces invaded Kyushu. To keep this picture up to date, new information was filled in as rapidly as it was received from the various intelligence sources.

1. ESTIMATE OF THE ENEMY SITUATION

a. Development of Ground Strength:

(1) The initial estimates on Kyushu of 24 March and 25 April 1945 considered an initial enemy deployment of six (6) divisions but seriously forecast a potentially larger deployment to ten (10) divisions, viz:

> *"Although the Japanese obviously regard the Tokyo Plain as the ultimate decisive battle ground, it is apparent that Kyushu is considered a critical sector on their planned Empire Battle Position. It is believed that plans will visualize assignment of about 6 combat divisions (plus 2 depot divisions) to garrison Kyushu initially and that they are prepared to expend up to 10 divisions, all they can tactically employ in the area to insure its retention. Depot facilities to maintain such a force would have been established in Northern Kyushu.*

These divisions have since made their appearance, as predicted, and the end is not in sight. This threatening development, inherent in war, will affect our own troops basis and calls for special air missions. If this deployment is not checked it may grow to a point where we attack on a ratio of one (1) to one (1) which is not the receipt for victory. . . .

b. . . . .

c. Development of Command Structure:

Recent information suggests the grouping of mobile combat units under 2 Corps Headquarters, one in Southern and one in Northern Kyushu. However, it is possible that with the considerable increase in strength in Southern Kyushu, a third

43 *Downfall.*

44 *Olympic.*

Corps may eventually be formed.[45]

57 Corps: Established Headquarters at Takarabe (Miyakonojo Basin) during period April—June 1945. Believed responsible for defense of Southern Kyushu, i.e., of area south of central mountain mass.

56 Corps: Headquarters established at Iizuka (east of Fukuoka) during June 1945. Probably responsible for defense of area north of central mountain mass.

d. Organization of Volunteer Defense Units:

Information received since publication of the G-2 Estimate of 24 April 1945, indicates that the Japanese have materially accelerated their mobilization program and are now tapping a hitherto unused source of reinforcement, i.e., they are augmenting Army and Navy Units with large numbers of Volunteer Home Defense Units composed largely of partially trained reservists, over and above increments inducted into the actual armed forces. It has been estimated that approximately 125,000 of these are available in Southern Kyushu and approximately 450,000 in Northern Kyushu. Note the recent assignment of Major Generals as Commanders of 26 of the 51 Regimental Recruiting Districts in Japan, all of which were formerly commanded by Colonels, and of Lieutenant Generals of 10 other similar but generally more populous districts, suggests the grouping of these units of odd size and composition into divisions and possibly even into Corps. It is equally significant that these new Commanders have had recent front-line field commands. Credence is lent this theory by the recent sudden appearance in various parts of Japan of new divisions numbered in the 200 and 300 series whereas the highest previously identified divisional number was 161. Two of these new and unusually high numbered divisions, the 206th and 212th have appeared in Kyushu. . . . (Plate No. 119)

e. Tactical Significance of Defiles:

(1) The rapidity with which the Japanese are

45 The estimated arrivals of enemy reinforcements are shown in the following table:

	30 April			31 May			30 June			20 July		
Tactical Area	Divs	Brigs	Tk Regts	Divs	Brigs	Tk Regts	Divs	Brigs	Tk Regts	Divs	Brigs	Tk Regts
Southern Kyushu	86	1 Pcht	—	86	1 Pcht	—	86	1 Pcht	37 (b)	86	1 Pcht	37 (b)
					3 Amph		25	3 Amph	40 (b)	25	3 Amph	40 (b)
							77			77		
							212 (a)			212 (a)		
										156		
										154		
										146		
Total	1	1	0	1	2	0	4	2	2	7	2	2
Northern Kyushu	57	—	18	57	—	18	57	—	18	57	126	18
	6 Dep*	—	—	6 Dep	—	—	6 Dep	—	—	6 Dep	—	—
	56 Dep	—	—	56 Dep	—	—	56 Dep	—	—	56 Dep	—	—
		—	—	206 (a)	—	—	206 (a)	—	—	206 (a)	—	—
Total	3	0	1	4	0	1	4	0	1	4	1	1
Unlocated								96 IM**			96 IM	
											122 IM	
Total	0	0	0	0	0	0	0	1	0	0	2	0
Aggregate (c)	4	1	1	5	2	1	8	3	3	11	5	3

(a) Probably composed of Volunteer Defense Units.

(b) Believed elements of u/i Tank Brigade, Code Name "To."

(c) Does not include: u/i Unit, Code name "Yaku." Believed to be in Southern Kyushu. May be one of the numbered units listed hereon, u/i Unit, Code name "Ho" believed to be in Northern Kyushu. Character unknown.

* Dep refers to Depot.

** IM refers to Independent Mixed Brigade.

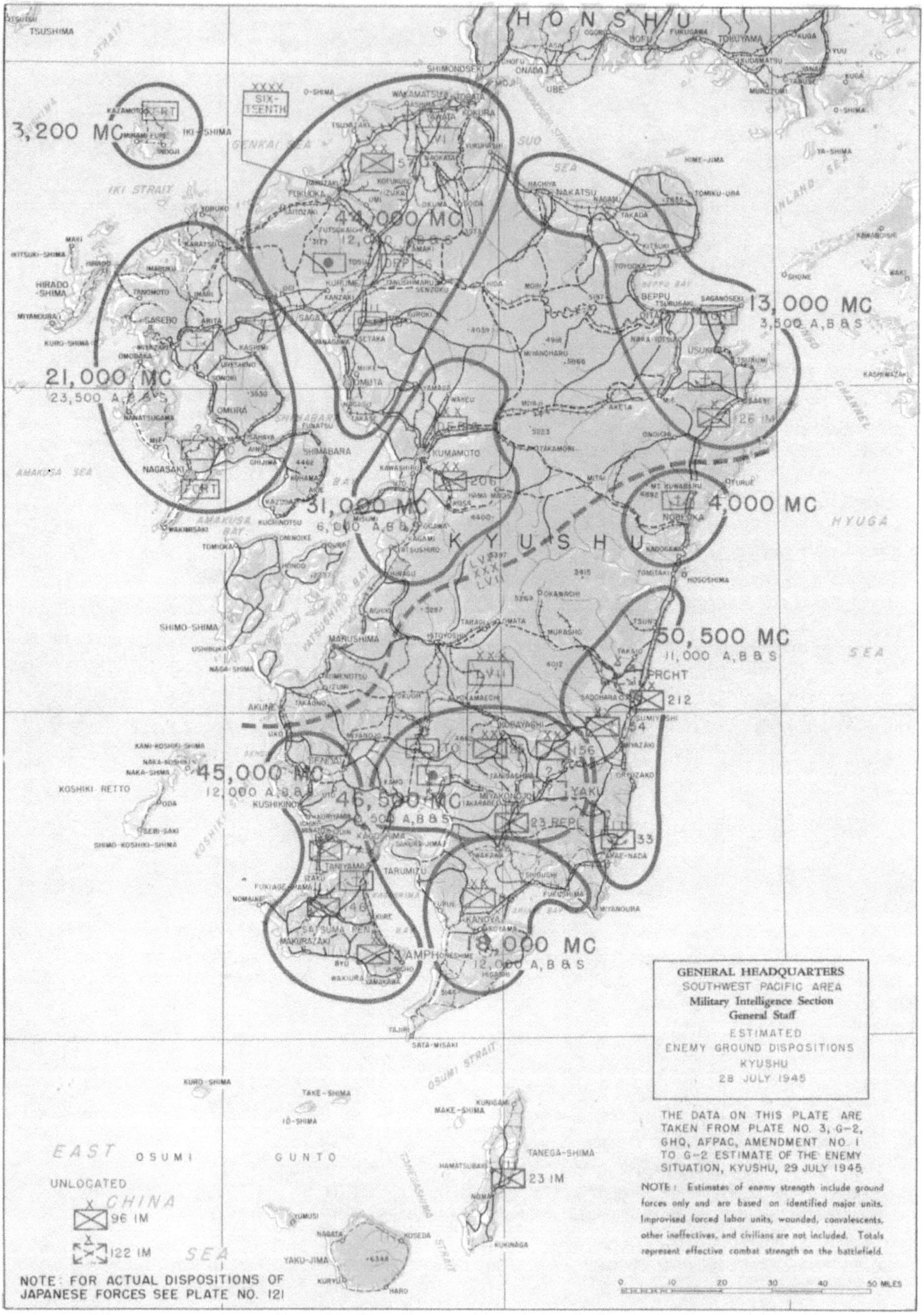

PLATE NO. 119
Estimated Enemy Ground Dispositions on Kyushu, 28 July 1945

effecting their concentration in Southern Kyushu is undoubtedly influenced by the continued availability at full capacity of the two overland roads and railroad routes from Northern to Southern Kyushu. Although some of the incoming units may have arrived by sea and one (the 212th Division) was probably recruited locally, it is probable that 3 to 4 Divisions which are believed to have been activated by Depots in Northern Kyushu and Southwest Honshu utilized one or both of the overland routes. Also, in view of the fact that these troops must be maintained principally by the Kurume and Kumamoto Depots, the roads and rail lines have undoubtedly become vital supply lines. Concern of the Japanese as to the continuity of these overland lines of communications and recognition of their vulnerability to air bombardment is emphasized by recent initiation of Special AA Defense measures....

(2) Attention is invited to the following paragraphs of the G-2 Estimates of 25 April 1945:

Par. 1 e (4)

"These factors (detailed discussion of critical defiles traversed by east and west coast road and railroad routes from Northern to Southern Kyushu and the part-way alternate route Yatsushiro—Kokubu)— lead to the conclusion that the blocking of a limited number of defiles, i.e., destruction of critical bridges, tunnels, cuts, and fills, by concentrated aerial bombing and vigorous maintenance of interdiction by both air and naval bombardment will render it exceedingly difficult for the Japanese either to reinforce Southern Kyushu with elements of the Northern Kyushu Garrison, or to supply and maintain their forces which are in Southern Kyushu prior to the blocking of their only two lines of overland communication."

Par. 2 a (2) (c)

"Effective interdiction of both coastal routes and the alternate route from Yatsushiro through concentrated aerial bombing and/or naval gunfire would reduce overland reinforcement from Northern Kyushu to a mere trickle as long as maintained. Laborious passage of blocked defiles and long overland marches would be required.... It is probable that the Japanese may resort to night barge movemement. An ample supply of small coastal craft is available and their tactics have long stressed the use of this means of troops transport.... Effective air and sea control should limit if not prevent the use of this method...."

Par. 2 a (2) (d)

"Reinforcement from Honshu: Troops from this source would probably be moved into Northern Kyushu via the Shimonoseki Tunnel or by night overwater movement; however, their subsequent movement to Southern Kyushu would be subject to the same restrictions that apply to Northern Kyushu garrison. Overwater movement across the Bungo Channel direct to Nobeoka or points south thereof should be easily prevented by air and/or naval action."

2. CONCLUSIONS:

a. The rate and probable continuity of Japanese reinforcements into the Kyushu area are changing the tactical and strategical situation sharply.

At least six (6) additional major units have been picked up in June/July; it is obvious that they are coming in from adjacent areas over lines of communication, that have apparantly not been seriously affected by air strikes.

There is a strong likelihood that additional major units will enter the area before target date; we are engaged in a race against time by which the ratio of attack-effort vis-a-vis defense capacity is perilously balanced.

b. The Japanese have correctly estimated Southern Kyushu as a probable invasion objective, and have hastened their preparations to defend it.

c. They have fully recognized the precarious nature of the land and sea routes by which they must concentrate and support their forces in Southern Kyushu. They are vigorously exploiting available time to complete the deployment and supply lines of strong forces in the area before they are deprived of the full use of their limited lines of communication.

d. Since April 1945, enemy strength in Southern Kyushu has grown from approximately 80,000 troops including in mobile combat the equivalent of about 2 Infantry Divisions to an estimated 206,000 including 7 divisions and 2 to 3 brigades, plus Naval, Air-Ground, and Base and Service Troops. This rapid expansion within a few weeks' time, supply of this large concentration of troops, and the movement of defensive material has undoubtedly so strained the capacity of all existing lines of communication that any major interruption thereof would seriously reduce the effectiveness of the enemy's preparations. It is also probable that some of the new units identifed in Southern Kyushu are not yet fully assembled and that at least one division (the 212th) which was probably formed of local volunteers has not yet been completely equipped.

e. The assumption that enemy strengths will remain divided in North and South (Kyushu) compartments is no longer tenable.

f. The number of enemy major units rapidly tend to balance our attack units.

g. The trend of reinforcements from North to South (Kyushu) is unmistakable.

h. Massing in present attack sectors is apparent.

i. Unless the use of these routes is restricted by air and/or naval action as suggested in Pars. 1 e (4) and 2 a (2) (c), and (d), G-2 Estimate of 25 April, enemy forces in Southern Kyushu may be still further augmented until our planned local superiority is overcome, and the Japanese will enjoy complete freedom of action in organizing the area and in completing their preparations for defense.[46]

The Planned Defense of Kyushu

Increasing air raids by American bombers against southern Kyushu and Shikoku in June 1945 gave the Japanese another reason to believe that these areas would be the focal point for the initial invasion of the Homeland. Accordingly, Imperial General Headquarters hastened operational preparations in those areas and gave urgent priority to transporting and accumulating supplies to support the newly organized ground forces of southern Japan.[47]

Shifting the emphasis of defense to Kyushu necessitated a postponement of military preparations in the Tokyo—Yokohama area. The Japanese feared that if the invasion of the Kanto region should be initiated immediately after the Kyushu attack, or if Honshu should be assaulted directly, an adequate defense could not be made since most of the Japanese air and naval power would already have been committed to Kyushu. For this reason, Imperial General Headquarters were reluctant to transfer ground strength from the capital city area until the very last moment. Although air and sea preparations were being pushed in Kyushu, and supplies and munitions were being moved southward, the Japanese held up the scheduled commitment of troops to that area pending further developments. They assumed correctly, however, that the first aim of the United States would be the annihilation of the Japanese forces on the southern Kyushu front and the occupation of strategic air bases and harbors in Miyazaki and Kagoshima Prefectures.

Following the invasion of Tanega Shima, where a fighter base would be installed, it was considered probable that the American forces would direct their main attack against the Shibushi area on the Ariake Bay front and the Sumiyoshi coast on the Miyazaki Plain. A secondary attack was expected on Fukiagehama on the Satsuma Peninsula.

46 G-2, GHQ, AFPAC, "Amendment No. 1 to G-2 Estimate of the Enemy Situation with Respect to Kyushu," 29 Jul 45.

47 The strategists at Imperial General Headquarters believed that, if they could succeed in inflicting unacceptable losses on the United States in the Kyushu operation, convince the American people of the huge sacrifices involved in an amphibious invasion of Japan, and make them aware of the determined fighting spirit of the Japanese army and civilian population, they might be able to postpone, if not escape altogether, a crucial battle in the Kanto area. In this way, they hoped to gain time and grasp an opportunity which would lead to the termination of hostilities on more favorable terms than those which unconditional surrender offered. Statements by Lt. Gen. Seizo Arisue, 10 May 49, and Lt. Gen. Torashiro Kawabe, Deputy Chief, Army General Staff, 13 Jun 49, Interrogation Files, G-2 Historical Section, GHQ, FEC.

Since these three areas were geographically cut off from each other, it was believed that the Americans would seek to divide and contain the Japanese units therein. The Japanese, therefore, estimated that the first attack would be against either the Miyazaki Plain or the Satsuma Peninsula, or both. The main force of the assault would then in all likelihood be directed at Ariake Bay while other elements would attempt to break through the mouth of Kagoshima Bay. Simultaneously with these attacks by ground forces, an airborne landing would possibly be made on the airfields around Kanoya and Miyakonojo. It was also thought probable that the Tosa Plain of southern Shikoku would be invaded to destroy the launching places of Japanese special-attack planes and, at the same time, establish American fighter bases.

It was anticipated that the United States would employ about 15 divisions, with 10 or 12 divisions attacking southern Kyushu. Two divisions, according to Japanese estimates, would be committed to Shikoku and the rest to strategic reserve. The initial landing force, which was expected to be sufficiently strong to defeat a Japanese counterattack, would probably comprise at least 3 divisions in the main sector and about 2 divisions on the other fronts.[48]

To counter the expected United States invasion of southern Kyushu, plans were made to attack the main body of the landing force on the sea and in the coastal landing areas. The entire air strength in the Kyushu area would be employed against the invasion convoys. Reconnaissance was to be conducted on a day and night basis along a 600-mile radius of the Homeland by 140 naval reconnaissance planes augmented by army air reconnaissance elements. Secret airfields would be utilized by suicidal special-attack units to escape bombing raids. To conserve air strength for employment against the landing forces, carrier task forces supporting the invasion would be attacked by only a few hundred planes and then only when it became unmistakable that a full-scale invasion was in progress. The bulk of the air fleet of 10,500 planes, mostly of the small special-attack type, would be launched against the warships and transports in the crucial invasion area. Japanese plans called for these planes to be completely expended within a ten-day period in a supreme effort to repel the invasion forces. By the end of June 1945, 8,000 of the necessary craft were already available; the other 2,500 were expected to be finished by Fall.[49]

The Japanese Navy had little power left to oppose an Allied assault. Its remaining forces consisted of a variety of small underseas special-attack craft and a few heavy surface units that had been hidden from Allied attack. The total destroyer strength of nineteen operational ships was to be distributed along the islands of the Inland Sea. As soon as the invasion convoy entered its anchorage, the destroyers would attack in co-ordination with air force planes and surface and underwater special-attack units. About thirty-four submarines of all types were available for patrol duty and harassment of Allied convoy and supply lines.

The surface special-attack forces, in conjunction with air and other naval units, were earmarked for strong mass assaults on American transports in the invasion area. Midget submarines (Koryu) would strike at the convoy before and after its entrance into the anchorge. "Human torpedoes" (Kaiten) would execute

48 Japanese First Demobilization Bureau Report, *Homeland Operations Record*, Supplement to Vol III "Essentials of Sixteenth Area Army Ketsu Operation Plan," Sep 46, G-2 Historical Section, GHQ, FEC.

49 Japanese First Demobilization Bureau Report, *Homeland Air Operations Record*, Dec 46, G-2 Historical Section, GHQ, FEC.

PLATE NO. 120
Japanese Ground Dispositions on Kyushu, 18 August 1945

similar attacks and, in addition, strike at the fleet forces supporting the bombardment while small motorboats (Shinyo) would be employed in night attacks.[50]

Land preparations on Kyushu in June and July were largely inadequate and an Allied assault at that time would have found Japan's defenses generally incomplete. In their planned policy of playing for time, the Japanese were aided by the strong resistance on Okinawa, which they felt would further delay the invasion date. Utilizing this needed interval, the Japanese implemented their operational plans and hastened to complete all defenses and to coordinate all activities for the objective of "annihilating the enemy on the beaches through offensive action."[51] Every effort was bent toward this goal and it was estimated that all steps to meet the invasion would be completed by November.

Responsibility for the defense of the Kyushu area was assigned to the Sixteenth Area Army with headquarters at Fukuoka. Under its command were the Fortieth, Fifty-sixth, and Fifty-seventh Armies (each the equivalent of an American corps), one independent line-combat division, four independent coastal-combat brigades, and miscellaneous service units. (Plate No. 120)

By mid-July 1945, the Fortieth Army was deployed along the Satsuma Peninsula in southwest Kyushu below Sendai. The Fifty-sixth Army occupied northern Kyushu, with most of its strength distributed along the northwestern coast from Moji to Sasebo. The Fifty-seventh Army was responsible for the defense of the southeast and was concentrated in the Miyazaki Plain region. The total strength assigned to the immediate defense of Kyushu amounted to approximately 7 line-combat divisions, independent mixed brigades, and 3 tank brigades.[52] Three divisions and 2 tank brigades were kept in mobile reserve in south central Kyushu to be moved at an instant's notice and concentrated at the points of amphibious invasion.

The Japanese tactical plan called for a rigid defense of the beaches which would be reinforced as rapidly as units could be sent from other parts of Kyushu and the western regions of Honshu. The ridges immediately to the rear of the anticipated landing areas were fortified by a network of large caves and tunnels covered by a series of strong artillery and mortar emplacements and manned by large forces of coastal-combat ground troops. These troops would meet the first assault waves and, together with the line-combat divisions, would endeavor to hold all positions until the arrival of the mobile reinforcements at the beach battleground. If the primary target of the Allies could be ascertained before the landings, then the main Japanese forces would rush to that point. If the Allied target should be unclear or if all points should be invaded simultaneously, then the main enemy strength would be committed to Ariake Bay, while delaying actions were fought on other fronts. In all cases, the decisive engagement would be waged in the beach areas with all available forces moving toward the southern Kyushu coasts. The whole object of the Japanese was to repel the invaders before they could put their heavy armor and artillery ashore.

The Japanese realized that their defense plans contained numerous weak points but their situa-

50 Japanese Second Demobilization Bureau Report, *Preparations for Counterattacks against Enemy Invasion Landings*, G-2 Historical Section, GHQ, FEC.

51 Japanese First Demobilization Bureau Report, *Homeland Operations Record*, Vol III, G-2 Historical Section, GHQ, FEC.

52 Ibid.

ORGANIZATION of FORCES for "CORONET"

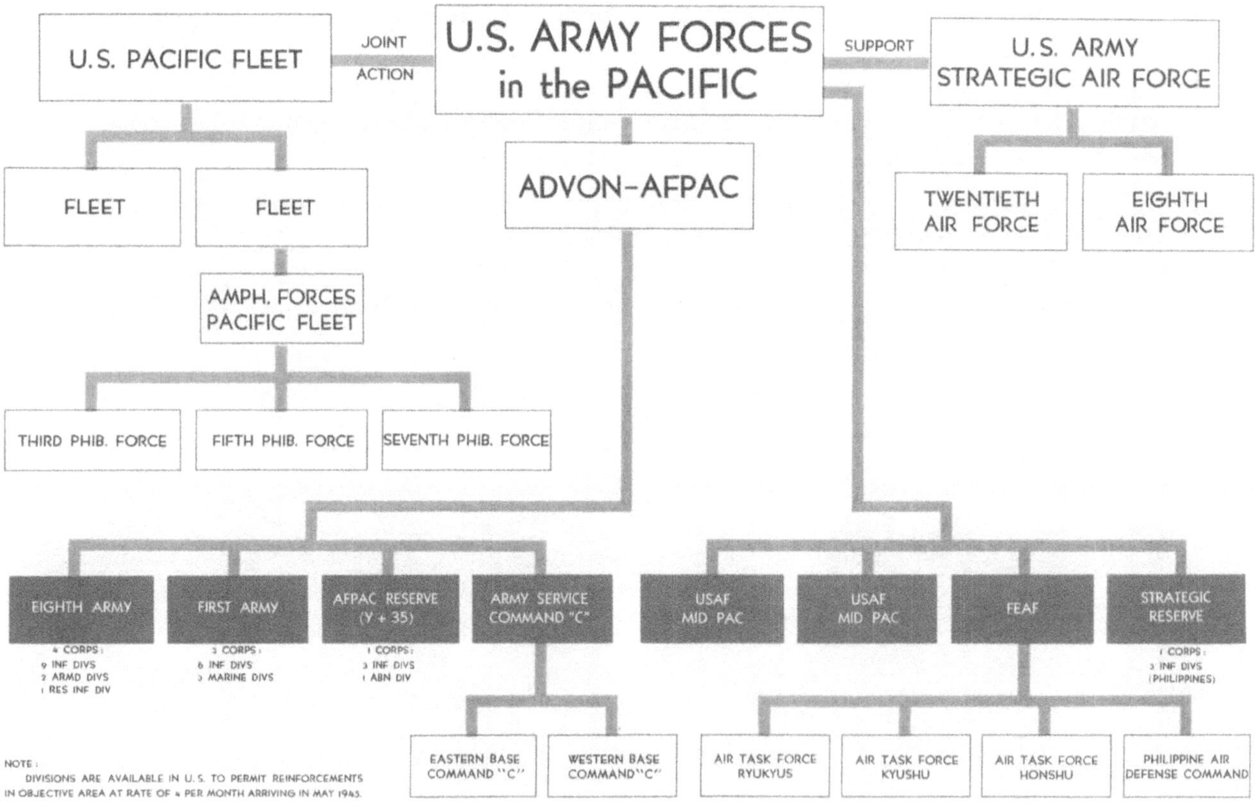

PLATE NO. 121
Organization of Forces for "Coronet"

tion was desperate and they felt compelled to use desperate measures.

American Plans for the Invasion of Honshu—Operation "Coronet"

Four months after the first American troops had fought their way ashore on Kyushu, the next and decisive amphibious operation against Japan would be launched. "Coronet," the invasion of the Kanto Plain area of Honshu, was scheduled for 1 March 1945.[53]

Two armies, the First and the Eighth, were charged with the second mammoth assault against the heartland of Japan. Their immediate mission was to destroy all opposition and occupy the Tokyo—Yokohama area. General MacArthur would exercise personal command of the landing forces and direct the ground operations on the mainland. With him would be the advance echelon of his General Headquarters to act as the Army Group Headquarters in the field.

Operation "Coronet" would be supported by all army and navy forces in the Pacific and, in addition, would be augmented by numerous combat units redeployed from Europe. Air, naval, and logistic support was outlined along the same general pattern as "Olympic" except on an enlarged scale. The command relationships established between General MacArthur, Admiral Nimitz and General Spaatz were to continue. (Plate No. 121)

The initial landings would be staged by 10 reinforced infantry divisions, 3 marine divisions, and 2 armored divisions. Launched from the Philippines and Central Pacific bases, the attacking forces would be constantly protected by ships and planes of the Pacific Fleet as well as by land-based aircraft. Thirty days after the initial assault, each army would be reinforced by a corps of 3 divisions. Five days after this reinforcement, an airborne division and an AFPAC Reserve Corps of 3 divisions would be made available. These 25 divisions were to seize the Kanto Plain, including the general areas of Tokyo and Yokohama, and then carry out any additional operations necessary to break Japanese resistance. Strategic reserve for the entire operation would consist of a corps of 3 divisions located in the Philippines and a sufficient number of divisions from the United States to permit reinforcement at the rate of 4 per month.

The amphibious assault against Honshu would be preceded by heavy blows of Allied naval and air forces against the Japanese Empire. Carrier planes from the Pacific Fleet would co-operate with the Strategic Air Force in carrying out repeated attacks against vital areas of the Japanese Home Islands to strangle land and sea communication and wipe out selected targets ashore. Land-based planes from newly won fields in Kyushu together with fighters and bombers from Okinawa would continue to range over the Empire and the Asiatic coast, destroying any remaining enemy aircraft, shattering land communications, and reducing defensive installations. All air attacks would be intensified as the landing date approached, culminating in an all-out effort, co-ordinated with naval bombardment, during the last fifteen days.

At the same time, subsidiary actions in other theaters of operations would be aimed at containing Japanese air and ground forces. The China Theater would conduct neutralizing air and ground attacks against the enemy on the Asiatic mainland. The Southeast Asia Command would launch similar operations in the

53 GHQ, AFPAC, *Staff Study "Coronet,"* 15 Aug 45. This staff study was published as a matter of interest only. It set forth the first draft of the plan formulated prior to the cessation of hostilities, and was still in an incomplete form.

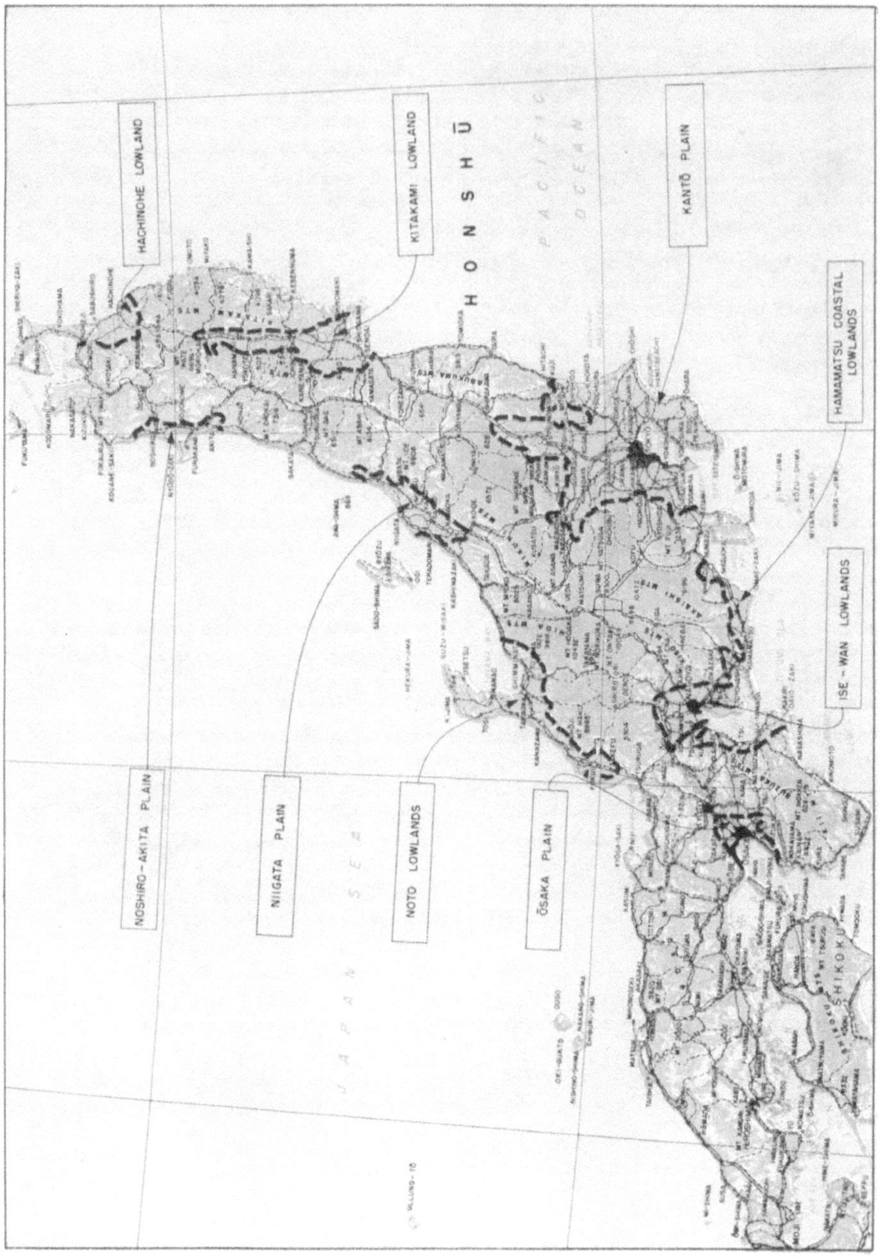

PLATE NO. 122
Honshu

PLATE NO. 123

Landing Beaches, Roads, Railroads, and Critical Defiles, Kanto Plain southern areas. Naval and air forces based in the Aleutians would be called upon to lend general support wherever possible. All plans were directed toward the successful execution of the greatest amphibious operation ever planned.

Geographical Considerations

The decisive operation in the projected campaign to bring about the final collapse of Japan was the invasion of Honshu. (Plate No. 122) The total defeat of Japan's armies in the core of the Empire was the over-all primary objective. In the event that the campaign in the Kanto Plain did not prove to be the last battle, the secondary objective would be achieved—secure positions from which to continue air, ground, and amphibious operations in the main islands of Japan.

The choice of the Kanto Plain for the final campaign in the Japanese Homeland had several distinct advantages. Firstly, that area offered a wide choice of suitable landing beaches—a cardinal consideration in any amphibious operation. (Plate No. 123) As the largest lowland region in the Japanese Home Islands, the Kanto Plain would enable the Allies to capitalize readily on their superiority in mechanization and armor. Logistic requirements to support the large forces involved necessitated good port facilities; the western shores of Tokyo Bay offered the best in Japan. In addition, the Kanto area served as the political and communications center for the Japanese Empire, containing approximately 50 per cent of Japan's war industry.[54]

Geographically, the Kanto Plain extends approximately 80 miles north and south and 70 miles east and west, covering between 5,000 and 6,000 square miles. The north and west sides are bordered by the mountainous masses of central and northern Honshu, which rise sharply to heights of 1,000—8,000 feet. Its eastern side is bounded by the Pacific Ocean and its southern side, by the waters of Tokyo and Sagami Bays.[55] The outer reaches of the Plain have three principal landing areas, each fronted with long sandy beaches, generally well suited to amphibious operations. These are the Kashima and Kujukuri beaches on the east coast, and Sagami beach at the head of Sagami Bay. From each of these landing points, fairly good routes lead into the Kanto Plain. Landings made in other regions would require overland movements through narrow, easily defended corridors.[56]

The capital of Tokyo is the hub of a widespread network of roads and railroads radiating outward along the Kanto Plain to northern, western, and southwestern Honshu. Numerous transverse roads and railroads within the Plain provide good routes of travel to or from Tokyo. Although these routes were important factors to the Japanese in reinforcement potential, they were relatively vulnerable to air attack. The destruction or blockade of the exposed bridges, tunnels, and defiles would seriously disrupt the transportation system and impede commercial and industrial activities. In addition, because of the widespread electrification of the railroads, effective bombing of dams, hydro-electric plants, and power stations would be catastrophic to the operation of the rail transport system.[57]

54 Joint Staff Planners, "An Outline Plan for the Invasion of the Kanto (Tokyo) Plain," Enclosure "A," 15 May 45, pp. 1–2 (TS).

55 AGS, G-2, GHQ, SWPA, Terrain Study No. 132, *Tokyo and Kwanto Plain*, 30 May 45.

56 Joint Staff Planners, "An Outline Plan for the Invasion of the Kanto (Tokyo) Plain," Enclosure "B," 15 May 45 (TS).

57 Ibid.

Employment of "Coronet" Ground Forces

The American landings on Honshu were to strike along the center of the east coast. The forces of the Eighth Army under General Eichelberger would constitute a western attack force to seize beachheads at the top of Sagami Bay. (Plate No. 124) From their initial positions, these troops would fan out to the north and east, taking the western shores of Tokyo Bay as far north as Yokohama. The armored elements would drive beyond Tokyo to the Kumagaya—Koga area to cut off Japanese reinforcement routes. If necessary, General Eichelberger would turn back his armor to strike at Tokyo from the rear. At the same time, other units under his command would complete the seizure of Yokohama.[58]

The veteran First Army of Gen. Courtney H. Hodges, redeployed to the Pacific from the battlefields of Europe, would strike at the Kujukuri beaches about fifty miles east of Tokyo, provide protection for the northeastern flank, and then strike out to the west and south to clear the eastern shores of Tokyo and Sagami Bays. One spearhead would advance directly toward Tokyo to destroy all hostile forces there in preparation for the establishment of air, naval, and supply facilities in the vicinity of the Japanese capital.

Only American troops would be engaged initially in central Honshu, but plans were made for the use of Australian, Canadian, British, and French divisions in subsequent stages of the campaign. They would be employed in case Japanese resistance should continue even after the heart of their Homeland was in American hands.[59]

Enemy Plans for the Defense of Honshu

By mid-June 1945, Japanese plans for the vital campaign in Honshu envisioned an all-out battle on the main beaches leading to the Kanto Plain. Kashima-nada, Kujukuri-hama, and the head of Sagami Bay were regarded as the crucial areas to be held at all costs against an amphibious invasion. Of the three beaches, Kujukuri-hama was selected as the area where the prime initial effort of the Japanese forces would be concentrated.

The general strategy called for the coastal-combat divisions to send their entire resources against the American assault head-on, with the underlying objective of merging all lines into an interlocking and continuously fluid struggle in which American air, artillery, and naval gunfire would be seriously hampered in choice of targets. It was felt that this was the only possible way to neutralize the tremendous air and sea superiority of the Allies.

Regular line-combat divisions, in prepared defenses, would take up the fight alongside and in the immediate rear of the coastal-combat troops. Other forces would reinforce the area of initial contact as fast as they could be moved into position. Without waiting to mass their strength, they would plunge immediately into the battle lines to be committed on a narrow front in great depth against a thin Allied beachhead. Japanese strategy was directed toward giving the landing forces no respite or opportunity to gain a firm foothold. It was not

---

58 GHQ, AFPAC, *Staff Study "Coronet,"* 15 Aug 45.

59 Ibid. CINCAFPAC Radio No. CX 28768 to WARCOS, 27 Jul 45, G-3, GHQ Admin 381/107–7 (TS). An earlier version of "Coronet" which is mentioned in General of the Army George C. Marshall's biennial report to the Secretary of War of 1 September 1945 called for an invasion by three armies: the First, the Eighth, and the Tenth; it was later decided not to use the Tenth Army in this operation.

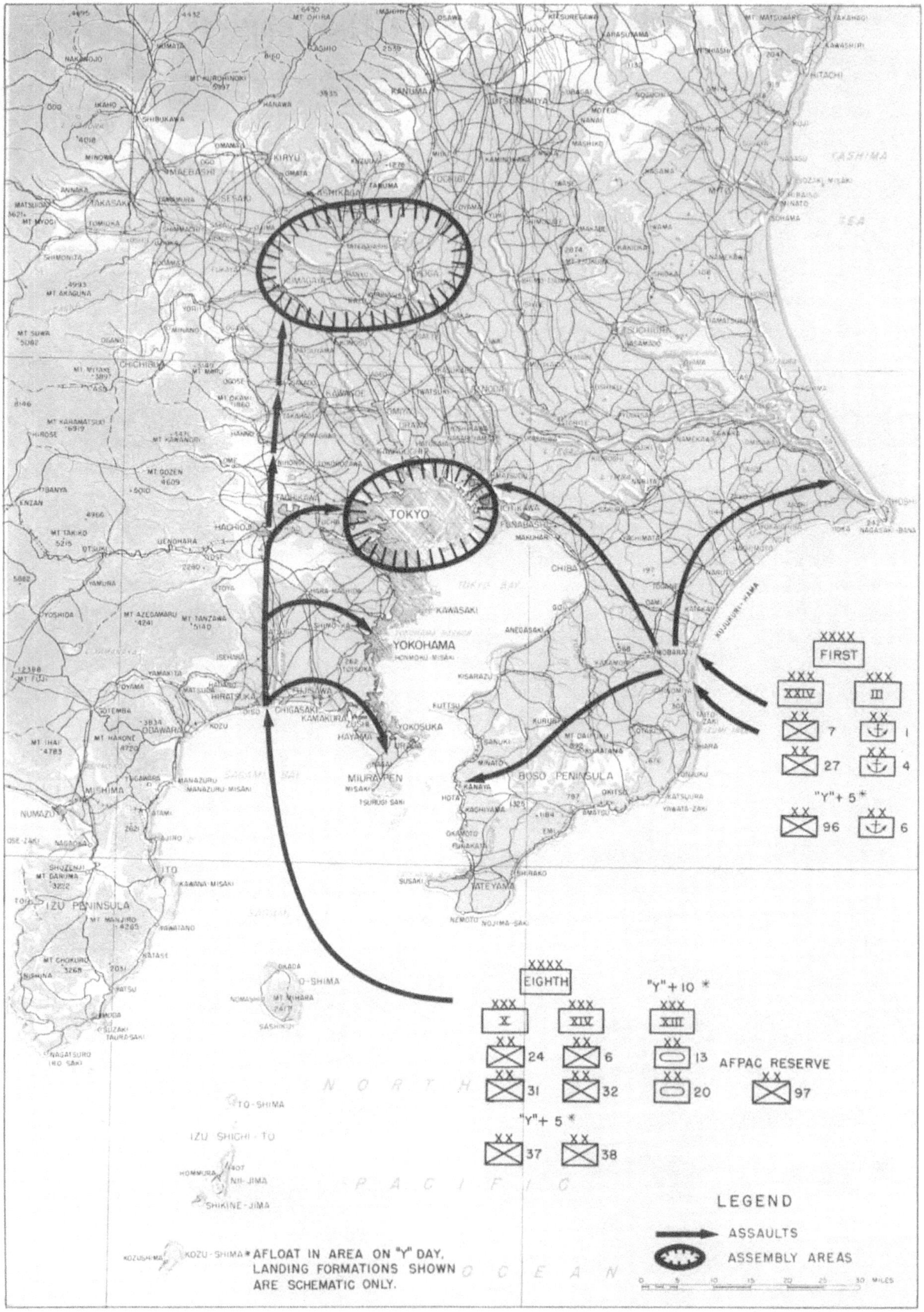

PLATE NO. 124

"Coronet," the Invasion of Honshu

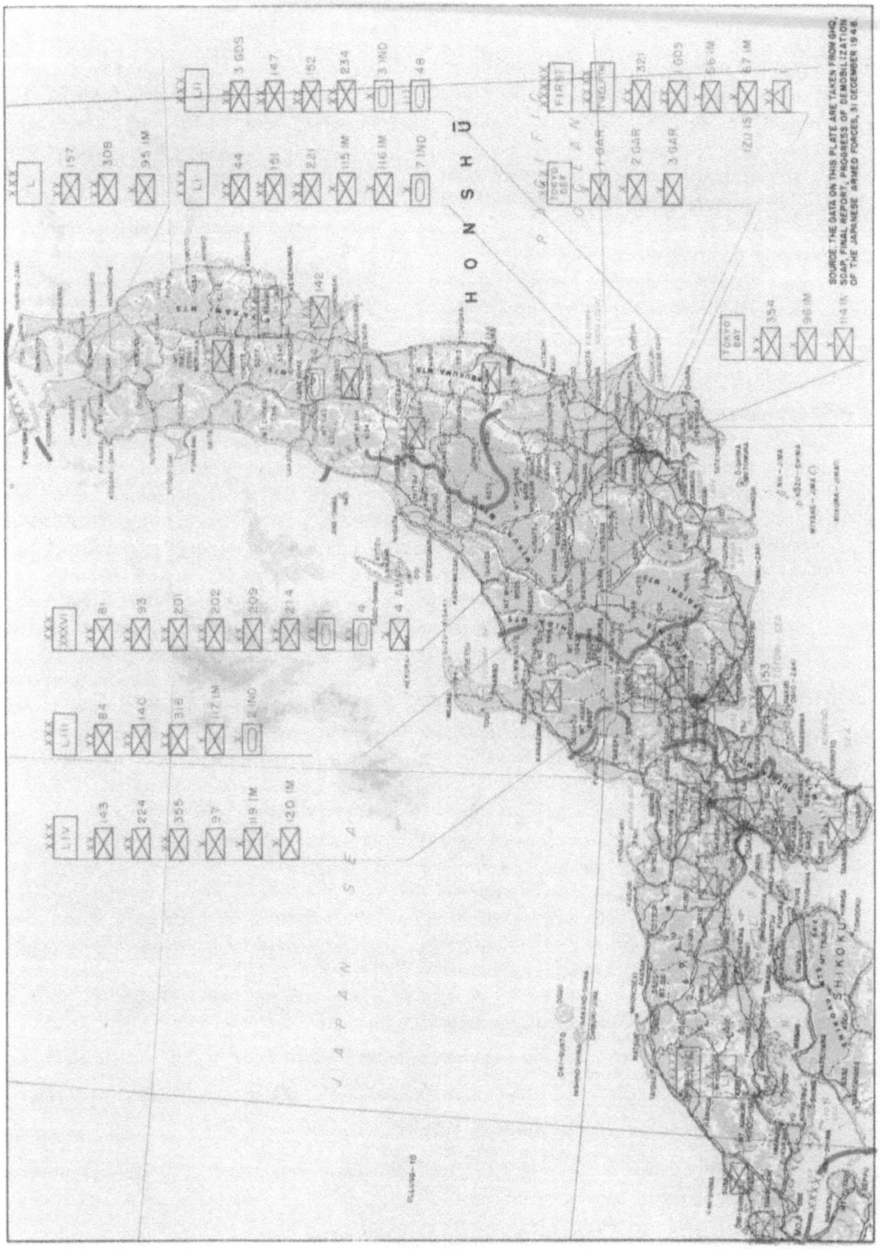

PLATE NO. 125
Japanese Ground Dispositions on Honshu, 18 August 1945

planned to keep any sizeable reserves in the central Kanto Plain region. If the Japanese could not hold the beaches and prevent a landing of heavy weapons and equipment, they felt that their last hope of a successful defense of Honshu would be irretrievably lost.[60]

The total ground combat strength in the general Kanto Plain region, including the shorelines of the three key areas, consisted of 18 infantry divisions, 7 independent mixed brigades, 2 armored divisions, and 3 tank brigades. Of the infantry divisions, 11 were line-combat while the other 7 were made up of the specially organized coastal-combat troops. In addition, a force of division strength was organized to make a last-ditch stand in the heart of the city of Tokyo and around the Imperial Palace.[61]

No definite provisions were made for the employment of air power in the event Honshu was assaulted in the spring of 1946, since it was anticipated that the Kyushu campaign would have consumed Japan's entire remaining air force. Special-attack surface craft were to be utilized to the fullest extent.

By July 1945, the Japanese had deployed an army at each of the three probable battlefronts —the Fifty-first Army at Kashima-nada; the Fifty-second Army at Kujukuri-hama; and the Fifty-third Army at Sagami Bay. (Plate No. 125) The Thirty-sixth Army, consisting of 6 line-combat infantry divisions and 2 armored divisions, was scheduled to move into Kujukuri-hama and be in position, fully equipped, by January 1946. Regardless of where the main force of the anticipated invasion might actually strike, the Japanese had determined to repel first any American troops coming ashore at Kujukuri-hama. Accordingly, Honshu defense plans directed that the bulk of Japan's fighting forces be massed at this point. Only after the action at Kujukuri-hama was completed would the Japanese offensive move to other battlefronts. The Tokyo Bay Group and the Yokosuka Naval District Group were responsible for holding the Boso Peninsula and the Miura Peninsula to prevent Allied penetration of Tokyo Bay.[62]

The Tokyo Defense Army had been formed late in June 1945 to bolster the weak defenses of the capital. This force, consisting of three infantry brigades and two independent engineer battalions, was to be reinforced immediately after the battle was joined. Two or three divisions and several tank and artillery units were to move in from other parts of Honshu to help secure the strategic central area of Tokyo and delay its capture as long as possible. It was also planned to construct a labyrinth of underground fortifications manned and supplied to resist for more than a year.[63]

Whether or not these desperate but extensive defense measures would have made an invasion prohibitively expensive in American lives is a matter for speculation. It is reasonable to assume, however, that Operation "Downfall" could have been successfully concluded only after a hard and bitter struggle with no quarter asked or given. It was fortunate for both sides that Japan realized the wisdom of surrender and that the Allied plan eventually executed was not "Downfall" but "Blacklist"—a peaceful occupation without gunfire, without further destruction, and without bloodshed.

---

60 Japanese First Demobilization Bureau Report, *Homeland Operations Record,* Vol. II; also statement by Gen. Kawabe, 13 Jun 49, Interrogation Files, G-2 Historical Section, GHQ, FEC.

61 Japanese First Demobilization Bureau Report, *Homeland Operations Record,* Vol. II, Attached Maps Nos. 1–3, G-2 Historical Section, GHQ, FEC.

62 Ibid. Also statement by Col. Hiroshi Fuwa, Operations Staff Officer, Twelfth Area Army, 24 May 49, Interrogation Files, G-2 Historical Section, GHQ, FEC.

63 Imperial General Headquarters, Army Order No. 1353, Directive No. 2513, 20 Jun 45, G-2 Historical Section, GHQ, FEC.

CHAPTER XIV
JAPAN'S SURRENDER

Continued Pressure on Japan

By the end of June 1945, United States forces had advanced their Pacific battle line thousands of miles from Australia and Pearl Harbor to reach the very threshold of the Japanese Homeland. They had overcome an enemy who fought with fierce tenacity and had solved unprecedented problems of logistics and enormous distance as they progressively occupied the coasts of New Guinea and New Britain, secured the strategic islands of the Solomons, Admiralties, Marianas, and Palaus, established airfields on Iwo Jima, moved into the Halmaheras, swept through the entire Philippines, and stood poised on Okinawa, the last military barrier to Japan Proper. (Plate No. 126)

Allied power dominated the land, sky, and sea of the western Pacific. General MacArthur's divisions had retaken vast island territories seized by Japan's armies at the outbreak of war and were now preparing to invade Japan itself. Huge formations of American Superfortresses pounded military and industrial targets on the Japanese mainland with increasing power. The U. S. Pacific Fleet had progressively cleared the ocean of Japanese warships in successive battles which stretched from the waters of Midway to the East China Sea and had bottled the decimated remnants of the Imperial Navy within their base ports. Even in its own Inland Sea and Tokyo Bay, the enemy fleet found neither respite nor refuge as fast American carriers navigated freely off the shores of Honshu and sent their bombing planes to hammer the great anchorages at Kure and Yokosuka. The time was ripe to hurl the whole might of the Allies against the defenses of Kyushu as the first step in Operation " Downfall."

In American hands, Kyushu could accommodate forty groups of the Far East Air Forces and provide unlimited opportunities for the use of air power against the military heart of Japan. In preparation for the main operation, " Coronet," planes from Kyushu could bomb every important target in Honshu, Korea, eastern Manchuria, and northern China. An additional forty air groups based in the Marianas, Iwo Jima, and Okinawa would magnify the potential force of destruction. The planes of these eighty air groups could drop 100,000 tons of bombs in September 1945 and 170,000 tons in January 1946. It was estimated that in March 1946 the projected date of the Honshu invasion, at least 220,000 tons of explosives could be released over the enemy's four main islands. In a single month, therefore, the industrial targets of Japan, contained in about one-tenth the area in which German targets were located, would be saturated by almost one-fourth the total bomb tonnage dropped on the Germans during the entire twelve months of 1944.[1]

With the approach of summer, the general air and naval offensive against Japan was

1 Allied Air Forces had unloosed 900,000 tons of explosives over German targets during 1944. CG USAAF Radio No. C-19585 to WARCOS, 17 Jun 45, C/S GHQ, WD No. 1053 (TS)

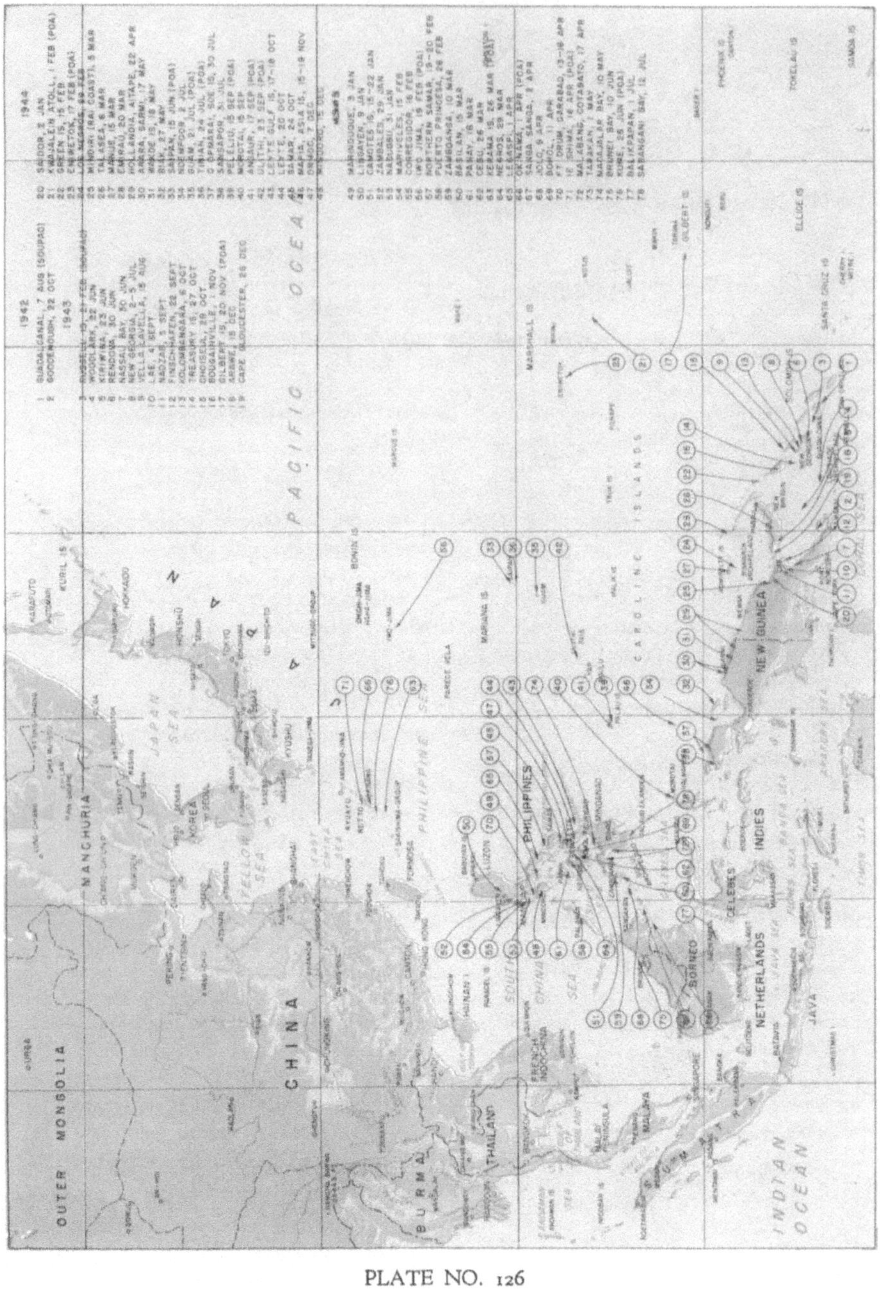

PLATE NO. 126
Allied Landings, August 1942 to August 1945

intensified to pave the way for the planned invasion of Kyushu. From the middle of May, when fighters based on the island of Ie Shima first attacked targets on southern Japan, the scale of co-ordinated air raids by the Fifth and Seventh Air Forces rose steadily, reaching a peak previously unknown in the Pacific War.[2]

On Okinawa, all organized Japanese resistance was ended by 21 June, and within two weeks fighters and bombers of the Fifth and Seventh Air Forces began their powerful assaults against Kyushu, neutralizing enemy air strength, severing lines of communication, and isolating the island from the rest of Japan.

Japanese targets in China also received their share of Allied attacks. Shanghai experienced its first large-scale aerial bombardment on 17–18 July, when the Seventh Air Force sent more than 200 Liberators, Mitchells, Invaders, and Thunderbolts from Okinawa over the great enemy-held industrial center in a two-day demonstration of air power. While the Seventh Air Force maintained its raids against Shanghai, the Fifth and Thirteenth Air Forces struck from bases in the Philippines to hit Formosa, Amoy, Swatow, Canton, and Hong Kong. The long-range bombers from the Marianas maintained a continuous shuttle over Japan itself, reducing its great industrial cities to ashes and rubble.[3]

Naval Pre-Invasion Operations

In July, carrier planes from Admiral Halsey's powerful Third Fleet contributed to the ceaseless strikes against the Japanese capital and its surrounding airfields. After supporting the Okinawa operation, the fast carriers of Vice Adm. John S. McCain's Task Force 38 departed from Leyte Gulf on 1 July and proceeded northward toward Japan. Arriving within striking range of Tokyo on 10 July, the armada launched fighter and bomber sweeps against military installations in the metropolitan area and blasted the targets with bombs and rockets. (Plate No. 127) More than 1,000 carrier-based planes were employed as the relentless assault continued virtually unopposed throughout the day. It was the greatest massing of U. S. naval air power against the Japanese since the beginning of the Pacific War. Simultaneously with the carrier assaults, between 500 and 600 Marianas-based B-29's made their deepest penetration of Japan to that time, in destructive raids against the war factories of the enemy's home islands.[4]

In a direct challenge to Japan's remaining air and naval strength, the Third Fleet on 14 July approached to within a few thousand yards of the enemy mainland off the steel plant city of Kamaishi and, in the first direct naval bombardment of the Homeland, fired thundering salvos into shore targets. Then, steaming 250 miles to the north, the mighty dreadnoughts and carriers on 14–15 July struck installations in northern Honshu and southern Hokkaido. Moving southward again, the Third Fleet was augmented by a carrier task force of the British Pacific Fleet and on 17 July carried out the first joint American-British bombardment of Japan. More than 2,000 tons of shells were fired into the coastal area at Hitachi, northeast of Tokyo. The next day over 1,500 United States and British carrier planes climaxed the shore attacks with the greatest carrier strike in history against the

---

2 Gen. Henry H. Arnold, "Third Report of the Commanding General of the Army Air Forces to the Secretary of War," 12 Nov 45.

3 G-3, GHQ, SWPA, *Monthly Summary of Operations*, Jul 45.

4 ONI, USN, *Office of Naval Intelligence Weekly*, Vol. IV, No. 28. 11 Jul 45, pp. 2156–2160. Hereinafter cited as: O. N. I. Weekly.

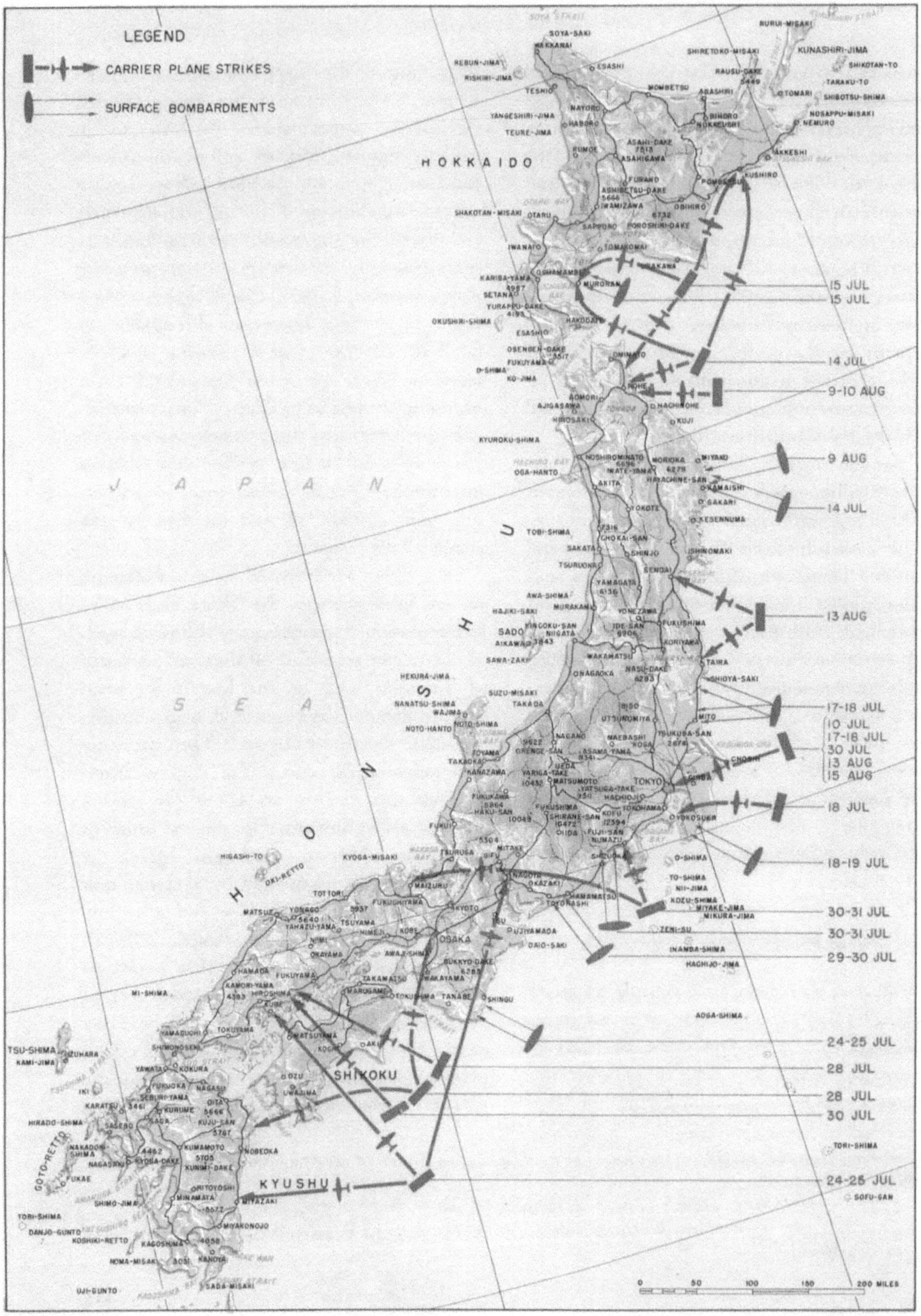

PLATE NO. 127

Third Fleet Pre-Invasion Operations against Japan

Tokyo area.

Following eight days of continuous assault on enemy airfields, shipping, and industrial targets, the Third Fleet, together with its accompanying British fleet units, turned its attention to eliminating whatever was left of the Japanese Navy. A heavy attack was launched on 18 July, when hundreds of carrier planes concentrated on the enemy warships, including the battleship *Nagato*, which were anchored at Yokosuka naval base. At the same time, a light detached force of the Third Fleet bombarded military installations at Nojima Saki during the night of 18–19 July.

Five days later, on 24 and 25 July, extensive air raids were launched against Kure and the Inland Sea area by the combined American-British naval force. This sixth carrier strike against the Japanese Homeland since 10 July was followed by another raid on 28 July. Seventh Air Force Liberators from bases on Okinawa joined the attack on Japan's remaining naval units in Empire waters by blasting the anchored warships at Kure. On 30 July, the Tokyo area was again pounded by aircraft from the fast carriers, while battleships poured more than 1,000 tons of shells into the key port, rail center, and industrial city of Hamamatsu on the east coast of central Honshu. Bad weather delayed the naval onslaught for the first few days of August, but on the 9th and 10th, northern Honshu was again subjected to air and sea attacks. The final heavy blow was delivered against Tokyo on 13 August, with carrier planes raking airfields and other military installations as primary targets.[5] The suspension of hostilities early on the morning of the 15th found some carrier planes already airborne for an attack on Tokyo, but only one wave hit the objective area; a second wave was recalled before reaching its targets.[6]

Resistance by the Japanese to this stinging nine-tailed lash of Allied naval and air power was scattered and ineffective; the enemy sought to conserve his few remaining planes and ships for the expected invasion. In Asia, meanwhile, several Japanese ground divisions were being steadily deployed to defend the important industrial regions of northern China after the Soviet Union on 5 April had announced its intention not to renew its existing neutrality treaty with Japan.[7]

The Potsdam Declaration

Against the background of final military preparations for the invasion of Japan, international negotiations were under way which were ultimately to make the projected operations "Olympic" and "Coronet" unnecessary. On 17 July 1945, the President of the United States, the Prime Minister of Great Britain, and the Premier of the Union of Soviet Socialist Republics met in a series of conferences at Potsdam, Germany, and discussed, among other things, the acceleration of the campaign against Japan. One result of this tripartite conference was that the Soviet Union finally agreed to enter the Pacific war. Another equally outstanding product of the Potsdam conferences was the Potsdam Declaration. President Truman and Prime Minister Atlee, with the concurrence of the President of the National Government of China, issued a final ultimatum

---

5 G-3, GHQ, AFPAC, Memo, "Third Fleet Operations in Jul 45," 1 Jul 45, G-3, GHQ Admin 381/134–9 (TS); Adm. Ernest J. King, "Third Official Report to the Secretary of the Navy," 8 Dec 45; William F. Halsey and J. Bryan III, *Admiral Halsey's Story* (New York, 1947), pp. 258–265.

6 Halsey and Bryan, op. cit. pp. 270–271.

7 JCS Radio No. WX 15811 to CINCAFPAC, CINCPOA, CG CHINA, 13 Jun 45, G-3, GHQ Admin 381/81–6P (TS).

to the Japanese Government that gave Japan the choice of surrender or destruction. Set forth in powerful words of warning, the Potsdam Declaration read in part:

. . . The prodigious land, sea and air forces of the United States, the British Empire and of China, many times reinforced by their armies and air fleets from the west, are poised to strike the final blows upon Japan. This military power is sustained and inspired by the determination of all the Allied Nations to prosecute the war against Japan until she ceases to resist.

The result of the futile and senseless German resistance to the might of the aroused free peoples of the world stands forth in awful clarity as an example to the people of Japan. The might that now converges on Japan is immeasurably greater than that which, when applied to the resisting Nazis, necessarily laid waste the lands, the industry and the method of life of the whole German people. The full application of our military powers, backed by our resolve, will mean the inevitable and complete destruction of the Japanese armed forces and just as inevitably the utter destruction of the Japanese homeland.

The time has come for Japan to decide whether she will continue to be controlled by those self-willed militaristic advisers whose unintelligent calculations have brought the Empire of Japan to the threshold of annihilation, or whether she will follow the path of reason . . . .

After listing seven terms under which the Allies would accept the Japanese capitulation, the declaration continued:

We call upon the Government of Japan to proclaim now the unconditional surrender of all Japanese armed forces, and to provide proper and adequate assurances of their good faith in such action. The alternative for Japan is prompt and utter destruction.[8]

The decision rested with Japan. Would it be "prompt and utter destruction," or surrender according to the plans outlined at Cairo[9] and Potsdam, which accorded Japan an opportunity to refit herself for membership in a world of peaceful nations?

"Blacklist" Plan

In anticipation of Japan's possible surrender, it now became necessary to accelerate the preparation of plans for a peaceful entry into the enemy's homeland. The course of events had given the strategic control of the Pacific War to the United States, and as the conflict progressed it became the nation chiefly responsible for the conduct of all operations dealing with Japan. Although the broad policies of occupation were agreed upon by the major Allied governments in accordance with the United Nations Charter, the United States would execute these policies, provide and regulate the main occupation forces, and designate their commander. Such control also would enable the institution of a strong centralized administration to avoid dividing Japan into national zones of independent responsibility as had been done with Germany.[10]

In accordance with instructions from the Joint Chiefs of Staff received early in May 1945, General MacArthur had immediately

8 U. S. Department of State, *The Axis in Defeat, A Collection of Documents on American Policy Toward Germany and Japan*, pp. 27–29. Hereinafter cited as: *The Axis in Defeat.*

9 The Cairo Conference in Egypt held on 22–26 November 1943 by President Roosevelt, Prime Minister Churchill, and Generalissimo Chiang Kai-shek, was for the purpose of forming an agreement upon future military operations against Japan. The heads of the three governments resolved to bring unrelenting pressure against Japan, to take away from Japan all the Pacific Islands which she had seized or occupied, to expel the Japanese from all territories which had been taken by violence, and to restore the independence of Korea. *The Axis in Defeat*, "Cairo Declaration," pp. 4–5.

10 State-War-Navy Co-ordinating Committee Memo, 11 Aug 45, G-3, GHQ Admin (TS); WARCOS Radio No. WAR 48342 to CINCAFPAC, 12 Aug 45, C/S GHQ, WD No. 1118 (TS); WARCOS Radio No. W48672 to CINCAFPAC, 13 Aug 45, C/S GHQ S, WD No. 1119 (TS).

directed his staff to prepare a plan for a possible peaceful occupation. The first edition of this plan, designated "Blacklist," was published on 16 July and presented four days later at Guam for comparison with a concurrent plan for occupation termed "Campus" which was being formulated by Admiral Nimitz.[11] General MacArthur's proposals were based on the assumption that it would be his responsibility to impose surrender terms upon all elements of the Japanese military forces within Japan and that he also would be responsible for enforcing the demands of Allied commanders in other areas.

The final edition of "Blacklist" called for the progressive and orderly occupation in strength of an estimated fourteen major areas in Japan and from three to six areas in Korea so that the Allies could exercise unhampered control of the various phases of administration.[12] These operations would employ 22 divisions and 2 regimental combat teams, together with air and naval elements, and would utilize all United States forces immediately available in the Pacific. (Plate No. 128) Additional forces from outside the theater would be requisitioned if occupational duties in Formosa or in China were required. "Blacklist" Plan provided for the maximum use of existing Japanese political and administrative organizations since these agencies exerted an effective control over the population and could obviously be employed to good advantage by the Allies. If the functioning governmental machinery were to be completely swept away, the difficulties of orderly direction would be enormously multiplied, demanding the use of greater numbers of occupation forces.[13]

The preliminary opinions of the Joint Chiefs of Staff regarding the initial phase of the occupation inclined generally toward large-scale independent naval landings as envisioned in the Pacific Fleet Headquarters plan "Campus."[14] "Campus," the naval counterpart of "Blacklist," provided for entry into Japan by United States Army forces only after independently operating advance naval units had made an emergency occupation of Tokyo Bay and seized possession of key positions on shore including, if practicable, an operational airfield in the vicinity of each principal anchorage.[15]

General MacArthur felt that this concept was strategically unwise and dangerous. Although he agreed that, immediately after capitulation, the United States Fleet should move forward, seize control of Japan's Homeland waters, take positions off critical localities, and begin mine sweeping operations, he thought that these steps should introduce immediate landings by strong, co-ordinated ground and air forces of the army, fully prepared to overcome any potential opposition. General MacArthur believed that naval forces were not designed to effect the preliminary occupation of a hostile country whose ground divisions were still intact and contended that the occupation of large land areas involved operations which were fundamentally and basically a mission of the army. The occupation, he maintained, should proceed along sound tactical lines with each

---

11 CINCAFPAC Radio No. C-25892 to CINCPAC ADV, 16 Jul 45, G-3, GHQ Admin 381/61 (TS).

12 GHQ, APFAC, *Basic Outline Plan for "Blacklist" Operations to Occupy Japan Proper and Korea after Surrender or Collapse*, 3rd edition, 8 Aug 45, G-3, GHQ Planning File (TS). Hereinafter cited as: *Blacklist.*

13 Ibid; CINCAFPAC Radio No. C-28793 to WARCOS, 27 Jul 45, C/S GHQ, WD No. 1088 (TS).

14 JCS Radio No. BG-412V357 to CINCAFPAC, CINCPAC, 27 Jul 45, JCS and CCS Papers 7, G-3, GHQ Ex (TS).

15 CINCPAC and CINCPOA *Joint Staff Study "Campus,"* 9 Aug 45, G-3, GHQ Admin Planning Files (TS). Hereinafter cited as: *Campus.*

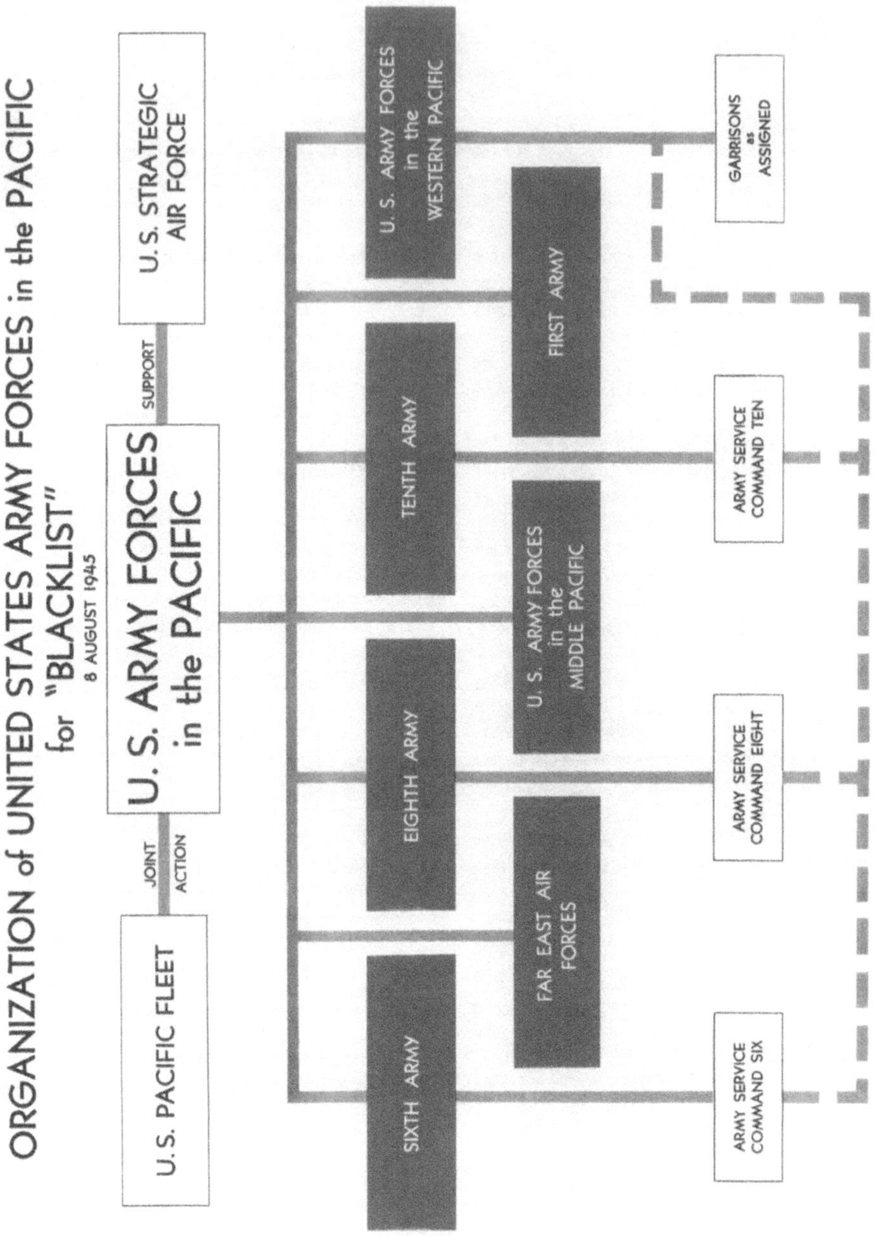

PLATE NO. 128
"Blacklist" Organization of Forces

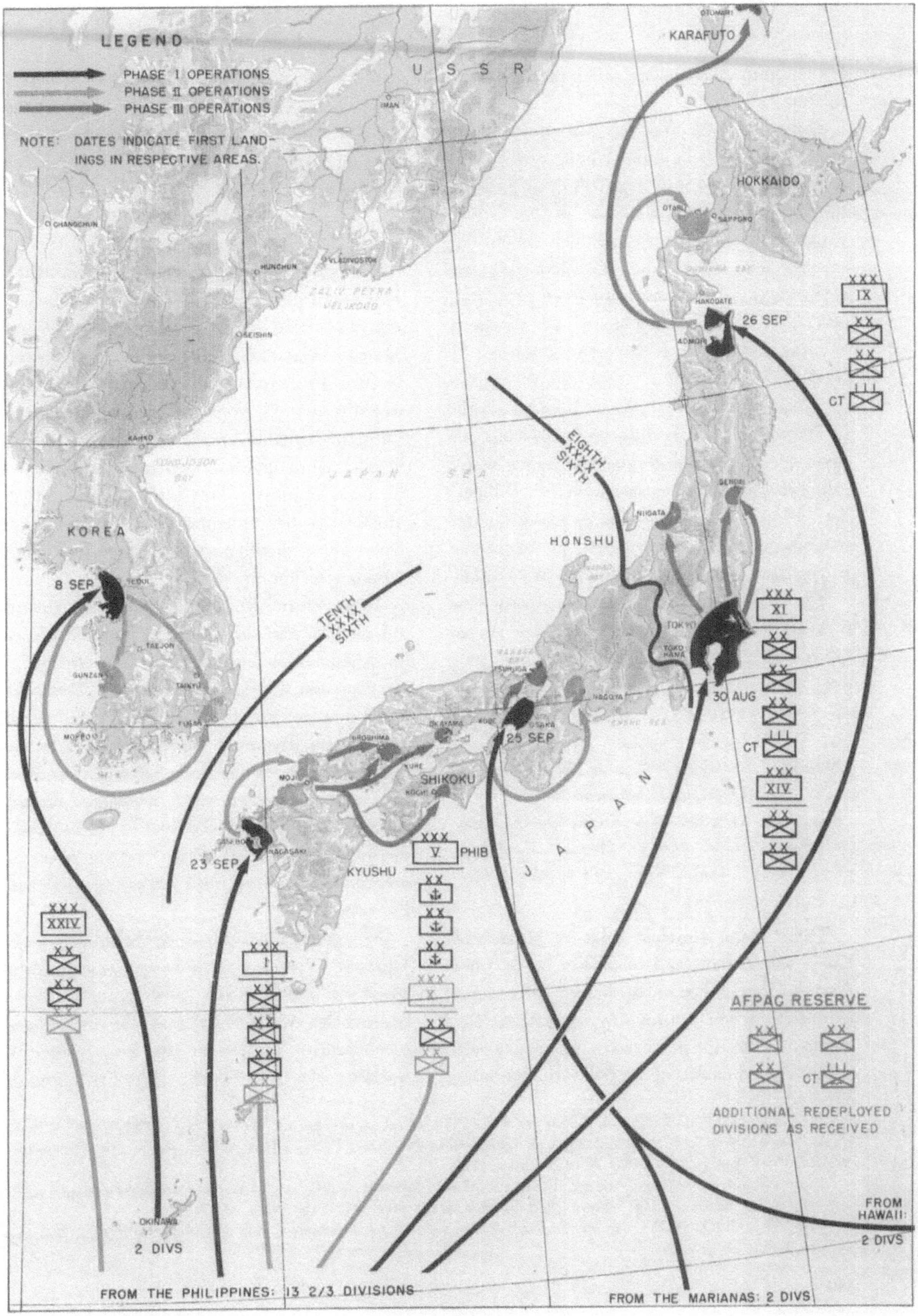

PLATE NO. 129
Basic Plan for the Occupation of Japan

branch of the service performing its appropriate mission.

General MacArthur felt also that, in the event landings by light naval units were authorized, army troops should go ashore at the same time to implement the display of force. Occupation by a weak Allied force might encourage opposition from dissident Japanese elements among the bomb-shattered population and lead to grave repercussions. In a radio to Washington, General MacArthur declared, "I hold the firm belief . . . that sound military judgment dictates that the occupation should be effected in force in order to impose our will upon the enemy and to avoid incidents which might develop serious proportions."[16] "Blacklist" Plan therefore provided for a co-ordinated movement of ground, naval, and air forces and a gradual but firmly regulated occupation.

The final edition of "Blacklist" issued on 8 August was divided into three main phases of successive occupation, viz: (Plate No. 129)

Phase I: *Kanto Plain, Nagasaki-Sasebo, Kobe-Osaka-Kyoto, Seoul (Korea), and Aomori-Ominato.*

Phase II: *Shimonoseki-Fukuoka, Nagoya, Sapporo (Hokkaido), and Fusan (Korea).*

Phase III: *Hiroshima-Kure, Kochi (Shikoku), Okayama, Tsuruga, Otomari (Karafuto), Sendai, Niigata, and Gunzan-Zenshu (Korea).*

These major strategic areas of Japan and Korea would be seized to isolate Japan from Asia, to immobilize enemy-armed forces, and to initiate action against any recalcitrant elements. Thus, the projected occupations would permit close direction of the political, economic, and military institutions of the two countries. Other areas would be occupied as deemed necessary by army commanders to accomplish their missions.[17]

The Final Blows

To lend additional force to the terms of the Potsdam ultimatum, the air and naval offensive was stepped up with even greater power. B-29's from the Marianas, supported by fighters based on Iwo Jima, averaged 1,200 sorties a week over the enemy's homeland, while planes from Okinawa airfields ripped his positions on the Asiatic mainland and destroyed what remained of his shipping. The Third Fleet and its attached British units meanwhile roamed Japanese waters, shelling coastal cities and shore targets with impunity.[18]

In an effort to minimize the losses among the civilian population and to counteract false propaganda concerning Allied aims spread by the Japanese High Command, the Twentieth Air Force and the Far East Air Forces on 28 July began dropping warning leaflets to announce seventy-two hours in advance the names of the cities marked for destruction. In addition to notifying all civilians to flee to safety, the leaflets advised them to "restore peace by demanding new and good leaders who will end the war."[19]

As a direct consequence of the failure of the Japanese Government to accept promptly the terms of the Potsdam proclamation, Japan became the victim of the most destructive and revolutionary weapon in the long history of warfare—the atom bomb. The first bomb of

---

16 CINCAFPAC Radio No. C-29035 to WARCOS, 26 Jul 45, C/S GHQ, WD No. 1092 (TS); CINCAFPAC Radio No. C-28810 to JCS, 27 Jul 45, C/S GHQ, WD No. 1090 (TS); CINCAFPAC Radio No. C-30508 to WARCOS, 3 Aug 45, C/S GHQ, WD No. 1099 (TS).

17 *Blacklist*, 3rd edition, 8 Aug 45. The final edition of the naval counterpart, "Campus," was dated 9 August 1945.

18 Adm. Ernest J. King, "Third Official Report to the Secretary of the Navy," 8 Dec 45.

19 G-3, GHQ, SWPA, *Monthly Summary of Operations*, "Air Summary," Aug 45; O. N. I. Weekly, Vol IV, No. 31, 1 Aug 45, p. 2293.

this type ever used against an enemy was released early on 6 August from an American Superfortress over the military base city of Hiroshima and exploded with incomparable and devastating force. The city was almost completely and uniformly leveled.

On 7 August President Truman electrified the world with a broadcast statement which declared:

> *Sixteen hours ago an American airplane dropped one bomb on Hiroshima, an important Japanese army base. That bomb had more power than 20,000 tons of TNT.... With this bomb we have now added a new and revolutionary increase in destruction to supplement the growing power of our armed forces....*
>
> *It is an atomic bomb. It is a harnessing of the basic power of the universe....*
>
> *We are now prepared to obliterate more rapidly and completely every productive enterprise the Japanese have above ground in any city. We shall destroy their docks, their factories, and their communications. Let there be no mistake; we shall completely destroy Japan's power to make war.*
>
> *It was to spare the Japanese people from utter destruction that the ultimatum of July 26 was issued at Potsdam.*[20]

With the echo of this cataclysmic explosion reverberating around the world, another staggering blow befell the Japanese. The Soviet Union on 8 August declared war on Japan and hastily sent troops against the Japanese Kwantung Army in Manchuria. The Japanese were now assailed from every side.

On 9 August a second atomic bomb destroyed the city of Nagasaki amid a cloud of dust and debris that rose 50,000 feet and was visible for more than 175 miles.[20] The two bombs which fell on Hiroshima and Nagasaki were dropped by the 509th Composite Bomb Group based on Tinian. The selection of Nagasaki as the second objective of the atomic bomb was caused by unfavorable weather conditions. After circling for fifty minutes above the smoke-obscured city of Kokura, which was the primary target, the bombing plane flew on to drop the bomb over Nagasaki, the alternate target.[21]

The advent of the atomic bomb coming on the heels of a long series of paralyzing military disasters, hastened the surrender which was already being intensively deliberated by Japan's leaders. By 10 August Japan had had enough; she recognized her situation as hopeless. After much internal struggle and argument, the Japanese Government instructed its Minister to Switzerland to advise the United States through the Swiss Government that the terms of the Potsdam ultimatum would be accepted if Japan's national polity could be preserved. The Japanese note read in part:

> *...the Japanese Government several weeks ago asked the Soviet Government, with which neutral relations then prevailed, to render good offices in restoring peace vis-a-vis the enemy power. Unfortunately, these efforts in the interest of peace having failed, the Japanese Government in conformity with the august wish of His Majesty to restore the general peace and desiring to put an end to the untold sufferings entailed by war as quickly as possible, have decided upon the following:*
>
> *The Japanese Government are ready to accept the terms enumerated in the joint declaration which was issued at Potsdam on July 26th, 1945, by the heads of the Governments of the United States, Great Britain, and China, and later subscribed by the Soviet Government with the understanding that the said declaration does not comprise any demand which prejudices the prerogatives of His Majesty as a Sovereign Ruler.*
>
> *The Japanese Government sincerely hope that this understanding is warranted and desire keenly that an explicit indication to that effect will be speedily forthcoming.*[22]

---

20 Gen. George C. Marshall, "Third Official Report to the Secretary of War," 1 Jul 43–30 Jun 45.

21 USSBS, *Strategic Air Operation of Very Heavy Bombardment in the War against Japan*, (*Twelfth Air Force*), p. 18.

22 *The Axis in Defeat*, pp. 29–30

On 11 August, the United States, acting on behalf of the United Nations, transmitted a reply which stated:

> *... From the moment of surrender the authority of the Emperor and the Japanese Government to rule the state shall be subject to the Supreme Commander of the Allied Powers who will take such steps as he deems proper to effectuate the surrender terms.*
>
> *The Emperor will be required to authorize and ensure the signature by the Government of Japan and the Japanese Imperial General Headquarters of the surrender terms necessary to carry out the provisions of the Potsdam Declaration, and shall issue his commands to all the Japanese military, naval, and air authorities and to all the forces under their control wherever located to cease active operations and to surrender their arms, and to issue such other orders as the Supreme Commander may require to give effect to the surrender terms.*
>
> *Immediately upon the surrender, the Japanese Government shall transport prisoners of war and civilian internees to places of safety as directed, where they can quickly be placed aboard Allied transports.*
>
> *The ultimate form of government of Japan shall, in accordance with the Potsdam Declaration, be established by the freely expressed will of the Japanese people.*
>
> *The armed forces of the Allied Powers will remain in Japan until the purposes set forth in the Potsdam Declaration are achieved.*[23]

While the Japanese Government pondered the Allied answer, President Truman, on 12 August, directed the Strategic Air Force to cease its attacks.[24] The Far East Air Forces and the Allied Fleet in Japanese waters, however, continued their steady pounding. When no reply was received from the Japanese by 13 August, the Strategic Air Force was ordered to renew its operations[25] and on the same day 1,000 carrier planes from the Third Fleet made their final raid on Tokyo.

Never before in history had one nation been the target of such concentrated air power. (Plate No. 130) In the last fifteen days of the war, the Fifth and Seventh Air Forces flew 6,372 sorties against Kyushu alone. Forty-nine per cent of this devastating effort was directed at manufacturing areas and docks. The remaining percentage was divided among enemy shipping, air installations, and lines of communication. Thus, with a deafening crescendo of blasting bombs, the Far East Air Forces culminated their blows against Japan. During the last seven and one-half months of the war their planes had destroyed or badly damaged 2,846,932 tons of shipping and 1,375 enemy aircraft, dropped 100,000 tons of bombs, and flown over 150,000 sorties.[26]

The end of the war in Europe had not only released additional ground, air, and naval forces for the war against Japan but it had also enabled the Soviet Union to mass its forces for an attack upon Manchuria and northern China. The veteran armies of General MacArthur were poised and ready for an invasion of Kyushu and Honshu. The warning to surrender or be destroyed had not been composed of idle words. Stark and ruinous defeat was already a frightening certainty for the Japanese.

Japan Capitulates

While Japan considered its final acceptance of the Allied terms, preparations for the progressive occupation of its cities and military possessions were already completed. The formal directive for the occupation of Japan, Korea, and the China coast was issued by the Joint Chiefs of Staff on 11 August. Arrange-

23 Ibid, pp. 30–31

24 WARCOS Radio No. WAR 47899 to CINCAFPAC, 12 Aug 45, C/S GHQ, WD No. 1117 (TS).

25 WARCOS Radio No. WAR 48689 to CINCAFPAC, CG USASTAF, 13 Aug 45, C/S GHQ, WD No. 1120 (TS).

26 G-3, GHQ, SWPA, *Monthly Summary of Operations*, "Air Summary," Aug 45.

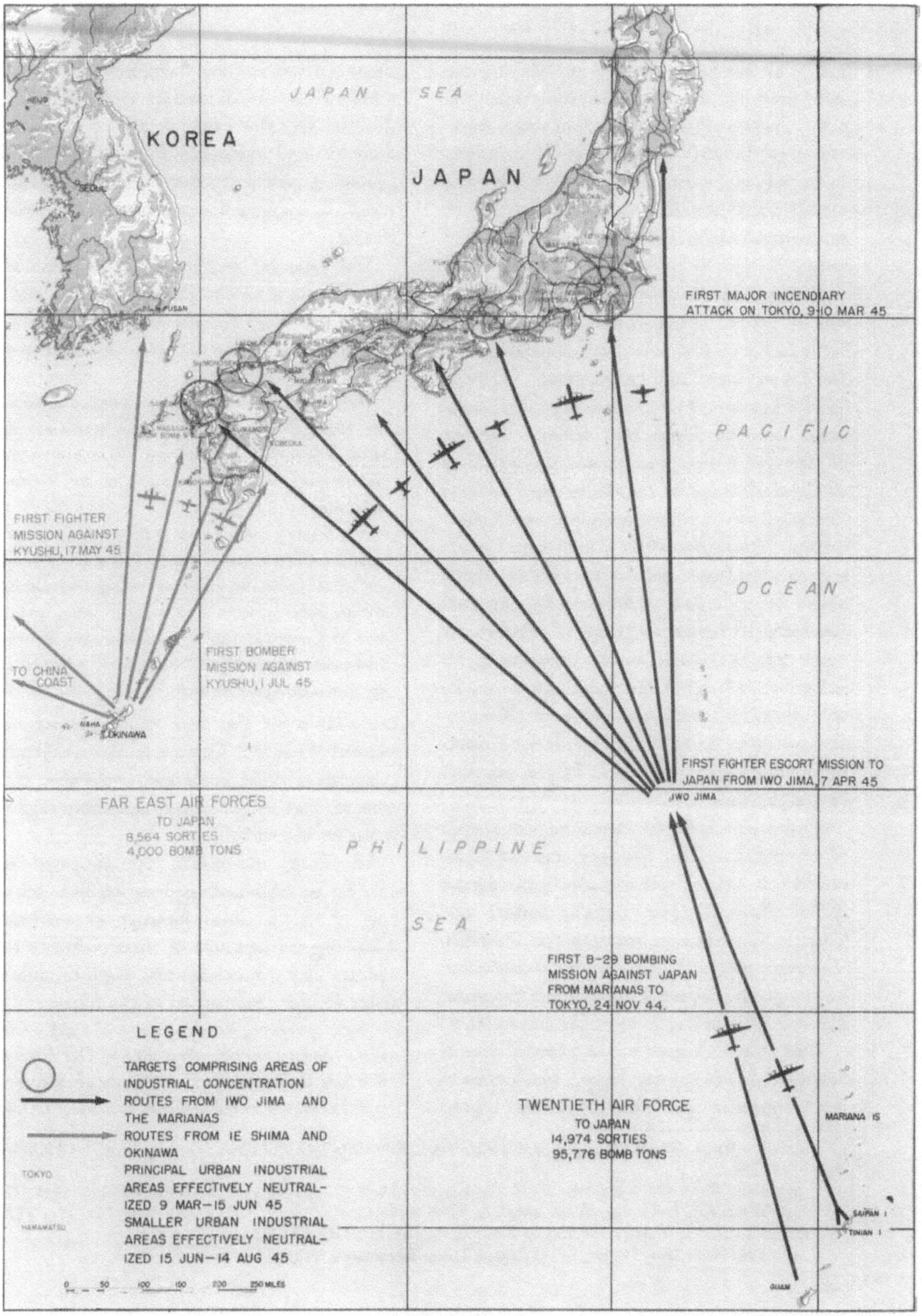

PLATE NO. 130
Aerial Bombardment of Japan

ments for the peaceful entry of Allied forces were patterned as far as practicable upon the actual invasion plans. The immediate objectives were the early introduction of occupying forces into major strategic areas, the control of critical ports, port facilities, and airfields, and the demobilization and disarmament of enemy troops.[27]

First priority was given to the prompt occupation of Japan, second to the consolidation of Keijo in Korea, and third to the operations on the China coast and in Formosa. General MacArthur was to assume responsibility for the forces entering Japan and Korea. General Wedemeyer was assigned operational control of the forces landing on the China coast and was instructed to co-ordinate his plans with Generalissimo Chiang Kai-shek. Naval forces entering ports in Japan and China were to remain under the command of Admiral Nimitz until General MacArthur and General Wedemeyer could take over these areas. Japanese forces in Southeast Asia were earmarked for surrender to Admiral Mountbatten, those in China to Generalissimo Chiang, and those in the Russian area of operations to the Soviet High Command in the Far East.[28]

Final authority for the execution of the terms of surrender and for the occupation of Japan would rest with a Supreme Commander for the Allied Powers. The British, Soviet, and Chinese Governments concurred in President Truman's proposal that General MacArthur be designated Supreme Commander to assume the over-all administration of the surrender.[29]

The 15th of August was an eventful date in history. It was the day Japan's notification of final surrender was received in the United States; it was the day President Truman announced the end of conflict in the Pacific; it was the day the Emperor of Japan made a dramatic and unexampled broadcast to his people; it was the day that General MacArthur was made Supreme Commander for the Allied Powers.

The Japanese acceptance of the terms laid down by the Allies at Potsdam read in part:

His Majesty the Emperor has issued an Imperial Rescript regarding Japan's acceptance of the provisions of the Potsdam declaration:

His Majesty the Emperor is prepared to authorize and ensure the signature by his Government and the Imperial General Headquarters of the necessary terms for carrying out the provisions of the Potsdam declaration.

His Majesty is also prepared to issue his commands to all the military, naval, and air authorities of Japan and all the forces under their control wherever located to cease active operations, to surrender arms, and to issue such orders as may be required by the Supreme Commander of the Allied Forces for the execution of the above-mentioned terms.[30]

As soon as the note of acceptance was received, President Truman announced Japan's capitulation to the world and, at the same time, ordered that all offensive operations against Japan be suspended.

In Japan, meanwhile, the Emperor was making an unprecedented nation-wide broadcast of the Imperial Rescript of surrender, informing the Japanese of their country's first military defeat and exhorting them to unite in peace for the construction of the future:

After pondering deeply the general trend of the world situation and the actual state of Our Empire, We have decided to effect a settlement of the present crisis by resort to an extraordinary measure. To Our

27 JCS Radio No. WAR 47945 to CINCAFPAC, CINCPAC, CG CHINA, 11 Aug 45, C/S GHQ, WD No. 1112 (TS).

28 Ibid; WARCOS Radio No. WAR 48342 to CINCAFPAC, 12 Aug 45, C/S GHQ, WD No. 1118 (TS).

29 WARCOS Radio No. WAR 47838 to CINCAFPAC, 11 Aug 45, C/S GHQ, WD No. 1111 (TS); WARCOS Radio No. W48672 to CINCAFPAC, 13 Aug 45, C/S GHQ S, WD No. 1119 (TS).

30 *The Axis in Defeat*, p. 32; Tokyo *Nippon Times*, September 9, 1945.

good and loyal subjects, we hereby convey Our will.

We have commanded Our Government to communicate to the Governments of the United States, Great Britain, China and the Soviet Union that Our Empire accepts the terms of their Joint Declaration....

Hostilities have now continued for nearly four years. Despite the gallant fighting of the Officers and Men of Our Army and Navy, the diligence and assiduity of Our servants of State, and the devoted service of Our hundred million subjects—despite the best efforts of all—the war has not necessarily developed in our favor, and the general world situation also is not to Japan's advantage ....

Should we continue to fight, the ultimate result would be not only the obliteration of the race but the extinction of human civilization. Then, how should We be able to save the millions of Our subjects and make atonement to the hallowed spirits of Our Imperial Ancestors? That is why We have commanded the Imperial Government to comply with the terms of the Joint Declaration of the Powers....

The suffering and hardship which Our nation yet must undergo will certainly be great. We are keenly aware of the innermost feelings of all ye, Our subjects. However, it is according to the dictates of time and fate that We have resolved, by enduring the unendurable and bearing the unbearable, to pave the way for a grand peace for all generations to come.

Since it has been possible to preserve the structure of the Imperial State, We shall always be with ye, Our good and royal subjects, placing Our trust in your sincerity and integrity. Beware most strictly of any outburst of emotion which may engender needless complications, and refrain from fraternal contention and strife which may create confusion, lead ye astray and cause ye to lose the confidence of the world. Let the nation continue as one family from generation to generation with unwavering faith in the imperishability of our divine land and ever mindful of its heavy burden of responsibility and the long road ahead. Turn your full strength to the task of building a new future. Cultivate the ways of rectitude, foster nobility of spirit, and work with resolution so that ye may enhance the innate glory of the Imperial State and keep pace with the progress of the world. We charge ye, Our loyal subjects, to carry out faithfully Our will.[31]

To initiate the steps towards the implementation of the surrender terms, President Truman announced the appointment of General MacArthur as the Supreme Commander for the Allied Powers and dispatched the following orders to the Japanese Government:

You are to proceed as follows:

Direct prompt cessation of hostilities by Japanese forces, informing the Supreme Commander for the Allied Powers of the effective date and hour of such cessation.

Send emissaries at once to the Supreme Commander for the Allied Powers with information of the disposition of the Japanese forces and commanders, and fully empowered to make any arrangements directed by the Supreme Commander for the Allied Powers to enable him and his accompaying forces to arrive at the place designated by him to receive the formal surrender.

For the purpose of receiving such surrender and carrying it into effect, General of the Army Douglas MacArthur has been designated as Supreme Commander for the Allied Powers, and he will notify the Japanese Government of the time, place, and other details of the formal surrender.[32]

From the War Department in Washington, the Army Chief of Staff dispatched a message to General MacArthur which read, "Your directive as Supreme Commander for the Allied Powers is effective with the receipt of this message."[33]

Preparations for Surrender

General MacArthur received the announce-

31 Japanese *Imperial Rescript*, 15 Aug 45, G-2 Historical Section, GHQ, FEC.

32 *The Axis in Defeat*, pp. 32–33.

33 WARCOS Radio No. 1408 to CINCAFPAC, 15 Aug 45, C/S GHQ, WD No. 1125.

ment of Japan's capitulation with an expression of deep gratitude:

> *I thank a merciful God that this mighty struggle is about to end. I shall at once take steps to stop hostilities and further bloodshed. The magnificent men and women who have fought so nobly to victory can now return to their homes in due course and resume their civil pursuits. They have been good soldiers in war. May they be equally good citizens in peace.*[34]

The air waves crackled with urgent radio messages between Manila Headquarters and Tokyo. General MacArthur notified the Emperor and the Japanese Government on 15 August that he had been designated Supreme Commander for the Allied Powers and was authorized to arrange for the cessation of hostilities at the earliest practicable date.[35] Accordingly, he directed that the Japanese forces terminate hostilities immediately and that he be notified at once of the effective date and hour of such termination. He further directed that Japan send to Manila on 17 August "a competent representative empowered to receive in the name of the Emperor of Japan, the Japanese Imperial Government, and the Japanese Imperial General Headquarters certain requirements for carrying into effect the terms of surrender."

General MacArthur's stipulations to the Japanese Government included specific instructions regarding the journey of the Japanese representatives to Manila. The emissaries were to leave Sata Misaki, on the southern tip of Kyushu, on the morning of 17 August. They were instructed to travel in a Douglas DC-3-type transport plane, painted white and marked with green crosses on the wings and fuselage, and to fly under Allied escort to an airdrome on Ie Shima in the Ryukus. From there the Japanese would be transported to Manila in a United States plane. The code designation chosen for communication between the Japanese plane and United States forces was the symbolic word "Bataan."[37]

On the evening of the 16th, General MacArthur was notified that the Emperor of Japan had issued an order at 1600 that day commanding the entire armed forces of his nation to halt their fighting immediately.[38] The wide dispersion and the disrupted communications of the Japanese forces, however, made the rapid and complete implementation of such an order exceedingly difficult. As the Japanese apologetically explained, the Imperial order would take approximately two to twelve days to reach Japanese forces throughout the Pacific and Asiatic areas, "but whether and when the order will be received by the first line it is difficult to foresee."[39] The radio also stated that members of the Imperial family were being sent to Japan's numerous theaters of operations as personal representatives of the Emperor to expedite and insure full compliance with the Imperial order to cease hostilities.

The departure of the delegates for the Manila negotiations, the Japanese continued, would be slightly delayed "as it is impossible for us to arrange for the flight of our representatives on 17 August due to the scarcity of time allowed us."[40] The radio added, however, that preparations were being made with all possible speed and that General MacArthur

34 GHQ, AFPAC, Press Release, 15 Aug 45.

35 SCAP Radio No. Z-500 to Japanese Emperor, Japanese Imperial Government, and Japanese Imperial General Headquarters, 15 Aug 45, G-3, GHQ Admin Files.

36 SCAP Radio No. Z-501 to Japanese Emperor et al, 15 Aug 45, G-3, GHQ Admin Files.

37 Ibid.

38 Japanese Government Radio to SCAP, 16 Aug 45, G-3, GHQ Admin Files.

39 Ibid.

40 Ibid.

would be immediately informed of the re-scheduled flight date.[41] A second message from the Japanese Government on the 16th described the tentative itineraries of the Imperial emissaries who were being dispatched by air to the various fronts.[42]

General MacArthur's Headquarters assured the Japanese that their intended measures were satisfactory and promised that every precaution would be taken to ensure the safety of the Emperor's representatives on their missions. Further instructions were issued regarding the type of plane to be used in sending the Japanese to Manila, but authorization was given to change the type of plane if necessary.[43]

Another communication from the Japanese on the 16th asked for clarification of the phrase, "certain requirements for carrying into effect the terms of surrender."[44] General MacArthur replied that the signing of the surrender terms would not be among the tasks of the Japanese representatives despatched to Manila.[45]

On 16 August, Japan's leaders announced that their delegates had been selected and would leave Tokyo for Manila on 19 August.[46] It was now only a matter of days before the long-awaited moment of final surrender would become a reality.

The Manila Conference

Headed by Lt. Gen. Torashiro Kawabe, Vice-Chief of the Army General Staff, the sixteen-man Japanese delegation[47] on the morn-

41 General MacArthur's message concerning the Manila mission was not received in Japan until the morning of the 16th. The change-over in cabinets at this time was another important reason for the inability of the Japanese Government to arrange for the dispatch of the desired emissaries by 17 August. The Suzuki Cabinet had resigned on the afternoon of the 15th and the succeeding Higashikuni Cabinet was not installed until the afternoon of the 17th. The problem of selection of the proper officials during this critical interim period, together with the task of preparing the numerous documents of information required by the Supreme Commander, necessitated the request for more time. Statement by Shunichi Matsumoto, former Japanese Vice-Minister of Foreign Affairs, G-2 Historical Section, GHQ, FEC.

42 Japanese Imperial GHQ Radio No. 2 to SCAP, 16 Aug 45, G-3, GHQ Admin Files. General Prince Haruhito Kanin was sent to Saigon and Singapore; General Prince Yasuhiko Asaka to Nanking and Peiping; and Lt. Col. Prince Tsuneyoshi Takeda to Hsinking (Changchun) in Manchuria.

43 SCAP Radio No. Z-502 to Japanese Imperial GHQ, 17 Aug 45, G-3, GHQ Admin Files.

44 Japanese Imperial GHQ Radio No. 4 to SCAP, 17 Aug 45, G-3, GHQ Admin Files.

45 SCAP Radio No. Z-504 to Japanese Imperial GHQ, 17 Aug 45, G-3, GHQ Admin Files.

46 Japanese Imperial GHQ Radio No. 9 to SCAP, 18 Aug 45, G-3, GHQ Admin Files.

47 Other Japanese delegates included the following:

Mr. Katsuo Okazaki	Chief, Research Bureau, Ministry of Foreign Affairs.
Maj. Gen. Masakazu Amano	Chief, Operations Section, Army General Staff.
R. Adm. Ichiro Yokoyama	Member, Naval General Staff.
Mr. Shuichi Mizota	Secretary to the Navy Minister.
Col. Arata Yamamoto	Member, Army General Staff.
Capt. Toshikazu Ohmae	Chief, Operations Section, Naval General Staff.
Capt. Hidemi Yoshida	Member, Navy Ministry.
Mr. Kazuma Sugita	Secretary to the Navy Minister.
Mr. Morio Yukawa	Member, Minister of Foreign Affairs.
Comdr. Yoshimori Terai	Member, Naval General Staff.
Lt. Col. Masao Matsuda	Member, War Ministry.
Lt. Col. Kiyoshi Minami	Member, War Ministry.
Lt. Col. Morio Takakura	Member, War Ministry.
2nd Lt. Sadao Otake	Member, Army General Staff.
2nd Lt. Harumi Takeuchi	Member, Army General Staff.

GHQ, AFPAC, "List of Japanese Emissaries to Manila," 21 Aug 45, G-3, GHQ Admin Files.

ing of 19 August boarded two white, green-crossed, disarmed Navy medium bombers and departed secretly from Kisarazu Airdrome, on the eastern shore of Tokyo Bay.[48] After landing at Ie Shima, according to General MacArthur's instructions, the Japanese passengers were immediately transferred to a U. S. Army transport plane and put down on Nichols Field south of Manila at about 1800 that same day.

On hand to meet the Japanese envoys as they emerged from the plane was a party of linguist officers headed by General Willoughby, General MacArthur's wartime director of intelligence. Following the necessary introductions and identifications, the Japanese were taken immediately to temporary quarters on Manila's Dewey Boulevard to await the meetings scheduled for that evening.[49]

Less than three hours after their arrival, the sixteen-man Japanese delegation was led by General Willoughby to the first of two conferences held that night with members of General MacArthur's staff. General MacArthur himself was not present. As the solemn procession moved from Dewey Boulevard through the battered and war-torn streets of Manila and up the broad steps of the City Hall, the stony-faced Japanese officers in their beribboned gray-green uniforms, with their peculiarly peaked caps, and with their two-handed Samurai swords dangling from their waists almost to the ground, made a grim and curious picture. Shortly after 2100, the Japanese and American representatives entered General Chamberlin's office and sat down facing each other across the long, black table of the map-covered conference room.[50]

The meetings continued through the night of the 19th and into the next day.[51] As General Sutherland led the discussions, linguists busily scanned, translated, and photostated the various reports, maps, and charts which the Japanese had brought with them. Allied Translator and Interpreter Section personnel worked throughout the night to put General MacArthur's requirements into accurate Japanese before morning. It was a matter of vital importance that all documents be capably and correctly translated so that arrangements for surrender could be completed with a minimum of misunderstanding and a maximum of speed.

The conference proceeded smoothly and all

48 Careful planning preceded the actual flight, according to Japanese officers concerned with the arrangements. "In planning the course of the airplane taking the Japanese delegates to Manila . . . precautions had to be taken to prevent interference by certain elements of the Japanese Air Force who, it was rumored, were planning to attack and destroy the plane. To insure the safe arrival of the delegates, the course to be taken by their plane was kept secret, and it was planned to fly a decoy plane on a direct route to Kyushu. At 0611 on 19 August, a Douglas-type naval transport plane, bearing the delegates, left Haneda, arriving at Kisarazu 14 minutes later. Here the group boarded two medium bombers, leaving Kisarazu at 0707. The planes proceeded southeast for approximately 100 nautical miles and, once outside the fighter-plane range, headed for Tanegashima. At 1115 they were met just south of Tanegashima by two American Army fighter units, comprising two B-24's and twelve P-38's. . . . They arrived safely on Ie Jima at 1240. . . ." Combined statement of Rear Adm. Ichiro Yokoyama, Navy Ministry; Capt. Toshikazu Ohmae, Naval General Staff; Capt. Hidemi Yoshida, Navy Ministry; Cmdr. Yoshimoto Terai, Naval General Staff, 21 Nov 49, G-2 Historical Section, GHQ, FEC.

49 GHQ, AFPAC, Press Release, 19 Aug 45.

50 GHQ, AFPAC, Press Release, 19 Aug 45. During the conference, the Japanese and American representatives sat in the following order: Capt. Hidemi Yoshida opposite Maj. Gen. Lester Whitlock, ACofS, G-4; Capt. Toshikazu Ohmae oppsite Maj. Gen. Richard J. Marshall, DCofS; Rear Adm. Ichiro Yokoyama opposite Rear Adm. Forest P. Sherman, ACofS for Planning for CINCPAC; Lt. Gen. Torashiro Kawabe opposite Lt. Gen. Richard K. Sutherland, CofS; Mr. Katsuo Okasaki opposite Maj. Gen. Stephen J. Chamberlin, ACofS, G-3; Maj. Gen. Masakazu Amano opposite Maj. Gen. Charles A. Willoughby, ACofS, G-2; and Lt. Col. Masao Matsuda opposite Brig. Gen. Donald R. Hutchinson, CofS, FEAF. Interpreters sat at both ends of the table.

51 Ibid.

major problems were resolved satisfactorily. Results of the negotiations made it advisable to modify some of the original concepts on the problem of occupation. Based upon the full co-operation of the Japanese Government and Imperial General Headquarters, the new modifications provided for gradual occupation of designated areas after the Japanese had disarmed the local troops. No direct de-militarization was to be carried out by Allied personnel; the Japanese were to control the disarmament and demobilization of their own armed forces under Allied supervision.[52]

General Kawabe expressed his belief that the Japanese would faithfully carry out all Allied demands, but because of the unpredictable reactions of the Japanese civilian and army elements he requested that Japan be given an additional period of preparation before the actual steps of occupation were taken. General Sutherland allowed three extra days. The target date for the initial landings was postponed from 25 August to 28 August. The arrival of the advance unit at Atsugi Airfield was scheduled for 26 August.

At the close of the conference, General Kawabe was handed the documents containing the "Requirements of the Supreme Commander for the Allied Powers."[53] These directives stipulated General MacArthur's requirements concerning the arrival of the first echelons of the Allied forces, the formal surrender ceremony, and the subsequent reception of the occupation forces. Also given to General Kawabe were a draft of the Imperial Proclamation by which the Emperor would accept the terms of the Potsdam Declaration and command his subjects to cease hostilities, a copy of General Order No. 1 by which Imperial General Headquarters would direct all military and naval commanders to lay down their arms and surrender their units to designated Allied commanders, and lastly the Instrument of Surrender itself which would later be signed on board an American battleship in Tokyo Bay.

The Manila Conference was over. The Japanese delegation left at 1300 on 20 August and started back to Japan along the same route by which it had come. The homeward trip, however. was marred by an accident which caused a few anxious moments to the bearers of the surrender documents. The plane carrying the key emissaries had to make a forced landing on a beach near Hamamatsu, and it was not until seven hours after their scheduled time of return that the members of the mission were able to report the results of the Manila Conference to their waiting Premier.[54] It now remained for Japan to prepare itself to carry out

52 10th Information and Historical Service, HQ Eighth Army, *Occupational Monograph of the United States Eighth Army in Japan*, Vol I, Aug 45—Jan 46, pp. 11–12. Hereinafter cited as: *Monograph of Eighth Army in Japan*. G-2, GHQ, familiar with the structure of the Japanese armed forces and in control of the linguist service, was designated to supervise the Japanese plans for demobilization and disarmament.

53 GHQ, SCAP, *Requirements of the Supreme Commander for the Allied Powers Presented to the Japanese Representatives at Manila, P. I.* 19 Aug 45 (TS).

54 On the first leg of their return trip in an American transport plane, the delegates arrived at Ie Jima at 1745 on the 20th. Since one of the Japanese planes waiting there to take them back to Japan had developed trouble, it was decided to split the mission into two groups and leave one group behind on Ie Jima until its plane was repaired. The members of the first group left Ie Jima at 1840 and were escorted by twelve P-38's to the vicinity of Tanegashima. From there their plane proceeded along the coast from Shionomisaki until 2345 when the fuel supply became exhausted and they were compelled to make a forced landing on a beach approximately three kilometers east of the mouth of the Tenryu River. Except for Katsuo Okazaki, who suffered slight injuries, there were no casualties.

By foot and then by truck the delegates finally arrived at the evacuated headquarters of the Hamamatsu Air School at 0330 on the 21st. At 0700 they left Hamamatsu Airfield in an Army heavy bomber, arriving at Chofu Airfield at 0800. The second plane left Ie Jima at 0830 on the 21st and arrived safely at Kisarazu at 1400. Combined statement of Ichiro Yokoyama, Toshikazu Ohmae, Hidemi Yoshida, Yoshimori Terai, G-2 Historical Section, GHQ, FEC.

the provisions of surrender and to accept a peaceful military occupation of the Homeland by Allied forces.

General MacArthur anticipated that, subject to weather conditions which would permit the necessary air and naval operations, the Instrument of Surrender would be signed within ten days. "It is my earnest hope," he announced after the departure of Japan's representatives from Manila, "that pending the formal accomplishment of the Instrument of Surrender, armistice conditions may prevail on every front and that bloodless surrender may be effectuated."[55]

Reorganization of AFPAC

Simultaneously with President Truman's announcement of Japan's acceptance of the Potsdam Declaration, a thorough reorganization of AFPAC forces was effected in preparation for the forthcoming Allied mission in Japan. On 15 August, General MacArthur ordered sweeping changes to strengthen Sixth and Eighth Armies and XXIV Corps (then with Tenth Army in the Ryukyus) for their imminent duties of occupation. (Plate No. 131)

The 11th Airborne Division had moved 800 miles by air with full combat equipment from Luzon to Okinawa in a record time of forty-four hours and had already passed to the control of General Eichelberger's Eighth Army. Eighth Army, which would institute the occupation unassisted until 22 September, was also given control of XI Corps with the Americal, 1st Cavalry, and 43rd Divisions and the 112th Regimental Combat Team, IX Corps with the 77th and 81st Divisions and the 158th Regimental Combat Team, and the 27th Division on Okinawa. In addition Eighth Army included XIV Corps, the 31st Division and the 368th Regimental Combat Team, taken from X Corps, and the 40th Division, taken from Sixth Army.[56]

General Krueger's Sixth Army was meanwhile increased by the addition of V Amphibious Corps, with its 2nd, 3rd, and 5th Marine Divisions located in Saipan, Oahu, Guam, and Hawaii. General Krueger also assumed command of X Corps, with the 24th and 41st Divisions; the 6th Division from Eighth Army, and the 98th Division from AFMIDPAC were also transferred to Sixth Army.

XXIV Corps on Okinawa passed from Tenth Army to the direct control of AFPAC to operate independently as the occupation force in Korea, south of the 38 degree parallel.

General MacArthur assigned the responsibility for the security of the Ryukyus to Army Service Command I (ASCOM-I) and directed AFWESPAC to assume combat responsibilities in the Southwest Pacific Area. Lt. Gen. Wilhelm D. Styer, Commanding General of AFWESPAC, established two commands to maintain security in the Philippines: the Luzon Area Command and the Southern Islands Area command.[57] The former SWPA commands, Allied Land Forces, Allied Naval Forces, and Allied Air Forces, were to be abolished with the signing of the surrender terms, at which time control of the southern portion of the Southwest Pacific Area would pass to the British.[58]

Initiation of "Blacklist"

As the day of formal surrender drew near, all available troop transports of the Far East

55 GHQ, AFPAC, Press Release, 20 Aug 45.

56 G-3, GHQ, SWPA-AFPAC, *Monthly Summary of Operations*, Aug 45.

57 Ibid.

58 G-3, GHQ, SCAP-AFPAC, *Operations Report* No. 4, 1 Sep 45, G-3, GHQ, SWPA *Journal* (S).

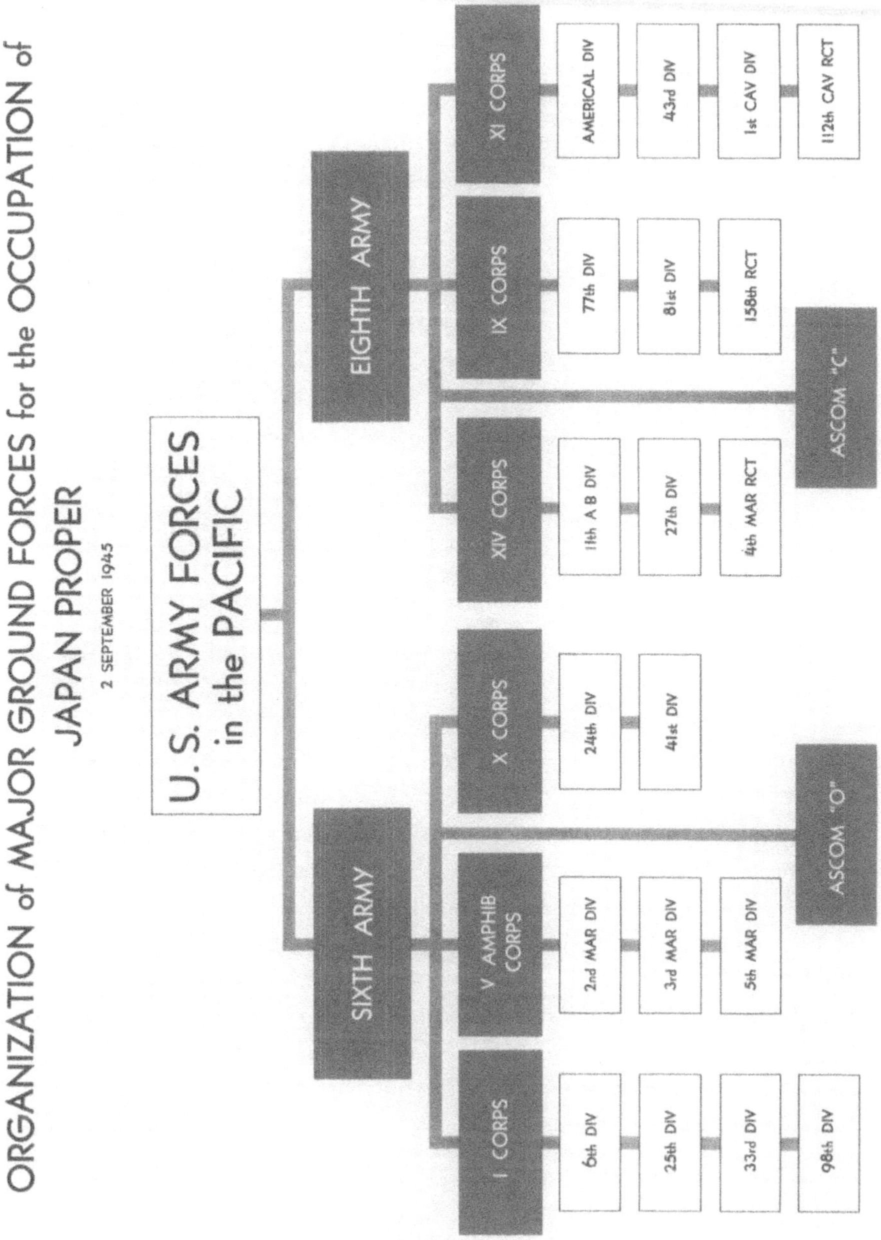

PLATE NO. 131

Organization of Major Ground Forces for the Occupation of Japan Proper

Air Forces and dozens of the huge Skytrains and Skymasters of the Pacific Air Transport Command were massed at Okinawa to airlift the first occupation forces to Japan in the greatest aerial movement of the Pacific War. On 26 August, General Eichelberger transferred the Eighth Army Command Post from the eastern coastal plain of Leyte to Okinawa and prepared to lead the vanguard forces of the 11th Airborne and 27th Infantry Divisions onto Japanese soil. At this critical juncture, however, a typhoon raging through the Japanese Home Islands caused a delay in Japan's final preparations to receive the occupation forces and resulted in a two-day postponement of the preliminary landings, originally scheduled for 26 August.

The first American landings in Japan were made at 0900 on 28 August by a small airborne advance party of 150 communications experts and engineers. Deplaning at the large navy airfield at Atsugi, some twenty miles southwest of Tokyo, the daring little group fell immediately to the task of setting up the communications and other operational facilities for the swarms of four-engined planes that would bring the 11th Airborne Division to establish the American airhead in the Atsugi area. This advance group was followed three hours later by thirty-eight troop transports carrying protective combat forces and necessary supplies of gasoline, oil, and other equipment.[59]

The main phase of the airborne operation began at dawn on 30 August. The first plane, bearing a regular forty-man load, touched the runway at 0600. Practically every three minutes thereafter throughout the day, American planes landed on the huge Japanese airfield, gliding down with clockwork precision and without a single mishap. By evening, 4,200 combat-equipped troops of the 11th Airborne were on the ground and strategically deployed to protect the airhead against any eventuality.[60]

It was a great, though calculated, military gamble. The American elements, outnumbered by thousands to one, were landing in a hostile country where huge numbers of enemy soldiers still had access to their arms. The occupation plan was predicated upon the ability of the Emperor to maintain psychological control over his people and to quell any recalcitrant elements. It was doubtful that the majority of the Japanese people would disobey the Imperial command to surrender peaceably, but the possibility that certain dissident extremists would forcibly oppose the occupation despite all orders had to be carefully considered.[61]

In view of the unpredictable reactions of the Japanese troops, General Eichelberger flew in to Atsugi early the first day to take personal command of the situation and to make preparations for the arrival of the Supreme Com-

---

59 G-3, GHQ, SWPA-AFPAC, *Monthly Summary of Operations*, Aug 45.

60 *Monograph of Eighth Army in Japan*, p. 20.

61 Although the American forces arriving on the 28th met with no incidents, the days before their landings had seen several outbreaks of violence among Japanese army and navy troops. On the night of the 14th, certain members of the Imperial Guard Division in Tokyo tried to seize and isolate the Palace grounds and prevent the Imperial Rescript from being broadcast. The attempted revolt was subdued by the following morning, however, and most of the rebellious participants committed suicide. On the 14th, also, the residence of the Prime Minister and the President of the Privy Council were attacked and burned.

Atsugi Airfield, the site chosen for General MacArthur's landing, was the scene of a five-day series of disorders during which naval Kamikaze pilots dropped leaflets over Tokyo denouncing the Emperor's advisors and exhorting the people not to surrender. It was not until 26 August, two days before the arrival of General Eichelberger's advance units, that the recalcitrant elements were brought under control. Court Record for the International Military Tribunal for the Far East, pp. 31194–31198 and 29322–29326; Court Martial Decision, Captain Yasuma Kosono, IJM, 16 Oct 45, G-2 Historical Section, GHQ, FEC.

mander. Fortunately, all apprehensions proved groundless. There was no trouble; not a sign of resistance was apparent. The Japanese had stationed picked and trusted troops around the field and along the main roads of egress to provide security for the greatly outnumbered American soldiers.[62] General MacArthur's calculated risk had been well taken.

Shortly after 1400 a famous C-54—the name "Bataan" in large letters on its nose—circled the field and glided in for a landing. From it stepped General MacArthur, accompanied by General Sutherland and his other staff officers. The Supreme Commander's first words to General Eichelberger and the men of Eighth Army and the 11th Airborne Division who greeted him were:

> *From Melbourne to Tokyo is a long road. It has been a long and hard road, but this looks like the payoff. The surrender plans are going splendidly and completely according to previous arrangements. In all outlying areas, fighting has practically ceased.*
>
> *In this area a week ago, there were 300,000 troops which have been disarmed and demobilized. The Japanese seem to be acting in complete good faith. There is every hope of the success of the capitulation without undue friction and without unnecessary bloodshed.*

The entire operation proceeded smoothly. General MacArthur paused momentarily to inspect the airfield, then stepped into a waiting automobile for the drive to Yokohama. Thousands of Japanese troops were posted along the fifteen miles of road from Atsugi to Yokohama, to guard the route of the Allied motor cavalcade as it proceeded to the temporary SCAP Headquarters in Japan's great seaport city.[63]

Meanwhile, Admiral Halsey's Third Fleet, including the British warships, steamed into Japan's coastal waters and anchored in Sagami Bay on the 27th. Japanese naval officers met the incoming fleet units to receive instructions for the safe entry of the Allied Fleet into Tokyo Bay in accordance with General MacArthur's surrender directives. On 29 August, the Third Fleet moved forward into Tokyo Bay to prepare for the landings at Yokosuka.

As the 11th Airborne poured into Atsugi Airfield, troops of the 4th Marine Regimental Combat Team, 6th Marine Division, went ashore at Yokosuka Naval Base on the west bank of Tokyo Bay below Yokohama. Immediately after the airborne landing, elements of the 188th Parachute Glider Regiment sped to Yokohama to take control of the huge dock area. Other patrols of the airborne unit fanned out to the south, and contacted troops of the 4th Marines, whose landing was also completed without incident. The Marine Regiment passed to the control of the Commanding General, Eighth Army immediately after it disembarked.[64]

The intensive preparation and excitement that attended these first landings on the Japanese mainland did not interfere with the mission of affording relief and rescue to the unfortunate Allied personnel already inside Japan as internees or prisoners. Despite the bad weather that delayed the occupation operation, units of the Far East Air Forces and planes from the Third Fleet continued their surveillance missions. On 25 August they began dropping relief supplies of food, medicine, and clothing to Allied soldiers and civilians in prisoner-of-war and internment camps throughout the main islands.

While the advance echelon of the occupation forces was still on Okinawa, "mercy teams" were organized to accompany the first elements of Eighth Army Headquarters. Immediately

62 Robert L. Eichelberger and Milton MacKaye, "Our Bloody Jungle Road to Tokyo," Part 7, *The Saturday Evening Post* (September 24, 1949), p. 163.

63 *Monograph of Eighth Army in Japan*, p. 20.

64 G-3, GHQ, SWPA-AFPAC, *Monthly Summary of Operations*, Aug 45.

after the initial landings, these teams established contact with the Swiss and Swedish Legations, the International Red Cross, the United States Navy, and the Japanese Liaison Office and rushed to expedite the release and evacuation, where necessary, of the thousands of Allied internees.[65]

By 31 August the 511th Parachute Infantry Regiment had joined the 188th Paragliders in the Yokohama area and established contact with the 4th Marines at Yokosuka. The 187th Parachute Infantry Regiment had consolidated the Atsugi airhead to secure it as a base for subsequent air activities by the Far East Air Forces and to protect the incoming 27th Infantry Division, also to be airborne from Okinawa.[66]

The Reconnaisance Troop of the 11th Airborne Division made a subsidiary airlift operation on 1 September, flying from Atsugi to Kisarazu Airfield. This airfield was occupied without incident. On the morning of 2 September, the 1st Cavalry Division began landing at Yokohama as the surrender ceremonies took place in Tokyo Bay. With the exception of Tokyo itself, most of the strategic areas along the shores of Tokyo Bay had by then been secured by Allied forces.[67]

Tokyo Bay: 2 September 1945

Japan's formal capitulation to the Allies climaxed a week of historic events as the initial steps of the occupation program went into effect. The surrender ceremony took place aboard the Third Fleet flagship, *U. S. S. Missouri*, on the misty morning of Sunday, 2 September 1945. As the *Missouri* lay majestically at anchor in the calm waters of Tokyo Bay, convoys of large and small vessels formed a tight cordon around the surrender ship, while army and navy planes maintained a protective vigil overhead. This was the objective toward which the Allies had long been striving—the unconditional surrender of the previously undefeated military forces of Japan and the final end to conflict in World War II.

The decks of the *Missouri* that morning were crowded with the representatives of the various United Nations that had participated in the Pacific War. Outstanding among the Americans flanking General MacArthur were Admirals Nimitz and Halsey, and General Wainwright who had recently been released from a Manchurian internment camp, flown to Manila, and then brought aboard to witness the occasion. Present also were the veteran staff members who had fought with General MacArthur since the early dark days of Melbourne and Port Moresby.

Shortly before 0900 Tokyo time, a launch from the mainland pulled alongside the great United States warship and the emissaries of defeated Japan climbed silently and glumly aboard. The Japanese delegation included two representatives empowered to sign the Instrument of Surrender, Mamoru Shigemitsu, Minister of Foreign Affairs and Gen. Yoshijiro Umezu of the Imperial General Staff, in addition to three representatives from the Foreign Office, three representatives from the Army, and three representatives from the Navy.[68]

As Supreme Commander for the Allied

65 *Monograph of Eighth Army in Japan*, p. 20; *Adm. Halsey's Story*, p. 274.

66 G-3, GHQ, SWPA-AFPAC, *Monthly Summary of Operations*, Aug 45.

67 Op. cit. Sep 45.

68 The representatives of the Foreign Office were Katsuo Okazaki, Shunichi Kase, and Saburo Ota; the representatives of the Army were Lt. Gen. Shuichi Miyazaki, Lt. Gen. Yatsuji Nagai, and Col. Ichiji Sugita; the representatives of the Navy were Rear Adm. Sadatoshi Tomioka, Rear Adm. Ichiro Yokoyama, and Capt. Katsuo Shiba. Tokyo *Asahi Shimbun*, September 3, 1945, p. 1 and Tokyo *Yomiuri-Hochi*, September 3, 1945, p. 1.

Powers, General MacArthur presided over the epoch-making ceremony, and with the following words he inaugurated the proceedings which would ring down the curtain of war in the Pacific:

We are gathered here, representatives of the major warring powers, to conclude a solemn agreement whereby peace may be restored. The issues, involving divergent ideals and ideologies, have been determined on the battlefields of the world and hence are not for our discussion or debate. Nor is it for us here to meet, representing as we do a majority of the people of the earth, in a spirit of distrust, malice or hatred. But rather it is for us, both victors and vanquished, to rise to that higher dignity which alone befits the sacred purposes we are about to serve, committing all our peoples unreservedly to faithful compliance with the understandings they are here formally to assume.

It is my earnest hope, and indeed the hope of all mankind, that from this solemn occasion a better world shall emerge out of the blood and carnage of the past—a world dedicated to the dignity of man and the fulfillment of his most cherished wish for freedom, tolerance and justice.

The terms and conditions upon which surrender of the Japanese Imperial Forces is here to be given and accepted are contained in the instrument of surrender now before you . . . .[69]

The Supreme Commander then invited the two Japanese plenipotentiaries to sign the duplicate surrender documents: Foreign Minister Shigemitsu, on behalf of the Emperor and the Japanese Government, and General Umezu, for the Japanese Imperial General Headquarters. He then called forward two famous former prisoners of the Japanese to stand behind him while he himself affixed his signature to the formal acceptance of the surrender: Gen. Jonathan M. Wainwright, hero of Bataan and Corregidor and Lt. Gen. Sir Arthur E. Percival, who had been forced to yield the British stronghold at Singapore.

General MacArthur was followed in turn by Admiral Nimitz, who signed on behalf of the United States, and by the representatives of the other United Nations present: Gen. Hsu Yung-Chang for China, Adm. Sir Bruce Fraser for the United Kingdom, Lt. Gen. Kuzma N. Derevyanko for the Union of Soviet Socialist Republics, Gen. Sir Thomas A. Blamey for Australia, Col. L. Moore-Cosgrave for Canada, Gen. Jacques P. LeClerc for France, Adm. Conrad E. L. Helfrich for the Netherlands, and Air Vice-Marshall Leonard M. Isitt for New Zealand.

The Instrument of Surrender was completely signed within twenty minutes. (Plate No. 132) The first signature of the Japanese delegation was affixed at 0904; General MacArthur wrote his name at 0910; and the last of the Allied representatives signed at 0920. The Japanese envoys then received their copy of the surrender document, bowed stiffly and departed for Tokyo. Simultaneously, hundreds of army and navy planes roared low over the *Missouri* in one last display of massed air might.

In signing the Instrument of Surrender, the Japanese bound themselves to accept the provisions of the Potsdam Declaration, to surrender unconditionally their armed forces wherever located, to liberate all internees and prisoners of war, and to carry out all orders issued by the Supreme Commander for the Allied Powers to effectuate the terms of surrender.

On that same eventful day, the Supreme Commander broadcast a report to the people of the United States. Having been associated with Pacific events since the Russo-Japanese war, General MacArthur was able to speak with the authority of long experience to forecast a future for Japan:

We stand in Tokyo today reminiscent of our

69 GHQ, AFPAC, Press Release, "Remarks by the Supreme Commander for the Allied Powers, Surrender Ceremony, Tokyo Bay," 2 Sep 45.

INSTRUMENT OF SURRENDER

We, acting by command of and in behalf of the Emperor of Japan, the Japanese Government and the Japanese Imperial General Headquarters, hereby accept the provisions set forth in the declaration issued by the heads of the Governments of the United States, China and Great Britain on 26 July 1945, at Potsdam, and subsequently adhered to by the Union of Soviet Socialist Republics, which four powers are hereafter referred to as the Allied Powers.

We hereby proclaim the unconditional surrender to the Allied Powers of the Japanese Imperial General Headquarters and of all Japanese armed forces and all armed forces under Japanese control wherever situated.

We hereby command all Japanese forces wherever situated and the Japanese people to cease hostilities forthwith, to preserve and save from damage all ships, aircraft, and military and civil property and to comply with all requirements which may be imposed by the Supreme Commander for the Allied Powers or by agencies of the Japanese Government at his direction.

We hereby command the Japanese Imperial General Headquarters to issue at once orders to the Commanders of all Japanese forces and all forces under Japanese control wherever situated to surrender unconditionally themselves and all forces under their control.

We hereby command all civil, military and naval officials to obey and enforce all proclamations, orders and directives deemed by the Supreme Commander for the Allied Powers to be proper to effectuate this surrender and issued by him or under his authority and we direct all such officials to remain at their posts and to continue to perform their non-combatant duties unless specifically relieved by him or under his authority.

We hereby undertake for the Emperor, the Japanese Government and their successors to carry out the provisions of the Potsdam Declaration in good faith, and to issue whatever orders and take whatever action may be required by the Supreme Commander for the Allied Powers or by any other designated representative of the Allied Powers for the purpose of giving effect to that Declaration.

We hereby command the Japanese Imperial Government and the Japanese Imperial General Headquarters at once to liberate all allied prisoners of war and civilian internees now under Japanese control and to provide for their protection, care, maintenance and immediate transportation to places as directed.

The authority of the Emperor and the Japanese Government to rule the state shall be subject to the Supreme Commander for the Allied Powers who will take such steps as he deems proper to effectuate these terms of surrender.

PLATE NO. 132

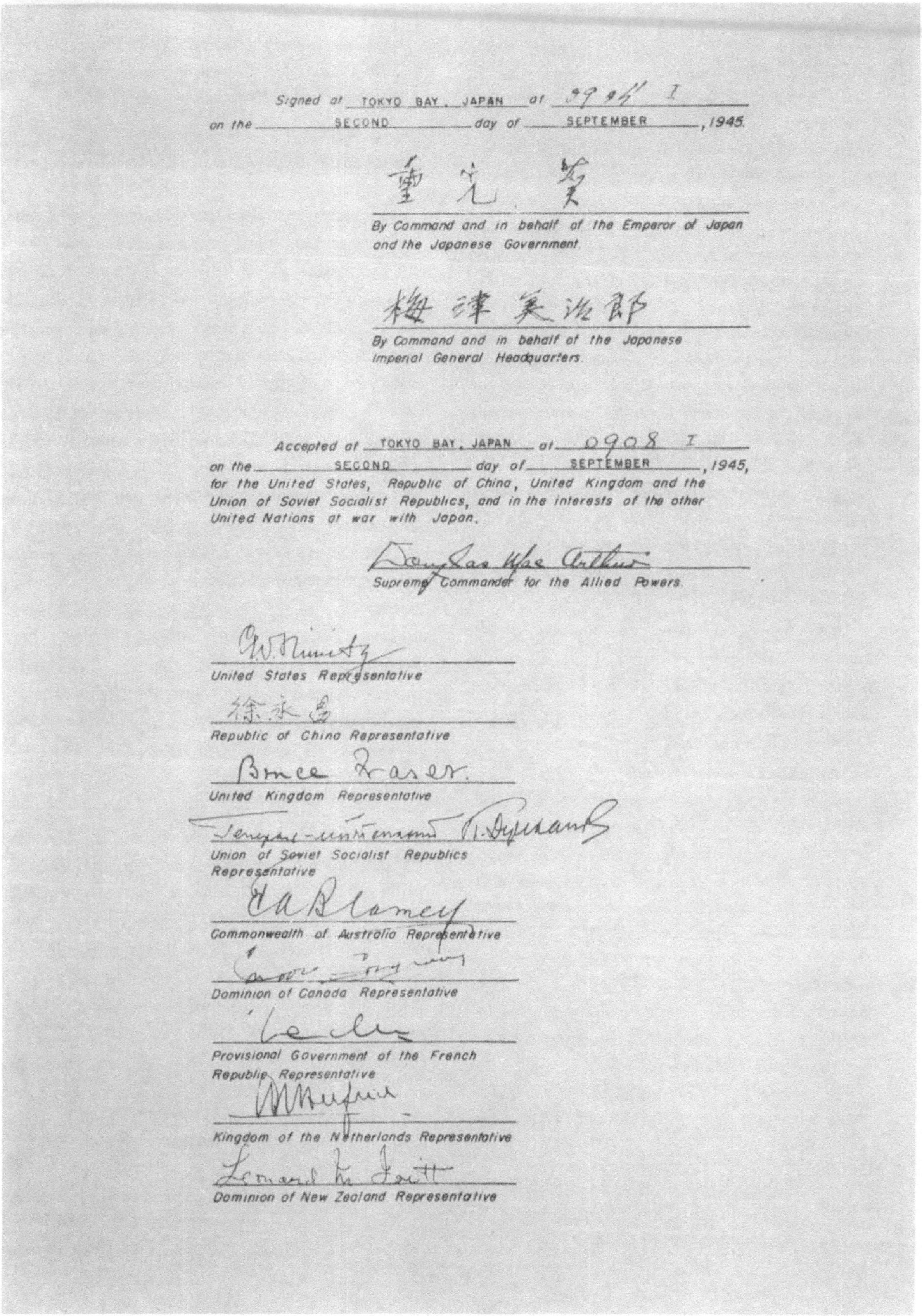

Signed at TOKYO BAY, JAPAN at 0904 I
on the SECOND day of SEPTEMBER, 1945.

重光葵

By Command and in behalf of the Emperor of Japan and the Japanese Government.

梅津美治郎

By Command and in behalf of the Japanese Imperial General Headquarters.

Accepted at TOKYO BAY, JAPAN at 0908 I
on the SECOND day of SEPTEMBER, 1945,
for the United States, Republic of China, United Kingdom and the Union of Soviet Socialist Republics, and in the interests of the other United Nations at war with Japan.

Supreme Commander for the Allied Powers.

United States Representative

徐永昌

Republic of China Representative

United Kingdom Representative

Union of Soviet Socialist Republics Representative

Commonwealth of Australia Representative

Dominion of Canada Representative

Provisional Government of the French Republic Representative

Kingdom of the Netherlands Representative

Dominion of New Zealand Representative

Surrender Document

countryman, Commodore Perry, ninety-two years ago. His purpose was to bring to Japan an era of enlightenment and progress by lifting the veil of isolation to the friendship, trade and commerce of the world. But, alas, the knowledge thereby gained of Western science was forged into an instrument of oppression and human enslavement. Freedom of expression, freedom of action, even freedom of thought were denied through supervision of liberal education, through appeal to superstition and through the application of force. We are committed by the Potsdam Declaration of Principles to see that the Japanese people are liberated from this condition of slavery. It is my purpose to implement this commitment just as rapidly as the armed forces are demobilized and other essential steps taken to neutralize the war potential. The energy of the Japanese race, if properly directed, will enable expansion vertically rather than horizontally. If the talents of the race are turned into constructive channels, the country can lift itself from its present deplorable state into a position of dignity . . . .[70]

Immediately following the signing of the surrender articles, the Imperial Proclamation of capitulation was issued. The Proclamation, the draft of which had been given to General Kawabe at Manila, read as follows:

Accepting the terms set forth in the Declaration issued by the heads of the Governments of the United States, Great Britain and China on July 26th 1945 at Potsdam and subsequently adhered to by the Union of Soviet Socialist Republics, We have commanded the Japanese Imperial Government and the Japanese Imperial General Headquarters to sign on Our behalf the instrument of surrender presented by the Supreme Commander for the Allied Powers and to issue General Orders to the Military and Naval forces in accordance with the direction of the Supreme Commander for the Allied Powers.

We command all Our people forthwith to cease hostilities, to lay down their arms and faithfully to carry out all the provisions of the Instrument of Surrender and the General Orders issued by the Japanese Imperial Government and the Japanese Imperial General Headquarters hereunder.[71]

Surrender throughout the SWPA Areas

In accordance with the Emperor's proclamation, Japanese army commanders took steps to surrender the millions of their forces in overseas areas. (Chart) Offensive action by U. S. troops had been suspended on 15 August with the announcement of Japan's acceptance of the Potsdam Declaration. All fighting did not cease, however, since it took many days, and in some cases weeks, for the official word of surrender to be carried along the badly disrupted Japanese communication channels.

Various devices were employed by the American commanders to transmit the news of final defeat to the dispersed and isolated enemy troops. Plane-strewn leaflets, loudspeaker broadcasts, strategically placed signboards, prisoner-of-war volunteers—all helped in persuading reluctant Japanese to submit themselves peaceably in conformity with the Imperial Rescript.

General MacArthur ordered General Styer, commanding the Army Forces of the Western Pacific, to receive the surrender of the Japanese units remaining in the Philippines. Rounding up the remaining enemy forces, however, was not a simple task. The remnants of the enemy scattered throughout the islands were split into a number of independent groups, all of which were operating from the comparative security of the mountainous terrain of the interior. Although these troops doggedly attempted to continue fighting, their resistance was disorganized and relatively ineffective. Malnutrition,

70 GHQ, SCAP, Press Release, "Text of Speech by the Supreme Commander for the Allied Powers Broadcast to the American People on V-J Day," 2 Sep 45.

71 *The Axis in Defeat*, pp. 33–34.

ESTIMATES OF JAPANESE STRENGTH AS OF AUGUST 1945

AREA	G-2 ESTIMATE OF ENEMY GROUND FORCES[1]	JAPANESE STRENGTH ESTIMATES			
		ARMY[2]	NAVY[3]	CIVILIANS[4]	TOTAL
KURILES	85,000	50,000	1,621	18,515	70,136
KARAFUTO	30,000	20,000	1,328	373,223	394,551
RYUKYUS	80,000	40,882	9,766	9,964	60,612
FORMOSA	260,000	128,080	46,713	307,147	481,940
BONINS	20,000	14,996	7,735	5	22,736
KOREA	365,000	274,200	29,431	712,583	1,016,214
MANCHURIA	615,000	760,000	1,185	998,815	1,760,000
CHINA	1,010,000	1,049,700	63,755	428,518	1,541,973
INDO-CHINA	90,000	90,370	8,914	6,900	106,184
THAILAND	50,000	106,000	3,051	5,300	114,351
BURMA	70,000	70,350	1,372	11	71,733
MALAYA-ANDAMAN SEA	100,000	95,581	36,473	17,036	149,090
SUMATRA	75,000	59,480	4,984	4,300	68,764
JAVA	30,000	40,360	15,180	10,000	65,540
LESSER SUNDAS, TIMOR	25,000	17,500	4,238	450	22,188
BORNEO	20,000	24,580	10,879	7,000	42,459
CELEBES, MOLUCCAS	79,000	17,650	6,518	5,050	29,218
PHILIPPINES	25,000	97,300[2]	36,151[3]	17,651	151,102
NEW GUINEA	20,000	30,230	7,159	269	37,658
BISMARCKS	55,000	57,530	30,854	0	88,384
SOLOMONS	15,000	12,330	16,729	0	29,059
MANDATES	95,000	48,644	44,178	18,845	111,667
TOTALS	3,214,000	3,157,683	406,890	2,941,636	6,516,959

1 Source: G-2, GHQ, SWPA, *Monthly Summary of Enemy Dispositions*, 31 July 1945, Plate 1.

2 Figures include civilians attached to the Army. In the case of the Philippines, approximately 4,200 civilians were included in the army strength figure. Source: Japanese First Demobilization Bureau Report, G-2 Historical Section, GHQ, FEC.

3 Figures include civilians attached to the Navy. In the case of the Philippines, approximately 20,365 civilians were included in the navy strength figure. Source: Japanese Second Demobilization Bureau Report, Historical Section, G-2, GHQ, FEC.

4 Source: Japanese Foreign Office Report, G-2 Historical Section, GHQ, FEC.

The G-2 estimates of enemy ground forces in the last phases of the war are in fair agreement with the figures for repatriated army and navy personnel subsequently developed by the Japanese Demobilization Bureaus, although air and navy remnants and civilians introduced unidentified variables. In areas of actual combat such as the Philippines and New Guinea where the G-2 estimates were concerned only with effective combat strength, the total number of army personnel including civilians attached to the army which were present in the area naturally exceeded the number of effectives estimated by G-2.

ESTIMATES OF ENEMY EFFECTIVE GROUND STRENGTH AND REPATRIATION TOTALS: AUGUST 1945

AREA	G-2 ESTIMATE OF ENEMY[1] GROUND FORCES	PERSONS REPATRIATED AND AWAITIING REPATRIATION[2]			
		ARMED FORCES		CIVILIANS	TOTAL
		ARMY	NAVY		
MANCHURIA	615,000	679,057	1,092	1,270,330	1,950,479
CHINA	1,010,000	1,001,384	42,414	446,272	1,490,070
KOREA	365,000	215,405	23,237	651,328	889,970
FORMOSA	260,000	127,170	43,161	318,086	488,417
OKINAWA	80,000	46,360	9,913	6,141	62,414
BONINS	20,000	14,379	8,995	5	23,379
PHILIPPINES	22,000[3]	99,673[4]	8,414[4]	30,802[4]	138,889[4]
INEFFECTIVES	60,000	DISTRIBUTED TROUGHOUT COMBAT AREAS[5]			
BORNEO	20,000	16,049	6,088	3,422	25,559
NEW GUINEA CELEBES AND LESSER SUNDAS	115,000	113,545	33,560	5,636	152,741
RABAUL	55,000	64,356	31,356	74	95,795
BOUGAINVILLE	15,000	10,447	10,377	511	21,335
CENTRAL PACIFIC	95,000	58,493	39,652	18,818	116,963
JAVA	30,000	23,302	5,844	2,636	31,782
MALAY-SUMATRA	175,000	183,021	49,312	24,977	257,310
THAILAND-BURMA	120,000	177,437	2,744	4,600	184,781
FRENCH INDO-CHINA	90,000	86,095	8,144	4,695	98,934
TOTALS	3,147,000	2,916,173	324,312	2,788,333	6,028,828

1 Figures from G-2, GHQ, SWPA, Monthly Summary of Enemy Dispositions, 31 Jul 45 (PLATE 1).
2 Japanese Foreign Office as of 31 Dec 46.
3 G-2, GHQ, SWPA, Supplement to Monthly Summary of Enemy Dispositions, 23 Aug 45.
4 LETTER, HQ PHILRYCOM to Brig. Gen. Willoughby, 23 Jan 47.
5 There was a strong pre-war colony of Japanese in the Davao area of Mindanao (25,000) and in the Visayas and on Luzon (10–15,000).

The G-2 estimate of enemy ground forces in the Philippines-Bonins-Okinawa area in the last phases of the campaign are in fair agreement with the repatriation figures subsequently developed by the Japanese Foreign Office and the embarkation authorities in Manila, although air and navy remnants and civilians introduced unidentified variables.

The 205,000 effective combat troops in the Philippines in December 1944 had been reduced to 95,000 in less than four months, by April 1945, and had shrunk to 22,000 in August 1945, the balance were in hospitals, were convalescents, or were in static, isolated pockets of resistance scattered throughout the combat area. The classification of ineffectives is based on the known intelligence factor that there is a fairly steady ratio of killed to wounded. In the United States Army it was roughly one (1) to four (4) with a high percentage returning to the battlefield. In the Japanese Army, the ratio of recovered wounded was set lower because of callous handling by a fanatical and cold-blooded enemy who frequently killed his wounded when forced to retreat; wounded were not credited as "effective" on the battlefield for any particular date and period. In July—August G-2 estimated 22,000 effectives; 12,715 prisoners of war; 8–12,000 air and navy remnants; 50–60,000 ineffectives; 20–30,000 civilians of the pre-war Japanese colony in the Philippines, or an aggregate of 133,000 Japanese. The repatriation figures for January 1947 show an aggregate of 138,889.

disease, and improperly treated wounds had sapped their strength and taken a high toll in lives. On Luzon, the largest group of Japanese was concentrated around Kiangan, northeast of Baguio, fighting stubbornly in what had come to be called the "Yamashita Pocket." In this sector General Yamashita still held forth in command of the remaining forces of the Fourteenth Area Army, now reduced to a conglomeration of sick and wounded military elements, civilian employees, and refugees totaling about 40,000 men. In the Sierra Madre Mountains east of the Cagayan Valley, in the Zambales Mountains of Bataan, and in the dense areas of southern Luzon other thousands continued to struggle for survival.[72] The jungles of Negros and Cebu and the wilds of Zamboanga and Mindanao provided temporary sanctuary for additional Japanese who roamed the islands of the South.[73]

In British New Guinea, V-J Day found the Australian 6th Division pushing some 8,000 of the enemy out of the Wewak area southward across the Sepik River. In Bougainville, an estimated 13,000 enemy troops were compressed into the southern coastal areas by the Australian 3rd Division, while another group of about 6,000 held out along the Buin Road in the south. In New Britain, approximately 47,000 Japanese were gathered at the tip of the Gazelle Peninsula. In Borneo most of the enemy units had retreated into the hills and jungles leaving the Australian 7th Division to occupy the cities along the coasts.[74]

In most cases, steps toward surrender took place as soon as the local Japanese commanders heard the Emperor's proclamation by radio. On Mindanao, the commanding officer of 4,000 air force personnel in the mountains about 30 miles east of Valencia in central Bukidnon Province began immediate negotiations to turn over his troops.[75] Rear Adm. Takasue Furuse, in command of 1,500 Japanese naval personnel in the Infanta area of southern Luzon, initiated surrender talks with a guerrilla organization.[76] Japanese forces on Bataan agreed to turn over their arms on 1 September while the enemy commander in the Cagayan Valley, Maj. Gen. Shintaro Yuguchi, signified that he would surrender his remaining units as soon as he received word from General Yamashita.[77] Colonel Matsui, commander of Japanese forces in the southern part of the Cagayan Valley, arranged to turn over his men to the U. S. 27th Division on 2 September. Another enemy force, numbering more than 2,300, located farther north in the Dummun River and Capisayan District, sent word that it would give itself up to the Americans during the period of 2–6 September.[78]

The first direct contact made by United States forces with General Yamashita came on 26 August. Prior to that time it had been strongly suspected but not definitely known

72 In speaking of the problem of physical survival of the Japanese forces at this time, Lt. Gen. Akira Muto, Chief of Staff, Fourteenth Area Army said: "Staff officers and the rest of the personnel somewhere secured three meals a day, but the quantity of food decreased gradually and the illness increased. . . . According to the report of the Chief of the Administration Department, we could somehow get by the whole of August, but thereafter there was nothing on which we could rely. Each section of the Headquarters formed parties to forage afar for tubers and edible grasses." "Memoirs of Lt. Gen. Akira Muto," p. 58, G-2 Historical Section, GHQ, FEC.

73 G-3, GHQ, SWPA-AFPAC, *Monthly Summary of Operations*, Aug 45.

74 Ibid. These figures were soon to be swelled by refugee Japanese colonists, civilian employees, service elements, and port and shipping detachments.

75 G-2, GHQ, SWPA-AFPAC, *Daily Summary* No. 1236, 22/23 Aug 45.

76 G-2, GHQ, SWPA-AFPAC, *Daily Summary* No. 1243, 30 Aug 45.

77 Ibid.

78 G-2, GHQ, SWPA-AFPAC, *Daily Summary* No. 1245, 1 Sep 45.

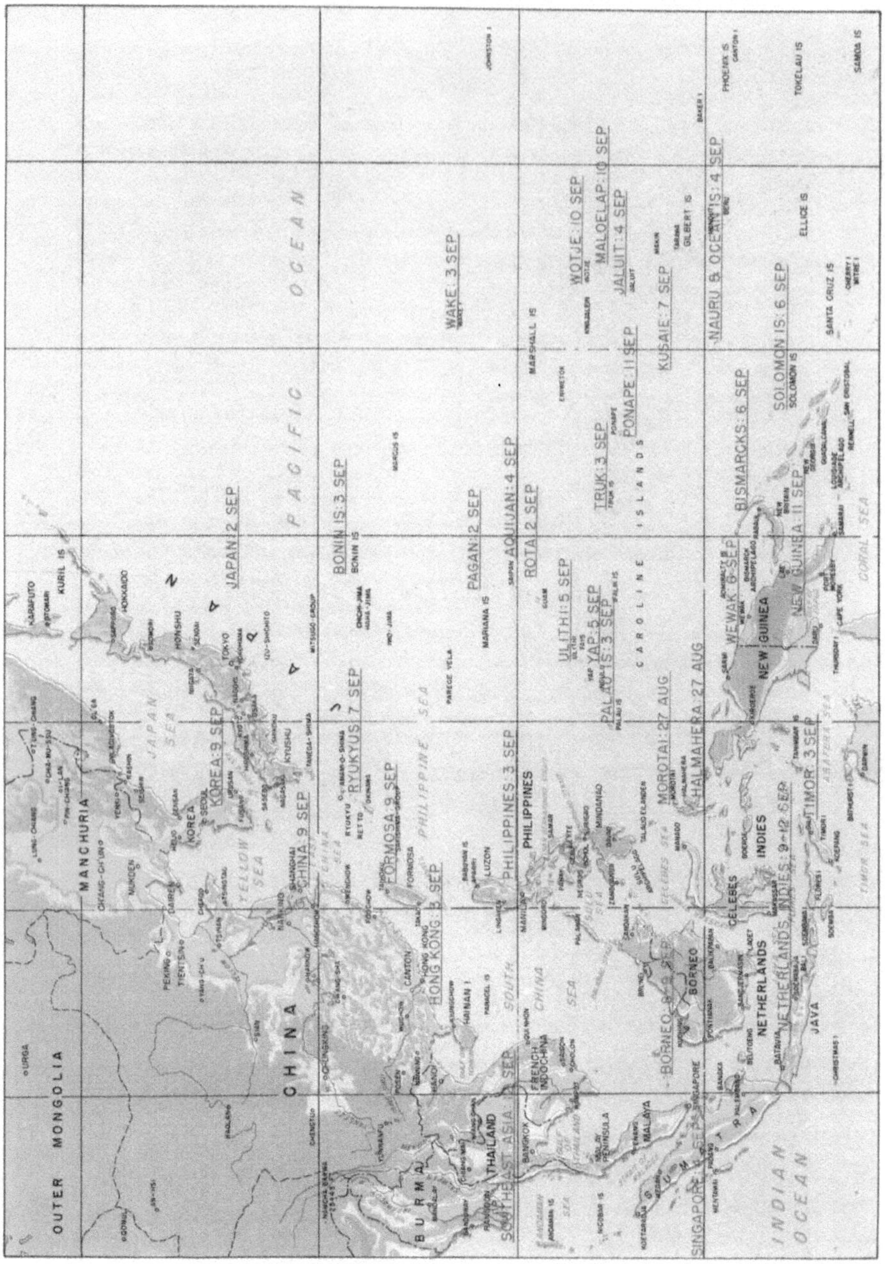

PLATE NO. 133
Japanese Surrender throughout the Pacific

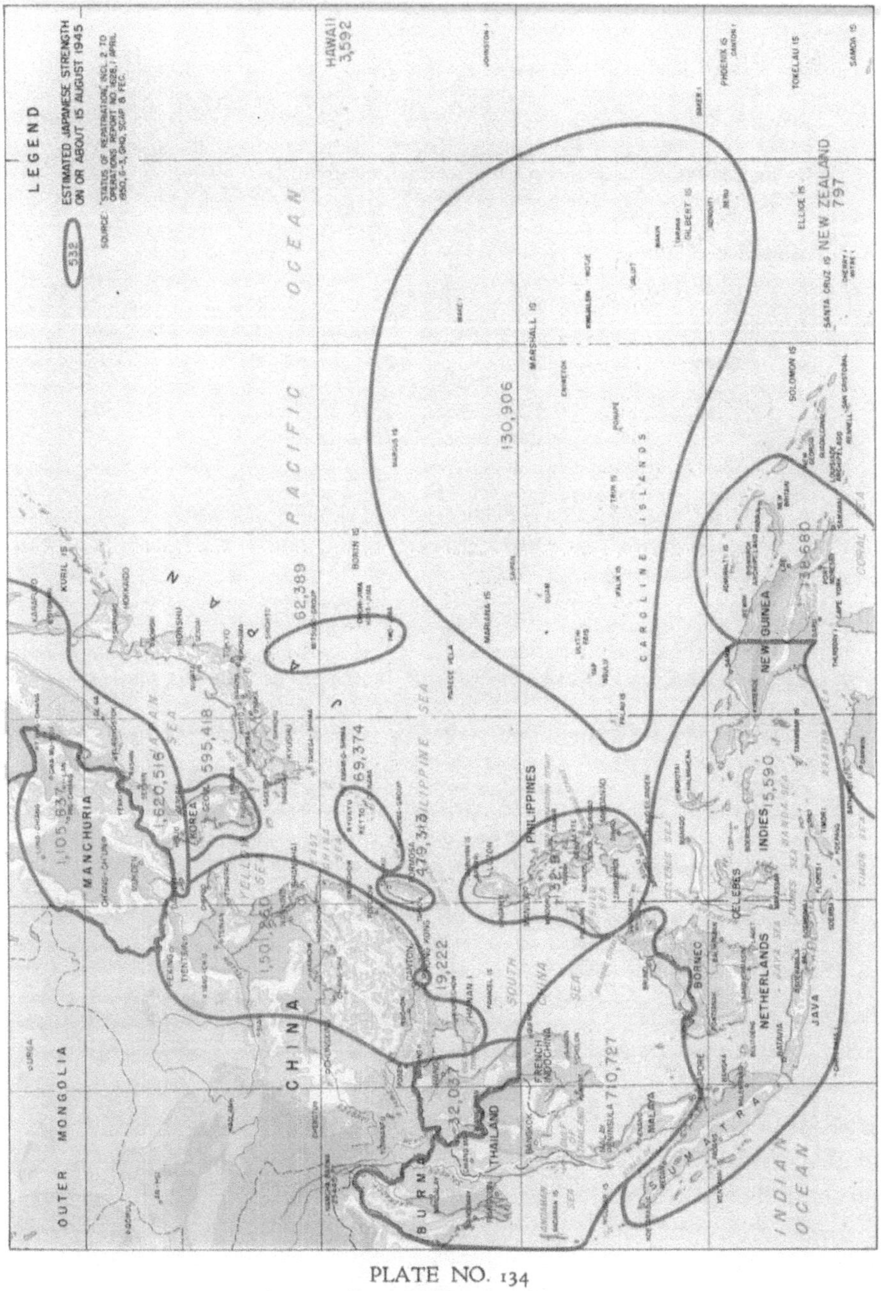

PLATE NO. 134
Japanese Strength Overseas, August 1945

that General Yamashita had remained with his troops in northern Luzon. Although attempts had been made to determine the general's whereabouts, his exact location had not been discovered. Filipino guerrilla reports to the effect that the Japanese general was present on the island were substantiated when a captured American pilot, who had been forced to bail out over enemy lines and who was subsequently released by the Japanese, later returned by air to point out the location of General Yamashita's headquarters.[79]

A message, signed by Maj. Gen. William H. Gill, commander of the 32nd Division, was dropped on 24 August requesting that contact be established to discuss surrender terms. At the same time, liaison parties were sent out by the 128th Infantry Regiment, 32nd Division, which by the middle of August had fought to within a few miles of General Yamashita's last stronghold. A second letter was dropped the following day directing General Yamashita to send a representative for the purpose of arranging the surrender of the Japanese forces which he controlled.[80]

The Japanese representative arrived at a 32nd Division outpost near Kiangan on 26 August with a letter from General Yamashita addressed to General Gill.[81] Acknowledging the receipt of the two previous messages, General Yamashita stated that although he had received and transmitted an order to cease hostilities, he was without authority to surrender his troops until he was formally notified of the signing of the general surrender in Tokyo.

Further exchanges of messages, aided by an air-ground radio at the Japanese headquarters, indicated General Yamashita's willingness to send representatives to Baguio in about five days to make preliminary negotiations for the surrender of his 40,000 men in northwestern Luzon, more than one-third of whom, he declared, were either sick or wounded. This proposal was rejected, however, because of the excessive delay it would cause before the final terms could be agreed upon.

General Yamashita finally radioed to a liaison plane circling above his headquarters that he would meet American parties in Kiangan on 2 September, and that he would be ready to proceed to Baguio to sign surrender papers for the Japanese military forces in the Philippines.[82]

At 0900 on 2 September General Yamashita, together with his Chief of Staff, Lt. Gen. Akira Muto, and other staff officers delivered themselves to an escort of the 32nd Division near Kiangan. A second party, consisting of naval personnel under Vice Adm. Denhichi Okochi, also appeared on the scene to negotiate the surrender of the naval personnel in that area. From Kiangan the Japanese were transported by motor vehicle to Bagabag Airfield where a C-47 transport plane waited to take them to Baguio.[83]

General Styler, who had attended the general surrender ceremony in Tokyo, flew back to the Philippines on the morning of 3 September, accompanied by General Wainwright and General Percival. At noon the American and Japanese officers gathered in the former residence of the U. S. High Commissioner at Camp John Hay, Baguio. General Styler opened the conference with a terse statement of purpose: "We are gathered here today to consummate the surrender of the Japanese

79 GHQ, AFPAC, Press Release, 27 Aug 45.

80 Ibid.

81 Ibid.

82 Ibid.

83 32nd Inf Div Radio No. ES-32 to PRO, GHQ, AFPAC, 3 Sep 45, G-3, GHQ *Journal* (R).

armed forces in the Philippines. I shall ask my Chief of Staff, General Leavey, to commence the proceedings."[84]

Maj. Gen. Edmond H. Leavey directed that the instrument of surrender be read and then the necessary documents were signed by General Yamashita and Admiral Okochi at 1210 on the 3rd. General Leavy thereupon affixed his own signature, after which he announced the official acceptance of the surrender and the conclusion of the surrender ceremonies. Three hours later the Japanese delegates were flown to Nielson Airfield in Manila and then sent immediately to New Bilibid Prison, sixteen miles south of the city, for internment.

With General Yamashita's capitulation, subordinate commanders began turning over their forces in rapidly increasing numbers. On Cebu, 2,900 surrendered; another 1,400 gave themselves up on Negros; local Japanese commanders on Mindanao began negotiations to surrender 4,000 troops in Davao Province, 2,000 in the Agusan Valley, and over 1,000 air force personnel in the mountains northwest of Davao.[85]

The long and brutal fighting in the Philippines was finished. An army of close to half a million Japanese had been overpowered, divided into ineffective remnants, and virtually annihilated. So completely cut off and separated had the various enemy units become that, for another year after the general surrender, patrols of the United States and the Philippine Armies continued to bring in straggling troops who had been unaware that their country had capitulated. On tiny Corregidor Island, twenty Japanese who had been hiding in a great jungle-filled ravine along the steep south shore cliffs did not surrender until 1 January 1946.

In other Pacific areas, the first large-scale surrender of Japanese forces came on 27 August, when Lt. Gen. Yoshio Ishii surrendered Morotai and Halmahera to the U. S. 93rd Division. Capitulation of Japanese units in the remaining parts of the Asiatic-Pacific Theater followed in quick succession. (Plate No. 133) In the Marianas, the Japanese commanders on Rota and Pagan Islands relinquished their commands almost simultaneously with the Tokyo Bay ceremony. While General Yamashita was yielding his Philippine Islands forces on 3 September, other Japanese garrisons were being surrendered in the Bonins, on Wake Island, in the Palaus, and at Truk in the Carolines, as well as at Hong Kong on the South China coast.[86]

Japanese commanding officers on Aquijan (Agiguan) Islands in the Marianas, and on Jaluit Atoll in the Marshalls, surrendered on 4 September, as did the Japanese on Ocean and Nauru Islands, to the west of the Gilberts. The same day Japanese forces at Singapore surrendered to the British aboard a warship in the harbor. On 5 September, Ulithi Atoll and the Yap garrison in the Carolines capitulated. The over-all surrender of Japanese forces in the Solomons and Bismarcks and in the Wewak area of New Guinea was signed on 6 September by Gen. Hitoshi Imamura on a British warship off Rabaul, once the center of Japanese power in the South Pacific.

Japanese naval and army forces in the Ryukyus officially capitulated on 7 September at the headquarters of Gen. Joseph E. Stilwell's Tenth Army on Okinawa. On the same day the enemy forces on the island of Kusaie gave up. In Borneo, Japanese forces in the southeast surrendered at Balikpapan on the 8th,

84 Stanley A. Frankel, *The 37th Infantry Division in World War II*, "Division Commander's Account of Yamashita's Surrender," pp. 377–378.

85 G-2, GHQ, SWPA-AFPAC, *Daily Summary* No. 1248 (with Encl No. 3), 4 Sep 46.

86 G-3, GHQ, SCAP-AFPAC, *Daily Situation Report* No. 17, 14 Sep 45.

while those in the northwest followed suit at Brunei Bay the following day.[87]

The official capitulation of all the Japanese in the China Theater took place on the morning of 9 September at the Central Military Academy in Nanking. The surrender by Yasuji Okamura, Commander of Japanese Expeditionary forces in China, included Japanese forces in China, Formosa, and northern Indo-China but not in Manchuria. That same afternoon at the Government Central Building in Seoul, Korea, Lt. Gen. Yoshio Kotsuki, commanding the Japanese Seventeenth Area Army, and Vice Adm. Gizaburo Yamaguchi, Commandant of the Chinkai Guard District, together with the Japanese Governor-General, Nobuyuki Abe, surrendered all Japanese forces in Korea south of the 38-degree parallel to Lt. Gen. John R. Hodge, Commanding General of the XXIV Corps and Admiral Kincaid, Commander of the U. S. Seventh Fleet. Forces north of the 38-degree line were to be surrendered to the Russian commander.[88]

On the same day at Morotai, General Blamey, Commander-in-Chief of the Australian Military Forces, accepted from Lt. Gen. Fusataro Teshima, commanding the Japanese Second Army, the surrender of all enemy forces remaining in Borneo and the Netherlands East Indies east of Lombok Island. On 10 September, Wotje Atoll and Maloelap in the Marshalls capitulated. The next day Timor just north of Australia, Ponape in the Carolines, and the remaining Japanese forces on the New Guinea mainland surrendered. Formal surrender of all southeast Asia, the Netherlands East Indies, and the various islands to the east was signed on the 12th at Singapore by Lt. Gen. Seishiro Itagaki, who was in command of the forces at Singapore and deputy commander of the Southern Area Command.[89]

"The Guns are Silent"

The signing of the surrender aboard the *Missouri* on 2 September marked the end of the "long and hard road" to Tokyo. Japan was defeated; her armies had surrendered and her cities were being occupied. World War II was over. (Plate No. 134)

As the Allied forces were disembarking from the hundreds of ships that lined Tokyo Bay to take up their duties of occupation, General MacArthur broadcast the announcement of peace to the world. His words expressed the thoughts of one who had felt both the bitterness of defeat and the gratification of victory, who knew well the horrors of warfare and the futility of attempting to achieve peace by resort to the sword.

> ***Today the guns are silent. A great tragedy has ended. A great victory has been won. . . .***
>
> ***As I look back upon the long, tortuous trail from those grim days of Bataan and Corregidor, when an entire world lived in fear, when Democracy was on the defensive everywhere, when modern civilization trembled in the balance, I thank a merciful God that he has given us the faith, the courage and the power from which to mold victory. We have known the bitterness of defeat and the exultation of triumph, and from both we have learned there can be no turning back. We must go forward to preserve in Peace what we won in War.***
>
> ***A new era is upon us. Even the lesson of Victory itself brings with it profound concern, both for our future security and the survival of civilization. The destructiveness of the War potential,***

87 Ibid; O.N.I. Weekly, Vol IV, No. 37, 12 Sep 45, pp. 2561–64.

88 O. N. I. Weekly, Vol IV, No. 37, 12 Sep 45, p. 2566.

89 Ibid, p. 2567.

through progressive advances in scientific discovery, has in fact now reached a point which revises the traditional concept of War.

Men since the beginning of time have sought peace. Various methods through the ages have been attempted to devise an international process to prevent or settle disputes between nations. From the very start, workable methods were found insofar as individual citizens were concerned but the mechanics of an instrumentality of large international scope have never been successful. Military Alliances, Balances of Power, Leagues of Nations all in turn failed leaving the only path to be by way of the crucible of war. The utter destructiveness of war now blots out this alternative. We have had our last chance. If we do not devise some greater and more equitable system Armageddon will be at our door. The problem basically is theological and involves a spiritual recrudescence and improvement of human character that will synchronize with our almost matchless advance in science, art, literature and all material and cultural developments of the past two thousand years. It must be of the spirit if we are to save the flesh. . . .

REPORTS OF GENERAL MacARTHUR

VOL I: The Campaigns of MacArthur in the Pacific
VOL I: Supplement: MacArthur in Japan: The Occupation, Military Phase
VOL II: Japanese Operations in the Southwest Pacific Area

". . . This report has been prepared by the General Staff to serve as a background for, and introduction to the detailed operational histories of the various tactical commands involved.

The pressure of other duties having prevented my personal participation in its preparation, it has been entrusted by me to that magnificent staff group which actually conducted the staff work during the progress of the campaigns. They speak with that sincere and accurate knowledge which is possessed only by those who have personally participated in the operations which they record . . ."

Preface by General Douglas MacArthur.

Senior Commanders: South West Pacific Areas

Gen W. Krueger: Sixth Army; Lt Gen R. L. Eichelberger: Eighth Army; Gen Sir Thomas Blamey: Aust. Imp. Forces; Lt Gen G. C. Kenny: AAF; Adm T. H. Kinkaid, USN: Seventh Fleet

The General Staff: GHQ: South West Pacific Area

Lt Gen R. K. Sutherland, CofS; Maj Gen R. J. Marshall, D CofS; Maj Gen C. P. Stivers, G–1: Maj Gen C. A. Willoughby, G–2; Maj Gen S. J. Chamberlain, G–3; Maj Gen L. J. Whitlock, G–4; Maj Gen S. B. Akin, CSO; Maj Gen W. F. Marquat, AAO; Maj Gen H. J. Casey, CE; Brig Gen B. M. Fitch, AG; Brig Gen L. A. Diller, PRO.

Editor in Chief

Maj Gen Charles A. Willoughby, G–2

Senior Editors

Col E. H. F. Svensson, G–2; Gordon W. Prange PhD; Mr. Stewart Thorn

Associate and Contributing Editors

Brig Gen H. E. Eastwood, G–4; Col F. H. Wilson, G–2; Col R. L. Ring, G–2; Col W. J. Niederpruem, G–3; Lt Col M. K. Schiffman, G–2; Maj J. M. Roberts, G–3; Capt J. C. Bateman, G–2; Capt Mary Guyette, G–2; Capt John L. Moore, G–2; Lt Stanley Falk, G–2; Mr. Jerome Forrest; Mr. Kenneth W. Myers; Miss Joan Corrigan.

Translation–Interrogation–Production

Lt Col W. H. Brown, G–2; Louis W. Doll, PhD; Capt E. B. Ryckaert, G–2; Capt K. J. Knapp, Jr., G–2; Lt Y. G. Kanegai, G–2; Mr. James J. Wickel; Mr. John Shelton, ATIS; Mr. Norman Sparnon, ATIS; SFC H. Y. Uno, G–2; Mr. K. Takeuchi; Mr. S. Wada.

INDEX

☆ U.S. GOVERNMENT PRINTING OFFICE: 1994 360-573

TOKYO
YOKOHAMA
HONSHU
SHIKOKU
KYUSHU
NAGASAKI
KAGOSHIMA
MIYAZAKI
SHIMONOSEKI
HIROSHIMA
OSAKA
KYOTO
NAGOYA
MATSUE
KANAZAWA
TAEGU
FUSAN
MOKPO
SAISHU
YELLOW SEA
TSINGTAO
TUNGSHAN
NANKING
SHANGHAI
HANKOW
WUCHANG
NANCHANG
CHANGSHA
HENGYANG
KWEILIN
KUNMING
CANTON
HONGKONG
MACAU
SWATOW
AMOY
FOOCHOW
WENCHOW
KIENNINGFU
NANCHOW
C H I N A
CHUNGKING
CHENGTU
LANCHOW
EAST CHINA SEA
YAKUSHIMA
TANEGASHIMA
AMAMI OSHIMA
OKINAWA IS
RYUKYU RETTO
SAKISHIMA GROUP
TAIHOKU
FORMOSA
TAKAO
TAITO
OSHIMA
AOGASHIMA
TORIJIMA
MUKO JIMA
CHICHI JIMA
HAHA JIMA
BONIN IS
KITA IWO JIMA
IWO JIMA
MINAMI IWO JIMA
PHILIPPINE SEA
BATAN IS
MAUG I
AGRIJAN I
PAGAN I
MARIANAS IS
SAIPAN I
TINIAN I
ROTA I
GUAM I
HANOI
KIUNGCHOW
HAINAN I
LUANGPRABANG
QUANGTRI
THAILAND
BANGKOK
INDO
CHINA
SAIGON
PHANRANG
GULF OF SIAM
LONGXUYEN
APARRI
LUZON
LINGAYEN
MANILA
PHILIPPINE IS
MINDORO
SAMAR
LEYTE
PANAY
CEBU
NEGROS
PALAWAN
BOHOL
SURIGAO
SULU SEA
MINDANAO
DAVAO
PALAU IS
ULITHI IS
FAIS IS
YAP I
FARAULEP IS
WOLEAI IS
M I C R
C A R O L I N E I
NORTH BORNEO
JESSELTON
SANDAKAN
TAWAO
SULU ARCHIPELAGO
BRUNEI
KOTA BHARU
MALAYA
KUALA LUMPUR
MALACCA
SINGAPORE
NATOENA IS
ANAMBAS IS
TARAKAN
CELEBES SEA
TALAUD IS
MOROTAI I
HALMAHERA
MANADO
GORONTALO
TERNATE
WAIGEO
M E L A N
EQUATOR
KUCHING
SAMARINDA
BORNEO
BANGKA
PALEMBANG
BILLITON I
SUMATRA
PADANG
DONGGALA
MOLUCCA SEA
SOELA IS
MANOKWARI
SCHOUTEN IS
BIAK
SARMI
HOLLANDIA
AITAPE
WEWAK
ADMIRALTY IS
MUSSAU I
BISMARCK ARCHIPELAGO
RABAUL
CELEBES
BANDJERMASIN
KOTA BAROE
BOEROE I
AMBOINA
KAI IS
AROE IS
NEW GUINEA
PAPUA
LAE
SALAMAUA
NEW BRITAIN
NETHERLANDS INDIES
MAKASSAR
JAVA SEA
BANDA SEA
TELOEKBETOENG
BATAVIA
BANDOENG
SEMARANG
SOERABAJA
JAVA
BALI
LOMBOK
FLORES SEA
FLORES I
ALOR IS
WETAR I
DILI
TIMOR
KOEPANG
TANIMBAR IS
ARAFURA SEA
MERAUKE
DARU
GULF OF PAPUA
TORRES STRAIT
PORT MORESBY
SAMARAI
CHRISTMAS I
MELVILLE I
BATHURST I
THURSDAY I
C YORK
TIMOR SEA
DARWIN
INDIAN
OCEAN
GULF OF CARPENTARIA
BIRDUM
DALY WATERS
WYNDHAM
TURKEY CREEK
DERBY
BROOME
HALLS CREEK
WOLLAL
BURKETOWN
NORMANTON
COOKTOWN
CAIRNS
INNISFAIL
GREAT BARRIER REEF
CHARLESTON
TOWNSVILLE
CLONCURRY
POWELLS CREEK
JOANNA SPR
BARROW I
PORT HEDLAND
ROEBOURNE
MARBLE BAR
NORTH WEST CAPE
BARROW CREEK
TEATREE WELL
ALICE SPRINGS
WINTON
LONGREACH
ST LAWRENCE
ROCKHAMPTON
CARNARVON
DIRK HARTOGS I
A U S T R A L I A
CHARLEVILLE
OODNADATTA
NORTHAMPTON
GERALDTON
YALGOO
SANDSTONE
LAVERTON
MALCOLM
STRANGWAYS SPR
120°
20°

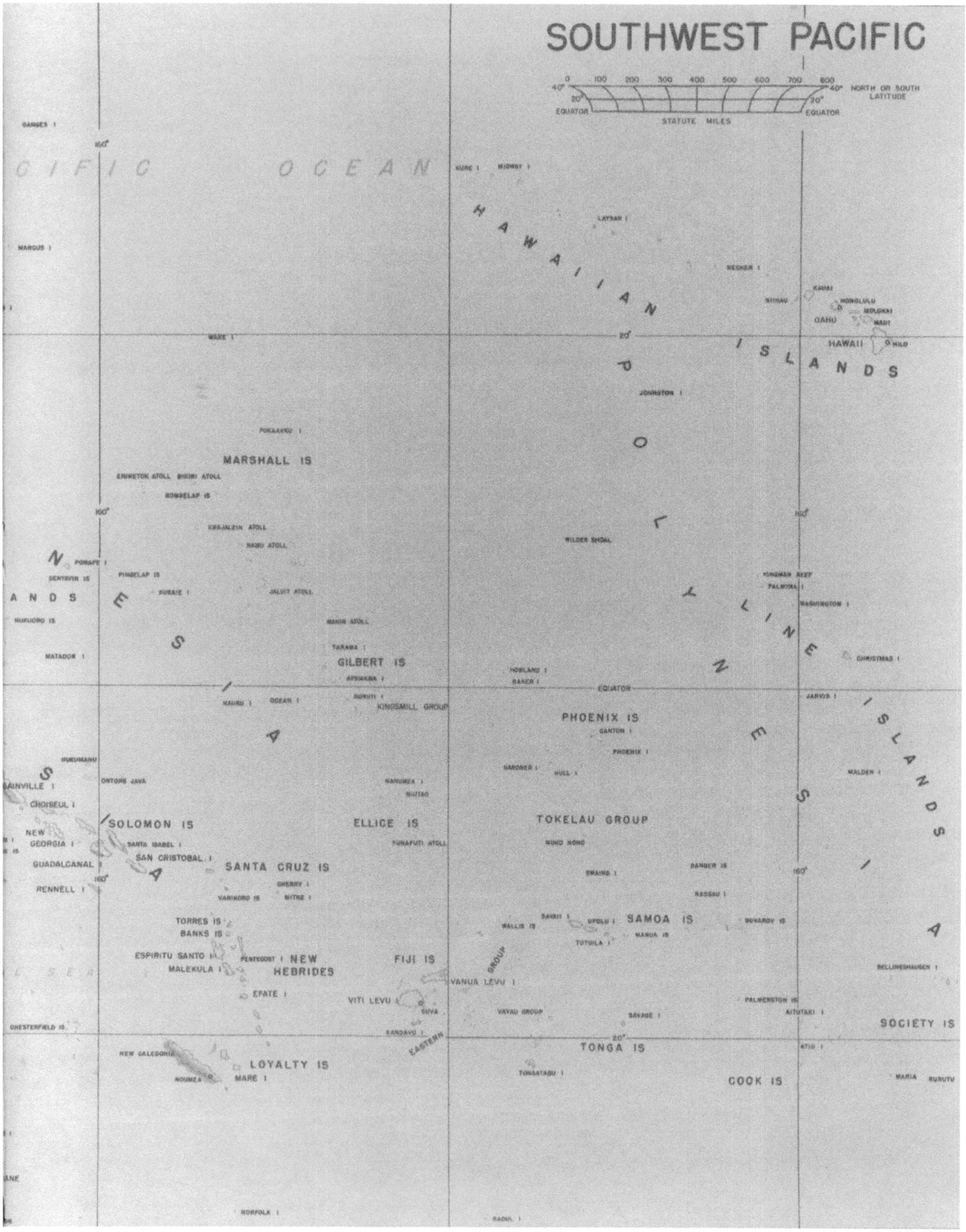

SOUTHWEST PACIFIC
0 100 200 300 400 500 600 700 800
NORTH OR SOUTH LATITUDE
EQUATOR
STATUTE MILES
PACIFIC OCEAN
HAWAIIAN ISLANDS
OAHU
HAWAII
POLYNESIA
MARSHALL IS
GILBERT IS
KINGSMILL GROUP
LINE ISLANDS
PHOENIX IS
SOLOMON IS
ELLICE IS
TOKELAU GROUP
SANTA CRUZ IS
SAMOA IS
NEW HEBRIDES
FIJI IS
VITI LEVU
VANUA LEVU
TONGA IS
LOYALTY IS
MARE I
COOK IS
SOCIETY IS
CHOISEUL I
NEW GEORGIA I
GUADALCANAL I
RENNELL I
SAN CRISTOBAL I
TORRES IS
BANKS IS
ESPIRITU SANTO I
MALEKULA I
EPATE I

Made in United States
Cleveland, OH
05 November 2026

47673398R00278